Fundamental Theories of Physics

Volume 225

The international monograph series "Fundamental Theories of Physics" aims to stretch the boundaries of mainstream physics by clarifying and developing the theoretical and conceptual framework of physics and by applying it to a wide range of interdisciplinary scientific fields. Original contributions in well-established fields such as Quantum Physics, Relativity Theory, Cosmology, Quantum Field Theory, Statistical Mechanics and Nonlinear Dynamics are welcome. The series also provides a forum for non-conventional approaches to these fields. Publications should present new and promising ideas, with prospects for their further development, and carefully show how they connect to conventional views of the topic. Although the aim of this series is to go beyond established mainstream physics, a high profile and open-minded Editorial Board will evaluate all contributions carefully to ensure a high scientific standard.

Andrei Barvinsky · Alexander Kamenshchik
Editors

Open Issues in Gravitation and Cosmology

Original Contributions, Essays, and Recollections in Honor

of Alexei Starobinsky

Springer

Editors
Andrei Barvinsky
P. N. Lebedev Physical Institute
of the Russian Academy of Sciences
Moscow, Russia

Alexander Kamenshchik
Department of Physics
and Astronomy "A. Righi"
University of Bologna
Bologna, Italy

ISSN 0168-1222 ISSN 2365-6425 (electronic)
Fundamental Theories of Physics
ISBN 978-3-032-15702-7 ISBN 978-3-032-15703-4 (eBook)
https://doi.org/10.1007/978-3-032-15703-4

This Springer imprint is published by the registered company Springer Nature Switzerland AG
The registered company address is: Gewerbestrasse 11, 6330 Cham, Switzerland

Preface

With the publication of this collection of essays, we are commemorating the life and work of Alexei Alexandrovich Starobinsky, who passed away in December 2023. It is hard to imagine the progress of quantum gravity and modern precision cosmology without referencing his scientific biography, which has been so tightly intertwined with the history of these branches of physics for the last 50 years. One need only mention his pioneering contribution to such fields as black hole physics, quantum field theory in curved spacetime, theoretical foundations of rapidly developing gravitational wave astronomy, and finally, cosmological inflation theory, which describes the formation of the large-scale structure of the Universe with unprecedented accuracy.

Another part of the Alexei Starobinsky phenomenon is that, despite the fact that he would rarely supervise students at academic institutions, he was able to bring to the international scientific stage generations of talented physicists who were inspired by his striking and unfading creative power. Indeed, at the international conferences he would attend, the majority of speakers would be people who had been driven by his scientific charisma and were implementing his ideas and methods. The collection of papers in this book is the work of those people—his former students, coauthors, friends and physicists for whom he served as a moving spirit behind their work. This collection includes original research papers, reviews and personal memories concerning open issues in gravity theory and cosmology related to the scientific legacy of Alexei Starobinsky.

Starobinsky's scientific biography began with his graduation from the Physics Department of the Lomonosov Moscow State University, after which he began working with his supervisor, the famous physicist Yakov Borisovich Zeldovich. Zeldovich, whose interests ranged from nuclear physics to astrophysics and relativistic cosmology, was one of the great pillars of Soviet physics, and one could say that Starobinsky inherited the greatness of his teacher. Like Zeldovich, he skipped the stage of scientific habilitation by being directly elected to the Russian Academy of Sciences. It is worth mentioning in passing that Zeldovich never even got his Ph.D., such exceptions from the need to 'get the admission ticket to the temple of science' (in the famous words of Freeman Dyson) being very rare, but still possible, in that epoch.

The election of Starobinsky to the Russian Academy of Sciences however, would take place much, much later. In the early 1970s the joint paper by Starobinsky and Zeldovich[1] was of a seminal nature by predicting the isotropization effect of quantum particle creation in an anisotropic expanding Universe. The next Starobinsky paper dealt with the amplification effect for waves reflecting from a rotating black hole, now called "super-radiance", and contained the hypothesis of the spontaneous radiation from black holes—the famous thermal black hole evaporation phenomenon suggested a year later by Stephen Hawking and marking the beginning of a new era in the history of quantum gravity theory.

Then A. Starobinsky studied the quasi-exponential stage of the cosmological expansion. This work[2] laid the foundations of the cosmological inflation scenario and its observational signatures in the spectrum of relic gravitational radiation, the tensor part of the cosmic microwave background. This was the very beginning of the inflation paradigm conceived by A. A. Starobinsky more than ten years ahead of the actual experimental discovery of the dominant scalar part of CMB in the early 1990s by the COBE satellite. This discovery marked the beginning of the precision cosmology and it was followed by a series of experimentally triumphant missions of CMB studies. The dynamical origin of this quasi-de Sitter expansion stage with the subsequent transition to the power-law Friedmann expansion was derived by A. Starobinsky in his arguably most famous and influential paper,[3] where he derived a singularity-free cosmological scenario from the effective equations generated by the conformal anomaly of quantum matter. A simplified version of these equations underlies the Starobinsky R^2-model of modified gravity theory, one of the most popular inflationary models. Further studies were undertaken on cosmological perturbations and their relation to the parameters of CMB[4,5,6]. Exit from inflation and a subsequent transition to a hot universe stage, which are necessary for the formation

[1] Zeldovich, Ya.B., Starobinskii, A.A.: Particle Production and Vacuum Polarization in an Anisotropic Gravitational Field. JETP **34**, 1159–1412 (1972).

[2] Starobinskii, A.A.: Spectrum of relict gravitational radiation and the early state of the universe. JETP Lett. **30**, 682–685 (1979).

[3] Starobinsky, A.A.: A new type of isotropic cosmological models without singularity. Phys. Lett. B **91**, 99–102 (1980).

[4] Starobinsky, A.A.: Dynamics of phase transition in the new inflationary universe scenario and generation of perturbations. Phys. Lett. B **117**, 175–178 (1982).

[5] Starobinsky, A.A.: The Perturbation Spectrum Evolving from a Nonsingular Initially De-Sitter Cosmology and the Microwave Background Anisotropy. Sov. Astron. Lett. **9**, 302 (1983).

[6] Starobinsky, A.A.: Multicomponent de Sitter (Inflationary) Stages and the Generation of Perturbations. JETP Lett. **42**, 152–155 (1985).

of the observable large scale structure of the universe, were studied by A. Starobinsky and his coauthors in their pioneering papers on reheating theory.[7,8]

The success of inflation theory in explaining the problems of the Standard Big Bang scenario would still leave open the issue of the origin of inflation itself—the problem of quantum creation of the Universe, which is the subject of quantum cosmology. There were two fundamental suggestions for the quantum state of the early Universe, the tunneling and no-boundary cosmological wavefunctions—two special solutions of the Wheeler-DeWitt equation. A. Starobinsky, together with Ya. Zeldovich,[9] was among the pioneers of the tunneling version of the quantum state signifying the ontologically attractive notion of the birth of the Universe from "nothing"—its origin from the classically forbidden state of the gravitational field, described by the Euclidean signature spacetime metric.

A. Starobinsky made also a great contribution to quantum field theory in curved spacetime by suggesting the stochastic approach to the description of long-wavelength field modes in the inflationary universe.[10,11] His hypothesis that inflaton quantum dynamics can be modelled by a stochastic classical field, evolving according to the Langevin equation, turned out to be very productive and went in parallel with the development of very popular stochastic methods in quantum field theory. The idea that the stochastic force arises as a collective effect of averaging over ultraviolet modes has led A. Starobinsky to the formulation of the Fokker-Planck equation, which allows one to perform nonperturbative resummation of field-theoretical series and ascend to such popular research directions as eternal inflation and the notion of Nature as the realm of multiverse cosmology.

While the origin of the Universe has quantum nature, the observable large-scale structure of the Universe is essentially classical. To explain this classicalization phenomenon, A. Starobinsky and his coauthors combined the quantum decoherence study with the squeezed state formalism.[12,13] A crucial observation here was that when neglecting the decaying mode of cosmological perturbations, one arrives at the classicalization of quantum observables, the effect especially characteristic of cosmological expansion phenomenon.

---

[7] Kofman, L., Linde, A.D., Starobinsky, A.A.: Reheating after inflation. Phys. Rev. Lett. **73**, 3195–3198 (1994).

[8] Kofman, L., Linde, A.D., Starobinsky, A.A.: Towards the theory of reheating after inflation. Phys. Rev. D **56**, 3258–3295 (1997).

[9] Zeldovich, Ya.B., Starobinsky, A.A.: Quantum creation of a universe in a nontrivial topology. Sov. Astron. Lett. **10**, 135–137 (1984).

[10] Starobinsky, A.A.: Stochastic De Sitter (Inflationary) Stage in the Early Universe. Lect. Notes Phys. **246**, 107–126 (1986).

[11] Starobinsky, A.A., Yokoyama, J.: Equilibrium state of a selfinteracting scalar field in the De Sitter background. Phys. Rev. D **50**, 6357–6368 (1994).

[12] Polarski, D., Starobinsky, A.A.: Semiclassicality and decoherence of cosmological perturbations. Class. Quant. Grav. **13**, 377–392 (1996).

[13] Kiefer, C., Polarski, D., Starobinsky, A.A.: Quantum to classical transition for fluctuations in the early universe. Int. J. Mod. Phys. D **7**, 455–462 (1998).

Cosmic accelerated expansion of the modern Universe was discovered at the end of 1990s. This stimulated searches for the dark energy mechanism responsible for such a phenomenon, including the development of the techniques that would allow one to confront observational data with theoretical predictions. A new diagnostic method for dark energy, the so-called "statefinder" one, was invented by A. Starobinsky and his coauthors[14] and turned out to be much more refined than the other methods. The discovery of cosmic acceleration also gave rise to a search for modified gravity theories. A very thought-provoking and influential paper[15] on such a modification in the form of a scalar-tensor theory incorporated, as functions of the redshift, the parameters of the luminosity distance and the density perturbation in the dustlike matter component of the model. A. Starobinsky also suggested a modified $f(R)$-theory of gravity generalizing his R^2-model[16], which is likely to unify inflation, Standard Cosmology and dark energy scenario, simultaneously being compatible with modern observational data.

The progress of precision cosmology would always be associated with the need for a fundamental link between the observable cosmological parameters and particle physics data. This became especially intriguing within the context of the so-called Higgs inflation model prior to the discovery of Higgs boson and its mass measurement. Such a link was suggested by A. Starobinsky and his coauthors[17] in the form of a certain relation between this mass and the observable red tilt $n_s = 0.96$ of the inflationary CMB spectrum. The further renormalization group improvement of this relation[18] predicted approximately 135 GeV for the magnitude of the Higgs mass, which was only 10 GeV higher than the actual value measured later by LHC. This accuracy inspired a great interest in the Higgs inflation model and resulted in its numerous quantitative and qualitative modifications.

This brief overview certainly cannot fully convey the scholarly grandeur of A. Starobinsky and his prophetic gift as a scientist. His famous R^2 and $f(R)$ gravity theories still remain the best fit models for CMB observations during the 45 years since his seminal works mentioned above (see the footnotes 3 and 4). Undoubtedly, his other breakthrough revelations, like the light of extinct star, are still in store for us. We need only mention his predictions[19] for possible scenarios of the future of the Universe in light of recent cosmological discoveries.

---

[14] Sahni, V., Saini, T.D., Starobinsky, A.A., Alam, U.: Statefinder—a new geometrical diagnostic of dark energy. JETP Lett. **77**, 201–206 (2003).

[15] Boisseau, B., Esposito-Farese, G., Polarski, D., Starobinsky, A.A.: Reconstruction of a scalar tensor theory of gravity in an accelerating universe. Phys. Rev. Lett. **85**, 2236 (2000).

[16] Starobinsky, A.A.: Disappearing cosmological constant in f(R) gravity. JETP Lett. **86**, 157–163 (2007).

[17] Barvinsky, A.O., Kamenshchik, A.Yu., Starobinsky, A.A.: Inflation scenario via the Standard Model Higgs boson and LHC. JCAP **11**, 021 (2008).

[18] Barvinsky, A.O., Kamenshchik, A.Yu., Kiefer, C., Starobinsky, A.A., Steinwachs, C.: Asymptotic freedom in inflationary cosmology with a non-minimally coupled Higgs field. JCAP **12**, 003 (2009).

[19] Starobinsky, A.A.: Future and origin of our universe: Modern view. Grav. Cosmol. **6**, 157–163 (2000).

This book consists of seven parts containing 33 essays associated with various aspects of A. Starobinsky's scientific legacy. It could not have been created without the positive response from theoretical physicists to our proposal to publish this commemorative collection of research papers and personal recollections, and we are thankful to all of them. We are also grateful to Chris Caron, Publishing Editor of Springer, for his help at the early stage of the project of this book, to Angela Lahee, Lisa Scalone, Ravi Vengadachalam and Umamagesh Perumal for their attention, benevolence and tolerance in the progression of this project and to Alison Wearing for her invaluable assistance in literary work on this collection.

From various viewpoints, this book is truly universal and international. It is universal by the subject which forms its content, the theory of our Universe whose cosmological evolution has an existential role for all of us. It is also international by virtue of its editors and authors from 18 countries, united both by the noble spirit of the international scientific community and by their readiness to express their respect to their outstanding colleague, Alexei Alexandrovich Starobinsky.

Moscow, Russia
Bologna, Italy
October 2025

Andrei Barvinsky
Alexander Kamenshchik

Contents

Inflation

Early Universe and Quantum Cosmology

Modified Gravity Models

Effects of Gravity Theory

Quantum Gravity and Quantum Field Theory in Curved

Inflation

On Legacy of Starobinsky Inflation

Sergei V. Ketov

Abstract Alexei Alexandrovich Starobinsky was one of the greatest cosmologists of all times, who made fundamental contributions to gravitational theory and cosmology based on geometrical ideas in physics, in the spirit of Einstein. One of his big achievements is the famous Starobinsky model of cosmological inflation in the early universe, proposed in 1979–1980. In this memorial paper, the Starobinsky inflation model is systematically reviewed from the modern perspective. Its deformation to include production of primordial black holes is proposed, and possible quantum corrections in the context of superstring theory and the Swampland Program are discussed. Starobinsky inflation also leads to the universal reheating mechanism for particle production after inflation.

1 Introduction

My research collaboration and personal encounters with Alexei Alexandrovich Starobinsky were both accidental and inevitable. In order to explain the apparent contradiction, I briefly recall where I came from.

I was raised in a tiny remote village in Western Siberia, in the former Soviet Union. Despite that, it was possible to get there by mail the high-quality journals for school children, known as the "Quant" and the "Pioneer", where advanced articles with exercises (problems) in physics and mathematics were published. Some of those problems included the names of their inventors, and among them were e.g., Ya.B. Zeldovich and A.N. Kolmogorov, just to mention a couple of famous scientists. There were regular competitions in physics and mathematics also, organized by university teachers. After winning some of those competitions, I was sent to a high school specialized in physics and mathematics and later to a university without passing entrance exams because my teachers in the high school were university professors and I already completed the first-year university program in physics and

S. V. Ketov (✉)
Department of Physics, Tokyo Metropolitan University, Hachioji-shi, Tokyo, Japan
e-mail: ketov@tmu.ac.jp

A. Barvinsky and A. Kamenshchik (eds.), *Open Issues in Gravitation and Cosmology*, Fundamental Theories of Physics 225, https://doi.org/10.1007/978-3-032-15703-4_1

mathematics, being in the high school. In the middle of my university studies, I was sent to the Physical Institute (FIAN) of the Soviet Academy of Sciences in Moscow, known as the Lebedev Physical Institute now. Being attached to the Theory Laboratory headed by V. L. Ginzburg, another famous scientist, and attending numerous seminars there, I was thinking about a research topic for my graduation. There were two natural options because the Ginzburg Lab had the cosmology group with prominent scientists such as A.D. Sakharov, A.D. Linde and V.F. Mukhanov, and the quantum field theory group including other prominent scientists such as E.S. Fradkin, M.A. Vasiliev, I.A. Batalin and I.V. Tyutin. I have chosen to conduct my research in supersymmetric quantum field theory and string theory under the supervision of E.S. Fradkin because this topic had a solid mathematical foundation and promising connections to high energy particle physics unlike cosmology that lacked a reliable theoretical and experimental foundation in the late 70's and the early 80's of the last century. Much later, around 2010, being in Tokyo, I turned to theoretical cosmology of the early universe because my research path in formal high-energy physics naturally guided me there. At the same time, A.A. Starobinsky was a Visiting Professor at the University of Tokyo, and he approached me with a question about a realization of inflation in supergravity. Our research collaboration began with that question and resulted in joint papers proposing new supergravity-based realizations of inflation [1, 2]. Over the last 16 years my research group developed this subject by including spontaneous supersymmetry breaking, production of primordial black holes, dark energy, dark matter and gravitational waves [3–37].

In this memorial paper, Starobinsky inflation is reviewed without supersymmetry and supergravity for simplicity. It is needed and actual because legacy of Starobinsky inflation is still not settled since the groundbreaking first paper [38] appeared almost 45 years ago.

In Sect. 2, the Starobinsky model of inflation, its possible quantum origin and its predictions for observations and measurements of the cosmic microwave background (CMB) radiation are reviewed from the modern perspective. In Sect. 3, a phenomenological deformation of the Starobinsky inflation model in the framework of modified $F(R)$ gravity is given in order to accommodate production of primordial black holes (PBH) during inflation. In Sect. 4, the basic conjectures of Swampland Program in application to Starobinsky inflation are briefly discussed. An impact of the leading (closed) superstring correction to the low-energy gravitational effective action (EFT) on Starobinsky inflation is studied in Sect. 5. Section 6 is devoted to reheating after Starobinsky inflation. Section 7 is a conclusion where the future prospects are outlined.

2 Starobinsky Inflation Model

In this Section, the Starobinsky model of inflation is described from the current perspective (at present) without following the historical developments that were rather confusing.

The Starobinsky model of inflation is the generally covariant and non-perturbative modified gravity theory whose Lagrangian is a sum of two terms, the Einstein-Hilbert (EH) term and the term quadratic in the Ricci scalar curvature R with a positive coefficient. All the curvature-dependent terms beyond the EH one are irrelevant in the Solar system, while they may also be negligible during reheating after inflation in the weak-gravity regime. However, it is not the case during inflation in the high curvature regime where the R^2 term is the dominant contribution and the EH term can be ignored in the leading approximation.

Therefore, the Starobinsky model can be treated as the particular example of the modified $F(R)$ gravity theories that are geometrical because only gravitational interactions or General Relativity (GR) are used. A modified gravity action has the higher derivatives and generically suffers from Ostrogradsky instabilities and ghosts. However, in the most general modified gravity action, whose Lagrangian is quadratic in the spacetime curvature, the *only* ghost-free term is just given by R^2 with a positive coefficient, which immediately leads to the Starobinsky model with the action

$$S_{\text{Star.}} = \alpha \int d^4x\sqrt{-g}R^2 + \frac{M_{\text{Pl}}^2}{2}\int d^4x\sqrt{-g}R\ , \quad \alpha \equiv \frac{M_{\text{Pl}}^2}{12M^2}\ , \tag{1}$$

with just one free parameter α or M, where $M_{\text{Pl}} = 1/\sqrt{8\pi G_{\text{N}}} \approx 2.4 \times 10^{18}$ GeV, the spacetime signature is $(-,+,+,+,)$ and the natural units are used, $\hbar = c = 1$. The first term in this action is *scale invariant* because it does not have a dimensional coupling constant (the parameter α is dimensionless).

The origin of the R^2 term was initially proposed due to one-loop contributions of quantized matter fields in the EH gravity [38]. However, because the EH term was subleading during inflation, the alternative interpretation is also possible, namely, with the EH term originated from the scale-invariant gravity with matter after quantization. For instance, when starting from the scale-invariant action for gravity and a scalar field ϕ as [39–41]

$$S[g_{\mu\nu},\phi] = \int d^4x\sqrt{-g}\left[\alpha R^2 + \xi\phi^2 R - \frac{1}{2}(\partial\phi)^2 - \lambda\phi^4\right]\ , \tag{2}$$

one finds that it can undergo a phase transition (called dimensional transmutation) due to quantum corrections, known as the Coleman-Weinberg mechanism of spontaneous symmetry breaking [42]. It leads to the massive scalar field ϕ that may be identified with dilaton or Higgs field having a non-vanishing vacuum expectation value (VEV) in the effective action, as can be demonstrated in the one-loop perturbation theory [40, 41]. As a result, both the Planck mass and the EH term are generated with

$$\frac{1}{2}M_{\text{Pl}}^2 = \xi\,\langle\phi\rangle^2\ . \tag{3}$$

The alternative interpretation with the induced EH term is viable, though the action (2) is still nonrenormalisable and, hence, cannot be considered as the UV-completion of the Starobinsky modified gravity.

The background metric of a flat Friedman (early) universe is given by

$$ds^2 = -dt^2 + a^2 \left(dx_1^2 + dx_2^2 + dx_3^2\right) , \tag{4}$$

whose cosmic factor $a(t)$ is time-dependent. The action (1) with the metric (4) leads to equations of motion in the form

$$2H\ddot{H} - \left(\dot{H}\right)^2 + H^2 \left(6\dot{H} + M^2\right) = 0 , \quad H = \dot{a}/a , \tag{5}$$

known as the Starobinsky equation in the literature, where the dots stand for the time derivatives and $H(t)$ is Hubble function.

When searching for a solution to the Starobinsky equation in the form of left Painlevé series, $H(t) = \sum_{k=-\infty}^{k=p} c_k (t_0 - t)^k$, one finds the Hubble function (see e.g., Ref. [34])

$$\begin{aligned}\frac{H(t)}{M} &= \frac{M}{6}(t_0 - t) + \frac{1}{6M(t_0 - t)} - \frac{4}{9M^3(t_0 - t)^3} + \frac{146}{45M^5(t_0 - t)^5} \\ &\quad - \frac{11752}{315M^7(t_0 - t)^7} + O\left(M^{-9}(t_0 - t)^{-9}\right)\end{aligned} \tag{6}$$

valid for $M(t_0 - t) > 1$. This special solution is an attractor, while $R = 12H^2 + 6\dot{H}$.

Therefore, in the slow-roll (SR) approximation defined by $\left|\ddot{H}\right| \ll \left|H\dot{H}\right|$ and $\left|\dot{H}\right| \ll H^2$, the leading term in the Starobinsky solution reads

$$H(t) \approx \left(\frac{M^2}{6}\right)(t_0 - t) , \tag{7}$$

being entirely due to the R^2-term in the action. The attractor solution spontaneously breaks the scale invariance of the R^2-gravity and, therefore, implies the existence of the Nambu-Goldstone boson (called scalaron) that is the physical excitation of the higher-derivative Starobinsky gravity. It can be made manifest by rewriting the Starobinsky action to the quintessence form after the field redefinition (or the Legendre-Weyl transform) with [43]

$$\varphi = \sqrt{\frac{3}{2}} M_{\rm Pl} \ln F'(\chi) \quad \text{and} \quad g_{\mu\nu} \to \frac{2}{M_{\rm Pl}^2} F'(\chi) g_{\mu\nu} , \quad \chi = R. \tag{8}$$

It yields the standard quintessence action

$$S[g_{\mu\nu}, \varphi] = \frac{M_{\rm Pl}^2}{2} \int \mathrm{d}^4x\sqrt{-g}R - \int \mathrm{d}^4x\sqrt{-g}\left[\frac{1}{2}g^{\mu\nu}\partial_\mu\varphi\partial_\nu\varphi + V(\varphi)\right] \tag{9}$$

in terms of the canonical inflaton φ with the scalar potential

$$V(\varphi) = \frac{3}{4} M_{\text{Pl}}^2 M^2 \left[1 - \exp\left(-\sqrt{\frac{2}{3}} \varphi / M_{\text{Pl}} \right) \right]^2 . \tag{10}$$

This potential has the infinite plateau (for the large φ-field values of the order M_{Pl} and beyond) with the approximate shift symmetry of the inflaton field as the consequence of the scale invariance of the R^2 gravity or the approximate scale invariance of the action (1) in the large-curvature regime. The potential (10) also has the positive "cosmological constant" given by the first term in the square brackets, induced by the R^2 term in the action (1), which can be physically interpreted as the energy driving inflation. The scale of inflation is determined by the mass parameter M that is identified with the canonical inflaton mass. The universality class of the Starobinsky-like inflation models is determined by the critical parameter $\sqrt{2/3}$ in the exponential term [32].

The equivalent actions (1) and (9) are usually referred to the Jordan frame and the Einstein frame, respectively, in the literature. The approximate shift symmetry of the potential (10) required for proper duration of inflation arises as the consequence of the approximate scale invariance of the Starobinsky gravity. The R^2-term is unique among *all* the higher-derivative curvature terms because it is the only one that leads to a ghost-free scale-invariant action. Therefore, the R^2 term must be present in any viable model of inflation based on modified gravity. This conclusion becomes more transparent by using the *inverse* transformation from the quintessence action (with a potential V) in the Einstein frame to an $F(R)$ gravity action in the Jordan frame, having the parametric form [44]

$$R = \left(\frac{\sqrt{6}}{M_{\text{Pl}}} \frac{dV}{d\varphi} + \frac{4V}{M_{\text{Pl}}^2} \right) e^{\sqrt{\frac{2}{3}} \varphi / M_{\text{Pl}}} , \quad F = \left(\frac{\sqrt{6}}{M_{\text{Pl}}} \frac{dV}{d\varphi} + \frac{2V}{M_{\text{Pl}}^2} \right) e^{2\sqrt{\frac{2}{3}} \varphi / M_{\text{Pl}}} , \tag{11}$$

with the inflaton φ as the parameter. In the SR approximation, the first term in the brackets is negligible against the second term, which immediately implies $F(R) \sim R^2$.

The gravitational EFT during inflation does not have to be limited to the terms given in Eq. (1) but can also include the higher-order curvature terms. Those terms eliminate the infinite plateau in the inflaton potential (10) and may destabilize Starobinsky inflation. However, the fact that the Starobinsky model of inflation is in excellent agreement with the current CMB measurements (see below) implies those terms should also be sub-leading during Starobinsky inflation, which put restrictions on their (unknown) coefficients.

It is convenient to use the e-folds number N instead of time t, which are related by

$$N = \int_t^{t_0} H(\tilde{t}) d\tilde{t} , \tag{12}$$

and the co-moving wavenumber $k = 2\pi/\lambda$ as a scale. The scale k is related to e-folds N by the equation $d\ln k = -dN$.

The SR (running) parameters in the Einstein frame are defined by

$$\varepsilon_{\text{sr}}(\varphi) = \frac{M_{\text{Pl}}^2}{2}\left(\frac{V'}{V}\right)^2 \quad \text{and} \quad \eta_{\text{sr}}(\varphi) = M_{\text{Pl}}^2\left(\frac{V''}{V}\right) \tag{13}$$

in terms of the quintessence scalar potential V, where the primes denote the derivatives with respect to canonical inflaton φ. In the Jordan frame, one uses the Hubble flow functions,

$$\epsilon_H = -\frac{\dot{H}}{H^2}, \quad \eta_H = \epsilon_H - \frac{\dot{\epsilon}_H}{2\epsilon_H H}. \tag{14}$$

The amplitude of scalar perturbations at the horizon crossing with the pivot scale $k_* = 0.05\ \text{Mpc}^{-1}$ is known from CMB measurements (called WMAP or COBE normalization) as

$$A_s = \frac{V_*^3}{12\pi^2 M_{\text{Pl}}^6 (V_*')^2} = \frac{3M^2}{8\pi^2 M_{\text{Pl}}^2}\sinh^4\left(\frac{\varphi_*}{\sqrt{6}M_{\text{Pl}}}\right) \approx 2\cdot 10^{-9}\,, \tag{15}$$

where the subscript (*) refers to the CMB pivot scale in the case of Starobinsky inflation. Therefore, the known amplitude A_s determines the parameter M (or α) as well as the scale of inflation, $H_{\text{inf.}}$, in the Starobinsky model as

$$\frac{M}{M_{\text{Pl}}} \approx O(10^{-5}), \quad \alpha \approx O(10^9), \quad H \approx O(10^{14})\ \text{GeV}, \quad \frac{R}{M_{\text{Pl}}^2} \approx \frac{12H^2}{M_{\text{Pl}}^2} \approx 10^{-7}. \tag{16}$$

The higher-order curvature terms in the gravitational EFT beyond the Starobinsky model of inflation lead to power series in $H^2/M_{\text{Pl}}^2 \sim 10^{-8}$. Those terms are subleading during inflation unless they have very large coefficients. It is also worth noticing that the large value of α required by CMB does not speak in favor of generating the R^2-term by quantum matter contributions because a single quantized matter field contributes in the one-loop approximation about 10^{-3} to the α-parameter, so one needs about 10^{12} quantized matter fields in order to achieve the value of α needed.

The primordial spectrum $P_\zeta(k)$ of 3-dimensional scalar (density) perturbations $\zeta(x)$ in a flat Friedman universe is defined by the 2-point correlation function

$$\left\langle \frac{\delta\zeta(x)}{\zeta}\frac{\delta\zeta(y)}{\zeta}\right\rangle = \int \frac{d^3k}{k^3} e^{ik\cdot(x-y)} \frac{P_\zeta(k)}{P_0}\,, \tag{17}$$

where k is the comoving wavenumber (or scale). Similarly, one defines the primordial spectrum $P_t(k)$ of tensor perturbations, see e.g., Ref. [32] for details. The power spectra define the amplitudes $A_s(k)$ and $A_t(k)$, respectively.

Given the power spectra $P_\zeta(k)$ and $P_t(k)$, one defines the scalar tilt $n_s(k)$, its running parameter $\alpha_s(k)$, the tensor tilt $n_t(k)$ and its running parameter $\alpha_t(k)$ (all dimensionless) as

$$n_s = 1 + \frac{d\ln P_\zeta(k)}{d\ln k}\,, \quad \alpha_s = \frac{d^2\ln P_\zeta(k)}{(d\ln k)^2}\,, \quad n_t = \frac{d\ln P_t(k)}{d\ln k}\,, \quad \alpha_t = \frac{d^2\ln P_t(k)}{(d\ln k)^2}\,, \tag{18}$$

as well as the tensor-to-scalar ratio

$$r(k) = \frac{P_t}{P_\zeta} = 8\,|n_t|\,. \tag{19}$$

The Starobinsky model gives sharp predictions for the cosmological tilts of the CMB scalar and tensor power spectra. *In the leading order* with respect to the inverse e-folds N_* evaluated when perturbations left the horizon (at the horizon crossing), those predictions are [45, 46]

$$n_s \approx 1 - \frac{2}{N_*}\,, \quad \alpha_s \approx -\frac{2}{N_*^2}\,, \quad \alpha_t \approx -\frac{3}{N_*^3}\,, \quad r \approx \frac{12}{N_*^2}. \tag{20}$$

Therefore, tensor perturbations are suppressed with respect to scalar perturbations by the extra factor of N_*^{-1}. The theoretical predictions (20) are to be compared to the recent (Planck mission) CMB measurements [47, 48], which give

$$n_s \approx 0.9649 \pm 0.0042\ (68\%\text{CL}) \quad \text{and} \quad r < 0.032\ (95\%\text{CL})\,, \tag{21}$$

and fit the Starobinsky model predictions for

$$N_* = 56 \pm 8. \tag{22}$$

This prediction for the duration of inflation agrees with the calculations in Ref. [49] in the Jordan frame, based on the solution (6). The corresponding times for the end and the beginning of inflation are $M(t_0 - t_{\text{end}}) \approx 2.5$ and $M(t_0 - t_{\text{start}}) \approx 27.7$, respectively.

It is difficult to independently predict or measure the duration of inflation. However, excluding N_* from Eq. (20) yields the prediction of the Starobinsky model for the tensor-to-scalar ratio with high precision as

$$r \approx 3(1 - n_s)^2. \tag{23}$$

The Starobinsky model is sensitive to quantum ultra-violet (UV) corrections because of its high inflation scale and the inflaton field values near the Planck scale during inflation. Therefore, it is important to determine the UV-cutoff Λ_{UV} of the

Starobinsky model by studying scaling of inflaton scattering amplitudes with respect to energy, E/Λ_{UV}. A careful calculation yields [50]

$$\Lambda_{\text{UV}} = M_{\text{Pl}}. \tag{24}$$

Therefore, the predictions of the Starobinsky model for inflation and CMB make sense and the model itself can be considered as a trustable EFT after decoupling of heavy modes expected at the Planck scale [51].

It is worth mentioning that very little from this Section can be found in the original paper [38] that is the standard reference for Starobinsky inflation. The key point of Ref. [38] was a discovery that quantum-modified gravitational equations of motion can have attractor solutions suitable for describing inflation.

3 Deformation of the Starobinsky Inflation Model for Production of Primordial Black Holes at Smaller Scales

The Starobinsky model can be extended within modified $F(R)$ gravity in order to describe double slow-roll (SR) inflation with an ultra-slow-roll (USR) phase by engineering the function $F(R)$ leading to a near-inflection point in the inflaton potential below the scale of inflation [52, 53]. It results in large density perturbations whose gravitational collapse leads to production of primordial black holes (PBH). However, adding the near-inflection point and the USR phase requires fine-tuning of the model parameters [54], which often lowers the value of the CMB tilt n_s of scalar perturbations and thus leads to tension with CMB measurements, while large perturbations may also imply significant nonGaussianity and large quantum (loop) corrections that may invalidate classical single-field models of inflation with PBH production [55].

The phenomenological model [52, 53] is an extension of the singularity-free Appleby-Battye-Starobinsky (ABS) model [56] with other parameters whose fine-tuning leads to efficient production of PBH having asteroid-size masses (and atomic sizes) between 10^{16} g and 10^{20} g, exceeding the Hawking (black hole) evaporation limit of 10^{15} g. Those PBH may be part (or the whole) of the current dark matter (DM), while the model itself may be confirmed or falsified by detection or absence of the induced gravitational waves with frequencies in the 10^{-2} Hz range (see below).

The $F(R)$ gravity model [53] is defined by the action ($M_{\text{Pl}} = 1$)

$$S = \frac{1}{2}\int d^4x\sqrt{-g}\,F(R)\ , \tag{25}$$

whose F-function of the spacetime scalar curvature R reads

$$F(R) = (1 + g\tanh b)R + gE_{AB}\ln\left[\frac{\cosh\left(\frac{R}{E_{AB}} - b\right)}{\cosh(b)}\right] + \frac{R^2}{6M^2} + \delta\frac{R^4}{48M^6}\ , \tag{26}$$

where the first terms are known in the literature as the ABS model [56] with the Starobinsky mass $M \approx 1.3 \times 10^{-5}$ defining the scale of the first SR phase of inflation. The ABS parameter

$$E_{AB} = \frac{R_0}{2g\ln(1 + e^{2b})} \tag{27}$$

has the scale R_0 related to the second SR phase of inflation below the Starobinsky scale, while it differs from Ref. [56] where R_0 was associated with the dark energy scale. The other parameters g and b define the shape of the inflaton potential and have to be fine-tuned in order to engineer a near-inflection point. The last term in Eq. (26) may be considered as a quantum gravity correction (see Sect. 6) in order to get good (within 1σ) agreement with the measured CMB value of n_s. The function (26) obeys the no-ghost (stability) conditions, $F'(R) > 0$ and $F''(R) > 0$, for the relevant values of R, avoids singularities, and has the correct Newtonian limit.

To produce PBH, one needs a large enhancement of the power spectrum of scalar perturbations by seven orders of magnitude against the CMB spectrum. Then the parameters (R_0, g, b) should be fine-tuned as [53]

$$R_0 \approx 3.00M^2\ , \quad g \approx 2.25 \quad \text{and} \quad b \approx 2.89. \tag{28}$$

The numerically obtained profile of the inflaton potential $V(\phi)$ (in the Einstein frame) for some values of R_0 and δ is given in Fig. 1.

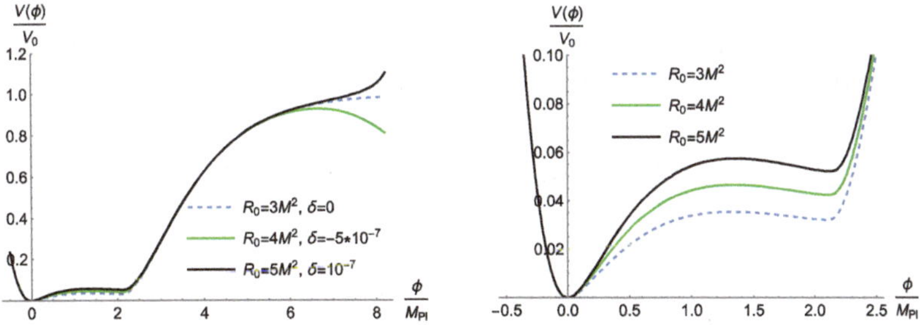

Fig. 1 The inflaton potential having two plateaus for $g = 2.25$ and $b = 2.89$ with $V_0 = \frac{3}{4}M^2$ (left). Zooming the potential for lower values of ϕ with a near-inflection point (right). The potential is unstable for negative values of δ, and has the infinite plateau for $\delta = 0$ leading to Starobinsky inflation

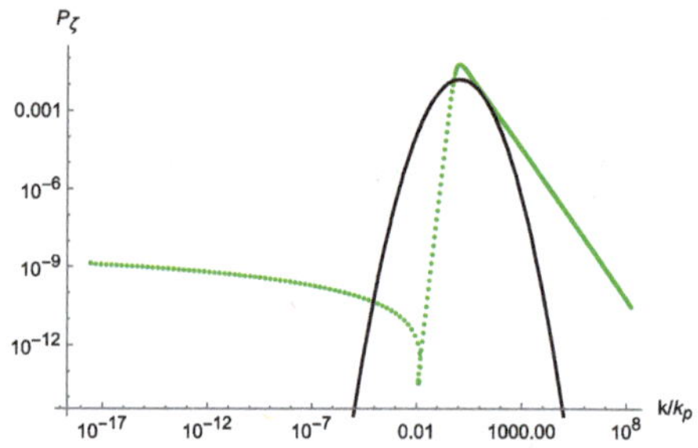

Fig. 2 The primordial power spectrum $P_\zeta(k)$ of scalar perturbations and its log-normal fit

The CMB observables with the model best fit are given by [53]

$$n_s \approx 0.965, \quad r \approx 0.0095, \quad M_{\rm PBH} \approx 1.0 \cdot 10^{20}\ \text{g}. \tag{29}$$

The PBH-producing peak in the scalar power-spectrum of Fig. 2 can be approximated by the log-normal fit [57, 58] as

$$P_\zeta^{\rm peak}(k) \approx \frac{A_\zeta}{\sqrt{2\pi}\,\Delta} \exp\left[\frac{-\ln^2(k/k_p)}{2\Delta^2}\right] \tag{30}$$

with the amplitude $A_\zeta \approx 0.06$ and the width $\Delta \approx 1.5$, where $k_p \approx 4.5 \cdot 10^{12}\ \text{Mpc}^{-1}$ is the location of the peak.

PBH production in the early universe leads to stochastic gravitational waves (GW) different from primordial GW caused by inflation. The PBH-induced GW density in the model also can be approximated by the lognormal fit as [57, 59]

$$\Omega_{\rm GW}(k) = \frac{A_{\rm GW}}{\sqrt{2\pi}\,\sigma_{\rm GW}} \exp\left[-\frac{\ln^2(k/k_p)}{2\sigma_{\rm GW}^2}\right], \tag{31}$$

with the amplitude $A_{\rm GW} \approx 5.6 \cdot 10^{-8}$ and the width $\sigma_{\rm GW} \approx \Delta/\sqrt{2} \approx 1.06$, where Δ is the width of the power spectrum in Fig. 2. The $\Omega_{\rm GW}^{\rm peak}(k)$ near the peak is roughly given by $10^{-6} P_\zeta^2(k)$.

The induced GW frequencies f_p are related to the PBH masses as [60]

$$f_p \approx 5.7 \left(\frac{M_\odot}{M_{\rm PBH}}\right)^{1/2} 10^{-9}\ \text{Hz}, \tag{32}$$

where the Sun mass is given by $M_\odot \approx 2 \cdot 10^{33}$ g. Given the PBH masses of 10^{20} g, it results in the GW frequency $f_p \approx 0.0255$ Hz that is higher than the GW frequencies between 3 and 400 nHz detected by NANOGrav [61]. A more specific comparison of the model predictions with future GW observations is possible by plotting the GW spectrum in the model against the expected sensitivity curves in the future space-based gravitational interferometers such as LISA [62], TianQin [63], Taiji [64] and

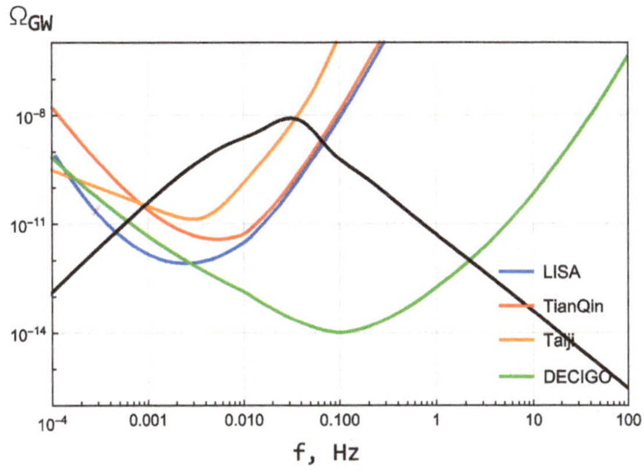

Fig. 3 The GW density induced by the power spectrum of scalar perturbations in the model (in black) against the expected sensitivity curves for the future space-based GW experiments (in color)

DECIGO [65], see Fig. 3. The space-based experiments are expected to be sensitive to stochastic GW in the frequencies between 10^{-3} Hz and 10^{-1} Hz, while part of the black curve in Fig. 3 from the model belongs to that frequency range also.

It was found in Ref. [59] by using the δN formalism that the relative size of quantum (loop) corrections to the power spectrum of scalar perturbations in the model is of the order 10^{-3} or less, so that the model of inflation and PBH production considered in this Section is not ruled out by the quantum loop corrections.

4 Starobinsky Model and Swampland Conjectures

The Starobinsky model is a *large-field* model whose inflaton field values are near the Planck scale during inflation. Large-field models of inflation are sensitive to quantum gravity that may ruin their consistency. Unfortunately, little is known about quantum gravity, not even its energy scale. Should the quantum gravity scale be the same as the Planck scale, then Eq. (24) implies robustness of Starobinsky inflation against quantum gravity corrections. However, should the quantum gravity scale be close to the Starobinsky inflation scale of 10^{14} GeV, then the predictions of Starobinsky inflation would not be reliable and the Starobinsky model could not be EFT, i.e. it would belong to Swampland [66].

A string theory derivation of the Starobinsky inflation is unknown, though it is possible to get the effective scalar potential close to that in Eq. (10) in the Einstein frame, see e.g., [67, 68].

In the absence of full theory of quantum gravity, one can employ a set of its expected features, which is known as the Swampland Program [69, 70]. The Swampland conjectures are all about consistency with quantum gravity (or UV completion) either in string theory or in general. The Swampland Program is aimed to discriminate between possible EFT.

To the end of this Section, the simplest Swampland conjectures are briefly discussed in relation to the Starobinsky model of inflation.

No global symmetries in quantum gravity:
This Swampland conjecture is about the absence of global (or rigid) symmetries in any fundamental theory of quantum gravity. It claims that only local (or gauge) symmetries are possible in quantum gravity. This conjecture is fully in line with fundamental principles of General Relativity but may appear unusual for particle physics based on the Standard Model near or below the electro-weak scale.

As regards the Starobinsky model, it has only approximate (global) scale invariance related to the R^2 term alone, which is obviously violated by the EH term and any other possible higher-order term with respect to the spacetime curvature in the gravitational EFT. Therefore, the Starobinsky model does not violate the no-global-symmetry conjecture.

Weak-gravity conjecture:
In its simplest version, this conjecture claims that gravity is always the weakest force, for example, in comparison to electromagnetic force. At first sight, Starobinsky inflation contradicts this conjecture because the Starobinsky inflation relies on the gravitational force as the main force driving inflation in the Jordan frame. However, in the Einstein frame, Starobinsky inflation is mainly due to the inflaton selfinteraction described by the scalar potential (10), i.e. due to the inflaton force (or inflaton exchange). Therefore, there is no contradiction between the Starobinsky model of inflation and the weak-gravity conjecture.

One may also argue that during inflation there were no sources of electromagnetic force because there were no charged particles yet (they appeared during reheating after inflation).

No-de-Sitter conjecture:
Claims that no de Sitter (dS) spacetime and no eternal inflation are possible in quantum gravity. Again, at first sight, it is in contradiction to Starobinsky inflation because the potential (10) has the infinite plateau. However, the infinite plateau in the Starobinsky model is destabilized by the higher-order terms with respect to the spacetime curvature, which are certainly present in any UV-completion of the Starobinsky action (1). A stronger version of the conjecture, excluding all locally flat or single-field inflaton potentials [71], is apparently in conflict with observations [72, 73].

Distance conjecture:
About possible EFT from string theory implies constraints on the inflaton range $\Delta\varphi$ during inflation and, hence, on the CMB tensor-to-scalar-ratio r, via the upper bound [74]

$$|\Delta\varphi| \leq M_{\rm Pl} \ln\left(\frac{M_{\rm Pl}}{H_{\rm inf}}\right) \approx \frac{M_{\rm Pl}}{2} \ln\left(\frac{2}{\pi^2 A_s r}\right) \tag{33}$$

that is complementary to the Lyth lower bound [75]

$$|\Delta\varphi| \geq O(1)\left(\frac{r}{0.01}\right)^{1/2} M_{\text{Pl}}. \tag{34}$$

Both constraints are in agreement with Starobinsky inflation that has $|\Delta\varphi| \approx 5.5 M_{\text{Pl}}$ but severely constrain it.

5 Quantum Correction from Superstring Theory

Superstring theory is a good candidate for the theory of quantum gravity, so it is natural to seek an UV completion of the Starobinsky model there. However, it turns out to be difficult because superstring theory is defined in 10 spacetime dimensions, whereas the Starobinsky model is defined in 4 space-time dimensions. Therefore, 6 spatial dimensions in superstring theory have to be hidden or compact, while the gravitational low-energy EFT in string theory is subject to large ambiguities related to spacetime metric redefinitions and compactification, see Refs. [76, 77] for the alternative proposals to hidden dimensions.

In the case of closed (type II) superstrings, the leading $(\alpha')^3$-correction beyond the EH term is given by the terms *quartic* in the space-time curvature, which were first derived by Grisaru and Zanon in 1985 from the vanishing four-loop renormalization group beta-function of the supersymmetric non-linear sigma model in two dimensions, describing propagation of a test superstring in the gravitational background [78]. The same quartic curvature terms arise from M-theory [79, 80] after dimensional reduction. The dependence of the gravitational EFT in string theory upon Ricci scalar curvature (and Ricci tensor also) is known to be ambiguous, see e.g., Ref. [81], because closed strings (generating gravity) are well defined only in Ricci-flat spacetimes by string scattering amplitudes (S-matrix).

By combining the Starobinsky model with the GZ correction one gets the Starobinsky-Grisaru-Zanon (SGZ) gravity defined by the action

$$S_{\text{SGZ}}[g] = \frac{M_{\text{Pl}}^2}{2}\int d^4x\sqrt{-g}\left(R + \frac{1}{6M^2}R^2 - \frac{72\gamma}{M^6}Z\right), \tag{35}$$

where the GZ superstring correction reads [78]

$$72Z = \left(R^{\mu\rho\sigma\nu}R_{\lambda\rho\sigma\tau} + \frac{1}{2}R^{\mu\nu\rho\sigma}R_{\lambda\tau\rho\sigma}\right)R_{\mu}{}^{\alpha\beta\lambda}R^{\tau}{}_{\alpha\beta\nu} \tag{36}$$

with the effective dimensionless (string) coupling constant $\gamma > 0$. The value of γ cannot be calculated from string theory because it depends upon a compactification from ten to four spacetime dimensions and unknown vacuum expectation value of string dilaton.

Due to the origin of SGZ gravity, the first two terms in the action (35) should be considered nonperturbatively but the last (quartic) curvature terms should be treated as a perturbation, in the first order with respect to the coupling constant γ. In particular, no Ostrogradsky ghosts arise.

In a flat Friedman universe (4) one finds [49]

$$Z = H^8 + 2H^6\dot{H} + \tfrac{11}{6}H^4\dot{H}^2 + \tfrac{2}{3}H^2\dot{H}^3 + \tfrac{1}{12}\dot{H}^4 , \tag{37}$$

and the SGZ gravity equations of motion are

$$\begin{aligned} & m^6H^2 + 6m^4H^2\dot{H} + 2m^4H\ddot{H} - m^4\dot{H}^2 \\ & - 12\gamma H^8 + 132\gamma H^6\dot{H} + 44\gamma H^5\ddot{H} + 138\gamma H^4\dot{H}^2 \\ & + 48\gamma H^3\dot{H}\ddot{H} + 28\gamma H^2\dot{H}^3 + 12\gamma H\dot{H}^2\ddot{H} - 3\gamma\dot{H}^4 = 0 , \end{aligned} \tag{38}$$

which extend the Starobinsky equation (5) by the γ-dependent terms. A solution to this equation in the first order with respect to the γ-parameter reads [49]

$$H(t) = H_0(t) + \gamma H_1(t) , \tag{39}$$

where

$$\begin{aligned} \frac{H_1(t)}{M} = & -\frac{1}{163296}M^7(t_0-t)^7 - \frac{2}{2835}M^5(t_0-t)^5 - \frac{391}{90720}M^3(t_0-t)^3 - \frac{9061}{306180}M(t_0-t) \\ & - \frac{127}{7776M(t_0-t)} - \frac{1931203}{5358150M^3(t_0-t)^3} + O(M^{-5}(t_0-t)^{-5}). \end{aligned} \tag{40}$$

Accordingly, the Ricci scalar is given by [49]

$$\begin{aligned} \frac{R}{M^2} = & \frac{M^2(t_0-t)^2}{3} - \frac{1}{3} - \frac{4}{9M^2(t_0-t)^2} + \frac{16}{5M^4(t_0-t)^4} - \frac{6908}{189M^6(t_0-t)^6} \\ & + \gamma\left[-\frac{M^8(t_0-t)^8}{40824} - \frac{151M^6(t_0-t)^6}{58320} + \frac{143M^4(t_0-t)^4}{122472} - \frac{4163M^2(t_0-t)^2}{81648} \right. \\ & - \frac{68713}{7144200} - \frac{5109281}{5143824M^2(t_0-t)^2} + \frac{17584432631}{964467000M^4(t_0-t)^4} \\ & \left. - \frac{75802186291}{300056400M^6(t_0-t)^6} + O(M^{-8}(t_0-t)^{-8}) \right]. \end{aligned} \tag{41}$$

It was found in Ref. [49] that demanding the absence of ghosts and negative energy fluxes implies the upper bound

$$\gamma \leq 1.12 \times 10^{-6}. \tag{42}$$

It is instructive to compare contributions of the GZ quantum correction to the CMB observables against the classical contributions beyond the leading order given by Eq. (20) within the possible range of N_* in Eq. (22). The subleading terms for the CMB observables in the Starobinsky inflation model are given by [82]

$$\begin{aligned} n_s &= 1 - \frac{2}{N_*} + \frac{2.4}{N_*^2} - \frac{\ln(2N_*)}{6N_*^2} + O\left(\frac{\ln 2N_*}{N_*^3}\right) , \\ \alpha_s &= -\frac{2}{N_*^2} + O\left(\frac{\ln 2N_*}{N_*^3}\right) , \\ r &= \frac{12}{N_*^2} + \frac{2\ln(2N_*)}{N_*^3} - \frac{56.76}{N_*^3} + O\left(N_*^{-4}\right) , \\ \alpha_t &= -\frac{3}{N_*^3} + O\left(N_*^{-4}\right) , \end{aligned} \tag{43}$$

where $\ln 2N_*/N_*^3 < 4.1 \cdot 10^{-5}$ and $N_*^{-4} < 2 \cdot 10^{-7}$. The subleading contribution to the scalar tilt n_s in the 3rd term above can increase the n_s-value by $1.0 \cdot 10^{-3}$, and the subleading contribution to the tensor-to-scalar-ratio r, given by the 3rd term above, can decrease the r-value by $5 \cdot 10^{-4}$. The subleading contributions are within the observational errors given in Eq. (21). On the other hand, in the first order with respect to the string parameter γ, the GZ quantum contributions to n_s and r are up to $+2.5 \cdot 10^{-4}$ and $-4.9 \cdot 10^{-5}$, respectively.

Therefore, the GZ quantum contributions to the CMB tilts are smaller than the subleading terms proportional to N_*^{-2} by one order of magnitude but may be of the same order of magnitude as the classical N_*^{-3} contributions. The same conclusion also applies to the running parameters α_s and α_t.

6 Universal Reheating After Starobinsky Inflation

Starobinsky inflation also implies the *universal reheating mechanism*, as an additional bonus, when matter is coupled to Starobinsky gravity in the Jordan frame [83]. Reheating was a transfer of energy from inflaton to ordinary particles when inflaton field coherently oscillated around the minimum of its potential in an expanding universe. Reheating took place after inflation but before the Big Bang Nucleosynthesis (BBN) and radiation domination. Before reheating, the leading channel of particle production can be due to a nonperturbative (broad) parametric resonance, known as preheating that is not considered here (see, however, Ref. [10] about preheating in the Starobinsky supergravity framework). The resonance disappeared when the inflaton field became sufficiently small, and it was replaced by perturbative decays. Reheating provided the initial conditions for leptogenesis and baryogenesis, DM abundance, baryon asymmetry, relic monopoles and gravitinos, and BBN. Both preheating and reheating are highly model-dependent, while Starobinsky gravity coupled to matter gives testable predictions.

The classical solution near the minimum of the inflaton scalar potential reads

$$a(t) \approx a_0 \left(\frac{t}{t_0}\right)^{2/3} \quad \text{and} \quad \varphi(t) \approx \left(\frac{M_{\rm Pl}}{3M}\right) \frac{\cos\left[M(t-t_0)\right]}{t-t_0} , \tag{44}$$

where M is the inflaton mass (Sect. 2). The time-dependent classical spacetime background leads to quantum production of particles with masses $m < M$. When particle production is included, the amplitude of inflaton oscillations decreases much faster via inflaton decay and the universe expansion as

$$\exp[-\frac{1}{2}(3H + \Gamma_{\varphi})t] , \tag{45}$$

where Γ_{φ} is the perturbative decay rate (see below). Accordingly, the inflaton equation of motion gets modified as

$$\ddot{\varphi} + (3H + \Gamma_{\varphi})\, \dot{\varphi} + V_{,\varphi} = 0. \tag{46}$$

The Starobinsky model (in the Jordan frame) with matter ψ is described by an action

$$S = \int d^4x \sqrt{-g_J}\, F_{\rm Star.}(R_J) + S_{\rm matter}(g^{\mu\nu}{}_J, \psi). \tag{47}$$

After the Legendre-Weyl transformation to the Einstein frame it takes the form

$$S = S_{\rm quintessence}(g_{\mu\nu}, \varphi) + S_{\rm matter}(g^{\mu\nu} e^{-\tilde{\kappa}\varphi}, \psi) , \tag{48}$$

where inflaton φ couples to *all* non-conformal terms and fields ψ due to universality of gravitational interaction, and $\tilde{\kappa} \sim M_{\rm Pl}^{-1}$. Therefore, Starobinsky inflation leads to the universal mechanism of particle production, see Ref. [83] for some specific examples. The decay rates are sensitive to spin and mass of created particles, while all of them are suppressed by the factors of $M_{\rm Pl}$. By construction, conformal couplings do not contribute to inflaton decay.

The perturbative decay rates of inflaton into a pair of scalars (s) and into a pair of spin-1/2 fermions (f) are given by [84, 85]

$$\Gamma_{\varphi \to ss} = \frac{M^3}{192\pi M_{\rm Pl}^2} \quad \text{and} \quad \Gamma_{\varphi \to ff} = \frac{M M_f^2}{48\pi M_{\rm Pl}^2} , \tag{49}$$

respectively. The energy transfers by the time $t_{\rm reh} \geq \left(\sum_{s,f} \Gamma_{s,f}\right)^{-1} = \Gamma_{\rm total}^{-1}$. One expects quick thermalization after inflation. i.e. a thermal bath filled with baryons, radiation and DM. The reheating temperature (the maximal temperature of primordial plasma) after Starobinsky inflation is given by [86]

$$T_{\rm reh} \propto \sqrt{\frac{M_{\rm Pl}\Gamma_{\rm total}}{(\#d.o.f.)^{1/2}}} \approx 10^9 \text{ GeV}\ , \tag{50}$$

where the effective # d.o.f. ($= g$) for the Standard Model is $g = 106.75$, and $g = 228.75$ for the Minimal Supersymmetric Standard Model. Once the universe is thermalized, the energy density of radiation $\rho_{\rm reh}$ is related to the reheating temperature as

$$\rho_{\rm reh} = \frac{\pi^2 g}{30} T^4_{\rm reh}. \tag{51}$$

The Starobinsky reheating may also lead to production of DM particles and right-handed Majorana neutrinos via scalaron decay, may explain the neutrino oscillations (via the standard seasaw mechanism) and the baryon asymmetry (via the standard leptogenesis and the related baryogenesis) also [83, 87].

The specific predictions of Starobinsky inflation for the cosmological tilts (Sects. 2 and 5) are dependent upon the e-folds number N_* that is also affected by the amount of reheating at the end of inflation. A combination of the Planck, BICEP2/Keck Array and BAO data implies $50 \leq N_* \leq 52$ at the 68% CL [88, 89], which is still consistent with (22) but also implies tight constraints on Starobinsky inflation.

7 Conclusion

The upcoming CMB measurements by CORE Collaboration [90], S4 Collaboration [91], LiteBIRD Collaboration [92], NASA PICO Collaboration [93], the Simons Observatory survey [94] and EUCLID Collaboration [95] are expected to probe the tensor-to-scalar ratio r in the range of 10^{-3} and improve the precision value of n_s. Should those measurements confirm Eq. (23), this would be a triumph of Starobinsky inflation. Should the disagreement in r be significant (say, two orders of magnitude), it would rule out Starobinsky inflation. Should the disagreement be small (say, within one order of magnitude), one may expect that due to the sub-leading corrections that may be either due to the sub-leading terms in the classical Starobinsky model or due to quantum gravity corrections to the gravitational EFT, or may have a different origin, say, due to reheating or new physics, see e.g., Ref. [96].

The Starobinsky model is just the simplest large-single-field model of inflation, which has solid theoretical motivation and well fits the current CMB measurements 45 years after the pioneering paper [38], which is a quite remarkable achievement. There exist no obstruction toward generalizations of single-field models to multi-field models of inflation preferred by supergravity and string theory. Should the primordial non-Gaussianity be detected, it would be another crucial test of Starobinsky inflation because a contribution of the nonlinear (NL) 3-point correlation function of primordial curvature perturbations to the amplitude of primordial non-Gaussianity is fixed by the Maldacena relation $f_{\rm NL} = \frac{5}{12}(1 - n_s)$ in all single-field models of inflation [97].

Acknowledgements The author (SVK) is grateful to Ignatios Antoniadis, Robert Brandenberger, Michael Duff, Gia Dvali and Dmitry Gorbunov for discussions. SVK was supported by Tokyo Metropolitan University, the Japanese Society for Promotion of Science under the grant No. 22K03624, and the World Premier International Research Center Initiative (MEXT, Japan).

Competing Interests The author declares no conflict of interest.

References

1. S.V. Ketov, A.A. Starobinsky, Embedding $(R + R^2)$-inflation into supergravity. Phys. Rev. D **83**, 063512 (2011)
2. S.V. Ketov, A.A. Starobinsky, Inflation and non-minimal scalar-curvature coupling in gravity and supergravity. JCAP **08**, 022 (2012)
3. S.J. Gates Jr., S.V. Ketov, Superstring-inspired supergravity as the universal source of inflation and quintessence. Phys. Lett. B **674**, 59–63 (2009)
4. S.V. Ketov, Scalar potential in F(R) supergravity. Class. Quant. Grav. **26**, 135006 (2009)
5. S.V. Ketov, $F(R)$ supergravity. AIP Conf. Proc. **1241**, 613–619 (2010)
6. S.V. Ketov, Chaotic inflation in F(R) supergravity. Phys. Lett. B **692**, 272–276 (2010)
7. S.V. Ketov, N. Watanabe, Cosmological properties of a generic $\mathcal{R}^2$-supergravity. JCAP **03**, 011 (2011)
8. S.V. Ketov, N. Watanabe, Cosmological constant in F(R) supergravity. Phys. Lett. B **705**, 410–413 (2011)
9. S.V. Ketov, Supergravity and early universe: the meeting point of cosmology and high-energy physics. Int. J. Mod. Phys. A **28**, 1330021 (2013)
10. S.V. Ketov, S. Tsujikawa, Consistency of inflation and preheating in F(R) supergravity. Phys. Rev. D **86**, 023529 (2012)
11. S.V. Ketov, N. Watanabe, Dark energy in modified supergravity. Mod. Phys. Lett. A **27**, 1250225 (2012)
12. S.V. Ketov, On the supersymmetrization of inflation in f(R) gravity. PTEP **2013** (2013), 123B04 (2013)
13. S.V. Ketov, T. Terada, Old-minimal supergravity models of inflation. JHEP **12**, 040 (2013)
14. S.V. Ketov, Starobinsky model in $N = 2$ supergravity. Phys. Rev. D **89**, 085042 (2014)
15. S.V. Ketov, T. Terada, Inflation in supergravity with a single chiral superfield. Phys. Lett. B **736**, 272–277 (2014)
16. S.V. Ketov, T. Terada, Generic scalar potentials for inflation in supergravity with a single chiral superfield. JHEP **12**, 062 (2014)
17. A. Addazi, S.V. Ketov, Energy conditions in Starobinsky supergravity. JCAP **03**, 061 (2017)
18. A. Addazi, S.V. Ketov, M.Y. Khlopov, Gravitino and Polonyi production in supergravity. Eur. Phys. J. C **78**, 642 (2018)
19. Y. Aldabergenov, S.V. Ketov, Removing instability of inflation in Polonyi-Starobinsky supergravity by adding FI term. Mod. Phys. Lett. A **33**, 1850032 (2018)
20. A. Addazi, A. Marciano, S.V. Ketov, M.Y. Khlopov, Physics of superheavy dark matter in supergravity. Int. J. Mod. Phys. D **27**, 1841011 (2018)
21. S.V. Ketov, M.Y. Khlopov, Extending Starobinsky inflationary model in gravity and supergravity. Bled Workshops Phys. **19**, 148–163 (2018)
22. H. Abe, Y. Aldabergenov, S. Aoki, S.V. Ketov, Polonyi-Starobinsky supergravity with inflaton in a massive vector multiplet with DBI and FI terms. Class. Quant. Grav. **36**, 075012 (2019)
23. S.V. Ketov, Y. Aldabergenov, Inflation, dark energy, and dark matter in supergravity. Astron. Nachr. **340**, 126–130 (2019)

24. S.V. Ketov, M.Y. Khlopov, Cosmological probes of supersymmetric field theory models at superhigh energy scales. Symmetry **11**, 511 (2019)
25. Y. Aldabergenov, A. Chatrabhuti, S.V. Ketov, Generalized dilaton-axion models of inflation, de Sitter Vacua and spontaneous SUSY breaking in supergravity. Eur. Phys. J. C **79**, 713 (2019)
26. S.V. Ketov, Inflation, dark energy and dark matter in supergravity, Meeting of the Division of Particles and Fields of the American Physical Society (2019). arXiv:1909.05599
27. S.V. Ketov, On the equivalence of Starobinsky and Higgs inflationary models in gravity and supergravity. J. Phys. A **53**, 084001 (2020)
28. Y. Aldabergenov, S. Aoki, S.V. Ketov, Minimal Starobinsky supergravity coupled to a dilaton-axion superfield. Phys. Rev. D **101**, 075012 (2020)
29. S.V. Ketov, Supergravity as the dark side of the universe. Int. J. Mod. Phys. A **35**, 2040038 (2020)
30. Y. Aldabergenov, A. Addazi, S.V. Ketov, Primordial black holes from modified supergravity. Eur. Phys. J. C **80**, 917 (2020)
31. Y. Aldabergenov, A. Addazi, S.V. Ketov, Testing primordial black holes as dark matter in supergravity from gravitational waves. Phys. Lett. B **814**, 136069 (2021)
32. S.V. Ketov, Multi-field versus single-field in the supergravity models of inflation and primordial black holes. Universe **7**, 115 (2021)
33. Y. Aldabergenov, A. Addazi, S.V. Ketov, Inflation, SUSY breaking, and primordial black holes in modified supergravity coupled to chiral matter. Eur. Phys. J. C **82**, 681 (2022)
34. S.V. Ketov, E.O. Pozdeeva, S.Y. Vernov, On the superstring-inspired quantum correction to the Starobinsky model of inflation. JCAP **12**, 032 (2022)
35. S.V. Ketov, Inflationary cosmology from supergravity, in *Handbook of Quantum Gravity* (Springer, Singapore, 2023). https://doi.org/10.1007/978-981-19-3079-9_51-1
36. R. Ishikawa, S.V. Ketov, Improved model of large-field inflation with primordial black hole production in Starobinsky-like supergravity. Class. Quant. Grav. **41**, 195014 (2024)
37. D. Frolovsky, S.V. Ketov, Dilaton-Axion modular inflation in supergravity. Int. J. Mod. Phys. D **33**, 2340008 (2024)
38. A.A. Starobinsky, New type of isotropic cosmological models without singularity. Phys. Lett. B **91**, 99–102 (1980)
39. F. Cooper, G. Venturi, Cosmology and broken scale invariance. Phys. Rev. D **24**, 3338 (1981)
40. I.L. Buchbinder, Mechanism for introduction of Einstein gravitation. Sov. Phys. J. **29**, 220–224 (1986)
41. M.B. Einhorn, D.R.T. Jones, Induced gravity I: real scalar field. JHEP Ser. **01**, 019 (2016)
42. S.R. Coleman, E.J. Weinberg, Radiative corrections as the origin of spontaneous symmetry breaking. Phys. Rev. D **7**, 1888–1910 (1973)
43. K.-I. Maeda, Towards the Einstein-Hilbert action via conformal transformation. Phys. Rev. D **39**, 3159 (1989)
44. S.V. Ketov, N. Watanabe, The f(R) gravity function of the Linde quintessence. Phys. Lett. B **741**, 242–245 (2015)
45. V.F. Mukhanov, G.V. Chibisov, Quantum fluctuations and a nonsingular universe. JETP Lett. **33**, 532–535 (1981)
46. V.F. Mukhanov, H.A. Feldman, R.H. Brandenberger, Theory of cosmological perturbations. Part 1. Classical perturbations. Part 2. Quantum theory of perturbations. Part 3. Extensions. Phys. Rept. **215**, 203–333 (1992)
47. **BICEP, Keck** Collaboration, P.A.R. Ade, et al., Improved constraints on primordial gravitational waves using Planck, WMAP, and BICEP/Keck observations through the 2018 observing season. Phys. Rev. Lett. **127**, 151301 (2021)
48. M. Tristram et al., Improved limits on the tensor-to-scalar ratio using BICEP and Planck data. Phys. Rev. D **105**, 083524 (2022)
49. S. Toyama, S.V. Ketov, Starobinsky inflation beyond the leading order. Phys. Rev. D **110**, 063552 (2024)
50. M.P. Hertzberg, On inflation with non-minimal coupling. JHEP **11**, 023 (2010)
51. S.V. Ketov, Starobinsky inflation and swampland conjectures. Russ. Phys. J. **68**, 1 (2025)

52. D. Frolovsky, S.V. Ketov, S. Saburov, Formation of primordial black holes after Starobinsky inflation. Mod. Phys. Lett. A **37**, 2250135 (2022)
53. S. Saburov, S.V. Ketov, Improved model of primordial black hole formation after Starobinsky inflation. Universe **9**, 323 (2023)
54. S.R. Geller, W. Qin, E. McDonough, D.I. Kaiser, Primordial black holes from multifield inflation with nonminimal couplings. Phys. Rev. D **106**, 063535 (2022)
55. J. Kristiano, J. Yokoyama, Constraining primordial black hole formation from single-field inflation. Phys. Rev. Lett. **132**, 221003 (2024)
56. S.A. Appleby, R.A. Battye, A.A. Starobinsky, Curing singularities in cosmological evolution of F(R) gravity. JCAP **06**, 005 (2010)
57. S. Pi, M. Sasaki, Gravitational waves induced by scalar perturbations with a lognormal peak. JCAP **09**, 037 (2020)
58. D. Frolovsky, S.V. Ketov, Fitting power spectrum of scalar perturbations for primordial black hole production during inflation. Astron. **2**, 47–57 (2023)
59. S. Saburov, S.V. Ketov, Quantum loop corrections in the modified gravity model of Starobinsky inflation with primordial black hole production. Universe **10**, 354 (2024)
60. V. De Luca, G. Franciolini, A. Riotto, NANOGrav data hints at primordial black holes as dark matter. Phys. Rev. Lett. **126**, 041303 (2021)
61. **NANOGrav** Collaboration, G. Agazie, et al., The NANOGrav 15 yr data set: evidence for a gravitational-wave background. Astrophys. J. Lett. **951**, L8 (2023)
62. **LISA** Collaboration, P. Amaro-Seoane, et al., Laser interferometer space antenna (2017). http://arxiv.org/abs/1702.00786
63. **TianQin** Collaboration, J. Luo, et al., TianQin: a space-borne gravitational wave detector. Class. Quant. Grav. **33**, 035010 (2016)
64. X. Gong, et al., Descope of the ALIA mission. J. Phys. Conf. Ser. **610**, 012011 (2015)
65. H. Kudoh, A. Taruya, T. Hiramatsu, Y. Himemoto, Detecting a gravitational-wave background with next-generation space interferometers. Phys. Rev. D **73**, 064006 (2006)
66. D. Lüst, J. Masias, B. Muntz, M. Scalisi, Starobinsky inflation in the swampland. JHEP **07**, 186 (2024)
67. R. Blumenhagen, A. Font, M. Fuchs, D. Herschmann, E. Plauschinn, Towards axionic Starobinsky-like inflation in string theory. Phys. Lett. B **746**, 217–222 (2015)
68. M. Brinkmann, M. Cicoli, P. Zito, Starobinsky inflation from string theory. JHEP **09**, 038 (2023)
69. E. Palti, The swampland: introduction and review. Fortsch. Phys. **67**, 1900037 (2019)
70. M. van Beest, J. Calderón-Infante, D. Mirfendereski, I. Valenzuela, Lectures on the swampland program in string compactifications. Phys. Rept. **989**, 1–50 (2022)
71. G. Obied, H. Ooguri, L. Spodyneiko, C. Vafa, De Sitter space and the swampland (2018). http://arxiv.org/abs/1806.08362
72. A. Kehagias, A. Riotto, A note on Inflation and the Swampland. Fortsch. Phys. **66**, 1800052 (2018)
73. F. Denef, A. Hebecker, T. Wrase, de Sitter swampland conjecture and the Higgs potential. Phys. Rev. D **98**, 086004 (2018)
74. M. Scalisi, I. Valenzuela, Swampland distance conjecture, inflation and α-attractors. JHEP **08**, 160 (2019)
75. D.H. Lyth, What would we learn by detecting a gravitational wave signal in the cosmic microwave background anisotropy? Phys. Rev. Lett. **78**, 1861–1863 (1997)
76. S.V. Ketov, H. Nakada, Inflation from $(R + \gamma R^n - 2\Lambda)$ gravity in higher dimensions. Phys. Rev. D Ser. **95**, 103507 (2017)
77. H. Nakada, S.V. Ketov, Inflation from higher dimensions. Phys. Rev. D **96**, 123530 (2017)
78. M.T. Grisaru, D. Zanon, σ model superstring corrections to the Einstein-Hilbert action. Phys. Lett. B **177**, 347–351 (1986)
79. M.B. Green, M. Gutperle, P. Vanhove, One loop in eleven-dimensions. Phys. Lett. B **409**, 177–184 (1997)

80. R. Blumenhagen, N. Cribiori, A. Gligovic, A. Paraskevopoulou, Emergence of R^4-terms in M-theory. JHEP 07, **018** (2024)
81. S.V. Ketov, *Quantum Nonlinear Sigma Models: From Quantum Field Theory to Supersymmetry, Conformal Field Theory, Black Holes and Strings* (Springer, Berlin, 2000)
82. E. Bianchi, M. Gamonal, Primordial power spectrum at N3LO in effective theories of inflation. Phys. Rev. D **110**, 104032 (2024)
83. D.S. Gorbunov, A.G. Panin, Scalaron the mighty: producing dark matter and baryon asymmetry at reheating. Phys. Lett. B **700**, 157–162 (2011)
84. A.A. Starobinsky, Nonsingular model of the universe with the quantum gravitational de Sitter stage and its observational consequences, in *The Proceedings of the Second Seminar on Quantum Gravity*, Moscow (1981)
85. A. Vilenkin, Classical and quantum cosmology of the Starobinsky inflationary model. Phys. Rev. D **32**, 2511 (1985)
86. A.A. Starobinsky, The perturbation spectrum evolving from a nonsingular initially de-sitter cosmology and the microwave background anisotropy. Sov. Astron. Lett. **9**, 302 (1983)
87. H. Jeong, K. Kamada, A.A. Starobinsky, L. Yokoyama, Reheating process in the R^2 inflationary model with the baryogenesis scenario. JCAP **11**, 023 (2023)
88. J. Ellis, M.A.G. Garcia, D.V. Nanopoulos, K.A. Olive, Calculations of inflaton decays and reheating: with applications to no-scale inflation models. JCAP **07**, 050 (2015)
89. J. Ellis, M.A.G. Garcia, D.V. Nanopoulos, K.A. Olive, M. Peloso, Post-inflationary gravitino production revisited. JCAP **03**, 008 (2016)
90. **CORE** Collaboration, F. Finelli, et al., Exploring cosmic origins with CORE: Inflation. JCAP **04**, 016 (2018)
91. **CMB-S4** Collaboration, K. Abazajian, et al., CMB-S4: forecasting constraints on primordial gravitational waves. Astrophys. J. **926**, 54 (2022)
92. **LiteBIRD** Collaboration, D. Paoletti, The LiteBIRD mission. PoS ICHEP **085** (2022)
93. **NASA PICO** Collaboration, S. Hanany, et al., PICO: probe of inflation and cosmic origins (2019). arXiv:1902.10541
94. **Simons Observatory** Collaboration, P. Ade, et al., The Simons observatory: science goals and forecasts. JCAP **02**, 056 (2019)
95. **Euclid** Collaboration, S. Ilić, et al., Euclid preparation. XV. Forecasting cosmological constraints for the Euclid and CMB joint analysis. Astron. Astrophys. **657**, A91 (2022)
96. Y. Aldabergenov, R. Ishikawa, S.V. Ketov, S.I. Kruglov, Beyond Starobinsky inflation. Phys. Rev. D **98**, 083511 (2018)
97. J.M. Maldacena, Non-Gaussian features of primordial fluctuations in single field inflationary models. JHEP **05**, 013 (2003)

Non-local Starobinsky Inflation and Beyond

Alexey S. Koshelev

Abstract Celebrated Starobinsky inflation model based on the R^2 modification of the Einstein gravity has served for around 45 years as the best fit to all known observational data. This model has come from a thorough consideration of the gravity quantization and a subsequent phenomenon of the trace anomaly. Further theoretical developments in constructing quantum gravity signal an inevitability of non-locality in the fundamental physics. It is thus natural to expect that the original local Starobinsky model gets modified. Its elegant analytic embedding into a string field theory inspired non-local gravity model raises hopes to have a UV complete model of the cosmic inflation. The fresh controversy regarding inflationary models raised by the new Atacama Cosmology Telescope data if confirmed may also prompt for certain modifications of the original Starobinsky model which can be easily addressed in a non-local framework. In this note we overview the non-local Starobinsky inflation and interesting further developments in this direction.

1 Introduction

After decades of developing and studying of the cosmic inflation it is very natural to say that it is an extremely neat idea. The paradigm of inflation can resolve problems of the Big Bang cosmology such as the horizon problem, flatness problem, homogeneity problem and others [1–3]. While there are (very) many models of inflation [4] Starobinsky model of inflation [5] takes its special place. Being an application of the trace anomaly [6] to the questions of the early Universe cosmology it nicely implemented the cosmic inflation with just one extra parameter compared to the Einstein's general relativity (GR) perfectly matching all the observational data [7, 8] for around 45 years. The latter statement is right now being challenged by the new Atacama Cosmology Telescope (ACT) dataset [9, 10] and this will be touched later on in the present note.

A. S. Koshelev (✉)
School of Physical Science and Technology, ShanghaiTech University, Shanghai, China
e-mail: askoshelev@shanghaitech.edu.cn

A. Barvinsky and A. Kamenshchik (eds.), *Open Issues in Gravitation and Cosmology*, Fundamental Theories of Physics 225, https://doi.org/10.1007/978-3-032-15703-4_2

From the theoretical perspective of constructing a self-consistent gravity theory the challenge existed from the very beginning though. Neither GR nor R^2 modification of it (even upon inclusion of the Weyl tensor squared counterterm [11, 12]) are healthy quantum gravity candidates. The former is non-renormalizable while the latter is non-unitary due to an appearance of the Ostrogradski ghost [13]. A way to circumvent this trouble may be in the use of infinite-derivative gravity model. It is obvious that any gravity modification in UV is essentially a theory with higher derivatives as it is clear from an EFT of gravity approach [14]. Analytic infinite derivative (AID) gravity [15] proposal provides a statement that a healthy unitary model arises only when an infinite tower of derivatives is present. An asymptotically safe gravity program [16] shares this property in its construction. Both approaches provide a hopeful way towards a consistent quantum gravity theory and feature operator functions of the d'Alembertian operator (also named form-factors) which have intriguing mathematical similarity to those arising in string field theory (SFT) [17].

Starobinsky inflation is essentially a solution of local R^2 gravity. There have been several attempts in the last decades of embedding this or somewhat modified solution into more fundamental and presumably UV complete theories such as string theory and supergravity, for example [18, 19]. A major breakthrough was achieved in joining the success of the Starobinsky model and novel theoretical development when it was understood that the Starobinsky solution can be exactly analytically embedded into an SFT motivated infinite derivative renormalizable and ghost-free AID gravity model [20, 21]. This raises an importance of the Starobinsky model even higher. The corresponding generalization is named non-local Starobinsky inflation.

In this note we will try to review in a brief but a concise way the non-local Starobinsky inflation which was intensively developed thanks to the unforgettable and extremely fruitful collaboration with Alexei Starobinsky himself. Further developments focusing on their current status and open questions are to be discussed and outlined as well.

2 Original Starobinsky Inflation Model

Starobinsky model originates from the idea of utilizing the conformal (or trace) anomaly. The latter can be phrased in short as an appearance of the non-zero trace of the stress-energy tensor of conformal fields at 1-loop level. The corresponding schematic equation reads

$$M_P^2 R = g^{\mu\nu}\langle T_{\mu\nu}\rangle = aF + bG + c\Box R\,, \tag{1}$$

where one has the Weyl tensor square term $F = C_{\mu\nu\rho\sigma}C^{\mu\nu\rho\sigma}$, the Gauss-Bonnet term $G = R_{\mu\nu\rho\sigma}R^{\mu\nu\rho\sigma} - 4R_{\mu\nu}R^{\mu\nu} + R^2$ with $R_{\mu\nu\rho\sigma}$ being Riemann tensor, $R_{\mu\nu}$ the Ricci tensor and $R = g^{\alpha\beta}R^{\mu}_{\alpha\mu\beta}$ the Ricci scalar. Here a, b, c are dimensionless coefficients that depend on the number of conformal fields, and $\Box = \nabla^\mu\nabla_\mu$ is the covariant d'Alembertian with ∇_μ being a covariant derivative. M_P as usual is the Planck mass.

In a regime when the last term in (1) dominates one can find an inflationary solution for the Universe evolution in the form of the Friedmann-Lemaître-Robertson-Walker (FLRW) metric describing a nearly de Sitter (dS) expansion lasting for a long enough time with a graceful exit. It is the Starobinsky solution. It can be obtained from the following effective action

$$S_{R^2} = \int d^4x\sqrt{-g}\left(\frac{M_P^2}{2}R + \frac{M_P^2}{12M^2}R^2\right) . \tag{2}$$

Here M is the mass parameter which becomes the mass of a propagating scalar in this model, named scalaron.

To find an inflationary solution one can solve for example the trace of the generalized Einstein equations

$$\Box R = M^2 R , \tag{3}$$

perturbatively in the slow-roll parameter $\varepsilon \equiv -\dot{H}/H^2$ where H is the Hubble rate. Hereafter we assume a spatially flat FLRW metric which has the following line element $ds^2 = -dt^2 + a^2(t)d\mathbf{x}^2$. Here t is the cosmic time and $a(t)$ is the scale factor. $H \equiv \dot{a}/a$ and the dot is a derivative with respect to the cosmic time. The slow-roll parameter ε should be small in order to have a lasting long enough Universe expansion. The nearly dS phase of inflation is given by the following expressions:

$$a(t) \approx a_0(t_s - t)^{-1/6}\, e^{-M^2(t_s-t)^2/12} , \tag{4}$$

$$H = \frac{\dot{a}}{a} = \frac{M^2(t_s - t)}{6} + \frac{1}{6(t_s - t)} + \cdots , \tag{5}$$

$$R = 6(\dot{H} + 2H^2) = \frac{M^4(t_s - t)^2}{3} - \frac{M^2}{3} + \frac{4}{3(t_s - t)^2} + \cdots , \tag{6}$$

where t_s is the time of the end of inflation when the graceful exit should happen. It is when $|\dot{H}| \sim H^2$. Parameter t_s plays a role of the integration constant. The above formulae are valid when $M(t_s - t) \gg 1$. The inflation happens when $t < t_s$, while for $t > t_s$ there is a graceful exit to the subsequent power-law expansion stage with $a(t) \propto t^{2/3}$ accompanied by small oscillations. Note that the local R^2 inflation never enters the exact dS phase with a constant R as it is not a solution to the trace Eq. (3). One can deduce from the above relations that a slow-roll inflation implies $\varepsilon \approx M^2/(6H^2) \ll 1$. This in turn means that in order to have a long-lasting inflation one must have $M \ll M_P$ making the coefficient in front of the R^2 term in action (2) very large. This explicitly demonstrates that this model allows for a cosmic inflation. On top of constructing this solution the original paper positively answered the question of stability of this configuration. Moreover it was shown that this is a past attractor meaning a favorable answer for the absence for the fine-tuning of the initial data.

Once a background solution is determined one can quite straightforwardly even though sometimes lengthy deduce various inflationary observables, such as power spectra of perturbations of different spins, corresponding spectral indices, ratio of different power spectra, higher order correlation functions, i.e. non-Gaussian correlations, etc.

Power spectra are essentially 2-point correlation functions of quantized perturbation fields of a given spin. One has to identify canonically normalized perturbative degrees of freedom. For a spin-zero scalar it is named a Mukhanov-Sasaki variable. Linear perturbations can in many relevant situations be obtained in general without resorting to a slow-roll approximation. Quantization is performed around the background and one can easily implement it as long as the background is a nearly dS stage of evolution when the slow-roll approximation is valid. The respective propagators in this approximation in the Jordan frame appear to be covariant Klein-Gordon equations for a field of given spin on a background solution. That is, for a scalar and a spin-2 field

$$(\Box - M^2)\varphi = 0 \text{ and } \left(\Box - \frac{R}{6}\right) h^{\perp}_{\mu\nu} = 0\,, \tag{7}$$

where the $\perp$ sign designates the transverse and traceless part of the metric perturbations. One should take into account that during the inflation $R \gg M$ and thus we can consider the scalar excitation to be nearly massless. For a tensor excitation the term $R/6$ restores a massless graviton as appears after careful Cristoffel symbols contractions. Since the background is not an exact dS space-time, masses of perturbations become slowly varying. For linear perturbations no mixing of different spins arises. Curvature perturbations obtained upon proper normalization play a special role in the local R^2 (or in any $f(R)$) model as they stay constant in the super-Hubble limit. This results in a nearly scale invariant power spectrum P_R of curvature perturbations.

The only new parameter of the Starobinsky model is the scalaron mass M which can be fixed by the observed characteristics of the Fourier power spectrum of curvature perturbations. In order to do so the number of e-foldings during the time of inflation should be fixed. The latter is determined as

$$N = \int_{t_i}^{t_e} H dt \approx 1/(2\varepsilon)\,, \tag{8}$$

where t_i and t_e are the initial and the end time for inflation respectively. N is known to be around 70 so that the number of e-foldings from the time when curvature perturbations cross the Hubble horizon to the end of inflation is in the range of $N_k = 50 - 60$. The latter constraint comes from Baryon Acoustic Oscillations observations. Consequently one can compute the scalar spectral index, $n_s = \frac{d\ln P_R}{d\ln k}$, which is the tilt of the power spectrum P_R, the ratio of the tensor to the scalar power spectra, $r = \frac{2P_T}{P_R}$, where 2 is due to two polarizations of the graviton, and the tensor tilt, $n_t = \frac{d\ln P_T}{d\ln k}$, at the Hubble horizon exit. In the local R^2 model these parameters become

$$n_s = 1 - \frac{2}{N}, \quad r = \frac{12}{N^2}, \quad r = -8n_t\,. \tag{9}$$

The last relation above implies the negative tensor tilt $n_t < 0$. This property has been regarded as a test for R^2 inflation.

3 AID Gravity Model

Here we sketch some steps in constructing AID gravity model in order to emphasize its importance and introduce relevant terms and notations. The model in question is described by the following action

$$S = \int d^4x\sqrt{-g}\left[\frac{M_P^2}{2}R + \frac{\lambda}{2}R\mathcal{F}(\Box/\mathcal{M}^2)R + \frac{\lambda}{2}C_{\mu\nu\rho\sigma}\mathcal{F}_C(\Box/\mathcal{M}^2)C^{\mu\nu\rho\sigma}\right]. \tag{10}$$

Here $\mathcal{F}(\Box)$ and $\mathcal{F}_C(\Box)$ are operator functions, or form-factors which depend only on the covariant d'Alembertian operator and are analytic at zero. The latter is required to have a smooth GR limit. This limit is tracked by the dimensionless parameter λ. Analyticity at zero implies that those form-factors can be written as Taylor series expansions at zero with constant coefficients.

A new mass scale $\mathcal{M}$ is a non-locality scale which tells at which energies new corrections begin to become important. Its value is of a great importance but it cannot be fixed right away. Only two things are obvious. This scale is less or equal M_P as otherwise these extra terms will not help to resolve non-renormalizability problems of GR. Also, this scale is not small as we nowhere observe it up to several TeV at least. We will moreover see in the next Sections that it should be well above the scale of inflation. Unless important, we will put it to a unit hereafter.

This action has a separate motivation without any relation to other more fundamental models like strings. Namely this action was proven [22] to be the most general action capturing the quadratic variation behavior in any general gravity model around Minkowski space-time in four dimensions. To be more precise this statement means the following. Let one start from any gravity theory built as an analytic at zero functional of the Riemann tensor, metric and covariant derivatives, and let one compute a second variation of the corresponding action around a Minkowski background. Then one will arrive to the Lagrangian for linear perturbations which equivalently can be derived from an action given by (10). This construction can be generalized to (anti-)dS space-times and to higher dimensions with minimal modifications to the above action.

We thus can study the most general propagator in a gravity theory starting from action (10). In doing so for the transverse and traceless part of the metric tensor perturbation one can get

$$S_{s-2} = \int d^4x \sqrt{-\eta}\frac{1}{2}h^{\perp}_{\mu\nu}\Box\left[M_P^2 + 6\lambda\Box\mathcal{F}_C(\Box)\right]h^{\perp\mu\nu}\,. \tag{11}$$

Here η is the determinant of the Minkowski metric $\eta_{\mu\nu}$, metric perturbations are defined as $g_{\mu\nu} = \eta_{\mu\nu} + h_{\mu\nu}$ and as before the $\perp$ sign designates its transverse and traceless part. The $\Box$ factor corresponds to a massless graviton in the spectrum. Further it is clear that if $\mathcal{F}_C(\Box) \neq 0$ and is a finite degree polynomial then it will result in Ostrogradski ghosts in the spectrum. Stelle theory corresponds to a situation when $\mathcal{F}_C(\Box) = 1$.

There is however a way to circumvent this problem if $\mathcal{F}_C(\Box)$ is a non-polynomial function. The idea is to require that the combination in brackets in action (11) is an exponent of an entire function. Namely

$$M_P^2 + 2\lambda\Box\mathcal{F}_C(\Box) = M_P^2 e^{2\omega(\Box)}\,, \tag{12}$$

where $\omega(\Box)$ is some entire function. Indeed, an entire function is regular on the whole complex plane and therefore being exponentiated it is a new entire function which has no zeros. Thus its inverse does not have poles and does not create new physical excitations. The corresponding $\mathcal{F}_C(\Box)$ is simply

$$\mathcal{F}_C(\Box) = M_P^2\frac{e^{2\omega(\Box)} - 1}{2\lambda\Box}\,. \tag{13}$$

This is a function analytic at zero as long as we require $\omega(0) = 0$.

From here it is straightforward to conclude that in order to avoid ghosts one should either stay in GR which is non-renormalizable or include an infinite derivative form-factor $\mathcal{F}_C(\Box)$. So, logically, the only possibility to construct a renormalizable and unitary gravity is to have an AID gravity.

Going further and extracting a canonical propagating scalar degree of freedom one comes for the corresponding second order action

$$S_{s-0} = \int d^4x \sqrt{-\eta}\frac{1}{2}\varphi\frac{\left[6\lambda\Box\mathcal{F}(\Box) - M_P^2\right]\left[M_P^2 + 2\lambda\mathcal{F}_C(\Box)\right]}{6\lambda\left[\mathcal{F}(\Box) + \frac{1}{3}\mathcal{F}_c\Box\right]}\varphi\,. \tag{14}$$

The denominator is the result of a resolution of constraints and normalization of the field φ. An invertability of such an operator implies that either it is again an exponent of some entire function or a constant, or it is identically zero. The latter means that we do not have a propagating scalar at all. A simple algebra in the case when a scalar is present and all the conditions of invertability of operators are taken into account reveals that

$$\mathcal{F}(\Box) = M_P^2\frac{e^{2\omega(\Box)}(\Box/M^2 - 1) + 1}{6\lambda\Box}\,. \tag{15}$$

Here parameter M appears exactly like in the local R^2 action such that $\lambda\mathcal{F}(0) = M_P^2/(6M^2)$. Note that an absence of a scalar corresponds to $M = \infty$. We thus see that $\mathcal{F}(\Box)$ is also an infinite derivative form-factor determined by the same entire function $\omega(\Box)$.

The described setup guarantees that action (10) describes perturbatively around Minkowski background in four dimensions only a standard graviton with two polarizations and perhaps a scalar field of mass M. Thus it is ghost-free. Moreover, given a potential power of form-factors in suppressing loop integrals for large momenta one can show that this theory can be easily made power-counting renormalizable. While determining the relevant function $\omega(\Box)$ is not a trivial question in general some of its properties are established during various studies. The most important for the construction of quantum gravity is that function $e^{\omega(\Box)}$ should have a polynomial asymptotics along the positive real axis. This guarantees an absence of the strong coupling regime in the model. Such an example was constructed in [23].

At this stage we can put forward a proposal that action (10) can be a sufficient generalization of GR to build a self-consistent quantum gravity theory. Then we can depart from an initial motivation of describing only linear fluctuations around Minkowski background and pose a question which other solutions can be present. We are going to demonstrate that in particular Starobinsky inflation can be an exact solution in this model. We note that inflation is happening at energies well below the Planck scale. Thus inflation will not help fixing a really high energy behavior of function $\omega(\Box)$ but it can be sensitive to the low-energy expansion of the form-factor.

An interesting phenomenon arises for any background solution different from Minkowski space-time for action (10). As long as we fix form-factors to have a good healthy theory in Minkowski background we lose this luxury around any other solution. Corresponding new excitations however are unusual and can be shown to have complex masses squared for reasonable parameter value ranges and they come in complex conjugate pairs in the spectrum of perturbations.

To illustrate this statement we can consider a scalar field example as follows

$$S = \int d^4x \left(\frac{1}{2}\varphi(\Box - m^2)f(\Box)\varphi - W(\varphi)\right), \tag{16}$$

where potential $W(\varphi)$ contains terms of degree three and higher. In a trivial vacuum $\varphi = 0$ we have only one excitation of mass m as long as $f(\Box) = e^{2\sigma(\Box)}$ for some entire function $\sigma(\Box)$. However, for any other background (not obligatory a constant vacuum) we have for excitations

$$S = \int d^4x \frac{1}{2}\chi\left[(\Box - m^2)f(\Box) - W''(\varphi)\right]\chi\,. \tag{17}$$

Here it is obvious that an expression in brackets cannot be again an exponent of some entire function except a situation when $W''(\varphi) = 0$ on the background solution. This however can virtually never be as long as the background is not a constant. If $W''(\varphi)$ is approximately a constant we can rewrite an expression in brackets using the Weierstrass product valid for entire functions

$$(\Box - m^2)f(\Box) - W''(\varphi) = e^{2\tilde{\sigma}(\Box)}\prod_i(\Box - \mu_i^2)^{n_i}\,,$$

where μ_i^2 are roots satisfying $(\mu_i^2 - m^2)f(\mu_i^2) - W''(\varphi) = 0$ and $\tilde{\sigma}(\Box)$ is yet another entire function. n_i are root multiplicities which we take to be a unit for simplicity. A solution for χ becomes

$$\chi = \sum_i \chi_i \text{ where } (\Box - \mu_i^2)\chi_i = 0\,. \tag{18}$$

Having in mind that an original action contains only physical parameters which are expected to be real, roots μ_i^2 are either real or complex conjugate. Half of new extra real poles will become definitely ghosts according to the Ostrogradski statement. These are better to be avoided. Complex conjugate masses are more subtle though. Consider some pair of complex conjugate roots μ^2 and $\bar{\mu}^2$. A Lagrangian reproducing their propagators is

$$L = \frac{1}{2}\chi(\Box - \mu^2)\chi + \frac{1}{2}\bar{\chi}(\Box - \bar{\mu}^2)\bar{\chi}\,.$$

Here bar indicates complex conjugation. It is not a standard $U(1)$ symmetric complex field but the combination can be real as long as solutions for χ and $\bar{\chi}$ have precisely complex conjugate initial conditions. An interesting point is that in an attempt of rewriting the latter Lagrangian in pure real quantities one yields a non-diagonal quadratic form. That is, we cannot have either a clear definition of mass or kinetic energy. As a consequence, we cannot certainly attribute a label of a tachyon or a ghost to such degrees of freedom. They are somewhat special and we name them Background Induced States (BIS) [24].

An important fact about BIS-s is that they actually can be stable on a non-trivial background. Indeed, first of all it is obvious that complex masses will generate unstable excitations in a flat space-time. An imaginary part of a mass will contribute to a real positive exponential with respect to time in one of the solutions. This gets changed in a dS background which is of a primary importance for inflation. Given that the $\Box$ operator is covariant and the dS space-time is characterized by the Hubble rate H we get the following parabola-shaped region of stable masses

$$\mathrm{Im}(\mu^2)^2 < 9H^2\mathrm{Re}(\mu^2)\,. \tag{19}$$

We may not be able to find roots at first, but we may be able to invert a problem and study which values of roots are good for some given effect.

The same happens for perturbations around some background different from Minkowski space-time in action (10). While it is difficult to write a general second order variation it is possible to do so around a nearly (anti-)dS space-times. Note that an exact (anti-)dS cannot be a solution in this case unless an explicit cosmological constant is added to the action but then in contrast Minkowski space-time would not be an exact solution anymore. Also one can compute a second order variation around conformally flat solutions which satisfy (3). The latter will be shown to be indeed exact background solutions up to two extra conditions on form-factors.

These new states will be shown to play an interesting role during inflation and provide possible observational signatures.

4 Non-local Starobinsky Inflation

Starobinsky inflation can be exactly embedded in the AID gravity as a background solution. Hence the name non-local Starobinsky inflation. Skipping the details which can be found in [20] we note that the embedding can happen thanks to the fact that equations of motion for action (10) admit any conformally flat solution of a local R^2 model which features Eq. (3) as long as two algebraic conditions on the form-factor $\mathcal{F}(\Box)$ are satisfied. Namely

$$\mathcal{F}'(M^2) = 0 \text{ and } 6\lambda M^2 \mathcal{F}(M^2) = M_P^2 \,. \tag{20}$$

Here prime is a derivative with respect to an argument. For the form-factor (15) this implies $\mathcal{F}(0) = \mathcal{F}(M^2)$ and $\omega(M^2) = 0$. This is essentially new information about the theory compared to the construction in the previous Section.

Therefore, we can explicitly and exactly embed the Starobinsky solution in AID gravity. Since it is the same solution, the background retains all the properties it had in a local case. It is obvious, however, that perturbations can and will be different as the underlying model is very much different. Namely equations for free perturbations get modified as follows

$$\frac{\mathcal{W}(\Box)}{\mathcal{F}(\Box) + \frac{1}{3}\mathcal{F}_C(\Box)}(\Box - M^2)\varphi = 0 \text{ and } \mathcal{P}(\Box)\left(\Box - \frac{R}{6}\right) h^{\perp}_{\mu\nu} = 0\,. \tag{21}$$

Here $\mathcal{W}(\Box)$ operator in a nearly dS limit is given by

$$\mathcal{W}(\Box) = 3\mathcal{F}(\Box) + (R + 3M^2)\frac{\mathcal{F}(\Box) - \mathcal{F}(M^2)}{\Box - M^2} + \mathcal{F}_C(\Box)\,,$$

and it is an entire function as one can see no singularities using a Taylor expansion around $\Box = M^2$. $\mathcal{P}(\Box)$ operator in a nearly dS limit is given by

$$\mathcal{P}(\Box) = M_P^2 + 2\lambda \mathcal{F}(0) R + 2\lambda \left(\Box - \frac{R}{3}\right) \mathcal{F}_C\left(\Box + \frac{R}{3}\right),$$

where the first term containing M_P^2 is suppressed by the second one on the Starobinsky solution during inflation.

Newly appeared operators $\mathcal{W}(\Box)$ and $\mathcal{P}(\Box)$ give rise to BIS-s. For the parameter values in the system there are no new real poles. Those states should not be simply ignored. There are two ways though to argue that in principle they can be neglected. The first one is to see that their dynamics is heavily suppressed due to the fact that all the d'Alembertian operators entering those operators are suppressed by the non-locality scale $\mathcal{M}$. In the meantime we understand that the $\mathcal{M} \to \infty$ limit restores a standard Starobinsky inflation which simply works. We thus conclude that in general a hierarchy $M \ll H \ll \mathcal{M}$ should be realized and therefore can expect those new excitations to be irrelevant. The second way is to draw an expectation that those excitations are classically stable and thus exponentially decaying following the arguments in the previous Section and condition (19). This is something one cannot easily trace from the shape of a form-factor though as finding roots of even a given function is not an easy task. However, if it is the case then the respective dynamics of those excitations will not affect the power-spectra much. The last statement requires clarifications which we will skip until Sect. 5 and subsequent analysis. There models generalizing the present consideration and using Einstein frame will be discussed. Obviously the second way is the more systematic approach as it does not depend on initial conditions for BIS-s.

Continuing by presuming that the effect of new excitations is negligible for the power-spectra we can see that the new factor in the first equation in (21) will give a numeric factor of 3. This is clear from an already obtained condition on $\mathcal{F}(\Box)$ that $\mathcal{F}(M^2) = \mathcal{F}(0)$. This factor of 3 contributes to the normalization of the scalar excitation and respective scalar power-spectrum exactly as in a local Starobinsky inflation. Thus the scalar power-spectrum will remain the same

$$P_R = P_{R\,local}\,. \tag{22}$$

Obviously, the scalar spectral index remains intact either.

However, the situation is a bit less trivial for the operator $\mathcal{P}(\Box)$ as upon evaluation at $\Box = R/6$, the value dictated by the massless graviton pole factor, a non-trivial factor $\mathcal{P}(R/6)$ is retained. We name it non-trivial as it would be a unit in the local case without a Weyl tensor square term. This produces a differently normalized tensor power-spectrum

$$P_T = \mathcal{P}(R/6)P_{T\,local}\,. \tag{23}$$

The latter in turn results in a modified value of the tensor to scalar ratio r which becomes

$$r = \mathcal{P}(R/6)r_{local}\,. \tag{24}$$

Obviously, the tensor spectral index remains intact analogously to the scalar one.

Thus in the leading slow-roll approximation untouched scalar power-spectrum and both scalar and tensor spectral indices in a combination with a modified tensor power-spectrum together with a modified tensor to scalar ratio can be signatures of the non-local Starobinsky inflation. Note that in the above results one can restore the non-locality scale in order to emphasize the underlying physics. This scale would enter the argument of the $\mathcal{F}_C(\Box)$ form-factor inside of the operator $\mathcal{P}(\Box)$.

One can try to modify the model further to make operators $\mathcal{W}(\Box)$ and $\mathcal{P}(\Box)$ to generate no new solutions exactly. While not in principle impossible, coming with an explicit construction appears to be a very difficult problem. One can somewhat alleviate a problem for tensor excitations by assuming that $\mathcal{F}_C(\Box)$ would also depend on the scalar curvature such that $\mathcal{F}_C(R, \Box)$ [25]. With this one can achieve that there will be only graviton on any constant curvature background which will be a resolution for an exact (anti-)dS space-time. However this approach has two important issues. First, it will work only around an exact dS and will not help around an approximate dS already. That is, it is not an absolute cure for an inflation which is a nearly dS background, and not an exact one. Second, analogous change for the $\mathcal{F}(\Box)$ form-factor will modify the Starobinsky solution on the background in the first place. This will create further difficulties in considering an inflationary solution. In general modifying action (10) such that it will be ghost-free around any background is an important, interesting and an extremely tough task to be done.

Third and higher variations of the model Lagrangian reveal non-Gaussian corrections which in the case of three-point correlation functions are characterized by the non-linearity parameter f_{NL} which is proportional to the bi-spectrum $A(k_1, k_2, k_3)$ as $(\sum_i k_i^2) f_{NL} \sim A(k_1, k_2, k_3)$ with k_i being the comoving wave-vectors. Corresponding computation is fairly long and involved [26] as it requires to use the next to leading expansion in the slow-roll parameter. However, many technical aspects remain as they are in standard models mainly because the background still has a slow-roll behavior. Skipping all the computational details we lay down the results for the non-linearity parameter for the triple scalar non-Gaussianities

$$\begin{aligned}
f_{NL}^{sq} &\approx \frac{5}{12}(1-n_s) - 23\varepsilon^2\left[e^{\omega\left(\frac{R}{4\mathcal{M}^2}\right)} - 1\right] - \frac{4R}{\mathcal{M}^2}\varepsilon^3 e^{\omega\left(\frac{R}{4\mathcal{M}^2}\right)}\omega'\left(\frac{R}{4\mathcal{M}_s^2}\right), \\
f_{NL}^{equiv} &\approx \frac{5}{12}(1-n_s) - 49\varepsilon^2\left[e^{\omega\left(\frac{R}{4\mathcal{M}^2}\right)} - 1\right] - \frac{9R}{\mathcal{M}^2}\varepsilon^3 e^{\omega\left(\frac{R}{4\mathcal{M}^2}\right)}\omega'\left(\frac{R}{4\mathcal{M}_s^2}\right), \\
f_{NL}^{ortho} &\approx \frac{5}{12}(1-n_s) - 43\varepsilon^2\left[e^{\omega\left(\frac{R}{4\mathcal{M}^2}\right)} - 1\right] - \frac{22R}{3\mathcal{M}^2}\varepsilon^3 e^{\omega\left(\frac{R}{4\mathcal{M}^2}\right)}\omega'\left(\frac{R}{4\mathcal{M}_s^2}\right),
\end{aligned} \tag{25}$$

where prime is a derivative with respect to an argument and we have restored the non-locality scale for more physical clarity.

Here the three limits are: squeezed with $k_1 = 0$ and $k_2 = k_3$, equilateral with $k_1 = k_2 = k_3$, and orthogonal with $k_1 = k_2 = k_3/2$. We see that the so-called Maldacena consistency relation [27] is restored as long as $\omega(\Box) \equiv 0$ or analogously when $\mathcal{M} \to \infty$. However, all the limits get corrections in our case. The primary reason for this is the new perturbative degrees of freedom hidden in the operator $\mathcal{W}(\Box)$. An analogous effect even though not computed yet is expected when tensor or mixed scalar-tensor non-Gaussianities are considered since the operator $\mathcal{P}(\Box)$ will give rise to new perturbative degrees of freedom as well.

We note here that the underlying computations were performed around an exact Starobinsky solution which allowed to use on top of the slow-roll approximation relation (3) and conditions (20). This simplified greatly the technical part leading to a possibility to account all values of $\mathcal{M}$ irregardless of the relation $H \ll \mathcal{M}$ and even compute ε^3 corrections. Moreover, one can analogously compute next in slow-roll correction to the tilt of the tensor power-spectrum leading to a potentially positive value for n_t as long as $\mathcal{M}$ is larger but not much larger than M. The latter does not contradict a presence for the stability condition (19) but $H \ll \mathcal{M}$ relation is more anticipated theoretically.

With this in mind we remind the hierarchy of scales in the problem once again

$$M \ll H \overset{?}{\ll} \mathcal{M} \lesssim M_P\,, \tag{26}$$

where we put a question mark over the second relation as it in principle can be changed to $<$ without an essential trouble and with more possible observational signatures. Note also that the first relation is true in the Jordan frame while in the Einstein frame we have $M = H/2$.

5 Beyond Starobinsky Inflation in AID Gravity

So far we have considered Starobinsky inflation embedding in the AID gravity action (10), corresponding properties and observational signatures. This study was possible greatly due to a possibility of an exact embedding and subsequent significant simplifications despite the fact that the model is heavily involved technically. However, there are also limitations in physics which can be studied. Namely, thinking about some standard inflationary model one may have in mind a picture with a nearly flat scalar field potential. This is possible thanks to the trick of a conformal transformation of the metric $g_{\mu\nu} = e^{2\Omega(x)}\hat{g}_{\mu\nu}$ with some function $\Omega(x)$. This can decouple the Einstein gravity and the scalar degree of freedom, hence leading to the Einstein frame, in any $f(R)$ gravity. Upon such a transformation the local Starobinsky model in particular leads to a standard scalar field minimally coupled to gravity with a potential

$$V(\phi) = \frac{3}{4}M^2 M_P^2 \left(1 - e^{2\varphi/(\sqrt{6}M_P)}\right)^2\,. \tag{27}$$

Note that the mass term is inside the above potential.

Here two comments are in order. The first one is that in the case of higher derivatives and the Weyl tensor present in action (10) doing a conformal transform is either useless or mathematically incorrect. Indeed, the d'Alembertian operator is not conformally invariant. That means that its transform even though possible to write will lead to higher and higher powers of the scalar field with extra derivative operators in all instances. Moreover, a would be reasonable transformation which can replace the term $R\mathcal{F}(\Box)R$ with a scalar field structure with derivatives would need to use an inverse of the operator $\mathcal{F}(\Box)$. While we know how to construct an invertible operator with infinitely many derivatives, we have already learned that it is not $\mathcal{F}(\Box)$ but rather some further combination which must be an exponent of an entire function in order to avoid ghosts in the Jordan frame. This does not exclude that some modification of either the model or the transform may lead to interesting results. However, even if we succeed in some reformulation of the action (10) we will end up with some perhaps more intuitive Lagrangian which is still realizing the same original Starobinsky solution. A wish to break free from this limitation leads to the second comment.

The second comment is that it looks much simpler and more transparent to study different models of inflation in the language of the Einstein frame. That is, one can for example modify the scalar potential with easy and thus study inflationary backgrounds different from Starobinsky solution. However, following the previous comment we cannot easily connect Jordan to Einstein frames in case of higher derivatives present in the model. This prompts to consider a new construction without appealing to the conformal transformation much. To do so we emphasize that the main characteristic feature appeared in the higher derivative model in the Jordan frame is modified higher derivative equations for perturbations (21) which lead to new perturbative degrees of freedom or BIS-s. This phenomenon can be re-created in an Einstein frame model when a scalar field has a higher derivative kinetic term for example. The following action can be considered [26]

$$S = \int d^4x\sqrt{-g}\left[\frac{M_P^2}{2}R + \frac{1}{2}\varphi(\Box - M^2)\mathcal{F}(\Box/\mathcal{M}^2)^2\varphi - W(\varphi)\right], \tag{28}$$

where $W(\varphi)$ contains only interaction terms with the powers of field grater than two. As usual we assume that the form-factor is a function analytic at zero to have a well-defined IR limit, that is $\mathcal{F}(\Box) = \sum_{n\geq 0} f_n\Box^n$ and $f_0 = 1$ to preserve a canonical normalization of the scalar field. Again we put $\mathcal{M} = 1$ in most formulae. Minkowski vacuum corresponds to $\varphi = 0$ and it will have no extra excitations as long as $\mathcal{F}(\Box) = e^{2\sigma(\Box)}$ where $\sigma(\Box)$ is an entire function. Note that here the mass term is separate from the interaction potential while the full potential which is the minus scalar field Lagrangian evaluated for $\Box = 0$ is $V = \frac{1}{2}M^2\varphi^2 + W(\varphi)$. The mass term is written separately as it participates in the propagator construction around the Minkowski vacuum and should also be scaled by the form-factor as long as the kinetic term is. However, even if we adjust $V(\varphi)$ to be the Starobinsky potential the system will not have exactly the same solution. Another note is that the above invertible operator $\mathcal{F}(\Box)$ allows us to canonically normalize the field φ as follows

$$\chi = \mathcal{F}(\Box)\varphi\,,\ \varphi = \mathcal{F}(\Box)^{-1}\chi \equiv \mathcal{G}(\Box)\chi \equiv \tilde{\chi}\,, \tag{29}$$

and we can rewrite action (28) as

$$S = \int d^4x \sqrt{-g} \left[\frac{M_P^2}{2} R + \frac{1}{2} \chi \left(\Box - m^2 \right) \chi - W(\tilde{\chi}) \right] . \tag{30}$$

This change of variable is correct because it does not introduce new degrees of freedom. When the non-localities are now in the interaction potential and the kinetic term is kept canonical a similarity with SFT is even more apparent.

We stress that the model formulated above is not a transformation of AID gravity given by action (10). It is a different model which we will name AID scalar-tensor theory and which features the same property of higher derivatives and as we will see gives rise to BIS-s. Moreover it offers a possibility to study different shapes of the potentials. Also, even though we could we did not introduce an analog of the Weyl tensor dependent term. Its presence if it is quadratic in the Weyl tensor will not change any conformally flat background solution. Neither it will significantly affect an analysis of scalar perturbations on which we will focus the most.

6 Inflation in AID Scalar-Tensor Theories

A scalar-tensor model (30) presented in the previous Section resembles a standard model of inflation despite having higher derivatives. However, these derivative operators make a lot of difference and forbid in general deriving an analog of Mukhanov-Sasaki equation unless an extra approximation $M \ll \mathcal{M}$ is taken into consideration. While such a relation is not unnatural, we recall that in the case of AID gravity we could circumvent its direct use thanks to presence of a simplifying ansatz (3) valid on the solution of interest. In fact we would be also forced to use some extra condition given that there is no simplifying ansatz which would be the case if model (10) would contain higher curvature corrections.

As anticipated we will treat model (30) perturbatively using the slow-roll approximation. Two relevant slow-roll parameters are

$$\epsilon = -\frac{\dot{H}}{H^2}, \quad \eta = M_P^2 \frac{\tilde{V}''}{\tilde{V}} \bigg|_{\mathcal{G}(\Box)=1} . \tag{31}$$

Here we use another notation for the first slow-roll parameter because technically the Hubble parameter H is different in Jordan and Einstein frames. Prime here is the derivative with respect to an argument and $\tilde{V} = \frac{1}{2} M^2 \chi^2 + W(\tilde{\chi})$. To discuss the background solution it is enough to say that one can have nearly the same solution as in the local case because all the higher derivative corrections will be suppressed be the slow-roll parameters. It is not necessary yet to use a relation for masses.

It is however crucial to impose the mass hierarchy when deriving the perturbation equations because perturbations do not have to follow the slow-roll. Thus when derivatives act on perturbation fields we cannot say much about their behavior and a relation for masses helps to close the system of equations. Accounting the assumption

$M \ll \mathcal{M}$ we get the following single equation for perturbations

$$\left(\partial_0^2 + 2\mathcal{H}\partial_0 - \partial_i\partial^i\right)\frac{v}{a} + (3\eta - 6\epsilon)\,\mathcal{H}^2\frac{v}{a} + \left(3\mathcal{H}^2\eta - m^2a^2\right)\left(\mathcal{G}(\Box)^2 - 1\right)\frac{v}{a} = 0\,, \tag{32}$$

which appears to be a non-local Mukhanov-Sasaki equation. Here we use conformal time τ connected to the cosmic time t as $dt = a(\tau)d\tau$ and also $\mathcal{H} \equiv a'/a$ where hereafter prime or ∂_0 is a derivative with respect to the conformal time τ. In the local case when $\mathcal{G}(\Box) = 1$ this equation reduces to a familiar local Mukhanov-Sasaki equation

$$v'' - \partial_i\partial^i v - \frac{\nu^2 - \frac{1}{4}}{\tau^2}v = 0\,, \tag{33}$$

where $\nu^2 = \frac{9}{4} + 9\epsilon - 3\eta$. The quantity v in the local single-field inflation scenario is the canonically normalized variable related to curvature perturbation ζ as $v = z\zeta$ with $z = a\chi'/\mathcal{H} \approx a\sqrt{2\epsilon}$.

We note here following considerations in Sect. 3 that Eq. (32) even though being written as an equation for one function physics wise represents many degrees of freedom. In order to treat this equation using the Weierstrass product we have to rewrite it using only one and the same operator everywhere, say the d'Alembertian, and assure that all the coefficients are constants. Indeed, the equation in question can be rewritten as

$$\left[a^2\Box - (3\eta - 6\epsilon)\mathcal{H}^2 - \left(3\mathcal{H}^2\eta - m^2a^2\right)\left(\mathcal{G}(\Box/\mathcal{M}^2)^2 - 1\right)\right]\frac{v}{a} = 0\,. \tag{34}$$

In the nearly dS regime a relation $a^2 \sim \mathcal{H}^2 \sim 1/\tau^2$ allows to factor out the time-dependent coefficients a^2 and $\mathcal{H}^2$. Under an extra assumption that slow-roll parameters are constants we can resort to the Weierstrass product decomposition. We remind that we cannot just ignore higher derivatives as we do not have a slow-roll suppression for perturbations.

Following the recipe in Sect. 3 and decomposing the latter equation into factors $(\Box - \mu_i^2)$ and finding corresponding excitations v_i we get that each of it satisfies a Klein-Gordon equation in the dS space-time. Thus solutions are Hankel function of order $\nu_i = \sqrt{\frac{9}{4} - \frac{\mu_i^2}{H^2}}$. For our range of parameters we can easily judge that all masses but one are complex. The only real mass is similar to the standard massless inflaton with a small correction of order of $M/\mathcal{M}$ so following the same Eq. (33) as in the local case. We enumerate this excitation as v_0. This difference in mass even a slight one compared to the local case is in some contrast to the AID gravity model (10) where the mass of the scalaron got preserved following Eq. (21). The complex masses are not obligatory in the domain of stability. Intuitively we expect that if they are classically unstable then the system maybe in danger. Since we cannot easily tell where the masses fall on the complex plane, we promote this instead as a criterion for constraining the form-factor.

Quantization requires extra attention as we are going to qunatize not only a real mass excitations $v_0(\tau, x)$ but also complex conjugate masses pairs $v_i(\tau, x)$, $v_i^*(\tau, x)$. The following scheme is used

$$\hat{v}_0(\tau, x) = \int \frac{d^3k}{(2\pi)^{3/2}} \left\{\hat{a}_k^0 v_{0,k}(\tau) e^{i\mathbf{kx}} + \hat{a}_k^{0\dagger} v_{0,k}^*(\tau) e^{-i\mathbf{kx}}\right\}, \tag{35}$$

$$\hat{v}_i(\tau, x) = \int \frac{d^3k}{(2\pi)^{3/2}} \left\{\hat{a}_k^i v_{i,k}(\tau) e^{i\mathbf{kx}} + \hat{b}_k^{i\dagger} u_{i,k}(\tau) e^{-i\mathbf{kx}}\right\}, \tag{36}$$

$$\hat{v}_i^*(\tau, x) = \int \frac{d^3k}{(2\pi)^{3/2}} \left\{\hat{b}_k^i u_{i,k}^*(\tau) e^{i\mathbf{kx}} + \hat{a}_k^{i\dagger} v_{i,k}^*(\tau) e^{-i\mathbf{kx}}\right\}. \tag{37}$$

Here $v_{0,k}$, $v_{i,k}$, $u_{i,k}$ are the corresponding mode functions, and repeated indices do not imply summation. The commutation relations for the creation and annihilation operators of the 0-th and i-th excitations are

$$\begin{aligned}
&\left[\hat{a}_k^0, \hat{a}_{k'}^{0\dagger}\right] = \delta^{(3)}(k - k'), \\
&\left[\hat{a}_k^i, \hat{a}_{k'}^{j\dagger}\right] = \delta^{(3)}(k - k')\delta^{ij}, \quad \left[\hat{b}_k^i, \hat{b}_{k'}^{j\dagger}\right] = \delta^{(3)}(k - k')\delta^{ij},
\end{aligned} \tag{38}$$

with all other commutation relations vanishing. In this scheme we take care of the fact that in the limit when $\mu_i = \mu_i^*$, i.e. when the mass is real, we restore a quantization procedure for a standard $U(1)$ symmetric complex scalar field as described in details in [28]. Solutions to the mode functions $v_{0,k}(\tau)$, $v_{i,k}(\tau)$ and $u_{i,k}(\tau)$ can be written as

$$v_{0,k}(\tau) = \sqrt{-\tau}\left[\alpha_0 H_{\nu_0}^{(1)}(-k\tau) + \beta_0 H_{\nu_0}^{(2)}(-k\tau)\right], \tag{39}$$

$$v_{i,k}(\tau) = \sqrt{-\tau}\left[\alpha_i H_{\nu_i}^{(1)}(-k\tau) + \beta_i H_{\nu_i}^{(2)}(-k\tau)\right], \tag{40}$$

$$u_{i,k}(\tau) = \sqrt{-\tau}\left[\tilde{\alpha}_i H_{\nu_i}^{(1)}(-k\tau) + \tilde{\beta}_i H_{\nu_i}^{(2)}(-k\tau)\right]. \tag{41}$$

Here, $H_\nu^{(1,2)}(-k\tau)$ are Hankel functions of first and second kind respectively and α, β, α_i, β_i, $\tilde{\alpha}_i$ and $\tilde{\beta}_i$ are constants which can be determined by choosing the initial states of corresponding excitations. Assuming all of the excitations to start in Bunch-Davies initial state one yields

$$\alpha_0 = \alpha_i = e^{i\frac{\pi}{2}(\nu+\frac{1}{2})}\sqrt{\frac{\pi}{4}}, \quad \tilde{\beta}_i = e^{-i\frac{\pi}{2}(\nu+\frac{1}{2})}\sqrt{\frac{\pi}{4}}, \quad \beta_0 = \beta_i = \tilde{\alpha}_i = 0. \tag{42}$$

The total field $\hat{v}(\tau, x)$ can be expressed as

$$\hat{v}(\tau, x) = \hat{v}_0(\tau, x) + \sum_{i=1}^{N}\left(\hat{v}_i(\tau, x) + \hat{v}_i^*(\tau, x)\right), \tag{43}$$

where the summation is over N complex conjugate pairs of excitations with non-trivial initial conditions. The corresponding power-spectrum for $\hat{v}(\tau, x)$ according to commutation relations (38) is

$$P_v(k) = P_{v_0}(k) + \sum_{i=1}^{N} \left(P_{v_i}(k) + P_{v_i^*}(k)\right)$$
$$= v_{0,k}(\tau)v_{0,k}^*(\tau) + \sum_{i=1}^{N} \left(v_{i,k}(\tau)v_{i,k}^*(\tau) + u_{i,k}(\tau)u_{i,k}^*(\tau)\right) . \qquad (44)$$

Recalling that $\hat{v}_0(\tau, x) = z\zeta_0(\tau, x)$, $\hat{v}_i(\tau, x) = z\hat{\zeta}_i(\tau, x)$, and $\hat{v}_{i,k}^*(\tau, x) = z\hat{\zeta}_i^*(\tau, x)$ and using the relation $z \approx a\sqrt{2\epsilon}$ we can write the power-spectra of the Fourier modes $\zeta_{0,k}$, $\zeta_{i,k}$ and $\zeta_{i,k}^*$ as

$$\Delta_{\zeta_0}(k) = \frac{k^3}{2\pi^2}\frac{1}{z^2}P_{v,0}(k) = \frac{k^3}{2\pi^2}\frac{1}{2M_p^2a^2\epsilon}(-\tau)\frac{\pi}{4}|H_{\nu_0}^{(1)}(-k\tau)|^2 , \qquad (45)$$

$$\Delta_{\zeta_i}(k) = \frac{k^3}{2\pi^2}\frac{1}{z^2}P_{v_i}(k) = \frac{k^3}{2\pi^2}\frac{1}{2M_p^2a^2\epsilon}(-\tau)\frac{\pi}{4}|H_{\nu_i}^{(1)}(-k\tau)|^2 e^{i(\nu-\nu^*)} , \qquad (46)$$

$$\Delta_{\zeta_i^*}(k) = \frac{k^3}{2\pi^2}\frac{1}{z^2}P_{v_i^*}(k) = \frac{k^3}{2\pi^2}\frac{1}{2M_p^2a^2\epsilon}(-\tau)\frac{\pi}{4}|H_{\nu_i^*}^{(1)}(-k\tau)|^2 e^{i(\nu^*-\nu)} . \qquad (47)$$

The total scalar curvature power-spectrum is

$$\Delta_S = \Delta_{\zeta_0}(k) + \sum_{i=1}^{N} \left(\Delta_{\zeta_i}(k) + \Delta_{\zeta_i^*}(k)\right) . \qquad (48)$$

The power-spectrum of massless excitation (45) does not evolve after the horizon crossing which can be seen from the super-horizon behavior of the massless excitation using $\lim_{k\tau\to 0} H_{\nu_0}^{(1)}(-k\tau) \sim (-k\tau)^{\nu_0}$. To ensure that the massive excitations do not grow after the horizon crossing one should impose a stability constraint (19) on the complex masses. In terms of Hankel function order ν_i in the super-horizon limit where the behavior is $(-k\tau)^{3/2-\nu_i}$ this implies $Re(\nu_i) \leq 3/2$.

The scale dependence of the scalar power-spectrum can be easily written as

$$n_s - 1 = \frac{d\ln\Delta_S}{d\ln k} = \frac{d\ln\Delta_{\zeta_0}(k)}{d\ln k}$$
$$+ \frac{d}{d\ln k}\ln\left(1 + \frac{1}{\Delta_{\zeta_0}(k)}\sum_{i=1}^{N}\left(\Delta_{\zeta_i}(k) + \Delta_{\zeta_i^*}(k)\right)\right) . \qquad (49)$$

When μ_i are inside the mass parabola than $\frac{\Delta_{\zeta_i}(k)}{\Delta_{\zeta_0}(k)} \sim \frac{\Delta_{\zeta_i^*}(k)}{\Delta_{\zeta_0}(k)} \ll 1$ and we restore $n_s - 1 \simeq \frac{d \ln \Delta_{\zeta_0}(k)}{d \ln k} = 2\eta - 6\epsilon$. This can be shown by direct computation for some finite number N when masses μ_i are fixed to be inside the stability region.

In general it is very plausible that if masses are inside the mass parabola they do not affect the scalar power-spectrum or the spectral index. However, they can provide some shift compared to the value in a local model. To see this on an example of the spectral index we can compute it accounting the new corrections. An expression in terms of orders of Hankel functions and the number N_k of e-foldings during which the inflation was happening is

$$\begin{aligned} &n_s - 1 \approx 2\eta - 6\epsilon \\ &+2e^{-2\nu_0 N_k} \sum_{i=1}^{N} \left((\nu_0 - \nu_i) e^{2\nu_i N_k} e^{i(\nu_i - \nu_i^*)} + (\nu_0 - \nu_i^*) e^{2\nu_i^* N_k} e^{-i(\nu_i - \nu_i^*)} \right) . \end{aligned} \tag{50}$$

The correction can be shown to be positive and very small, being nearly zero if an imaginary part of a mass of the new excitation is near zero, i.e. definitely inside the stability domain. The correction is growing slowly and monotonically when the imaginary part of the mass is tending towards the boundary of the stability region. The correction grows steeper when this boundary is approached.

One can naturally ask whether this can explain the new ACT data. In principle yes but in order to achieve this we either should have a complex masses pair having a mass very close to the stability region or we should accumulate an effect of quite many complex mass pairs which then may not be that well tuned. While either scenario is not excluded we are quite cautious at the moment about new ACT data as the confidence level of just 2σ maybe be a bit premature to be really worried about.

Going further scalar non-Gaussianities can be computed in this setup. Computations are rather straightforward even though lengthy and tedious. Moreover, numerical evaluation is required to produce actual numbers. BIS-s play important role in amplifying non-Gaussianities and also lead to a violation of the Maldacena consistency relation. A somewhat surprising result here is that $f_{NL}^{squeezed}$ gets amplified to have large values already the half way to the border of the stability region for new excitations. Moreover, other limits lead to no significant amplification. This result despite looking different from the conclusion described earlier in AID gravity model can be explained by the fact that here we strictly work in a limit $M \ll \mathcal{M}$. We thus cannot wake up and get contributions of relevant operators and have only the squeezed limit different from a local scenario.

7 Conclusions and Outlook

In this note we have overviewed the non-local Starobinsky inflation which is an exact analytic embedding of an original Starobinsky model into an SFT inspired non-local gravity action. This gravity model having infinite-derivative operators paves the way towards a quantum gravity construction. Inflationary observables get modified and one can readily obtain deviations in the tensor to scalar ratio. Moreover an intriguing feature of having BIS-s in this model can in fact be a useful tool for generating corrections to other observable parameters such as scalar spectrum tilt n_s and generate large non-Gaussianities. New data from the ACT indicate that some corrections to the value of n_s compared to a local Starobinsky model can be present. Those can be explained in the presented framework but we stress that currently new data are lacking of a good confidence level benchmark.

Further a scalar-tensor variant of an AID inflationary model was discussed focusing on its greater generality with respect to the form of the potential. This model shares the same crucial feature of generating extra perturbative degrees of freedom and thus modifies results of local models. New excitations can be contained by requiring their classical stability and thus they do not play a destabilizing role. Note that in the present note we left aside questions regarding post-inflationary and induced gravitational waves in the framework of AID gravity which were also recently studied [24, 29]. An important open question in this class of models is a construction of a transformation mapping Jordan to Einstein frames.

Another open and very difficult question is to understand whether such models can avoid extra perturbative states around different backgrounds. Here we would like to reiterate that models of gravity with infinite number of derivatives is the only way to have a ghost-free propagator around Minkowski space-time beyond GR. Having this in mind an importance of such models is obvious. However, presence of BIS-s is not fully understood and avoiding them may solve many naive questions of stability in this framework. On the other hand, as it was demonstrated in this note such states can be benign and even can generate interesting observable signatures.

Other class of open questions are quantum field theory related and non-perturbative aspects of these constructions. Corresponding studies including a study of a 1-loop effective action, scattering amplitudes and a construction of a non-perturbative propagator are under investigation currently. There is a hope to answer these and many other questions in forthcoming publications.

Acknowledgements Primarily I would like to express my immeasurable gratitude to Alexei Starobinsky with whom I had the greatest pleasure to collaborate on several very interesting papers on the subject and who inspired and supported me throughout my career.
Also I would like to thank Slava Mukhanov on discussion about recent ACT data and their interpretation, and Abhishek Naskar for verification of some of the computations in this note.

References

1. A.H. Guth, The inflationary universe: a possible solution to the horizon and flatness problems. Phys. Rev. D **23**, 347–356 (1981). http://dx.doi.org/10.1103/PhysRevD.23.347
2. A.D. Linde, A new inflationary universe scenario: a possible solution of the horizon, flatness, homogeneity, isotropy and primordial monopole problems. Phys. Lett. B **108**, 389–393 (1982). http://dx.doi.org/10.1016/0370-2693(82)91219-9
3. D. Baumann, Inflation, in *Theoretical Advanced Study Institute in Elementary Particle Physics: Physics of the Large and the Small*, pp. 523–686 (2011). http://dx.doi.org/10.1142/9789814327183_0010. arXiv:0907.5424 [hep-th]
4. J. Martin, C. Ringeval, V. Vennin, Encyclopædia Inflationaris: Opiparous Edition. Phys. Dark Univ **5-6**, 75–235 (2014). http://dx.doi.org/10.1016/j.dark.2024.101653. arXiv:1303.3787 [astro-ph.CO]
5. A.A. Starobinsky, A new type of isotropic cosmological models without singularity. Phys. Lett. B **91**, 99–102 (1980). http://dx.doi.org/10.1016/0370-2693(80)90670-X
6. M.J. Duff, Twenty years of the Weyl anomaly. Class. Quant. Grav **11**, 1387–1404 (1994). http://dx.doi.org/10.1088/0264-9381/11/6/004. arXiv:hep-th/9308075
7. **Planck** Collaboration, Y. Akrami et al., Planck 2018 results. X. Constraints on inflation. Astron. Astrophys **641**, A10 (2020). http://dx.doi.org/10.1051/0004-6361/201833887. arXiv:1807.06211 [astro-ph.CO]
8. **Planck** Collaboration, N. Aghanim et al., Planck 2018 results. VI. Cosmological parameters. Astron. Astrophys **641**, A6 (2020). http://dx.doi.org/10.1051/0004-6361/201833910. arXiv:1807.06209 [astro-ph.CO]. [Erratum: Astron.Astrophys. 652, C4 (2021)]
9. **ACT** Collaboration, T. Louis et al., The Atacama Cosmology Telescope: DR6 Power Spectra, Likelihoods and ΛCDM Parameters. arXiv:2503.14452 [astro-ph.CO]
10. **ACT** Collaboration, E. Calabrese et al., The Atacama Cosmology Telescope: DR6 Constraints on Extended Cosmological Models. arXiv:2503.14454 [astro-ph.CO]
11. K.S. Stelle, Renormalization of higher derivative quantum gravity. Phys. Rev. D **16**, 953–969 (1977). http://dx.doi.org/10.1103/PhysRevD.16.953
12. K.S. Stelle, Classical gravity with higher derivatives. Gen. Rel. Grav **9**, 353–371 (1978). http://dx.doi.org/10.1007/BF00760427
13. M. Ostrogradsky, Mémoires sur les équations différentielles, relatives au problème des isopérimètres. Mem. Acad. St. Petersbourg **6**(4), 385–517 (1850)
14. M. Ruhdorfer, J. Serra, A. Weiler, Effective field theory of gravity to all orders. JHEP **05**, 083 (2020). http://dx.doi.org/10.1007/JHEP05(2020)083. arXiv:1908.08050 [hep-ph]
15. T. Biswas, A.S. Koshelev, A. Mazumdar, Consistent higher derivative gravitational theories with stable de Sitter and anti–de Sitter backgrounds. Phys. Rev. D **95**(4), 043533 (2017). http://dx.doi.org/10.1103/PhysRevD.95.043533. arXiv:1606.01250 [gr-qc]
16. B. Knorr, C. Ripken, F. Saueressig, Form factors in quantum gravity: contrasting non-local, ghost-free gravity and asymptotic safety. Nuovo Cim. C **45**(2), 28 (2022). http://dx.doi.org/10.1393/ncc/i2022-22028-5. arXiv:2111.12365 [hep-th]
17. I.Y. Arefeva, D.M. Belov, A.A. Giryavets, A.S. Koshelev, P.B. Medvedev, Noncommutative field theories and (super)string field theories, in *11th Jorge Andre Swieca Summer School on Particle and Fields*, pp. 1–163 (2001). http://dx.doi.org/10.1142/9789812777317_0001. arXiv:hep-th/0111208
18. R. Kallosh, A. Linde, On the present status of inflationary cosmology. arXiv:2505.13646 [hep-th]
19. A. Kehagias, A. Moradinezhad Dizgah, A. Riotto, Remarks on the Starobinsky model of inflation and its descendants. Phys. Rev. D **89**(4), 043527 (2014). http://dx.doi.org/10.1103/PhysRevD.89.043527. arXiv:1312.1155 [hep-th]
20. A.S. Koshelev, L. Modesto, L. Rachwal, A.A. Starobinsky, Occurrence of exact R^2 inflation in non-local UV-complete gravity. JHEP **11**, 067 (2016). http://dx.doi.org/10.1007/JHEP11(2016)067. arXiv:1604.03127 [hep-th]

21. A.S. Koshelev, K. Sravan Kumar, A.A. Starobinsky, R^2 inflation to probe non-perturbative quantum gravity. JHEP **03**, 071 (2018). http://dx.doi.org/10.1007/JHEP03(2018)071. arXiv:1711.08864 [hep-th]
22. T. Biswas, A.S. Koshelev, A. Mazumdar, Gravitational theories with stable (anti-)de Sitter backgrounds. Fundam. Theor. Phys **183**, 97–114 (2016). http://dx.doi.org/10.1007/978-3-319-31299-6_5. arXiv:1602.08475 [hep-th]
23. E.T. Tomboulis, Superrenormalizable gauge and gravitational theories. arXiv:hep-th/9702146
24. A. Addazi, A.S. Koshelev, S. Pi, A. Tokareva, Secondary gravitational waves in non-local Starobinsky inflation. arXiv:2408.04004 [gr-qc]
25. A.S. Koshelev, K.S. Kumar, A.A. Starobinsky, Generalized non-local R^2-like inflation. JHEP **07**, 146 (2023). http://dx.doi.org/10.1007/JHEP07(2023)146. arXiv:2209.02515 [hep-th]
26. A.S. Koshelev, K. Sravan Kumar, A. Mazumdar, A.A. Starobinsky, Non-Gaussianities and tensor-to-scalar ratio in non-local R^2-like inflation. JHEP **06**, 152 (2020). http://dx.doi.org/10.1007/JHEP06(2020)152. arXiv:2003.00629 [hep-th]
27. J.M. Maldacena, Non-Gaussian features of primordial fluctuations in single field inflationary models. JHEP **05**, 013 (2003). http://dx.doi.org/10.1088/1126-6708/2003/05/013. arXiv:astro-ph/0210603
28. N.N. Bogolyubov, D.V. Shirkov, *Introduction to the Theory of Quantized Fields*, vol. 3 (1959)
29. A.S. Koshelev, A.A. Starobinsky, A. Tokareva, Post-inflationary GW production in generic higher (infinite) derivative gravity. Phys. Lett. B **838**, 137686 (2023). http://dx.doi.org/10.1016/j.physletb.2023.137686. arXiv:2211.02070 [hep-th]

Alexei Starobinsky and Modern Cosmology

Andrei Linde

Abstract Alexei Starobinsky is one of the main authors of inflationary cosmology. Here I will discuss the Starobinsky model and its generalizations, including the theory of α-attractors. I will then describe the current status of these models in light of the latest observational results from ACT, SPT, and DESI.

1 Introduction

It's hard for me to believe that time runs so fast, and that Alexei (Alesha) Starobinsky and I met each other almost 60 years ago. In 1966, we became students at the Department of Physics of Moscow State University. For a while, our interests did not intersect. I was interested in quantum field theory and particle physics, working with Kirzhnits. Starobinsky had a wonderful start investigating particle production in external gravitational fields, in collaboration with Zeldovich. I recall telling my parents at the time that I was very interested in studying cosmology, but "I'm too old to pursue it."

And then things changed. Since the 1970s, I have been studying cosmological phase transitions, false vacuum decay, and the transfer of vacuum energy to matter energy, which later formed the basis of both old and new inflation. Meanwhile, Starobinsky studied quantum effects near the cosmological singularity and at the end of the 70's, he published two prolific papers. In the first one, he described the theory of creation of inflationary gravitational waves [1]. In the second paper, he developed the famous Starobinsky model of cosmic inflation, "A New Type of Isotropic Cosmological Models Without Singularity" [2].

What makes these papers especially interesting is that they were written before the inflationary theory was proposed by Alan Guth [3], and well before the development of the new inflation [4] and chaotic inflation [5].

A. Linde (✉)
Leinweber Institute for Theoretical Physics at Stanford, Stanford, CA, USA
e-mail: alinde@stanford.edu

A. Barvinsky and A. Kamenshchik (eds.), *Open Issues in Gravitation and Cosmology*, Fundamental Theories of Physics 225, https://doi.org/10.1007/978-3-032-15703-4_3

The Starobinsky model [2] was an instant sensation among Russian cosmologists back in 1980, but outside of Russia, it was soon overshadowed by many other inflationary models. Admittedly, he did not attempt to provide a solution to the homogeneity, isotropy, horizon, and flatness problems in these papers, which was the trademark of inflation. But the results of the theory of inflationary perturbations developed by Mukhanov and Chibisov [6] in the context of the Starobinsky model provided a perfect match to the WMAP and Planck data [7]. This model has required only some relatively minor modifications since its introduction 45 years ago, and it still remains one of the most popular inflationary models.

At that time, I was fully focused on the new inflationary scenario [4]. I invented it in early Summer 1981. I spent a few months trying to obtain permission for its publication, and gave a talk about it at the international conference in Moscow in October. The next day after my talk, I was asked to translate a talk by Stephen Hawking at the Sternberg Astronomical Institute. In his talk, Stephen was explaining why the scenario proposed by Guth does not work. And then he said that Linde recently proposed an improved version of this scenario, but it does not work either. For half an hour, I was translating into Russian a detailed explanation of why inflation cannot be improved. After that, I said that I translated, but I disagree. We had a great discussion, and he invited me to the Nuffield Workshop on inflation in the Summer of 1982 in Cambridge.

Preparing for the conference, I issued four new papers on new inflation. About a month before the Cambridge conference, another conference on cosmology took place in Tartu, Estonia. I gave a talk about new inflation there, and then... Starobinsky gave a talk claiming that inflationary perturbations in new inflation were too large, which effectively ruled out this theory.

Well, first, Hawking claimed that new inflation is dead, then Starobinsky also claimed that it is dead, but for a different reason. With friends like that... However, soon after I returned from Tartu to Moscow, I received a preprint by Hawking, which claimed that the amplitude of inflationary perturbations in new inflation is fine. That was getting really interesting.

The three-week-long Nuffield Workshop was the most significant conference of my life. This was the time when the status of inflationary cosmology was fully recognized. This included the status of the theory of inflationary perturbations in the context of the new inflation scenario. I will explain it by the words of Alesha Starobinsky from his message to me from 2021:

> I firmly remember that just after my arrival in Cambridge, I saw the first version of Steven's 1982 paper, and when I had a short meeting with him soon (probably just on the next day after arrival), I told him that I had got a different (and large for the Higgs-like value of λ) result for perturbations, and this would be presented in my talk. I didn't tell him what the result was actually (there was no blackboard nearby), and he didn't ask about it. He made no comment to me at all. However, as you know, in two or three days, a new version of his paper had appeared with the correct dependence on λ (up to a numerical coefficient for which he did not pretend). The latter coefficient was correctly and independently obtained in mine and Guth and Pi papers (though in terms of different gauge-invariant quantities, the

one which I used was $h = 2\mathcal{R}$ in the present notation). Thus, one may have different guesses to what extent my firm statement had affected Steven and had forced him to reconsider the first variant of his paper that he did very quickly, being a great scientist indeed.

This was a very stressful conference, where the new cosmological paradigm was developed. But we were young at that time. I remember a dinner at the house of Martin Rees, where I hypnotized Don Page and Dmitri Nanopoulos, but completely failed to hypnotize Hawking's secretary Judy Fella. And then I remember my surprise when Alesha suggested watching the famous movie "Apocalypse Now" by Francis Coppola. I asked why waste time watching movies during a conference like that? In reply, Alesha gave me a brief lecture about it and convinced me to go. I knew that he was a great physicist, but I did not expect him to be a movie expert. We went to see it, and yes, he was right: the movie was dark, shocking, and powerful.

Reflecting on that time, I recall how painful it was for me to hear from Alesha that my new inflation scenario, which I had been so proud of, had been ruled out. But without knowing it, I would not be bold enough to reconsider the basic principles of old and new inflation, and invent chaotic inflation [5], which did not rely on the standard hot Big Bang theory. Thank you, Alesha!

And then he invented a stochastic approach to inflation [8], which was instrumental in the development of the theory of eternal chaotic inflation [9–14]. Then, together with Kofman and Starobinsky, we developed the theory of preheating, which is a crucial component of the theory of matter creation after inflation [15–17].

But in this paper, I will mostly concentrate on the famous Starobinsky model [2], its development, some of its generalizations, including α-attractors, and the present status of these models in view of the recent observational data by ACT [18] and SPT [19].

2 Starobinsky Model and Its Different Versions

In most of the recent papers, the authors use the name "Starobinsky model" to describe, interchangeably, one of the two different models,

$$\frac{\mathcal{L}}{\sqrt{-g}} = \frac{R}{2} + \frac{R^2}{12M^2}\,, \tag{1}$$

or

$$\frac{\mathcal{L}}{\sqrt{-\tilde{g}}} = \frac{\tilde{R}}{2} - \frac{1}{2}\partial_\mu\varphi\partial^\mu\varphi - \frac{3M^2}{4}\left(1 - e^{-\sqrt{\frac{2}{3}}\varphi}\right)^2\,. \tag{2}$$

However, as we will discuss now, the relation between these two models is rather delicate. Moreover, the original version of the Starobinsky model [2] was somewhat different.

Instead of modifying the gravitational action as shown in (1), Starobinsky in [2] considered the Einstein equations with quantum corrections to the energy-momentum tensor,

$$\langle T_{ik}\rangle = \frac{1}{H_0^2}\left(R_i^l R_{kl} - \frac{2}{3}RR_{ik} - \frac{1}{2}g_{ik}R_{lm}R^{lm} + \frac{1}{4}g_{ik}R^2\right) \\ - \frac{1}{6M^2}\left(2R_{;i}^{;k}R_{kl} - 2g_{ik}R_{;l}^{;l} - 2RR_{ik} + \frac{1}{2}g_{ik}R^2\right). \quad (3)$$

Here $H_0^{-2} = \frac{k_2}{2880\pi^2}$, $M^{-2} = -\frac{k_3}{2880\pi^2}$, where k_i are the coefficients in the expression for the conformal anomaly. These numbers include the contributions of all particle species. A contribution of each type of particles is $O(1)$. Consistency of the model required $M \ll 1$, $H_0 \ll 1$, $H_0 \gtrsim 5M$ [6, 20]. During inflation, the Hubble constant decreased gradually, from H_0 to M, and then inflation ended.

According to the original version of the Starobinsky model, the universe at $t = 0$ was in an unstable de Sitter state with the Hubble constant H_0. Since de Sitter universe is non-singular, one should be able to continue this solution to $t \to -\infty$. However, an unstable de Sitter state could not survive from $t = -\infty$ to $t = 0$, so inflation could not begin in an unstable state at $t = 0$. Therefore, the original version of the Starobinsky model was incomplete.

Another problem was related to the amplitude of adiabatic perturbations in this model. To match the WMAP - Planck data, one would need to have $M \sim 1.3 \times 10^{-5}$, in Planck mass units $M_p = 1$ [6, 20, 21]. But to achieve it, one would need to have more than 10^{10} different particle species contributing to the conformal anomaly. Finding a realistic particle physics model with these properties is extremely difficult. Therefore in 1985, in our paper [22], Starobinsky abandoned the original idea of inflation due to radiative corrections and instead proposed using the action (1).

To explain the way from (3) to (1), one should note that the first term in the energy-momentum tensor (3), proportional to H_0^{-2}, originates from the conformal anomaly, whereas the second term, proportional to M^{-2}, could be either a result of a conformal anomaly, or of the term $\frac{R^2}{12M^2}$ in the action (1), or both [21]. Therefore, one can neglect all quantum corrections, thus removing the first term in (3), and interpret the second term in (3) as a result of introducing the term $\frac{R^2}{12M^2}$ in (1).

I discussed these developments here because the standard reference to the Starobinsky model brings the readers to the original paper [2], which considered inflation as a result of quantum corrections to the Einstein theory, which required an extremely large number of particle species. This may lead to a misunderstanding, for example, when one attempts to evaluate the Starobinsky model from the perspective of swampland constraints on the number of species [23]. That is why it is important to remember that in 1985 Starobinsky moved away from the idea of explaining inflation by quantum effects and started using the streamlined version of his model (1) which did not rely on the large number of particle species.

From the point of view of the cosmological predictions, the transition from (3) to (1) was relatively uneventful, since inflation in the Starobinsky model [2] requires

$H_0^2 \gg M^2$. Predictions of the original model [2] have a slight dependence on the value of H_0, but this dependence gradually disappears in the regime $H_0 \gg 5M$ [6, 20]. Under this condition, the observational consequences of the original version of the Starobinsky model [2] almost coincide with the predictions of the model (1) [6, 20]. A detailed evaluation of these predictions, including both cases with and without conformal anomaly, can be found in the paper by Kofman, Mukhanov, and Pogosyan [21].

From a more general perspective, this modification of the Starobinsky model was substantial. In the original model [2], inflation begins in a de Sitter state with the Hubble constant H_0, and $|R^2| = 12H_0^2 \ll 1$. This solution disappears when one removes the first term in Eq. (3). Thus, one could no longer argue that the universe was born in a dS state. However, inflationary solutions do exist [22, 24]. The Hubble constant H in the theory (1) satisfies equation

$$M^2 H^2 = -6\, H^2 \dot{H} + \dot{H}^2 - 2H\ddot{H} \ . \tag{4}$$

During the slow-roll inflationary regime, one can drop the last two terms and find

$$H = \frac{M^2}{6}(t_1 - t) \ . \tag{5}$$

Inflation ends at $t_1 - t = O(1/M)$. By analogy with the chaotic inflation scenario, one could argue that inflation begins at $-t \sim M^{-2}$, i.e., at the Planck boundary when $H^2 \sim |R| = O(1)$. This would provide a possible solution to the problem of initial conditions for this model. That is why Starobinsky compared his original model [2] with new inflation, where inflation may begin in an unstable dS state [4], and he compared its streamlined version (1) with chaotic inflation, which may begin at very large φ, close to the Planck density and curvature $R = O(1)$ [5].

However, the situation with initial conditions in the model (1) is a bit tricky because of the term $\frac{R^2}{12M^2}$. The gravitational action (1) can be represented as $\frac{R}{2}(1 + \frac{R}{6M^2})$. During inflation, the coefficient $(1 + \frac{R}{6M^2})$ for a long time remains nearly constant, playing the role of an effective value of M_p^2, which is equal to 1 at $R = 0$. But at large R, the effective value of M_p^2 increases as $\frac{R}{6M^2}$. For $R \gg M^2$, equation defining the Planck boundary $R \sim M_p^2 \sim \frac{R}{6M^2}$ does not have any solutions. This intuitive argument reveals a problem with the definition of the Planck boundary in the model (1). To analyze this issue properly, it is convenient to switch to the second formulation of the Starobinsky model given by (2).

A transition from (1) to (2) took several years. First of all, after the metric transformation $\tilde{g}_{\mu\nu} = (1 + \frac{\phi}{3M^2})g_{\mu\nu}$, the term R^2 disappears, and the model (1) transforms into a theory of a scalar field interacting with gravity [25]. Finally, a redefinition of the scalar field, making it canonically normalized [26–28], transforms this theory into (2). The transformation from (1) to (2) is often referred to as a transition from the Jordan frame to the Einstein frame.

In the Einstein frame, the definition of the Planck boundary is unambiguous: $\tilde{R} = O(1)$. Inflationary potential has a plateau with the height $V_0 = \frac{3}{4}M^2$. Inflation can only happen at the plateau with $V(\varphi) \leq \frac{3}{4}M^2 \sim 1.3 \times 10^{-10}$. Thus, inflation can never begin at the Planck boundary $\tilde{R} = 1$.

One may wonder whether inflation beginning at a density ten orders of magnitude below the Planck density is probable. To analyze it, one may calculate the probability of quantum creation of the Starobinsky universe "from nothing".

According to [29–31], the probability of quantum creation of a closed inflationary universe with potential $V(\phi)$ is given by

$$P \sim e^{-\frac{24\pi^2}{V}} . \tag{6}$$

In application to the theory (2), this yields [32]

$$P \sim e^{-\frac{32\pi^2}{M^2}} \sim 10^{-10^{12}} . \tag{7}$$

This makes the initial conditions for inflation in the Starobinsky model somewhat problematic.

To alleviate this problem, one could try to add to the model (1) a potential with a large energy density of the false vacuum. One could expect that this should increase V in (6) and therefore increase the probability. However, Vilenkin has found that this does not increase the final result $P \sim e^{-\frac{32\pi^2}{M^2}}$ [32].

One way to understand this surprising conclusion is to note that if we add a cosmological constant (false vacuum energy) to the Starobinsky model (1), then, in the Einstein frame, this would add to (2) not a constant but a term proportional to $e^{-2\sqrt{\frac{2}{3}}\varphi}$, which is too steep to support inflation. This explains why adding false vacuum energy to the model (1) does not increase the probability of inflation (7).

The simplest solution to this problem is to start not with the modified gravity model (1), but directly with the model (2), and add to it the chaotic inflation potential $\frac{m^2}{2}\chi^2$:

$$\frac{\mathcal{L}}{\sqrt{-\tilde{g}}} = \frac{\tilde{R}}{2} - \frac{1}{2}\partial_\mu\varphi\partial^\mu\varphi - \frac{3M^2}{4}\left(1 - e^{-\sqrt{\frac{2}{3}}\varphi}\right)^2 - \frac{1}{2}\partial_\mu\chi\partial^\mu\chi - \frac{m^2}{2}\chi^2 . \tag{8}$$

The potential of these two fields is shown in Fig. 1. The first stage of inflation in this scenario begins at $\chi \gg 1$, $\varphi \gg 1$. At that stage, the potential is dominated by the chaotic inflation potential $\frac{m^2}{2}\chi^2$. This stage may begin at $\frac{m^2}{2}\chi^2 = O(1)$. According to (6), the probability of such a process is not exponentially suppressed [29, 30]. During this first stage, the field χ moves down, but the field φ practically does not move. After a long stage of inflation, the scalar field χ falls to $\chi = 0$, and only then does inflation driven by the field φ begin, as shown in Fig. 1. This second stage of inflation is responsible for the formation of the large-scale structure of the observable universe. This scenario was described in [33] in application to α-attractor models, and its validity was confirmed by a detailed numerical investigation in [34].

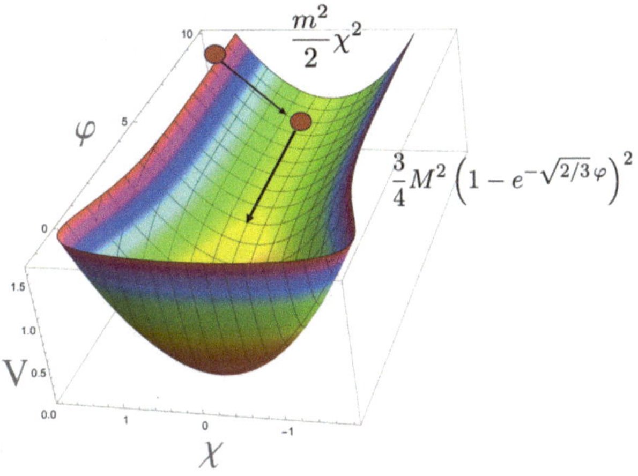

Fig. 1 Inflation in the Starobinsky model (2) combined with the chaotic inflation mode $\frac{m^2}{2}\chi^2$. The first stage of inflation begins at large χ, as in the simple chaotic inflation model. After the field χ rolls to $\chi = 0$, the second stage of inflation begins, which is described by the model (2). This solves the problem of initial conditions for inflation in the model (8)

The Starobinsky model predicts [6]

$$A_s = \frac{V_0 N_e^2}{18\pi^2}, \quad n_s = 1 - \frac{2}{N_e}, \quad r = \frac{12}{N_e^2}, \quad \text{where} \quad V_0 = \frac{3M^2}{4}. \tag{9}$$

The number of e-foldings N_e depends on the process of reheating after inflation. To make the theory complete, including the process of reheating, it is necessary to add matter fields to it. One can do it, for example, by adding the Lagrangian of the standard model either to (1) or to (2). If one adds the standard model to (2), it will be coupled to the field φ only via gravity. Meanwhile, if one adds the standard model to (1), the transformation to the Einstein frame will modify the standard model and induce some terms describing its interaction with the inflaton φ [35]. The resulting two models, originating from (1) and (2), and the process of reheating in these two models, differ from each other.

Another model, which provided a nearly perfect match to the Planck data, was the Higgs inflation model [36, 37]. The Higgs inflation model describes a scalar field with the standard Higgs potential, where the scalar field non-minimally is coupled to gravity, with the coupling $\sim \xi\phi^2 R$ and $\xi \gg 1$. One can make a transformation to the Einstein frame where this term disappears, and the potential of the canonically normalized Higgs at its large values approaches a plateau exponentially fast. This potential is very similar to the potential in (2), but it depends on $|\varphi|$, so it has two plateaus, at $\varphi \to \pm\infty$. For a discussion of initial conditions in this model, see [38].

The Starobinsky model and the Higgs inflation model differ significantly from each other, yet both predict the same values for n_s and r. There was some mystery surrounding it, so we sought to find a reason for this result and investigate whether other models may exhibit similar properties.

3 Conformal Attractors

The first class of a similar type that we were able to find had a rather unusual action [39]

$$\frac{\mathcal{L}}{\sqrt{-g}} = \frac{1}{2}\partial_\mu \chi \partial^\mu \chi + \frac{\chi^2}{12}R(g) - \frac{1}{2}\partial_\mu \phi \partial^\mu \phi - \frac{\phi^2}{12}R(g) - \frac{1}{36}F\,(\phi/\chi)\,(\phi^2 - \chi^2)^2 . \tag{10}$$

where F is an arbitrary function in terms of the variable $z = \phi/\chi$.

This theory is locally conformal invariant under the transformations

$$\tilde{g}_{\mu\nu} = e^{-2\sigma(x)} g_{\mu\nu}, \qquad \tilde{\chi} = e^{\sigma(x)}\chi, \qquad \tilde{\phi} = e^{\sigma(x)}\phi . \tag{11}$$

The field $\chi(x)$ is referred to as a conformal compensator. Its kinetic term has a negative sign, but this is not a problem because it can be removed from the theory by fixing the gauge symmetry (11). In particular, one can use the gauge $\chi^2 - \phi^2 = 6$ and resolve this constraint in terms of the fields $\chi = \sqrt{6}\cosh\frac{\varphi}{\sqrt{6}}$, $\phi = \sqrt{6}\sinh\frac{\varphi}{\sqrt{6}}$ and the field φ: $\frac{\phi}{\chi} = \tanh\frac{\varphi}{\sqrt{6}}$. Our action (10) becomes

$$\frac{\mathcal{L}}{\sqrt{-g}} = \frac{1}{2}R - \frac{1}{2}\partial_\mu \varphi \partial^\mu \varphi - F(\tanh\frac{\varphi}{\sqrt{6}}) . \tag{12}$$

Note that at large φ the potential approaches a plateau

$$V(\varphi) = V_0(1 - Ce^{-\sqrt{\frac{2}{3\alpha}}\varphi}) . \tag{13}$$

The constant C can be absorbed into a redefinition of the field φ. One can show that these models have the same predictions for n_s and r as the Starobinsky model for a very broad choice of the functions F. Moreover, for the special choice $F = \frac{27M^2}{(1+\chi/\phi)^2}$, the potential of this model exactly coincides with the potential of the Starobinsky model (2).

Models similar to (10) are often used in supergravity. By using this approach, we developed one of the first consistent supergravity generalizations of the Starobinsky model [39, 40].

Thus, we have a large class of models, which have a different origin, not related to modified gravity (1), but have very similar cosmological predictions. In this context, the model (2) is one of many models that equally well match the Planck data. Because of the conformal invariance of the models (10) and stability of predictions of these models with respect to large modifications of the function F, we called this class of models "conformal attractors" [39–41]. The next step, the theory of α-attractors, turned out to be even more interesting.

4 α-Attractors

The theory of α-attractors [39, 42–47] has its roots in the hyperbolic geometry of the moduli space [45], and in supergravity [48]. Here we will describe the phenomenological implications of two basic classes of single-field α-attractors, T-models and E-models. All of these models have supergravity generalizations.

4.1 T-Models

The simplest T-model is given by

$$\frac{\mathcal{L}}{\sqrt{-g}} = \frac{R}{2} - \frac{(\partial_\mu \phi)^2}{2\big(1 - \frac{\phi^2}{6\alpha}\big)^2} - V(\phi)\,. \tag{14}$$

Here $\phi(x)$ is the inflaton. In the limit $\alpha \to \infty$ the kinetic term becomes the canonical term $-\frac{(\partial_\mu \phi)^2}{2}$. One can recover the canonical normalization for any α by solving the equation $\frac{\partial \phi}{1-\frac{\phi^2}{6\alpha}} = \partial \varphi$. The solution is $\phi = \sqrt{6\alpha}\, \tanh \frac{\varphi}{\sqrt{6\alpha}}$. The full theory, in terms of the canonical variables φ, becomes a theory with a plateau potential

$$\frac{\mathcal{L}}{\sqrt{-g}} = \frac{R}{2} - \frac{(\partial_\mu \varphi)^2}{2} - V\big(\sqrt{6\alpha}\, \tanh \frac{\varphi}{\sqrt{6\alpha}}\big)\,. \tag{15}$$

At large φ, the potential can be represented as

$$V(\varphi) = V_0 - 2\sqrt{6\alpha}\, V_0'\, e^{-\sqrt{\frac{2}{3\alpha}}\varphi}\,. \tag{16}$$

Here $V_0 = V(\phi)|_{\phi=\sqrt{6\alpha}}$ is the height of the plateau potential, and $V_0' = \partial_\phi V|_{\phi=\sqrt{6\alpha}}$. The coefficient $2\sqrt{6\alpha}\, V_0'$ in front of the exponent can be absorbed into a shift of the field φ.

That is why all inflationary predictions in the regime with $e^{-\sqrt{\frac{2}{3\alpha}}\varphi} \ll 1$ are determined by two parameters, V_0 and α, and not by any other features of the potential $V(\phi)$:

$$A_s = \frac{V_0\, N_e^2}{18\pi^2 \alpha}\,, \qquad n_s = 1 - \frac{2}{N_e}\,, \qquad r = \frac{12\alpha}{N_e^2}\,. \tag{17}$$

Accuracy of these results increases with an increase of N_e and a decrease of α. These results are compatible with all presently available Planck/BICEP/Keck data.

The amplitude of inflationary perturbations in these models matches the Planck normalization $A_s \approx 2.01 \times 10^{-9}$ for $\frac{V_0}{\alpha} \sim 10^{-10}$, $N_e = 60$, or for $\frac{V_0}{\alpha} \sim 1.5 \times 10^{-10}$, $N_e = 50$. For the simplest model $V = \frac{m^2}{2}\phi^2$ one finds

$$V = 3m^2\alpha \tanh^2 \frac{\varphi}{\sqrt{6\alpha}} . \tag{18}$$

This simplest model is shown by the prominent vertical yellow band in Fig. 8 of the paper on inflation in the Planck2018 data release [7]. In this model, the condition $\frac{V_0}{\alpha} \sim 10^{-10}$ reads $m \sim 0.6 \times 10^{-5}$. The small magnitude of this parameter accounts for the small amplitude of perturbations $A_s \approx 2 \times 10^{-9}$. No other parameters are required to match the presently existing Planck/BICEP/Keck data in this model. If the inflationary gravitational waves are discovered, their amplitude can be accounted for by the choice of the parameter α in (17).

4.2 E-Models

The second family of α-attractors, called E-models ,is given by

$$\frac{\mathcal{L}}{\sqrt{-g}} = \frac{R}{2} - \frac{3\alpha}{4}\frac{(\partial\rho)^2}{\rho^2} - V(\rho) . \tag{19}$$

As before, one can go to canonical variables, which yields

$$\frac{\mathcal{L}}{\sqrt{-g}} = \frac{R}{2} - \frac{1}{2}(\partial\varphi)^2 - V(e^{-\sqrt{\frac{2}{3\alpha}}\varphi}). \tag{20}$$

We consider $V(\rho)$ not singular at $\rho = 0$, which has a Minkowski minimum, e.g. $V(\rho) = V_0(1-\rho)^{2n}$. In canonical variables, it becomes

$$V = V_0\Big(1 - e^{-\sqrt{\frac{2}{3\alpha}}\varphi}\Big)^{2n} . \tag{21}$$

For $\alpha = 1$ and $n = 1$, this potential coincides with the potential of the Starobinsky model [2]. In the small α limit, the predictions of the E-models coincide with the predictions of the T-models (17).

For a discussion of the problem of initial conditions in α-attractor models, see e.g. [33, 34, 49].

5 From Planck/BICEP/Keck to ACT, SPT and DESI

In the CMB White Paper [50] and LiteBIRD [51] figures, one can see only 3 viable targets for B-mode searches: the Starobinsky model [2], Higgs inflation [36, 37], and α-attractors [43].

With respect to the spectral index n_s, all of these models had a universal prediction $n_s = 1 - 2/N_e$. The typical value of N_e depends on the mechanism of reheating after inflation and is expected to be in the range of 47 to 60. For example, for $N_e = 60$ one finds $n_s = 0.967$.

The simplest reheating mechanisms in the Starobinsky model are not very efficient; therefore, one typically has $n_s \sim 0.962 - 0.963$. In the Higgs inflation model, reheating is very efficient because the Higgs field is a part of the standard model, which is why one may expect $n_s \sim 0.966$. Reheating in α-attractors is somewhat less constrained, depending on the specific model version. In particular, E-models introduced in [43] can reach $n_s \approx 0.97$ for $N_e = 60$, see Fig. 2. There is also a class of quintessential α-attractors, where the potential may not have a minimum. Such models may describe either a cosmological constant or dark energy, and they may have n_s higher than that in more conventional models by $\Delta n_s \sim 0.006$ [52].

The most recent constraint on n_s, based on the combination of all CMB-based data, including Planck. BICEP/Keck, ACT, and SPT, is $n_s = 0.9684 \pm 0.003$, which is compatible with the inflationary models mentioned above.

However, when the CMB data is combined with the results of DESI, the value of n_s increases considerably. In particular, CMB data, including ACT data and DESI DR1 results, yield $n_s = 0.9739 \pm 0.0034$ [18, 53]. The latest SPT data, in combination with all other CMB results plus DESI DR2, give $n_s = 0.9728 \pm 0.0027$ [19], which

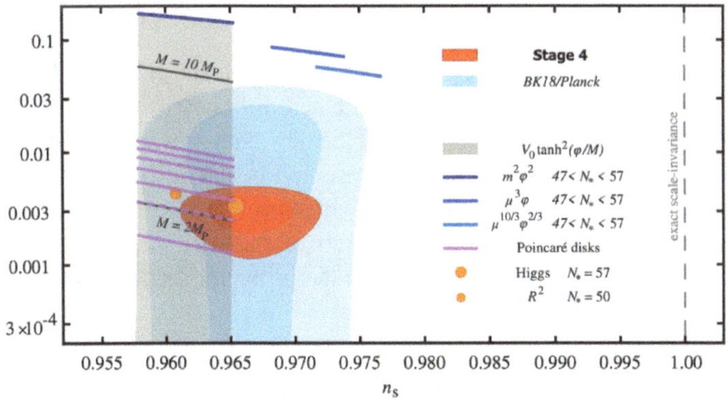

Fig. 2 The figure from "Snowmass2021 Cosmic Frontier: Cosmic Microwave Background Measurements White Paper" [50]. It shows the predictions of T-model α-attractors with unconstrained values of α (gray area), the predictions for $3\alpha = 7, 6, 5, 4, 3, 2, 1$ (purple lines), as well as Higgs inflation, R^2 inflation (red disks). The predictions are for $47 < N_e < 57$. Reprinted with permission from [50] ©2022, R. Flaugher for the CMB-S4 collaboration. All rights reserved

is consistent with the ACT-DESI results. As noted in [53], these results disfavor the Starobinsky model, the Higgs inflation model, and many α-attractors models at $\gtrsim 2\sigma$.

Thus, this result disfavored a large class of inflationary models, which had been favored by CMB data for more than a decade. Indeed, according to [18, 19, 53], the combination of CMB and DESI results eliminates all viable B-mode targets shown in figures by CMB-S4 and LiteBIRD collaborations.

However, the situation with the CMB-DESI related constraints is still highly uncertain. The authors of the ACT data release [18, 53] in their talks emphasized that this is "only 2σ". The discussion of these results in the SPT data release [19] was even more cautious and nuanced. In particular, they noticed that the constraints $n_s = 0.9739 \pm 0.0034$ [18, 53] and $n_s = 0.9728 \pm 0.0027$ [19] were obtained in the context of the ΛCDM theory, but in this context the CMB data are in about 2.8σ tension with the DESI DR2 data.

The simplest way to understand it is to note that the CMB data alone are fully consistent with ΛCDM, but the results of a combination of ACT and DESI DR2 are in more than 3σ tension with ΛCDM [54]. Thus, one may wonder whether it is feasible to disfavor the Starobinsky model at 2σ level by combining data that are inconsistent with each other at the more than 3σ level in the ΛCDM context. For a general discussion of combining mutually inconsistent data sets, see e.g. [55].

A more detailed discussion can be found in Section VII C "Evaluating the consistency of CMB and BAO data in ΛCDM" of the SPT data release [19]. The authors advise caution in interpreting CMB+DESI data results in ΛCDM.

A subsequent discussion of this issue in [56] demonstrates that the fit of ΛCDM to CMB data exhibits a high degree of correlation between n_s and the BAO parameters $r_d h$ and Ω_m. This correlation, and the tension in $r_d h$ and Ω_m between CMB data and DESI, can lead to significant shifts in n_s when the two are combined. The authors concluded: "Given the crucial role of n_s in discriminating between inflationary models, we urge caution in interpreting CMB+BAO constraints on n_s until the BAO-CMB tension is resolved."

Thus, for the time being, we may continue to use the Starobinsky model and Higgs inflation, and develop new, sophisticated versions of α-attractors [57, 58]. But it would be incorrect to simply discard the new data or wait until the dust settles. The first response to this situation was presented in [59], where we developed the non-minimal chaotic inflation model (30), which provides a good match to the ACT-DESI results. Shortly after that, approximately 50 papers on this issue were written, each offering its own approach to resolving the problem; see a recent review of this situation in [60].

In particular, there is a large variety of Starobinsky-like models, which can be modified to match the ACT-DESI data (see, e.g., [35, 61–66]). Many of these proposals consider general versions of $f(R)$ gravity. A detailed discussion of such models and many related references can be found in [65]. A particular model described there contains additional terms, such as R^3 and R^4. R^6 with different coefficients in the range between 10^{-5} and 10^{-9}. Such generalizations of the Starobinsky model can match the CMB+DESI data at the expense of introducing additional extremely small parameters.

E-model α attractors with $\alpha \sim 6$ are only in 1σ tension with the CMB-DESI related constraints for $N_e = 60$. The predictions of quintessential α-attractors may also move closer to the CMB-DESI constraints. More considerable modifications can be achieved in the two-field models, such as the hybrid α-attractors [67]. In this class of models, one has two basic attractor regimes. If $V_{\rm up}$ is large and $\phi_c \ll 1$, then the last stage of inflation occurs as in the original hybrid inflation scenario, with n_s very close to 1. On the other hand, if uplifting is small, one has the standard α-attractor result $n_s = 1 - \frac{2}{N_e}$, $r = \frac{12}{N_e^2}$. Thus, by changing the parameters of the model, one may interpolate between $n_s = 1 - \frac{2}{N_e}$ and the $n_s = 1$ [67, 68].

The full theory of this effect is rather nuanced, as it depends on the relations between many parameters of the α-attractor version of hybrid inflation models. For example, in certain cases, the process of spontaneous symmetry breaking can drive a short second stage of inflation, resulting in the production of large primordial black holes [69], which might even account for dark matter [70].

Once one switches to the two-field models, other non-trivial models may emerge, such as the revised version of the curvaton scenario proposed in [71].

6 Unity of Cosmological Attractors

To summarize the previous results: The Starobinsky model and the Higgs inflation model provide a very good match to the Planck data. There is a large class of the single-field models, such as conformal attractors and α-attractors, which generalize the Starobinsky model and the Higgs inflation model. All of these models have predictions that are stable with respect to strong modifications of the potential. In the previous section, we discussed a recent controversy that arises when combining CMB data and DESI results. If further investigations confirm the increase of n_s found in [18, 19, 53], then the stability of the predictions of the single-field models discussed above can make it difficult to modify them. We discussed some of such modifications in the previous section, including switching to the multi-field models.

Going forward, we should also consider more substantial modifications of these models. All of these models have different motivations and different structures. One may wonder whether there is something common to all of them, ensuring stability of their predictions, and whether this common element may be preserved?

Indeed, there is a common element that is often overlooked in discussions of the Starobinsky model and Higgs inflation. We all know that one can begin with (1) and end up with (2). However, we did not mention that halfway from one formulation to another, one deals with scalar fields with singular kinetic terms, similar to the kinetic term $\frac{3\alpha}{4} \frac{(\partial\rho)^2}{\rho^2}$ in the α-attractor action (19).

Exploring the Higgs inflation model, one starts with terms like $\phi^2 R$ and then switches to the Einstein frame. At the intermediate stages of calculations, one has a theory of a scalar field with a singular kinetic term of a similar type. Similar terms appear at the intermediate stage of calculations in conformal attractors. And finally, in α-attractors these singular terms become a central part of the theory.

Note that in all of these models, the pole is of order 2. Such singular terms have an especially interesting interpretation for complex scalar fields, in terms of hyperbolic geometry, famously illustrated in the artwork by Escher [45].

However, one may consider a more general situation, in the context of the so-called pole inflation models [44]. It is obtained by generalizing the α-attractor Eq. (19):

$$\frac{\mathcal{L}}{\sqrt{-g}} = \frac{R}{2} - \frac{a_q}{2} \frac{(\partial\rho)^2}{\rho^q} - V(\rho) . \tag{22}$$

Here, the pole of order q is at $\rho = 0$ and the residue at the pole is a_q. For $q = 2$, $a_2 = \frac{3\alpha}{2}$, this equation describes E-models of α-attractors, but here we consider general values of q. For $q \neq 2$ one can always rescale ρ to make $a_q = 1$. Just as in the theory of α-attractors, one can make a transformation to the canonical variables φ and find that the asymptotic behavior of the potential $V(\varphi)$ during inflation is determined only by $V(0)$ and the first derivative $\frac{dV(\rho)}{d\rho}|_{\rho=0}$.

The potentials in this family of models also have inflationary plateaus at large φ. For α-attractors, the plateau of the potential is reached exponentially, but for $q > 2$ the approach to the plateau is controlled by negative powers of φ:

$$V \sim V_0(1 - \frac{m^k}{\varphi^k} + \cdots) , \qquad n_s = 1 - \frac{2}{N_e}\frac{k+1}{k+2} , \qquad \text{where} \qquad k = \frac{2}{2-q} . \tag{23}$$

Here are two simple examples of the potentials of this type and their cosmological predictions:

$$V = V_0 \frac{\varphi^4}{m^4 + \varphi^4} , \qquad n_s = 1 - \frac{5}{3N_e} , \qquad r = \frac{4\, m^{\frac{4}{3}}}{(3N_e)^{\frac{5}{3}}} . \tag{24}$$

$$V = V_0 \frac{\varphi^2}{m^2 + \varphi^2} , \qquad n_s = 1 - \frac{3}{2N_e} , \qquad r = \frac{\sqrt{2}\, m}{N_e^{\frac{3}{2}}} . \tag{25}$$

Some of these models described in [46, 47, 72] have an interpretation in terms of Dp-brane inflation [73, 74]. However, they also appear in other contexts. For example, the quadratic model $V \sim \frac{\varphi^2}{m^2+\varphi^2}$ was first proposed in [75]. It was used in [76] as an example of a flattening mechanism for the φ^2 potential due to the inflaton interactions with heavy scalar fields. To explain the basic idea, consider a potential

$$V(\phi, \chi) = \phi^2\chi^2 + m^2(\chi - \chi_0)^2 . \tag{26}$$

If m^2 is sufficiently large, during inflation driven by the field ϕ, the field χ will track its instantaneous minimum, depending on ϕ. The effective potential of the field ϕ along the inflationary trajectory coincides with the potential (25), with $V_0 = m^2\chi_0^2$.

Potentials (23) can also be obtained in the context of polynomial α-attractors [77], where

$$\frac{\mathcal{L}}{\sqrt{-g}} = \frac{R}{2} - \frac{3\alpha}{2}\frac{(\partial\rho)^2}{\rho^2} - V_0\,\frac{\ln^k\rho}{\ln^k\rho+1}\,. \tag{27}$$

By the change of variables $\rho = e^{-\sqrt{\frac{2}{3\alpha}}\varphi}$, as in E-models (20), (21), this model is reduced to a model for a canonical scalar field with the potential

$$V = V_0\frac{|\varphi|^k}{\mu^k + |\varphi|^k}\,, \qquad \mu = \sqrt{\frac{3\alpha}{2}}\,. \tag{28}$$

Note that for any $k > 0$ the attractor values of n_s in (23) are greater than $1 - 2/N_e$, and in the limit $k \to 0$ one has $n_s \to 1 - 1/N_e$. This means, for example, that for $N_e = 55$, this class of models can account for n_s in the broad range $0.9636 < n_s < 0.982$.

Yet another example is provided by the simplest chaotic inflation model with non-minimal coupling to gravity [59]:

$$\frac{1}{\sqrt{-g}}\mathcal{L} = \tfrac{1}{2}(1+\phi)R - \tfrac{1}{2}(\partial\phi)^2 - \tfrac{1}{2}m^2\phi^2\,. \tag{29}$$

As in the Starobinsky model and in Higgs inflation, one can switch to the Einstein frame with a canonically normalized scalar field minimally coupled to gravity. In the large φ limit one has

$$V = \frac{m^2}{2}\left(1 - 8\varphi^{-2} + O(\phi^{-4})\right), \tag{30}$$

This model predicts, for $N_e = 60$,

$$n_s = 0.9733, \qquad r = 0.0094\,. \tag{31}$$

which matches the ACT and SPT constraints $n_s = 0.9739 \pm 0.0034$ [18, 53] and $n_s = 0.9728 \pm 0.0027$ [19]. Note that $r = 0.0094$ is almost 3 times higher than r in the Starobinsky and Higgs inflation model, so it should be easier to test this model by future observations.

An exact shape of the inflaton potential (30) of the non-minimal chaotic inflation model (29) is shown in Fig. 3. It is interesting and perhaps somewhat unexpected that this potential looks remarkably similar to that of the Starobinsky model. The only difference is that the Starobinsky model potential approaches the plateau exponentially fast, whereas the potential of the non-minimal chaotic inflation approaches the plateau more slowly, following the inverse power law (30). Both models are most economical, requiring only one parameter, but the Starobinsky model provides a good match to the Planck data, whereas the non-minimal chaotic inflation (29) matches the CMB-DESI related constraints [18, 19, 53].

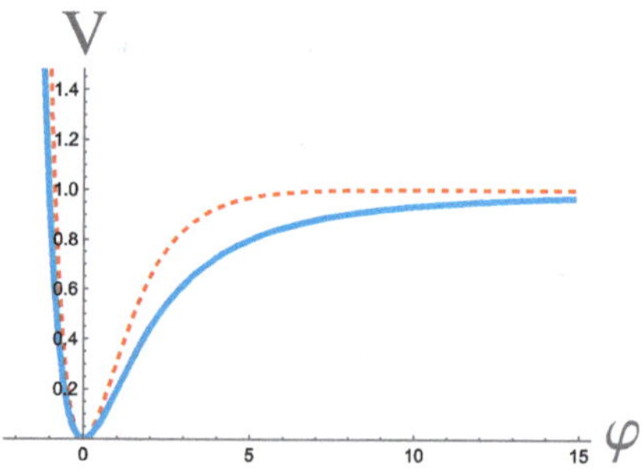

Fig. 3 The potential of the model (29) in the Einstein frame as a function of the canonical inflaton field φ. As one can see, it is similar to the potential in the Starobinsky model (red dashed line) and in the E-models of α-attractors [43], but it approaches the plateau more slowly

7 Discussion

In this paper, I have focused on a single topic, the Starobinsky model, because it is so popular, and this subject is particularly urgent in light of the recent CMB-DESI results. It is remarkable that a beautiful theoretical idea, proposed by Starobinsky 45 years ago, remains at the core of recent developments in both theoretical and observational cosmology. It is so unfortunate that we can no longer ask its author what he thinks about it now. I wish I could discuss his stochastic approach to inflation here, which was a stunning idea rediscovered and utilized by many. I remember how I was carrying in my backpack a draft of our paper with Lev Kofman and Alesha Starobinsky on reheating, first from Moscow to CERN, then to Stanford, until few yers later we had a chance to return to it, then completely re-write it, then throw this large paper to the garbage bin, and instead of it write a small paper on preheating. I still have on my computer a large draft of our paper with Kofman and Starobinsky on quantum cosmology, which has been left unchanged since 2009. There were many other urgent things to do, and we could certainly always call each other and return to it later.

I wish...

Acknowledgements I am grateful to R. Bond, R. Kallosh, D. Roest, and E. Silverstein for many related discussions. This work was supported by Leinweber Institute for Theoretical Physics at Stanford and by NSF Grant PHY-2310429.

References

1. A.A. Starobinsky, Spectrum of relict gravitational radiation and the early state of the universe. JETP Lett. **30**, 682 (1979)
2. A.A. Starobinsky, A new type of isotropic cosmological models without singularity. Phys. Lett. **91B**, 99 (1980). https://doi.org/10.1016/0370-2693(80)90670-X

3. A.H. Guth, The inflationary universe: a possible solution to the horizon and flatness problems. Phys. Rev. D **23**, 347 (1981). https://doi.org/10.1103/PhysRevD.23.347
4. A.D. Linde, A new inflationary universe scenario: a possible solution of the horizon, flatness, homogeneity, isotropy and primordial monopole problems. Phys. Lett. B **108**, 389 (1982). https://doi.org/10.1016/0370-2693(82)91219-9
5. A.D. Linde, Chaotic inflation. Phys. Lett. B **129**, 177 (1983). https://doi.org/10.1016/0370-2693(83)90837-7
6. V.F. Mukhanov, G.V. Chibisov, Quantum fluctuations and a nonsingular universe. JETP Lett. **33**, 532 (1981)
7. PLANCK collaboration, Planck 2018 results. X. Constraints on inflation. Astron. Astrophys **641**, A10 (2020). https://doi.org/10.1051/0004-6361/201833887. arXiv:1807.06211
8. A.A. Starobinsky, Stochastic de sitter (inflationary) stage in the early universe. Lect. Notes Phys. **246**, 107 (1986). https://doi.org/10.1007/3-540-16452-9_6
9. A.D. Linde, Eternal chaotic inflation. Mod. Phys. Lett. A **1**, 81 (1986). https://doi.org/10.1142/S0217732386000129
10. A.D. Linde, Eternally existing self-reproducing chaotic inflationary universe. Phys. Lett. B **175**, 395 (1986). https://doi.org/10.1016/0370-2693(86)90611-8
11. A.S. Goncharov, A.D. Linde, Global structure of inflationary universe. Sov. Phys. JETP **65**(4), 635 (1987). https://doi.org/10.1142/S0217751X87000211
12. A.S. Goncharov, A.D. Linde, V.F. Mukhanov, The global structure of the inflationary universe. Int. J. Mod. Phys **A2**, 561 (1987). https://doi.org/10.1142/S0217751X87000211
13. A.D. Linde, A. Mezhlumian, Stationary universe. Phys. Lett. B **307**, 25 (1993). https://doi.org/10.1016/0370-2693(93)90187-M. arXiv:gr-qc/9304015
14. A.D. Linde, D.A. Linde, A. Mezhlumian, From the Big Bang theory to the theory of a stationary universe. Phys. Rev. D **49**, 1783 (1994). https://doi.org/10.1103/PhysRevD.49.1783. arXiv:gr-qc/9306035
15. L. Kofman, A.D. Linde, A.A. Starobinsky, Reheating after inflation. Phys. Rev. Lett. **73**, 3195 (1994). https://doi.org/10.1103/PhysRevLett.73.3195. arXiv:hep-th/9405187
16. L. Kofman, A.D. Linde, A.A. Starobinsky, Towards the theory of reheating after inflation. Phys. Rev. D **56**, 3258 (1997). https://doi.org/10.1103/PhysRevD.56.3258. arXiv:hep-ph/9704452
17. P.B. Greene, L. Kofman, A.D. Linde, A.A. Starobinsky, Structure of resonance in preheating after inflation. Phys. Rev. D **56**, 6175 (1997). https://doi.org/10.1103/PhysRevD.56.6175. arXiv:hep-ph/9705347
18. ACT collaboration, The atacama cosmology telescope: DR6 power spectra, likelihoods and Λ CDM parameters. arXiv:2503.14452
19. SPT- 3G collaboration, SPT-3G D1: CMB temperature and polarization power spectra and cosmology from 2019 and 2020 observations of the SPT-3G Main field. arXiv:2506.20707
20. A.A. Starobinsky, The perturbation spectrum evolving from a nonsingular initially De-Sitter cosmology and the microwave background anisotropy. Sov. Astron. Lett. **9**, 302 (1983)
21. L.A. Kofman, V.F. Mukhanov, D.Y. Pogosian, Evolution of inhomogeneities in inflationary models in a theory of gravitation with higher derivatives. Sov. Phys. JETP **66**, 433 (1987)
22. L.A. Kofman, A.D. Linde, A.A. Starobinsky, Inflationary universe generated by the combined action of a scalar field and gravitational vacuum polarization. Phys. Lett. B **157**, 361 (1985). https://doi.org/10.1016/0370-2693(85)90381-8
23. D. Lüst, J. Masias, B. Muntz, M. Scalisi, Starobinsky inflation in the swampland. JHEP **07**, 186 (2024). https://doi.org/10.1007/JHEP07(2024)186. arXiv:2312.13210
24. J.D. Barrow, A.C. Ottewill, The stability of general relativistic cosmological theory. J. Phys. A **16**, 2757 (1983). https://doi.org/10.1088/0305-4470/16/12/022
25. B. Whitt, Fourth order gravity as general relativity plus matter. Phys. Lett. B **145**, 176 (1984). https://doi.org/10.1016/0370-2693(84)90332-0
26. J.D. Barrow, S. Cotsakis, Inflation and the conformal structure of higher order gravity theories. Phys. Lett. B **214**, 515 (1988). https://doi.org/10.1016/0370-2693(88)90110-4
27. K.-I. Maeda, Inflation as a transient attractor in R**2 cosmology. Phys. Rev. D **37**, 858 (1988). https://doi.org/10.1103/PhysRevD.37.858

28. D.H. Coule, M.B. Mijic, Quantum fluctuations and eternal inflation in the R^2 model. Int. J. Mod. Phys **A3**, 617 (1988). https://doi.org/10.1142/S0217751X88000266
29. A.D. Linde, Quantum creation of the inflationary universe. Lett. Nuovo Cim **39**, 401 (1984). https://doi.org/10.1007/BF02790571
30. A.D. Linde, The inflationary universe. Rept. Prog. Phys **47**, 925 (1984). https://doi.org/10.1088/0034-4885/47/8/002
31. A. Vilenkin, Quantum creation of universes. Phys. Rev. D **30**, 509 (1984). https://doi.org/10.1103/PhysRevD.30.509
32. A. Vilenkin, Classical and quantum cosmology of the Starobinsky inflationary model. Phys. Rev. D **32**, 2511 (1985). https://doi.org/10.1103/PhysRevD.32.2511
33. A. Linde, On the problem of initial conditions for inflation, in *Black Holes, Gravitational Waves and Spacetime Singularities Rome, Italy, May 9-12* (2017). http://inspirehep.net/record/1630432/files/arXiv:1710.04278.pdf. arXiv:1710.04278
34. M. Corman, W.E. East, Starting inflation from inhomogeneous initial conditions with momentum. JCAP **10**, 046 (2023). https://doi.org/10.1088/1475-7516/2023/10/046. arXiv:2212.04479
35. I. Antoniadis, J. Ellis, W. Ke, D.V. Nanopoulos, K.A. Olive, How accidental was inflation?. arXiv:2504.12283
36. D.S. Salopek, J.R. Bond, J.M. Bardeen, Designing density fluctuation spectra in inflation. Phys. Rev. D **40**, 1753 (1989). https://doi.org/10.1103/PhysRevD.40.1753
37. F.L. Bezrukov, M. Shaposhnikov, The standard model Higgs boson as the inflaton. Phys. Lett. B **659**, 703 (2008). https://doi.org/10.1016/j.physletb.2007.11.072. arXiv:0710.3755
38. A.O. Barvinsky, A.Y. Kamenshchik, Nonminimal Higgs inflation and initial conditions in cosmology (2023). https://doi.org/10.1007/978-981-19-3079-9_13-1. arXiv:2212.13077
39. R. Kallosh, A. Linde, Universality class in conformal inflation. JCAP **1307**, 002 (2013). https://doi.org/10.1088/1475-7516/2013/07/002. arXiv:1306.5220
40. R. Kallosh, A. Linde, Superconformal generalizations of the Starobinsky model. JCAP **06**, 028 (2013). https://doi.org/10.1088/1475-7516/2013/06/028. arXiv:1306.3214
41. R. Kallosh, A. Linde, Multi-field conformal cosmological attractors. JCAP **1312**, 006 (2013). https://doi.org/10.1088/1475-7516/2013/12/006. arXiv:1309.2015
42. S. Ferrara, R. Kallosh, A. Linde, M. Porrati, Minimal supergravity models of inflation. Phys. Rev. D **88**, 085038 (2013). https://doi.org/10.1103/PhysRevD.88.085038. arXiv:1307.7696
43. R. Kallosh, A. Linde, D. Roest, Superconformal inflationary α-attractors. JHEP **11**, 198 (2013). https://doi.org/10.1007/JHEP11(2013)198. arXiv:1311.0472
44. M. Galante, R. Kallosh, A. Linde, D. Roest, Unity of cosmological inflation attractors. Phys. Rev. Lett. **114**, 141302 (2015). https://doi.org/10.1103/PhysRevLett.114.141302. arXiv:1412.3797
45. R. Kallosh, A. Linde, Escher in the sky. C R Phys. **16**, 914 (2015). https://doi.org/10.1016/j.crhy.2015.07.004. arXiv:1503.06785
46. R. Kallosh, A. Linde, B-mode targets. Phys. Lett. B **798**, 134970 (2019). https://doi.org/10.1016/j.physletb.2019.134970. arXiv:1906.04729
47. R. Kallosh, A. Linde, CMB targets after the latest *Planck* data release. Phys. Rev. D **100**, 123523 (2019). https://doi.org/10.1103/PhysRevD.100.123523. arXiv:1909.04687
48. R. Kallosh, A. Linde, Attractors in supergravity. arXiv:2503.13682
49. J.J.M. Carrasco, R. Kallosh, A. Linde, Cosmological attractors and initial conditions for inflation. Phys. Rev. D **92**, 063519 (2015). https://doi.org/10.1103/PhysRevD.92.063519. arXiv:1506.00936
50. C.L. Chang et al., Snowmass2021 cosmic frontier: cosmic microwave background measurements white paper. arXiv:2203.07638
51. LiteBIRD collaboration, Probing cosmic inflation with the LiteBIRD cosmic microwave background polarization survey. PTEP **2023**, 042F01 (2023). https://doi.org/10.1093/ptep/ptac150. arXiv:2202.02773
52. Y. Akrami, R. Kallosh, A. Linde, V. Vardanyan, Dark energy, α-attractors, and large-scale structure surveys. JCAP **06**, 041 (2018). https://doi.org/10.1088/1475-7516/2018/06/041. arXiv:1712.09693

53. ACT collaboration, The Atacama cosmology telescope: DR6 constraints on extended cosmological models. arXiv:2503.14454
54. DESI collaboration, Cosmological implications of DESI DR2 BAO measurements in light of the latest ACT DR6 CMB data. arXiv:2504.18464
55. J. Luis Bernal, J.A. Peacock, Conservative cosmology: combining data with allowance for unknown systematics. JCAP **07**, 002 (2018). https://doi.org/10.1088/1475-7516/2018/07/002. arXiv:1803.04470
56. E.G.M. Ferreira, E. McDonough, L. Balkenhol, R. Kallosh, L. Knox, A. Linde, The BAO-CMB tension and implications for inflation. arXiv:2507.12459
57. R. Kallosh, A. Linde, $SL(2, \mathbb{Z})$ cosmological attractors. JCAP **04**, 045 (2025). https://doi.org/10.1088/1475-7516/2025/04/045. arXiv:2408.05203
58. J.J. Carrasco, R. Kallosh, A. Linde, D. Roest, Axion stabilization in modular cosmology. JCAP **07**, 090 (2025). https://doi.org/10.1088/1475-7516/2025/07/090. arXiv:2503.14904
59. R. Kallosh, A. Linde, D. Roest, A simple scenario for the last ACT. arXiv:2503.21030
60. R. Kallosh, A. Linde, On the present status of inflationary cosmology. arXiv:2505.13646
61. M. Drees, Y. Xu, Refined predictions for Starobinsky inflation and post-inflationary constraints in light of ACT. Phys. Lett. B **867**, 139612 (2025). https://doi.org/10.1016/j.physletb.2025.139612. arXiv:2504.20757
62. I.D. Gialamas, T. Katsoulas, K. Tamvakis, Keeping the relation between the Starobinsky model and no-scale supergravity ACTive. arXiv:2505.03608
63. A. Addazi, Y. Aldabergenov, S.V. Ketov, Curvature corrections to Starobinsky inflation can explain the ACT results. arXiv:2505.10305
64. D.S. Zharov, O.O. Sobol, S.I. Vilchinskii, ACT observations, reheating, and Starobinsky and Higgs inflation. Phys. Rev. D **112**, 023544 (2025). https://doi.org/10.1103/km3q-rm34
65. S.V. Ketov, E.O. Pozdeeva, S.Y. Vernov, Inflation in $F(R)$ gravity models revisited after ACT. arXiv:2508.08927
66. J. Ellis, M.A.G. Garcia, N. Nagata, D.V. Nanopoulos, K.A. Olive, Deformations of Starobinsky inflation in no-scale SU(5) and SO(10) GUTs. arXiv:2508.13279
67. R. Kallosh, A. Linde, Hybrid cosmological attractors. Phys. Rev. D **106**, 023522 (2022). https://doi.org/10.1103/PhysRevD.106.023522. arXiv:2204.02425
68. L. Iacconi, M. Bacchi, L.F. Guimarães, F.T. Falciano, Testing inflation on all scales: a case study with α-attractors. arXiv:2412.02544
69. M. Braglia, A. Linde, R. Kallosh, F. Finelli, Hybrid α-attractors, primordial black holes and gravitational wave backgrounds. JCAP **04**, 033 (2023). https://doi.org/10.1088/1475-7516/2023/04/033. arXiv:2211.14262
70. J. Garcia-Bellido, A.D. Linde, D. Wands, Density perturbations and black hole formation in hybrid inflation. Phys. Rev. D **54**, 6040 (1996). https://doi.org/10.1103/PhysRevD.54.6040. arXiv:astro-ph/9605094
71. C.T. Byrnes, M. Cortês, A.R. Liddle, The curvaton ACTs again. arXiv:2505.09682
72. R. Kallosh, A. Linde, Y. Yamada, Planck 2018 and Brane inflation revisited. JHEP **01**, 008 (2019). https://doi.org/10.1007/JHEP01(2019)008. arXiv:1811.01023
73. G.R. Dvali, S.H.H. Tye, Brane inflation. Phys. Lett. B **450**, 72 (1999). https://doi.org/10.1016/S0370-2693(99)00132-X. arXiv:hep-ph/9812483
74. S. Kachru, R. Kallosh, A.D. Linde, J.M. Maldacena, L.P. McAllister, S.P. Trivedi, Towards inflation in string theory. JCAP **0310**, 013 (2003). https://doi.org/10.1088/1475-7516/2003/10/013. arXiv:hep-th/0308055
75. E.D. Stewart, Mutated hybrid inflation. Phys. Lett. B **345**, 414 (1995). https://doi.org/10.1016/0370-2693(94)01646-T. arXiv:astro-ph/9407040
76. X. Dong, B. Horn, E. Silverstein, A. Westphal, Simple exercises to flatten your potential. Phys. Rev. D **84**, 026011 (2011). https://doi.org/10.1103/PhysRevD.84.026011. arXiv:1011.4521
77. R. Kallosh, A. Linde, Polynomial α-attractors. JCAP **04**, 017 (2022). https://doi.org/10.1088/1475-7516/2022/04/017. arXiv:2202.06492

Constant-Roll Inflation

Hayato Motohashi

Abstract Constant-roll inflation is a distinctive class of phenomenological inflationary models in which the inflaton's rate of roll remains constant. It provides an exact solution that is compatible with the latest observational constraints and offers a natural framework for enhancing the curvature power spectrum, which is relevant to the formation of primordial black holes. In this paper, I review constant-roll inflation in memory of Alexei Starobinsky.

1 Introduction

Alexei Starobinsky's pioneering work in cosmology, astrophysics, and gravitational physics has profoundly shaped our understanding of the universe. Among his numerous groundbreaking contributions, he played a pivotal role in developing the theory of the early universe. He proposed one of the first models of cosmic inflation [1–7], which is a cornerstone of modern cosmology, resolving fundamental issues in Big Bang cosmology–such as the horizon, flatness, and monopole problems–through a phase of quasi-exponential expansion of the universe. At the same time, it provides a mechanism for generating primordial perturbations [8–13], which act as the seeds for the large-scale structure of the universe. He was instrumental in the development of the δN formalism [10, 14–23] and pioneered the stochastic formalism [10, 24, 25] to explore generation and evolution of quantum fluctuations during inflation. He also made key contributions to the theory of reheating [26–28], the process by which the inflaton field oscillates and decays into Standard Model particles after inflation.

I had the privilege of collaborating with Alexei and spending valuable time working with him. From 2000 to 2019, he regularly visited the Research Center for the Early Universe (RESCEU) at the University of Tokyo, where I was a PhD student from 2008 to 2013. Alongside my PhD advisor, Jun'ichi Yokoyama, and Alexei, I explored topics in dark energy [29–38] and its connection to inflation [39]. Our discussions at RESCEU and at conferences around the world remain unforgettable,

H. Motohashi (✉)
Department of Physics, Tokyo Metropolitan University, Hachioji, Japan
e-mail: motohashi@tmu.ac.jp

A. Barvinsky and A. Kamenshchik (eds.), *Open Issues in Gravitation and Cosmology*, Fundamental Theories of Physics 225, https://doi.org/10.1007/978-3-032-15703-4_4

deepening both my understanding of fundamental physics and my appreciation for Alexei's unique approach and dedication to research.

Beyond our professional relationship, Alexei was warm and engaging on a personal level as well. Outside of physics, we sometimes enjoyed playing table tennis together in the university gym, where I was always impressed by his remarkably steady backhand.

Our collaboration continued after I completed my PhD, shifting focus to inflation [40–43]. We proposed a particular type of inflation, dubbing it 'constant-roll inflation', in which the slow-roll condition is replaced by a requirement that the inflaton maintains a constant rate of roll. This framework generalizes slow-roll inflation, provides an exact solution compatible with the latest observational constraints, and is also relevant to the formation of primordial black holes. In this paper, dedicated to the memory of Alexei, I shall review constant-roll inflation.

"We should work harder," Alexei once said during his banquet address at a conference a decade ago, mourning the untimely passing of his former PhD student Lev Kofman. Now, I feel the weight of his words more than ever. His research legacy continues to inspire us, and each reading of his papers reveals new insights. It is our responsibility to further advance cosmology and gravitational physics in directions that Alexei could not witness firsthand–but might have foreseen with his remarkable insight.

2 Constant-Roll Inflation

Constant-roll inflation is a phenomenological inflationary model, in which the slow-roll condition is replaced by a requirement that the inflaton's rate of roll remains constant. It allows for exact analytical solutions and exhibits unique theoretical properties. Our focus is on constant-roll models in single-field canonical inflation [40], which has been extensively studied [41, 44–53, 53–67]. Extensions beyond single-field or canonical inflation have also been explored [42, 43, 68–75].

2.1 Background Exact Solution

We consider a canonical single-field inflation model described by the action

$$S = \int d^4x \sqrt{-g} \left[\frac{M_{\rm Pl}^2}{2} R - \frac{1}{2} g^{\mu\nu} \partial_\mu \phi \partial_\nu \phi - V(\phi) \right], \tag{1}$$

where $M_{\rm Pl} \equiv (8\pi G)^{-1/2}$ is the reduced Planck mass. We work in a spatially flat Friedmann-Lemaître-Robertson-Walker (FLRW) background metric

$$ds^2 = -dt^2 + a^2\delta_{ij}dx^i dx^j, \tag{2}$$

where $a = a(t)$ is the scale factor. The equations of motion are then given by

$$3M_{\rm Pl}^2 H^2 = \frac{1}{2}\dot{\phi}^2 + V, \tag{3}$$
$$-2M_{\rm Pl}^2 \dot{H} = \dot{\phi}^2, \tag{4}$$
$$\ddot{\phi} + 3H\dot{\phi} + V' = 0, \tag{5}$$

where $H \equiv \dot{a}/a$ is the Hubble parameter, and dots and primes denote derivatives with respect to t and ϕ, respectively. Only two of (3)–(5) are independent.

For a given potential $V(\phi)$, one can solve the equations of motion (3)–(5) to obtain the evolution of the scale factor $a(t)$ and the inflaton field $\phi(t)$. On the other hand, one can require a particular type of inflationary evolution and find a potential that accommodates it. The constant-roll inflation is a representative example of the latter approach.

Constant-roll inflation is characterized by the following condition [44]

$$\ddot{\phi} = \beta H\dot{\phi}, \tag{6}$$

where β is a constant. This generalizes the slow-roll case, $\ddot{\phi}/(H\dot{\phi}) \ll 1$, and the ultra-slow-roll case with $\beta = -3$ [76, 77]. While [44] derived an approximate solution, it was later demonstrated in [40] that an exact solution satisfying the constant-roll condition can be obtained (see also [78–81] for earlier related work).

The key to deriving an exact solution lies in the Hamilton-Jacobi formalism [15, 82], where the Hubble parameter H is treated as a function of ϕ. This approach is valid as long as $t = t(\phi)$ is a single-valued function of ϕ. Substituting $\dot{H} = H'\dot{\phi}$ into (4), we find

$$-2M_{\rm Pl}^2 H' = \dot{\phi}. \tag{7}$$

Taking the time derivative of (7) and using the constant-roll condition (6), we obtain

$$H'' = -\frac{\beta}{2M_{\rm Pl}^2} H. \tag{8}$$

The general solution $H(\phi)$ to this equation is given by a linear combination of $\exp\left(\pm\sqrt{\frac{-\beta}{2}}\frac{\phi}{M_{\rm Pl}}\right)$. With this solution $H(\phi)$, we can construct the potential $V(\phi)$ from (3) and (7) as

$$V(\phi) = 3M_{\rm Pl}^2 H^2 - 2M_{\rm Pl}^4 H'^2. \tag{9}$$

Integrating (7), we obtain the time evolution of ϕ, allowing us to express H as a function of time and determine the scale factor $a(t)$.

We remark that the above derivation does not rely on the slow-roll approximation. The slow-roll parameters, defined as

$$\epsilon_1 \equiv -\frac{d \ln H}{d \ln a}, \qquad \epsilon_{n+1} \equiv \frac{d \ln \epsilon_n}{d \ln a}, \tag{10}$$

are not necessarily small in constant-roll inflation. In fact, in the asymptotic regime $\phi \simeq 0$, corresponding to $t \to \mp\infty$ for $\beta \gtrless 0$, they approach [40] [see also (28) below]

$$2\epsilon_1 = \epsilon_{2n+1} \simeq 2|\beta| a^{2\beta} \to 0, \qquad \epsilon_{2n} \to 2\beta. \tag{11}$$

Thus, constant-roll inflation provides a controlled deviation from slow-roll dynamics, offering a versatile framework for understanding inflationary behavior beyond the standard paradigm. While originally developed as a phenomenological model, the emergence of constant-roll models from string theory has been discussed recently [51].

2.2 *Stability and Duality*

Although the Hamilton-Jacobi formalism provides exact solutions for constant-roll inflation, these solutions are not necessarily stable. Hence, it is crucial to analyze their stability separately. The stability analysis reveals intriguing properties and constraints on constant-roll models for different ranges of the parameter β.

A remarkable feature of constant-roll inflation is the duality between background solutions with β and $\alpha \equiv -(3+\beta)$. As noted in [40], the effective inflaton mass squared asymptotically approaches $\alpha\beta M^2$ as $\phi \simeq 0$, and the general solution for the inflaton is the sum of two exponential functions, $e^{\alpha M t}$ and $e^{\beta M t}$, where M is the characteristic energy scale of inflation. This duality is analogous to the one found in inflation with a quadratic potential [83]. In fact, the constant-roll potential can be approximated by a quadratic form near its extremum, leading to the same duality.

It is important to note that the construction of the constant-roll solution provides only the latter solution, $e^{\beta M t}$, which is consistent with the condition (6). Although the Hamilton-Jacobi formalism produces only one solution, each constant-roll solution with a given β has a dual counterpart with α. The stability properties of these dual solutions can differ. The attractor behavior of the inflationary dynamics is governed by one of the two solutions, while the other serves as a non-attractor solution, making the constant-roll condition unstable for that branch.

It has been explicitly demonstrated that constant-roll inflation is unstable for certain parameter regions [52, 55, 56]. For $\beta < -3/2$, numerical studies show that $\ddot{\phi}/(H\dot{\phi})$ asymptotically approaches $-(3+\beta)$, indicating that the constant-roll analytical solution is not an attractor, including the ultra-slow-roll case with $\beta = -3$. Conversely, for $-3/2 < \beta$, the constant-roll solution is stable and acts as the attractor.

The duality has been further investigated and provided rich phenomenology in the analysis of primordial perturbations [83–86]. The duality in the quadratic potential was recently revisited in [87] as a logarithmic duality of curvature perturbations. Extensions of this concept to constant-roll inflation have clarified characteristic non-Gaussian tails [66, 67]. The implications of logarithmic non-Gaussianity for primordial black holes and induced gravitational waves have also been studied recently [88, 89].

2.3 Perturbations

Let us consider the gauge invariant curvature perturbation, ζ, related to the metric perturbation through $\delta g_{ij} = a^2(1-2\zeta)\delta_{ij}$ in a gauge $\delta\phi = 0$. The evolution of the mode function $v_k \equiv \sqrt{2}M_{\rm Pl}z\zeta_k$ with $z \equiv a\sqrt{\epsilon_1}$ is governed by the Mukhanov-Sasaki equation [12, 13]:

$$\frac{d^2 v_k}{d\tau^2} + \left(k^2 - \frac{1}{z}\frac{d^2 z}{d\tau^2}\right) v_k = 0, \tag{12}$$

where τ is the conformal time defined by $d\tau = dt/a$. The potential term $\frac{1}{z}\frac{d^2 z}{d\tau^2}$ is exactly expressed in terms of slow-roll parameters:

$$\frac{1}{z}\frac{d^2 z}{d\tau^2} = a^2 H^2 \left(2 - \epsilon_1 + \frac{3}{2}\epsilon_2 + \frac{1}{4}\epsilon_2^2 - \frac{1}{2}\epsilon_1\epsilon_2 + \frac{1}{2}\epsilon_2\epsilon_3\right). \tag{13}$$

The asymptotic expression of the slow-roll parameters (11) allows us to derive a simple relation

$$\frac{1}{z}\frac{d^2 z}{d\tau^2} \simeq \frac{\nu^2 - 1/4}{\tau^2}, \qquad \nu \equiv |\beta + 3/2|. \tag{14}$$

and then the spectral index n_s of the curvature power spectrum evaluated at horizon exit is given by [40]:

$$n_s - 1 = 3 - |2\beta + 3|. \tag{15}$$

Thus, the spectrum generated by constant-roll inflation is scale-invariant for $\beta = 0$ or -3, red-tilted for $\beta < -3$ or $\beta > 0$, and blue-tilted for $-3 < \beta < 0$. We can also see that the same spectral index is predicted by background solutions with β and $\alpha \equiv -(3+\beta)$. Below, we shall explore the properties of red-tilted and blue-tilted models in greater detail.

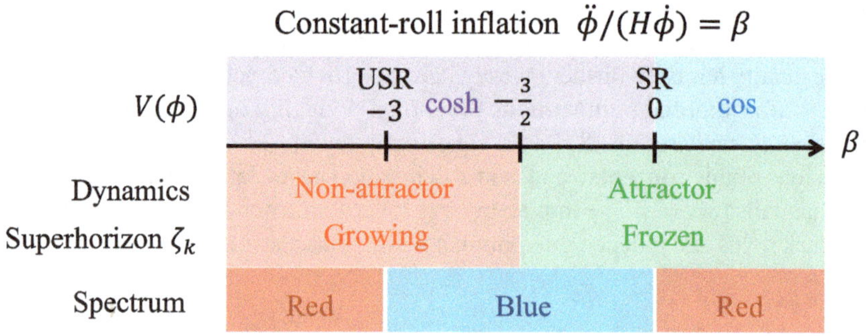

Fig. 1 Classification and properties of constant-roll models

An important aspect of constant-roll inflation is the behavior of the curvature perturbation ζ on superhorizon scales. The general solution of the Mukhanov-Sasaki equation (12) in this regime is given by

$$\zeta = A + B \int \frac{dt}{a^3 \epsilon_1}, \tag{16}$$

where A and B are constants. The first term remains constant, whereas the second term evolves in time. In standard slow-roll inflation, where ϵ_1 is nearly constant and small, this second term decays, causing the curvature perturbation to freeze on superhorizon scales. However, in more general case such as constant-roll inflation, the evolution of ϵ_1 can alter this behavior. Since the constant-roll condition (6) implies $\dot{\phi} \propto a^\beta$, it follows that $\epsilon_1 = \dot{\phi}^2/(2H^2) \simeq a^{2\beta}$. For $\beta < -3/2$, the second term in (16) grows, leading to a superhorizon growth of curvature perturbations. The ultra-slow-roll case ($\beta = -3$) is a well-known example of this phenomenon.

Constant-roll models can be classified based on the value of β, as illustrated in Fig. 1. Each regime exhibits distinct characteristics in terms of the potential shape, stability properties, superhorizon perturbation evolution, and spectral tilt.

In the following sections, we focus on attractor constant-roll models with $\beta > -3/2$. We examine the implications of a red-tilted spectrum for $\beta > 0$ and a blue-tilted spectrum for $-3/2 < \beta < 0$.

2.4 Red-Tilted Spectrum

Let us first consider the case of positive β. There are several potentials that maintain the constant-roll condition. Among them, a viable constant-roll potential that generates a red-tilted spectrum consistent with observational constraints is given by [40]

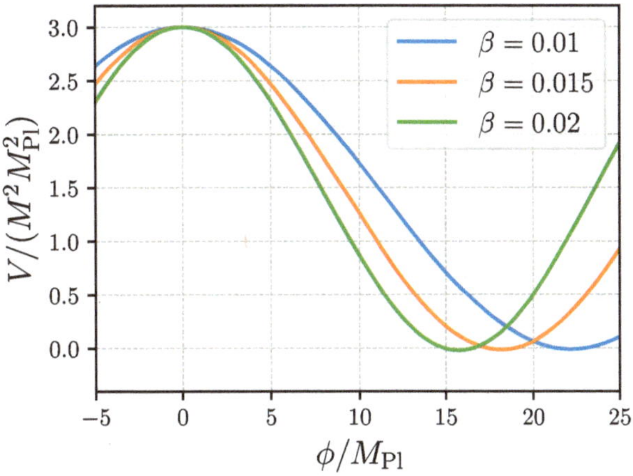

Fig. 2 Constant-roll potential (17) for $\beta > 0$

$$V(\phi) = 3M^2 M_{\rm Pl}^2 \left[1 - \frac{3+\beta}{6}\left\{1 - \cos\left(\sqrt{2\beta}\frac{\phi}{M_{\rm Pl}}\right)\right\}\right], \tag{17}$$

where M is the energy scale of inflation. The exact solutions for $a(t)$, $H(t)$ and $\phi(t)$ are given by

$$\phi = 2\sqrt{\frac{2}{\beta}} M_{\rm Pl} \arctan(e^{\beta M t}), \tag{18}$$

$$H = -M \tanh(\beta M t) = M \cos\left(\sqrt{\frac{\beta}{2}}\frac{\phi}{M_{\rm Pl}}\right), \tag{19}$$

$$a \propto \cosh^{-1/\beta}(\beta M t) = \sin^{1/\beta}\left(\sqrt{\frac{\beta}{2}}\frac{\phi}{M_{\rm Pl}}\right). \tag{20}$$

The potential is shown in Fig. 2 for representative parameters. In the following arguments we focus on the region $\phi \geq 0$. The inflaton asymptotically approaches the origin in the infinite past, and rolls down the potential in the positive direction as time progresses.

This potential resembles natural inflation [90] but includes an additional negative cosmological constant, $-M^2\beta$. The critical field value ϕ_c where $V = 0$ is given by

$$\phi_c = \frac{M_{\rm Pl}}{\sqrt{2\beta}} \arccos\left(1 - \frac{6}{3+\beta}\right) = M_{\rm Pl}\sqrt{\frac{2}{\beta}} \arcsin\sqrt{\frac{3}{3+\beta}}. \tag{21}$$

To ensure a graceful exit from constant-roll inflation, the potential must be truncated before reaching $V < 0$ at a cutoff field value ϕ_0, which satisfies $\phi_0 < \phi_c$. After ϕ_0

we assume a certain kind of phase transition that ends inflation. The cutoff can be chosen such that $\phi_0 \ll \phi_c$ or $\phi_0 \lesssim \phi_c$.

Let us denote ϕ_i as a field position where the generation of perturbations relevant to CMB scales occurs. This should occur sufficiently early, say 60 e-folds before inflation ends at ϕ_0. Consequently, regardless of whether the cutoff ϕ_0 is far from or close to the critical value ϕ_c, CMB scale perturbations are expected to be generated at $\phi = \phi_i \ll \phi_c$. In this region, the potential closely resembles that of quadratic hilltop inflation [91].

The number of e-folds counted from $\phi = \phi_c$ is given by

$$N \equiv \ln\left[\frac{a(\phi)}{a(\phi_c)}\right] = \frac{1}{\beta}\ln\left[\sqrt{\frac{3+\beta}{3}}\sin\left(\sqrt{\frac{\beta}{2}}\frac{\phi}{M_{\rm Pl}}\right)\right]. \tag{22}$$

Note that N is negative and increasing for $\phi < \phi_c$, meaning it does not measure the total duration of inflation but rather the e-folds until the critical field value ϕ_c where $V = 0$.

The CMB scale perturbations are generated at $\phi = \phi_i$ and inflation ends at some earlier point $\phi = \phi_0$, both of which are smaller than ϕ_c. While ϕ_c is fixed for a given β, ϕ_0 is not fixed and can be chosen any value smaller than ϕ_c in general. If one assumes that the CMB pivot scale exited the horizon 60 e-folds before the end of inflation, then N corresponding to the CMB scales satisfies $N \leq -60$ but not necessarily $N = -60$. Therefore, in addition to M and β, we have an additional parameter ϕ_i (or equivalently N) in the constant-roll model, and the prediction of CMB observables depends on these three parameters.

In [41], the slow-roll approximation was used to evaluate inflationary observables:

$$n_s - 1 \approx -6\epsilon_V + 2\eta_V, \tag{23}$$

$$r \approx 16\epsilon_V, \tag{24}$$

$$\frac{dn_s}{d\ln k} \approx 16\epsilon_V\eta_V - 24\epsilon_V^2 - 2\xi_V. \tag{25}$$

where the potential slow-roll parameters are defined as

$$\epsilon_V \equiv \frac{1}{2}\left(\frac{V'}{V}\right)^2, \quad \eta_V \equiv \frac{V''}{V}, \quad \xi_V \equiv \frac{V'V''}{V^2}, \tag{26}$$

and to constrain model parameters β and ϕ_i. It was found that $\beta \approx 0.015$ produces a red-tilted spectrum consistent with CMB constraints [40, 41].

We can also consider an approximation based on the Hubble-flow slow-roll parameters defined by (10). Since the Hubble parameter is expressed as

$$H = M\sqrt{1-h}, \qquad h \equiv \frac{3e^{2\beta N}}{3+\beta}, \tag{27}$$

the Hubble-flow slow-roll parameters (10) are expressed as

$$2\epsilon_1 = \epsilon_{2n+1} = \frac{2\beta h}{1-h}, \qquad \epsilon_{2n} = \frac{2\beta}{1-h}. \tag{28}$$

Applying the Hubble-flow slow-roll approximation yields

$$\Delta_\zeta^2 \approx \frac{H^2}{8\pi^2\epsilon_1} = \frac{M^2}{8\pi^2}\frac{(1-h)^2}{\beta h}, \tag{29}$$

$$n_s - 1 \approx -2\epsilon_1 - \epsilon_2 = -2\beta\frac{1+h}{1-h}, \tag{30}$$

$$r \approx 16\epsilon_1 = 16\beta\frac{h}{1-h}. \tag{31}$$

In general, the Hubble-flow slow-roll approximation works better than the potential slow-roll approximation as the latter assumes the approximate relations between the slow-roll parameters such as $\epsilon_1 \approx \epsilon_V$. The formulae (29)–(31) directly relate the observables to M, β, and N, the number of e-folds (22) counted from the critical field value $\phi = \phi_c$.

We see that there is a degeneracy between M and N since M only appears through the combination $M_s \equiv e^{-\beta N} M$ in (29)–(31). Therefore, the amplitude of the curvature power spectrum constrains the combination M_s rather than M [54].

As noted in [40], constant-roll inflation allows for a much smaller tensor-to-scalar ratio r than other scenarios due to the freedom in choosing N (or equivalently ϕ_i). By taking $|N|$ sufficiently large (or ϕ_i sufficiently small), r can be arbitrarily small. This topic was recently highlighted in [59], demonstrating that there is no lower bound on r for simple single-field inflation models.

Assuming the Hubble-flow slow-roll approximation, in Fig. 3, we compare the predicted spectral index and tensor-to-scalar ratio against constraints from Planck 2018, BICEP/Keck 2018, and BAO [92, 93]. The constant-roll inflation with $\beta \approx 0.015$ is consistent with the observational constraints. It covers the parameter region in a unique way and accommodates small r. While this result is based on the slow-roll approximation and in general deviates from numerical evaluations [50, 54], the qualitative behavior is robust.

The constant-roll inflation with a red-tilted spectrum serves as a simple, exactly solvable, and observationally viable model. It provides a unique prediction of the spectral index n_s and tensor-to-scalar ratio r, especially accommodating negligible r. Future observations will further constrain r, allowing this scenario to be tested in the near future.

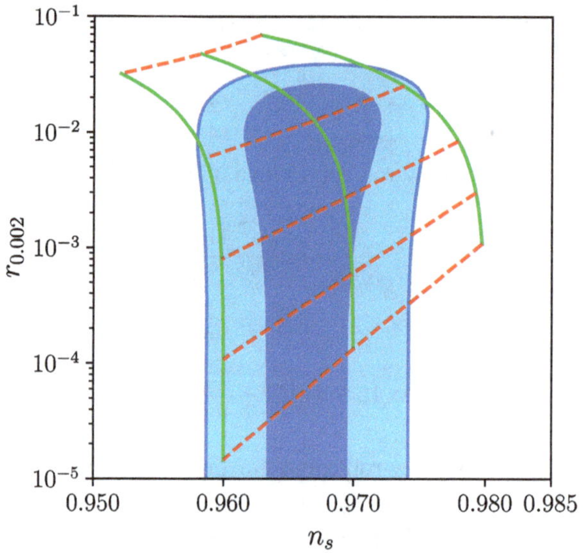

Fig. 3 Spectral index and tensor-to-scalar ratio predicted from constant-roll inflation assuming the Hubble-flow slow-roll approximation. $\beta = 0.010, 0.015, 0.020$ (green solid, from right to left) and $N = -60, -100, -150, -200, -250$ (red dashed, from top to bottom) are shown. Blue contours indicate the latest CMB constraints from Planck 2018, BICEP/Keck 2018, and BAO [92, 93], reproduced using the publicly available data and Python code released by the BICEP/Keck Collaboration (http://bicepkeck.org/bk18_2021_release.html)

2.5 Blue-Tilted Spectrum

Next, let us consider the case of negative β. In this case, the constant-roll potential is given by [40]

$$V(\phi) = 3M^2 M_{\rm Pl}^2 \left[1 - \frac{3+\beta}{6} \left\{ 1 - \cosh\left(\sqrt{2|\beta|} \frac{\phi}{M_{\rm Pl}} \right) \right\} \right], \tag{32}$$

and the corresponding exact solution for the background evolution is given by

$$\phi = M_{\rm Pl} \sqrt{\frac{2}{|\beta|}} \ln \left[\coth \left(\frac{|\beta|}{2} M t \right) \right], \tag{33}$$

$$H = M \coth(|\beta| M t) = M \cosh \left(\sqrt{\frac{|\beta|}{2}} \frac{\phi}{M_{\rm Pl}} \right), \tag{34}$$

$$a \propto \sinh^{1/|\beta|}(|\beta| M t) = \sinh^{-1/|\beta|} \left(\sqrt{\frac{|\beta|}{2}} \frac{\phi}{M_{\rm Pl}} \right). \tag{35}$$

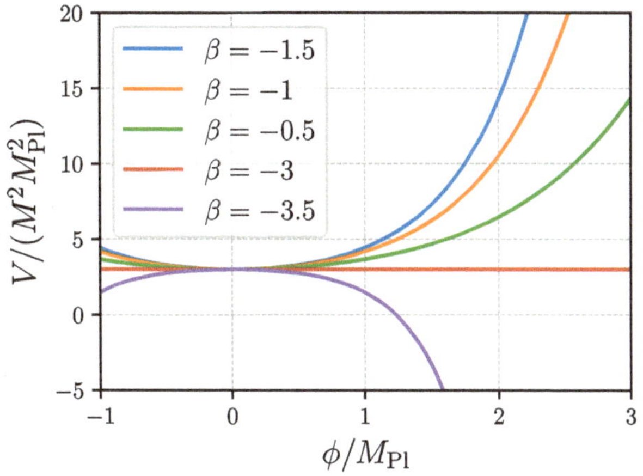

Fig. 4 Constant-roll potential (32) for $\beta < 0$

We show the potential in Fig. 4 for representative parameters. In the following arguments we focus on the region $\phi \geq 0$. For $\beta = -3$, the potential becomes constant, reducing to ultra-slow-roll inflation. The potential is convex for $-3 < \beta < 0$ and concave for $\beta < -3$. For the following, we focus on the regime $-3/2 < \beta < 0$, for which the above background solution is the attractor, and superhorizon curvature perturbations freeze. The inflaton monotonically rolls down the potential, asymptotically approaching the origin in the infinite future.

As seen from (15), this model generates a blue-tilted spectrum, making it particularly relevant to the formation of primordial black holes (PBHs) [94–97], which may form from the direct collapse of overdense regions in the early universe due to large primordial perturbations on small scales (for recent reviews, see [98–100]).

Constant-roll inflation provides a natural framework for PBH-generating models in connection with the no-go theorem of slow roll for PBH production in canonical single-field inflation [101]. This theorem states that to generate a peak in the curvature power spectrum relevant to PBH formation, a transient violation of the slow-roll condition is necessary. This is because the amplitude of the curvature power spectrum, constrained to $\mathcal{O}(10^{-9})$ on CMB scales, requires a significant enhancement to $\mathcal{O}(10^{-2})$ within 60 efolds, as depicted in Fig. 5.

The slow-roll violation necessary for PBH production is quantified by [101]

$$\frac{\Delta \ln \epsilon_1}{\Delta N} \lesssim -0.4. \tag{36}$$

where ΔN corresponds to the number of efolds between CMB and PBH scales, and $\Delta \ln \epsilon_1$ is the corresponding change in the slow-roll parameter. This bound corresponds to the scenario where the smallest mass (or equivalently the largest ΔN) PBHs constitute all the dark matter.

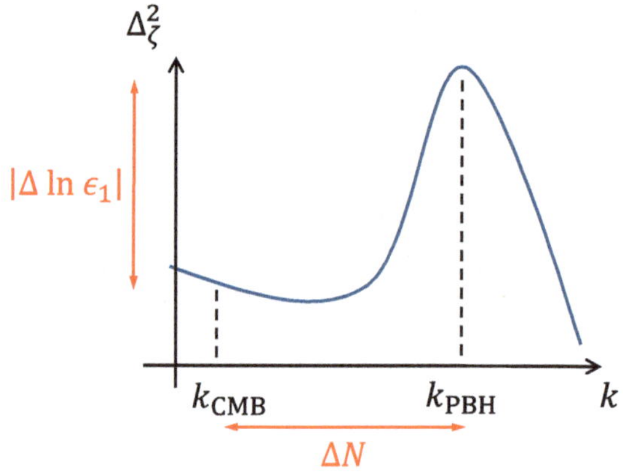

Fig. 5 Schematic representation of the no-go theorem of slow roll in PBH production

Stronger slow-roll violations are required for different PBH scenarios. For instance, if PBHs are responsible for the observed microlensing events towards the Galactic bulge, PBHs are roughly in the Earth-Jupiter mass range, $\mathcal{O}(10^{28})$ g, and constitute $\mathcal{O}(1)\%$ of dark matter [102]. The bound in this case tightens to $\Delta \ln \epsilon_1/\Delta N \lesssim -0.7$. On the other hand, if PBHs are responsible for gravitational wave events detected by LIGO-Virgo-KAGRA collaborations [103–106], the mass is $\mathcal{O}(10) M_\odot$ and the abundance is $\mathcal{O}(0.1)\%$ of dark matter [107]. The bound further strengthens to $\Delta \ln \epsilon_1/\Delta N \lesssim -1$.

Although the above estimations are based on various approximations carrying large uncertainties, they all enter into $\Delta \ln \epsilon_1/\Delta N$ logarithmically. Even orders of magnitude changes in the mass scale and power spectrum amplification would not qualitatively change this result.

These constraints inform viable inflationary models for PBH formation. The bound on $\Delta \ln \epsilon_1/\Delta N$ implies that the amplitude of the second slow-roll parameter, $\epsilon_2 = d \ln \epsilon_1/dN$, should be large. The simplest scenario would be to consider a constant but nonnegligible value of ϵ_2, which is where constant-roll models come into play. Such a constant-roll phase responsible for the significant enhancement of curvature power spectrum should be sandwiched between different phases of inflation, which are responsible for generating CMB scale perturbations and preventing overproduction of PBHs. Thus, we are interested in a transient phase of blue-tilted constant-roll inflation.

A special case $\epsilon_2 = -6$ corresponds to the ultra-slow-roll inflation with the flat plateau potential. The PBH production scenarios from transient ultra-slow-roll inflation or inflection models have been actively investigated [51, 61, 101, 108–129]. However, ultra-slow-roll inflation, despite its simplest constant potential, is not straightforward. In terms of slow-roll violation, ultra-slow-roll inflation represents an extreme case, with $\epsilon_2 = -6$, far exceeding the minimum violation (36), $\epsilon_2 \lesssim -0.4$, required for PBH production. It features a non-attractor phase where superhorizon

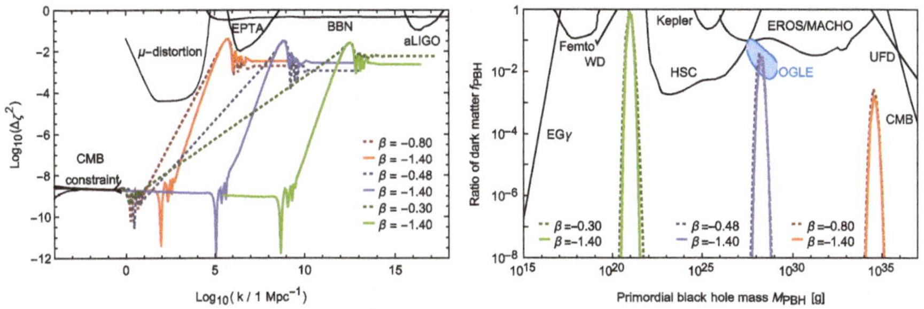

Fig. 6 Enhancement of the curvature power spectrum (left) and the resulting PBH abundance (right) from transient constant-roll inflation. Reprinted with permission from [58].

perturbations grow, leading to diffusion effects. These features require special care in analysis.

A more stable alternative is transient constant-roll inflation [51, 53, 57, 57, 58, 60–63, 130, 131] which naturally accommodates slow-roll violations necessary for PBH formation while maintaining an attractor behavior for $-3/2 < \beta < 0$. A simple piecewise model that realizes a transient constant-roll phase between two slow-roll phases is demonstrated in [58], where the spectral tilt of the enhancement can be controlled by the constant-roll parameter, providing greater flexibility and various PBH production scenarios, as illustrated in Fig. 6. The non-Gaussianity and one-loop corrections in such models have also been explored recently [62, 67]. Beyond exact constant-roll solution, approximate constant-roll behavior can be realized in potentials with bump-like features. In such models, a transient nearly-constant-roll phase leads to a localized enhancement of the power spectrum [57, 130].

A transient phase of blue-tilted constant-roll inflation provides a more flexible alternative to transient ultra-slow-roll scenarios. Its advantages make the model a compelling framework for PBH formation, providing both theoretical robustness and observational viability. Future studies, including refined non-Gaussianity analyses and constraints from gravitational wave experiments, will further test the predictions of this model.

3 Conclusion

Similar to many other models proposed by Starobinsky, constant-roll inflation provides a simple yet elegant framework with valuable insights, offering a rich playground for exploring inflationary dynamics beyond the standard slow-roll paradigm. It allows for an exact solution where the inflaton maintains a constant rate of roll. Depending on the parameter branches, constant-roll inflation can generate either a red-tilted spectrum of curvature perturbations, consistent with the latest observational constraints, or a blue-tilted spectrum conducive to the generation of PBHs. In both cases, future observations and theoretical developments will be essential for testing the model and deepening our understanding of the early universe.

Competing Interests The authors have no conflicts of interest to declare that are relevant to the content of this chapter.

Acknowledgements I am deeply grateful to Alexei A. Starobinsky for his collaborations, insightful discussions, and invaluable support over the years, as well as his friendship I will always cherish. I would also like to thank my collaborators, Wayne Hu, Ryoto Inui, Cristian Joana, Jerome Martin, Shinji Mukohyama, Michele Oliosi, Samuel Passaglia, Shi Pi, Teruaki Suyama, Yuichiro Tada, Jun'ichi Yokoyama, and Shuichiro Yokoyama for the fruitful collaborations on the topics discussed in this paper. This work was supported by JSPS KAKENHI Grant No. JP22K03639.

Appendix

In the main text, we discuss that the key to constructing exact solutions in constant-roll inflation is the Hamilton-Jacobi formalism. This approach is also useful for obtaining exact de Sitter constant-roll solutions in scalar-tensor gravity, as demonstrated in [43].

In this Appendix, we present additional examples of exact de Sitter solutions with a non-constant scalar field, including those beyond constant-roll models. These cases were originally discussed and considered for inclusion in [43] but were ultimately left out.

Let us consider the following scalar-tensor theory with a non-minimal coupling of a scalar field to gravity:

$$S = \int d^4x\sqrt{-g}\left[\frac{1}{2}F(\phi)R - \frac{1}{2}g^{\mu\nu}\partial_\mu\phi\partial_\nu\phi - U(\phi)\right]. \tag{37}$$

With the spatially flat FLRW background metric (2), the equations of motion for the action (37) are given by

$$3FH^2 = \frac{1}{2}\dot{\phi}^2 + U - 3H\dot{F}, \tag{38}$$

$$-2F\dot{H} = \dot{\phi}^2 + \ddot{F} - H\dot{F}, \tag{39}$$

$$\ddot{\phi} + 3H\dot{\phi} + U' - 3(\dot{H} + 2H^2)F' = 0. \tag{40}$$

We can derive the exact de Sitter solution with a constant scalar field as follows. Substituting constant profile $H = H_0$ and $\phi = \phi_0$ to (38) and (40), we obtain

$$3F(\phi_0)H_0^2 = U(\phi_0), \tag{41}$$

$$6F'(\phi_0)H_0^2 = U'(\phi_0). \tag{42}$$

Namely, ϕ_0 for the de Sitter solution is obtained as a root of the following algebraic equation:

$$\left(\frac{U}{F^2}\right)'\bigg|_{\phi=\phi_0} = 0, \tag{43}$$

and then H_0 is determined by (41) or (42).

Let us derive all the exact de Sitter solutions with a non-constant scalar field assuming $H = H_0$ and $\dot{\phi} \neq 0$, which is a more nontrivial task. We adopt a Hamilton-Jacobi formalism and regard $\dot{\phi}$ as a function of ϕ:

$$\dot{\phi} = A(\phi). \tag{44}$$

Then, (38) and (39) are rewritten as

$$A' + \frac{F'' + 1}{F'} = H_0, \tag{45}$$

$$U = 3H_0^2 F + 3H_0 F' A - \tfrac{1}{2} A^2. \tag{46}$$

Using the method of integrating factors, we can solve the differential equation (45) for $A(\phi)$ to obtain

$$A(\phi) = \frac{H_0}{F'} \exp\left(-\int \frac{d\phi}{F'}\right) \int d\phi F' \exp\left(\int \frac{d\phi}{F'}\right). \tag{47}$$

Plugging this into (46), we obtain the potential $U(\phi)$ that accommodates the exact de Sitter solutions. Note that this branch shows up only when $F' \neq 0$, i.e., when the scalar field non-minimally couples to gravity. The de Sitter branch with $\dot{\phi} \neq 0$ is not possible in the case of Einstein-Hilbert action with a canonical scalar field.

The de Sitter solution (47) includes both constant-roll and non-constant-roll solutions since it does not satisfy the constant-roll condition (6) in general. Indeed, we see from (47) that

$$\frac{\ddot{\phi}}{H_0 \dot{\phi}} = \frac{A'}{H_0} = 1 - \frac{F'' + 1}{F'^2} \exp\left(-\int \frac{d\phi}{F'}\right) \int d\phi F' \exp\left(\int \frac{d\phi}{F'}\right), \tag{48}$$

which is not constant in general. From (48), we can derive a simple criterion: The exact de Sitter solution (47) is also constant-roll one when δ is constant, which we define as

$$\delta = \frac{(F'' + 1)^2}{-F''' F' + F''(2F'' + 3) + 1}, \tag{49}$$

and $\delta = 1 - \beta$ holds when it is constant.

Now, let us focus on a particular subclass, the case of non-minimally coupled scalar field with

$$F(\phi) = F_0 - \xi \phi^2, \tag{50}$$

where F_0 and ξ are nonvanishing constants. In this case, we can obtain a simple solution, which turns out to be nothing but the constant-roll de Sitter solution obtained in [43], showing its uniqueness.

The solution (47) reads

$$A(\phi) = \frac{2\xi}{4\xi - 1} H_0 \phi + C H_0^{-\frac{1}{2\xi}+3} \phi^{\frac{1}{2\xi}-1}, \tag{51}$$

where C is a dimensionless integration constant. By plugging (51) into (46), we obtain the potential $U(\phi)$ as

$$\begin{aligned} U(\phi) = 3F_0 H_0^2 - \frac{\xi(6\xi-1)(16\xi-3)}{(4\xi-1)^2} H_0^2 \phi^2 \\ -\frac{1}{2} C^2 H_0^4 \left(\frac{\phi}{H_0}\right)^{\frac{1}{\xi}-2} - \frac{4C\xi(6\xi-1)}{4\xi-1} H_0^4 \left(\frac{\phi}{H_0}\right)^{\frac{1}{2\xi}}. \end{aligned} \tag{52}$$

Requiring the boundedness of the potential $U(\phi)$ prohibits solution with all $C \neq 0$. Namely, we can show that if $C \neq 0$, the potential (52) is unbounded from below for $\xi \neq 0$. The key is that the coefficient of the third term in (52) is always negative. First, when $\xi < 0$ or $\xi > \frac{1}{2}$, we have $U \to -\infty$ as $\phi \to 0$ due to the third term. Second, when $0 < \xi \leq \frac{1}{4}$, the third term has the largest positive exponent, and hence $U \to -\infty$ as $\phi \to \infty$. Finally, when $\frac{1}{4} \leq \xi \leq \frac{1}{2}$, the second term has the largest positive exponent and negative coefficient, and hence $U \to -\infty$ as $\phi \to \infty$.

For $C = 0$, the potential (52) is given by

$$U(\phi) = 3F_0 H_0^2 - \frac{\xi(6\xi-1)(16\xi-3)}{(4\xi-1)^2} H_0^2 \phi^2, \tag{53}$$

and integrating (51), the inflaton field $\phi(t)$ as a function of time is given by

$$\phi(t) \propto H_0 e^{\frac{2\xi}{4\xi-1} H_0 t}. \tag{54}$$

The potential (53) is bounded from below for $\xi \leq 0$ and for $\frac{1}{6} \leq \xi \leq \frac{3}{16}$. In both cases, this exact de Sitter solution with a massive non-minimally coupled scalar field is a late-time attractor.

Let us remark on the relation to the previous works. The solution with $\xi < 0$ was known long ago [78, 132], and a special case of the conformal coupling $\xi = \frac{1}{6}$ and constant potential was studied in [81]. The solutions of all cases fall into the constant-roll de Sitter solution derived in [43]. In [43], we started from the constant-roll condition (6) and found the de Sitter solution with the constant-roll parameter $\beta = \frac{2\xi}{4\xi-1}$ for a wider class of theory with $F(\phi) = \frac{\beta}{2(1-2\beta)}\phi^2 + c_1 \beta \phi^{\frac{1}{\beta}} + c_2$. On the other hand, in the above derivation, we start from the potential (46) with (47) that accommodates all the de Sitter solutions, and obtain the constant-roll solution. This shows that the constant-roll de Sitter solution found in [43] is the unique exact de Sitter solution in the case of a massive non-minimally coupled scalar field with (50) and bounded potential.

References

1. R. Brout, F. Englert and E. Gunzig, *The creation of the universe as a quantum phenomenon*. Ann. Phys. **115**, 78 (1978), https://doi.org/10.1016/0003-4916(78)90176-8
2. A. A. Starobinsky, *A New Type of Isotropic Cosmological Models Without Singularity*. Phys. Lett. B **91**, 99 (1980), https://doi.org/10.1016/0370-2693(80)90670-X
3. D. Kazanas, *Dynamics of the Universe and Spontaneous Symmetry Breaking*. Astrophys. J. Lett. **241**, L59 (1980), https://doi.org/10.1086/183361
4. K. Sato, *First-order phase transition of a vacuum and the expansion of the Universe*. Mon. Not. Roy. Astron. Soc. **195**, 467 (1981), https://doi.org/10.1093/mnras/195.3.467
5. A. H. Guth, *Inflationary universe: A possible solution to the horizon and flatness problems*. Phys. Rev. D **23**, 347 (1981), https://doi.org/10.1103/PhysRevD.23.347
6. A. Linde, *A new inflationary universe scenario: A possible solution of the horizon, flatness, homogeneity, isotropy and primordial monopole problems*. Phys. Lett. B **108**, 389 (1982), https://doi.org/10.1016/0370-2693(82)91219-9
7. A. Albrecht and P. J. Steinhardt, *Cosmology for grand unified theories with radiatively induced symmetry breaking*. Phys. Rev. Lett. **48**, 1220 (1982), https://doi.org/10.1103/PhysRevLett.48.1220
8. A.A. Starobinsky, Spectrum of relict gravitational radiation and the early state of the universe. JETP Lett. **30**, 682 (1979)
9. V.F. Mukhanov, G.V. Chibisov, Quantum Fluctuations and a Nonsingular Universe. JETP Lett. **33**, 532 (1981)
10. A. A. Starobinsky, *Dynamics of Phase Transition in the New Inflationary Universe Scenario and Generation of Perturbations*. Phys. Lett. B **117**, 175 (1982), https://doi.org/10.1016/0370-2693(82)90541-X
11. J. M. Bardeen, P. J. Steinhardt and M. S. Turner, *Spontaneous Creation of Almost Scale - Free Density Perturbations in an Inflationary Universe*. Phys. Rev. D **28**, 679 (1983), https://doi.org/10.1103/PhysRevD.28.679
12. V.F. Mukhanov, Gravitational Instability of the Universe Filled with a Scalar Field. JETP Lett. **41**, 493 (1985)
13. M. Sasaki, *Large Scale Quantum Fluctuations in the Inflationary Universe*. Prog. Theor. Phys. **76**, 1036 (1986), https://doi.org/10.1143/PTP.76.1036
14. E. Lifshitz and I. Khalatnikov, *About singularities of cosmological solutions of the gravitational equations. I*. Zh. Eksp. Teor. Fiz. **39**, 149 (1960), https://doi.org/10.1016/0370-2693(82)90541-X
15. D. S. Salopek and J. R. Bond, *Nonlinear evolution of long wavelength metric fluctuations in inflationary models*. Phys. Rev. D **42**, 3936 (1990), https://doi.org/10.1103/PhysRevD.42.3936
16. G. L. Comer, N. Deruelle, D. Langlois and J. Parry, *Growth or decay of cosmological inhomogeneities as a function of their equation of state*. Phys. Rev. D **49**, 2759 (1994), https://doi.org/10.1103/PhysRevD.49.2759
17. M. Sasaki and E. D. Stewart, *A General analytic formula for the spectral index of the density perturbations produced during inflation*. Prog. Theor. Phys. **95**, 71 (1996), https://doi.org/10.1143/PTP.95.71
18. M. Sasaki and T. Tanaka, *Superhorizon scale dynamics of multiscalar inflation*. Prog. Theor. Phys. **99**, 763 (1998), https://doi.org/10.1143/PTP.99.763

19. D. Wands, K. A. Malik, D. H. Lyth and A. R. Liddle, *A New approach to the evolution of cosmological perturbations on large scales*. Phys. Rev. D **62**, 043527 (2000), https://doi.org/10.1103/PhysRevD.62.043527
20. D. H. Lyth and D. Wands, *Conserved cosmological perturbations*. Phys. Rev. D **68**, 103515 (2003), https://doi.org/10.1103/PhysRevD.68.103515
21. G. I. Rigopoulos and E. P. S. Shellard, *The separate universe approach and the evolution of nonlinear superhorizon cosmological perturbations*. Phys. Rev. D **68**, 123518 (2003), https://doi.org/10.1103/PhysRevD.68.123518
22. D. H. Lyth, K. A. Malik and M. Sasaki, *A General proof of the conservation of the curvature perturbation*. JCAP **05**, 004 (2005), https://doi.org/10.1088/1475-7516/2005/05/004
23. D. H. Lyth and Y. Rodriguez, *The Inflationary prediction for primordial non-Gaussianity*. Phys. Rev. Lett. **95**, 121302 (2005), https://doi.org/10.1103/PhysRevLett.95.121302
24. A. A. Starobinsky, *STOCHASTIC DE SITTER (INFLATIONARY) STAGE IN THE EARLY UNIVERSE*. Lect. Notes Phys. **246**, 107 (1986), https://doi.org/10.1007/3-540-16452-9_6
25. A. A. Starobinsky and J. Yokoyama, *Equilibrium state of a selfinteracting scalar field in the De Sitter background*. Phys. Rev. D **50**, 6357 (1994), https://doi.org/10.1103/PhysRevD.50.6357
26. L. Kofman, A. D. Linde and A. A. Starobinsky, *Reheating after inflation*. Phys. Rev. Lett. **73**, 3195 (1994), https://doi.org/10.1103/PhysRevLett.73.3195
27. L. Kofman, A. D. Linde and A. A. Starobinsky, *Towards the theory of reheating after inflation*. Phys. Rev. D **56**, 3258 (1997), https://doi.org/10.1103/PhysRevD.56.3258
28. P. B. Greene, L. Kofman, A. D. Linde and A. A. Starobinsky, *Structure of resonance in preheating after inflation*. Phys. Rev. D **56**, 6175 (1997), https://doi.org/10.1103/PhysRevD.56.6175
29. H. Motohashi, A. A. Starobinsky and J. Yokoyama, *Analytic solution for matter density perturbations in a class of viable cosmological f(R) models*. Int. J. Mod. Phys. D **18**, 1731 (2009), https://doi.org/10.1142/S0218271809015278
30. H. Motohashi, A. A. Starobinsky and J. Yokoyama, *Phantom behaviour and growth index anomalous evolution in viable $f(R)$ gravity models*. 19th Workshop on General Relativity and Gravitation in Japan (2009)
31. H. Motohashi, A. A. Starobinsky and J. Yokoyama, *Phantom boundary crossing and anomalous growth index of fluctuations in viable f(R) models of cosmic acceleration*. Prog. Theor. Phys. **123**, 887 (2010), https://doi.org/10.1143/PTP.123.887
32. H. Motohashi, A. A. Starobinsky and J. Yokoyama, *Phantom behaviour and growth index anomalous evolution in viable f(R) gravity models*. arXiv: arXiv:1002.0462 (2010)
33. H. Motohashi, A. A. Starobinsky and J. Yokoyama, *Matter power spectrum in f(R) gravity with massive neutrinos*. Prog. Theor. Phys. **124**, 541 (2010), https://doi.org/10.1143/PTP.124.541
34. H. Motohashi, A. A. Starobinsky and J. Yokoyama, *f(R) Gravity and its Cosmological Implications*. Int. J. Mod. Phys. D **20**, 1347 (2011), https://doi.org/10.1142/S0218271811019529
35. H. Motohashi, A. A. Starobinsky and J. Yokoyama, *Future Oscillations around Phantom Divide in f(R) Gravity*. JCAP **06**, 006 (2011), https://doi.org/10.1088/1475-7516/2011/06/006
36. H. Motohashi, A. A. Starobinsky and J. Yokoyama, *Classification of Future Phantom-to-Normal Oscillations in $f(R)$ Gravity*. 20th Workshop on General Relativity and Gravitation in Japan, 286 (2011)
37. H. Motohashi, A. A. Starobinsky and J. Yokoyama, *f(R) cosmology and massive neutrinos*. Int. J. Mod. Phys. Conf. Ser. **10**, 35 (2012), https://doi.org/10.1142/S2010194512005739
38. H. Motohashi, A. A. Starobinsky and J. Yokoyama, *Cosmology Based on f(R) Gravity Admits 1 eV Sterile Neutrinos*. Phys. Rev. Lett. **110**, 121302 (2013), https://doi.org/10.1103/PhysRevLett.110.121302
39. H. Motohashi, A. A. Starobinsky and J. Yokoyama, *Joint description of inflation and dark energy in extended $f(R)$ gravity*. 21st Workshop on General Relativity and Gravitation in Japan, 34 (2012)

40. H. Motohashi, A. A. Starobinsky and J. Yokoyama, *Inflation with a constant rate of roll*. JCAP **09**, 018 (2015), https://doi.org/10.1088/1475-7516/2015/09/018
41. H. Motohashi and A. A. Starobinsky, *Constant-roll inflation: confrontation with recent observational data*. EPL **117**, 39001 (2017), https://doi.org/10.1209/0295-5075/117/39001
42. H. Motohashi and A. A. Starobinsky, $f(R)$ *constant-roll inflation*. Eur. Phys. J. C **77**, 538 (2017), https://doi.org/10.1140/epjc/s10052-017-5109-x
43. H. Motohashi and A. A. Starobinsky, *Constant-roll inflation in scalar-tensor gravity*. JCAP **11**, 025 (2019), https://doi.org/10.1088/1475-7516/2019/11/025
44. J. Martin, H. Motohashi and T. Suyama, *Ultra Slow-Roll Inflation and the non-Gaussianity Consistency Relation*. Phys. Rev. D **87**, 023514 (2013), https://doi.org/10.1103/PhysRevD.87.023514
45. Q. Gao and Y. Gong, *Reconstruction of extended inflationary potentials for attractors*. Eur. Phys. J. Plus **133**, 491 (2018), https://doi.org/10.1140/epjp/i2018-12324-3
46. F. Cicciarella, J. Mabillard and M. Pieroni, *New perspectives on constant-roll inflation*. JCAP **01**, 024 (2018), https://doi.org/10.1088/1475-7516/2018/01/024
47. L. Anguelova, P. Suranyi and L. C. R. Wijewardhana, *Systematics of Constant Roll Inflation*. JCAP **02**, 004 (2018), https://doi.org/10.1088/1475-7516/2018/02/004
48. A. Karam, L. Marzola, T. Pappas, A. Racioppi and K. Tamvakis, *Constant-Roll (Quasi-)Linear Inflation*. JCAP **05**, 011 (2018), https://doi.org/10.1088/1475-7516/2018/05/011
49. Z. Yi and Y. Gong, *On the constant-roll inflation*. JCAP **03**, 052 (2018), https://doi.org/10.1088/1475-7516/2018/03/052
50. Q. Gao, *The observational constraint on constant-roll inflation*. Sci. China Phys. Mech. Astron. **61**, 070411 (2018), https://doi.org/10.1007/s11433-018-9197-2
51. M. Cicoli, V. A. Diaz and F. G. Pedro, *Primordial Black Holes from String Inflation*. JCAP **06**, 034 (2018), https://doi.org/10.1088/1475-7516/2018/06/034
52. M. J. P. Morse and W. H. Kinney, *Large-*η *constant-roll inflation is never an attractor*. Phys. Rev. D **97**, 123519 (2018), https://doi.org/10.1103/PhysRevD.97.123519
53. D. Cruces, C. Germani and T. Prokopec, *Failure of the stochastic approach to inflation beyond slow-roll*. JCAP **03**, 048 (2019), https://doi.org/10.1088/1475-7516/2019/03/048
54. J. T. Galvez Ghersi, A. Zucca and A. V. Frolov, *Observational Constraints on Constant Roll Inflation*. JCAP **05**, 030 (2019), https://doi.org/10.1088/1475-7516/2019/05/030
55. Q. Gao, Y. Gong and Z. Yi, *On the constant-roll inflation with large and small* η_H. Universe **5**, 215 (2019), https://doi.org/10.3390/universe5110215
56. W.-C. Lin, M. J. P. Morse and W. H. Kinney, *Dynamical Analysis of Attractor Behavior in Constant Roll Inflation*. JCAP **09**, 063 (2019), https://doi.org/10.1088/1475-7516/2019/09/063
57. V. Atal, J. Cid, A. Escrivá and J. Garriga, *PBH in single field inflation: the effect of shape dispersion and non-Gaussianities*. JCAP **05**, 022 (2020), https://doi.org/10.1088/1475-7516/2020/05/022
58. H. Motohashi, S. Mukohyama and M. Oliosi, *Constant Roll and Primordial Black Holes*. JCAP **03**, 002 (2020), https://doi.org/10.1088/1475-7516/2020/03/002
59. N. K. Stein and W. H. Kinney, *Simple single-field inflation models with arbitrarily small tensor/scalar ratio*. JCAP **03**, 027 (2023), https://doi.org/10.1088/1475-7516/2023/03/027
60. M. W. Davies, P. Carrilho and D. J. Mulryne, *Non-Gaussianity in inflationary scenarios for primordial black holes*. JCAP **06**, 019 (2022), https://doi.org/10.1088/1475-7516/2022/06/019
61. A. Karam, N. Koivunen, E. Tomberg, V. Vaskonen and H. Veermäe, *Anatomy of single-field inflationary models for primordial black holes*. JCAP **03**, 013 (2023), https://doi.org/10.1088/1475-7516/2023/03/013
62. H. Motohashi and Y. Tada, *Squeezed bispectrum and one-loop corrections in transient constant-roll inflation*. JCAP **08**, 069 (2023), https://doi.org/10.1088/1475-7516/2023/08/069
63. S. S. Mishra, E. J. Copeland and A. M. Green, *Primordial black holes and stochastic inflation beyond slow roll. Part I. Noise matrix elements*. JCAP **09**, 005 (2023), https://doi.org/10.1088/1475-7516/2023/09/005

64. E. Tomberg, *Stochastic constant-roll inflation and primordial black holes*. Phys. Rev. D **108**, 043502 (2023), https://doi.org/10.1103/PhysRevD.108.043502
65. G. Tasinato, *Large $|\eta|$ approach to single field inflation*. Phys. Rev. D **108**, 043526 (2023), https://doi.org/10.1103/PhysRevD.108.043526
66. Y. Wang, Q. Gao, S. Gao and Y. Gong, *On the duality in constant-roll inflation*. arXiv: arXiv:2404.18548 (2024)
67. R. Inui, H. Motohashi, S. Pi, Y. Tada and S. Yokoyama, *Constant roll and non-Gaussian tail in light of logarithmic duality*. JCAP **02**, 042 (2025), https://doi.org/10.1088/1475-7516/2025/02/042
68. S. Hirano, T. Kobayashi and S. Yokoyama, *Ultra slow-roll G-inflation*. Phys. Rev. D **94**, 103515 (2016), https://doi.org/10.1103/PhysRevD.94.103515
69. A. Ito and J. Soda, *Anisotropic Constant-roll Inflation*. Eur. Phys. J. C **78**, 55 (2018), https://doi.org/10.1140/epjc/s10052-018-5534-5
70. A. Mohammadi, K. Saaidi and T. Golanbari, *Tachyon constant-roll inflation*. Phys. Rev. D **97**, 083006 (2018), https://doi.org/10.1103/PhysRevD.97.083006
71. Q. Gao, Y. Gong and Q. Fei, *Constant-roll tachyon inflation and observational constraints*. JCAP **05**, 005 (2018), https://doi.org/10.1088/1475-7516/2018/05/005
72. A. Mohammadi, K. Saaidi and H. Sheikhahmadi, *Constant-roll approach to non-canonical inflation*. Phys. Rev. D **100**, 083520 (2019), https://doi.org/10.1103/PhysRevD.100.083520
73. A. Mohammadi, T. Golanbari, S. Nasri and K. Saaidi, *Constant-roll brane inflation*. Phys. Rev. D **101**, 123537 (2020), https://doi.org/10.1103/PhysRevD.101.123537
74. M. Guerrero, D. Rubiera-Garcia and D. Saez-Chillon Gomez, *Constant roll inflation in multifield models*. Phys. Rev. D **102**, 123528 (2020), https://doi.org/10.1103/PhysRevD.102.123528
75. J. C. Garnica, L. G. Gomez, A. A. Navarro and Y. Rodriguez, *Constant-Roll Inflation in the Generalized SU(2) Proca Theory*. Annalen Phys. **534**, 2100453 (2022), https://doi.org/10.1002/andp.202100453
76. N. Tsamis and R. P. Woodard, *Improved estimates of cosmological perturbations*. Phys. Rev. D **69**, 084005 (2004), https://doi.org/10.1103/PhysRevD.69.084005
77. W. H. Kinney, *Horizon crossing and inflation with large eta*. Phys. Rev. D **72**, 023515 (2005), https://doi.org/10.1103/PhysRevD.72.023515
78. J. D. Barrow and K.-i. Maeda, *Extended inflationary universes*. Nucl. Phys. B **341**, 294 (1990), https://doi.org/10.1016/0550-3213(90)90272-F
79. J. D. Barrow, *Exact inflationary universes with potential minima*. Phys. Rev. D **49**, 3055 (1994), https://doi.org/10.1103/PhysRevD.49.3055
80. C. R. Contaldi, M. Peloso, L. Kofman and A. D. Linde, *Suppressing the lower multipoles in the CMB anisotropies*. JCAP **07**, 002 (2003), https://doi.org/10.1088/1475-7516/2003/07/002
81. L. Kofman and S. Mukohyama, *Rapid roll Inflation with Conformal Coupling*. Phys. Rev. D **77**, 043519 (2008), https://doi.org/10.1103/PhysRevD.77.043519
82. A. G. Muslimov, *On the Scalar Field Dynamics in a Spatially Flat Friedman Universe*. Class. Quant. Grav. **7**, 231 (1990), https://doi.org/10.1088/0264-9381/7/2/015
83. K. Tzirakis and W. H. Kinney, *Inflation over the hill*. Phys. Rev. D **75**, 123510 (2007), https://doi.org/10.1103/PhysRevD.75.123510
84. D. Wands, *Duality invariance of cosmological perturbation spectra*. Phys. Rev. D **60**, 023507 (1999), https://doi.org/10.1103/PhysRevD.60.023507
85. S. M. Leach and A. R. Liddle, *Inflationary perturbations near horizon crossing*. Phys. Rev. D **63**, 043508 (2001), https://doi.org/10.1103/PhysRevD.63.043508
86. A. A. Starobinsky, *Inflaton field potential producing the exactly flat spectrum of adiabatic perturbations*. JETP Lett. **82**, 169 (2005), https://doi.org/10.1134/1.2121807
87. S. Pi and M. Sasaki, *Logarithmic Duality of the Curvature Perturbation*. Phys. Rev. Lett. **131**, 011002 (2023), https://doi.org/10.1103/PhysRevLett.131.011002
88. R. Inui, C. Joana, H. Motohashi, S. Pi, Y. Tada and S. Yokoyama, *Primordial black holes and induced gravitational waves from logarithmic non-Gaussianity*. JCAP **03**, 021 (2025), https://doi.org/10.1088/1475-7516/2025/03/021

89. M. Shimada, A. Escrivá, D. Saito, K. Uehara and C.-M. Yoo, *Primordial black hole formation from type II fluctuations with primordial non-Gaussianity*. JCAP **02**, 018 (2025), https://doi.org/10.1088/1475-7516/2025/02/018
90. K. Freese, J. A. Frieman and A. V. Olinto, *Natural inflation with pseudo–Nambu-Goldstone bosons*. Phys. Rev. Lett. **65**, 3233 (1990), https://doi.org/10.1103/PhysRevLett.65.3233
91. L. Boubekeur and D. H. Lyth, *Hilltop inflation*. JCAP **07**, 010 (2005), https://doi.org/10.1088/1475-7516/2005/07/010
92. Planck Collaboration, Y. Akrami et al., *Planck 2018 results. X. Constraints on inflation*. Astron. Astrophys. **641**, A10 (2020), https://doi.org/10.1051/0004-6361/201833887
93. BICEP, Keck Collaboration, P. A. R. Ade et al., *Improved Constraints on Primordial Gravitational Waves using Planck, WMAP, and BICEP/Keck Observations through the 2018 Observing Season*. Phys. Rev. Lett. **127**, 151301 (2021), https://doi.org/10.1103/PhysRevLett.127.151301
94. Y.B. Zel'dovich, I.D. Novikov, The Hypothesis of Cores Retarded during Expansion and the Hot Cosmological Model. Soviet Astron. AJ **10**, 602 (1967)
95. S. Hawking, Gravitationally collapsed objects of very low mass. Mon. Not. Roy. Astron. Soc. **152**, 75 (1971)
96. B. J. Carr and S. W. Hawking, *Black holes in the early Universe*. Mon. Not. Roy. Astron. Soc. **168**, 399 (1974), https://doi.org/10.1093/mnras/168.2.399
97. B. J. Carr, *The Primordial black hole mass spectrum*. Astrophys. J. **201**, 1 (1975), https://doi.org/10.1086/153853
98. M. Sasaki, T. Suyama, T. Tanaka and S. Yokoyama, *Primordial black holes—perspectives in gravitational wave astronomy*. Class. Quant. Grav. **35**, 063001 (2018), https://doi.org/10.1088/1361-6382/aaa7b4
99. B. Carr, K. Kohri, Y. Sendouda and J. Yokoyama, *Constraints on primordial black holes*. Rept. Prog. Phys. **84**, 116902 (2021), https://doi.org/10.1088/1361-6633/ac1e31
100. A. Escrivá, F. Kuhnel and Y. Tada, *Primordial Black Holes*. arXiv: arXiv:2211.05767 (2022)
101. H. Motohashi and W. Hu, *Primordial Black Holes and Slow-Roll Violation*. Phys. Rev. D **96**, 063503 (2017), https://doi.org/10.1103/PhysRevD.96.063503
102. H. Niikura, M. Takada, S. Yokoyama, T. Sumi and S. Masaki, *Constraints on Earth-mass primordial black holes from OGLE 5-year microlensing events*. Phys. Rev. D **99**, 083503 (2019), https://doi.org/10.1103/PhysRevD.99.083503
103. LIGO Scientific, Virgo Collaboration, B. Abbott et al., *Observation of Gravitational Waves from a Binary Black Hole Merger*. Phys. Rev. Lett. **116**, 061102 (2016), https://doi.org/10.1103/PhysRevLett.116.061102
104. LIGO Scientific Collaboration, J. Aasi et al., *Advanced LIGO*. Class. Quant. Grav. **32**, 074001 (2015), https://doi.org/10.1088/0264-9381/32/7/074001
105. VIRGO Collaboration, F. Acernese et al., *Advanced Virgo: a second-generation interferometric gravitational wave detector*. Class. Quant. Grav. **32**, 024001 (2015), https://doi.org/10.1088/0264-9381/32/2/024001
106. KAGRA Collaboration, T. Akutsu et al., *Overview of KAGRA: Detector design and construction history*. PTEP **2021**, 05A101 (2021), https://doi.org/10.1093/ptep/ptaa125
107. M. Sasaki, T. Suyama, T. Tanaka and S. Yokoyama, *Primordial Black Hole Scenario for the Gravitational-Wave Event GW150914*. Phys. Rev. Lett. **117**, 061101 (2016), https://doi.org/10.1103/PhysRevLett.117.061101
108. J. Garcia-Bellido and E. Ruiz Morales, *Primordial black holes from single field models of inflation*. Phys. Dark Univ. **18**, 47 (2017), https://doi.org/10.1016/j.dark.2017.09.007
109. J. M. Ezquiaga, J. Garcia-Bellido and E. Ruiz Morales, *Primordial Black Hole production in Critical Higgs Inflation*. Phys. Lett. B **776**, 345 (2018), https://doi.org/10.1016/j.physletb.2017.11.039
110. C. Germani and T. Prokopec, *On primordial black holes from an inflection point*. Phys. Dark Univ. **18**, 6 (2017), https://doi.org/10.1016/j.dark.2017.09.001
111. G. Ballesteros and M. Taoso, *Primordial black hole dark matter from single field inflation*. Phys. Rev. D **97**, 023501 (2018), https://doi.org/10.1103/PhysRevD.97.023501

112. Y.-F. Cai, X. Chen, M. H. Namjoo, M. Sasaki, D.-G. Wang and Z. Wang, *Revisiting non-Gaussianity from non-attractor inflation models*. JCAP **05**, 012 (2018), https://doi.org/10.1088/1475-7516/2018/05/012
113. M. Biagetti, G. Franciolini, A. Kehagias and A. Riotto, *Primordial Black Holes from Inflation and Quantum Diffusion*. JCAP **07**, 032 (2018), https://doi.org/10.1088/1475-7516/2018/07/032
114. I. Dalianis, A. Kehagias and G. Tringas, *Primordial black holes from α-attractors*. JCAP **01**, 037 (2019), https://doi.org/10.1088/1475-7516/2019/01/037
115. C. Pattison, V. Vennin, H. Assadullahi and D. Wands, *The attractive behaviour of ultra-slow-roll inflation*. JCAP **08**, 048 (2018), https://doi.org/10.1088/1475-7516/2018/08/048
116. S. Rasanen and E. Tomberg, *Planck scale black hole dark matter from Higgs inflation*. JCAP **01**, 038 (2019), https://doi.org/10.1088/1475-7516/2019/01/038
117. C. T. Byrnes, P. S. Cole and S. P. Patil, *Steepest growth of the power spectrum and primordial black holes*. JCAP **06**, 028 (2019), https://doi.org/10.1088/1475-7516/2019/06/028
118. S. Passaglia, W. Hu and H. Motohashi, *Primordial black holes and local non-Gaussianity in canonical inflation*. Phys. Rev. D **99**, 043536 (2019), https://doi.org/10.1103/PhysRevD.99.043536
119. N. Bhaumik and R. K. Jain, *Primordial black holes dark matter from inflection point models of inflation and the effects of reheating*. JCAP **01**, 037 (2020), https://doi.org/10.1088/1475-7516/2020/01/037
120. D. Y. Cheong, S. M. Lee and S. C. Park, *Primordial black holes in Higgs-R^2 inflation as the whole of dark matter*. JCAP **01**, 032 (2021), https://doi.org/10.1088/1475-7516/2021/01/032
121. H. V. Ragavendra, P. Saha, L. Sriramkumar and J. Silk, *Primordial black holes and secondary gravitational waves from ultraslow roll and punctuated inflation*. Phys. Rev. D **103**, 083510 (2021), https://doi.org/10.1103/PhysRevD.103.083510
122. G. Tasinato, *An analytic approach to non-slow-roll inflation*. Phys. Rev. D **103**, 023535 (2021), https://doi.org/10.1103/PhysRevD.103.023535
123. D. G. Figueroa, S. Raatikainen, S. Rasanen and E. Tomberg, *Non-Gaussian Tail of the Curvature Perturbation in Stochastic Ultraslow-Roll Inflation: Implications for Primordial Black Hole Production*. Phys. Rev. Lett. **127**, 101302 (2021), https://doi.org/10.1103/PhysRevLett.127.101302
124. C. Pattison, V. Vennin, D. Wands and H. Assadullahi, *Ultra-slow-roll inflation with quantum diffusion*. JCAP **04**, 080 (2021), https://doi.org/10.1088/1475-7516/2021/04/080
125. K. Inomata, E. McDonough and W. Hu, *Primordial black holes arise when the inflaton falls*. Phys. Rev. D **104**, 123553 (2021), https://doi.org/10.1103/PhysRevD.104.123553
126. K. Inomata, E. McDonough and W. Hu, *Amplification of primordial perturbations from the rise or fall of the inflaton*. JCAP **02**, 031 (2022), https://doi.org/10.1088/1475-7516/2022/02/031
127. S. R. Geller, W. Qin, E. McDonough and D. I. Kaiser, *Primordial black holes from multifield inflation with nonminimal couplings*. Phys. Rev. D **106**, 063535 (2022), https://doi.org/10.1103/PhysRevD.106.063535
128. A. Caravano, G. Franciolini and S. Renaux-Petel, *Ultraslow-roll inflation on the lattice: Backreaction and nonlinear effects*. Phys. Rev. D **111**, 063518 (2025), https://doi.org/10.1103/PhysRevD.111.063518
129. T. Fujita, R. Kawaguchi, M. Sasaki and Y. Tada, *Dip and non-linearity in the curvature perturbation from inflation with a transient non-slow-roll stage*. arXiv: arXiv:2503.19744 (2025)
130. S. S. Mishra and V. Sahni, *Primordial Black Holes from a tiny bump/dip in the Inflaton potential*. JCAP **04**, 007 (2020), https://doi.org/10.1088/1475-7516/2020/04/007
131. O. Özsoy and Z. Lalak, *Primordial black holes as dark matter and gravitational waves from bumpy axion inflation*. JCAP **01**, 040 (2021), https://doi.org/10.1088/1475-7516/2021/01/040
132. V. Sahni and S. Habib, *Does inflationary particle production suggest Omega(m) less than 1?*. Phys. Rev. Lett. **81**, 1766 (1998), https://doi.org/10.1103/PhysRevLett.81.1766

A Complete Analysis of Inflation with Piecewise Quadratic Potential

Xinpeng Wang, Xiao-Han Ma, and Misao Sasaki

Abstract We conduct a thorough study of the comoving curvature perturbation $\mathcal{R}$ in single-field inflation with two stages, represented by a piecewise quadratic potential, where both the first and second derivatives are allowed to be discontinuous at the transition point. We calculate the evolution of $\mathcal{R}$ by combining the perturbative and non-perturbative methods consistently, and obtain the power spectrum and the non-Gaussian features in the probability distribution function. We find that both the spectrum and the statistics of $\mathcal{R}$ depend significantly on the second derivatives of the potential at both the first and second stages. Furthermore, we find a new parameter constructed from the potential parameters, which we call α, plays a decisive role in determining various features in the spectrum such as the amplitude, the slope, and the

X. Wang · X.-H. Ma · M. Sasaki (✉)
Kavli Institute for the Physics and Mathematics of the Universe (WPI), The University of Tokyo Institutes for Advanced Study, The University of Tokyo, Chiba, Japan
e-mail: misao.sasaki@apctp.org

X. Wang
e-mail: xinpeng.wang@ipmu.jp

X.-H. Ma
e-mail: mxh171554@mail.ustc.edu.cn

X. Wang
School of Physics Science and Engineering, Tongji University, Shanghai, China

X.-H. Ma
Deep Space Exploration Laboratory/School of Physical Sciences, University of Science and Technology of China, Hefei, Anhui, China

CAS Key Laboratory for Researches in Galaxies and Cosmology/Department of Astronomy, School of Astronomy and Space Science, University of Science and Technology of China, Hefei, Anhui, China

M. Sasaki
Center for Gravitational Physics and Quantum Information, Yukawa Institute for Theoretical Physics, Kyoto University, Kyoto, Japan

Leung Center for Cosmology and Particle Astrophysics, National Taiwan University, Taipei, Taiwan

Asia Pacific Center for Theoretical Physics, Pohang, Korea

A. Barvinsky and A. Kamenshchik (eds.), *Open Issues in Gravitation and Cosmology*, Fundamental Theories of Physics 225, https://doi.org/10.1007/978-3-032-15703-4_5

existence of a dip. In particular, we recover the typical k^4 growth of the spectrum in most cases, but the maximum growth rate of $k^5(\log k)^2$ can be obtained by fine-tuning the parameters. Then, using the δN formalism valid on superhorizon scales, we give fully nonlinear formulas for $\mathcal{R}$ in terms of the scalar field perturbation $\delta\phi$ and its time derivative. In passing, we point out the importance of the nonlinear evolution of $\delta\phi$ on superhorizon scales. Finally, using the Press-Schechter formalism for simplicity, we discuss the effect of the non-Gaussian tails of the probability distribution function on the primordial black hole formation.

1 Introduction

At first, let us say a few words about Alexei Starobinsky, as this article is dedicated to him. As everyone knows, Alexei Starobinsky was a giant in cosmology and general relativity. We have been constantly influenced by him and his work. In particular, focusing on cosmological perturbation theory, the present work is based on a number of his important achievements, not to mention his pioneering work on the δN formalism [99] and on the potential with a singular feature [100]. We were saddened by his death, but it is our honor that we can make a scientific contribution to this book in memory of Alexei Starobinsky.

Inflation, a quasi-exponential expansion stage before the radiation-dominated era, serves as a simple but effective phenomenological description of the very early universe. It was originally meant to solve several problems in the Big Bang cosmology, such as the initial singularity, horizon and flatness problems [59, 80, 98], though it is well-understood now that the most important prediction of inflation is that it explains the origin of all the structures in the universe. During inflation, the quantum vacuum fluctuations of a scalar field on microscopic scales are stretched beyond the Hubble radius and frozen on super Hubble scales. After the end of inflation, these fluctuations eventually turn into curvature perturbation and provide seeds for the inhomogeneous structures in the universe today.

Based on the concordance ΛCDM model, the Planck satellite [1] has put tight constraints on primordial perturbations on large scales, indicating they are nearly scale-invariant, adiabatic, and Gaussian. However, due to nonlinear astrophysical processes on small scales that erased primordial information, the properties of primordial perturbations on such scales are completely unknown.

The understanding of small scale primordial perturbations is not only crucial to establishing the full picture of inflation but also fascinating as it leads to rich phenomenology that could be observed in the near future, such as scalar-induced gravitational waves (SIGWs) and the formation of primordial black holes (PBHs). Both SIGWs and PBHs are associated with an enhancement of the primordial scalar power spectrum on small scales, while the large scale fluctuations remain nearly scale-invariant. There are numerous inflationary models in which the primordial scalar power spectrum is enhanced by various mechanisms, in single-field models

[15, 35, 54, 77, 87] and in multi-field models [3, 10, 13, 14, 16, 31, 33, 34, 38, 40, 43, 50, 52, 55, 56, 58, 63, 64, 73, 84, 90, 91, 96, 104, 111].

Let us focus on models of single-field inflation. These models can be broadly categorized into two types: those that modify the inflaton potential [7, 11, 18, 20, 22, 23, 29, 30, 37, 39, 47, 53, 65, 67, 68, 81, 85, 86, 89, 92, 100, 108, 109], leading to non-trivial evolution of the slow-roll parameters, and those that alter the kinetic term or introduce non-minimal couplings to gravity, resulting in modified equations of motion for the scalar perturbations [2, 8, 21, 25, 45, 46, 51, 57, 60, 69, 70, 72, 78, 79, 97, 103, 105–107, 110].

The inflection-point type or ultra-slow-roll (USR) scenario is a widely discussed approach to enhancing the power spectrum by modifying the inflaton potential with a feature that deviates the inflaton from the slow-roll attractor solution. However, the non-slow-roll phase cannot persist for too many e-folds; otherwise, inflation will either end prematurely or continue indefinitely. Therefore, in practice, a healthy and complete model of this type typically involves two phases of standard slow-roll inflation before and after the non-slow-roll phase. The simplest model of this kind is an inflationary model with a piecewise potential, composed of two segments of slow-roll potentials with different parameters. It has been extensively studied in single-field models with piecewise linear potentials, which are often called Starobinsky's linear potential model, that the enhancement of curvature perturbations exhibits distinct characteristics, such as dips and growth slopes in the power spectrum, as well as oscillatory behavior [15, 24, 39, 89, 100]. These features indicate that the analysis conducted using separate universe approach (leading order in the gradient expansion) needs further consideration [66, 77]. However, the piecewise linear potential implies that the non-trivial contribution from the second derivative of the potential, $\mathrm{d}^2 V(\phi)/\mathrm{d}\phi^2$, has been largely overlooked and remains unexplored in previous studies. On the other hand, it has been shown that a non-vanishing $\mathrm{d}^2 V(\phi)/\mathrm{d}\phi^2$ leads to an exponential tail in the probability density function (PDF) of curvature perturbations [5, 6, 12, 17–19, 89], due to a nonlinear relationship between the comoving curvature $\mathcal{R}$ and the field perturbation on flat slicing $\delta\phi$: $\mathcal{R} \propto -(1/\eta_V) \ln[1 + \eta_V \delta\phi]$, which could significantly affect PBH formation. Therefore, studying general piecewise potential models with non-zero second derivatives is of great interest and significance.

We thoroughly discuss the primordial power spectrum of curvature perturbations and non-Gaussianity within a general framework of single-field inflation with piecewise quadratic potentials. This paper is organized as follows: In Sect. 2, we introduce the piecewise quadratic potential and analytically solve for the background evolution in the most general case. Section 3 presents the analytical derivation of the power spectrum of curvature perturbations, along with a discussion of its scaling behavior and IR/UV features. In Sect. 4, we examine non-perturbative non-Gaussianity using the δN formalism. Finally, in Sect. 5, we explore the implications of non-Gaussianity for PBH formation.

2 The Background

We parameterize a piecewise potential in the vicinity of its junction point as $V_\star = V(\phi_\star)$ up to the second order in $\phi - \phi_\star$,

$$V = \begin{cases} V_\star\left(1 + \sqrt{2\epsilon_I}(\phi - \phi_\star) + \frac{1}{2}\eta_I(\phi - \phi_\star)^2\right) & \text{for } \phi \geq \phi_\star\,, \\ V_\star\left(1 + \sqrt{2\epsilon_{II}}(\phi - \phi_\star) + \frac{1}{2}\eta_{II}(\phi - \phi_\star)^2\right) & \text{for } \phi < \phi_\star\,. \end{cases} \tag{1}$$

In the scenario considered here, we do not have an upward step, that is, the case of $\Delta V = 0$ in [18, 19]. The slow-roll parameters we used in the following discussion are defined by $\epsilon_V \equiv M_p^2(V_{,\phi}/V)^2/2$ and $\eta_V \equiv M_p^2 V_{,\phi\phi}/V$, where M_p represents the reduced Planck mass. In the above, as well as in what follows, ϵ_X and η_X $(X = I,\ II)$ are the potential parameters, respectively, in the regions I $(\phi \geq \phi_\star)$ and II $(\phi \leq \phi_\star)$. As shown in Fig. 1, we consider four different piecewise potentials (Fig. 2).

The Friedmann equations and the scalar field equation on the flat Friedmann-Lemaître-Robertson-Walker (FLRW) background are given by

$$3M_p^2 H^2 = \frac{1}{2}\dot{\phi}^2 + V(\phi), \tag{2}$$

$$-\,2M_p^2 \dot{H} = \dot{\phi}^2, \tag{3}$$

$$\ddot{\phi} + 3H\dot{\phi} + V_{,\phi} = 0. \tag{4}$$

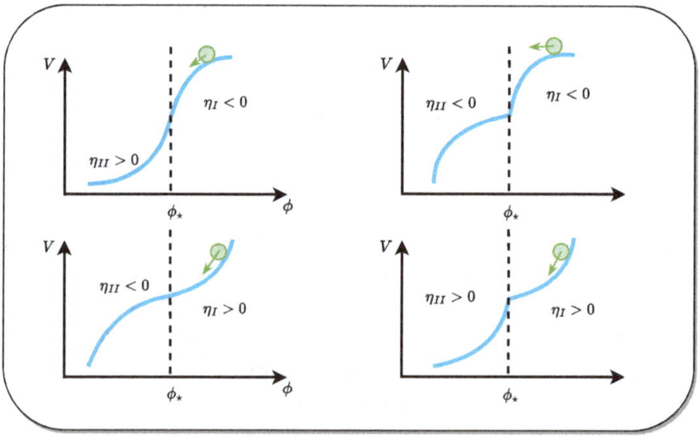

Fig. 1 A schematic diagram of the potentials of interest

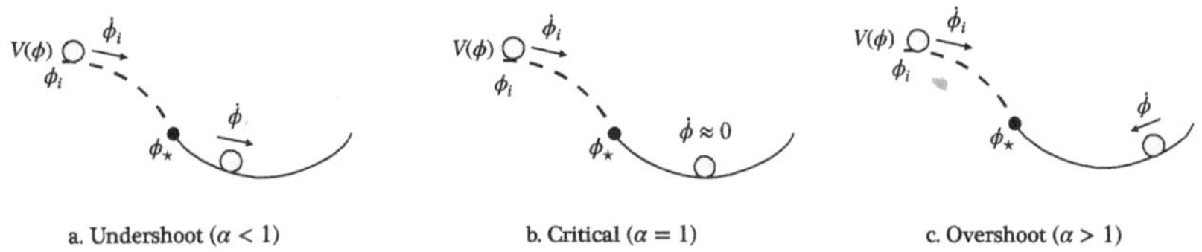

Fig. 2 Schematic plots for the potential and inflaton evolution when $\eta_{II} > 0$. In each plot, the hollow point on the left side shows the initial attractor state while the hollow point on the right side shows the final attractor state of the inflaton. Plot **a** shows the undershoot case that inflaton goes back to the attractor phase before reaching the potential minimum. Plot **c** shows the overshoot case that the inflaton goes back to the attractor phase after when it passes the potential minimum and makes a U-turn. Plot **b** shows the critical case for the eternal USR-like phase case, that the inflaton stops by the potential minimum

For simplicity, we adopt the reduced Planck units $M_{\rm p} = 1$ in the remaining part of the paper. We introduce the e-folding number $dn = Hdt$ as the time variable to solve the background field equation,

$$\frac{d^2\phi}{dn^2} + (3 - \epsilon_H)\frac{d\phi}{dn} + \frac{V_{,\phi}}{H^2} = 0\,, \tag{5}$$

where $\epsilon_H = -\dot{H}/H^2$, which is approximately equal to ϵ_V. Our primary interest lies in the local dynamics of the inflaton around the joint point. Inserting the potential (1) into the above and assuming $\epsilon_H \ll 1$, we have $H^2 = V_\star/3$, and the equation for ϕ may be written as

$$\frac{d^2\widetilde{\phi}_X}{dn^2} + 3\frac{d\widetilde{\phi}_X}{dn} + 3\eta_X\widetilde{\phi}_X = 0. \tag{6}$$

where $X = I,\ II$, and $\widetilde{\phi}_X \equiv \phi - \phi_\star + \sqrt{2\epsilon_X}/\eta_X$. Requiring the monotonicity of $\phi(n)$, we assume $\eta_X < 3/4$. Then, the general solution for ϕ is given by

$$\phi_X = C_{X,-}e^{-\lambda_{X,-}(n-n_\star)} + C_{X,+}e^{-\lambda_{X,+}(n-n_\star)} + \phi_{X,0}\,; \quad \phi_{X,0} = \phi_\star - \frac{\sqrt{2\epsilon_X}}{\eta_X}\,, \tag{7}$$

where $-\lambda_{X,\pm}$ are the characteristic roots of Eq. (6), given by

$$\lambda_{X,\pm} = \pm\nu_X + \frac{3}{2}\,; \quad \nu_X \equiv \sqrt{\frac{9}{4} - 3\eta_X}\,. \tag{8}$$

Sometimes it is convenient to express η_X in terms of ν_X,

$$\eta_X = \frac{1}{3}\left(\frac{3}{2} - \nu_X\right)\left(\frac{3}{2} + \nu_X\right) = \frac{1}{3}\lambda_{X+}\lambda_{X-}. \tag{9}$$

Given the initial condition ($\phi_i, \pi_i \equiv (\mathrm{d}\phi/\mathrm{d}n)_i$), the coefficients for the two linearly independent solutions can be determined as

$$C_{X,\pm} = \pm\frac{\pi_\star + \lambda_{X,\mp}\widetilde{\phi}_{X,\star}}{\lambda_{X,-} - \lambda_{X,+}}, \tag{10}$$

for the piecewise potential given by (1), with $\eta_X < 3/4$, $-\lambda_{X,+} = -\lambda_{X,-} - 2\nu_X < -\lambda_{X,-}$. Hence the solution with the characteristic exponent $-\lambda_{X,+}$ always decays faster than the other solution with $-\lambda_{X,-}$. The latter solution corresponds to the solution in the slow-roll limit, the so-called attractor solution, while we call the solution with $-\lambda_{X,+}$ the decaying solution. We focus on the initial condition for which the decaying solution has already disappeared and the solution is in the attractor regime before ϕ reaches $\phi_\star$. Namely, we assume $C_{I,+} = 0$. This gives $C_{I,-} = \sqrt{2\epsilon_I}/\eta_I$. Under this assumption, when ϕ reaches $\phi_\star$, we have

$$\pi_{\star,I} = -\lambda_{I,-}\frac{\sqrt{2\epsilon_I}}{\eta_I}\,. \tag{11}$$

Since π is continuous at $\phi_\star$, $\pi_{\star,I} = \pi_{\star,II} = \pi_\star$. This fixes the coefficients $C_{II,\pm}$,

$$C_{II,\pm}(\phi_\star, \pi_\star) = \pm\frac{3}{\nu_{II}}\left(\frac{\sqrt{2\epsilon_I}}{2\nu_I + 3} \mp \frac{\sqrt{2\epsilon_{II}}}{2\nu_{II} \pm 3}\right), \tag{12}$$

which determines the dynamic at the second stage. When $|C_{X,+}| > |C_{X,-}|$, the decay mode dominates soon after the transition, and thus leads to a period of non-attractor phase (See in Fig. 4). If the background dynamics can be described by only the decaying solution, the second slow-roll parameter reads

$$\eta_H \equiv \frac{\dot{\epsilon}_H}{H\epsilon_H} = -3 - 2\nu_{II}\,, \tag{13}$$

its value becomes $\eta_H \approx -6$ if $\eta_{II} \ll 1$, in which case the phase is called ultra-slow-roll (USR) inflation.

Typically, the relaxation period is short. However, re-writing (12) as

$$C_{II,-} = \frac{3}{\nu_{II}}\frac{\sqrt{2\epsilon_{II}}}{2\nu_{II} - 3}(\alpha - 1)\,, \tag{14}$$

where α is defined as

$$\alpha \equiv \frac{3 - 2\nu_{II}}{(2\nu_I + 3)R_\epsilon} = \frac{\lambda_{II,-}}{\lambda_{I,+}R_\epsilon}; \quad R_\epsilon = \sqrt{\epsilon_{II}/\epsilon_I}\,, \tag{15}$$

we see that $C_{II,-}$ vanishes for the parameters that satisfy $\alpha = 1$. In this case, the relaxation stage becomes infinitely long. It is worth mentioning that the magnitude of α depends linearly on η_{II} in the limit $\eta_{II} \ll 1$,

$$\alpha \approx \frac{\eta_{II}}{R_\epsilon(\nu_I + 3/2)} \quad \text{for } |\eta_{II}| \ll 1 \tag{16}$$

Thus $\alpha = 1$ may be realized for the value of η_{II} given by $\eta_{II} \approx (\nu_I + 3/2)R_\epsilon$ when $R_\epsilon \ll 1$. During the infinitely long period of approaching the asymptotic solution, the inflaton stays in a non-slow roll state (see in Fig. 3, the red line in the top panel).

We note that this eternal USR-like case is extremely unstable, that even a tiny deviation of the boundary condition at the joint point $(\phi_\star, \dot{\phi}_\star)$ or a slight time evolution of H can violate this situation, i.e. introduce a non-vanishing attractor mode that eventually dominates the background solution. Thus, in reality the parameters that satisfy $\alpha = 1$ can only lead to a relatively long USR-like phase. Nevertheless, it is of interest if we can tune the potential in such a way that it exactly leads to an eternal USR-like stage. We leave this issue for future study (Fig. 4).

The parameter α is also useful when considering the asymptotic behavior of ϕ. The sign of $C_{II,-}$ determines the sign of the asymptotic value of the momentum π, together with the sign of η_{II}, as the sign of $\pi(= d\phi/dn)$ is determined by that of the product $\lambda_{II,-}C_{II,-} \propto (\alpha - 1)$. If $\eta_{II} < 0, \alpha < 0$. Hence the inflaton approaches

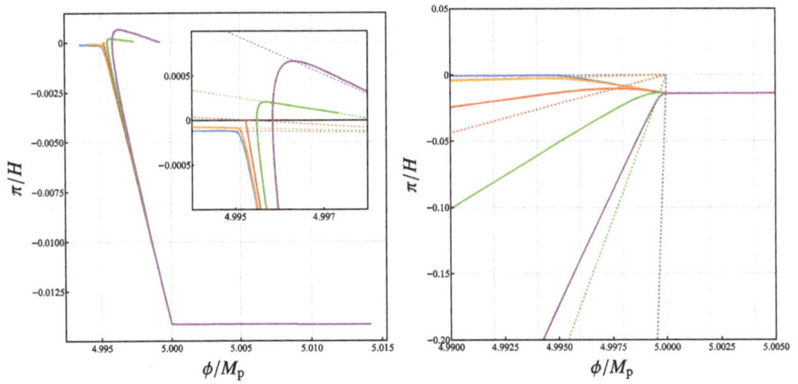

Fig. 3 **The background phase diagram** for a positive η_{II} (left panel) and negative η_{II} (right panel) starting from a slow roll attractor initial condition given during the first stage of inflation. We take $R_\epsilon = 10^{-2}$, $\eta_I = 10^{-3}$ and $\phi_\star = 5M_\text{p}$ in both figures, the solid lines are given by the numerical solution of the background equations, the dashed lines are the accordingly slow roll attractor solution of the second stage given analytically. In the left panel, we take $\eta_{II} = \eta_{II}^{r+} \times (e^{-2}, e^{-1}, 1, e^1, e^2)$ for the blue, orange, red, green and purple lines. In the right panel we take $\eta_{II} = \eta_{II}^{r-} \times (10^{-2}, 10^{-1}, 1, 10^1, 10^2)$ for the blue, orange, red, green and purple lines. We use $\eta_{II}^{r+} \approx 0.02969$ which is given by $\alpha = 0$ for the vanishing non-decay mode case, $\eta_{II}^{r-} = -4.3757$ which is given by $\xi_2 = 0$ defined in (44) for the vanishing second-order gradient expansion contribution to the curvature perturbation, which will be discussed in the next section

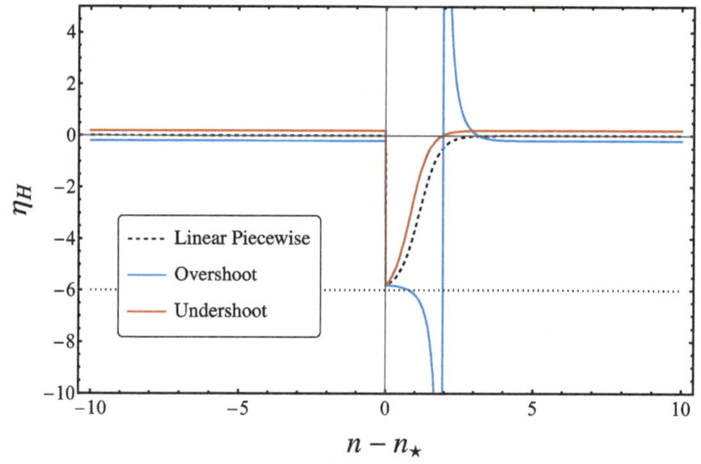

Fig. 4 The time evolution of η_H for different cases. $\eta_I = \eta_{II} = 0.1$, $\eta_I = \eta_{II} = -0.1$ are used for the blue curve and red curve separately. For the overshoot case which is shown by the blue curve, corresponding to $\alpha > 1$, η_H diverges because $\dot{\phi}$ hits 0 before the inflaton settles down to the attractor phase. The undershoot case is shown by the red curve, corresponding to $\alpha < 1$, which indicates that the inflaton enters the second attractor phase before it hits the potential minimum

the attractor solution smoothly. On the other hand, if $\eta_{II} > 0$, two kinds of behavior appear depending on the value of α. When (1) $\alpha < 1$, the attractor phase is realized before crossing the minimum of the potential $\phi_{X,0}$ (the blue and orange lines in the left panel of Fig. 3), while when (2) $\alpha > 1$, the attractor phase is realized after crossing $\phi_{X,0}$ (the purple and green lines in the left panel of Fig. 3). Namely, it enters the attractor phase after the moment it starts to roll back.

Here, it is worth mentioning that we assume that the end of inflation is in the attractor phase. If the attractor phase is realized after a U-turn, we may need a special mechanism to end inflation. We simply assume the existence of such a mechanism without discussing any specific models in this paper. If we would relax this attractor assumption, we would have to keep track of the time evolution of the curvature perturbation to determine its final spectrum, instead of taking the asymptotic future limit. As this would complicate our discussion significantly, we leave studies of such cases for the future.

In order to have some intuition about the power spectrum enhancement, we evaluate the ratio of power for the small scale modes that exit the horizon after the inflaton enters the attractor phase during the second stage to the large scale modes that exit the horizon during the first stage. Assuming small η limit, the power spectrum is nearly scale-invariant on both the long and short wavelength ends, and the ratio is thus given by

$$\frac{\mathcal{P}_{\mathcal{R}}(k_{UV})}{\mathcal{P}_{\mathcal{R}}(k_{IR})} \approx \frac{H^2/\epsilon_H\big|_{t=t_\star}}{H^2/\epsilon_H\big|_{t\gg t_\star}} = \frac{C_{I,-}^2\lambda_{I,-}^2}{C_{II,-}^2\lambda_{II,-}^2}$$
$$= \left(\frac{\lambda_{II,-}-\lambda_{II,+}}{\lambda_{II,-}-\lambda_{I,+}R_\epsilon}\right)^2$$
$$= \frac{1}{(1-\alpha)^2}\left(\frac{4\nu_{II}}{(3+2\nu_I)R_\epsilon}\right)^2, \tag{17}$$

which is a good approximation to the order of enhancement of the power spectrum due to the transition regardless of a small spectral index induced by nonzero η and ϵ. The upper expression easily degenerates to the standard slow roll version $\mathcal{P}_{\mathcal{R}}(k_{UV})/\mathcal{P}_{\mathcal{R}}(k_{IR}) = R_\epsilon$ when taking $\nu_{II} = \nu_I = 3/2$ (or $\eta_I, \eta_{II} = 0$, or equivalently $\lambda_{I,-} = \lambda_{II,-} = 0, \lambda_{I,+} = \lambda_{II,+} = 3$). However, α starts to determine the enhancement in cases that $R_\epsilon \neq 1$ (see an example in the next section Fig. 10).

3 The Curvature Power Spectrum

3.1 The Perturbation Equations

We consider the metric with scalar-type linear perturbations, together with the scalar field perturbation,

$$ds^2 = -(1+2A)dt^2 + 2aB_{,i}dx^i dt + a^2[1+2\mathcal{R}\delta_{ij} + 2E_{,ij}]dx^i dx^j, \tag{18}$$
$$\phi(\mathbf{x}, t) = \phi(t) + \delta\phi(\mathbf{x}, t) \tag{19}$$

Among various choices of gauges, the one most convenient for evaluating the quantum vacuum fluctuations is the flat slicing where we set $\mathcal{R} = 0$. Another is the slicing on which the fluctuation in the scalar field vanishes $\delta\phi = 0$, called the comoving slicing. The field fluctuation on flat slicing, denoted by $\delta\phi_f$ and the curvature perturbation on comoving slicing, denoted by $\mathcal{R}$ are related by

$$\mathcal{R} = -\frac{H}{\dot\phi}\delta\phi_f. \tag{20}$$

On flat slicing, the equation of motion for the Fourier mode of $\delta\phi_f$, denoted by $\delta\phi_k$, is given by

$$\delta\ddot\phi_k + 3H\delta\dot\phi_k + \left(\frac{k^2}{a^2} + V_{,\phi\phi}\right)\delta\phi_k = \frac{1}{a^3}\frac{d}{dt}\left(\frac{a^3}{H}\dot\phi^2\right)\delta\phi_k. \tag{21}$$

By introducing the conformal time $d\tau = dt/a$, the corresponding equation for the Fourier mode of $\mathcal{R}_c$, denoted by $\mathcal{R}_k$, becomes

$$\mathcal{R}_k'' + 2\frac{z'}{z}\mathcal{R}_k' + k^2\mathcal{R}_k = 0, \tag{22}$$

where $z \equiv ad\phi/dn$. Under the slow roll approximation $|\epsilon| \ll 1$, the dimensionless conformal time derivatives of z are given

$$\frac{z'}{z} = -\tau^{-1}\left(2\frac{d\phi}{dn} + \frac{V_{,\phi}}{H^2}\right), \tag{23}$$

$$\frac{z''}{z} = \tau^{-2}\left(2 - \frac{V_{,\phi\phi}}{H^2}\right). \tag{24}$$

It is worth mentioning that the first derivative of z depends explicitly on the background dynamics, while the second derivative does not.

In our model, z is a continuous function of the scale factor a,

$$z \simeq \begin{cases} a_\star\left(\dfrac{a}{a_\star}\right)^{\nu_I-\frac{1}{2}} \dfrac{3\sqrt{2\epsilon_I}}{\nu_I+3/2} & \text{for } \phi > \phi_\star, \\ a_\star\left(\dfrac{a}{a_\star}\right)^{\nu_{II}-\frac{1}{2}} \dfrac{3\sqrt{2\epsilon_{II}}}{2\nu_{II}}\left(1+\dfrac{2\nu_{II}-3}{(2\nu_I+3)R_\epsilon}\right) & \\ -a_\star\left(\dfrac{a}{a_\star}\right)^{-\nu_{II}-\frac{1}{2}} \dfrac{3\sqrt{2\epsilon_{II}}}{2\nu_{II}}\left(1-\dfrac{2\nu_{II}+3}{(2\nu_I+3)R_\epsilon}\right) & \text{for } \phi < \phi_\star. \end{cases} \tag{25}$$

Yet z'/z appears to have a discontinuity at $a_\star$. By taking the limit $\epsilon_H \ll 1$ and setting $\tau \approx -1/(aH_\star)$, the discontinuity of z'/z at $\phi_\star$ is found as

$$\frac{z'}{z} \simeq \begin{cases} \left(\dfrac{1}{2}-\nu_I\right)\tau_\star^{-1} & \phi-\phi_\star = 0^+, \\ \left[2 - R_\epsilon\left(\dfrac{3}{2}+\nu_I\right)\right]\tau_\star^{-1} & \phi-\phi_\star = 0^-. \end{cases} \tag{26}$$

As a result, the second derivative of $\mathcal{R}_k$ appears to be discontinuous at the joint point. To circumvent this potential issue/complexity, we introduce a new variable,

$$u = a\delta\phi = z\mathcal{R}, \tag{27}$$

and rewrite (22) as

$$u_k'' + \left(k^2 - \frac{z''}{z}\right)u_k = 0. \tag{28}$$

A convenient property of this form of the equation is that, as (24) tells us, z''/z depends only on the potential parameter,

$$\frac{z''}{z} \simeq \begin{cases} \tau^{-2}(2 - 3\eta_I) & \text{for } \phi > \phi_\star, \\ \tau^{-2}(2 - 3\eta_{II}) & \text{for } \phi < \phi_\star, \end{cases} \tag{29}$$

where $\epsilon_H \ll 1$ is assumed. Hence under this approximation, the general solution to (28) is analytically found as

$$u_{k,I} = C_I\sqrt{-k\tau}H^{(1)}_{\nu_I}(-k\tau) + D_I\sqrt{-k\tau}H^{(2)}_{\nu_I}(-k\tau) \quad \text{for } \phi > \phi_\star\,, \tag{30}$$

$$u_{k,II} = C_{II}\sqrt{-k\tau}H^{(1)}_{\nu_{II}}(-k\tau) + D_{II}\sqrt{-k\tau}H^{(2)}_{\nu_{II}}(-k\tau) \quad \text{for } \phi < \phi_\star\,. \tag{31}$$

where C_X and D_X $(X = I,\ II)$ are constants, and $H^{(1)}_\nu(x)$ and $H^{(2)}_\nu(x)$ are the Hankel functions of the first kind and second kind, respectively. In the limit of $-k\tau \to \infty$, the k-mode is deep inside the horizon. The Hankel function in this limit is

$$H^{(1)}_\nu(x \to \infty) \to \sqrt{\frac{2}{\pi x}}\exp\left(ix - \frac{i\pi}{4}(2\nu + 1)\right), \tag{32}$$

Thus we obtain the coefficients for the adiabatic vacuum as

$$C_I(k) = \frac{\sqrt{\pi}}{2\sqrt{k}}\exp\left[\frac{i\pi}{4}(2\nu + 1)\right], \quad D_I = 0\,, \tag{33}$$

where the phase is chosen for convenience. Note that ν is real in our model.

3.2 Matching at the Joint Point

Since $\mathcal{R}_k = u_k/z$ and its time derivative is continuous across the transition, we match $\mathcal{R}_k$ and $\mathcal{R}'_k$ at the junction point,

$$\frac{u_{\star,I}}{z_I} = \frac{u_{\star,II}}{z_{II}} \tag{34}$$

$$\frac{u'_{\star,I}}{z_I} - \frac{u_{\star,I}}{z_I}\frac{z'_I}{z_I} = \frac{u'_{\star,II}}{z_{II}} - \frac{u_{\star,II}}{z_{II}}\frac{z'_{II}}{z_{II}}\,. \tag{35}$$

For the sake of generality, we keep D_I as if it were non-vanishing, which enables us to obtain the power spectrum for a non-adiabatic vacuum initial condition. Denoting $x = -k\eta_\star$, the coefficients are given by

$$
\begin{aligned}
&C_{II} - D_{II} = \\
&-\frac{i\pi}{4}(C_I + D_I)\left(2xJ_{\nu_I^-}(x)J_{\nu_{II}}(x) - 2xJ_{\nu_{II}^-}(x)J_{\nu_I}(x) + (\lambda_{II}^+ - R_\epsilon\lambda_I^+)J_{\nu_{II}}(x)J_{\nu_I}(x)\right) \\
&-\frac{i\pi}{4}(C_I - D_I)\left(2xJ_{\nu_{II}^-}(x)Y_{\nu_I}(x) - 2xJ_{\nu_{II}}(x)Y_{\nu_I^-}(x) + (\lambda_{II}^+ - R_\epsilon\lambda_I^+)J_{\nu_{II}}(x)Y_{\nu_I}(x)\right),
\end{aligned} \tag{36}
$$

where J_ν and Y_ν are the Bessel functions of order ν, and we have introduced

$$
\begin{aligned}
\nu_I^\pm &= \pm 1 + \nu_I, \\
\nu_{II}^\pm &= \pm 1 + \nu_{II},
\end{aligned} \tag{37}
$$

for notational simplicity.

Now let us evaluate the final amplitude of the curvature perturbation, assuming that the end of the second stage is the end of inflation. We consider the late stage when $\tau \gg \tau_\star$ so that all the modes of interest are well outside the horizon, $-k\tau = (a_k H_k/aH) \to 0$. In this limit, we may use the asymptotic behavior of $H_\nu^{(1,2)}(x) \to \pm\Gamma(\nu)(2/x)^\nu/(i\pi)$ to simplify (31),

$$
u_{k,II}(-k\tau \to 0) = \sqrt{-k\tau}\,\Gamma(\nu_{II})\left(-\frac{2}{k\tau}\right)^{\nu_{II}} \frac{1}{i\pi}\,(C_{II} - D_{II}). \tag{38}
$$

Then, the curvature perturbation at the end of inflation is obtained as

$$
\begin{aligned}
\mathcal{P}_\mathcal{R}(k) &= \frac{k^3}{2\pi^2}|\mathcal{R}_k|^2 = \frac{k^3}{2\pi^2}\left|\frac{u_{k,II}}{z_{II}}\right|^2 \\
&\xrightarrow{-k\tau\to 0} \left(\frac{H_\star}{2\pi}\right)^2 \left(\frac{k/k_\star}{2}\right)^{3-2\nu_{II}} \left(\frac{\Gamma(\nu_{II})}{\sqrt{\pi}/2}\right)^2 \left|\frac{C_{II} - D_{II}}{\sqrt{\pi/k}/2}\right|^2 \left(\frac{2\nu_{II}}{3(1-\alpha)\sqrt{2\epsilon_{II}}}\right)^2.
\end{aligned} \tag{39}
$$

In the limit of no transition ($R_\epsilon = 1, \eta_{II} = \eta_I$), the above expression reduces to

$$
\mathcal{P}_\mathcal{R}(k) = \left(\frac{H_\star}{2\pi}\right)^2 \left(\frac{k/k_\star}{2}\right)^{3-2\nu_I} \left(\frac{\Gamma(\nu_I)}{\sqrt{\pi}/2}\frac{2\nu_I + 3}{6\sqrt{2\epsilon_I}}\right)^2 \xrightarrow{\eta_I \ll 1} \frac{H^2}{8\pi^2\epsilon_I}, \tag{40}
$$

which agrees with the standard slow-roll result. For ease in calculating the formalism δN after the background solution converges to the attractor phase, the asymptotic expression of $\delta\phi_k$ is given

$$
\mathcal{P}_{\delta\phi}(k) = \frac{k^3}{2\pi^2}\left|\frac{u_{k,II}(-k\tau \to 0)}{a}\right|^2 = \left(\frac{H}{2\pi}\right)^2 \left(\frac{k/(aH)}{2}\right)^{3-2\nu_{II}} \left(\frac{\Gamma(\nu_{II})}{\sqrt{\pi}/2}\right)^2 \left|\frac{C_{II} - D_{II}}{\sqrt{\pi/k}/2}\right|^2. \tag{41}
$$

The analytical result for $\mathcal{P}_\mathcal{R}(k)|_{\tau\to 0}$ under the constant H approximation is obtained by inserting (33) and (36) into (39). We find it matches the result obtained

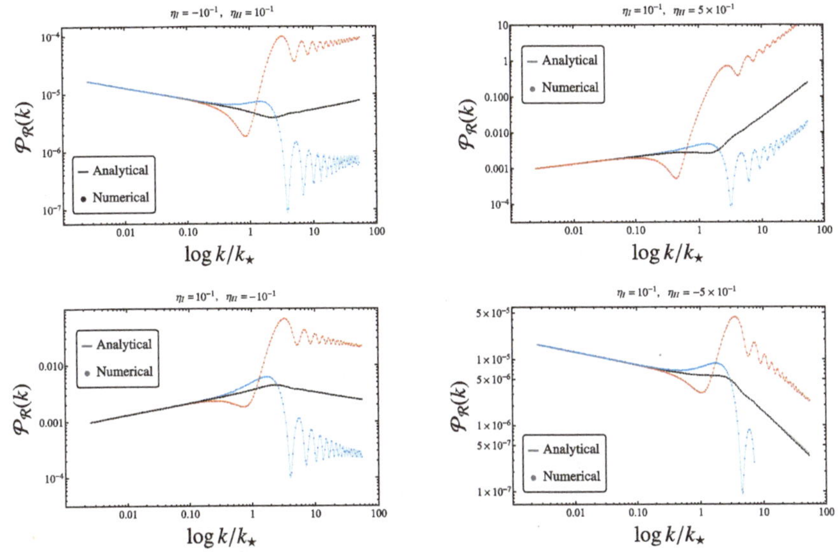

Fig. 5 The primordial power spectrum for the piecewise quadratic model. The positions of panels are arranged in accordance with those in Fig. 1, classified by the signs of η_I and η_{II}. The values of R_ϵ are $R_\epsilon^2 = 0.1$ (black), 0 (red), and 10 (blue)

by numerically solving the perturbation equations very well, as shown in Fig. 5. Using the analytical formula for the power spectrum, we plot several characteristic examples in Figs. 6 and 7. Both figures show the cases with different η for fixed R_ϵ.

In the figures above, the shape of the power spectrum varies with the parameter sets we used. In order to understand the features analytically, we divide our detailed discussion into the two parts, (1) Long wavelength modes that exit the horizon before $t_\star$, i.e. $k \ll k_\star$, (2) Short wavelength modes that exit the horizon after $t_\star$, i.e. $k \gg k_\star$.

3.3 The Long Wavelength Limit

Assuming that there was no stage with features prior to stage 1, we set $D_I = 0$. Then expanding the analytical expression for the power spectrum (39) to second order in $X = k/k_\star \ll 1$, we find

$$\mathcal{P}_\mathcal{R}(k;\, \tau_f) = \mathcal{A}_\mathcal{R}^L X^{3-2\nu_I} \left|1 + \xi_1 X^{2\nu_I} + \xi_2 X^2 + O(X^{2+2\nu_I})\right|^2, \tag{42}$$

where the amplitude $\mathcal{A}_\mathcal{R}^L$ is given by

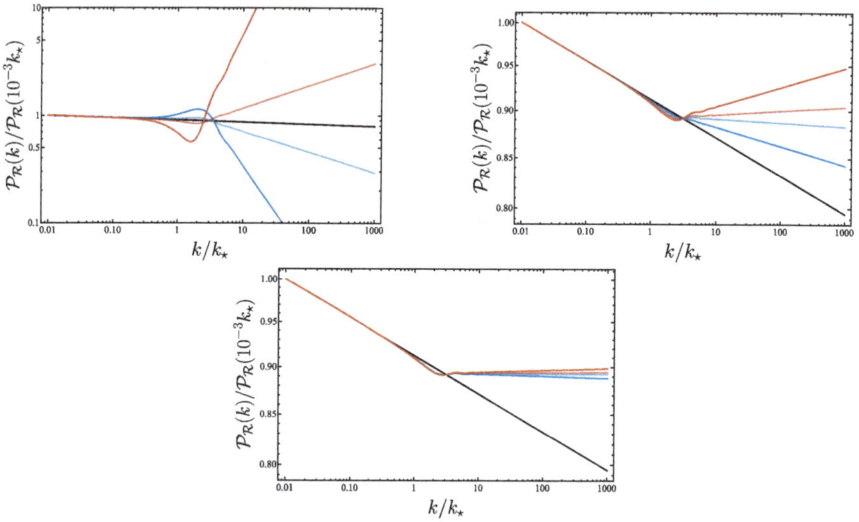

η_{II}	Top left	Top right	Bottom
Black	-10^{-2}	-10^{-2}	-10^{-2}
Dark blue	-5×10^{-1}	-5×10^{-3}	-5×10^{-4}
Light blue	-10^{-1}	-10^{-3}	-10^{-4}
Light red	10^{-1}	10^{-3}	10^{-4}
Dark red	5×10^{-1}	5×10^{-3}	5×10^{-4}

Fig. 6 Spectra for $R_\epsilon^2 = 1$, $\eta_I = -10^{-2}$ and for various values of η_{II} shown in the table

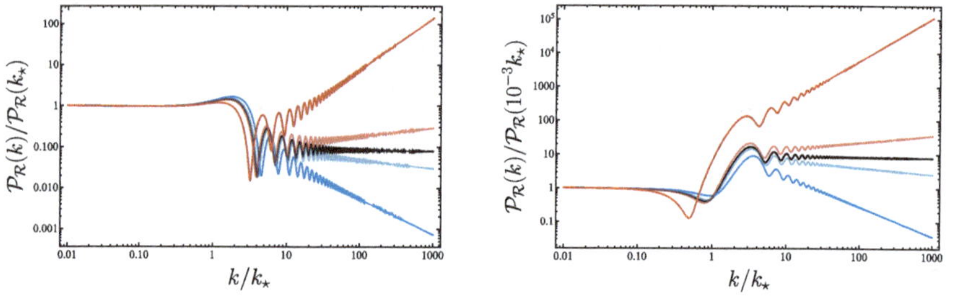

Fig. 7 Spectra for $\eta_I = -10^{-2}$ and for various values of η_{II}: $\eta_{II} = -10^{-2}$ (black), -5×10^{-1} (dark blue), -10^{-1} (light blue), 10^{-1} (light red), and 5×10^{-1} (dark red). The left panel is for $R_\epsilon^2 = 10$ and the right panel for $R_\epsilon^2 = 0.1$

$$\mathcal{A}^L_{\mathcal{R}} = \left(\frac{2^{-5/2+\nu_I}}{3}\frac{H_\star}{2\pi}\frac{\Gamma(\nu_I)}{\sqrt{\pi}/2}\frac{3+2\nu_I}{\sqrt{2\epsilon_I}}\right)^2, \tag{43}$$

and the coefficients ξ_1 and ξ_2 are given by

$$\xi_1 \simeq -4^{-\nu_I}\frac{\pi(\cot(\nu_I\pi)-i)}{\Gamma(1+\nu_I)\Gamma(\nu_I)}\left(1-\frac{4\nu_I}{2\nu_{II}-3+(2\nu_I+3)R_\epsilon}\right), \tag{44}$$

$$\xi_2 \simeq \frac{1}{4(\nu_{II}+1)(\nu_I-1)}\left(2+\nu_{II}-\nu_I-\frac{4(\nu_I+\nu_{II})}{2\nu_{II}-3+(2\nu_I+3)R_\epsilon}\right). \tag{45}$$

We note that the curvature perturbation approaches a constant in the superhorizon limit in general, but there are cases when the curvature perturbation evolves in time for a while on superhorizon scales if there is a feature in the potential. Therefore, the value of the curvature perturbation at (shortly after) the horizon crossing may not be the same as that at the end of inflation. This is the reason why we evaluate the power spectrum at the end of inflation by taking the limit $-k\tau \to 0$.

In the very long wavelength limit, $X = k/k_\star \to 0$, apparently the leading order term $X^{3/2-\nu_I}$ ($3/2-\nu_I \approx \eta_I$ for $\eta_I \ll 1$) dominates the power spectrum, obtaining a slightly tilted power spectrum due to the non-vanishing η_I,

$$\mathcal{P}^{IR}_{\mathcal{R}}(k) = \left(\frac{H_\star}{2\pi}\right)^2\left(\frac{k/k_\star}{2}\right)^{3-2\nu_I}\left(\frac{\Gamma(\nu_I)}{\sqrt{\pi}/2}\frac{2\nu_I+3}{6\sqrt{2\epsilon_I}}\right)^2. \tag{46}$$

Nevertheless, the next to leading order (NLO) terms may dominate the spectrum of near IR modes $k \lesssim k_\star$, unless both $|\xi_1|$ and $|\xi_2|$ happen to be much smaller than unity. Let us call the NLO terms, $\xi_1 X^{2\nu_I}$ and $\xi_2 X^2$, respectively, T1 and T2. When either T1 or T2 dominates, the scaling behavior of the near IR modes in the final power spectrum will be strongly tilted.

In order to look into this case in detail, let us first clarify the magnitude of T1 and T2. When $\nu_I < 1$, T1 is the term with a lower power-law index, and T2 for $\nu_I > 1$. The ratio of ξ_1 and ξ_2 reads

$$\frac{\xi_2}{\xi_1} = \frac{4^{\nu_I-1}\Gamma(\nu_I)\Gamma(1+\nu_I)}{\pi(1+\nu_{II})(\cot(\nu_I\pi)-i)}\left(1+\frac{1+\nu_{II}}{1-\nu_I}+\frac{4(\nu_I+\nu_{II})}{4\nu_I+(2\nu_{II}-3)(\alpha^{-1}-1)}\right). \tag{47}$$

If both η_I and η_{II} are small ($\ll 1$), we obtain $\xi_2/\xi_1 \approx -1.2i - 3.77\eta_I - 0.08i\eta_{II}$. We see that the effect of η_I on the ratio is much stronger than that of η_{II}. By inspecting the ratio numerically, we find the η_{II} dependence is small even if it is of order unity. Hence, when considering the parameter dependence of the ratio ξ_2/ξ_2, we may safely neglect the η_{II} contribution and focus on the η_I dependence.

We plot it in Fig. 8, where the limit $|\eta_{II}| \to 0$ is taken. As one can notice, there are two distinct values of η_I, i.e., $|\xi_2/\xi_1| = 1$ at $\eta_I = 5/12$ and $|\xi_2/\xi_1| = 0$ at $\eta_I = -7/12$, independent of the value of R_ϵ. For $\eta_I < -7/12$, the T2 contri-

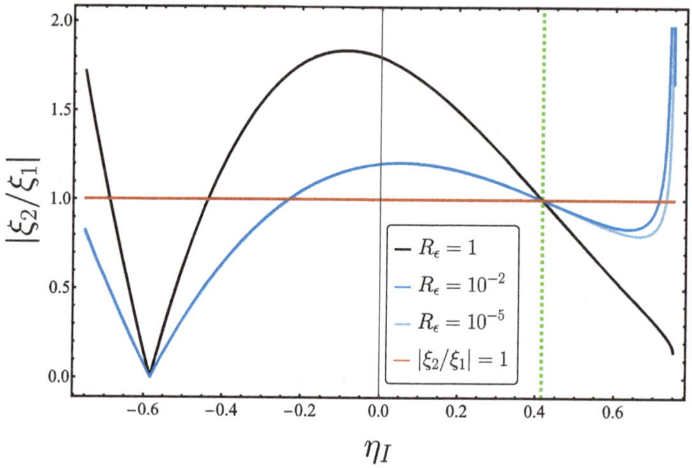

Fig. 8 The dependence of ξ_2/ξ_1 to η_I. For the black, blue and light blue solid lines, we take $R_\epsilon = 1, 10^{-2}, 10^{-5}$ to plot the expression given by (47). The green vertical line indicates $\eta_I = 5/12$, and the horizontal red line shows the $|\xi_2/\xi_1| = 1$, they divide the figure into four parts, for parameters lie in the top-left part, the ξ_2 term serves as next to leading order term with the biggest coefficient among the higher order terms, thus dominates the power spectrum for $|\alpha - 1| < 1$ cases. Whilst for parameters lie in the bottom right corner, the ξ_1 term dominates

bution increases as η_I decreases. For $-7/12 < \eta_I < 5/12$, the T2 contribution is maximized around $\eta_I = 0$. For $\eta_I > 5/12$, the T2 contribution may become dominant depending on the value of R_ϵ. This R_ϵ dependence can be seen by taking the limit $\eta_I \to 3/4$. We obtain $\xi_2/\xi_1 \to 7/20 - 1/(5R_\epsilon)$. This is smaller than unity for large R_ϵ, but exceeds unity for $R_\epsilon < 4/27$.

In the above, we ignored the possible higher-order contribution. To estimate them analytically is beyond the scope of the current paper, but as far as the range of parameters we consider is concerned, a good fit with the numerical results suggests that the higher orders are unimportant in most cases. Nevertheless, when $|\xi_2| \gg 1$, the higher order contribution may not be negligible near $k \sim k_\star$. In fact, the numerical result we obtained shows a slightly sharpened growth feature near $k_\star$ as depicted in Fig. 9.

Let us turn back to the case T2 dominates, and focus on the magnitude of ξ_2. As we have discussed, T2 dominance happens for a relatively small $\eta_I \sim O(0.1)$. By expanding ξ_2 in the limit η_I and η_{II} are small, we obtain

$$\xi_2 = -\frac{\eta_{II}(\eta_{II} - 2) + \alpha(6 + \eta_I(\eta_{II} - 2) - \eta_{II}^2)}{(1 - \alpha)(2\eta_I - 1)\eta_{II}(2\eta_{II} - 5)} \approx \frac{2(\eta_{II} + \alpha(\eta_I - 3))}{5\eta_{II}(1 - \alpha)}. \quad (48)$$

Apparently, the term $1 - \alpha$ in the denominator determines the behavior of the power spectrum for near-IR modes ($k \lesssim k_\star$). To be specific, T2 starts to dominate the full power spectrum for modes in the range,

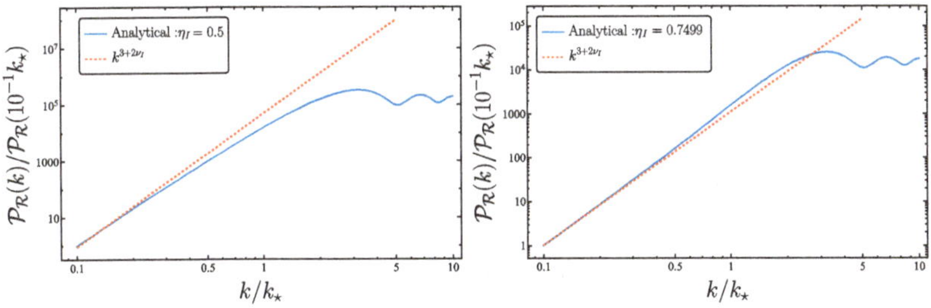

Fig. 9 The normal power spectrum V.S. sharpened power spectrum. We take $R_\epsilon = 10^{-5}$ for both graphs

$$1 > \frac{k}{k_\star} > \frac{1}{\sqrt{|\xi_2|}} \approx \sqrt{\left|\frac{5\eta_{II}(1-\alpha)}{2(\eta_{II}+\alpha(\eta_I-3))}\right|}, \tag{49}$$

when $|\xi_2| \gg 1$, which is realized either for $|\delta\alpha| \ll 1$ (where $\delta\alpha \equiv \alpha - 1$). In this case, the right inequality in (49) can easily be fulfilled even by very small k, since it simplifies to $k/k_\star \gtrsim \sqrt{5\eta_{II}|\delta\alpha|/6}$, where we used the fact that $\alpha \propto (3 - 2\nu_{II}) > 0$ implies $\eta_{II} > 0$ according to (15).

In the limit both η_I and η_{II} are very small, while R_ϵ is not so small, we have $\xi_2 = 2(R_\epsilon - 1)/(5R_\epsilon)$, which becomes smaller than -1 when $R_\epsilon < 2/7$. Note that (16) implies $|\alpha| \ll 1$ in this case. In this limit, T2 dominates since T1 is proportional to $X^{2\nu_I} \sim X^3$ while T2 to X^2. Thus, a dip may appear at $X = k/k_\star < 1$ when $R_\epsilon < 2/7$, and the spectrum is dominated by T2, hence $\mathcal{P}_\mathcal{R}(k) \propto k^4$ at $k/k_\star < 1$.

Another interesting finding is that the sign of ξ_2, which determines the existence or nonexistence of a dip in the power spectrum, depends on whether α is greater or smaller than unity. For the case $|\delta\alpha| = |\alpha - 1| \ll 1$, we obtain a simplified form of $\xi_2 \approx 6/(5\eta_{II}\delta\alpha) \gg 1$. Thus, the sign of ξ_2 directly depends on the sign of $\delta\alpha$.

We have $\xi_2 < 0$ if $\delta\alpha < 0$ $(\alpha < 1)$. In this case, a dip exists, and the position of the scale at which the dip appears is given by

$$\frac{k_{\rm dip}}{k_\star} = |\xi_2|^{1/2} = \sqrt{\left|\frac{(1-\alpha)(2\eta_I - 1)\eta_{II}(2\eta_{II} - 5)}{\eta_{II}(\Delta\eta_{II} - 2) + \alpha(6 + \eta_I(\eta_{II} - 2) - \eta_{II}^2)}\right|} \approx \sqrt{\left|\frac{5\eta_{II}(1-\alpha)}{2(\eta_{II}+\alpha(\eta_I-3))}\right|}. \tag{50}$$

On the other hand, if $\delta\alpha > 0$ $(\alpha > 1)$, we have $\xi_2 > 0$, and the dip ceases to exist (as we show in Fig. 10). It is worth noting here that since $\alpha > 1$ is the condition for the disappearance of a dip, it means that the attractor phase is realized only after the inflaton passes through the minimum, climbs up the potential, stops moving and makes a u-turn toward the minimum, as discussed in Sect. 2.

It is also worth mentioning that, in our model, the steepest growth of the power spectrum can exceed the "steepest growth" of k^4 if we tune the parameters. According

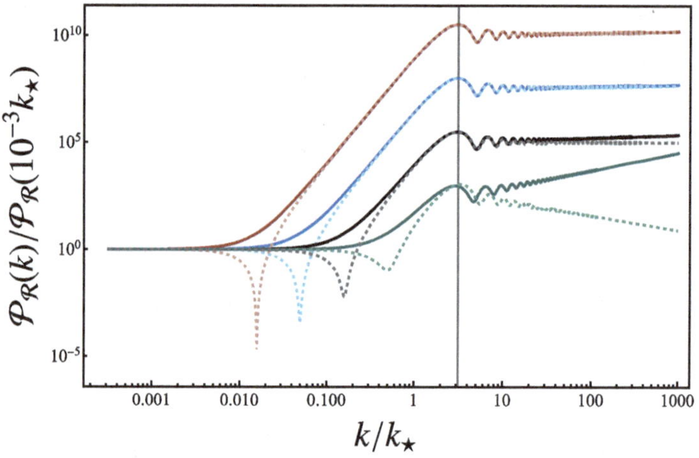

Fig. 10 **The enhancement of power spectrum for $\delta\alpha$ with different signs.** The figure is given by (39), by setting $R_{\epsilon} = 10^{-2}$, $\eta_I = -1 \times 10^{-10}$ and $D_I = 0$. For the red, blue, black, and green solid (dashed) lines, we take $\delta\alpha = +(-)10^{-2}, +(-)10^{-1}, +(-)1, +(-)10$ to show the alpha dependence of the shape of the power spectrum. The gray vertical line indicates $k/k_{\star} = \pi$, which is the analytical approximation of the peak position discussed in Sect. 3.4

to the (42), assuming both ξ_1 and ξ_2 are of the same order, the spectra index $n_{\star}$ near $k_{\star}$ is given by

$$n_{\star} = \min[3 + 2\nu_I, 7 - 2\nu_I]\,. \tag{51}$$

This takes the maximum value of 5 at $\nu_I = 1$ ($\eta_I = 5/12$) where both indices degenerate. When the index ν of the Hankel function takes an integer value, the solution near the origin starts to contain a logarithmic function. This gives rise to a logarithmic dependence of the spectral index in k. Thus the resulting spectral shape is becomes $k^5(\ln k)^2$.

One may then wonder if there could be a case when $\xi_2 = 0$ and $|\xi_1| > 1$ for $\nu_I > 1$. If this happens the spectral index would be given by $3 + 2\nu_I$ and it would exceed 5. In order to examine this possibility, we performed a perturbative analysis by setting $\xi_2 = 0$ and $\nu_I = 1 + \delta$ ($\delta > 0$). We found that ξ_1 can never exceed unity. Thus we conclude that the steepest growth rate in our model is $k^5(\ln k)^2$ [24] (Fig. 11).

As the appearance of a logarithmic function is a property of the Hankel function with an integer index, one might suspect that the $(\ln k)^2$ feature in the power spectrum could appear not only when $\nu_I = 1$ but when $\nu_I = 2, 3, \cdots$. However, Eq. (51) tells us that the T2 term always donimates, hence renders the logarithmic term subdominant.

It may be noted that our discussion of the steepest growth $k^5(\ln k)^2$ is restricted to the case of the adiabatic vacuum. As recently found in [32], for the α vacuum, a growth of k^6 is possible for some certain choice of the Bogoliubov coefficient. In our case, the α vacuum corresponds to the choice of non-vanishing D_I in Eq. (33).

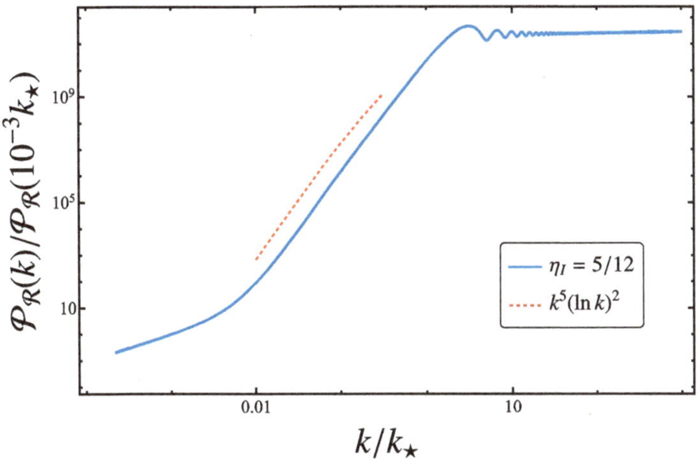

Fig. 11 The steepest growth of $k^5(\ln k)^2$. In this plot, we take $\eta_I = 5/12$, $R_\epsilon = 10^{-5}$

We leave the analysis for the α-vacuum in the generalized piecewise potential in the future work.

3.4 *The Short Wavelength Limit*

In the limit $X = k/k_\star \gg 1$, the final amplitude of the curvature perturbation can be expanded to second order as

$$\mathcal{P}_{\mathcal{R}}(k;\tau_f) = \mathcal{A}^S_{\mathcal{R}} \left| X^{\frac{3}{2}-\nu_{II}} + \Omega_1(X) X^{\frac{1}{2}-\nu_{II}} + \Omega_2(X) X^{-\frac{1}{2}-\nu_{II}} + O(X^{-\frac{3}{2}-\nu_{II}}) \right|^2, \tag{52}$$

where the amplitude $\mathcal{A}^S_{\mathcal{R}}$ is given by

$$\begin{aligned}\mathcal{A}^S_{\mathcal{R}} &= \left(\frac{2^{-5/2+\nu_{II}}}{3} \frac{H_\star}{2\pi} \frac{\Gamma(\nu_{II})}{\sqrt{\pi}/2} \frac{4\nu_{II}}{(1-\alpha)\sqrt{2\epsilon_{II}}} \right)^2 \\ &= \mathcal{A}^L_{\mathcal{R}} \left(\frac{2^{\nu_{II}}\Gamma(\nu_{II})}{2^{\nu_I}\Gamma(\nu_I)} \frac{4\nu_{II}}{(-3+2\nu_{II}) + R_\epsilon(3+2\nu_I)} \right)^2 .\end{aligned} \tag{53}$$

Thus the enhancement of the power spectrum in the short wavelength regime is estimated as

$$\frac{\mathcal{P}_{\mathcal{R}}(k_{UV})}{\mathcal{P}_{\mathcal{R}}(k_{IR})} = \frac{\mathcal{A}^S_{\mathcal{R}}}{\mathcal{A}^L_{\mathcal{R}}} = \left(\frac{2^{\nu_{II}}\Gamma(\nu_{II})}{2^{\nu_I}\Gamma(\nu_I)}\right)^2 \left(\frac{4\nu_{II}}{(-3+2\nu_{II}) + R_\epsilon(3+2\nu_I)}\right)^2$$
$$= \left(\frac{2^{\nu_{II}}\Gamma(\nu_{II})}{2^{\nu_I}\Gamma(\nu_I)}\right)^2 \frac{1}{(1-\alpha)^2} \left(\frac{4\nu_{II}}{(3+2\nu_I)R_\epsilon}\right)^2 . \qquad (54)$$

This slightly differs from the estimate (17) in the small η limit (i.e., both η_I and η_{II} are small), in which the factor involving the Gamma functions is absent. Here the expression is valid for any models with non-negligible η. The functions Ω_1 and Ω_2 in (52) are given by

$$\Omega_1 = \frac{1}{4}\left[\left(e^{2i(X-\nu_{II}\pi)} + i\right)(3+2\nu_I)(R_\epsilon - 1) - 2i(\nu_{II}^2 - \nu_I^2)\right], \qquad (55)$$

$$\Omega_2 = \frac{1}{16} i e^{2i(X-\nu_{II}\pi)}\left[(3+2\nu_1)(2\nu_I^2 + 2\nu_{II}^2 - 1)(R_\epsilon - 1) - 4(\nu_{II}^2 - \nu_I^2)\right]$$
$$+ \frac{1}{8}(\nu_{II}^2 - \nu_I^2)\left[(3+2\nu_1)(R_\epsilon - 1) - (\nu_{II}^2 - \nu_I^2)\right]. \qquad (56)$$

As they contain the oscillatory function e^{i2X}, and an oscillating feature appears in the power spectrum when $R_\epsilon \neq 1$, as seen in Figs. 5, 6 and 7.

In the regime of $R_\epsilon \ll 1$, Ω_1 and Ω_2 are approximately of the same order. Since ν_I and ν_{II} are both positive, the $\Omega_1 x^{1/2-\nu_{II}}$ term dominates the oscillation, with Ω_1 approximately given by

$$\Omega_1 \approx -\frac{(3+2\nu_I)}{4}\Big(1 + \exp\big[i\,(2X - 2\nu_{II})\big]\Big). \qquad (57)$$

The first peak gives the local maximum for modes $k > k_\star$. In the small η limit, by solving $d|\Omega_1|/dx = 0$, we find it is located at [89]

$$X_{\text{peak}} \approx \frac{\pi}{4}(1 + \nu_I + \nu_{II}) \sim \pi\,, \qquad (58)$$

as shown in Fig. 10. In the small wavelength limit, the leading order term in (52) implies a slightly tilted spectrum proportional to $k^{3-2\nu_{II}}$. Some examples of the case with significant tilts are given in Figs. 6 and 7.

To conclude this section, let us summarize the η dependence of the shape of the power spectrum. It appears in four ways:

- The slope of the far IR (UV) part of the power spectrum is determined by η_I (η_{II}) provided that it is much larger than ϵ (i.e., $\epsilon_X \ll |\eta_X|$ for $X = I,\ II$). It is proportional to $k^{3-2\nu_X}$.
- The amplitude of the enhancement is determined by parameter α defined by Eq. (15), which is a function of ϵ_X and η_X ($X = I,\ II$), that controls the state of inflaton at the end of the off-attractor stage. As given by (54), the small scale power spectrum is strongly enhanced when $|\alpha - 1| \ll 1$.

- The scaling behavior of the enhanced power spectrum for modes $k \lesssim k_\star$ is modified by η_I, resulting in the steepest growth of $k^5(\ln k)^2$ for $\nu_I = 1$ or equivalently $\eta_I = 5/12$.
- The dip at $k < k_*$ and the peak at $k > k_\star$ of the power spectrum depend also on the values of η_X. When $\alpha > 1$, which may happen if $\eta_{II} > 0$, the dip disappears. The peak position is well approximated by $k \approx \pi k_\star$ when both η_X ($X = I, II$) are small.

4 The Non-Gaussianity: δN Calculation

In Sect. 2, we have discussed the background evolution of inflaton with the potential with the form presented as (1). In this section, applying the δN formalism to the background solutions given there, we discuss the non-Gaussianity in this model.

The δN formalism is based on the separate universe approach that a comoving region greater than the Hubble horizon scale evolves like a homogeneous and isotropic universe at leading order in the spatial gradient expansion. It offers an intuitive and efficient method for linking the field fluctuations $\delta\phi$ with the curvature perturbation at the end of inflation by $\mathcal{R} = \delta N$. Here, N represents the e-folding number of a fiducial patch of the universe, calculated backward in time from the end of inflation as $N \equiv n_{\text{end}} - n$, where $\mathrm{d}n \equiv H\mathrm{d}t$, to an epoch when the scalar field ϕ and its velocity $\pi = -d\phi/dn$ takes given values, i.e., $N = N(\phi, \pi)$. The δN is then given by the difference in the e-folding number between the two different initial conditions $\delta N = N(\phi + \delta\phi, \pi + \delta\pi) - N(\phi, \pi)$, where $\delta\phi$ and $\delta\pi$ are given by the field fluctuations on the flat slicing at or after the horizon crossing.

Usually $\delta\phi$ approaches the adiabatic limit, $\delta\phi \propto \pi$ and $\delta\pi \to 0$. In this case, one can use the values of $\delta\phi$ at the horizon crossing and setting $\delta\pi = 0$ as the initial condition. This naturally introduces the scale-dependence of δN determined by the horizon crossing time of the comoving scale.

However, it is now well understood that there can be non-trivial evolution of the field fluctuations even after horizon crossing if there exists a feature in the potential that induces transitional behavior of the scalar field [66, 77], like the case we consider in the current paper. In such a case, it is known that the next leading order correction to $\delta\phi$ in gradient expansion must be taken into account in order to evaluate the curvature perturbation spectrum properly if we use the horizon crossing value of $\delta\phi$. An extension of the δN formalism that includes this correction has been done recently in [4].

In this paper, in order to avoid these issues, we apply the δN formula only at epoch sufficiently far outside the horizon when the scalar field perturbation has relaxed to the attractor limit at some late time $n = n_{\text{attr}} > n_\star$, under the assumption that the scalar field perturbation can still be treated in linear theory until $n = n_{\text{attr}}$. Then provided that the scalar field perturbation follows the classical equations of motion, using the fact that δN is independent of the time at which it is evaluated, one may pretend that one can evaluate δN at horizon-crossing $n = n_k$, not with the actual values of $\delta\phi$ and $\delta\pi$ at horizon-crossing, but with the values determined by those evolved backward in time from n_{attr} to n_k. With this prescription, we make sure that the non-Gaussian properties evaluated by using the δN formula are correct even if we take the initial slice to be the one at horizon-crossing.

In passing, we note that there is a subtlety in the evolution of $\delta\phi$ on superhorizon scales that has not been explicitly stated in most literature. Usually, the evolution of $\delta\phi$ is assumed to be linear. As in the case of the first or second stage of our model, this is true as long as the equation of motion is a linear second-order differential equation. However, if there is a non-trivial feature like the sudden change in the slope and/or the effective mass, its effect is the same as the case of a highly nonlinear potential for which the nonlinear corrections to the evolution of $\delta\phi$ is non-negligible. Thus if we evolve $\delta\phi$ from the first stage to the second stage on superhorizon scales, $\delta\phi$ at the first stage is perfectly Gaussian, but that at the second stage contains the nonlinear corrections due to the transition, hence is non-Gaussian in general. A detailed discussion on this point will be discussed in the subsection at the end of this section.

4.1 Analytical Formulas for δN

First, let us summarize the qualitative behavior of δN as one varies the comoving scale. Broadly speaking, they can be categorized into three cases depending on the horizon crossing time. Perturbations corresponding to larger scales (long-wavelength modes) exit the Hubble horizon earlier:

- Case A: Super-long-wavelength modes $n_\star - n_k \gg 1$. An approximate analytical expression for δN is given later by Eq. (74).
- Case B: Modes that exit the horizon shortly before the transition point $n_k \lesssim n_\star$. An approximate analytical expression is given by Eq. (76).
- Case C: Modes that exit the horizon after the transition point $n_k > n_\star$. An approximate analytical expression is given by Eq. (77).

Apparently, Case A and Case B differ from Case C in that the modes exit the horizon before the transition $n = n_\star$ for the former, while they exit the horizon after $n = n_\star$ for the latter. Thus, the modes undergo a two-stage evolution for Case A and Case B ($n_k < n_\star$), hence $N = N_I + N_{II}$, where $N_I = n_\star - n_k$ and $N_{II} = n_{\text{end}} - n_\star$. In contrast, for Case C ($n_k > n_\star$), there is only a single stage, $N = n_{\text{end}} - n_k$. The analytical results corresponding to these three cases can be obtained by applying the technique developed in [88]. A detailed numerical comparison is provided in Sect. 4.2. Below we recapitulate the formulas in [88].

By manipulating the background solution (7), we have

$$\frac{\pi_X + \lambda_{X,+}\widetilde{\phi}_X}{\pi_\star + \lambda_{X,+}\widetilde{\phi}_{X,\star}} = e^{-\lambda_{X,-}(n-n_\star)}\,, \tag{59}$$

$$\frac{\pi_X + \lambda_{X,-}\widetilde{\phi}_X}{\pi_\star + \lambda_{X,-}\widetilde{\phi}_{X,\star}} = e^{-\lambda_{X,+}(n-n_\star)}\,, \tag{60}$$

where $X = I$ or II, and $\widetilde{\phi}_X = \phi_X - \phi_{X,0}$ ($\phi_{X,0} = \phi_\star - \sqrt{2\epsilon_X}/\eta_X$) as before. Using either one of these expressions, we obtain $n - n_\star$ expressed in terms of $(\widetilde{\phi}, \pi)$ and $\pi_\star$ as

$$n - n_\star = -\frac{1}{\lambda_{X,\pm}} \ln\left(\frac{\pi_X + \lambda_{X,\mp}\widetilde{\phi}_X}{\pi_\star + \lambda_{X,\mp}\widetilde{\phi}_{X,\star}}\right). \tag{61}$$

Furthermore, by eliminating $n - n_\star$ from Eqs. (59) and (60), we can derive the relateion that determins $\pi_\star$ as a function of $(\widetilde{\phi}_X, \pi_X)$,

$$\lambda_{X,-} \ln\left(\frac{\pi_X + \lambda_{X,-}\widetilde{\phi}_X}{\pi_\star + \lambda_{X,-}\widetilde{\phi}_{X,\star}}\right) = \lambda_{X,+} \ln\left(\frac{\pi_X + \lambda_{X,+}\widetilde{\phi}_X}{\pi_\star + \lambda_{X,+}\widetilde{\phi}_{X,\star}}\right). \tag{62}$$

This equality ensures that the result in Eq. (61) is consistent regardless of whether the plus or minus sign is used, leading to the *logarithmic duality* [88].

Following the procedure in [88], the calculation of the δN formalism is straightforward by subtracting Eq. (61) from a perturbed version where $\widetilde{\phi}_X \to \widetilde{\phi}_X + \delta\phi_X$, $\pi_X \to \pi_X + \delta\pi_X$, and $\pi_\star(\phi, \pi) \to \pi_\star + \delta\pi_\star$.[1] We obtain

$$\delta(n - n_\star) = -\frac{1}{\lambda_{X,\pm}} \ln\left[1 + \frac{\delta\pi_X + \lambda_{X,\mp}\delta\phi_X}{\pi_X + \lambda_{X,\mp}\widetilde{\phi}_X}\right] + \frac{1}{\lambda_{X,\pm}} \ln\left[1 + \frac{\delta\pi_\star}{\pi_\star + \lambda_{X,\mp}\widetilde{\phi}_{X,\star}}\right], \tag{63}$$

where $\delta\pi_\star(\phi, \pi)$ is determined by the perturbed version of Eq. (62),

$$\begin{aligned} &\left(1 + \frac{\delta\pi_\star}{\pi_\star + \lambda_{X,-}\widetilde{\phi}_{X,\star}}\right)^{\lambda_{X,-}} \left(1 + \frac{\delta\pi_\star}{\pi_\star + \lambda_{X,+}\widetilde{\phi}_{X,\star}}\right)^{-\lambda_{X,+}} \\ &= \left(1 + \frac{\delta\pi_X + \lambda_{X,+}\delta\phi_X}{\pi_X + \lambda_{X,+}\widetilde{\phi}_X}\right)^{-\lambda_{X,+}} \left(1 + \frac{\delta\pi_X + \lambda_{X,-}\delta\phi_X}{\pi_X + \lambda_{X,-}\widetilde{\phi}_X}\right)^{\lambda_{X,-}}, \end{aligned} \tag{64}$$

which also guarantees the equivalence between the plus and minus signs of Eq. (63).

For $n_k < n_\star$, the evolution at the second stage $n_\star < n$ is totally determined by the initial condition at $n = n_\star$, namely, $(\phi_\star, \pi_\star)$. Thus for $n_k < n_\star$, setting $n = n_k$ and $X = I$ in (63) gives $N_I = n_\star - n_k$, and setting $n = n_{\text{end}}$ and $X = II$ gives $N_{II} = n_{\text{end}} - n_\star$. On the other hand, for $n_k > n_\star$, N is given by setting $n = n_k$ and

[1] Note that $\phi_\star$ is merely a parameter of the model that determines the transition point. Hence it is not perturbed.

$X = II$, and replacing $n_\star$ by $n_{\rm end}$ in (63). Perturbing thus obtained expressions, we obtain the curvature perturbation at the end of inflation.

For the long-wavelength modes that exit the horizon before the transition (i.e., Case A and B) the comoving curvature at the end of inflation is given by

$$\mathcal{R} \equiv \delta N = \delta N_I + \delta N_{II}\,, \tag{65}$$

where

$$\begin{aligned}\delta N_I &= \delta(n_\star - n_k)\\ &= \frac{1}{\lambda_{I,\mp}}\left[\ln\left(1+\frac{\delta\pi_I + \lambda_{I,\pm}\delta\phi_I}{\pi_I + \lambda_{I,\pm}\tilde{\phi}_I}\right) - \ln\left(1+\frac{\delta\pi_\star}{\pi_\star + \lambda_{I,\pm}\tilde{\phi}_{I,\star}}\right)\right], \end{aligned}\tag{66}$$

$$\begin{aligned}\delta N_{II} &= \delta(n_{\rm end} - n_\star)\\ &= \frac{1}{\lambda_{II,\mp}}\left[\ln\left(1+\frac{\delta\pi_\star}{\pi_\star + \lambda_{II,\pm}\tilde{\phi}_{II,\star}}\right) - \ln\left(1+\frac{\delta\pi_{\rm end}}{\pi_{\rm end} + \lambda_{II,\pm}\tilde{\phi}_{\rm end}}\right)\right]. \end{aligned}\tag{67}$$

In the above, $\delta\pi_\star$ is a function of $(\delta\phi, \delta\pi)$ implicitly determined by Eq. (64) and $\delta\pi_{\rm end}$ is by

$$\begin{aligned}&\left(1+\frac{\delta\pi_\star}{\pi_\star + \lambda_{II,-}\tilde{\phi}_{II,\star}}\right)^{\lambda_{II,-}}\left(1+\frac{\delta\pi_\star}{\pi_\star + \lambda_{II,+}\tilde{\phi}_{II,\star}}\right)^{-\lambda_{II,+}}\\ &\quad= \left(1+\frac{\delta\pi_{\rm end}}{\pi_{\rm end} + \lambda_{II,+}\tilde{\phi}_{\rm end}}\right)^{-\lambda_{II,+}}\left(1+\frac{\delta\pi_{\rm end}}{\pi_{\rm end} + \lambda_{II,-}\tilde{\phi}_{\rm end}}\right)^{\lambda_{II,-}}. \end{aligned}\tag{68}$$

For the short-wavelength modes that exit the horizon after the transition (i.e., Case C), the curvature perturbation at the end of inflation is given by

$$\begin{aligned}\mathcal{R} &= \delta N = \delta N_{II} = \delta(n_{\rm end} - n_k)\\ &= \frac{1}{\lambda_{II,\mp}}\left[\ln\left(1+\frac{\delta\pi_{II} + \lambda_{II,\pm}\delta\phi_{II}}{\pi_{II} + \lambda_{II,\pm}\tilde{\phi}_{II}}\right) - \ln\left(1+\frac{\delta\pi_{\rm end}}{\pi_{\rm end} + \lambda_{II,\pm}\tilde{\phi}_{\rm end}}\right)\right], \end{aligned}\tag{69}$$

where $\delta\pi_{\rm end}$ is implicitly determined by

$$\begin{aligned}&\left(1+\frac{\delta\pi_{II} + \lambda_{II,-}\delta\phi_{II}}{\pi_{II} + \lambda_{II,-}\tilde{\phi}_{II}}\right)^{\lambda_{II,-}}\left(1+\frac{\delta\pi_{II} + \lambda_{II,+}\delta\phi_{II}}{\pi_{II} + \lambda_{II,+}\tilde{\phi}_{II}}\right)^{-\lambda_{II,+}}\\ &\quad= \left(1+\frac{\delta\pi_{\rm end}}{\pi_{\rm end} + \lambda_{II,+}\tilde{\phi}_{\rm end}}\right)^{-\lambda_{II,+}}\left(1+\frac{\delta\pi_{\rm end}}{\pi_{\rm end} + \lambda_{II,-}\tilde{\phi}_{\rm end}}\right)^{\lambda_{II,-}}. \end{aligned}\tag{70}$$

Below we more closely examine the non-Gaussian nature of the curvature perturbation for Case A, Case B and Case C separately.

Case A: $k \ll k_\star$

For the modes that exit the horizon well before the transition, the field fluctuation already settles down to the attractor phase when the transition occurs. Then, as discussed in [88], if the inflaton is already in the attractor phase at $\phi = \phi_\star$ as well as at the end of inflation $\phi = \phi_{\rm end}$, Eqs. (64) and (68) can be simplified as

$$1 + \frac{\delta\pi_\star}{\pi_\star + \lambda_{I,-}\widetilde{\phi}_{I,\star}} \simeq \left(1 + \frac{\delta\pi_I + \lambda_{I,-}\delta\phi_I}{\pi_I + \lambda_{I,-}\widetilde{\phi}_I}\right)\left(1 + \frac{\delta\pi_I + \lambda_{I,+}\delta\phi_I}{\pi_I + \lambda_{I,+}\widetilde{\phi}_I}\right)^{-\frac{\lambda_{I,+}}{\lambda_{I,-}}}, \tag{71}$$

$$1 + \frac{\delta\pi_{\rm end}}{\pi_{\rm end} + \lambda_{II,-}\widetilde{\phi}_{II,{\rm end}}} \simeq \left(1 + \frac{\delta\pi_\star}{\pi_\star + \lambda_{II,-}\widetilde{\phi}_{II,\star}}\right)\left(1 + \frac{\delta\pi_\star}{\pi_\star + \lambda_{II,+}\widetilde{\phi}_{II,\star}}\right)^{-\frac{\lambda_{II,+}}{\lambda_{II,-}}}. \tag{72}$$

Note that $\delta\pi_\star \to 0$ and $\delta_{\rm end} \to 0$ in the attractor limit. But the left-hand sides of the above equations are non-trivial because $\pi_\star + \lambda_{I,-}\widetilde{\phi}_{I,\star} \to 0$ and $\pi_{\rm end} + \lambda_{II,-}\widetilde{\phi}_{II,{\rm end}} \to 0$.

Substituting the above into (65), we obtain

$$\mathcal{R} = \frac{1}{\lambda_{I,-}} \ln\left[1 + \frac{\delta\pi_I + \lambda_{I,+}\delta\phi_I}{\pi_I + \lambda_{I,+}\widetilde{\phi}_I}\right] + \frac{1}{\lambda_{II,-}} \ln\left[1 + \frac{\delta\pi_\star}{\pi_\star + \lambda_{II,+}\widetilde{\phi}_{II,\star}}\right]. \tag{73}$$

Furthermore, under the assumption that the background is in the attractor phase initially at the first stage, we have $\pi_I + \lambda_{I,+}\widetilde{\phi}_I \simeq (\lambda_{I,+} - \lambda_{I,-})\widetilde{\phi}_I$ and $\lambda_{II,+}\widetilde{\phi}_{II,\star} + \pi_\star \simeq \lambda_{II,+}\widetilde{\phi}_{II,\star} - \lambda_{I,-}\widetilde{\phi}_{I,\star}$. Thus except for an extremely fine-tuned case $\lambda_{II,+}\widetilde{\phi}_{II,\star} - \lambda_{I,-}\widetilde{\phi}_{I,\star} = 0$, the contribution of the second logarithmic term is negligible. In the end, the curvature perturbation for the super-long-wavelength modes is given by

$$\mathcal{R} \simeq \frac{1}{\lambda_{I,-}} \ln\left[1 + \frac{\delta\pi_I + \lambda_{I,+}\delta\phi_I}{\pi_I + \lambda_{I,+}\widetilde{\phi}_I}\right] \quad \text{for } k \ll k_\star\,. \tag{74}$$

In the limit $|\lambda_{I,-}| \ll 1$, this reduces to the standard slow-roll result.

Case B: $k \lesssim k_\star$

The modes that exit the Hubble horizon shortly before the transition point show highly non-trivial behavior. It is not immediately apparent that the general expression (65) conceals some non-trivial features for those modes. The non-trivial scale dependence of $\delta\pi_\star(\delta\phi, \delta\pi; \phi, \pi)$ is implicitly embedded in (62).

To analytically clarify this non-trivial dependence, we use the approximation that $-\lambda_{I,-}(n - n_\star) \ll 1$. This enables us to approximate $e^{-\lambda_{I,-}(n-n_\star)} \approx 1$. Then from (7) and its n derivative ($\pi = {\rm d}\phi/{\rm d}n$), we have

$$\lambda_{I,+}\widetilde{\phi}_I + \pi_I \simeq \lambda_{X,+}\widetilde{\phi}_{I,\star} + \pi_\star\,. \tag{75}$$

In the limit $\eta_I \to 0$, we have $\lambda_{I,+} \to 3$, this reduces to the familiar relation for an ultra-slow-roll trajectory $3\phi + \pi = 3\phi_\star + \pi_\star$. Substituting the perturbed (75) and (72) into (65), we find $\delta N_I \approx 0$ and obtain

$$\mathcal{R} \simeq \frac{1}{\lambda_{II,-}} \ln\left[1 + \frac{\delta\pi_\star}{\pi_\star + \lambda_{II,+}\widetilde{\phi}_{II,\star}}\right] \simeq \frac{1}{\lambda_{II,-}} \ln\left[1 + \frac{\lambda_{I,+}\delta\phi_I + \delta\pi_I}{\pi_\star + \lambda_{II,+}\widetilde{\phi}_{II,\star}}\right] \quad \text{for } k \lesssim k_\star\,. \tag{76}$$

Thus, the main contribution is from the off-attractor behavior at the beginning of the second stage. There exists no USR term in the δN result, although (75) gives very similar trajectories in phase space.

We mention that the formula (76) differs from those reported in [18, 19]. Specifically, the first term in (2.34) of [18] and in (6) of [19] should be absent. This slightly modifies the remaining equations but not in an essential way as the main focus of these papers is on the effect of the upward step that leads to a cutoff in the probability distribution function of the curvature perturbation.

Case C: $k > k_\star$

For the modes that exit the horizon after the transition, the calculations are simpler, as we only need to compute $\delta N_{II} = \delta(n_{\text{end}} - n)$. We have

$$\begin{aligned}\mathcal{R} &= \frac{1}{\lambda_{II,-}} \ln\left[1 + \frac{\lambda_{II,+}\delta\phi_{II} + \delta\pi_{II}}{\pi_{II} + \lambda_{II,+}\widetilde{\phi}_{II}}\right] - \frac{1}{\lambda_{II,-}} \ln\left[1 + \frac{\delta\pi_{\text{end}}}{\pi_{\text{end}} + \lambda_{II,+}\widetilde{\phi}_{\text{end}}}\right] \\ &\simeq \frac{1}{\lambda_{II,-}} \ln\left[1 + \frac{\lambda_{II,+}\delta\phi_{II} + \delta\pi_{II}}{\pi_{II} + \lambda_{II,+}\widetilde{\phi}_{II}}\right] \quad \text{for } k > k_\star\,,\end{aligned} \tag{77}$$

where again we assumed that the inflaton is already in the attractor phase at the end of inflation.

We now have fully nonlinear δN results for the curvature perturbation $\mathcal{R}$ in our piecewise potential model, $\mathcal{R} = \delta N(\delta\phi, \delta\pi; \phi, \pi)$ where $\delta\phi$ and $\delta\pi$ are normally Gaussian random variables, but δN is a nonlinear function of $(\delta\phi, \delta\pi)$. Although there exists $\delta\pi$-dependence in $\delta N(\delta\phi, \delta\pi)$, the variance of $\delta\pi$ is negligible compared to that of $\delta\phi$ in the combination $\delta\pi + \lambda_+\delta\phi$ if $|\eta| \ll 1$. In such cases, we can further simplify the result by setting $\mathcal{R} = \delta N(\delta\phi, \delta\pi = 0)$. It is easy to check that the modes $k \ll k_\star$ and $k \gg k_\star$ follow the standard slow-roll result. The case given by (76) for the modes $k \lesssim k_\star$ provides an interesting example in which the curvature perturbation is not determined by η_I at the first stage but by η_{II} at the second stage. Ignoring $\delta\pi_I$ we have

$$\mathcal{R} \simeq \frac{1}{\lambda_{II,-}} \ln\left[1 + \frac{(3 - \eta_I)\delta\phi_I}{\pi_\star + \lambda_{II,+}\sqrt{2\epsilon_{II}}/\eta_{II}}\right], \tag{78}$$

where we have assumed $|\eta_I| \ll 1$ for simplicity, but η_{II} is not necessarily small. Using $\pi_\star$ given by (11) and the fact $|\eta_I| \ll 1$, we have $\pi_\star \simeq -\sqrt{2\epsilon_I}$. Hence the above equation can be concisely expressed as

$$\mathcal{R} = \frac{1}{\gamma} \ln\left[1 + \gamma \mathcal{R}_{\rm G}\right] ; \quad \mathcal{R}_{\rm G} \equiv \frac{(3-\eta_I)\delta\phi_I}{\lambda_{II,-}(\lambda_{II,+}\sqrt{2\epsilon_{II}}/\eta_{II} - \lambda_{I,-}\sqrt{2\epsilon_I}/\eta_I)} \tag{79}$$

where $\gamma = \lambda_{II,-}$ and $\mathcal{R}_{\rm G}$ is the Gaussian part of the curvature perturbation. Expanding the logarithmic function, one finds

$$\mathcal{R} = \mathcal{R}_{\rm G} + \frac{3}{5} f_{\rm NL} \mathcal{R}_{\rm G}^2 + \cdots ; \quad f_{\rm NL} = -\frac{5}{6}\gamma . \tag{80}$$

Since $\gamma = \lambda_{II,-} = (3 - \sqrt{9 - 12\eta_{II}})/2$, the non-Gaussian parameter $f_{\rm NL}$ can be large if η_{II} is large and negative.

As in the case of (78) above, when one of the logarithmic functions dominates $\mathcal{R}$, one can always express the curvature perturbation in the form of (79), except for the form of $\mathcal{R}_{\rm G}$. In such a case, using the probability conservation $\mathbb{P}[\mathcal{R}]{\rm d}\mathcal{R} = \mathbb{P}[\mathcal{R}_{\rm G}]{\rm d}\mathcal{R}_{\rm G}$, we can easily get the probability distribution function (PDF) of $\mathcal{R}$,

$$\mathbb{P}[\mathcal{R}] = \frac{1}{\sqrt{2\pi}\sigma_{\mathcal{R}_{\rm G}}} \exp\left[\gamma\mathcal{R} - \frac{(e^{\gamma\mathcal{R}} - 1)^2}{2\gamma^2\sigma_{\mathcal{R}_{\rm G}}^2}\right] , \tag{81}$$

where $\sigma_{\mathcal{R}_{\rm G}}^2$ is the variance of the Gaussian part of the PDF, $\mathbb{P}[\mathcal{R}_{\rm G}]$,

$$\sigma_{\mathcal{R}_{\rm G}}^2 = \int \mathcal{R}_{\rm G}^2 \mathbb{P}[\mathcal{R}_{\rm G}] d\mathcal{R}_{\rm G} . \tag{82}$$

In particular, there always appears an exponential tail in the PDF at large positive $\mathcal{R}$ if $\beta < 0$ and at large negative $\mathcal{R}$ if $\beta > 0$.

4.2 Numerical Algorithm and Comparison with Analytical Formulas

To check the robustness of our analytical approximations (73), (76), and (77), we numerically evaluated the exact δN formula (65) for Case A, Case B and Case C. Here we first describe the numerical algorithm we adopted.

As mentioned in Sect. 4, there are three situations where perturbation modes exit the Hubble horizon at different times. However, the consistency of the analytical approximations throughout the entire transition from the first slow-roll stage to the second has not been fully clear yet. In particular, for the modes crossing the horizon very near the transition, the field fluctuation determines whether the horizon crossing occurs before or after the transition. Thus the fluctuation significantly affects the corresponding background trajectory, leading to fully non-perturbative phenomena.

The key to analytically determining the nonlinear formula for $\mathcal{R}(\delta\phi)$ in (65) is to explicitly express $\delta\pi_\star(\phi, \pi)$ and $\delta\pi_{\rm end}(\pi_\star)$. These quantities are determined by perturbing the *logarithmic duality* (64) or (68), which typically requires numerical evaluations. The special cases in which it can be algebraically solved occur when $\lambda_{X,+}/\lambda_{X,-} = m/n$, where $m, n \in \mathbb{Z}$, $m > |n|$, and $\max(m, m-n) < 4$. The details about solvable cases are given in [88]. One option is to solve (64) or (68) numerically and substitute the result into the expression for $\mathcal{R}$. However, here we adopt a more straightforward approach.

The algorithm is outlined in the following steps:

- Step 1: Write down the analytical solution of the equation of motion (EOM) (6) for both stages, using the initial conditions $(\phi_{\rm i}, \pi_{\rm i})$ for the first stage $X = I$ and $(\phi_{\rm i}, \pi_{\rm i})$ for the second stage $X = II$:

$$\begin{aligned}\phi_I(n) = \phi_\star - \frac{\sqrt{2\epsilon_I}}{\eta_I} - \frac{\pi_{\rm i} + \lambda_{I,+}(\phi_{\rm i} - \phi_\star + \sqrt{2\epsilon_I}/\eta_I)}{\lambda_{I,-} - \lambda_{I,+}} e^{-\lambda_{I,-}(n-n_{\rm i})}\\ + \frac{\pi_{\rm i} + \lambda_{I,-}(\phi_{\rm i} - \phi_\star + \sqrt{2\epsilon_I}/\eta_I)}{\lambda_{I,-} - \lambda_{I,+}} e^{-\lambda_{I,+}(n-n_{\rm i})}\,; \quad n < n_\star\,,\end{aligned} \tag{83}$$

$$\begin{aligned}\phi_{II}(n) = \phi_\star - \frac{\sqrt{2\epsilon_{II}}}{\eta_{II}} - \frac{\pi_{\rm i} + \lambda_{II,+}(\phi_{\rm i} - \phi_\star + \sqrt{2\epsilon_{II}}/\eta_I)}{\lambda_{II,-} - \lambda_{II,+}} e^{-\lambda_{II,-}(n-n_{\rm i})}\\ + \frac{\pi_{\rm i} + \lambda_{II,-}(\phi_{\rm i} - \phi_\star + \sqrt{2\epsilon_{II}}/\eta_{II})}{\lambda_{II,-} - \lambda_{II,+}} e^{-\lambda_{II,+}(n-n_{\rm i})}\,; \quad n > n_\star\,.\end{aligned} \tag{84}$$

- Step 2: Choose an initial condition $(\phi_{\rm i}, \pi_{\rm i})$ at the first stage that ensures the background evolution is physically valid. Then, evaluate $\pi_\star \equiv {\rm d}\phi_I(n_\star)/{\rm d}n$ and the e-folding number from $\phi_{\rm i}$ to $\phi_\star$: $N_I = n_\star - n_{\rm i}$.
- Step 3: Given the end of inflation $\phi_{\rm end}$, solve for the e-folding number $N_{II} = n_{\rm end} - n_\star$ from $\phi_\star$ to $\phi_{\rm end}$, using $(\phi_\star, \pi_\star)$ as the initial condition for (84).
- Step 4: Sum the total e-folding number $N = N_I + N_{II}$ for the background solution.
- Step 5: Calculate the new total e-folding number by introducing a different initial condition $(\phi_{\rm i} + \delta\phi, \pi_{\rm i} + \delta\pi)$. The key point here is to check whether the perturbed $\phi_{\rm i} + \delta\phi$ is greater or less than $\phi_\star$. If $\phi_{\rm i} + \delta\phi > \phi_\star$, then we repeat Steps 1 to 3. On the other hand, if $\phi_{\rm i} + \delta\phi < \phi_\star$, we obtain the total e-folding number by solving $N = N_{II} = n_{\rm end} - n_{\rm i}$.
- Step 6: By repeating the above, numerically establish the dependence of the total e-folding number as a function of the initial condition, $\delta N(\phi_{\rm i}, \pi_{\rm i}; \delta\phi, \delta\pi)$.

By employing this algorithm, we can generate a contour plot of numerically computed δN for a set of initial conditions $(\phi_{\rm i} + \delta\phi, \pi_{\rm i} + \delta\pi)$.

We select the same first slow-roll parameter, $\epsilon_I = \epsilon_{II} = 10^{-10}$, to isolate and highlight the effect of the second slow-roll parameter η_V. With $\epsilon_V = 10^{-10}$, a typ-

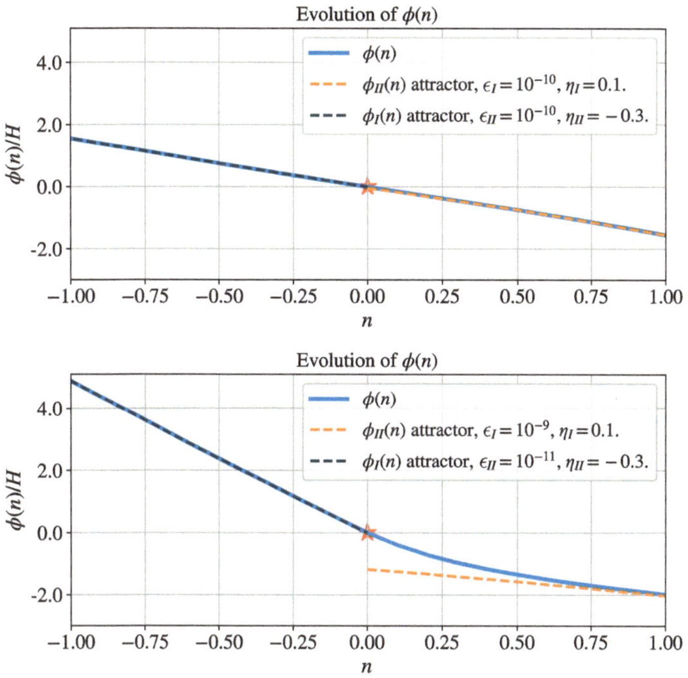

Fig. 12 The evolution of $\phi(n)$ is governed by Eqs. (83) and (84). We select $\pi_\star = 1.03576\sqrt{2\epsilon_I}$ for the blue line. The red star marks the joint point at $(n = 0, \phi = 0)$

ical inflationary energy scale of $H/M_\mathrm{p} = 10^{-5}$ can enhance the primordial power spectrum to $\mathcal{O}(10^{-2})$. The background evolution of $\phi(n)$ is illustrated in Fig. 12. The blue line represents the analytical results from (83) and (84), with $\phi_\star = 0$ and $\pi_\star = 1.03576\sqrt{2\epsilon_I}$. The dashed orange and black lines indicate the attractor solution. Here, we set $n_\star = 0$.

As shown in Fig. 12, $\phi(n)$ has already converged to the first-stage attractor solution when $n_\mathrm{i} - n_\star = -5$. We select (ϕ, π) at $n_\mathrm{i} - n_\star = -5$ to illustrate Case A; see Fig. 13. The orange line represents the analytical result from (74) for Case A and matches the numerical δN result well. We also observe a deviation from linearity at $\delta N \approx \mathcal{O}(1)$, which may be important for PBH formation.

To study the modes of most interest, Cases B and C, we choose $n_\mathrm{i} - n_\star = 0.5$ for a complete numerical demonstration, as shown in Fig. 14. At $n = n_\mathrm{i}$, the background solution yields $\phi_\mathrm{i} = 0.75H$ and $\pi_\mathrm{i} = 1.543H$ (represented by the solid blue line in Fig. 12). This makes it easy for the quantum fluctuation of $\delta\phi$ to shift the initial data, $(\phi_\mathrm{i} + \delta\phi, \pi_\mathrm{i} + \delta\pi)$, beyond the transition point. This mechanism highlights the non-perturbative phenomena observed in our model. As shown by the contour lines of δN in Fig. 14, the intervals are non-uniform along the $\delta\phi$ direction, which directly reflects the nonlinearity observed in the top panel of Fig. 15.

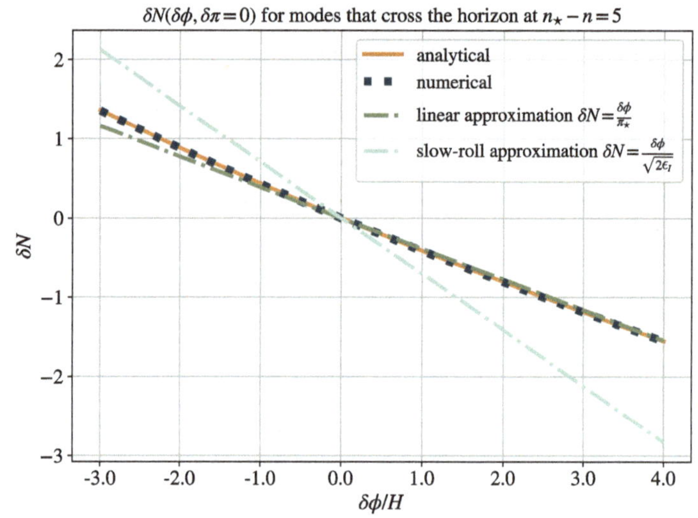

Fig. 13 We consider the point $(\phi_{\rm i}, \pi_{\rm i})$ on the blue line in Fig. 12, 5 e-folds before the inflaton reaches the joint point, and evaluate the numerical δN. The parameters are set as $\epsilon_I = \epsilon_{II} = 10^{-10}$, $\eta_I = 0.1$, and $\eta_{II} = -0.3$

It is worth noting that $\delta\pi$ also contributes to δN, as indicated by the slope of the contour lines. We typically ignore the contribution of $\delta\pi$ because δN is valid when the modes are superhorizon, i.e., several e-folds after horizon crossing. The $\delta\pi$ term decays exponentially and can therefore be ignored.

In the case when there is a sharp transition feature like in the current model, another noteworthy feature appears in the PDF. Consider the mode that exits the Hubble horizon at the transition point, $k = k_\star$. As in our model, let us assume that the background inflaton rolls down toward smaller values, $d\phi/dn < 0$. Then the field exits the horizon *before* the transition if $\delta\phi > 0$, while it exits the horizon *after* the transition if $\delta\phi < 0$. This implies that depending on the sign of $\delta\phi$, one has to use the different formulas of $\mathcal{R}$. For example, (76) for $\delta\phi < 0$ and (77) for $\delta\phi > 0$. This further implies that the formula to be used depends on the sign and the magnitude of $\delta\phi$. For example, for a mode that leaves the horizon slightly before the transition, $k = k_\star - \Delta k$ $(\Delta k > 0)$, (76) should be used for $\delta\phi > -\epsilon$ $(\epsilon > 0)$ while (77) should be used for $\delta\phi < -\epsilon$. We illustrate this interesting nonlinear effect in Fig. 15, which are obtained numerically by using the complete expression of $\mathcal{R}$, (65).

Here it is important to recall that the evaluation of the curvature perturbation by the δN formula does not depend on the time at which one evaluates the field fluctuations as long as the scale is superhorizon and the evolution is classical. Thus the above argument is independent of the precise definition of the horizon crossing time. In fact, one can evaluate δN at any time after horizon crossing. We will come back to this point in the next subsection.

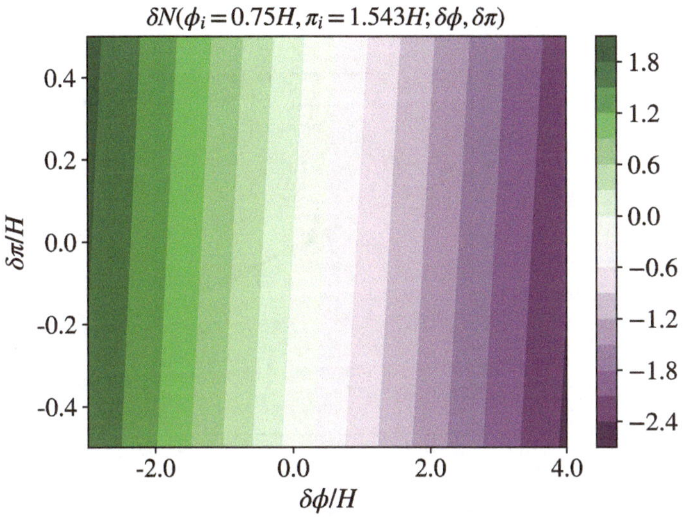

Fig. 14 Contour plot of $\delta N(\delta\phi, \delta\pi)$ for modes crossing the Hubble horizon 5 e-folds before the joint point. The parameters are set as $\epsilon_I = \epsilon_{II} = 10^{-10}$, $\eta_I = 0.1$, and $\eta_{II} = -0.3$. We set $\delta\pi = 0$ to obtain the numerical dashed line in the top panel of Fig. 15

4.3 Nonlinear Evolution of Superhorizon Modes

In this subsection, we discuss the conservation of the comoving curvature $\mathcal{R}$ in the nonlinear regime. Although it is already proved in [82] generically, we find it useful to consider our case explicitly.

Let us focus on the super-long wavelength (SLWL) modes, say, on the scales measured by the cosmic microwave background [1]. We know from the observational data that the slow-roll inflation is valid on those scales. The SLWL modes are frozen, and the spatial gradients can be ignored soon after horizon crossing. Thus, we can evaluate δN soon after horizon crossing. Namely, we use (74) with $\delta\phi$ inferred from linear perturbation theory.

Now, let us evaluate δN of the SLWL modes after the epoch of transition, $n > n_\star$. Then we should use (77). But the formula looks completely different from (74). So if one would naively solve for $\delta\phi$ by using the linear perturbation equation, one would obtain a completely wrong answer. This seems a contradiction as one would expect that $\delta\phi$ would follow the linear evolution equation since the scalar field evolution is linear already at the background level.

The answer is that the evolution of $\delta\phi$ is actually highly nonlinear on superhorizon scales when there is a sharp feature in the potential. Let us explain the origin of the nonlinearity. To do so, we first write down the solution of ϕ by neglecting the spatial-gradient terms at two different times $n = n_1$ and n_2, where n_1 belongs to the first

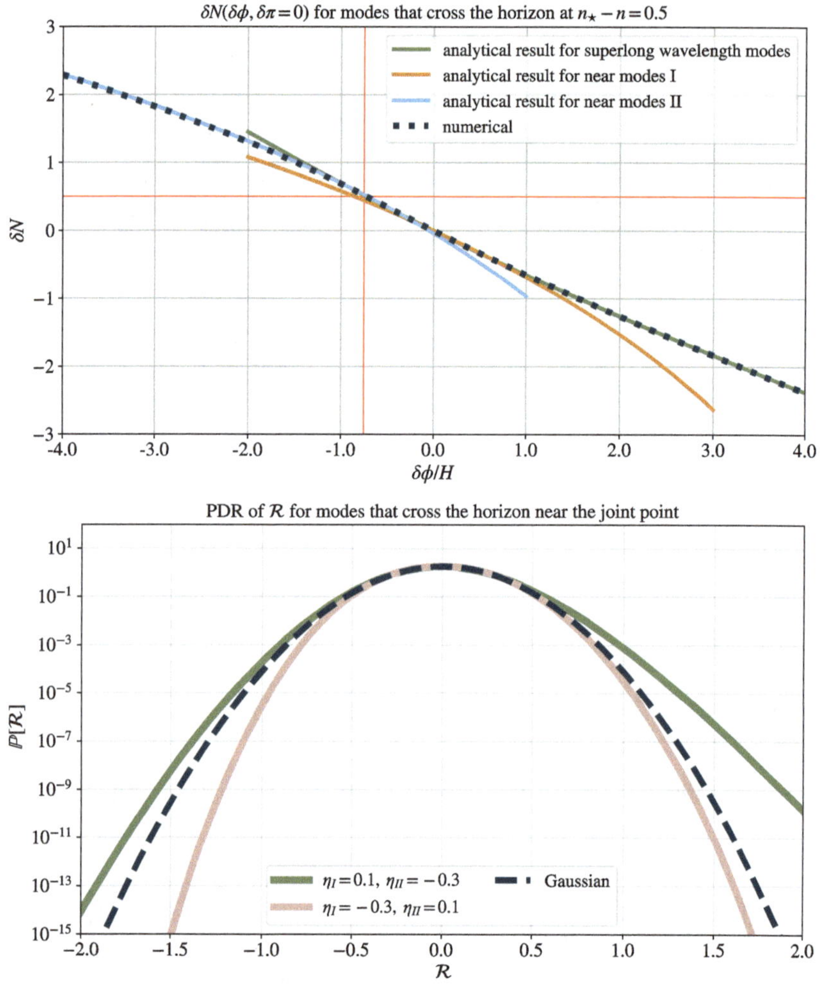

Fig. 15 **Top panel:** We demonstrate the nonlinear behavior of the curvature perturbation (65) for modes crossing the horizon near the joint point $\phi = \phi_\star$. We choose parameters as $\epsilon_I = \epsilon_{II} = 10^{-10}$, $\eta_I = 0.1$, and $\eta_{II} = -0.3$ for demonstration. The black dots denote numerical δN results. The modes crossing the horizon before and after the joint point are represented on the right and left sides of the vertical thin red solid line, respectively. The blue line represents the approximate result of (77), the orange line represents (76), and the green line represents (73). **Bottom panel:** The nonlinearity of $\mathcal{R}(\delta\phi)$ causes $\mathbb{P}[\mathcal{R}]$ to deviate from a Gaussian distribution. The PDF of the green line is given by the fully nonlinear relation demonstrated in the top panel. Changing the parameters leads to the pink one. There are always exponential tails in the PDF; however, the exponential tail on the positive $\mathcal{R}$ side is determined by η_{II}, while the exponential tail on the negative side is determined by η_I

stage and n_2 to the second stage. Assuming that the system is in the attractor regime both at $n = n_1$ and n_2, we have

$$\begin{aligned}\widetilde{\phi_I}(n_1) &= C_{I,-}\exp(-\lambda_{I,-}(n_1 - n_\star)),\\ \widetilde{\phi_{II}}(n_2) &= C_{II,-}\exp(-\lambda_{II,-}(n_2 - n_\star)),\end{aligned} \tag{85}$$

where the time when $\phi = \phi_\star$ is given by

$$n_\star = n_1 + \frac{1}{\lambda_{I,-}}\ln\left(\frac{\widetilde{\phi}_{I\star}}{\widetilde{\phi_I}(n_1)}\right). \tag{86}$$

As in the case of $\phi_\star$ being the end of inflation, $n_\star$ is a nonlinear function of the initial condition $\phi(n_1)$. Inserting the expression of $n_\star$ into $\widetilde{\phi_{II}}(n_2)$, we clearly see that the relation between $\widetilde{\phi_{II}}(n_2)$ and $\widetilde{\phi_I}(n_1)$ is nonlinear. Setting $\widetilde{\phi_I}(n_1) \to \widetilde{\phi_I}(n_1) + \delta\phi(n_1)$ and $\widetilde{\phi_{II}}(n_2) \to \widetilde{\phi_{II}}(n_2) + \delta\phi(n_2)$, we obtain $\delta\phi(n_2)$ as a highly nonlinear function of $\delta\phi(n_1)$,

$$\delta\phi(n_2) = \widetilde{\phi_{II}}(n_2)\left\{\left(\frac{\widetilde{\phi}_{I\star}}{\widetilde{\phi_I}(n_1)}\right)^{\lambda_{II,-}/\lambda_{I,-}} - \left(\frac{\widetilde{\phi}_{I\star}}{\widetilde{\phi_I}(n_1) + \delta\phi(n_1)}\right)^{\lambda_{II,-}/\lambda_{I,-}}\right\}. \tag{87}$$

Thus, assuming $\delta\phi$ is Gaussian at horizon crossing does not necessarily mean that it stays Gaussian afterward. As expected, if we take this nonlinearity into account in the evolution of $\delta\phi$, $\mathcal{R}$ on superhorizon scales for the SLWL modes remain conserved.

In passing, we mention its possible implication to the recent dispute over the loop effects on the curvature perturbation on superhorizon scales [9, 26–28, 36, 41, 42, 44, 48, 49, 61, 62, 71, 74–76, 83, 93–95, 101, 102]. As the field perturbation $\delta\phi$ in our computations corresponds to the one on flat slicing in perturbation theory, it implies that the conservation of the curvature perturbation cannot be guaranteed even at the classical level unless the nonlinear corrections to $\delta\phi$ are properly taken into account, if the perturbation is evolved in the flat slicing.

5 Primordial Black Holes from Non-Gaussian Tails

In this section, we consider the effect of non-Gaussianity on the abundance of primordial black holes (PBHs) in our piecewise quadratic potential model. Since detailed discussions on the formation criteria are beyond the scope of the present paper, for the sake of simplicity, we adopt the Press-Schechter formalism in which the threshold for the PBH formation is given by a critical amplitude of the density contract smoothed over a given comoving scale at the time of the horizon re-entry after inflation. Further, we ignore the nonlinearity arising from the Einstein equation and set the threshold in the framework of linear perturbation theory.

In linear theory, the density contrast $\delta \equiv \rho/\bar{\rho} - 1$ in the comoving slicing is related to the comoving curvature perturbation as

$$\delta = -\frac{2}{3}\frac{3+3\omega}{5+3\omega}\left(\frac{1}{aH}\right)^2 \nabla^2 \mathcal{R}, \tag{88}$$

where $w = 1/3$ during the radiation-dominated era, which we assume in the following. If we consider a comoving scale R_s, (88) at the time of horizon re-entry, $R_s = aH$, gives

$$\delta \approx \frac{4}{9}\mathcal{R}, \tag{89}$$

where we have assumed that the smoothing on the scale R_s effectively singles out the amplitude of the mode $k = 1/R_s$ in Fourier space. Then it follows that the threshold of δ is simply related to that of $\mathcal{R}$ by $\delta_{\text{th}} = 9/4\mathcal{R}_{\text{th}}$. Namely, we have the linear relation between the thresholds of the density contrast threshold and the comoving curvature perturbation. Furthermore, assuming that we evaluate δN when the field fluctuation is still Gaussian, the Gaussian part of the curvature perturbation is linearly related to the field fluctuation by $\mathcal{R}_{\text{G}} = -\delta\phi/(d\phi/dn) = \delta\phi/\pi$ as in the standard slow-roll case.

Under the above approximations, the abundance of PBHs at the time of formation on the comoving scale R_s is determined by the threshold $\mathcal{R}_{\text{th}}(R_s)$ as

$$\begin{aligned}
\beta_{\text{PBH}}(R_s) &\equiv \frac{\rho_{\text{PBH}}(R_s)}{\rho_{\text{tot}}} = \int_{\mathcal{R}_{\text{th}}(R_s)}^{+\infty} \mathbb{P}[\mathcal{R}]\text{d}\mathcal{R} = \int_{\mathcal{R}_{\text{G,th}}(R_s)}^{+\infty} \mathbb{P}[\mathcal{R}_{\text{G}}]\text{d}\mathcal{R}_{\text{G}} \\
&= \int_{\delta\phi_{\text{th}}(R_s)}^{+\infty} \frac{1}{\sqrt{2\pi}\sigma_{\delta\phi}(R_s)} \exp\left[-\frac{\delta\phi^2}{2\sigma_{\delta\phi}^2(R_s)}\right] \text{d}\delta\phi \\
&= \text{erfc}\left(\frac{\delta\phi_{\text{th}}(R_s)}{\sqrt{2}\sigma_{\delta\phi}(R_s)}\right),
\end{aligned} \tag{90}$$

where the probability conservation $\mathbb{P}[\mathcal{R}]\text{d}\mathcal{R} = \mathbb{P}[\mathcal{R}_{\text{G}}]\text{d}\mathcal{R}_{\text{G}}$ and the relation between the Gaussian variables $\mathcal{R}_{\text{G}} = \delta\phi/\pi$ have been used, where the Gaussianity of $\delta\phi$ is assumed. Denoting $\mathcal{R}$ as a function of $\mathcal{R}_{\text{G}}$ by $\mathcal{R} = F(\mathcal{R}_{\text{G}})$, the threshold for each variable is related as $\delta\phi_{\text{th}} = \pi\mathcal{R}_{\text{G,th}} = \pi F^{-1}(\mathcal{R}_{\text{th}})$, where F^{-1} is the inverse function of F.

Now, we discuss the dependence of the PBH abundance on the degree of non-Gaussianity. As an interesting specific example for the function F, let us consider the curvature perturbation for the modes that cross the horizon just before the transition, given by (79). The threshold for $\delta\phi$ is determined by that for $\mathcal{R}$ as

$$\delta\phi_{\text{th}} = \frac{1 - e^{\lambda_{II,-}\mathcal{R}_{\text{th}}}}{3}\left(\sqrt{2\epsilon_I} - \frac{\lambda_{II,+}}{\eta_{II}}\sqrt{2\epsilon_{II}}\right), \tag{91}$$

where we have taken the limit $\eta_I \to 0$ as its effect is unimportant. In comparison, if we would ignore the non-Gaussianity, the threshold for $\delta\phi$ would become

$$\delta\phi_{\mathrm{th}}^{\mathrm{G}} = -\frac{\lambda_{II,-}}{3}\mathcal{R}_{\mathrm{th}}\left(\sqrt{2\epsilon_I} - \frac{\lambda_{II,+}}{\eta_{II}}\sqrt{2\epsilon_{II}}\right), \tag{92}$$

Here, combining the above two equations, we find the effect of non-Gaussianity on the threshold value of $\delta\phi$ as

$$\delta\phi_{\mathrm{th}} = \frac{1 - e^{\lambda_{II,-}\mathcal{R}_{\mathrm{th}}}}{-\lambda_{II,-}\mathcal{R}_{\mathrm{th}}}\delta\phi_{\mathrm{th}}^{\mathrm{G}} = \left(1 + \frac{\eta_{II}\mathcal{R}_{\mathrm{th}}}{2} + \frac{(\eta_{II}\mathcal{R}_{\mathrm{th}})^2}{3} + O(\eta_{II}^3)\right)\delta\phi_{\mathrm{th}}^{\mathrm{G}}, \tag{93}$$

for a given threshold value of $\mathcal{R}_{\mathrm{th}}$, where the second equality is for $|\eta_{II}| \ll 1$.

Let us denote the abundance of PBH that fully includes non-Gaussianity by $\beta_{\mathrm{PBH}} = \beta_{\mathrm{PBH}}(\delta\phi_{\mathrm{th}})$, while that ignores non-Gaussianity by $\beta_{\mathrm{PBH}}^{\mathrm{G}} = \beta_{\mathrm{PBH}}(\delta\phi_{\mathrm{th}}^{\mathrm{G}})$. For definiteness, we fix the curvature perturbation threshold to unity, $\mathcal{R}_{\mathrm{th}} = 1$ (which corresponds to $\delta_{\mathrm{th}} = 4/9 = 0.444\cdots$). The ratio $\beta_{\mathrm{PBH}}/\beta_{\mathrm{PBH}}^{\mathrm{G}}$ as a function of $\delta\phi_{\mathrm{th}}$ is shown in the upper panel of Fig. 16. We can see from the figure that the modification of the abundance caused by the nonlinearity grows while the threshold grows. Interestingly, for the same absolute value of η_{II}, the suppression caused by a positive η_{II} is always more significant than the enhancement caused by a heavy tail when η_{II} is negative. This result can be understood from the right-hand side of the second equality in (93), i.e., the increase in $\delta\phi_{\mathrm{th}}$ is accumulative for $\eta_{II} > 0$ while the decrease in $\delta\phi_{\mathrm{th}}$ is not for $\eta_{II} < 0$. The lower panel of Fig. 16 shows the η_{II}-dependence of the ratio $\beta_{\mathrm{PBH}}/\beta_{\mathrm{PBH}}^{\mathrm{G}}$. One can also see that the suppression is more significant than the enhancement for the same absolute value of η_{II}.

6 Conclusion and Discussion

In this paper, we thoroughly studied a two-stage inflationary model with piecewise quadratic potential. Allowing the discontinuity in both the first and second derivatives of the potential, we explored distinct features in the primordial curvature power spectrum and the non-Gaussianity in the probability distribution function (PDF) of the curvature perturbation.

We obtained an analytic expression for the curvature perturbation power spectrum for the general initial condition by solving the perturbation equation until the end of inflation. Then for the modes $k < k_\star$ with the adiabatic vacuum initial condition, we obtained a general expression valid up to next-to-leading order corrections in $k/k_\star$. It is known that the amplitude of the curvature perturbation that crosses the horizon before the transition, namely, $k < k_\star$ where $k_\star$ is the comoving wavenumber that crosses the horizon at the transition, can be strongly enhanced on superhorizon scales. From the expression we obtained, we clarified the condition under which the

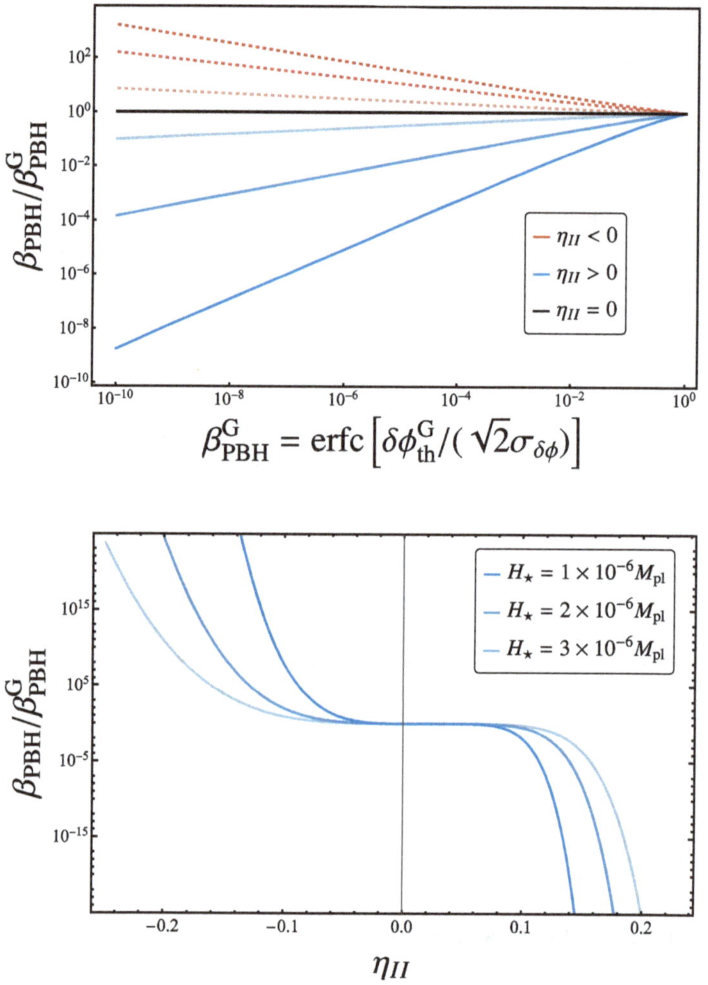

Fig. 16 **The ratio** $\beta_{\mathrm{PBH}}/\beta^{\mathrm{G}}_{\mathrm{PBH}}$**.** Top panel: For the red dashed (blue solid) lines from dark to light, we take $\eta_{II} = -(+)0.5, -(+)0.3, -(+)0.1$. As a reference, we plot the black solid line with $\eta_{II} = 0$ where the β_{PBH} overlaps with $\beta^{\mathrm{G}}_{\mathrm{PBH}}$. Bottom panel: The η_{II}-dependence of the PBH abundance

commonly observed k^4 growth appears and the fine-tuned condition under which the maximum growth rate of $k^5(\ln k)^2$ is realized [24].

The amplitude of the small scale power spectrum (i.e., at $k > k_\star$) relative to the large scale one is found to be proportional to $(\alpha - 1)^{-2}$ where α is a function of the first and second derivatives of the potential at both stages, first introduced in this paper, and $\alpha - 1$ is a measure of inflaton's attractor component at the beginning of the second stage. If $\alpha = 1$, the attractor component at the second stage vanishes, and the enhancement becomes infinite. Of course, this divergence is an artifact due to our approximation, which will disappear in realistic situations.

We also observed that there often appears a dip in the power spectrum, and its position is sensitive to the values of the second derivative of the potential (i.e., the η_V parameter) at both stages. The dip disappears when $\alpha > 1$, which corresponds to the case where the potential at the second stage is convex and the inflaton does not reach the attractor phase when it passes the minimum of the potential.

After clarifying many features of the curvature perturbation power spectrum, we discussed the non-Gaussianity. Adopting the δN formalism, we obtained the formula for the comoving curvature perturbation $\mathcal{R}$ as a function of the field perturbation $\delta\phi$ in flat slicing. The formula was originally derived in [88]. Here we performed a more detailed analysis of non-Gaussian properties. Specifically, we presented approximate analytical expressions for the probability distribution function of $\mathcal{R}$ in three cases: (1) the modes leaving the horizon well before the transition, (2) the modes exiting just before the transition, and (3) the modes exiting after the transition. In doing so, we clearly identified for the first time the scale dependence in the shape of the PDF as the motion of the inflaton varies in time from the initial slow-roll stage to the non-slow-roll stage right after transition and settles down to the final slow-roll stage. In particular, assuming Gaussian $\delta\phi$, we demonstrated that the modes exiting near the transition produce an asymmetric tail for the PDF of $\mathcal{R}$, with the negative tail determined by η_I and the positive tail determined by η_{II}.

Finally, we discussed the dependence of the primordial black hole (PBH) abundance on the shape of the PDF tails. In our model, we found that the decisive parameter is the second derivative of the potential at the second stage, η_{II}. We explored the η_{II}-dependence using the Press-Schetcher formalism and found that the PBH formation is strongly suppressed for $\eta_{II} > 0$ while enhanced for $\eta_{II} < 0$, under the assumption that the RMS variation of the field perturbation $\sigma_{\delta\phi}$ is below the threshold for PBH formation in the Gaussian case. For the same absolute values of η_{II}, suppression is found to be more prominent than enhancement.

Acknowledgements We would like to thank Yi-Fu Cai, Xin-Chen He, Shi Pi, Xiaoding Wang, and Ying-Li Zhang for useful discussions. This work is supported in part by JSPS KAKENHI Grant Nos. JP20H05853 and JP24K00624, National Key R&D Program of China (2021YFC2203100), by NSFC (12261131497), by CAS young interdisciplinary innovation team (JCTD-2022-20), by 111 Project (B23042), by Fundamental Research Funds for Central Universities, and by CSC Innovation Talent Funds. Kavli IPMU is supported by World Premier International Research Center Initiative (WPI), MEXT, Japan. XW and XHM gratefully acknowledge the hospitality and support of the Kavli Institute of Physics and Mathematics of the Universe (Kavli IPMU), the University of Tokyo, during their visit when the work has been done. We would also like to acknowledge the valuable input and discussions from the IBS CTPU-CGA, Tokyo Tech, USTC 2024 summer school held at Tateyama, Dynamics of Primordial Black Hole Formation II (DPBHF2) workshop held at Nagoya University, and Special Seminar at the University of Science and Technology of China, which greatly contributed to the completion of this work.

References

1. Y. Akrami et al., Planck 2018 results. X. Constraints on inflation. Astron. Astrophys. **641**, A10 (2020)
2. Y. Aldabergenov, A. Addazi, S.V. Ketov, Primordial black holes from modified supergravity. Eur. Phys. J. C **80**(10), 917 (2020)
3. L. Anguelova, On primordial black holes from rapid turns in two-field models. JCAP **06**, 004 (2021)
4. D. Artigas, S. Pi, T. Tanaka, Extended δN formalism (2024)
5. V. Atal, J. Cid, A. Escrivà, J. Garriga, PBH in single field inflation: the effect of shape dispersion and non-Gaussianities. JCAP **05**, 022 (2020)
6. V. Atal, J. Garriga, A. Marcos-Caballero, Primordial black hole formation with non-Gaussian curvature perturbations. JCAP **09**, 073 (2019)
7. S. Balaji, J. Silk, W. Yi-Peng, Induced gravitational waves from the cosmic coincidence. JCAP **06**(06), 008 (2022)
8. G. Ballesteros, J.B. Jimenez, M. Pieroni, Black hole formation from a general quadratic action for inflationary primordial fluctuations. JCAP **06**, 016 (2019)
9. G. Ballesteros, J.G. Egea, One-loop power spectrum in ultra slow-roll inflation and implications for primordial black hole dark matter. JCAP **07**, 052 (2024)
10. S. Bhattacharya, I. Zavala, Sharp turns in axion monodromy: primordial black holes and gravitational waves. JCAP **04**, 065 (2023)
11. N. Bhaumik, R.K. Jain, Primordial black holes dark matter from inflection point models of inflation and the effects of reheating. JCAP **01**, 037 (2020)
12. M. Biagetti, G. Franciolini, A. Kehagias, A. Riotto, Primordial black holes from inflation and quantum diffusion. JCAP **07**, 032 (2018)
13. M. Braglia, D.K. Hazra, F. Finelli, G.F. Smoot, L. Sriramkumar, A.A. Starobinsky, Generating PBHs and small-scale GWs in two-field models of inflation. JCAP **08**, 001 (2020)
14. M. Braglia, D.K. Hazra, L. Sriramkumar, F. Finelli, Generating primordial features at large scales in two field models of inflation. JCAP **08**, 025 (2020)
15. C.T. Byrnes, P.S. Cole, S.P. Patil, Steepest growth of the power spectrum and primordial black holes. JCAP **06**, 028 (2019)
16. R.-G. Cai, C. Chen, F. Chengjie, Primordial black holes and stochastic gravitational wave background from inflation with a noncanonical spectator field. Phys. Rev. D **104**(8), 083537 (2021)
17. Y.-F. Cai, X. Chen, M.H. Namjoo, M. Sasaki, D.-G. Wang, Z. Wang, Revisiting non-Gaussianity from non-attractor inflation models. JCAP **05**, 012 (2018)
18. Y.-F. Cai, X.-H. Ma, M. Sasaki, D.-G. Wang, Z. Zhou, Highly non-Gaussian tails and primordial black holes from single-field inflation. JCAP **12**, 034 (2022)
19. Y.-F. Cai, X.-H. Ma, M. Sasaki, D.-G. Wang, Z. Zhou, One small step for an inflaton, one giant leap for inflation: a novel non-Gaussian tail and primordial black holes. Phys. Lett. B **834**, 137461 (2022)
20. Y. Cai, M. Zhu, Y.-S. Piao, Primordial black holes from null energy condition violation during inflation. Phys. Rev. Lett. **133**(2), 021001 (2024)
21. D.D. Canko, I.D. Gialamas, G.P. Kodaxis, A simple $F(\mathcal{R}, \phi)$ deformation of Starobinsky inflationary model. Eur. Phys. J. C **80**(5), 458 (2020)
22. A. Caravano, G. Franciolini, S. Renaux-Petel, Ultra-slow-roll inflation on the lattice: backreaction and nonlinear effects (2024)
23. A. Caravano, K. Inomata, S. Renaux-Petel, The inflationary butterfly effect: non-perturbative dynamics from small-scale features (2024)
24. P. Carrilho, K.A. Malik, D.J. Mulryne, Dissecting the growth of the power spectrum for primordial black holes. Phys. Rev. D **100**(10), 103529 (2019)
25. P. Chen, S. Koh, G. Tumurtushaa, Primordial black holes and induced gravitational waves from inflation in the Horndeski theory of gravity (2021)

26. S.-L. Cheng, D.-S. Lee, K.-W. Ng, Power spectrum of primordial perturbations during ultra-slow-roll inflation with back reaction effects. Phys. Lett. B **827**, 136956 (2022)
27. S.-L. Cheng, D.-S. Lee, K.-W. Ng, Primordial perturbations from ultra-slow-roll single-field inflation with quantum loop effects. JCAP **03**, 008 (2024)
28. S. Choudhury, M.R. Gangopadhyay, M. Sami, No-go for the formation of heavy mass primordial black holes in single field inflation. Eur. Phys. J. C **84**(9), 884 (2024)
29. S. Choudhury, A. Karde, S. Panda, M. Sami, Scalar induced gravity waves from ultra slow-roll Galileon inflation (2023)
30. S. Choudhury, A. Karde, S. Panda, M. Sami, Realisation of the ultra-slow roll phase in Galileon inflation and PBH overproduction (2024)
31. P. Christodoulidis, J.-O. Gong, Enhanced power spectra from multi-field inflation (2023)
32. M. Cielo, G. Mangano, O. Pisanti, D. Wands, Steepest growth in the primordial power spectrum from excited states at a sudden transition (2024)
33. S. Clesse, J. García-Bellido, Massive primordial black holes from hybrid inflation as dark matter and the seeds of galaxies. Phys. Rev. D **92**(2), 023524 (2015)
34. E. Cotner, A. Kusenko, M. Sasaki, V. Takhistov, Analytic description of primordial black hole formation from scalar field fragmentation. JCAP **10**, 077 (2019)
35. I. Dalianis, A. Kehagias, G. Tringas, Primordial black holes from α-attractors. JCAP **01**, 037 (2019)
36. M.W. Davies, L. Iacconi, D.J. Mulryne, Numerical 1-loop correction from a potential yielding ultra-slow-roll dynamics. JCAP **04**, 050 (2024)
37. H. Di, Y. Gong, Primordial black holes and second order gravitational waves from ultra-slow-roll inflation. JCAP **07**, 007 (2018)
38. E. Dimastrogiovanni, M. Fasiello, A. Papageorgiou, A novel PBH production mechanism from non-Abelian gauge fields during inflation (2024)
39. G. Domènech, G. Vargas, T. Vargas, An exact model for enhancing/suppressing primordial fluctuations (2023)
40. Y. Ema, K. Mukaida, J. van de Vis, Higgs inflation as nonlinear sigma model and scalaron as its σ-meson. JHEP **11**, 011 (2020)
41. H. Firouzjahi, One-loop corrections in power spectrum in single field inflation. JCAP **10**, 006 (2023)
42. H. Firouzjahi, Revisiting loop corrections in single field ultraslow-roll inflation. Phys. Rev. D **109**(4), 043514 (2024)
43. P.H. Frampton, M. Kawasaki, F. Takahashi, T.T. Yanagida, Primordial black holes as all dark matter. JCAP **04**, 023 (2010)
44. G. Franciolini, A. Iovino, Jr., M. Taoso, A. Urbano, Perturbativity in the presence of ultraslow-roll dynamics. Phys. Rev. D **109**(12), 123550 (2024)
45. D. Frolovsky, S.V. Ketov, S. Saburov, Formation of primordial black holes after Starobinsky inflation. Mod. Phys. Lett. A **37**(21), 2250135 (2022)
46. F. Chengjie, W. Puxun, Yu. Hongwei, Primordial black holes from inflation with nonminimal derivative coupling. Phys. Rev. D **100**(6), 063532 (2019)
47. F. Chengjie, W. Puxun, Yu. Hongwei, Primordial black holes and oscillating gravitational waves in slow-roll and slow-climb inflation with an intermediate noninflationary phase. Phys. Rev. D **102**(4), 043527 (2020)
48. J. Fumagalli, Absence of one-loop effects on large scales from small scales in non-slow-roll dynamics (2023)
49. J. Fumagalli, Absence of one-loop effects on large scales from small scales in non-slow-roll dynamics. Part 2. Quartic interactions and consistency relations. JHEP **01**, 108 (2025)
50. J. Fumagalli, S. Renaux-Petel, J.W. Ronayne, L.T. Witkowski, Turning in the landscape: a new mechanism for generating primordial black holes. Phys. Lett. B **841**, 137921 (2023)
51. Q. Gao, Primordial black holes and secondary gravitational waves from chaotic inflation. Sci. China Phys. Mech. Astron. **64**(8), 280411 (2021)
52. J. Garcia-Bellido, A.D. Linde, D. Wands, Density perturbations and black hole formation in hybrid inflation. Phys. Rev. D **54**, 6040–6058 (1996)

53. J. Garcia-Bellido, M. Peloso, C. Unal, Gravitational waves at interferometer scales and primordial black holes in axion inflation. JCAP **12**, 031 (2016)
54. J. Garcia-Bellido, E.R. Morales, Primordial black holes from single field models of inflation. Phys. Dark Univ. **18**, 47–54 (2017)
55. M. Giovannini, Secondary graviton spectra and waterfall-like fields. Phys. Rev. D **82**, 083523 (2010)
56. C. Gordon, D. Wands, B.A. Bassett, R. Maartens, Adiabatic and entropy perturbations from inflation. Phys. Rev. D **63**, 023506 (2000)
57. A. Gundhi, S.V. Ketov, C.F. Steinwachs, Primordial black hole dark matter in dilaton-extended two-field Starobinsky inflation. Phys. Rev. D **103**(8), 083518 (2021)
58. A. Gundhi, C.F. Steinwachs, Scalaron-Higgs inflation. Nucl. Phys. B **954**, 114989 (2020)
59. A.H. Guth, The inflationary universe: a possible solution to the horizon and flatness problems. Phys. Rev. D **23**, 347–356 (1981)
60. S. Heydari, K. Karami, Primordial black holes in nonminimal derivative coupling inflation with quartic potential and reheating consideration. Eur. Phys. J. C **82**(1), 83 (2022)
61. L. Iacconi, D. Mulryne, D. Seery, Loop corrections in the separate universe picture. JCAP **06**, 062 (2024)
62. K. Inomata, Superhorizon curvature perturbations are protected against one-loop corrections. Phys. Rev. Lett. **133**(14), 141001 (2024)
63. K. Inomata, M. Kawasaki, K. Mukaida, Y. Tada, T.T. Yanagida, Inflationary primordial black holes as all dark matter. Phys. Rev. D **96**(4), 043504 (2017)
64. K. Inomata, M. Kawasaki, K. Mukaida, T.T. Yanagida, Double inflation as a single origin of primordial black holes for all dark matter and LIGO observations. Phys. Rev. D **97**(4), 043514 (2018)
65. P. Ivanov, P. Naselsky, I. Novikov, Inflation and primordial black holes as dark matter. Phys. Rev. D **50**, 7173–7178 (1994)
66. J.H.P. Jackson, H. Assadullahi, A.D. Gow, K. Koyama, V. Vennin, D. Wands, The separate-universe approach and sudden transitions during inflation. JCAP **05**, 053 (2024)
67. A. Karam, N. Koivunen, E. Tomberg, A. Racioppi, H. Veermäe, Primordial black holes and inflation from double-well potentials. JCAP **09**, 002 (2023)
68. A. Karam, N. Koivunen, E. Tomberg, V. Vaskonen, H. Veermäe, Anatomy of single-field inflationary models for primordial black holes. JCAP **03**, 013 (2023)
69. A. Karam, T. Pappas, K. Tamvakis, Nonminimal Coleman-Weinberg inflation with an R^2 term. JCAP **02**, 006 (2019)
70. S. Karydas, E. Papantonopoulos, E.N. Saridakis, Successful Higgs inflation from combined nonminimal and derivative couplings. Phys. Rev. D **104**(2), 023530 (2021)
71. R. Kawaguchi, S. Tsujikawa, Y. Yamada, Proving the absence of large one-loop corrections to the power spectrum of curvature perturbations in transient ultra-slow-roll inflation within the path-integral approach (2024)
72. S. Kawai, J. Kim, Primordial black holes from Gauss-Bonnet-corrected single field inflation. Phys. Rev. D **104**(8), 083545 (2021)
73. M. Kawasaki, N. Sugiyama, T. Yanagida, Primordial black hole formation in a double inflation model in supergravity. Phys. Rev. D **57**, 6050–6056 (1998)
74. J. Kristiano, J. Yokoyama, Comparing sharp and smooth transitions of the second slow-roll parameter in single-field inflation. JCAP **10**, 036 (2024)
75. J. Kristiano, J. Yokoyama, Constraining primordial black hole formation from single-field inflation. Phys. Rev. Lett. **132**(22), 221003 (2024)
76. J. Kristiano, J. Yokoyama, Note on the bispectrum and one-loop corrections in single-field inflation with primordial black hole formation. Phys. Rev. D **109**(10), 103541 (2024)
77. S.M. Leach, M. Sasaki, D. Wands, A.R. Liddle, Enhancement of superhorizon scale inflationary curvature perturbations. Phys. Rev. D **64**, 023512 (2001)
78. J. Lin, Q. Gao, Y. Gong, L. Yizhou, C. Zhang, F. Zhang, Primordial black holes and secondary gravitational waves from k and G inflation. Phys. Rev. D **101**(10), 103515 (2020)

79. J. Lin, S. Gao, Y. Gong, L. Yizhou, Z. Wang, F. Zhang, Primordial black holes and scalar induced gravitational waves from Higgs inflation with noncanonical kinetic term. Phys. Rev. D **107**(4), 043517 (2023)
80. A.D. Linde, A new inflationary universe scenario: a possible solution of the horizon, flatness, homogeneity, isotropy and primordial monopole problems. Phys. Lett. B **108**, 389–393 (1982)
81. J. Liu, Z.-K. Guo, R.-G. Cai, Analytical approximation of the scalar spectrum in the ultraslow-roll inflationary models. Phys. Rev. D **101**(8), 083535 (2020)
82. D.H. Lyth, K.A. Malik, M. Sasaki, A General proof of the conservation of the curvature perturbation. JCAP **05**, 004 (2005)
83. S. Maity, H.V. Ragavendra, S.K. Sethi, L. Sriramkumar, Loop contributions to the scalar power spectrum due to quartic order action in ultra slow roll inflation. JCAP **05**, 046 (2024)
84. D.-S. Meng, C. Yuan, Q.-G. Huang, Primordial black holes generated by the non-minimal spectator field. Sci. China Phys. Mech. Astron. **66**(8), 280411 (2023)
85. S.S. Mishra, V. Sahni, Primordial black holes from a tiny bump/dip in the inflaton potential. JCAP **04**, 007 (2020)
86. K.-W. Ng, W. Yi-Peng, Constant-rate inflation: primordial black holes from conformal weight transitions. JHEP **11**, 076 (2021)
87. O. Özsoy, G. Tasinato, On the slope of the curvature power spectrum in non-attractor inflation. JCAP **04**, 048 (2020)
88. S. Pi, M. Sasaki, Logarithmic duality of the curvature perturbation. Phys. Rev. Lett. **131**(1), 011002 (2023)
89. S. Pi, J. Wang, Primordial black hole formation in Starobinsky's linear potential model. JCAP **06**, 018 (2023)
90. S. Pi, Y. Zhang, Q.-G. Huang, M. Sasaki, Scalaron from R^2-gravity as a heavy field. JCAP **05**, 042 (2018)
91. W. Qin, S.R. Geller, S. Balaji, E. McDonough, D.I. Kaiser, Planck constraints and gravitational wave forecasts for primordial black hole dark matter seeded by multifield inflation. Phys. Rev. D **108**(4), 043508 (2023)
92. H.V. Ragavendra, P. Saha, L. Sriramkumar, J. Silk, Primordial black holes and secondary gravitational waves from ultraslow roll and punctuated inflation. Phys. Rev. D **103**(8), 083510 (2021)
93. A. Riotto, The primordial black hole formation from single-field inflation is still not ruled out (2023)
94. A. Riotto, The primordial black hole formation from single-field inflation is not ruled out (2023)
95. S. Saburov, S.V. Ketov, Quantum loop corrections in the modified gravity model of Starobinsky inflation with primordial black hole production. Universe **10**(9), 354 (2024)
96. M. Sasaki, T. Tanaka, Superhorizon scale dynamics of multiscalar inflation. Prog. Theor. Phys. **99**, 763–782 (1998)
97. M. Solbi, K. Karami, Primordial black holes and induced gravitational waves in k-inflation. JCAP **08**, 056 (2021)
98. A.A. Starobinsky, A new type of isotropic cosmological models without singularity. Phys. Lett. B **91**, 99–102 (1980)
99. A.A. Starobinsky, Multicomponent de Sitter (inflationary) stages and the generation of perturbations. JETP Lett. **42**, 152–155 (1985)
100. A.A. Starobinsky, Spectrum of adiabatic perturbations in the universe when there are singularities in the inflation potential. JETP Lett. **55**, 489–494 (1992)
101. Y. Tada, T. Terada, J. Tokuda, Cancellation of quantum corrections on the soft curvature perturbations. JHEP **01**, 105 (2024)
102. G. Tasinato, Large $|\eta|$ approach to single field inflation. Phys. Rev. D **108**(4), 043526 (2023)
103. Z. Teimoori, K. Rezazadeh, M.A. Rasheed, K. Karami, Mechanism of primordial black holes production and secondary gravitational waves in α-attractor Galileon inflationary scenario. JCAP **10** (2021)

104. X. Wang, Y. Zhang, M. Sasaki, Enhanced curvature perturbation and primordial black hole formation in two-stage inflation with a break. JCAP **07**, 076 (2024)
105. Z. Yi, Primordial black holes and scalar-induced gravitational waves from the generalized Brans-Dicke theory. JCAP **03**, 048 (2023)
106. Z. Yi, Q. Gao, Y. Gong, Z. Zhu, Primordial black holes and scalar-induced secondary gravitational waves from inflationary models with a noncanonical kinetic term. Phys. Rev. D **103**(6), 063534 (2021)
107. Z. Yi, Y. Gong, B. Wang, Z. Zhu, Primordial black holes and secondary gravitational waves from the Higgs field. Phys. Rev. D **103**(6), 063535 (2021)
108. J. Yokoyama, Chaotic new inflation and formation of primordial black holes. Phys. Rev. D **58**, 083510 (1998)
109. R. Zhai, Yu. Hongwei, W. Puxun, Growth of power spectrum due to decrease of sound speed during inflation. Phys. Rev. D **106**(2), 023517 (2022)
110. F. Zhang, Primordial black holes and scalar induced gravitational waves from the E model with a Gauss-Bonnet term. Phys. Rev. D **105**(6), 063539 (2022)
111. S. Pi and M. Sasaki, Primordial black hole formation in nonminimal curvaton scenarios, Phys. Rev. D **108**, L101301 (2023)

Bounce Solutions with Quantum Vacuum Effects of Massive Fields and Subsequent Starobinsky Inflation

Wagno Cesar e Silva, Samuel W. P. Oliveira, and Ilya L. Shapiro

Abstract We extend the previous work about the cosmological solutions with a bounce without modifications of gravity or introducing an extra scalar field. The main finding was that the bounce is possible in the initially contracting Universe filled with matter. After a strong contraction, matter gains the equation of state close to the one of radiation, such that the effect of matter on the evolution of the FLRW metric disappears at the classical level. However, this effect comes back owing to the quantum trace anomaly in the matter/radiation sector. In the present contribution, we explore the weak impact of massive fields on the anomaly-driven bounce solution and discuss the role of the vacuum terms. The masses are assumed small and regarded as small perturbations, which enables using trace anomaly even in this case. On the other hand, by adding the R^2 term to the action, we arrive at the model with the trans-Planckian bounce and subsequent Starobinsky inflation. In such a framework, using the numerical analysis, we consider three scenarios providing bounce solutions.

1 Remembering Alexei Alexandrovich Starobinsky

Before we present our contribution to this volume dedicated to the memory of Professor Starobinsky, let us tell a little bit about the history of contacts and collaboration with him by one of the authors of this paper—Ilya Shapiro, on whose behalf these reminiscences are conveyed below.

Alexei Alexandrovich was the person who greatly impressed many people who had professional and personal contacts with him and I was not an exception. We

W. C. e Silva
Centro Brasileiro de Pesquisas Físicas, Rio de Janeiro, RJ, Brazil

S. W. P. Oliveira · I. L. Shapiro
PPGCosmo, Universidade Federal do Espírito Santo, Vitória, ES, Brazil

I. L. Shapiro (✉)
Departamento de Física, ICE, Universidade Federal de Juiz de Fora, Campus Universitário, Juiz de Fora, MG, Brazil
e-mail: ilyashapiro2003@ufjf.br

A. Barvinsky and A. Kamenshchik (eds.), *Open Issues in Gravitation and Cosmology*, Fundamental Theories of Physics 225, https://doi.org/10.1007/978-3-032-15703-4_6

briefly met with him before on several occasions, but the first time we had a serious conversation was at the meeting in Paris in 2006, devoted to the 25 years of inflation. I am not a professional cosmologist and did not know well "who is who" in this community. The organizers invited me because of our papers on the model of inflation based on the conformal anomaly—something which was done originally by Fischetti, Hartle and Hu [1]. We added a few new details and some people got interested, but not the main cosmologists present at this meeting. At some point, Starobinsky addressed me and explained a few points which we did not understand well. This proved very useful for us since we got much better idea about the relation of anomaly induced inflation and the Starobinsky model. In general, some discussions in this meeting were hot. What impressed me was that nobody ever complained if Starobinsky was saying something. It looked as if he was one of a very few participants who had this privilege.

Next time we met in 2008, when Starobinsky came to Juiz de Fora for the first time for a small meeting which we organized as a continuation of the larger conference in João Pessoa. I have an impression that because of this visit Starobinsky started to read about Minas Gerais and came to be interested in the works of the artist of the XVIII—XIX centuries, called Aleijadinho. This name belongs to an outstanding painter and sculptor, unfortunately not very much known out of Brazil. Next time, Alexei Alexandrovich came for a few days and asked to organize his visit to Ouro Preto or to Tiradentes, the historic cities where one can see some Aleijadinho's works. In the next few years, we went to the region of Tiradentes, which is closer to Juiz de Fora, and everybody liked it a lot. Every time, in the next few visits, Starobinsky was contracting a particular driver to visit the town called Congonhas, where one can see the main sculptures of Aleijadinho: the set of Twelve Prophets sculptures made of a soapstone between 1800 and 1805, and painted sculptures in the famous Stations of the Cross scenes, carved between 1796 and 1799. When we came to the same place with Alexei Alexandrovich later on, on the way from Ouro Preto, he gave us detailed and very interesting lecture about Aleijadinho and about these sculptures, which he knew at the professional level and esteemed very high. Another detail was that, when we were walking in the Ouro Preto (old-style city, former capital of Minas Gerais state), it looked like he knew every corner and every direction, as if he would have a GPS in his head. This and many other things of this sort impressed a lot all of our company, including me.

On top of these excursions, every time we had a lot of discussions about physics. We started with quantum effects in the early Universe but, this time, it was difficult to overcome some differences. Anyway, at least for me, these discussions were very instructive. At some point we decided to change the subject and I asked Starobinsky to look at the calculation of equivalence between $f(R)$ gravity and metric-scalar theory. I made this small calculation from scratch, without basic knowledge of the literature. He looked at the calculations and immediately said that the same method can be applied to many other theories. We decided to offer this problem to one of my best students Felipe Salles. In the next months, with Felipe we met some mathematical difficulties and asked former Juiz de Fora student, Davi Rodrigues, to assist us. Eventually, after two years of work and a lot of discussions we could finish a paper with four authors [2], based on this idea. Needless to say that me and also our young collaborators were very excited to work under the problem suggested by Starobinsky.

After this, our long discussions about quantum effects and higher derivatives started to produce some results. Eventually, we did two works. The first one was about the transition between stable and unstable phases of the anomaly-induced inflation [3]. This was our main idea in the mentioned works of 2003, but we did not know well how to connect it to cosmology because there was no reliable model of quantum corrections in the transition period. At some point, Alexei Alexandrovich suggested to simplify things and explained to us how the approximation works in this case. This lesson was very useful for our group, and the approach appeared to be operational in some other models. The second work emerged from numerous discussions of how to deal with higher derivative models in cosmology. The conventional approach (mainly in quantum theory) was the one by Simon [4]. In our discussions, we always agreed that it is not a perfect thing, but at least enables us to ignore the problems. And at some point we decided to put it to practice, for the first time, that resulted in [5].

Even after more than a year after Alexei Alexandrovich passed away, it is difficult for me to accept this. Every time we have a doubt about cosmology or many other issues, the first idea is to ask the main expert in the field and it is painful to remember that he is not available. We miss not only his knowledge, constructive criticism and advices, but also his outstanding personality. What is true is that Prof. Starobinsky left a strong imprint on many people worldwide, including in Brazil and in our small group. And many people, including us, will remember him again and again.

2 Introduction

The Standard Cosmological Model [6–8] is a very successful theory but leaves significant theoretical and observational challenges. Some of the issues may be addressed by introducing the cosmic inflation phase in the early stage of the evolution of the Universe [9–11]. However, certain conceptual limitations of this framework persist unresolved. In particular, the existence of an initial singularity is unavoidable in the framework of GR [12, 13], and even inflation may be insufficient to solve this problem. This fact may be regarded as a strong motivation to modify GR, either through introducing the scalar fields, modifying the gravitational action, or introducing some exotic forms of "matter" with an unusual equation of state (see, e.g., [14]).

In this framework, bouncing models emerge as an interesting alternative to avoid singularities and, in some cases, provide an explanation for the primordial density perturbations. There are many different models of cosmological bounce, where the contraction of the Universe precedes its expansion. The systematic study of these models attracted a lot of interest in recent decades (see, e.g., the reviews [15, 16]). In particular, the possibility of achieving nonsingular bouncing solutions within the framework of the anomaly-induced effective action was explored in [17] by two of the present authors (earlier, the bounce in such a model was discussed in [18]). In a subsequent work [19], an analytic solution describing the cosmological bounce was found without additional *ad hoc* assumptions, i.e., without modifying the action of gravity or introducing a scalar field, or accounting for vacuum quantum effects.

The scheme of bounce owing to the quantum trace anomaly in the radiation sector [17] is quite simple, as it is based on the one-loop approximation.[1] The last means, the violation of local conformal symmetry is restricted to the terms linear in the scale parameter $\sigma = \log\, a(t)$. The model does not require *ad hoc* modifications of gravity and ignores vacuum quantum effects which may be relevant, in principle. On the other hand, this model of bounce leaves several open questions. The list starts with the numerical discrepancy between ignoring the vacuum (purely gravitational, i.e. metric-dependent) part of anomaly and the fact that the bounce requires the trans-Planckian energy density at the point of minimum of $a(t)$ with vanishing $H = 0$. The way out consists of looking beyond the basic QED framework and invoking the possible non-perturbative effects, i.e., including the higher orders in $\sigma(t)$. The second problem concerns the analysis of cosmic perturbations which may be critical for the consistency of the bounce model [16]. Finally, there are several smaller problems which we shall address in the present contribution. The first one is to check how the small masses of the quantum fields affect the bounce solution and its stability in both vacuum- and radiation-dependent anomaly-induced cases. In what follows, we deal with this problem in a way similar to what was done in the modified Starobinsky model of inflation [3, 22]. Another question is what happens to the bounce solution if we add the R^2 term with a large coefficient to the classical action of gravity, as it is effectively done in the Starobinsky model [9]. If the bounce is not critically affected by such an addition, one can assume that the vacuum contributions to the anomaly are irrelevant compared to the dominating R^2-term. In this way, we arrive at the model with cosmological bounce at very high energies and Starobinsky inflation occurs later on, after a significant expansion of the Universe.

The paper is organized as follows. Section 3 is a brief survey of the previous work [19] where the cosmological bounce solution based on the trace anomaly in the radiation sector of the theory has been found. In Sect. 4, we review the approach of [22, 23], which permits to use the anomaly-induced effective action in the case of massive fields. This approach is extended to the presence of radiation and is used a basis to explore the dynamics of the scale factor in the FLRW (Friedmann-Lemaitre-Robertson-Walker) cosmological metric. Section 5 describes the stability analysis of the bounce solution under linear perturbations of the conformal factor, taking into account the effects of massive fields. In Sect. 6, we present the numerical results for the semiclassical bounce and Sect. 7 describes the extended version of the radiation-dominated bounce with an additional R^2-term, which may have a subsequent phase of the Starobinsky inflation. Finally, in the last section, we draw our conclusions.

[1] It is worth mentioning that matter-bouncing cosmological models, including with the discussion of primordial black holes and induced gravitational waves production, were elaborated in the [20] (partially based on the techniques of [21]). However, these models are not based on the quantum corrections.

3 Bounce Solution from the Trace Anomaly

Let us start with a brief survey of the previous work. The model considered in [19] consists solely of the radiation sector from the anomaly-induced action of the metric $g_{\mu\nu} = \bar{g}_{\mu\nu} \exp \sigma(t)$, with the flat fiducial metric $\bar{g}_{\mu\nu} = \eta_{\mu\nu}$ at $a(t) = a_0$. The gravitational action may include the cosmological term. We use the signature $(+, -, -, -)$ for the Minkowski metric and adopt the natural system of units, where $c = 1$ and $\hbar = 1$. The action with the anomaly-induced term is [24, 25]

$$\Gamma = -\frac{1}{16\pi G} \int d^4x \sqrt{-g} \, (R + 2\Lambda) - \frac{\beta g^2}{4} \int d^4x \sqrt{-\bar{g}} \, \bar{F}^2 \sigma. \tag{1}$$

The bars denote quantities defined using the fiducial metric, e.g., $\bar{F}^2 = \bar{g}^{\mu\alpha} \bar{g}^{\nu\beta} F_{\mu\nu} F_{\alpha\beta}$, and βg^4 is the one-loop beta function for the square of the gauge coupling g^2,

$$\beta = -\frac{2}{(4\pi)^2} \Big(\frac{11}{3} C_1 - \frac{1}{6} N_{cs} - \frac{4}{3} N_f \Big). \tag{2}$$

Here N_{cs} and N_f are the numbers of complex scalars and fermions coupled to the given vector field, and C_1 is the Casimir operator of the corresponding gauge group.

The anomalous term in (1) produces the modified Friedmann equation,

$$H^2 = \frac{a_0^4}{a^4} \left[\frac{\rho_{r0}}{M_P^2} + \frac{2\pi \beta g^2 \bar{F}^2}{3 M_P^2} \log\Big(\frac{a}{a_0}\Big) \right] + \frac{\Lambda}{3}, \tag{3}$$

where $M_P = 1/\sqrt{G}$ is the Planck mass and ρ_{r0} is a radiation energy density at $a(t) = a_0$. The expression (3) admits vanishing Hubble parameter for a negative β and enables one to estimate the energy density of the radiation at the bounce point, when the scale factor assumes its minimum value. Thus, the bounce occurs owing to the equilibrium between the classical radiation term and the quantum correction in the radiation-gravitational sector. In the analysis of [19], the cosmological constant is supposed to satisfy $0 < \Lambda \ll M_P^2$, in this case its presence does not change the qualitative result stated above.

4 Anomaly-Induced Action of Massive Fields

Since the typical energy scale in the region of the bounce is, presumably, very high, all matter fields can safely be regarded massless. Then, relevant quantum effects near a bounce point may be described by the effects of massless fields in the zero-order approximation. On the other hand, it is interesting to check whether the masses of quantum fields may affect the bounce solution described in the previous section (see

[17] for more details), including the stability in the contraction phase. The bounce occurs in the far UV when the masses of real fields filling the Universe are completely irrelevant because the equation of state is one of radiation. Thus, we need to evaluate the effect of masses in the quantum corrections to the terms in the action (1). Since the masses are relatively small, the most useful technique is the conformal approach developed in [22, 23].

Let us start with a brief discussion of deriving the conformal anomaly and the anomaly-induced effective action in the theory with small masses of the fields. It is worth noting that there are many possible definitions of an anomaly for the classically non-conformal theories (see, e.g., [26] and references therein). Our approach is focused on effective action and treats the masses of the quantum fields as small perturbations [27].

In the high-energy regime of our interest, we can assume that only scalar and fermion fields are massive. Let us denote these masses by m_s and m_f respectively. To get rid of these dimensional parameters, one can use the approach that can be called conformal Stückelberg trick. The masses of the fields, the Newton constant, and the cosmological constant are replaced by the powers of a new auxiliary scalar field χ, i.e.,

$$m_{s,f} \to \frac{m_{s,f}}{M}\chi, \qquad \frac{1}{16\pi G}R \to \frac{M_P^2}{16\pi M^2}\big[R\chi^2 + 6(\partial\chi)^2\big], \qquad \Lambda \to \frac{\Lambda}{M^2}\chi^2, \tag{4}$$

where M is a new dimensional parameter. Applying this procedure to the Einstein-Hilbert term provides the local conformal invariance in all sectors of the theory. The new field χ is auxiliary and should not be quantized. It is assumed that at some point we replace back $\chi \to M$, but this will be done only after we derive the effective action.

To guarantee the local conformal invariance, we have to assume that χ transforms as

$$\chi = \bar{\chi}\, e^{-\sigma(x)}, \tag{5}$$

while the metric and matter fields transform as usual, i.e.,

$$g_{\mu\nu} = \bar{g}_{\mu\nu}\, e^{2\sigma(x)}, \qquad \varphi = \bar{\varphi}\, e^{-\sigma(x)}, \qquad \psi = \bar{\psi}\, e^{-3\sigma(x)/2}, \qquad A_\mu = \bar{A}_\mu. \tag{6}$$

The relevant part of the classical action that is a subject of renormalization is formed by the vacuum sector, including the Einstein-Hilbert and the cosmological constant terms, four-derivative terms, plus the matter sector, with a radiation, scalar and fermionic terms,

$$S = \int d^4x\sqrt{-g}\,\Big\{-\frac{1}{16\pi G}\big(R+2\Lambda\big) + a_1C^2 + a_2E_4 + a_3\Box R\Big\} + S_{\text{matter}}\,. \tag{7}$$

Here a_i are dimensionless coefficients, $C^2 = R^2_{\alpha\beta\mu\nu} - 2R^2_{\alpha\beta} + \frac{1}{3}R^2$ is the square of the Weyl tensor and $E_4 = R^2_{\mu\nu\alpha\beta} - 4R^2_{\alpha\beta} + R^2$ is the integrand of the topological Gauss-Bonnet term. Under the replacement trick (4) the first two terms become conformal invariant and the same for the matter action.

At the quantum level, the local conformal symmetry is violated by the corresponding anomaly. Then, the effective action of the background fields can be derived using the approach of the anomaly-induced action (see, e.g., [28] for a detailed introduction). As we have already mentioned, χ has to be treated as a background field, similar to the metric. How does the trace anomaly change under the procedure (4)? The gauge invariance prevents mass-dependent contributions to the radiation terms thus the last term in the expression (1) does not get modified. Thus, the conformal anomaly takes the form elaborated in the previous works [22, 23],

$$\langle \mathcal{T} \rangle = -\Big\{ wC^2 + \frac{1}{4}\beta g^2 F^2 + bE_4 + c\Box R + \frac{M_P^2}{4\pi M^2}\tilde{f}\Big[R\chi^2 + 6(\partial\chi)^2\Big] + \frac{M_P^4}{4\pi M^4}\tilde{g}\chi^4\Big\}. \tag{8}$$

The coefficients w, b and c are the semiclassical one-loop beta functions for the higher-derivative terms in the vacuum sector. These beta functions are well-known to depend on the number of fields of different spins (see, e.g., [29] or the textbook [28]),

$$\begin{aligned} w &= \frac{1}{120(4\pi)^2}\big(N_s + 6N_f + 12N_v\big), \\ b &= -\frac{1}{360(4\pi)^2}\big(N_s + 11N_f + 62N_v\big), \\ c &= \frac{1}{180(4\pi)^2}\big(N_s + 6N_f - 18N_v\big). \end{aligned} \tag{9}$$

On the other hand, the beta functions for the Einstein-Hilbert and cosmological terms are proportional, respectively, to the dimensionless quantities [23]

$$\tilde{f} = \frac{1}{3\pi M_P^2}\sum_{\text{fermions}} N_f m_f^2, \qquad \tilde{g} = \frac{1}{4\pi M_P^2\Lambda}\Big(\sum_{\text{scalars}} N_s m_s^4 - 4\sum_{\text{fermions}} N_f m_f^4\Big), \tag{10}$$

where N_s and N_f are multiplicities of scalars and fermions with the corresponding masses.

Since the F^2-part and the terms that depend on χ have the same conformal properties as the square of the Weyl tensor, such contributions can be incorporated into a conformally invariant structure $Y(g_{\mu\nu}, A_\mu, \chi)$, and the integration procedure of the anomaly is reduced to the usual case [24, 30]. In this way, after obtaining the anomaly-induced effective action Γ_{ind} and fixing the conformal unitary gauge

$\bar{\chi}\, e^{-\sigma} = M$, one can write complete effective action for massive fields, including the vacuum sector, in the form

$$\Gamma = S_{vac} + \Gamma_{\rm ind}[\chi \to M] = S_{HD} + S_c[\bar{g}_{\mu\nu}, M] - \frac{3c+2b}{36}\int d^4x\sqrt{-g}\,R^2$$
$$+ \int d^4x\sqrt{-\bar{g}}\,\Big\{\sigma\Big(w\bar{C}^2 - \frac{1}{4}\beta g^2\bar{F}^2\Big) + b\sigma\Big(\bar{E} - \frac{2}{3}\bar{\Box}\bar{R}\Big) + 2b\sigma\,\bar{\Delta}_4\sigma$$
$$- \frac{e^{2\sigma}}{16\pi G}\Big[\bar{R} + 6(\partial\bar{\sigma})^2\Big](1 - \tilde{f}\sigma) - \frac{\Lambda\, e^{4\sigma}}{8\pi G}(1 - \tilde{g}\sigma)\Big\}, \tag{11}$$

where $\Delta_4 = e^{-4\sigma}\,\bar{\Delta}_4$ is the Paneitz operator [31, 32],

$$\Delta_4 = \Box^2 + 2R^{\mu\nu}\nabla_\mu\nabla_\nu - \frac{2}{3}R\Box + \frac{1}{3}(\nabla^\mu R)\nabla_\mu. \tag{12}$$

In Eq. (11), the functional $S_c[\bar{g}_{\mu\nu}, M]$ is an "integration constant" that, in the massless case and with the homogeneous and isotropic background metric, does not contribute to the dynamics of the conformal factor. In principle, the same result cannot be guaranteed since this term is not conformally invariant. This means, (11) should be treated as an approximate expression. What is the physical sense of this approximation? It is easy to note that (11) is a local version of the renormalization-group improved classical action of the theory, where the global parameter of metric rescaling is replaced by the local conformal factor. Thus, the approximations we made are essentially the same as in the case of the renormalization group based on the Minimal Subtraction scheme. The comparison with explicit expressions shows that this approximation is reliable in the UV [28, 33], when the masses of quantum fields can be regarded as small corrections compared to the energy scale of the curvature tensor or the density of matter. In the case of a bounce based on the action (1) this is exactly what we need.

Our purpose is to evaluate the effect of the mass terms related to the coefficients $\tilde{f}$ and $\tilde{g}$ in Eq. (10), to the two scenarios that provide nonsingular bouncing solutions: (i) when the complete anomaly-induced corrections are taken into account, in particular the higher-derivative sector. This is the case studied in [17], for massless fields; (ii) the model (1) described by the mixing of radiation and gravity in the effective action. In this case, the higher-derivative terms are neglected, and the bounce is due to the contribution of the radiation sector. (iii) The same as the previous, but with an extra R^2 term.

4.1 Dynamic Equations and De Sitter-Like Solutions

Consider the spacetime described by the FLRW metric,

$$ds^2 = N^2(t)dt^2 - a^2(t)\Big(\frac{dr^2}{1-kr^2} + r^2 d\theta^2 + r^2 \sin^2\theta\, d\varphi^2\Big), \tag{13}$$

where $a(t) = e^{\sigma(t)}$ is the scale factor of the universe, $N(t)$ is the lapse function and k describes the spatial curvature. Since our interest is to study the dynamics of the scale factor, $N(t)$ is employed as an auxiliary function to obtain the second Friedmann equation.

By introducing the conformal time $d\eta = dt/a(t)$, Eq. (13) admits the conformal transformation (6) with the fiducial metric

$$\bar{g}_{\mu\nu} = \text{diag}\Big(N^2(\eta),\, -\frac{1}{1-kr^2},\, -r^2,\, -r^2\sin^2\theta\Big). \tag{14}$$

Using this metric in the complete action (11), we derive the modified first Friedmann equation by applying the variational principle with respect to the scale factor $a(\eta)$. This procedure is equivalent to taking the trace of the modified Einstein equations with the terms from the higher derivative and radiation sectors. The results for the traces can be presented in terms of $\sigma(t) = \ln a(t)$, taking $N = 1$ and changing to the physical time t,

$$T_{EH} + T_{HD} = T_{rad}\,, \tag{15}$$

where the traces of the equations of motion for the corresponding terms in the action are

$$\begin{aligned}
T_{EH} &= -\frac{3M_P^2}{4\pi}\Big[(\ddot{\sigma} + 2\dot{\sigma}^2 + ke^{-2\sigma})(1 - \tilde{f}\sigma) - \frac{\tilde{f}}{2}(\dot{\sigma}^2 + ke^{-2\sigma}) - \frac{2\Lambda}{3}\Big(1 - \tilde{g}\sigma - \frac{\tilde{g}}{4}\Big)\Big], \\
T_{HD} &= 6c\Big[\ddddot{\sigma} + 7\dddot{\sigma}\dot{\sigma} + 4\ddot{\sigma}^2 + 4\Big(3 - \frac{b}{c}\Big)\ddot{\sigma}\dot{\sigma}^2 - \frac{4b}{c}\dot{\sigma}^4 - 2k\Big(1 + \frac{2b}{c}\Big)(\dot{\sigma}^2 + \ddot{\sigma})e^{-2\sigma}\Big], \\
T_{rad} &= -\frac{\beta F^2}{4}e^{-4\sigma},
\end{aligned} \tag{16}$$

with $k = 0$ or $k = \pm 1$ for different space geometries. We derive the 00-component, ρ_i, using the conservation law for T_{EH}, T_{HD} and T_{rad},[2]

$$\frac{d\rho_i}{da^3} + \frac{4}{3}\frac{\rho_i}{a^3} = \frac{T_i}{3a^3}, \tag{17}$$

whose general solution is given by

---

[2] The same equation can be obtained by varying the action (11) with respect to the lapse function $N(t)$ and assuming $N = 1$ after that. See more details in [25, 34, 35].

$$\rho_i(a) = C(a)\, a^{-4}, \quad \text{with} \quad \frac{dC}{dt} = T_i\, a^3\, \dot{a}. \tag{18}$$

After integrating the expression (18) for the components (16) and collecting the results from each sector, we arrive at the following equation

$$\rho_{EH} + \rho_{HD} = \rho_{rad}, \tag{19}$$

where

$$\begin{aligned}
\rho_{EH} &= \frac{3M_P^2}{8\pi}\left[\left(\dot{\sigma}^2 + k\, e^{-2\sigma}\right)(1 - \tilde{f}\sigma) - \frac{\Lambda}{3}(1 - \tilde{g}\sigma)\right],\\
\rho_{HD} &= 6b\dot{\sigma}^4 + 6k\,(2b + c)\dot{\sigma}^2\, e^{-2\sigma} + 3c\left(\ddot{\sigma}^2 - 2\,\dddot{\sigma}\,\dot{\sigma} - 6\ddot{\sigma}\,\dot{\sigma}^2\right),\\
\rho_{rad} &= \frac{1}{4}(\rho_{r0} + \beta \bar{F}^2\sigma)\, e^{-4\sigma}.
\end{aligned} \tag{20}$$

In the case of massless matter fields, i.e., $\tilde{f} = \tilde{g} = 0$, with $k = 0$, and omitting the radiation term, Eq. (19) reduces to a biquadratic equation for the Hubble parameter, $H = \dot{\sigma} = const$, with the following dS-like solutions (obtained in [22] from the trace term)

$$H_{\text{dS}} = \pm \frac{M_P}{\sqrt{-32\pi b}}\left(1 \pm \sqrt{1 + \frac{64\pi b}{3}\,\frac{\Lambda}{M_P^2}}\right)^{1/2}. \tag{21}$$

The solutions with the positive H correspond to expanding, and the ones with the negative H, to contracting Universes. Let us note that, for $\Lambda \ll M_P^2$, the $(+,-)$ version of (21) gives the classical de Sitter solution with H_1, while the $(+,+)$ solution H_2 is close to the exponential one obtained in [9, 36, 37]. This solution corresponds to the equilibrium between the quantum anomaly-induced part and the classical part, i.e.,

$$H_1 = \sqrt{\frac{\Lambda}{3}} \quad \text{and} \quad H_2 = \frac{M_P}{\sqrt{-16\pi b}}. \tag{22}$$

On the other hand, even in the case of massive matter fields, we can still find an approximate solution for $\sigma(t) = \ln a(t)$. This is achieved by replacing [22],

$$M_P^2 \;\rightarrow\; \tilde{M}_P^2 = M_P^2\big(1 - \tilde{f}\sigma\big), \qquad \Lambda \;\rightarrow\; \tilde{\Lambda} = \Lambda\big(1 - \tilde{g}\sigma\big), \tag{23}$$

in the solutions for the massless case. From the second solution in (22), one can easily integrate $\tilde{H} = H_0\,(1 - \tilde{f}\sigma)^{\frac{1}{2}}$, under the initial condition $\sigma(0) = 0$, to obtain, in the absence of a cosmological constant and assuming $\tilde{g} = 0$,

$$\sigma(t) = H_2 t - \frac{H_2^2}{4}\tilde{f} t^2. \tag{24}$$

The parabola (24) reproduces the numerical solution [23] with the precision of 10^{-6}. Unlike the basic conformal case, this approximate solution for the conformal factor leads to a running Hubble parameter that decreases linearly with time and smoothly transforms into the usual [9] Starobinsky inflationary solution [3].

5 Stability Analysis

As the first contribution, in this section, we consider the bounce solution in the theory where the bounce is produced by vacuum anomaly-induced terms. We extend the previous consideration of the stability [22] to the contraction phase of the bounce solution and also include, for the sake of generality, the case of a theory with nonzero cosmological constant and its logarithmic (i.e., proportional to $\sigma(t)$-parameter) running according to the effective action (11). Indeed, according to the well-known results [38] we can assume that the cosmological constant is small, even in the very early Universe. Another relevant detail is that the physically justified way to perturb the cosmological solution is to consider the variation of $\sigma(t)$ and *not* the variation of $a(t)$.

This issue was discussed in detail in [22] and the interested reader may look at this reference. This difference is also described in the mathematical literature [39]. Anyway, it is worthwhile to explain it in more detail. Let us consider the asymptotic stability of the solution $a(t) = \exp\big[\sigma(t)\big]$, perturbing either $a(t)$ or $\sigma(t)$. For the sake of simplicity, consider exponential time dependence, i.e.,

$$a(t) = e^{H_0 t} + \delta a(t) \qquad \text{and} \qquad \sigma(t) = H_0 t + \delta\sigma(t)\,, \tag{25}$$

which obviously means

$$\delta a(t) = e^{H_0 t}\delta\sigma(t)\,. \tag{26}$$

This formula explains the difference. If the eigenvalues of the characteristic equation for the $\sigma(t)$ are all negative, then the solution is stable in this variable. On the other hand, if at least one of these eigenvalues is larger than $-H_0$, the same solution is unstable in the variable $a(t)$. It is important that, for the non-exponential form of $a(t)$, the different may still take place, but only for finite intervals of time, not asymptotically. Still, this may be relevant because the mentioned interval may be greater than the duration of the corresponding epoch of the evolution of the Universe.

To explore the asymptotic stability conditions in the framework described above, let us perform a linear perturbation of the conformal factor, $\sigma(t) \to \sigma(t) + x(t)$, in the Friedmann equation (19), neglecting the radiation part and assuming $k = 0$. This leads to the following equation for $x(t)$,

$$\dot{\sigma}\dddot{x} - (\ddot{\sigma} - 3\dot{\sigma}^2)\ddot{x} + \left(\dddot{\sigma} + 6\ddot{\sigma}\dot{\sigma} - \frac{4b}{c}\dot{\sigma}^3\right)\dot{x}$$
$$- \frac{M_P^2}{16\pi c}\left[2\dot{\sigma}(1 - \tilde{f}\sigma)\,\dot{x} - \left(\dot{\sigma}^2\tilde{f} - \frac{\tilde{g}}{3}\Lambda\right)x\right] = 0. \tag{27}$$

In the first approximation, one can consider that the perturbations have frequencies much greater than the Hubble parameter characterizing the dynamics of the background solution. This means, we can consider $\tilde{H} = \dot{\sigma} = constant$ on a sufficiently small interval of time $(t_i, t_i + \Delta t_i)$. We note that even using this approximation, evaluating the stability analytically is a challenging task when mass contributions are taken into account. For the simplest case with $\Lambda = \tilde{g} = 0$, this implies that $(1 - \tilde{f}\sigma) = 1$ and $\tilde{H} = H_0$. In this case, the equation for the perturbation $x(t)$ reduces to a simpler form,

$$\dddot{x} + 3H_0\ddot{x} - \frac{2b}{c}H_0^2\,\dot{x} - \frac{b}{c}\,H_0^3\,\tilde{f}\,x = 0, \tag{28}$$

whose solution can be written as

$$x(t) = C_0\,e^{\alpha t}\cos(\beta t + \varphi) + C_1\,e^{\lambda_3 t}, \tag{29}$$

where $\lambda_{1/2} = \alpha \pm i\beta$ and λ_3 are the roots of the polynomial equation corresponding to (28). In the expansion phase, the asymptotic stability corresponds to a negative real part of $\lambda_{1/2}$, i.e., $\alpha < 0$. The explicit form of α, β, and λ_3 is

$$\begin{aligned}
\alpha &= -H_0\left[1 - \frac{1}{3\cdot 2^{2/3}}\left(\frac{Z_1}{Z_3} - \frac{Z_3}{2^{2/3}}\right)\right],\\
\beta &= H_0\left[\frac{1}{\sqrt{3}\cdot 2^{2/3}}\left(\frac{Z_1}{Z_3} + \frac{Z_3}{2^{2/3}}\right)\right],\\
\lambda_3 &= -H_0\left[1 + \frac{2^{1/3}}{3}\left(\frac{Z_1}{Z_3} - \frac{Z_3}{2^{2/3}}\right)\right],
\end{aligned} \tag{30}$$

where we used notations

$$Z_1 = -3\left(3 + \frac{2b}{c}\right), \qquad Z_2 = -54\left[1 + \frac{b}{c}\left(1 - \frac{\tilde{f}}{2}\right)\right],$$
$$\text{and} \quad Z_3 = \left[Z_2 + \sqrt{4\,Z_1^3 + Z_2^2}\right]^{1/3}.$$

As in the inflationary case [22, 23], it is possible to classify the bounce models with respect to the particle contents which affects the coefficients (9). The classification is based on the sign of coefficient[3] c. E.g., for the particle content of the

[3] We assume the value that emerges in all regularizations except dimensional and covariant Pauli-Villars [40], also see the recent discussion of this issue in conformal quantum gravity [41].

Minimum Supersymmetric Standard Model (MSSM), with $N_{s,f,v} = (104, 32, 12)$, and the parameter $\tilde{f} = 10^{-3}$ [22], we can solve numerically the polynomial equation for (28) and obtain the results:

$$\lambda_{1/2} \approx -\big[1.50 \pm i\,(3.57)\big]H_0 \qquad \text{and} \qquad \lambda_3 \approx -5.00 \times 10^{-4} H_0. \tag{31}$$

In this case, stability is verified for the expansion phase, i.e., for $H_0 > 0$, as discussed in previous works [22]. On the other hand, the same formulas demonstrate the instabilities for the contraction phase when $H_0 < 0$. These values for $\lambda_{1/2}$ coincide with the analytical result for the massless case obtained in [17], where the contraction phase was also analyzed. This happens because the effect of $\tilde{f}$ is only significant for λ_3. From the numerical side, to get a small magnitude of the roots $\lambda_{1/2}$, it is necessary to have $\tilde{f} \geq 0.1$. This contradicts the limit on this parameter imposed above and also appears to be a non-realistic condition because it means that, according to Eq. (10), the spectrum of the underlying semiclassical GUT-like model includes particles with the almost-Planck magnitude of masses.

For the particle content corresponding to the Minimal Standard Model (MSM), where $N_{s,f,v} = (4, 24, 12)$, the oscillatory part becomes negligible, and the set of eigenvalues consists of positive and negative real roots. This feature signals an unstable solution for both the expansion and contraction phases.

To complete our numerical analysis, consider the non-zero values of Λ and $\tilde{g}$. Starting from the expression (27), and assuming the approximation of $\tilde{H}$ as constant, we can obtain

$$\tilde{H}\dddot{x} + 3\tilde{H}^2\ddot{x} - \frac{4b}{c}\tilde{H}^3\dot{x} - \frac{M_P^2}{16\pi c}\left[2\tilde{H}\dot{x} - \left(\tilde{H}^2\tilde{f} - \frac{\tilde{g}}{3}\Lambda\right)x\right] = 0, \tag{32}$$

where we use that $\big(1 - \tilde{f}\sigma\big) = 1$ and $\big(1 - \tilde{g}\sigma\big) = 1$. Here, $\tilde{H}$ assumes the standard structure of (21). Finally, considering the MSSM particle content and the estimates of $\tilde{g} = 10^{-2}$ and $\Lambda = 10^{-12}$ [42], we arrive at

$$\lambda_{1/2} \approx 1.46 \pm i\,(3.47) \qquad \text{and} \qquad \lambda_3 \approx 4.85 \times 10^{-4}, \qquad \text{for } \tilde{H} < 0. \tag{33}$$

This result shows that for nonzero $\tilde{g}$ and Λ, the solution corresponding to the contraction phase remains unstable under small perturbations. This means the nonzero cosmological constant and its running do not change the stability properties of the bounce solution.

6 Numerical Results for the Bounce Solutions

In this section, we present the numerical results for the scale factor and the Hubble parameter in the two bounce scenarios, (i) and (ii), mentioned in Sect. 4. To this end, we use the expression for the trace equation (15) together with the constraint

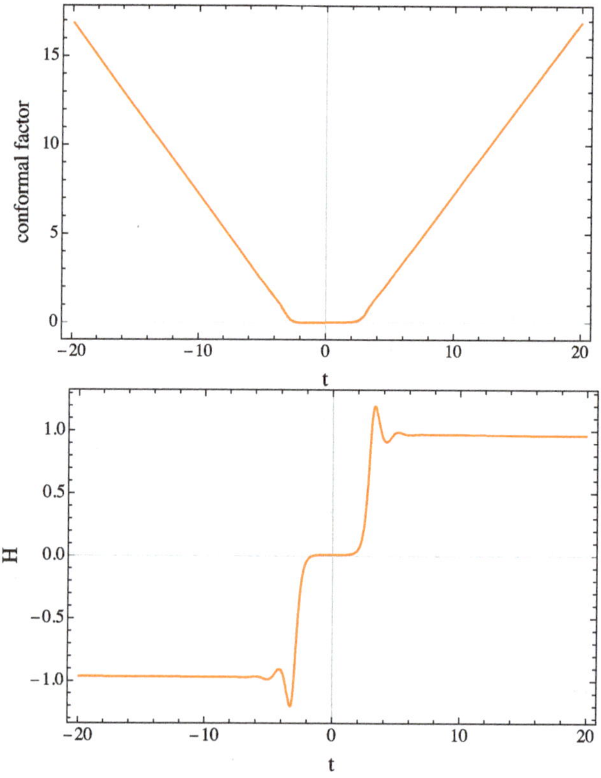

Fig. 1 Numerical solutions for the conformal factor $\sigma(t)$ in scenario (i). Here, we assumed the initial conditions: $\sigma(0) = 0,\ \dot{\sigma}(0) = -10^{-2}\tilde{H},\ \ddot{\sigma}(0) = 0,\ \dddot{\sigma}(0) = 0$. The bottom plot shows the Hubble parameter $H(t)$

equation (17), which provides the energy balance and can be useful for setting the initial conditions, especially in the second scenario (ii), based on the radiation effects. The analysis is numerical and all the plots shown in this section were generated using Mathematica [43].

For the first scenario, without loss of generality, we consider $T_{EH} + T_{HD} = 0$ with $k = 0$. The solutions obtained in this case are presented in Fig. 1, where we adopt Planck units for time t. Similar to the massless case studied in [17], it is also possible to find bouncing solutions in the present context for specific initial conditions. Moreover, such solutions persist under small variations of the initial conditions.

Figure 1 contains only the configuration with $\tilde{f} = 10^{-3}$, $\tilde{g} = 10^{-2}$ and $\Lambda = 10^{-8}$. We tried some other configurations, especially (a) $\tilde{f} = 0$, $\tilde{g} = 10^{-2}$ and $\Lambda = 10^{-8}$ and (b) $\tilde{f} = 10^{-3}$, $\tilde{g} = 0$ and $\Lambda = 10^{-8}$. All these three configurations provide very similar behavior for the conformal factor and the same is true for other sets of parameters. Indeed, the mentioned values of $\tilde{f}$, Λ, and $\tilde{g}$ are huge compared to

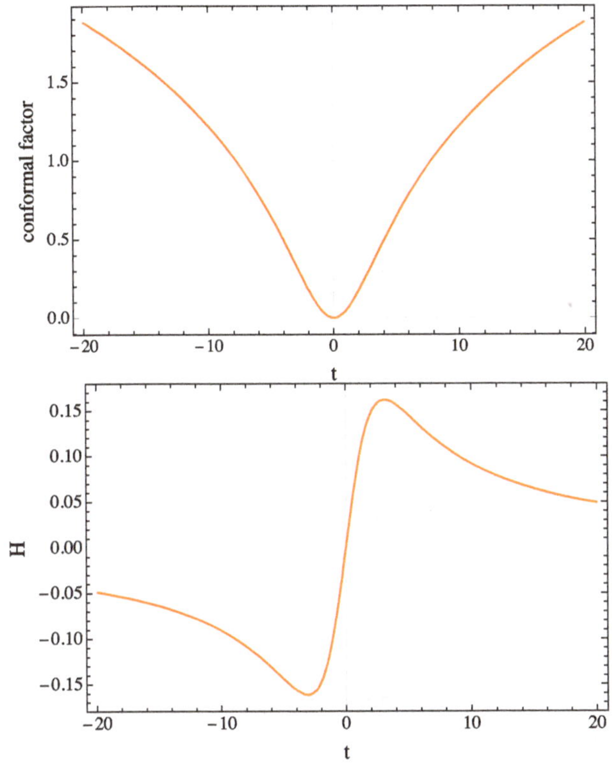

Fig. 2 Numerical solutions for the conformal factor $\sigma(t)$ in scenario (ii), where the higher-derivative terms are neglected. We assumed the value $\beta g^2 \bar{F}^2 = 0.1$ in the Planck units and the initial conditions $\sigma(0) = 0$ and $\dot{\sigma}(0) = -10^{-1}\tilde{H}$. The bottom plot shows the Hubble parameter $H(t)$

the physically acceptable ones. For instance, the well-motivated bound for the first parameter is $\tilde{f} < 10^{-6}$, which means the masses of the quantum particles do not exceed $M_X \sim 10^{16}$ GeV. The values we tried are far above this bound and still do not produce significant changes. The same is true for the parameters Λ and $\tilde{g}$. Thus, we conclude that, in general, the effects of the masses of quantum fields do not bring relevant changes to the bounce point in the scenario (i). Compared to the massless case, there are only small differences in the shapes of the plots, but no qualitative changes. To be precise, minimally visible changes are observed only in the presence of the nonzero cosmological constant Λ and at times far from the bounce point, $t < -50$ and $t > 50$. From the physical side, the negligible effect of masses should be expected since the closer one gets to the transition point, the higher the typical energy becomes, and the less relevant the mass effects of the fields are.

In the scenario (ii), the equation for the trace is given by $T_{EH} - T_{rad} = 0$, where we also assume $k = 0$. Figure 2 shows the corresponding solutions, with $\tilde{f} = 10^{-3}$, $\tilde{g} = 10^{-2}$ and $\Lambda = 10^{-8}$. In this case, we find that the quantum effects of massive fields are even more subtle than in the first scenario. The solutions with $\Lambda = 0$ and $\tilde{g} = 0$ do not exhibit any visible difference between the massive and massless cases, such that there is no reason to include the corresponding plots in our presentation. To get a visible (still very small) difference with respect to the massless case, it is necessary to postulate a large magnitude of the cosmological constant, which contradicts the standard assumptions [38].

7 The Radiation-Dominated Bounce with the R^2-Term

Let us, finally, present a third scenario providing a bounce solutions. The action we shall consider here is composed of the Einstein-Hilbert term, including the cosmological constant, the R^2-term, and the radiation part induced by the anomaly,

$$\Gamma_{R^2} = \int d^4x\sqrt{-g}\Big[-\frac{1}{16\pi G}(R+2\Lambda) + a_4 R^2\Big] - \frac{\beta g^2}{4}\int d^4x\sqrt{-\bar{g}}\,\bar{F}^2\sigma, \tag{34}$$

where $a_4 \approx 5 \times 10^8$ is the huge coefficient of the R^2-term required in the Starobinsky model [9, 44]. Taking a direct variational derivative of the action or, equivalently, from the trace of the modified Friedmann equations, we arrive at the following expression describing the dynamics of the conformal factor,

$$\ddddot{\sigma} + 7\dddot{\sigma}\dot{\sigma} + 4\ddot{\sigma}^2 + 12\ddot{\sigma}\dot{\sigma}^2 + \frac{M_P^2}{48\pi a_4}\Big(\dot{\sigma}^2 + \frac{\ddot{\sigma}}{2} - \frac{\Lambda}{3}\Big) - \frac{\beta g^2}{288\,a_4}\bar{F}^2 e^{-4\sigma} = 0. \tag{35}$$

Without the radiation term, we can easily reduce the order of Eq. (35) and arrive at the same results as in [9], close to what we described in the first sections. In particular, assuming $\dot{\sigma} = H = constant$, there is the classical de Sitter solution $H_1 = \sqrt{\Lambda/3}$. In comparison with the scenario (i), the higher-derivative sector of (34), composed only of R^2, does not generate a term of the type $\dot{\sigma}^4$ in the equation for the trace, and this makes it more difficult to obtain a nontrivial analytical solution in the de Sitter case.

On the other hand, we can explore the solution space of the scenario (iii) numerically by varying the parameters and initial conditions and, in this way, find the bounce-type solutions, as presented in Fig. 3. The lower extreme curve represents the situation with a very small coefficient a_4, approaching the well-known solution from [19], where $a_4 = 0$. While the upper extreme curve corresponds to the case with $a_4 \approx 5 \times 10^8$. The region between these two extreme solutions contains a set of possible bounce solutions with different values for the parameters a_4, $\beta g^2 \bar{F}^2$, and Λ, as well as variations in the initial conditions. Additionally, the smaller plot provides a zoom into the region around the bounce point, i.e., in the higher energy region.

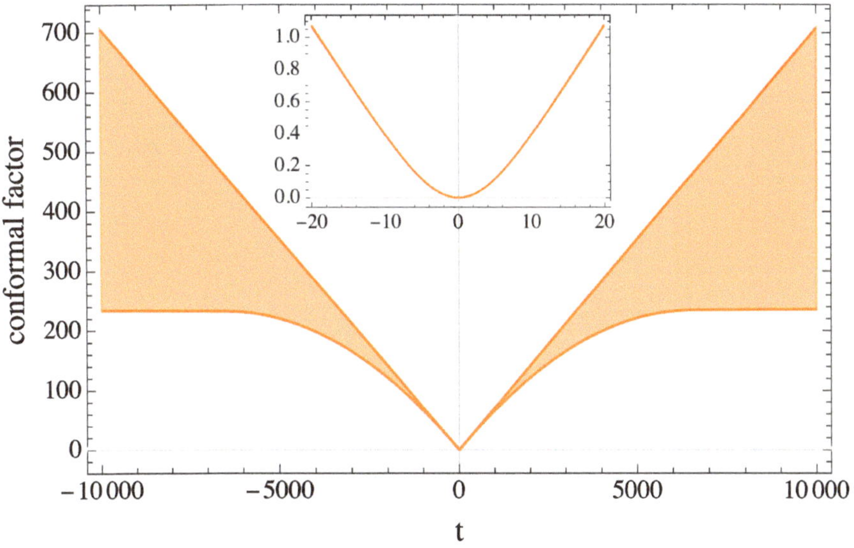

Fig. 3 Set of numerical solutions for the conformal factor $\sigma(t)$ in scenario (iii), which includes the contribution from the R^2-term. The extreme curves are obtained by assuming the initial conditions: $\sigma(0) = 0,\ \dot{\sigma}(0) = 10^{-4},\ \ddot{\sigma}(0) = 10^{-2},\ \dddot{\sigma}(0) = 0$. The smaller plot shows a zoom in the region around the bounce point, for the interval $-20 \leq t \leq 20$ in Planck units

Note that the difference in the bounce solutions appears only at the lower energies, but still above the Planck scale. In this case, we observe that such solutions arise even without the radiation term provided that fine-tuning of the initial conditions are implemented.

However, it is important to mention that the bounce solutions that do not take into account the radiation contribution are more sensitive to changes in the initial conditions, indicating possible dynamic instability. For example, consider the numerical results for the conformal factor, using the two configurations for $\beta g^2 \bar{F}^2$ and Λ, shown in Fig. 4. Here, it is easy to see that the bounce is possible only in the presence of the radiation term. This solution holds for both zero and nonzero cosmological constant cases. For a systematic study of dynamic instability in this system, it is necessary to implement cosmological perturbations, and this is an issue that we intend to address in future work.

8 Conclusions and Discussions

We explored several particular problems related to the bounce solutions in the gravity theory with quantum corrections. Assuming that the Universe is in the initial contracting phase, it has to come to the instant when the typical distance becomes very small and, therefore, the typical energy scale is very high. At this point, the masses

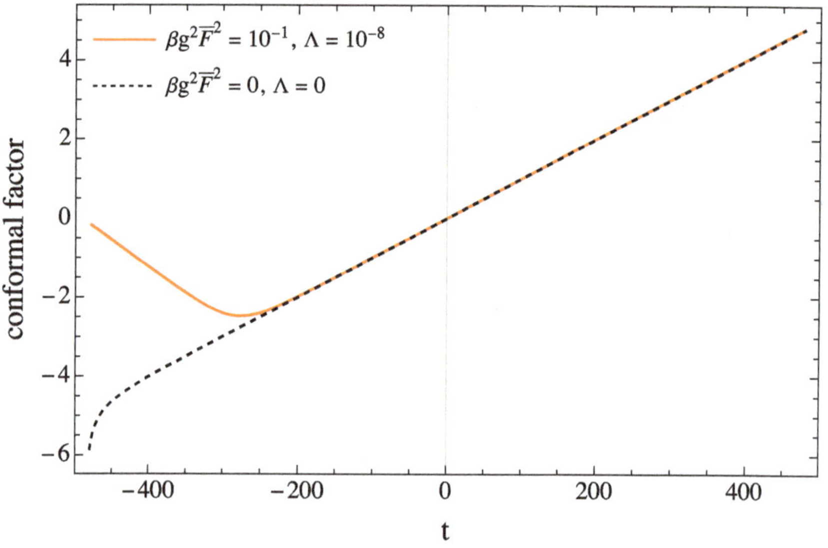

Fig. 4 Numerical solutions for the conformal factor $\sigma(t)$ in scenario involving a large coefficient for the R^2-term ($a_4 \approx 5 \times 10^8$). We assumed the initial conditions: $\sigma(0) = 0$, $\dot{\sigma}(0) = 10^{-2}$, $\ddot{\sigma}(0) = 0$, $\dddot{\sigma}(0) = 0$, and the interval $-480 \leq t \leq 480$ in the Planck units

of the particle contents of the Universe become irrelevant and the effective equation of state is the one of radiation. On the other hand, the quantum effects of effectively massless fields may be described by the trace anomaly. The anomaly-induced effective action may generate a cosmological bounce either by taking into account the vacuum (purely metric-dependent) effects (model i) or changing the equation of state of the radiation (model ii).

In the first place, we considered the effect of small masses of the quantum fields on the anomaly-induced bounce. In both cases of the models (i) and (ii), the masses do not produce strong changes on the bounce solutions. The second issue is the stability of the model (i) that was explored for the first time. It turns out that all versions of this model manifest instabilities in the contracting phase. This output means that model (i) cannot be seriously considered as a prospective model to describe cosmological bounce, at least in the form this model was formulated in [17, 18]. Finally, we considered adding the R^2-term to the most elegant and simplest bounce model (ii). The main problem of this model is that in the perturbative regime [19], the bounce occurs in the deep trans-Planckian regime, where we are not allowed to ignore vacuum quantum (and even the quantum gravity) effects. The possible way out is to add the R^2 classical term with a numerically big coefficient, such that the vacuum effects are shadowed by this dominating term. We have found that, in this case, there is a bounce solution. However, the stability properties in the presence of the dominating R^2 term are difficult to establish because the investigation of this issue requires an extensive numerical analysis for a variety of initial conditions. We hope to complete this study in a future work.

Acknowledgements The authors are very grateful to Patrick Peter for discussions and to Nicolas Bertini for the important assistance with the preparation of plots presented in Sect. 7. W.C.S. is grateful to *Conselho Nacional de Desenvolvimento Científico e Tecnológico* (CNPq—Brazil) for supporting his postdoctoral project under the PCI grant 302494/2024-3. S.W.P.O expresses his gratitude to *Fundação Coordenação de Aperfeiçoamento de Pessoal de Nível Superior* (CAPES-Brazil) for supporting his PhD project. The work of I.Sh. is partially supported by CNPq under the grant 305122/2023-1.

References

1. M.V. Fischetti, J.B. Hartle, B.L. Hu, Quantum effects in the early universe. I. Influence of trace anomalies on homogeneous, isotropic, classical geometries. Phys. Rev. **D20**, 1757 (1979)
2. D.C. Rodrigues, F. de O. Salles, I.L. Shapiro, A.A. Starobinsky, Auxiliary fields representation for modified gravity models. Phys. Rev. **D83**, 084028 (2011), arXiv:1101.5028
3. T.d.P. Netto, A.M. Pelinson, I.L. Shapiro, A.A. Starobinsky, From stable to unstable anomaly-induced inflation. Eur. Phys. J. **C76**, 544 (2016), arXiv:1509.08882
4. J.Z. Simon, Higher-derivative Lagrangians, nonlocality, problems, and solutions. Phys. Rev. D **41**, 3720 (1990)
5. A.R. Romero Castellanos, F. Sobreira, I.L. Shapiro, A.A. Starobinsky, On higher derivative corrections to the $R + R^2$ inflationary model. JCAP **12**, 007 (2018), arXiv:1810.07787
6. P.J.E. Peebles, *Principles of Physical Cosmology* (Princeton University Press, 1993)
7. S. Weinberg, *Cosmology* (Oxford University Press Inc., New York, 2008)
8. Yu. Baryshev, P. Teerikorpi, *Fundamental Questions of Practical Cosmology* (Springer, Dordrecht Heidelberg London New York, 2012)
9. A.A. Starobinski, A new type of isotropic cosmological models without singularity. Phys. Lett. B **91**, 99 (1980)
10. A.H. Guth, The inflationary Universe: a possible solution to the horizon and flatness problems. Phys. Rev. D **23**, 347 (1981)
11. A.D. Linde, A new inflationary Universe scenario: a possible solution of the horizon, flatness, homogeneity, isotropy and primordial monopole problems. Phys. Lett. B **108**, 389 (1982)
12. R. Penrose, Gravitational collapse and space-time singularities. Phys. Rev. Lett. **14**, 57 (1965). Gravitational collapse: the role of general relativity. Riv. Nuovo Cimento **1**, 252 (1969)
13. S.W. Hawking, Occurrence of singularities in open universes. Phys. Rev. Lett. **15**, 689 (1965); Singularities in the Universe. Phys. Rev. Lett. **17**, 443 (1966)
14. P. Coles, F. Lucchin, *Cosmology: The Origin and Evolution of Cosmic Structure*, 2nd edn. (Wiley-VCH, 2002)
15. M. Novello, S.E.P. Bergliaffa, Bouncing cosmologies. Phys. Rept. **463**, 127 (2008), arXiv:0802.1634
16. D. Battefeld, P. Peter, A critical review of classical bouncing cosmologies. Phys. Rept. **571**, 1 (2015), arXiv:1406.2790
17. W.C. e Silva, I.L. Shapiro, Bounce and stability in the early cosmology with anomaly-induced corrections. Symmetry **13**, 50 (2021), arXiv:2012.10554
18. J.C. Fabris, A.M. Pelinson, I.L. Shapiro, Anomaly induced effective action for gravity and inflation. Grav. Cosmol. **6**, 59 (2000), arXiv:gr-qc/9810032
19. W.C. e Silva, I.L. Shapiro, Semiclassical bounce with strong minimal assumptions. Phys. Rev. **D110**, 043540 (2024), arXiv:2402.18785

20. T. Papanikolaou, S. Banerjee, Y.F. Cai, S. Capozziello, E.N. Saridakis, Primordial black holes and induced gravitational waves in non-singular matter bouncing cosmology. JCAP **06**, 066 (2024), arXiv:2404.03779
21. S. Banerjee, T. Papanikolaou, E.N. Saridakis, Constraining $F(R)$ bouncing cosmologies through primordial black holes. Phys. Rev. D **106**, 124012 (2022), arXiv:2206.01150
22. A.M. Pelinson, I.L. Shapiro, F.I. Takakura, On the stability of the anomaly-induced inflation. Nucl. Phys. B **648**, 417 (2003), arXiv:hep-ph/0208184
23. I.L. Shapiro, J. Solà, Massive fields temper anomaly-induced inflation: the clue to graceful exit? Phys. Lett. B **530**, 10 (2002), arXiv:hep-ph/0104182
24. R.J. Riegert, A non-local action for the trace anomaly. Phys. Lett. B **134**, 56 (1984)
25. A.M. Pelinson, I.L. Shapiro, On the scaling rules for the anomaly-induced effective action of metric and electromagnetic field. Phys. Lett. B **694**, 467 (2011), arXiv:1005.1313
26. R. Ferrero, S.A. Franchino-Viñas, M.B. Fröb, W.C.C. Lima, Universal definition of the non-conformal trace anomaly. Phys. Rev. Lett. **132**, 071601 (2024), arXiv:2312.07666
27. I.L. Shapiro, Effective action of vacuum: semiclassical approach. Class. Quant. Grav. **25**, 103001 (2008), arXiv:0801.0216
28. I.L. Buchbinder, I.L. Shapiro, *Introduction to Quantum Field Theory with Applications to Quantum Gravity* (Oxford University Press, 2021)
29. N.D. Birrell, P.C.W. Davies, *Quantum Fields in Curved Space* (Cambridge University Press, Cambridge, 1982)
30. E.S. Fradkin, A.A. Tseytlin, Conformal anomaly in Weyl theory and anomaly free superconformal theories. Phys. Lett. B **134**, 187 (1984)
31. E.S. Fradkin, A.A. Tseytlin, Asymptotic freedom on extended conformal supergravities, Phys. Lett. **B110**, 117 (1982); One-loop beta function in conformal supergravities. Nucl. Phys. **B203**, 157 (1982)
32. S. Paneitz, A quartic conformally covariant differential operator for arbitrary pseudo Riemannian manifolds, MIT preprint - 1983; SIGMA **4**, 036 (2008), arXiv:0803.4331
33. E.V. Gorbar, I.L. Shapiro, Renormalization group and decoupling in curved space. JHEP **02**, 021 (2003), arXiv:hep-ph/0210388
34. D.F. Carneiro, E.A. Freitas, B. Gonçalves, A.G. de Lima, I.L. Shapiro, On useful conformal tranformations in general relativity. Grav. Cosm. **40**, 305 (2004); arXiv:gr-qc/0412113
35. I.L. Shapiro, *Primer in Tensor Analysis and Relativity* (Springer, NY, 2019)
36. S.G. Mamaev, V.M. Mostepanenko, Isotropic cosmological models determined by vacuum quantum effects. Sov. Phys. JETP **51**, 9 (1980)
37. E.S. Fradkin, A.A. Tseytlin, One loop effective potential in gauged $O(4)$ supergravity and the problem of the Λ term. Nucl. Phys. B **234**, 472 (1984)
38. S.A. Bludman, M.A. Ruderman, Induced cosmological constant expected above the phase transition restoring the broken symmetry. Phys. Rev. Lett. **38**, 255 (1977)
39. L. Cesari, *Asymptotic Behavior and Stability Problems in Ordinary Differential Equations* (Springer, 1963)
40. M. Asorey, E.V. Gorbar, I.L. Shapiro, Universality and ambiguities of the conformal anomaly. Class. Quant. Grav. **21**, 163 (2004), arXiv:hep-th/0307187
41. A.O. Barvinsky, G.H.S. Camargo, A.E. Kalugin, N. Ohta, I.L. Shapiro, On the local term in the anomaly-induced action of Weyl quantum gravity. Phys. Rev. D **108**, 086018 (2023), arXiv:2308.05251
42. A.M. Pelinson, Revisiting the modified Starobinsky model with cosmological constant. Int. J. Mod. Phys. D **18**, 1355 (2009), arXiv:0903.1970
43. Wolfram Research, Inc., *Mathematica,* (Version 12.0, Champaign, IL, 2019)
44. A.A. Starobinsky, The perturbation spectrum evolving from a nonsingular initially de-Sitter cosmology and the microwave background anisotropy. Sov. Astron. Lett. **9**, 302 (1983)

Early Universe and Quantum Cosmology

Induced Gravity and the Creation of the Universe from "nothing"

Victor A. Berezin and Inna D. Ivanova

Abstract A phenomenological description of particle creation in cosmological spacetimes is considered. We demonstrate that, contrary to previous conclusions, homogeneous and isotropic cosmological models can produce particles when back reaction is properly considered, potentially explaining gravitating dark matter and dark energy. We introduce the "gravitating mirages"–terms in the energy-momentum tensor that gravitate but are not conventional matter–and show that our model is equivalent to $f(R) \propto R^{\frac{3}{2}}$ gravity in a specific gauge. By analyzing the cosmological field equations using conformally invariant variables, we identify three distinct vacuum states and prove that transitions from vacuum to non-vacuum states are possible for two types of vacua and only dust could be born immediately after the transition.

1 Inrtoduction

According to Sakharov, the gravitational field is not a fundamental field existing independently, but rather the tension of quantum vacua from all other quantized fields [1]. This approach is known as induced gravity, which we adopt in this study.

In the early 1970s, researchers like L. Parker, S. Fulling, Y.B. Zel'dovich, A.A. Starobinsky, A. A. Grib, V. M. Mostepanenko, S. G. Mamaev and many other groups made pioneering discoveries about particle creation in curved spacetimes [2–4]. They found that particle production is closely related to the Weyl tensor and trace anomaly. Consequently, homogeneous and isotropic cosmological models with a zero Weyl tensor do not produce particles.

V. A. Berezin
Institute for Nuclear Research of the Russian Academy of Sciences, Moscow, Russia
e-mail: berezin@inr.ac.ru

I. D. Ivanova (✉)
Ben-Gurion University of the Negev, Beer-Sheva, Israel
e-mail: pc_mouse@mail.ru

A. Barvinsky and A. Kamenshchik (eds.), *Open Issues in Gravitation and Cosmology*, Fundamental Theories of Physics 225, https://doi.org/10.1007/978-3-032-15703-4_7

However, the problem of cosmological particle creation remained unresolved because the back reaction of created particles and their creation process on the spacetime metric were not fully explained.

The back reaction problem is particularly complex. Solving quantum field equations requires boundary conditions that depend on the spacetime structure, which originates from gravitational equations that need an energy-momentum tensor, which in turn depends on quantum field averages.

To circumvent these obvious obstacles, we propose a phenomenological description of particle creation. We use the action for an ideal fluid in Euler variables, where the particle conservation law is replaced by the particle creation law [5]. This method enables us to study particle creation at the classical level while accounting for the back reaction.

When the back reaction of particle creation on spacetime is properly considered, the previous conclusion that homogeneous and isotropic models cannot produce particles is overturned. Our model further suggests that particle creation could be responsible for gravitating dark matter and possibly explaining dark energy.

Particle production is a quantum process, but classical description becomes possible when external fields are strong enough and particle separation approaches their Compton length. This scenario allows approximation of particles as condensed matter within a quasi-classical description related to an effective action that may be complex. The imaginary component of effective action is common in quantum field theory, typically arising from logarithmic structures in UV form factors and indicating potential particle production by external fields (see e.g. [6–9]). For such production to occur, the energy of created particles must be less than the external field intensity. Considering these principles, we modified our model to include an imaginary component in the action, mirroring effective actions from quantum field theory.

2 Construction of the Lagrangian

In classical hydrodynamics, two types of dynamical variables exist: Lagrangian and Eulerian. Lagrangian coordinates are tied to individual particle motion, where each particle's world line is subject to variation when applying the principle of least action. These coordinates are not suitable for describing creation or annihilation processes. Therefore, the Euler formalism is preferred, using fields that describe the average characteristics of the medium. The corresponding action integral in Eulerian variables was developed by J.R. Ray [10]:

$$S_{\mathrm{m}} = -\int \varepsilon(X, n)\sqrt{-g}\, d^4x + \int \lambda_0(u_\mu u^\mu - 1)\sqrt{-g}\, d^4x + \\ + \int \lambda_1 (n u^\mu)_{;\mu}\sqrt{-g}\, d^4x + \int \lambda_2 X_{,\mu} u^\mu \sqrt{-g}\, d^4x, \tag{1}$$

where g is the determinant of the metric tensor $g_{\mu\nu}(x)$. We will be working in the framework of Riemannian geometry and using the units with $\hbar = c = 1$.

The dynamical variables in this formalism in addition to the metric include: the number density n, the four-velocity vector of fluid flow u^μ, and an auxiliary field X for world-line enumeration, while $\lambda_0(x)$, $\lambda_1(x)$ and $\lambda_2(x)$ are Lagrange multipliers. The energy density ε depends only on n and X. The particle conservation law is explicitly incorporated into the action through the constraint with the Lagrange multiplier λ_1.

Variation of this action with respect to Lagrange multipliers yields specific constraints:

$$u^\sigma u_\sigma - 1 = 0,$$

makes the vector field u^μ similar to the four-velocity of particles,

$$(nu^\sigma)_{;\sigma} = 0,$$

manifests the particle conservation law, and

$$X_{,\mu} u^\mu = 0,$$

links the vector field u^μ to the particle trajectories and enumerates them.

For a transition to a model with a variable number of particles, presented for the first time in the article [5], one needs only to replace the term in the original action, which, when varied with respect to λ_1, gives the particle conservation law, with a law of particle creation:

$$(n\, u^\mu)_{;\mu} = \Phi(inv),$$

where function Φ depends on the invariants of the fields responsible for the creation process.

It can be demonstrated that the left-hand side of the creation law, when multiplied by $\sqrt{-g}$, is invariant under local conformal transformation.

The local conformal transformation is defined as follows:

$$ds^2 = g_{\mu\nu} dx^\mu dx^\nu = \Omega^2 d\hat{s}^2 = \Omega^2(x) \hat{g}_{\mu\nu} dx^\mu dx^\nu,$$

where " ^" means "transformed", and $\Omega(x)$ is called "the conformal factor". Evidently, by definitions,

$$g_{\mu\nu} = \Omega^2\, \widehat{g}_{\mu\nu}, \quad g^{\mu\nu} = \frac{1}{\Omega^2}\, \widehat{g}^{\mu\nu}, \quad \sqrt{-g} = \Omega^4 \sqrt{-\widehat{g}}, \quad n = \frac{\hat{n}}{\Omega^3}, \quad u^\mu = \frac{\hat{u}^\mu}{\Omega}.$$

When for the rate of particle production we have:

$$(n\, u^\mu)_{;\mu} \sqrt{-g} = \left(n\, u^\mu \sqrt{-g}\right)_{,\mu} = \left(\frac{\hat{n}}{\Omega^3} \frac{\hat{u}^\mu}{\Omega}\, \Omega^4 \sqrt{-\widehat{g}}\right)_{,\mu} = \left(\hat{n}\, \hat{u}^\mu \sqrt{-\hat{g}}\right)_{,\mu}.$$

Consequently, the quantity $\Phi\sqrt{-g}$ is also conformally invariant. This condition imposes quite strict restrictions on the possible types of sources.

We consider a source function Φ of the following form:

$$\Phi = \alpha C^2 + \beta\left(\varphi\Box\varphi - \frac{1}{6}\varphi^2 R + \Lambda\varphi^4\right) + \varepsilon_1(\varphi, n),$$

where $\Box$ is the Laplace–Beltrami operator, $\alpha,\ \beta,\ \Lambda$ are some constants, R is a scalar curvature, $C^2 = C_{\mu\nu\lambda\sigma}\, C^{\mu\nu\lambda\sigma}$ is a square of Weyl tensor.

We propose that particles are created by quantum fluctuations of a scalar field, with this process being induced by spacetime curvature. The creation law Φ should contain a pure geometrical contribution causing vacuum polarization. Only one candidate exists (up to quadratic curvatures), the square of the Weyl tensor. Assuming the scalar field's usual transformation under the local conformal transformation,

$$\varphi = \frac{\widehat{\varphi}}{\Omega},$$

we have substantiated the presence of the second term in Φ.

Our particles are mass shell quanta of the scalar field, capable of annihilating back to virtual particles at a rate dependent on particle number density. Therefore, there should be a corresponding contribution to the creation law: $\varepsilon_1(\varphi, n)$.

The matter action integral S_m now reads as follows:

$$S_{\text{m}} = -\int \varepsilon(X, \varphi, n)\sqrt{-g}\, d^4x + \int \lambda_0(u_\mu u^\mu - 1)\sqrt{-g}\, d^4x + \\ + \int \lambda_1\left((nu^\mu)_{;\mu} - \Phi\right)\sqrt{-g}\, d^4x + \int \lambda_2 X_{,\mu}u^\mu\sqrt{-g}\, d^4x,$$

Note that $\varepsilon = \varepsilon(X, \varphi, n)$.

We consider a situation in which the gravity action is conformally invariant. This case is to some extent equivalent to induced gravity, where only matter action exists, since the Lagrangian multiplier λ_1 is defined up to a constant. We adopt this perspective and obtain an example of Sakharov's induced gravity [1]:

$$S_m = S_{tot}\ .$$

The variation of the total action integral is conformally invariant on the equations of motion. It is straightforward to prove that the terms with Lagrange multipliers in our action S_m are conformally invariant. The condition of conformal invariance in the variation of the first term provides the structure of energy density:

$$\frac{\delta}{\delta\Omega}\left(\int \varepsilon\sqrt{-g}\,d^4x\right) = 0 \Rightarrow \varphi\frac{\partial\varepsilon}{\partial\varphi} + 3n\frac{\partial\varepsilon}{\partial n} = 4\,\varepsilon \Rightarrow \tag{2}$$

$$\varepsilon = F\,(x)\,\varphi^4, \quad x = \frac{n}{\varphi^3}. \tag{3}$$

The term ϵ_1 in the creation law has a similar form:

$$\varepsilon_1 = F_1\,(x)\,\varphi^4,$$

where F and F_1 are some functions of x, $F_1(0) = 0$, since the φ^4-term is already present in the conservation law. While $F(x)$ is determined from the equations of motion, $F_1(x)$ is given in the Lagrangian.

From the formula (3) it follows an interesting connection between the equation of state and the scalar field. For example, for dust we find that the energy density is proportional to the scalar field and for radiation the energy density does not depend on φ.

3 Gravitating Mirages

The variation of the action by metric tensor yields the energy-momentum tensor:

$$S_m = -\frac{1}{2}\int T^{\mu\nu}\,\delta g_{\mu\nu}\,\sqrt{-g}\,d^4x.$$

It is convenient to divide S_m into two parts, $S[n]$ and $S[\varphi]$, the last one being independent on the hydrodynamical variables,

$$S[\varphi] = -\int\left\{\alpha\lambda_1 C^2 + \beta\lambda_1\left(\varphi\Box\varphi - \frac{1}{6}\varphi^2 R + \Lambda\varphi^4\right)\right\}\sqrt{-g}\,d^4x.$$

For homogeneous and isotropic cosmological models, we are considering in this paper, we don't need to add the contribution from C^2 to the energy momentum tensor and the equations of motion. Taking this into account, for the energy-momentum tensor we have:

$$\begin{aligned} T^{\mu\nu}[\varphi] &= \lambda_1\beta\Lambda\,\varphi^4\,g^{\mu\nu} - \beta\,(\lambda_1\varphi)_{,\sigma}\,\varphi^{,\sigma}\,g^{\mu\nu} + \beta\,(\lambda_1\varphi)^{,\mu}\,\varphi^{,\nu} + \\ &+ \beta\,(\lambda_1\varphi)^{,\nu}\,\varphi^{,\mu} + \frac{\beta}{3}\left\{\lambda_1\varphi^2\,G^{\mu\nu} - \left(\lambda_1\varphi^2\right)^{;\mu;\nu} + g^{\mu\nu}\,\Box\left(\lambda_1\varphi^2\right)\right\}, \end{aligned} \tag{4}$$

$$T^{\mu\nu}[n] = \left(\varepsilon + p + n\,\lambda_1\frac{\partial\varepsilon_1}{\partial n}\right)u^\mu\,u^\nu - g^{\mu\nu}\left(p + \lambda_1\left(n\,\frac{\partial\varepsilon_1}{\partial n} - F_1\,\varphi^4\right)\right),$$

where $p = -\varepsilon + n\,\frac{d\varepsilon}{dn}$ is the hydrodynamic pressure and $G^{\mu\nu} = R^{\mu\nu} - \frac{1}{2}Rg^{\mu\nu}$ is Einstein tensor. Because of the induced gravity:

$$T^{\mu\nu}[n] + T^{\mu\nu}[\varphi] = 0,$$

which in turn means that $T = T^{\mu\nu}g_{\mu\nu} = 0$. This condition is equivalent to the equation (3):

$$T = \varepsilon - 3p + 4\beta\,\lambda_1\,\Lambda\,\varphi^4 - \frac{\beta}{3}\,\lambda_1\,\varphi^2\,R + \beta\,\varphi\,\Box\,(\lambda_1\varphi) + \beta\,\lambda_1\varphi\,\Box\varphi + \\ +\lambda_1\varphi\,\frac{\partial\varepsilon_1}{\partial\varphi} = \varepsilon - 3p - \varphi\,\frac{\partial\varepsilon}{\partial\varphi} = 0, \quad (5)$$

where in the second equality the equation of motion for φ was used:

$$\delta\varphi:\quad \beta\left(\lambda_1\Box\varphi + \Box\,(\lambda_1\varphi) + 4\lambda_1\,\Lambda\varphi^3 - \frac{1}{3}\lambda_1\,\varphi\,R\right) + \lambda_1\,\frac{\partial\varepsilon_1}{\partial\varphi} = -\frac{\partial\varepsilon}{\partial\varphi}. \quad (6)$$

We'll call the term ε_1 in the creation law "gravitating mirages" [11]. It contributes to the energy-momentum tensor through the Lagrange multiplier λ_1. While $F(x)$ describes actual particles, $F_1(x)$ can exhibit different behavior and represents something that gravitates but it is not the matter at all. These mirages represent the gravitational back-reaction of particle creation on spacetime geometry, forming an invisible component of the gravitating system.

The linear terms in $F_1(x)$ suggest invisible dust, which could potentially explain dark matter.

Terms proportional to $x^{\frac{4}{3}}$ in $F_1(x)$ represent an invisible thermal bath, which might explain discrepancies between the visible cosmic microwave background and total gravitating thermal content.

4 Cosmology

We examine cosmological solutions, specifically homogeneous and isotropic spacetimes described by the Robertson-Walker metric:

$$ds^2 = dt^2 - a^2(t)dl^2 = a^2(\eta)(d\eta^2 - dl^2), \quad (7)$$

$$dl^2 = \gamma_{ij}dx^i dx^j = \frac{dr^2}{1 - kr^2} + r^2(d\theta^2 + \sin^2\theta d\varphi^2), \quad (k = 0, \pm 1),$$

here t is the cosmological time and η is the conformal time. For the closed cosmological model $k = +1$, while $k = -1$ stands for the open model and $k = 0$ for the spatially flat one.

For any scale factor $a(t)$ the Weyl tensor is identically zero,

$$C^{\mu}_{\ \nu\lambda\sigma} \equiv 0.$$

Due to the high level of the symmetry we assume that all dynamic variables except the metric depend only on t and $u^{\mu} = \delta^{\mu}_{0}$ so the constraint for the λ_0 is automatically satisfied. Since $T_1^1 = T_2^2 = T_3^3$ we can use only T^{00} and T. From the constraint for λ_2 we get:

$$\dot{X} = 0 \quad \Rightarrow \quad X = const, \tag{8}$$

the dot denotes the derivative with respect to t .

It turns out that equations of motion can be expressed in terms of conformally invariant functions x, $\lambda_1(\eta)$ and $f = \varphi a$ and conformal time η:

$$\begin{aligned}
\lambda_1' f f' + \lambda_1 \left(f'^2 + k f^2 + \Lambda f^4\right) + \frac{f^4}{\beta}\left(F + \lambda_1 F_1\right) = 0,\\
f\lambda_1'' + 2\lambda_1' f' + 2\lambda_1 \left(f'' + kf + 2\Lambda f^3\right) +\\
+\frac{f^3}{\beta}\left(4F - 3xF_x + \lambda_1\left(4F_1 - 3xF_{1x}\right)\right) = 0,\\
\beta f\left(f'' + kf + \Lambda f^3\right) + f^4 F_1 = \left(x f^3\right)',\\
\left(F_x + \lambda_1 F_{1x}\right) f + \lambda_1' = 0,
\end{aligned}$$

here prime denotes the derivative with respect to η, $F_x = \frac{dF}{dx}$.

The scale factor a does not enter these equations as it should, because it actually plays the role of the conformal factor. Consequently, the only trace of geometry is the type of a cosmological model encoded in the value of k.

For some problems it is more convenient to switch from the variable x to the variable $N = xf^3$:

$$\begin{aligned}
\lambda_1' f f' + \lambda_1 \left(f'^2 + k f^2 + \Lambda f^4\right) + \frac{f^4}{\beta}\left(F + \lambda_1 F_1\right) = 0,\\
f\lambda_1'' + 2\lambda_1' f' + 2\lambda_1 \left(f'' + kf + 2\Lambda f^3\right) +\\
+\frac{1}{\beta}\left(4f^3 F - 3NF_x + \lambda_1\left(4f^3 F_1 - 3NF_{1x}\right)\right) = 0,\\
\beta f\left(f'' + kf + \Lambda f^3\right) + f^4 F_1 = N',\\
\left(F_x + \lambda_1 F_{1x}\right) f + \lambda_1' = 0,
\end{aligned}$$

where F, F_1, F_x, F_{1x} are considered as functions of the combination $\frac{N}{f^3}$.

As was indicated in the introduction, we want to consider a modification of our model where the action has an imaginary part by analogy with the effective action.

Since the imaginary part of the effective action is associated with the creation of particles, it makes sense to make the Lagrange multiplier λ_1 complex:

$$\lambda_1 = \tilde{\lambda}_1 + i\,\approx{\lambda}_1$$

The equations for $\tilde{\lambda}_1$ are the same as before. The equations with $\approx{\lambda}_1$ are as follows:

$$\approx{\lambda}_1' + f\,\approx{\lambda}_1 F_{1x} = 0, \tag{9}$$

$$N' = \beta f\left\{f'' + kf + \Lambda f^3\right\} + F_1 f^4, \tag{10}$$

$$f\approx{\lambda}_1'' + 2\approx{\lambda}_1' f'' + 2\approx{\lambda}_1\left(f'' + kf + 2\Lambda f^3\right) = -\frac{f^3}{\beta}\approx{\lambda}_1\left(4F_1 - 3xF_{1x}\right), \tag{11}$$

$$f\,\approx{\lambda}_1' f' + \approx{\lambda}_1\left\{f'^2 + kf^2 + \Lambda f^4\right\} + \frac{f^4}{\beta}\approx{\lambda}_1 F_1 = 0. \tag{12}$$

5 Gauge Fixing

The conformal invariance of the gravity action and, by extension, the equations of motion admit an arbitrary gauge choice.

If we found the specific solution for the set of dynamical variables $\{\widehat{a}, \widehat{n}, \widehat{\varphi}\}$, then the general solution is: $\{a, n, \varphi\}$, where $a = \widehat{a}\,\Omega$, $\widehat{n} = n\,\Omega^3$, $\widehat{\varphi} = \varphi\,\Omega$ with a arbitrary smooth function $\Omega(t)$.

There are three important cases. In the first one we can set $a = 1$ everywhere except at $a = 0$. This does not imply a static model. In such a gauge, all previous equations remain unchanged but with f replaced by φ.

The second case is setting the derivative of λ_1 equal to one. This reveals the nonlocal nature of the Lagrange multiplier λ_1,

$$\frac{d\lambda_1}{dt} = \frac{1}{a}\frac{d\lambda_1}{d\eta} = 1 \quad \Rightarrow \quad \lambda_1 = t + \text{const} = \int a(\eta)d\eta,$$

and indicates its relationship with the nonlocal part of the trace anomaly.

The third one is the so-called "GR-gauge":

$$\beta\,\lambda_1\varphi^2 = -\frac{3}{\kappa}, \quad \kappa = \frac{8\pi G}{c^4},$$

which restores the cosmological Friedmann equations, enabling comparison with observations based on General Relativity.

This gauge also allows us to draw analogy with $f(R)$ gravity. In order to show this, let us consider how does the part of the action $S[\varphi]$ look like under the condition $\lambda_1 = -\frac{3}{\kappa\beta\varphi^2}$:

$$S[\varphi] = -\frac{1}{2\kappa}\int\left\{R - \frac{6}{\varphi^2}\,\varphi_{,\mu}\,\varphi^{,\mu} - 6\Lambda\varphi^2\right\}\sqrt{-g}\,d^4x. \tag{13}$$

Using the replacement: $\widetilde{\varphi} = 2\sqrt{3}\,ln\varphi$ the action (13) can be brought to the form:

$$S[\varphi] = -\frac{1}{2\kappa}\int\left(R - \frac{1}{2}\,g^{\mu\nu}\,\widetilde{\varphi}_\mu\widetilde{\varphi}_\nu - \widetilde{V}(\widetilde{\varphi})\right)\sqrt{-g}\,d^4x, \tag{14}$$

$$\widetilde{V}(\widetilde{\varphi}) = 6\Lambda\,e^{\frac{\widetilde{\varphi}}{\sqrt{3}}},$$

which is in turn equivalent to $f(R) \propto R^{\frac{3}{2}}$ gravity.

6 Vacua

By "vacua" we understand the solutions with no particles and no particle creation. Thus,

$$N = 0, \quad N' = 0.$$

Let us introduce the constants: $F(0) = \sigma$, $F_x(0) = \mu_1$, $F_{1x}(0) = \gamma_1$, and we remember that $F_1(0) = 0$. Our set of equations becomes:

$$\begin{aligned} \lambda_1' + f\,(\mu_1 + \lambda_1\,\gamma_1) &= 0,\\ f\left\{f'' + k\,f + \Lambda\,f^3\right\} &= 0,\\ \sigma f^4 + \beta\left\{(\lambda_1 f)'\,f' + \lambda_1\,kf^2 + \lambda_1\,\Lambda f^4\right\} &= 0, \end{aligned}$$

where, as in the previous Sect. 5, we consider λ_1 to be complex: $\lambda_1 = \widetilde{\lambda}_1 + i\,\widetilde{\widetilde{\lambda}}_1$. We did not include in this set the equation obtained by variation $\delta\varphi$, since it is the differential consequence of all other equations.

There are three types of vacuum solutions:

- Vacuum #1: $f(\eta) \equiv 0$, $\widetilde{\lambda}_1 = const$, $\widetilde{\widetilde{\lambda}}_1 = const$;
- Vacuum #2:

$$f = f_0 = \pm\sqrt{-\frac{k}{\Lambda}}, \quad \widetilde{\lambda}_1(\eta) = \left(\widetilde{\lambda}_1(0) + \frac{\mu_1}{\gamma_1}\right)e^{-f_0\gamma_1\eta} - \frac{\mu_1}{\gamma_1},$$

$$\sigma = 0, \quad \widetilde{\widetilde{\lambda}}_1(\eta) = \widetilde{\widetilde{\lambda}}_1(0)\,e^{-f_0\gamma_1\eta};$$

- Vacuum #3: $\widetilde{\lambda}_1 = -\frac{\mu_1}{\gamma_1}$, $\widetilde{\widetilde{\lambda}}_1 = 0$, $\sigma = \frac{1}{2}\frac{\beta\mu_1}{\gamma_1}\Lambda$.

Vacuum #3 does not require a special value for λ_1. In contrast, the value of λ_1 determines the characteristics of the particles created. A special value of σ means the choice of the solution (since σ is just the integration constant). For vacuum #3 , the function f satisfies the equation:

$$f'^2 + U(f) = 0, \quad U(f) = kf^2 + \frac{1}{2}\Lambda f^4,$$

with the solutions:

- $k = 0,\ \Lambda < 0$: $f^2 = -\frac{2}{\Lambda}\frac{1}{(\eta-\eta_0)^2}$;
- $k = 0,\ \Lambda = 0$: $f = const$;
- $k = -1,\ \Lambda > 0$: $f^2 = \frac{2}{\Lambda}\frac{1}{ch^2(\eta-\eta_0)}$;
- $k = -1,\ \Lambda < 0$: $f^2 = -\frac{2}{\Lambda}\frac{1}{sh^2(\eta-\eta_0)}$;
- $k = -1,\ \Lambda = 0$: $f = e^{\pm(\eta-\eta_0)}$;
- $k = 1,\ \Lambda < 0$: $f^2 = -\frac{2}{\Lambda}\frac{1}{sin^2(\eta-\eta_0)}$.

Note that if $\gamma_1 = 0$, then vacuum #3 exists only under the condition that $\mu_1 = 0$. In that case:

$$\widetilde{\lambda}_1 = const, \quad \widetilde{\widetilde{\lambda}}_1 = 0, \quad \sigma = -\frac{1}{2}\beta\Lambda\widetilde{\lambda}_1.$$

The Table 1 shows whether a particular vacuum exists depending on the value of the parameters. Here "+" means that the vacuum exists, "−" means that it does not exists, "≡" means that it coincides with another vacuum.

There cannot be classical transitions between different vacua because, as it will be explained below, the functions $\lambda_1(\eta),\ f(\eta),\ a(\eta)$ must be continuous under the transition.

There is however one case of the quantum tunneling between two of them. For $k = 1,\ \Lambda < 0$ vacuum #1 may undergo the quantum tunneling into the vacuum #3

Table 1 Dependence of the existence of vacuums on the parameters Λ and k

k	Λ	#1	#2	#3
0	> 0	+	≡ #1	−
0	< 0	+	≡ #1	+
0	$= 0$	+	+	+
1	> 0	+	−	−
1	< 0	+	+	+
1	$= 0$	+	≡ #1	−
−1	> 0	+	+	+
−1	< 0	+	−	+
−1	$= 0$	+	≡ #1	+

with $\lambda_1 = const$, thus determining the value of μ_1. Such a process begins at the past infinity of the imaginary time $i\eta$. Since the period of the imaginary time is infinite, the temperature after this transition is zero.

7 Transition from Vacuum to Non-vacuum

When deriving the equations of motion, the dynamic variable n and therefore N were considered continuous, but from a physical point of view it is discrete, since it is the total number of particles. Thus, the dependence of N on time should look like this:

$$N = N_0\,\theta(\eta - \eta_0) + (N_1 - N_0)\,\theta(\eta - \eta_1) + (N_2 - N_1)\,\theta(\eta - \eta_2) + \ldots,$$

where N_i and η_i are some constants. When differentiating this expression, delta functions arise:

$$N' = N_0\,\delta(\eta - \eta_0) + (N_1 - N_0)\,\delta(\eta - \eta_1) + (N_2 - N_1)\,\delta(\eta - \eta_2) + \ldots$$

Let us consider the transition from vacuum to non-vacuum that occurs on the hypersurface: $\eta = \eta_0$. The functions $f(\eta)$, $a(\eta)$, $\widetilde{\lambda}_1(\eta)$ and $\overset{\approx}{\lambda}_1(\eta)$ should be continuous on the hypersurface in order to avoid the occurrence of undefined functions, but their first derivatives in general may contain jumps, and the second may contain a delta function. From the restriction of the equations of motion to the hypersurface we obtain:

$$f[\widetilde{\lambda}_1'\, f'] + \widetilde{\lambda}_1[f'^2] + \frac{f^4}{\beta}\left[F + \widetilde{\lambda}_1\, F_1\right] = 0, \tag{15}$$

$$f[\widetilde{\lambda}_1''] + 2[\widetilde{\lambda}_1' f'] + 2\widetilde{\lambda}_1[f''] + \frac{1}{\beta}\left[4Ff^3 - 3NF_x + \widetilde{\lambda}_1\left(4F_1 f^3 - 3NF_{1x}\right)\right] = 0, \tag{16}$$

$$\delta:\ f\,[\widetilde{\lambda}_1'] = -2\widetilde{\lambda}_1\,[f'], \tag{17}$$

$$\beta f[f''] + f^4[F_1] = 0, \tag{18}$$

$$\delta:\ \beta f[f'] = N_0, \tag{19}$$

$$[\widetilde{\lambda}_1'] = -\left[F_x + \widetilde{\lambda}_1 F_{1x}\right] f, \tag{20}$$

$$f[\overset{\approx}{\lambda}_1'\, f'] + \overset{\approx}{\lambda}_1[f'^2] + \frac{f^4}{\beta}\overset{\approx}{\lambda}_1\,[F_1] = 0, \tag{21}$$

$$f[\overset{\approx}{\lambda}_1''] + 2[\overset{\approx}{\lambda}_1' f'] + 2\overset{\approx}{\lambda}_1[f''] + \frac{1}{\beta}\overset{\approx}{\lambda}_1\left[4F_1 f^3 - 3NF_{1x}\right] = 0, \tag{22}$$

$$\delta:\ f\,[\overset{\approx}{\lambda}_1'] = -2\overset{\approx}{\lambda}_1\,[f'], \tag{23}$$

$$[\overset{\approx}{\lambda}_1'] = -\overset{\approx}{\lambda}_1 f\,[F_{1x}], \tag{24}$$

where square brackets denote the jump of the corresponding quantity on the hypersurface, for example: $[F] = F - F(vac)$. The symbol δ marks the equations that were obtained by matching delta functions.

In order to have the exit from vacuum #1, it is necessary that the value of $f[f']$ remains finite as a tends to zero. Then from equation (18) it follows that the value of F_1 must contain a term proportional to x^2. This agrees with the result that will be obtained further.

For vacuum #3: $\overset{\approx}{\lambda}_1 = 0$, therefore regardless of the form of the function $F_1(x)$ from equations (23, 24) we obtain $[f'] = 0$ and as a consequence of (19) we will have $N_0 = 0$. Thus, there is no exit from vacuum #3.

From the equations (19,23), we get:

$$[f'] = \frac{N_0}{\beta f}, \quad [\overset{\approx}{\lambda}{}_1'] = -\frac{2\overset{\approx}{\lambda}_1}{\beta f^2} N_0 .$$

Since in all vacuums either $f'(vac) = 0$ or $\overset{\approx}{\lambda}{}_1'(vac) = 0$, then

$$[\overset{\approx}{\lambda}{}_1' f'] = \overset{\approx}{\lambda}{}_1' f' = \left(\overset{\approx}{\lambda}{}_1'(vac) + [\overset{\approx}{\lambda}{}_1']\right)\left(f'(vac) + [f']\right) =$$

$$= -\frac{2\overset{\approx}{\lambda}_1}{\beta f^2} N_0\, f'(vac) + \frac{N_0}{\beta f}\overset{\approx}{\lambda}{}_1'(vac) - \frac{2\overset{\approx}{\lambda}_1}{\beta^2 f^3} N_0^2,$$

$$[f'^2] = [f']\left(f' + f'(vac)\right) = \frac{N_0}{\beta f}\left(2f'(vac) + \frac{N_0}{\beta f}\right)$$

Using the relations above from the equation (21) we get:

$$[F_1] = \frac{1}{\beta}\frac{N_0^2}{f^6} - \frac{N_0}{f^3}\frac{\overset{\approx}{\lambda}{}_1'(vac)}{\overset{\approx}{\lambda}_1 f}. \tag{25}$$

The only vacuum for which $\overset{\approx}{\lambda}{}_1'(vac) \neq 0$ is vacuum #2, for which we get:

$$[F_1] = \frac{1}{\beta} x^2 + \gamma_1 x. \tag{26}$$

The function F_1 is given in the Lagrangian and should not depend on the type of vacuum from which the transition to particle production occurs. That is, the formula means that the values of the dynamic variables for a specific vacuum solution depend on the initially specified parameter gamma, and not vice versa. At first glance, this means that an exit is only possible from vacuum #2 with non-zero γ_1, but in fact, since for vacuum #1 f tends to zero, the contribution from the linear term in F_1 turns out to be suppressed compared to the quadratic one, that is, the form (26) of the function F_1 is compatible with an exit from the first vacuum.

Next, let us find out the form of the function F. Substituting $\overset{\approx}{\lambda}{}_1'$ from (9) into (12) and assuming $\overset{\approx}{\lambda}_1 \neq 0$, we get:

$$- f^2 F_{1x} f' + f'^2 + kf^2 + \Lambda f^4 + \frac{f^4}{\beta} F_1 = 0,$$

Similarly we transform the analogous equation for $\widetilde{\lambda}_1$:

$$- f^2 (F_x + \widetilde{\lambda}_1 F_{1x}) f' + \widetilde{\lambda}_1 (f'^2 + kf^2 + \Lambda f^4) + \frac{f^4}{\beta} \left(F + \widetilde{\lambda}_1 F_1 \right) = 0.$$

Combining these two equations we get:

$$F_x f' = \frac{f^2}{\beta} F,$$

For the vacuums #1 and #2 : $f'(vac) = 0$, then $f' = [f'] = \frac{N_0}{\beta f}$ for the transition, therefore:

$$F_x x = F \Rightarrow F = \mu_1 x,$$

that is, when leaving a vacuum, only dust is born.

8 Discussion and Conclusion

Modified theory of particle creation is considered in the framework of induced gravity, where a perfect fluid action is altered to include explicit particle creation rather than conservation. The model accounts for the back reaction on spacetime from both created particles and the creation process itself, incorporating the trace anomaly and enabling particle creation in homogeneous and isotropic cosmology.

It turns out that conformal invariance of the "creation law" imposes quite strict restrictions on the possible types of sources. A key discovery is the emergence of "gravitating mirages"–terms that appear in the energy-momentum tensor which, while not being conventional matter, have gravitational effects and may explain dark matter and a part of the cosmic microwave background.

The cosmological field equations were rewritten using only conformal invariant variables, where the scale factor drops out of the equations. A "GR-gauge" is introduced to compare with the observational data, showing our model to be equivalent to $f(R)$ gravity when $f(R) \propto R^{3/2}$.

It turns out that there are three types of vacua by which we understand the solutions with no particles and no particle creation. Classical transitions between different vacua are forbidden. There is however one case of the quantum tunneling between two of them. Such a process begins at the past infinity of the imaginary time. Thus, since the period of the imaginary time during the tunneling is infinite, the temperature after this transition is zero.

We analyzed the phase transition of the system from a vacuum state to a non-vacuum one. It was shown that the exit from vacuum #3 is impossible, and the existence of such a transition for vacuums #1 and #2 limits the form of the functions F and F_1. So F_1 contains only quadratic and linear terms, and F only linear, that is, immediately after exiting the vacuum, only dust is born.

Competing Interests The authors have no conflicts of interest to declare that are relevant to the content of this chapter.

References

1. A.D. Sakharov, Vacuum quantum fluctuations in curved space and the theory of gravitation. Dokl. Akad. Nauk Ser. Fiz. **177**, 70–71 (1967)
2. L. Parker, S.A. Fulling, Quantized matter fields and the avoidance of singularities in general relativity. Phys. Rev. D **7**, 2357–2374 (1973)
3. A.A. Grib, S.G. Mamayev, V.M. Mostepanenko, Particle creation from vacuum in homogeneus isotropic models of the Universe. Gen. Relativ. Gravit. **7**, 535–547 (1976)
4. Y.B. Zeldovich, A.A. Starobinskii, Rate of particle production in gravitational fields. JETP Lett. **26**, 252–255 (1977)
5. V.A. Berezin, Unusual hydrodynamics. Int. J. Mod. Phys. A **2**, 1591–1615 (1987)
6. J.S. Schwinger, On Gauge Invariance and Vacuum Polarization. Phys. Rev. **82**, 664–679 (1951)
7. L. Parker, Particle creation in expanding universes. Phys. Rev. Lett. **21**, 562–564 (1968)
8. P. Candelas, D.J. Raine, General-relativistic quantum field theory: An exactly soluble model. Phys. Rev. D **12**, 965–974 (1975)
9. E. Mottola, Particle creation in de Sitter space. Phys. Rev. D **31**, 754–766 (1985)
10. J.R. Ray, Lagrangian Density for Perfect Fluids in General Relativity. J. Math. Phys. **13**, 1451–1453 (1972)
11. V.A. Berezin, I.D. Ivanova, Conformally invariant gravity and gravitating Mirages. Universe. **10**, 147 (2024)

On Quantum Creation of a Toroidal Universe

Alan H. Guth and Alexander Vilenkin

Abstract We consider the quantum creation of a universe with flat spatial sections and the topology of a 3-torus, taking into account the effect of Casimir energy. We show that the corresponding instantons are singular. Since these instantons describe universes originating in a state of infinite energy, we argue that they cannot be interpreted as quantum creation from 'nothing'. If quantum corrections to the energy-momentum tensor are neglected, the spacetime of the toroidal universe reduces to de Sitter space with appropriate periodic identifications. Contrary to previous claims in the literature, this spacetime is geodesically incomplete. We argue that this spacetime describes a classical universe originating at a singularity, and not a quantum origin. We conclude that the quantum creation of a toroidal universe *from nothing* cannot be described in the context of semiclassical quantum gravity—it is either impossible, or it depends essentially on Planck-scale physics. We therefore see no reasonable way to compare the probability of creation of a toroidal universe, if it is possible at all, with that of a spherical universe.

1 Introduction

Both of us met Alexei Starobinsky for the first time in the summer of 1982, at the three-week-long Nuffield Workshop on the Early Universe, in Cambridge, UK, organized by Stephen Hawking and Gary Gibbons. Alexei and AHG were both working on the very exciting topic of calculating the density perturbations arising from inflation. AV was working on another very important cosmological problem, cosmic strings. Talking with Alexei was difficult, due to his stutter, but it was immediately clear

A. H. Guth (✉)
Center for Theoretical Physics–A Leinweber Institute, Laboratory for Nuclear Science, and Department of Physics, Massachusetts Institute of Technology, Cambridge, MA, USA
e-mail: guth@ctp.mit.edu

A. Vilenkin
Institute of Cosmology, Department of Physics and Astronomy, Tufts University, Medford, MA, USA
e-mail: vilenkin@cosmos.phy.tufts.edu

A. Barvinsky and A. Kamenshchik (eds.), *Open Issues in Gravitation and Cosmology*, Fundamental Theories of Physics 225, https://doi.org/10.1007/978-3-032-15703-4_8

that patient listening was richly rewarded. Alexei's understanding of physics, and his ability to explain it, were very impressive. Unfortunately, after Nuffield neither of us had the opportunity to meet Alexei very frequently, but his work had a big influence on both of us throughout our careers. We also have great respect for the enormous courage that Alexei showed in pushing back against government interference with the Russian Academy of Sciences, as part of the "July 1 Club" and beyond. Alexei was a very positive influence both scientifically and politically, so his passing is a great loss. We are grateful for the opportunity to contribute to this volume in his honor.

This paper concerns the possible quantum creation of the universe, a topic that was very important in Alexei's work. Our discussion starts with the observation that inflationary spacetimes are known to be past-incomplete [1]. This indicates that the inflating region of such spacetimes must have a past boundary, and raises the question of what determined the initial conditions at that boundary. Even though it may not be essential for the observational predictions of inflationary models, this is an important question of principle, and without resolving it our understanding of cosmology remains incomplete.

Perhaps the most promising approach to this problem is based on quantum cosmology, which suggests that a spatially compact universe can spontaneously nucleate out of 'nothing', where 'nothing' refers to a state with no classical space, time, or matter [2–7]. Most of the early work on quantum cosmology focused on spherical Friedmann-Robertson-Walker (FRW) universes. Zeldovich and Starobinsky [7] (hereafter ZS) considered the creation of a spatially flat universe with toroidal topology, discussing both quantum creation and the possibility that the universe was "'born classically' from a singularity." They made no attempt to compare the probabilities for the classical or quantum creation of toroidal universes, or the creation of spherical universes. The quantum creation of toroidal universes was further discussed by Coule and Martin [8], and by Linde [9], who argued that the probability of quantum creation of a toroidal universe is much greater than that of a spherical universe. In the present paper we re-examine this issue and argue that the toroidal universe describes a classical universe originating at a singularity, and not a quantum origin. We argue that the quantum creation of a toroidal universe from nothing cannot be described in the context of semiclassical quantum gravity, and therefore it is either impossible or is essentially dependent on Planck-scale physics. Since we do not interpret the quantum treatments that have been given for toroidal universes as describing creation from nothing, we see no foundation for comparing the probability for creation of toroidal universes with that of a spherical universe.

In quantum cosmology the universe is described by a single wave function ψ. This wave function is defined on the space of all possible 3-geometries and all matter field configurations, called superspace. The role of the Schrödinger equation for ψ is played by the Wheeler-DeWitt (WDW) equation

$$\mathcal{H}\psi = 0, \tag{1}$$

where $\mathcal{H}$ is the Hamiltonian operator.

In ordinary quantum mechanics, the wave function of a system is found by solving the Schrödinger equation with boundary conditions determined by the physical setup external to the system. But since there is nothing external to the universe, it appears that the boundary conditions for the WDW equation should be postulated as an independent physical law. Several possible forms of this law have been discussed in the literature; the most developed proposals are the Hartle-Hawking [3] and the tunneling [10, 11] boundary conditions.[1]

The tunneling boundary condition requires that ψ should include only outgoing waves at the boundary of superspace, except for the part of the boundary corresponding to vanishing 3-geometries. More precisely, incoming waves are allowed only on the part of the boundary corresponding to a vanishing 3-geometry which is nonsingular, in the sense that it can be described as a slice through a nonsingular four-geometry.[2] This is supplemented by the regularity condition, requiring that ψ remains finite everywhere, including the boundary. (See Refs. [10, 11] for more details.) Thus, the probability flux enters superspace through 3-geometries of vanishing size and leaves it through the rest of the boundary, corresponding to singular or infinitely large universes. The resulting wave function can be interpreted as describing a universe originating at zero size, that is, from 'nothing'.

The predictions resulting from the tunneling proposals have mostly been studied assuming a slowly varying potential $V(\phi)$, where ϕ is a scalar field. The wave function then predicts the nucleation with ϕ near the maximum of $V(\phi)$, which is just the right initial condition for inflation [10, 11].

The Hartle-Hawking wave function [3] is defined as a path integral over Euclidean histories bounded by a given 3-geometry and matter field configuration. This wave function has also been interpreted as describing quantum nucleation from nothing. In contrast to the tunneling wave function, the Hartle-Hawking wave function predicts that the universe is most likely to nucleate with the lowest value of $V(\phi)$ [3], which is not favorable for inflation. Various proposals have been advanced, however, to argue that the Hartle-Hawking wave function might nonetheless be compatible with inflation.[3]

It should be noted that, at the present level of understanding, both the Hartle-Hawking and the tunneling boundary conditions can only be implemented in the semi-classical approximation. The analyses of toroidal universes in Refs. [7–9] were based

---

[1] For early work closely related to the tunneling proposal, see Refs. [12, 4–7]. For discussions of the path integral approach to quantum cosmology, see [36–44]. In the present paper we use the WDW formalism of quan- tum cosmology, and will not be concerned with the subtleties of the path integral.

[2] A simple example of such a region is the slicing of a four-sphere along latitude lines near a pole—as the latitude approaches the pole, the slices are three-spheres with radii approaching zero and curvature approaching infinity, although the four-geometry remains completely regular. This condition is motivated by the idea that quantum nucleation of the universe should be described by a non-singular instanton. See Sect. III of Ref. [11] for more details.

[3] In 2002, Hawking and Hertog [13] introduced a "top-down approach," a form of the weak anthropic principle, to solve this problem. Alternatively, the Hartle-Hawking wave function might evolve into eternal inflation, which is believed to generically lead to predictions that are independent of the initial state [14, 15]. For a more recent discussion of these issues, see Maldacena, Ref. [16].

on the tunneling approach; hence we shall adopt the tunneling boundary condition in the present paper.

The paper is organized as follows. In the next section we review quantum cosmology of a spherical universe with a positive vacuum energy density. We first consider a minisuperspace model with a single degree of freedom (the radius of the universe), and then a perturbative superspace, including an infinite number of degrees of freedom of a quantum field. Quantum cosmology of a toroidal universe is discussed in the rest of the paper. A universe with a positive vacuum energy density is considered in Sect. 3 and the effects of including the Casimir energy are discussed in Sect. 4. Our conclusions are summarized and discussed in Sect. 5.

2 Quantum Creation of a Spherical Universe

2.1 A Minisuperspace Model

To illustrate the concept of quantum creation from nothing, we consider a simple model of a spherical universe with a constant vacuum energy density $\rho_v > 0$. We shall discuss this model in some detail, in order to compare it with the toroidal universe model analyzed in subsequent sections. The metric of our model is that of a closed FRW universe,

$$ds^2 = dt^2 - a^2(t)d\Omega_3^2, \tag{2}$$

where $d\Omega_3^2$ is the metric on a unit 3-sphere, the scale factor $a(t)$ satisfies

$$\dot{a}^2 + 1 = H_v^2 a^2, \tag{3}$$

and

$$H_v = (\rho_v/3)^{1/2}, \tag{4}$$

where we use Planck units with $8\pi G = 1$. The solution of Eq. (3) is de Sitter space,

$$a(t) = H_v^{-1}\cosh(H_v t). \tag{5}$$

It describes a universe contracting for $t < 0$, bouncing at $t = 0$ at radius $a = H_v^{-1}$, and expanding for $t > 0$.

In the quantum theory, the universe is described by the wave function $\psi(a)$, satisfying the Wheeler-DeWitt equation

$$\mathcal{H}\psi = 0. \tag{6}$$

The action for our model is

$$S = \int d^4x\sqrt{-g}\left[\frac{1}{2}R - \rho_v\right], \tag{7}$$

where R is the Ricci scalar, which for the metric of Eq. (2) is given by

$$R = 6(1 + \dot{a}^2 + a\ddot{a})/a^2. \tag{8}$$

Integrating over space and integrating $\ddot{a}$ by parts,

$$S = 3\Omega \int dt\,(1 - \dot{a}^2 - H_v^2 a^2)\,a, \tag{9}$$

where $\Omega = 2\pi^2$ is the volume of a unit 3-sphere. Then

$$p_a = -6\Omega a\dot{a} \tag{10}$$

is the momentum conjugate to a, and the Hamiltonian is

$$\mathcal{H} = -\frac{1}{12\Omega a}\left[p_a^2 + U(a)\right], \tag{11}$$

where

$$U(a) = (6\Omega)^2(1 - H_v^2 a^2)\,a^2. \tag{12}$$

With the replacement $p_a \to -i\hbar d/da$, the WDW equation takes the form

$$\left[\hbar^2\frac{d^2}{da^2} - U(a)\right]\psi(a) = 0. \tag{13}$$

Although we have set $\hbar \equiv 1$, we have written out $\hbar$ in Eq. (13) as a bookkeeping device, to make the WKB approximation more explicit. We have ignored the ordering ambiguity of the factors p_a and $1/a$, which is unimportant in the semiclassical approximation.

The WKB solutions of Eq. (13) in the classically allowed region, $a > H_v^{-1}$, are

$$\psi_\pm(a) = [p(a)]^{-1/2}\exp\left[\mp\frac{i}{\hbar}\int_{H_v^{-1}}^{a} p(a')da'\right], \tag{14}$$

where

$$p(a) \equiv [-U(a)]^{1/2}. \tag{15}$$

$\psi_+(a)$ and $\psi_-(a)$ correspond respectively to the expanding and contracting branches of the classical solution (5). (Note that the sign of p_a^2 in the Hamiltonian is the opposite from the usual situation, which affects the distinction between expansion and contraction.) The under-barrier, $a < H_v^{-1}$, solutions, $\tilde{\psi}_+(a)$ and $\tilde{\psi}_-(a)$,

$$\tilde{\psi}_{\pm}(a) = |p(a)|^{-1/2} \exp\left[\mp\frac{1}{\hbar}\int_a^{H_v^{-1}} |p(a')|da'\right] \tag{16}$$

respectively describe the branches of the wave function that are exponentially growing and decreasing as a function of a.

In the quantum creation scenario, the universe tunnels through the classically forbidden range from $a = 0$ to $a = H_v^{-1}$, and then evolves along the expanding branch of the de Sitter solution (5). The tunneling boundary condition requires that only the expanding (outgoing) component $\psi_+(a)$ should be present at $a > H_v^{-1}$. The amplitudes of the growing and decaying components $\tilde{\psi}_{\pm}(a)$ in the under-barrier region $0 < a < H_v^{-1}$ can be found from the WKB matching conditions. One finds that these components have comparable magnitudes at the classical turning point $a = H_v^{-1}$. The semiclassical nucleation probability can be estimated as

$$\mathcal{P} \sim e^{-\frac{1}{\hbar}|S_E|}, \tag{17}$$

where

$$|S_E| = 2\int_0^{H_v^{-1}} |p(a)|da = 8\pi^2/H_v^2. \tag{18}$$

is the Euclidean action of the instanton.

The tunneling can formally be described as evolution in Euclidean time, $\tau = -it$; then Eqs. (2) and (5) become

$$ds^2 = d\tau^2 + a^2(\tau)d\Omega_3^2, \tag{19}$$

$$a(\tau) = H_v^{-1}\cos(H_v\tau), \tag{20}$$

where we have changed the overall sign of the metric in Eq. (19), in order to conform to the standard convention. The quantity S_E in Eq. (17) is the Euclidean action of this instanton solution.

The Euclidean metric of Eqs. (19) and (20) is that of a 4-sphere of radius H_v^{-1}. It cannot be interpreted as describing evolution in ordinary time, but it is still meaningful to talk about the initial and final states (3-geometries) and about the succession of intermediate states corresponding to the 'most probable escape path' [17, 18]. In our model, the intermediate configurations are 3-spheres of radius a and 3-volume $\mathcal{V}(a) = 2\pi^2 a^3$. At the moment of nucleation the universe has the geometry of a 3-sphere of radius $a = H_v^{-1}$, and the initial state 'prior' to the tunneling corresponds to a vanishing radius, $a \to 0$. The total energy of matter also vanishes in this limit,

$$E(a) = 2\pi^2\rho_v a^3 \to 0 \quad (a \to 0). \tag{21}$$

This state with no classical space, time, or matter is what we are referring to as 'nothing'.

The semiclassical treatment is justified when $\rho_v \ll 1$, so that $H_v^{-1} \gg 1$ and $S_E \gg 1$. An important point to note is that even though the universe tunnels from zero size, this semiclassical description is not sensitive to Planck-scale physics. In particular, the quadratic and higher order curvature corrections are negligible in this case. The curvature radius of the instanton is H_v^{-1} and is everywhere large compared to 1.

In models involving a spatially homogeneous scalar field ϕ with a potential $V(\phi)$, the same kind of instanton describes tunneling to different extrema ϕ_j of the potential; the corresponding probabilities are given by $\mathcal{P}_j \propto \exp(-24\pi^2/V_j)$, where $V_j = V(\phi_j)$ and we assume that $V_j > 0$. If $V(\phi)$ is bounded from above, the most probable initial state is at the highest maximum of the potential. For a slowly varying potential, $V'(\phi) \ll V(\phi)$, approximate WKB solutions of the WDW equation for the wave function $\psi(a, \phi)$ have been found in Refs. [10] and [11]. The resulting probability distribution for the initial values of ϕ at nucleation is

$$\mathcal{P}(\phi) \sim \exp(-24\pi^2/V(\phi)). \tag{22}$$

The quantity in the exponent of Eq. (22) can be thought of as the Euclidean action of an approximate instanton solution, where the scalar field is held fixed at a given value ϕ.

2.2 *Perturbative Superspace*

Linear perturbations about spherical minisuperspace have been studied in Refs. [19–23]. The time-evolution of such perturbations is equivalent to the evolution in the quantum field theory of de Sitter space, but the tunneling boundary conditions for the cosmological wave function should uniquely specify the quantum states of the gravitational and matter fields. Refs. [19–23] collectively show that for a free scalar field of arbitrary mass, with a non-minimal $R\phi^2$ coupling to gravity, the tunneling condition uniquely specifies the full quantum state. With a plausible assumption to connect the behavior of the wave function for small field values to its behavior for large field values, this state is found to be the de Sitter-invariant (Bunch-Davies) state. Here we shall review the argument, using a conformally coupled massless field as an example. Having in mind application to models other than spherical de Sitter in subsequent sections, we shall keep the discussion as general as possible.

Including a free scalar field with non-minimal coupling to gravity, the action becomes

$$S = \int d^4x\sqrt{-g}\left[\frac{1}{2}R - \rho_v - \frac{1}{2}(\nabla\phi)^2 - \frac{1}{2}m^2\phi^2 - \frac{1}{2}\xi R\phi^2\right], \tag{23}$$

where conformal coupling corresponds to $\xi = \frac{1}{6}$. Introducing the definition $\phi(\mathbf{x}, t) \equiv \sqrt{6}\,\chi(\mathbf{x}, t)/a(t)$, using Eq. (8) for R, and integrating $\ddot{a}$ and $(\nabla\chi)^2$ by parts, one finds that the action for a massless conformally coupled field can be written as

$$S = \int d^4x\sqrt{-g}\,\frac{3}{a^4}\left[-a^2\dot{a}^2 + \frac{U(a)}{(6\Omega)^2} + a^2\dot{\chi}^2 + \chi\nabla^2\chi - \chi^2\right]. \tag{24}$$

Small excitations of the field about its vacuum value $\chi = 0$ can be expanded as

$$\chi(\mathbf{x}, t) = \sqrt{\Omega}\sum_{n,\ell,m} f_{n\ell m}(t)Q_{n\ell m}(\mathbf{x}), \tag{25}$$

where $Q_{n\ell m}$ are the harmonics on the 3-sphere, with $n = 1, 2, \ldots; \ell = 0, \ldots, n-1$; and $m = -\ell, -\ell+1, \ldots, \ell$. The harmonics are normalized by

$$\int d\Omega_3\, Q^*_{n'\ell'm'}(\mathbf{x})Q^*_{n\ell m}(\mathbf{x}) = \delta_{n'n}\delta_{\ell'\ell}\delta_{m'm}, \tag{26}$$

and satisfy

$$\nabla^2 Q_{n\ell m}(\mathbf{x}) \equiv g^{ij}\partial_i\partial_j Q_{n\ell m}(\mathbf{x}) = -(n^2-1)Q_{n\ell m}(\mathbf{x}). \tag{27}$$

The factor $\sqrt{\Omega}$, along with the factor of $\sqrt{6}$ in the definition of χ, are introduced so that the canonical momenta conjugate to the $f_{n\ell m}$ and to a appear in the Hamiltonian with coefficients of the same magnitude. Hereafter, the labels $\{n, \ell, m\}$ will be denoted simply by n. With all modes of the scalar field included, the wave function becomes a function of an infinite number of variables, $\psi(a, \{f_n\})$, and the action can be written as

$$S = \int dt\,\frac{3\Omega}{a}\left\{-a^2\dot{a}^2 + \frac{U(a)}{(6\Omega)^2} + \sum_n\left[a^2\dot{f}_n^2 - \omega_n^2 f_n^2\right]\right\}, \tag{28}$$

where

$$\omega_n^2 = n^2. \tag{29}$$

The canonical momentum conjugate to f_n is then

$$p_{f_n} = 6\Omega a\dot{f}_n, \tag{30}$$

and the Hamiltonian is then

$$\mathcal{H} = \frac{1}{12\Omega a}\left\{-p_a^2 - U(a) + \sum_n\left[p_{f_n}^2 + (6\Omega)^2\omega_n^2 f_n^2\right]\right\}. \tag{31}$$

The WDW equation takes the form

$$\left[\left(\hbar^2\frac{\partial^2}{\partial a^2} - U(a)\right) + \sum_n\left(-\hbar^2\frac{\partial^2}{\partial f_n^2} + (6\Omega)^2\omega_n^2 f_n^2\right)\right]\psi = 0\,. \tag{32}$$

This equation can be solved by treating f_n in lowest order approximation and by using the WKB approximation, writing the WDW wave function as[4]

$$\psi = e^{iS/\hbar}. \tag{33}$$

Writing the action as a power series in the f_n's:

$$S(a, \{f_n\}) = S_0(a) + \frac{1}{2}\sum_n S_n(a) f_n^2 + \mathcal{O}(f_n^3). \tag{34}$$

$S_0(a)$ and $S_n(a)$ are functions to be determined.

We substitute Eq. (34) into the WDW equation (32), neglect terms of order $\mathcal{O}(\hbar)$ and $\mathcal{O}(f_n^4)$, and then set each term in the power series in f_n equal to zero, obtaining the relations

$$\left(S_0{}'\right)^2 + U(a) = 0, \tag{35}$$

$$S_0{}' S_n{}' - S_n^2 - (6\Omega)^2 \omega_n^2 = 0, \tag{36}$$

where primes stand for derivatives with respect to a. Equation (35) is the Hamilton-Jacobi equation.

For our de Sitter model, the WDW potential $U(a)$ is given by Eq. (12), but the following analysis will apply to an arbitrary form of $U(a)$. We shall assume that the potential has a turning point a_0, where $U(a_0) = 0$, separating the classically allowed region at $a > a_0$ and classically forbidden region at $a < a_0$. In the classically allowed (cl) range, $U(a) < 0$, and we have

$$S_0^{\mathrm{cl}\,\prime} = -\sqrt{-U(a)}, \tag{37}$$

where the sign of the square root is selected by the tunneling boundary condition. We introduce the new variable

$$\eta(a) \equiv 6\Omega \int_{a_0}^{a} \frac{da'}{\sqrt{-U(a')}} = \int_0^t \frac{dt'}{a(t')}, \tag{38}$$

where the last equality, identifying η as the conformal time, follows from the classical equation $da/dt = \partial\mathcal{H}/\partial p_a$ and the fact that $\mathcal{H} = 0$. We can then rewrite Eq. (36) as

$$6\Omega \frac{dS_n^{\mathrm{cl}}}{d\eta} + S_n^{\mathrm{cl}\,2} + (6\Omega)^2 \omega_n^2 = 0. \tag{39}$$

[4] The WDW equation (32) can also be solved by separation of variables, but the method we use here is better suited for studying a wide class of possible quantum states. The relation between the two methods has been discussed in Ref. [21].

Equation (39) is a Riccati equation; it can be solved by the ansatz

$$S_n^{\rm cl} = 6\Omega \frac{\dot{u}_n^{\rm cl}}{u_n^{\rm cl}}, \tag{40}$$

where a dot now stands for a derivative with respect to η. Substituting this into Eq. (39), we obtain a linear equation for $u_n^{\rm cl}$:

$$\ddot{u}_n^{\rm cl} + \omega_n^2 u_n^{\rm cl} = 0. \tag{41}$$

The general solution of Eq. (41) is

$$u_n^{\rm cl} \propto e^{i\omega_n\eta} - B_n^{\rm cl} e^{-i\omega_n\eta}, \tag{42}$$

where $B_n^{\rm cl} = \text{const}$. Thus, we obtain

$$S_n^{\rm cl} = 6i\Omega\omega_n \frac{1 + B_n^{\rm cl} e^{-2i\omega_n\eta}}{1 - B_n^{\rm cl} e^{-2i\omega_n\eta}}. \tag{43}$$

The quantum state of the scalar field is completely specified by the choice of the functions $u_n(\eta)$, which play the role of the 'negative energy' mode functions, or equivalently, by the choice of the constants $B_n^{\rm cl}$.

In the under-barrier region, $U(a) > 0$, and in this case we have two relevant solutions to Eq. (35),

$$S_0^{\pm\prime} = \mp i\sqrt{U(a)}, \tag{44}$$

corresponding to the solutions $\tilde{\psi}_\pm(a)$ in Eq. (16). The full wave function of Eq. (33) will be generalized to be the sum of two terms, one of the form $e^{iS^+/\hbar}$, and one of the form $e^{iS^-/\hbar}$. In analogy to Eq. (38), we introduce the new variable

$$\zeta(a) \equiv 6\Omega \int_a^{a_0} \frac{da'}{\sqrt{U(a')}}. \tag{45}$$

Equation (36) can then be rewritten as

$$\pm 6i\Omega \frac{dS_n^\pm}{d\zeta} - S_n^{\pm 2} - (6\Omega)^2\omega_n^2 = 0. \tag{46}$$

The ansatz

$$S_n^\pm(\zeta) = \mp \frac{6\Omega i}{u_n^\pm(\zeta)} \frac{du_n^\pm(\zeta)}{d\zeta} \tag{47}$$

leads to the equation

$$\frac{d^2 u_n^\pm(\zeta)}{d\zeta^2} = \omega_n^2 u_n^\pm(\zeta). \tag{48}$$

We can write the general solution as

$$u_n^+(\zeta) \propto e^{\omega_n \zeta} - B_n^+ e^{-\omega_n \zeta}, \tag{49}$$

$$u_n^-(\zeta) \propto e^{-\omega_n \zeta} - B_n^- e^{\omega_n \zeta}, \tag{50}$$

which leads finally to the solution

$$S_n^{\pm}(\zeta) = 6i\Omega\omega_n \left[\frac{1 + B_n^{\pm} e^{\mp 2\omega_n \zeta}}{1 - B_n^{\pm} e^{\mp 2\omega_n \zeta}} \right]. \tag{51}$$

It was shown in Refs. [20, 23], following the WKB matching procedure developed in Ref. [17], that the f_n dependence of the wave function at the turning point, $\eta = \zeta = 0$, must be the same in all three branches—$e^{iS^{\rm cl}(a)/\hbar}$, $e^{iS^{+}(\zeta)/\hbar}$, and $e^{iS^{-}(\zeta)/\hbar}$. Thus, $B_n^{\rm cl} = B_n^+ = B_n^-$ for all n.

The constants B_n should in principle be determined by the outgoing-wave and regularity conditions at the boundaries of superspace, where $a \to 0$ or $a \to \infty$, or where $f_n \to \pm\infty$. This, however, cannot be implemented within the perturbative approach, since the expansion (34) breaks down at large values of f_n. Instead, we follow Refs. [20, 23] and require that the wave function does not grow towards large f_n, that is,

$$\text{Im}\, S_n(a) > 0 \text{ (for all } n \text{ and for all } a). \tag{52}$$

It appears that this is the best one can do to represent the boundary condition $\psi(|f_n| \to \infty) < \infty$.

Using Eq. (43), it is easily verified that in the classically allowed range the regularity condition (52) is equivalent to

$$|B_n| < 1. \tag{53}$$

In the under-barrier region, the corresponding condition is

$$|B_n| < e^{\mp 2\omega_n \zeta}. \tag{54}$$

Since $\zeta > 0$, the decreasing branch of the wave function (which corresponds to the lower sign in the exponent) does not contribute any additional constraints. The strongest constraint on the quantum state of the field comes from the growing branch:

$$|B_n| < e^{-2\omega_n \zeta_m}, \tag{55}$$

where ζ_m is the maximum value of ζ.

The above analysis applies to any closed FRW model with an arbitrary WDW potential $U(a)$. Turning now to a spherical de Sitter universe, we have $a_0 = H_v^{-1}$, and Eq. (45) gives

$$\zeta(a) = \cosh^{-1}\left(\frac{1}{H_v a}\right). \tag{56}$$

Hence $\zeta_m = \zeta(a{=}0) = \infty$, and the constraint (55) determines the values of B_n uniquely: $B_n = 0$ for all n. This selects the conformal vacuum, where the (negative-frequency) mode functions $u_n(\eta)$ are conformally related to the corresponding Minkowski space mode functions. For the de Sitter model this is the same as the Bunch-Davies state.

3 A Vacuum-Energy-Dominated Toroidal Universe

We now turn to quantum cosmology of a spatially flat universe having topology of a 3-torus. The metric ansatz is

$$ds^2 = dt^2 - a^2(t)d\mathbf{x}^2, \tag{57}$$

with the coordinates x^i taking values in the range $0 \le x^i \le 1$ and the surfaces $x^i = 0$ and $x^i = 1$ identified. That is, $(x^1 = 0, x^2, x^3)$ is identified with $(x^1 = 1, x^2, x^3)$, with similar identifications for x^2 and x^3. Assuming a positive vacuum energy density ρ_v, and no other relevant forms of energy, the classical Friedmann equation for $a(t)$ is

$$\dot{a}^2 = H_v^2 a^2 \tag{58}$$

with H_v from Eq. (4), and the solution is

$$a(t) = a_0 \exp(H_v t). \tag{59}$$

The universe begins having zero size as $t \to -\infty$, and Zeldovich and Starobinsky (ZS) interpret this solution as describing the quantum, or perhaps classical, creation of the universe. In contrast to a spherical universe, there is no barrier to penetrate, so there is no tunneling suppression factor. Coule and Martin [8], and also Linde [9], have suggested that this absence of suppression implies that nucleation of a toroidal universe is much more probable than that of a spherical universe.

It is also claimed in Refs. [7] and [9] that the toroidal universe with a scale factor (59) is geodesically complete. We, however, dispute this claim. We will give an argument below to show directly that the spacetime is geodesically incomplete, but we first point out that geodesic completeness would contradict a theorem proven by Arvind Borde and us. In Ref. [1], we showed that a spacetime must be geodesically incomplete if it contains at least one past-directed null or non-comoving timelike geodesic on which the average Hubble expansion rate H is positive. The theorem uses a generalized definition of H which is meaningful for any spacetime, but requires specifying the velocity vector of "comoving particles" at all points along the past-directed geodesic. The theorem does not require these "comoving particles" to exist, as long as the expansion of these "comoving particles" is used to define H. To apply the theorem to the toroidal universe of Eq. (57) (or to any Friedmann-Robertson-Walker spacetime, with or without periodic identifications), we can globally choose the "comoving particles" to be at rest in the standard coordinates (x^i = const), in which case the definition reduces to the usual $H = \dot{a}/a$. Thus, for the toroidal universe the value of H along any past-directed timelike or null geodesic would be H_v at all points, so the average would be H_v. Since $H_v > 0$, the theorem would imply geodesic incompleteness.[5]

After discussing the case with only vacuum energy, ZS point out that the energy-momentum tensor of quantum fields in a toroidal universe generally has a nonzero expectation value. In particular, they considered the effect of the Casimir energy density and found that it can, depending on the particle content, result in a classically forbidden range at small values of a. The corresponding tunneling action is finite, and ZS interpreted this as indicating that the tunneling can describe the quantum birth of a toroidal universe with Casimir energy. Below we will argue that the geodesic incompleteness causes this analysis to be outside the framework of quantum cosmology that we described in Sect. 2, resulting at best in a description with far less predictivity.

3.1 Past-Incompleteness

It is not difficult to show directly that the toroidal model of Eqs. (57) and (59) is actually past-incomplete. We consider first the covering space, described by the metric (57) without toroidal identification. This covering space is simply half of de Sitter space and is known (see, for example, Hawking and Ellis [25], pp. 125–126) to be past geodesically incomplete: all past-directed geodesics, except 'comoving' timelike geodesics x^i = const, reach $t = -\infty$ in a finite proper time. If we now consider the toroidal spacetime, we see that this statement is unchanged. Whenever a geodesic reaches a surface $x^i = 1$, it is continued at the identified point on the surface

[5] The theorem proven in Ref. [1] was recently disputed by Ref. [24], but those criticisms were based on a misunderstanding of the theorem. In Ref. [1], we concluded that "null and timelike geodesics are, in general, past-incomplete in inflationary models, whether or not energy conditions hold, provided only that the averaged expansion condition $H_{\text{av}} > 0$ holds along these past-directed geodesics." By H_{av} along a past-directed geodesic, we meant the average of H along the *whole* past-directed geodesic. We proved a finite bound on the integral of H along a null or non-comoving past-directed geodesic, so if the length of the geodesic were infinite, then H_{av} would have to be zero. In Ref. [24], however, the authors restated this condition (see Theorem 1 of that paper) as the mere requirement that the average value of H along any finite segment of the geodesic has to be strictly positive, which leads to a false statement to which they constructed counter-examples. For the case discussed here, $H = H_v$ at all points, so the average over the *whole* of each geodesic equals H_v, so the theorem of Ref. [1] shows that these geodesics cannot have infinite length.

$x^i = 0$ and vice versa. For a past-directed timelike geodesic starting at some time t_0 with a given 4-velocity, the coordinate time t as a function of proper time τ is clearly the same as it would be in the covering space. Since, in the covering space, all past-directed timelike geodesics except for the comoving ones reach $t = -\infty$ at a finite value of τ, the same will be true for the toroidal universe, implying that the spacetime is geodesically incomplete. The same argument can be used for past-directed null geodesics, using the affine parameter instead of proper time.

Thus, the past boundary defined by $t \to -\infty$ exhibits geodesic incompleteness, and is therefore singular in the sense of Ref. [25]. The classical toroidal model therefore describes a universe starting at a singularity and does not seem to address the problem of the initial conditions. For example, it does not determine the states of quantum fields, as we will show in Sect. 3.3.

3.2 Minisuperspace Quantization

To develop a quantum minisuperspace description of the toroidal model, all we need to do is to set

$$\Omega = \int_0^1 d^3x = 1, \tag{60}$$

and to replace Eqs. (8), (9), and (12) by

$$R = 6(\dot{a}^2 + a\ddot{a})/a^2, \tag{61}$$

$$S = -3\int dt\,(\dot{a}^2 + H_v^2 a^2)\,a, \tag{62}$$

and

$$U(a) = -36H_v^2 a^4. \tag{63}$$

Thus the WDW equation becomes

$$\left[\frac{d^2}{da^2} + (6H_v)^2 a^4\right]\psi = 0. \tag{64}$$

The general solution of this equation is a linear superposition of Hankel functions of the first and second kind [9],

$$\psi(a) = \sqrt{a}\left[\alpha H_{1/6}^{(1)}\left(2H_v a^3\right) + \beta H_{1/6}^{(2)}\left(2H_v a^3\right)\right], \tag{65}$$

where α and β are constants.

For $a \gg H_v^{-1/3}$, the two terms in Eq. (65) have a WKB form,

$$\psi(a) \propto e^{\pm 2i H_v a^3}, \tag{66}$$

with upper and lower signs corresponding to the first and second term, respectively. The outgoing wave as $a \to \infty$ corresponds to the "-" sign; hence we set $\alpha = 0$. Then, for $a \ll H_v^{-1/3}$ we have, up to a normalization constant,

$$\psi \approx c_1 + c_2 H_v^{1/3} a + c_3 H_v^2 a^6, \tag{67}$$

where the c_i's are constants of order 1. We see that ψ approaches a constant as $a \to 0$.

Coule and Martin [8] and Linde [9] note that the wave function (67) does not oscillate; hence the semiclassical description breaks down at small a.

The wave function (65) does not satisfy the tunneling boundary conditions as they were formulated in Sect. 2. It includes only the outgoing wave as $a \to \infty$, as desired, but the boundary at $a = 0$ corresponds to a torus of vanishing size, which cannot be obtained as a slice of a regular 4-geometry. Thus, the tunneling boundary condition does not allow quantum nucleation of a toroidal universe, at least not in the minisuperspace model considered in Refs. [7–9].

The authors of Refs. [7–9] nonetheless interpret the wave function for this model as a tunneling wave function, which amounts to dropping the condition that the probability flux must enter superspace through non-singular configurations. In the next subsection we shall consider applying this method in the context of perturbative superspace, showing that it leads to an almost complete loss of predictivity.

3.3 Perturbative Superspace

The perturbative superspace analysis of a toroidal universe, still considering only the vacuum-energy-dominated case, can now be performed along the lines of Sect. 2.2. The scalar field can now be represented as

$$\phi(\mathbf{x}, t) = \frac{1}{a(t)} \sum_{\mathbf{k}} f_{\mathbf{k}}(t) e^{i\mathbf{kx}}, \tag{68}$$

where the components of the vector $\mathbf{k}$ take values

$$k_i = 2\pi n_i, \tag{69}$$

where the n_i are integers. The WDW equation has the same form as Eq. (32) with

$$U(a) = -(6H_v a^2)^2, \tag{70}$$

$\Omega = 1$, and

$$\omega_n^2 = 4\pi^2(n_1^2 + n_2^2 + n_3^2). \tag{71}$$

Equations (33)–(43) still apply with the same substitutions. The main difference is that the WDW potential $U(a)$ is now negative for all values of a, and thus the entire range of a is classically allowed. The only constraint from the regularity condition is therefore given by Eq. (53), which means that the quantum state of the field is largely unconstrained.

In Refs. [22, 23] it was shown, for a massive non-minimally coupled field in a de Sitter minisuperspace model, that the regularity condition is satisfied in the entire classical range if it is satisfied at any one point in this range. In the toroidal model the entire range of a is classically allowed, and since the functions $S_n(a)$ are determined by the first order differential equation (36), it is clear that we can construct an infinite set of solutions consistent with the regularity condition: we just pick a point $a = a_1$ and any value $S_{n,1}$ with Im $S_{n,1} \geq 0$, and then solve Eq. (36) with that initial condition.

Thus, unlike the spherical case, in the vacuum-energy-dominated toroidal case the quantum state of the scalar field is not determined. As is the case for classical universe models, the singularity of the origin of the toroidal universe is associated with a lack of predictivity. Anything can collapse into a singularity, and since classical general relativity is time-reversal invariant, anything can emerge from a singularity. A singularity could produce a Friedmann-Robertson-Walker universe that matches our observed universe, but it could also produce *anything else.* For example, ZS considered an extension of the model allowing different expansion laws in the three directions of the torus,

$$ds^2 = dt^2 - A_1^2(t)dx^2 - A_2^2(t)dy^2 - A_3^2(t)dz^2, \tag{72}$$

and found that the classical solution has a Kasner-type singularity with $A_i(t) \propto t^{k_i}$ as $t \to 0$, where the Kasner indices k_i satisfy $k_1 + k_2 + k_3 = k_1^2 + k_2^2 + k_3^2 = 1$. A quantum version of this model can easily be constructed and solutions satisfying the outgoing wave condition can be found. There is an infinity of solutions corresponding to different values of k_i, and the outgoing wave condition does not give preference to any of them. In fact, any classical cosmological model beginning at a singularity can be recast in the form of a WKB wave function, but this does not bring us any closer to determining the initial state of the universe.

4 Tunneling with Casimir Energy

Let us now consider the effect of Casimir energy on the quantum dynamics of a toroidal universe. Zeldovich and Starobinsky (ZS), citing Refs. [12, 26–29], state that for massless conformally coupled fields in a toroidal universe described by the metric (57) with the stated identifications, the Casimir contribution to the energy-momentum tensor has the form

$$T_\nu^{(C)\mu} = -\frac{C}{a^4}\text{diag}(1, -1/3, -1/3, -1/3). \tag{73}$$

Note that it is traceless and covariantly conserved: $T_{\nu}^{(C)\nu} = 0$, $T_{\mu;\nu}^{(C)\nu} = 0$. ZS give the coefficient C for a real scalar field as $C_0 = 0.8375$, for a vector field $C_1 = 2C_0$, and for a chiral spin-1/2 field with periodic boundary conditions $C_{1/2} = -C_0$. In general, for non-interacting conformal fields,

$$C = (N_b - N_f)C_0, \tag{74}$$

where $N_b = N_0 + 2N_1$ and $N_f = 2N_{1/2}$ are the numbers of bosonic and fermionic spin degrees of freedom, with N_0, $N_{1/2}$, and N_1 being respectively the numbers of (real) scalar, chiral spinor, and vector fields in the model. ZS note that the Casimir energy vanishes in supersymmetric models, where $N_b = N_f$. On the other hand, for a spin-1/2 field with antiperiodic boundary conditions, as studied in Ref. [30], ZS state that $C_{1/2} = 0.3914$, so there is no cancellation.[6]

In a general FRW spacetime, the expectation value of the energy-momentum tensor includes, in addition to the Casimir term (73), contributions depending on the time derivatives of the scale factor $a(t)$, associated with the trace anomaly. In this paper, however, we will follow ZS and include only the Casimir contribution.

4.1 Minisuperspace Quantization

ZS assume that the Casimir energy density is nonzero, either because the particle theory is not supersymmetric, or due to antiperiodic boundary conditions for fermions, so that

$$\rho_C = -\frac{C}{a^4} \tag{75}$$

with $C > 0$. Then the Friedmann equation takes the form

$$\left(\frac{\dot{a}}{a}\right)^2 = H_v^2 - \frac{C}{3a^4}. \tag{76}$$

This has a 'bouncing' solution

$$a(t) = a_{\min} \cosh^{1/2}(2H_v t). \tag{77}$$

[6] The numerical value of C_0 is given slightly differently in the papers ZS refer to. In particular, Ref. [28] gives it as 0.833, but the paper also contains an analytic formula, found as a special case of Eq. (4), which becomes $C_0 = \frac{\pi^2}{90} + \frac{\zeta(3)}{2\pi} + \frac{\pi}{6}$, which evaluates as 0.8246. These numbers are all close to each other, but in any case the numerical value is not relevant to our calculations. We also note that these Casimir energy calculations apply to a static universe. Saharian and Mkhitaryan [31] carried out the corresponding calculation for a power-law expanding universe, with $a(t) \propto t^p$, and found, as one would expect, that it agrees with the static calculation in the limit of small a.

with

$$a_{\min} = \left(\frac{C}{3H_v^2}\right)^{1/4}. \tag{78}$$

The range $0 < a < a_{\min}$ is classically forbidden, due to the negative Casimir energy, but the universe can tunnel through this barrier.

To find the semiclassical tunneling amplitude, we write the Lagrangian of the model as

$$\mathcal{L} = -3a\dot{a}^2 - 3H_v^2 a^3 + \frac{C}{a}, \tag{79}$$

which leads to a Hamiltonian

$$\mathcal{H} = -\frac{p_a^2}{12a} + 3a^3 H_v^2 - \frac{C}{a}, \tag{80}$$

where the momentum conjugate to the scale factor a is

$$p_a = -6\dot{a}a = -6H_v \left(a^4 - a_{\min}^4\right)^{1/2}. \tag{81}$$

The under-barrier Euclidean action is

$$|S_E| = 2\int_0^{a_{\min}} |p_a| da = A(C^3/H_v^2)^{1/4}, \tag{82}$$

where

$$A = 4\,(3)^{1/4} \int_0^1 dx(1 - x^4)^{1/2}. \tag{83}$$

The integral in the last formula equals $\sqrt{\pi}\Gamma(1/4)/6\Gamma(3/4)$, and we obtain

$$A = \left(\frac{16\pi^2}{27}\right)^{1/4} \frac{\Gamma(1/4)}{\Gamma(3/4)} \approx 4.6, \tag{84}$$

in agreement with ZS.

Equation (82) can be compared with the tunneling action (18) for a spherical universe. The tunneling action is smaller for the toroidal universe as long as

$$H_v < \left(\frac{8\pi^2}{A}\right)^{2/3} \frac{1}{\sqrt{C}} \approx \frac{6.65}{\sqrt{C}}. \tag{85}$$

The constant C can be estimated as

$$C \lesssim N, \tag{86}$$

where N is the number of quantum fields in the model. Assuming that N is not excessively large, this seems to suggest that the tunneling probability (17) is much greater for a toroidal universe than for a spherical one, as long as H_v is well below the Planck scale. This is a potentially important conclusion, although ZS do not emphasize it in their paper.

However, the wave function for this toroidal model cannot satisfy the tunneling boundary condition for the same reason as before. In fact, the configurations as $a \to 0$ in this case are even more singular than for a pure de Sitter model. The total Casimir energy in a toroidal universe is

$$E(a) = \rho_C a^3 = -C/a, \quad (87)$$

which approaches $-\infty$ as $a \to 0$. So the tunneling here is not from 'nothing', but rather from a state with an infinite expectation value of the total energy of matter. This situation should be contrasted with the nucleation of a spherical universe, in which case the total energy of matter vanishes in the limit $a \to 0$. (Note that for $H_v \ll 1$, Casimir and other quantum corrections to the energy density are negligible in the spherical case.)

The divergence of energy of a toroidal universe at $a \to 0$ is reflected in the behavior of the instanton describing the tunneling. Analytic continuation of Eq. (77) gives

$$a(\tau) = a_{\text{min}} \sin^{1/2}(2H_v\tau), \quad (88)$$

where the origin for the Euclidean time τ is chosen so that $\tau = 0$ at $a = 0$. At small values of τ,

$$a(\tau) \propto \tau^{1/2}, \quad (89)$$

resulting in a curvature singularity. Singular instantons in quantum cosmology have been discussed by Hawking and Turok [32], but it was argued in Ref. [33] that such instantons are not actually stationary points of the Euclidean action, since they fail to satisfy the equations of motion at the singular point. Reference [33] also points out that if such singular instantons were treated as valid stationary points, then it would be possible to construct a singular instanton which would lead to the rapid decay of flat space.[7] Furthermore, as in the pure de Sitter case, the singularity leads to the lack of predictivity, as we shall now discuss.

---

[7] It has been shown in Refs. [34] and [35] that the Hawking-Turok instanton can be obtained as a $4D$ section of a non-singular instanton in 5 dimensions. Here we are concerned only with cosmology in $4D$, so this $5D$ extension is not relevant for our discussion.

4.2 Perturbative Superspace

As before, we consider a massless conformally coupled field with a mode expansion (68). The WDW equation for the perturbative superspace model has the form of Eq. (32) with

$$U(a) = (6H_v)^2(a_{\min}^4 - a^4), \tag{90}$$

$\Omega = 1$, $a_{\min}$ from Eq. (78) and ω_n from Eq. (71). The turning point of the potential is at $a = a_{\min}$, and the range $0 < a < a_{\min}$ is classically forbidden. The analysis of Sect. 2.2 still applies, so the regularity condition imposes the following constraints on the constants B_n:

$$|B_n| \leq e^{-2\omega_n \zeta_m}, \tag{91}$$

where ζ_m is the maximum value of the Euclidean conformal time ζ.

From Eq. (45), ζ_m is given by

$$\zeta_m = \frac{1}{H_v}\int_0^{a_{\min}} \frac{da}{\sqrt{a_{\min}^4 - a^4}} = \left(\frac{3}{CH_v^2}\right)^{1/4} F, \tag{92}$$

where

$$F = \int_0^1 \frac{dx}{\sqrt{1-x^4}} = \frac{\sqrt{\pi}\,\Gamma(1/4)}{4\,\Gamma(3/4)} \approx 1.31. \tag{93}$$

For $H_v \ll 1$ and C not too large, we have $\zeta_m \gg 1$. Then the bound (91) restricts the values of B_n to a narrow range around zero, but it does not determine them completely. The width of the allowed range of B_n decreases exponentially towards larger values of ω_n, so the allowed states may be observationally indistinguishable from the Bunch-Davies state. But, as a matter of principle, the quantum state should be uniquely determined by the boundary conditions. Note also that a semiclassical regime with $CH_v^2 \gg 1$ could in principle exist. Then we could have $\zeta_m \ll 1$ and a large uncertainty in B_n for a large range of values of n.

If we consider the limit in which the Casimir energy approaches zero from above, $C \to 0+$, Eq. (92) implies that $\zeta_m \to \infty$, which means that the quantum state is uniquely determined. However, in Sect. 3.3, we analyzed the case $C = 0$, and found that the quantum state is definitely not determined uniquely. While limits are sometimes discontinuous, we suggest that this peculiar behavior is more likely just another symptom of incorrectly treating the singular instanton as if it is a legitimate stationary point of the action.

5 Conclusions

The quantum creation of toroidal universes has been studied in a number of previous papers, including Zeldovich and Starobinsky (ZS) [7], Coule and Martin [8], and Linde [9]. In this paper we reach conclusions that differ from any of these papers.

Two of these papers, ZS and Linde, indicate that a toroidal universe—a spacetime foliated with flat equal-time slices with periodic identifications in all three directions, with the energy density dominated by vacuum energy—is geodesically complete. Here we argued that this conclusion is not true. To see this, we described the toroidal universe as de Sitter space in a spatially flat coordinate system, with periodic identifications. In the full de Sitter space, the flat slicing covers only half the manifold, and any non-comoving backward-going timelike trajectory reaches $t = -\infty$ in a finite amount of proper time. The periodic identifications cannot change that. A similar statement holds for all backward-going lightlike trajectories, with proper time replaced by the affine parameter. This lack of geodesic completeness means that classically the cosmological model begins with a singularity. The evolution is not unique, unless one fixes initial conditions that control what comes out of the singularity.

The presence of the initial singularity is highly relevant, because since the early work of Friedmann and Lemaître in the 1920's, we have had mathematical models of a classical universe that begins with a singularity. But such models have not been considered serious candidates for a theory of the origin of the universe, largely because of their lack of predictivity. Essentially anything can emerge from a singularity.

In this paper we argued that the toroidal universe describes nothing more than a classical universe originating at a singularity, and not a quantum origin of the universe. Our conclusion is that the quantum creation of a toroidal universe from nothing cannot be described in the context of semiclassical quantum gravity. If it is possible at all, it depends on Planck-scale physics which is currently out of reach.

To support this conclusion, we began by reviewing the formalism for the quantum creation of spherical universes, and then considered the model of a toroidal universe dominated by vacuum energy alone. We gave a more detailed discussion of the geodesic incompleteness of this model, and then discussed its quantization, first in the context of a "minisuperspace" model in which the scale factor a is treated as the only degree of freedom. We constructed the Wheeler-DeWitt equation and its WKB solution, showing explicitly that the singular behavior of the wave function $\psi(a)$ as $a \to 0$ places it outside the formalism developed for the spherical case. We then went on to study a superspace extension, with a massless conformally coupled scalar field, treated perturbatively. We showed, as expected, that the quantum state of the scalar field is not determined by the Wheeler-DeWitt equation plus boundary conditions. We argued that a valid quantum tunelling formalism should completely determine the quantum state, as we had found for the case of tunneling to a spherical universe.

Following the treatment by Zeldovich and Starobinsky, we then considered the possibility of including Casimir energy, in addition to vacuum energy. This addition to the equations results in a classically forbidden region of scale factors, from $a = 0$ to some $a = a_{\text{min}}$. This situation has a much closer resemblance to the spherical universe tunneling calculation, and leads to a WKB estimate of a tunneling rate which could, depending on parameters, make it much more probable than tunneling to a spherical universe. However, there are clear problems that lead us to believe that this calculation is not valid. First, the instanton describing the tunneling is singular at $a = 0$, with the total energy of the universe actually diverging as $a \to 0$. The corresponding instanton therefore has a curvature singularity at $a = 0$. The curvature singularity implies that the instanton does not satisfy the equations of motion at $a = 0$, which in turn means that it is not a valid stationary point of the Euclidean action. In Ref. [33] it was shown that if singular instantons of this type were treated as valid stationary points, then a singular instanton could be constructed that would describe the rapid decay of flat space.

As in our treatment of the vacuum-energy-only model, we extended our treatment of Casimir energy to a perturbative treatment of a model that includes a massless conformally coupled scalar field. As in the vacuum-energy-only case, we found that the quantum state of the scalar field was not determined by the Wheeler-DeWitt equation plus boundary conditions.

Thus, we are convinced that there is no valid reason to believe that toroidal universes should dominate the quantum creation of universes from nothing, as has sometimes been claimed. To the contrary, the quantum creation of a toroidal universe from nothing is not possible within the semiclassical quantum gravity description. Such quantum creation is either completely impossible, or perhaps it could take place through Planck-scale processes beyond our current understanding.

Acknowledgements A.H.G. is supported in part by the U.S. Department of Energy under Grant DE-SC0012567.

References

1. A. Borde, A.H. Guth, A. Vilenkin, Inflationary space-times are incomplete in past directions. Phys. Rev. Lett. **90**, 151301 (2003). arXiv:gr-qc/0110012
2. A. Vilenkin, Creation of universes from nothing. Phys. Lett. B **117**, 25 (1982)
3. J.B. Hartle, S.W. Hawking, Wave function of the universe. Phys. Rev. D **28**, 2960 (1983)
4. A.D. Linde, Quantum creation of the inflationary universe. Lett. Nuovo Cim. **39**, 401 (1984)
5. V.A. Rubakov, Quantum mechanics in the tunneling universe. Phys. Lett. B **148**, 280 (1984)
6. A. Vilenkin, Quantum creation of universes. Phys. Rev. D **30**, 509 (1984)
7. Y.B. Zeldovich, A.A. Starobinsky, Quantum creation of a universe in a nontrivial topology. Sov. Astron. Lett. **10**, 135 (1984)
8. D.H. Coule, J. Martin, Quantum cosmology and open universes. Phys. Rev. D **61**, 063501 (2000). arXiv:gr-qc/9905056
9. A.D. Linde, Creation of a compact topologically nontrivial inflationary universe. JCAP **0410**, 004 (2004). arXiv:hep-th/0408164
10. A. Vilenkin, Boundary conditions in quantum cosmology. Phys. Rev. D **33**, 3560 (1986)
11. A. Vilenkin, Quantum cosmology and the initial state of the universe. Phys. Rev. D **37**, 888 (1988)

12. B.S. DeWitt, Quantum theory of gravity. 1. The canonical theory. Phys. Rev. 160, 1113 (1967)
13. S.W. Hawking, T. Hertog, Why does inflation start at the top of the hill? Phys. Rev. D **66**, 123509 (2002). (hep-th)
14. A.D. Linde, D.A. Linde, A. Mezhlumian, From the Big Bang theory to the theory of a stationary universe. Phys. Rev. D 49, 1783–1826 (1994). arXiv:gr-qc/9306035 [gr-qc]
15. A. De Simone, A.H. Guth, M.P. Salem, A. Vilenkin, Predicting the cosmological constant with the scale-factor cutoff measure. Phys. Rev. D 78, 063520 (2008). arXiv:0805.2173 [hep-th]. (See especially Appendix A)
16. J. Maldacena, *Comments on the no boundary wavefunction and slow roll inflation (hep-th)*
17. T. Banks, C.M. Bender, T.T. Wu, Coupled anharmonic oscillators. 1. Equal mass case. Phys. Rev. D **8**, 3346 (1973)
18. A. Vilenkin, S. Winitzki, Detection of particles under potential barrier. Phys. Rev. D **50**, 5409 (1994). arXiv:hep-th/9404163
19. J.J. Halliwell, S.W. Hawking, The origin of structure in the universe. Phys. Rev. D **31**, 1777 (1985)
20. T. Vachaspati, A. Vilenkin, On the uniqueness of the tunneling wave function of the universe. Phys. Rev. D **37**, 898 (1988)
21. J. Hong, A. Vilenkin, S. Winitzki, Particle creation in a tunneling universe. Phys. Rev. D **68**, 023520 (2003). arXiv:gr-qc/0210034
22. J. Hong, A. Vilenkin, S. Winitzki, Creation of massive particles in a tunneling universe. Phys. Rev. D **68**, 023521 (2003). arXiv:gr-qc/0305025
23. S.J. Wang, M. Yamada, A. Vilenkin, Constraints on non-minimal coupling from quantum cosmology. JCAP **08**, 025 (2019)
24. J.E. Lesnefsky, D.A. Easson, P.C.W. Davies, Past-completeness of inflationary spacetimes. Phys. Rev. D **107**(4), 044024 (2023). (arXiv:2207.00955 [gr-qc])
25. S.W. Hawking, G.F.R. Ellis, *The Large Scale Structure of Space-Time* (Cambridge University Press, Cambridge, 1973)
26. A.A. Starobinsky, in Classical and Quantum Gravitation Theory [in Russian] (abstracts, 4th All-Union Gravitation Conf., Minsk, July 1976), Inst. Phys. Beloruss. Acad. Sci., Minsk (1976), p. 110
27. S.G. Mamaev, N.N. Trunov, Dependence of the vacuum expectation values of the energy momentum tensor on the geometry and topology of the manifold. Theor. Math. Phys. 38, 228 (1979). (Teor. Mat. Fiz. 38, 345 (1979))
28. S.G. Mamaev, N.N. Trunov, Vacuum expectation values of the energy momentum tensor of quantized fields on manifolds with different topologies and geometries. II, Sov. Phys. J. 22, 966 (1980). (Ivz. Vyssh. Uchebn. Zaved., Fizika, No. 9, 51 (1979))
29. R. Banach, J.S. Dowker, The vacuum stress tensor for automorphic fields on some flat space-times. J. Phys. A **12**, 2545 (1979)
30. L.H. Ford, Vacuum polarization in a nonsimply connected space-time. Phys. Rev. D **21**, 933 (1980)
31. A.A. Saharian, A.L. Mkhitaryan, Vacuum fluctuations and topological Casimir effect in Friedmann-Robertson-Walker cosmologies with compact dimensions. Eur. Phys. J. C **66**, 295 (2010). (hep-th)
32. S.W. Hawking, N. Turok, Open inflation without false vacua. Phys. Lett. B **425**, 25 (1998). (arXiv:hep-th/9802030)
33. A. Vilenkin, Singular instantons and creation of open universes. Phys. Rev. D **57**, 7069 (1998). arXiv:hep-th/9803084
34. J. Garriga, *Smooth 'creation' of an open universe in five-dimensions*. arXiv:hep-th/9804106
35. J.J. Blanco-Pillado, H.S. Ramadhan, B. Shlaer, Bubbles from Nothing. JCAP **1201**, 045 (2012). (arXiv:1104.5229 [gr-qc])
36. J. Feldbrugge, J. L. Lehners and N. Turok, Lorentzian Quantum Cosmology. Phys. Rev. D **95**(10), 103508 (2017). https://doi.org/10.1103/PhysRevD.95.103508 (arXiv:1703.02076 [hep-th])

37. J. Feldbrugge, J. L. Lehners and N. Turok, No smooth beginning for spacetime. Phys. Rev. Lett. **119**(17), 171301 (2017) https://doi.org/10.1103/PhysRevLett.119.171301 (arXiv:1705.00192 [hep-th])
38. J. L. Lehners, No smooth beginning for spacetime. Int. J. Mod. Phys. D **28**(02), 1930005 (2018). https://doi.org/10.1142/9789811258251_0005 (arXiv:1705.05340 [gr-qc])
39. J. Diaz Dorronsoro, J. J. Halliwell, J. B. Hartle, T. Hertog and O. Janssen, Real no-boundary wave function in Lorentzian quantum cosmology. Phys. Rev. D **96**(4), 043505 (2017). https://doi.org/10.1103/PhysRevD.96.043505 (arXiv:1705.05340 [gr-qc])
40. J. Feldbrugge, J. L. Lehners and N. Turok, Inconsistencies of the New No-Boundary Proposal. Universe **4**(10), 100 (2018). https://doi.org/10.10.3390/universe4100100 (arXiv: 1805.01609 [hep-th])
41. J. Feldbrugge, J. L. Lehners and N. Turok, No rescue for the no boundary proposal: Pointers to the future of quantum cosmology. Phys. Rev. D **97**(2), 023509 (2018) https://doi.org/10.1103/PhysRevD.97.023509 (arXiv:1708.05104 [hep-th])
42. J. Diaz Dorronsoro, J. J. Halliwell, J. B. Hartle, T. Hertog, O. Janssen and Y. Vreys, Damped perturbations in the no-boundary state. Phys. Rev. Lett. **121**(8), 081302 (2018). https://doi.org/10.1103/PhysRevLett.121.081302 (arXiv: 1804.01102 [gr-qc])
43. A. Vilenkin and M. Yamada, Tunneling wave function of the universe.Phys. Rev. D **98**(6), 066003 (2018). https://doi.org/10.1103/PhysRevD.98.066003(arXiv: 1808.02032 [gr-qc])
44. A. Vilenkin and M. Yamada, Tunneling wave function of the universe II: the backreaction problem. Phys. Rev. D **99**(6), 066010 (2019). https://doi.org/10.1103/PhysRevD.99.066010 (arXiv: 1812.08084 [gr-qc])

From a Quantum World to Our Classical Universe

Claus Kiefer

Abstract Modern cosmological theories invoke the idea that all structure in the Universe originates from quantum fluctuations. Understanding the quantum-to-classical transition for these fluctuations is of central importance not only for the foundations of quantum theory, but also for observational astronomy. In my contribution, I review the essential features of this transition, emphasizing in particular the role of Alexei Starobinsky.

1 Introduction

Astronomical observations show the presence of structure in the Universe. Visible and dark matter are grouped into galaxies and clusters of galaxies. An impressive example is displayed in Fig. 1. The photo was made in 12.5 h of exposure with the James Webb Space Telescope. It shows members of a distant galaxy cluster at about 4.6 billion lightyears away. In addition, one observes luminous arcs that are generated by the dark matter dominated mass of this cluster via gravitational lensing and that originate from a much more distant galaxy at about 9.5 billion lightyears.

An understanding of the origin of structure must be based on an understanding of gravity. In fact, much work was and is devoted to doing (mostly numerical) calculations using general relativity or its Newtonian limit in order to understand the evolution of structure from initial conditions.

But what are these initial conditions? The perhaps most promising candidate for a theory of the early Universe is *inflation*, the idea that the Universe underwent a quasi-exponential expansion at a very early time, but still much later than the Planck era. Alexei Starobinsky, in the memory of whom this article is written, was one of the pioneers of this idea.

C. Kiefer (✉)
Faculty of Mathematics and Natural Sciences, Institute for Theoretical Physics,
University of Cologne, Cologne, Germany
e-mail: kiefer@thp.uni-koeln.de

A. Barvinsky and A. Kamenshchik (eds.), *Open Issues in Gravitation and Cosmology*,
Fundamental Theories of Physics 225, https://doi.org/10.1007/978-3-032-15703-4_9

Fig. 1 Webb's First Deep Field. This view of the early Universe was achieved with the NIRCam instrument of the James Webb Spacer Telescope in 2022. Image Credit: NASA, ESA, CSA, STcI, NIRCam

A main feature of inflation is that structure has emerged from initial *quantum* fluctuations of the spacetime metric and a hypothetical inflation field. But this raises the fundamental question of how the transition from quantum to classical behaviour happened and which physical, potentially observational, consequences this transition ensued. It was this question that initiated my collaboration with Alexei and our colleague David Polarski. The perhaps most exciting aspect in our work was (and still is) the insight that an understanding of the foundations of quantum theory and in particular the formation of entanglement between quantum states is needed to understand the origin of galaxies and galaxy clusters.

In this article, I will focus on our work on the decoherence of the primordial fluctuations. I have enjoyed working with Alexei also on other topics, such as Higgs inflation and the cosmological constant, which I will leave out here.

My article is organized as follows. In Sect. 2, I will focus on the quantum issues in the inflationary scenario and state the problem of the quantum-to-classical transition. This problem is dealt with (and solved, in my opinion) by invoking the process of decoherence, which is familiar from ordinary quantum theory, where it was experimentally confirmed. Section 3 gives a brief introduction to decoherence with emphasis on its relevance for quantum cosmology. Section 4 then explains the decoherence of the primordial fluctuations and its consequences for, in particular, their entropy. Finally, Sect. 5 focuses on the relevance of these investigations for the arrow of time in our Universe.

2 The Inflationary Universe

Alexei considered an exponential expansion of the early Universe already in [1]. He was mainly motivated by symmetry arguments and explored the idea that the "universe was in a maximum symmetrical quantum state before the beginning of the classical Friedman expansion". This brief statement involves the ideas that the Universe early on (before the radiation-dominated phase) was in a quantum state and that this state had maximum symmetry. Since he assumed this maximum symmetry state to correspond to a spacetime solution different from Minkowski space, he is led to considering a quantum state in de Sitter space.

The only information from such a hypothetical early state, so the argument, can come from gravitons (corresponding to gravitational radiation in the semiclassical limit). Detection of such primordial gravitons could even provide information about a fundamental physical theory describing the symmetric state. Alexei concluded his paper with the words:

> It is remarkable that observation of the spectrum [of gravitons] would not only allow us to determine the initial state of the universe, but also to measure experimentally the s constant[1] which is fundamental for a unified theory of interactions.

So far, primordial gravitons have not been detected, but their observation at least by indirect means belongs to the central goals of work on inflation. The primordial gravitons, together with the primordial density fluctuations, also play a central role in our research on decoherence reviewed below.

Ideas about an initial state of maximum symmetry are, in fact, old and can be traced back to the pre-Socratics. Anaximander, who lived in the sixth century BC in Miletus, postulated the *apeiron* as the source of all things.[2] He also used symmetry arguments to conclude that the Earth must be round and freely floaring in empty space.[3] Anaxagoras, another Greek philosopher, who lived in the fifth century BC partly in Athens, entertained the idea of a homogeneous initial state out of which all constituents of the later cosmos would emerge.

To implement his ideas of maximum symmetry, Alexei considered a model of exponential expansion that is based on one-loop quantum-gravitational corrections to the Einstein field equations [4]. This model is now called *Starobinsky inflation*

---

[1] In [1], this constant is defined as $s = H_0 l_P$, where H_0 is the Hubble parameter of inflation, and l_P is the Planck length.

[2] Cf. the remarks in [2], in particular footnote 1.

[3] Cf. the excellent discussion in [3].

and constitutes one of the best-fit models to observations of the PLANCK satellite and other missions. A selection by observation is of utmost importance, since here exist many models of inflation, even if one restricts attention to the observationally preferred slow-roll single field models [5].

A major issue in inflation is the calculation of the power spectrum for scalar and tensor fluctuations in de Sitter space; see, for example, [6] for an extensive review and references. Besides his already mentioned work on the tensor (graviton) fluctuations, Alexei also presented an early calculation of the adiabatic amplitude of adiabatic scalar fluctuations generated during inflation [7].

The standard picture for the evolution of the fluctuations is shown in Fig. 2 and explained in the caption. The essential point is that a causal evolution is possible—a fluctuation with a certain wavelength λ starts from within the Hubble scale during inflation, leaves the Hubble scale during this phase and later re-enters the Hubble scale during the radiation (or matter-) dominated phase. The power spectra after re-entry then serve as initial conditions for the formation of structure as seen, for example, in Fig. 1.

How do we describe the emergence and evolution of quantum fluctuations during an inflationary phase? From a fundamental point of view, we should start with a

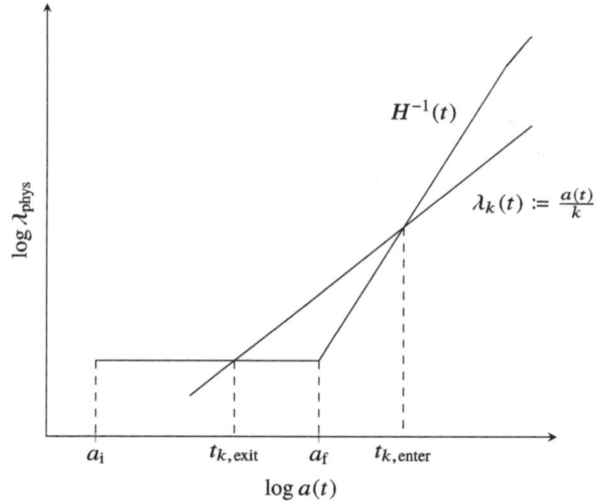

Fig. 2 Time development of a physical scale $\lambda_k(t) := a(t)/k$, where $a(t)$ is the scale factor of a Friedmann universe and k is the dimensionless wave number, and the Hubble scale $H^{-1}(t)$. During an inflationary phase, $H^{-1}(t)$ remains approximately constant. After the end of inflation (a_f), the Hubble scale $H^{-1}(t)$ increases faster than any physical scale. Therefore the scale described by λ_k, which has left the Hubble scale at time $t_{k,\mathrm{exit}}$, enters the Hubble scale again at $t_{k,\mathrm{enter}}$ in the radiation- (or matter-) dominated phase. This figure was produced by using real data from the PLANCK satellite and using the astropy package in python, see www.astropy.org/acknowledging. I am grateful to Yi-Fan Wang for producing this figure. Reprinted with permission from [8].

quantum theory of gravity [8]. Such a theory is not yet available in complete form, but there exist various approaches on which a discussion can be based. One of the most suitable approach for our purpose is quantum geometrodynamics, with the Wheeler–DeWitt equation playing the central role [8, 9]. In quantum geometrodynamics, it is possible to perform a consistent semiclassical approximation scheme, which is sufficient for the treatment of the inflationary fluctuations. Symbolically, one can write the total quantum state of gravity Ψ, which is a solution of the Wheeler–DeWitt equation, in the form

$$\Psi \propto \exp(\mathrm{i} m_{\mathrm{P}}^2 S_0[\text{gravity}])\psi[\text{gravity, matter}], \tag{1}$$

where m_{P} denotes the Planck mass. Performing an expansion with respect to the Planck mass, one finds that S_0 obeys the Hamilton–Jacobi equation of gravity (providing a semiclassical background allowing, in particular, the introduction of a time parameter) and that ψ satisfies a (functional) Schrödinger equation for quantum fields on that background [8, 10]. Higher orders in the Planck-mass expansion yield quantum-gravitational correction terms to the Schrödinger equation [11]. For the discussion below, we will not need these correction terms; they are important for searches of genuine quantum-gravitational effects in cosmology such as the modification of the CMB anisotropy spectrum at large scales [10, 12].

The quantum fluctuations in inflation can be described by the "Mukhanov–Sasaki variable", which is a gauge-invariant combination of the scale factor a, the homogeneous part of the inflaton field, and their perturbations [6]. For our discussion of decoherence, it is not necessary to employ the full gauge-invariant formalism; it is sufficient to describe them by a massless scalar field ϕ in a Friedmann–Lemaître universe (here chosen to be spatially flat). This formalism can encompass both scalar and tensor modes. Since the fluctuations are small, there mutual interaction is negligible, so they can be treated as being independent of each other and can be characterized by their (dimensionless) wave number k (absolute value of the vector $\mathbf{k}$) or their wavelength $\lambda_k = a/k$. Because of this, it is appropriate to go to Fourier variables,

$$y_k := a\phi_k = y^*_{-k}. \tag{2}$$

The Fourier modes are complex, but for our purpose it is sufficient to focus attention on one real mode y_k (in the following skipping the index k).

In the semiclassical approximation to quantum gravity, each mode can be characterized by a wave function ψ obeying a quantum-mechanical Schrödinger equation separately for each mode,

$$\mathrm{i}\hbar \frac{\partial \psi(y, t)}{\partial t} = \hat{H}\psi(y, t), \tag{3}$$

where the Hamiltonian $\hat{H}$ is given by

$$\hat{H} = \frac{1}{2}\left(p^2 + k^2 y^2 + \frac{2\dot{a}}{a} yp\right); \tag{4}$$

p denotes the momentum conjugate to y. The first two terms are standard harmonic oscillator terms, while the last term describes the coupling of the mode to the evolving universe described by $a(t)$. It is well known from quantum optics that this last term leads to a squeezing of the quantum state (characterized by increasing variance of either y or p and decreasing variance of the respective other variable). The squeezing formalism for the cosmic perturbations was first presented in [13]. Taking into account both the real and the imaginary part of y_k in (2) leads to a two-mode squeezed state.

In order to solve the Schrödinger equation (3) for the modes, we need to impose an initial condition. This may be straightforward to do in quantum optics, but is far from trivial in cosmology. The simplest, and most commonly used, initial condition is the adiabatic vacuum state already used by Alexei in [1]. In the Schrödinger picture, it is given by the mode wave function

$$\psi_0(y) = \left(\frac{2k}{\pi}\right)^{1/4} \exp\left(-ky^2\right). \tag{5}$$

This corresponds to the usual harmonic oscillator ground state. With this initial state and the Hamiltonian (4), the solution of (3) remains of Gaussian form and can be written as

$$\psi(y,t) = \left(\frac{2\Omega_{\mathrm{R}}(t)}{\pi}\right)^{1/4} \exp\left(-\Omega(t)y^2\right), \quad \Omega := \Omega_{\mathrm{R}} + \mathrm{i}\Omega_{\mathrm{I}}. \tag{6}$$

The dynamics leads to a squeezing of the vacuum state and can be parametrized as

$$\Omega_{\mathrm{R}} = \frac{k}{\cosh 2r + \cos 2\varphi \sinh 2r}, \quad \Omega_{\mathrm{I}} = -\Omega_{\mathrm{R}} \sin 2\varphi \sinh 2r, \tag{7}$$

with $\Omega_{\mathrm{R}} = k$ and $\Omega_{\mathrm{I}} = 0$ for the initial state. Here, r is called squeezing parameter, and φ is called squeezing angle. Figure 3 shows the variances (resp. contours of the Wigner function) for the initial vacuum state (blue circle) and the resulting squeezed vacuum state.

For eternal inflation, one finds asymptotically $r \to \infty$ and $\varphi \to 0$, that is, the major axis with size $\mathrm{e}^r \to \infty$ of the ellipse becomes aligned to the y-axis, and the minor axis becomes arbitrarily small. Since, however, inflation is not eternal, this limit is not obtained, and we are left with a finite, but large value of the squeezing parameter r.

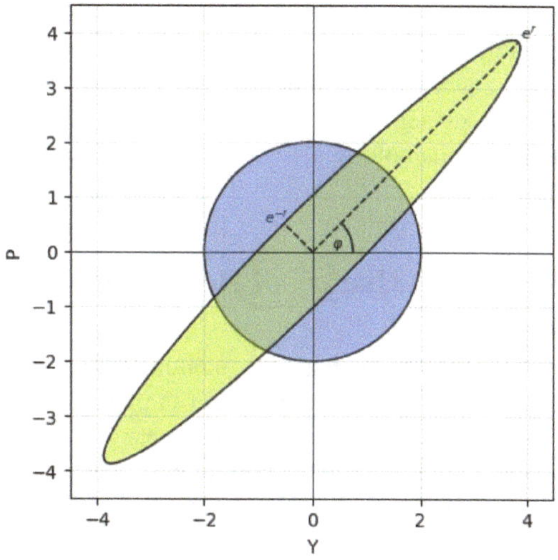

Fig. 3 Squeezed vacuum state (ellipse in green) compared with the vacuum state (circle in blue)

For concrete cases of evolution laws $a(t)$, the parameters can be given explicitly (at least numerically). In the simplest case of de Sitter exponential evolution,

$$a(t) = a_0 \exp(H_{\mathrm{I}} t),$$

one finds [14]

$$\sinh r = \frac{a H_{\mathrm{I}}}{2k}, \quad \cos 2\varphi = \tanh r.$$

For $r \gg 1$, this gives $\mathrm{e}^r \approx a H_{\mathrm{I}}/k$, so the squeezing parameter scales approximately with the logarithm of the scsale factor. For the largest cosmological scales, one has $r \approx 120$ and $\varphi \approx 0$. In this limit, the major axis of the ellipse seen in Fig. 3 will then be aligned along the y-axis.

So far, we have seen that the primordial fluctuations are in a (two-mode) squeezed quantum state. But how, then, can one understand the transition to a state describing classical stochastic fluctuations as they are used when analysing the CMB anisotropy spectrum and needed to understand the structure shown in Fig. 1?

A first step in this direction was achieved in the seminal paper [14]. There it was shown that quantum expectation values calculated from (6) cannot be distinguished from classical stochastic expectation values in the limit $r \to \infty$ (corresponding to neglecting what is called, in the Heisenberg picture, the "decaying mode"). For illustration, the analogy of a simple quantum-mechanical example is presented in [15].

But this is not yet enough. As is well known, squeezed states are very sensitive to interactions with other degrees of freedom (their 'natural environment') [16]. This feature is especially pronounced for states with a broad variance in position (this role is here played by the variable y), because most realistic interactions are in position, not in momentum. This leads to the process of decoherence, which decides which variables do assume classical properties and which do not. Decoherence as applied to the squeezed quantum state (6) was first discussed in [17], but before we turn to this, we present a brief history of decoherence with an eye towards applications in quantum cosmology.

3 A Brief History of Decoherence

In 1927, the formalism of quantum theory as it is used today was more or less complete. A central role therein is played by states (wave functions) obeying the superposition principle. The same year 1927 also witnessed the beginning of the still ongoing debate about the correct interpretation of the theory and, in particular, about the meaning of the wave functions. In 1929, Nevill Mott addressed the important question why a particle described by a spherically-symmetric wave function does not lead to a spherical structure in a cloud chamber, but instead to a trajectory [18]. To cite from that paper:

> In the theory of radioactive disintegration, as presented by Gamow, the α-particle is represented by a spherical wave which slowly leaks out of the nucleus. On the other hand, the α-particle, once emerged, has particle-like properties, the most striking being the ray tracks that it forms in a Wilson cloud chamber. It is a little difficult to picture how it is that an outgoing spherical wave can produce a straight line.

Mott's important insight was the observation that the quantum nature of the many atoms in the cloud chamber must be taken adequately into account when dealing with their interaction with the α-particle. In this, it is of the utmost importance to note that the total wave function is not a wave function in ordinary three-dimensional space (as e.g. classical fields are), but in the high-dimensional configuration space of α-particle *plus* cloud chamber atoms. In Mott's words:

> The difficulty that we have in picturing how it is that a spherical wave can produce a straight track arises from our tendency to picture the wave as existing in ordinary three dimensional space, whereas we are really dealing with wave functions in the multispace formed by the co-ordinates both of the particle and of every atom in the Wilson chamber.

Mott then went on to show that the full quantum state can be written as a superposition of track-like states for the α-particle, with each of these states correlated with a corresponding state of the chamber atoms. In todays's language (which goes back to Schrödinger's work in the 1930s) this total state is called an *entangled state*. Mott then argued (as was common in those days and happens sometimes even today) that this superposition can be replaced by a statistical ensemble with Born probabilities for the various components. What is observed in the cloud chamber is then one component of this total state with a probability given by the Born rule.

From a heuristic and pragmatic point of view, such an attitude may be understandable, but it is in conflict with the formalism of quantum theory. A superposition of states is fundamentally and observationally different from a statistical *ensemble* as myriads of interference experiments have demonstrated and continue to demonstrate. An electronic spin-state pointing in x-direction, for example, can be written as a superposition of spin states in $+z$- and $-z$-direction, but it is mathematically and experimentally different form an ensemble of spin states pointing into the $+z$- and $-z$-directions. Other examples can be found in particle physics where e.g. the superposition of a K-meson and its antiparticle results in a *new* particle (K-long or K-short) and not in a mixture of the original particles. The same holds for neutrino oscillations.

So something important must be missing in Mott's analysis. The missing important ingredient is what is today called the process of *decoherence*. It was first discussed in a seminal paper by H.-Dieter Zeh in 1970 [20], not yet using the word decoherence. Zeh showed that it is inconsistent to treat macroscopic quantum systems as isolated systems because they become strongly entangled with their environment.[4] These systems thus do *not* obey a Schrödinger equation, and their time evolution is *not* unitary.

Why has it taken so long from Mott's to Zeh's paper? The main reason is perhaps the prevalence of the Copenhagen interpretation of quantum theory after 1930 where classical concepts are postulated from the outset and where thus such questions were ignored. They were also considered irrelevant because quantum theory worked successfully at a pragmatic level; experiments able to test foundational issues were only possible after the advent of the laser and other tools.

The origin of Zeh's work lies in certain puzzling features of nuclear physics described by him, for example, in his essay [22]. Since his line of reasoning reflects certains ideas that are relevant for quantum cosmology, I shall spend some words on it.

In nuclear physics, there are many examples where a many-nuclei system is described by an entangled wave function Ψ obeying the stationary Schrödinger equation

$$H\Psi = E\Psi. \tag{8}$$

Nevertheless, for a calculation of the spectra there exist successful methods that make use of time-*dependent* single-nucleon wave functions (e.g. using the 'Hartree–Fock approximation'). These spectra also show features of deformed nuclei (e.g. Coriolis-force type of effects), in spite of the total state Ψ in (8) being in an angular momentum eigenstate. In fact, the state Ψ can be written as a superposition of states with different orientations in the form

$$\Psi = \int \mathrm{D}\Omega\, f(\phi, \theta, \chi) U(\phi, \theta, \chi) \Phi(\mathbf{x}_1, \ldots, \mathbf{x}_n), \tag{9}$$

[4] Here and below 'environment' means degrees of freedom whose coupling to the system cannot be neglected, but which are not "under control". A standard example is a dust particle in interaction with air molecules or photons [21].

where $U(\phi, \theta, \chi)$ denotes the unitary transformation for a rotation with angles ϕ, θ, χ (e.g. the Euler angles), $D\Omega$ is the volume element in this rotation space, and Φ is, for example, the product of many asymmetric single-particle wave functions $\phi_i(\mathbf{x}_i)$ (or generally a determinant wave function). The wave function Φ is then a 'deformed' state, which cannot be an eigenstate of (8), because such eigenstates are rotation-invariant angular-momentum eigenstates, exhibiting rotational symmetry instead of definite orientation. Only the superposition over all deformed nuclei in the form of (9) is then an (approximate) solution to the Schrödinger equation (8). This total state Ψ is strongly *entangled* with respect to the many single-nuclei states.

This counterintuitive situation (full symmetric state from a superposition of many states exhibiting a different direction) led Zeh in the 1960s to a weird speculation. In his own words ([22], p. 4):

> So one may say that the individual nucleons "observe" an apparent asymmetry in spite of the symmetric global superposition of all orientations. ...This analogy led me to the weird speculation about a nucleus that is big enough to contain a complex subsystem which may resemble a registration device or even a conscious observer. ...If the nucleons in the deformed nucleus dynamically feel a definite orientation in spite of the global superposition, would an internal observer then not similarly have to become "aware of" a *certain* measurement result?

We know at least since the famous work of Einstein, Podolsky, and Rosen (EPR) that entanglement is a central feature of quantum theory [23]. It is of great importance to see that entanglement is not only responsible for genuine quantum aspects (as in the EPR gedanken experiment), but also for the classical *appearance* of our world. The key to understanding this is the above mentioned process of decoherence. Let me give a brief explanation (see [16, 24] for details).

Following original ideas by John von Neumann, we consider an interaction between a system (S) and an apparatus (A), see Fig. 4. Both are described by quantum theory.

If the states of the measured system that are to be distinguished by the apparatus are denoted by $|n\rangle$ (e.g. spin up and spin down), a simple interaction Hamiltonian can be formulated as follows:

$$H_{\text{int}} = \sum_n |n\rangle\langle n| \otimes \hat{A}_n . \tag{10}$$

The operators $\hat{A}_n$ act on the states of the apparatus and must depend on the 'quantum number' n in order to describe a measurement. Equation (10) describes an idealized situation in which the apparatus becomes correlated with the system state without

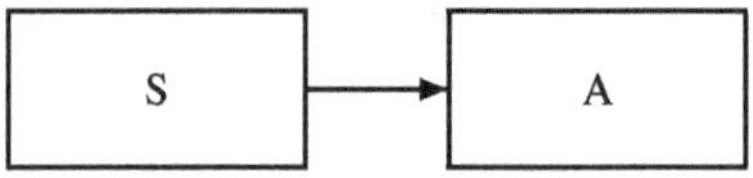

Fig. 4 Original form of the von Neumann measurement model

changing the latter. There is thus no disturbance of the system by the apparatus—on the contrary, the apparatus is disturbed by the system (in order to yield a measurement result).

If the measured system is initially in the state $|n\rangle$ and the apparatus in some initial state $|\Phi_0\rangle$, the evolution according to the Schrödinger equation with the Hamiltonian (10) reads

$$|n\rangle|\Phi_0\rangle \xrightarrow{t} \exp\left(-\mathrm{i}H_{\mathrm{int}}t\right)|n\rangle|\Phi_0\rangle = |n\rangle \exp\left(-\mathrm{i}\hat{A}_n t\right)|\Phi_0\rangle$$
$$=: |n\rangle|\Phi_n(t)\rangle. \tag{11}$$

The resulting apparatus states $|\Phi_n(t)\rangle$ are called *pointer states*. A process analogous to (11) can also be formulated in classical physics. The essential quantum features enter when one considers as initial state a *superposition* of different eigenstates of the measured 'observable' (such as a superposition of spin up and spin down). The linearity of time evolution then leads immediately to

$$\left(\sum_n c_n|n\rangle\right)|\Phi_0\rangle \xrightarrow{t} \sum_n c_n|n\rangle|\Phi_n(t)\rangle. \tag{12}$$

This state describes a superposition of macroscopic measurement results (as in the gedanken experiment with Schrödinger's cat). To avoid such a bizarre non-classical state, and to avoid the apparent conflict with experience, von Neumann introduced a dynamical collapse of the wave function as a new law that violates the universality of the Schrödinger equation. This alleged collapse should then select one component with probability $|c_n|^2$ (according to the Born rule).

The crucial idea to avoid such Schrödinger-cat type of states and to arrive at a definite measurement result (more precisely, the *appearance* of a definite result), the unavoidable interaction with the environment must be taken into account [20]. Environmental degrees of freedom usually occur in big numbers (e.g. in the case of scattering with photons or air molecules [21]) and are not under control (Fig. 5).

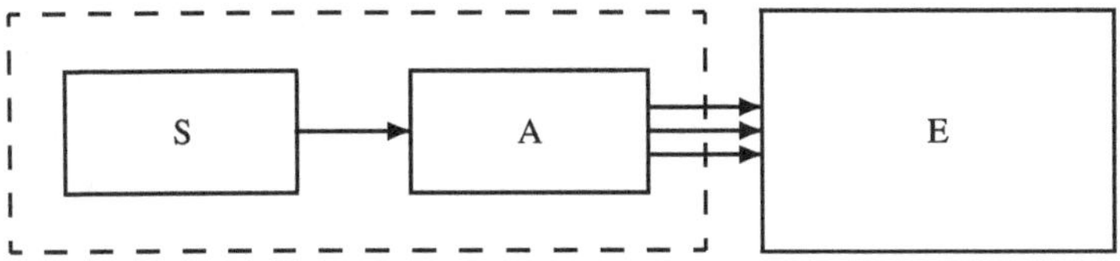

Fig. 5 Realistic extension of the von Neumann measurement model including the environment. Classical properties emerge through the unavoidable, irreversible interaction of the apparatus with the environment

Assuming that the interaction of the apparatus with the environment can again be described by a Hamiltonian of the form (10), one arrives at a gigantic superposition now *including* the environment, according to

$$\left(\sum_n c_n|n\rangle|\Phi_n\rangle\right)|E_0\rangle \quad \stackrel{t}{\longrightarrow} \quad \sum_n c_n|n\rangle|\Phi_n\rangle|E_n\rangle. \tag{13}$$

At first sight it seems as if the situation with macroscopic superpositions did even get worse. The important observation is, however, that most of the many environmental degrees of freedom are inaccessible. They thus have to be integrated out from the full state (13) in order to describe what can be seen at the apparatus. The result is a reduced density matrix for system plus apparatus, which contains all the information that can be retrieved from the two,

$$\rho_{\mathrm{SA}} \approx \sum_n |c_n|^2 |n\rangle\langle n| \otimes |\Phi_n\rangle\langle\Phi_n| \quad \text{if} \quad \langle E_n|E_m\rangle \approx \delta_{nm}. \tag{14}$$

In the last step we have used that, under realistic conditions, different environmental states are orthogonal to each other. Equation (14) is approximately identical to the density matrix of an ensemble of measurement results $|n\rangle|\Phi_n\rangle$. The system and apparatus thus seem to be in one of the states $|n\rangle$ and $|\Phi_n\rangle$, given by the probability $|c_n|^2$. This is also called an *apparent collapse*. In a universally valid quantum theory, the interferences described by (14) still exist—they are present in the entanglement between S and A with the environmental degrees of freedom, but are no longer recognizable in S and A alone. Or, in the words of [21]: "The interferences exist, but they are not *there*."

By now, the process of decoherence has been successfully tested in a variety of experiments; see, for example, the corresponding contributions to [24]. Further developments in the general formalism have led, for example, to the concept of *Quantum Darwinism* [25]. This refers to the fact that an abundance of many copies of the pointer states (the classical states distinguished by the interaction with the environment) are imprinted in the environment, so information about those pointer states can be obtained from observing the environment.

Decoherence originates from the entanglement between a system with its environment. If gravitational degrees of freedom are also described in quantum terms (as we assume here), they also take part in decoherence. In an early simple model, it was shown how a quantum metric in Newtonian approximation becomes decohered by the interaction with gas molecules [26]. How decoherence can in principle be applied to a full quantum theory of gravity and, in particular, to quantum cosmology, was first suggested in [27]. The first concrete calculations in the context of the Wheeler–DeWitt equation of canonical quantum gravity were presented in [28], see also [29], where anlogous calculations are given for quantum electrodynamics. In these papers, it was shown how the scale factor of an approximately homogeneous and isotropy universe as well as the homogeneous part of an inflaton are decohered by quantum fluctuations of gravity and matter. To give one concrete example of a later calculation [30], the reduced density matrix for the scale factor a after integrating out the many degrees of freedom of primordial gravitational waves was found to read

$$\rho_0(a, a') \xrightarrow{t} \rho_0(a, a') \exp\left(-C H_{\mathrm{I}}^3 a(a - a')^2\right) \ , \ C > 0, \tag{15}$$

where H_{I} is the Hubble scale of inflation. The interpretation of this result is that the homogeneous background, even if quantized, assumes classical properties quickly during inflation, since non-diagonal terms in the density matrix become tiny for increasing a; see also the illuminating discussion in [31].

While these quantum gravity considerations may seem mainly of conceptual interest, the quantum-to-classical transition for the primordial fluctuations in an inflationary universe is of more direct relevance because it leads to observational consequences. It is here where our collaboration starts [17].

4 Decoherence of Primordial Fluctuations

As we have reviewed above, an initial adiabatic ground state for metric and inflaton perturbations becomes highly squeezed during inflation, see (6) and (7). The squeezing appears in the momentum p canonically conjugate to the field amplitude y, that is, the resulting quantum state is very broad in y. As is well known from quantum optics, broad states in position are highly sensitive to an influence by environmental degrees of freedom [16]. The reason is that usual interaction terms in the Hamiltonian invoke a coupling in position, not in momentum. This sensitivity motivated us in [17] to study possible decoherence effects arising from an interaction with environmental degrees of freedom.

But where can such degrees of freedom come from? There are many possibilities. The perhaps most natural one is to take into account the interaction between modes of different wavelength, which arises from the non-linear interaction in the full theory and which is neglected in the approximation used in Sect. 2. Other options are interactions with degrees of freedom from fundamental (as yet unknown) theories that lead, for example, to string-effective actions containing the coupling of many different fields. Fortunately, the details of such interactions are not needed for understanding the major features of decoherence. The only general input is the fact that interactions in quantum field theory are usually local interactions coupling field amplitudes (and not their canonical momenta). With this, a quite general formalism for the decoherence of primordial fluctuations can be developed [32, 33].

The simplest case is an ideal interaction in the spirit already discussed by von Neumann. In this, there is not direct perturbation of the system, but only the formation of entanglement between the system and the interacting degrees of freedom. It can be described by the following modification of the reduced density matrix:

$$\rho_0(y, y') \longrightarrow \rho_\xi(y, y') = \rho_0(y, y') \exp\left(-\frac{\xi}{2}(y - y')^2\right), \tag{16}$$

where ξ denotes a phenomenological parameter that contains the details of the interaction and that can be calculated from the parameters and the form of a given interaction Hamiltonian.

In order to describe a realistic decoherence condition for the primordial fluctuations, one has to assume that ξ is much bigger than the parameter Ω_R appearing in (6) and given in (7),

$$\xi \gg \Omega_R \approx k e^{-2r}.$$

Otherwise, (16) does not describe a decrease in off-diagonal density-matrix terms, as is required for decoherence. It turns out that the decoherence time scale t_d for the modes during inflation is (apart from prefactors) given by the Hubble scale H_I^{-1} of inflation, $t_d \sim H_I^{-1}$ [32]. Although the system describing the modes is not a chaotic system (but only classically unstable), there is nevertheless a close analogy to a chaotic system: the Hubble parameter H_I is analogous to the Lyapunov parameter. The decoherence time can also be calculated from the rate of de-separation, that is, the rate at which originally unentangled states become entangled [15].

Correlations between y and p can easily been recognized from the contour of the Wigner function (the Wigner ellipse). In the case of Gaussian systems, as is the case here, the Wigner function remains positive, in spite of the presence of quantum features. If the modes are in a pure state, see (6), the Wigner ellipse has the form displayed in Fig. 3. In the limit of large r, its major axis α is $\alpha \approx e^r$, and its minor axis is $\beta \approx e^{-r}$. In the case of interaction with an environment, the modes are in a mixed state given by (16). For large r, the axes of the Wigner ellipse are then given by

$$\alpha \approx e^r, \quad \beta \approx \sqrt{\frac{\xi}{k}} \gg e^{-r}.$$

This corresponds to the situation displayed in Fig. 6.

The interaction of the primordial modes with other degrees of freedom can, of course, not be so strong as to be in conflict with observations. This means, in particular, that the correlation between y and p cannot be completely washed out, because otherwise the acoustic peaks in the CMB anisotropy spectrum observed by the PLANCK satellite and other missions[5] would not be present [32]. In the formalism, this correlation condition is implemented by

$$\frac{\xi}{k} \ll e^{2r}. \tag{17}$$

[5] See https://sci.esa.int/web/planck/-/51555-planck-power-spectrum-of-temperature-fluctuations-in-the-cosmic-microwave-background.

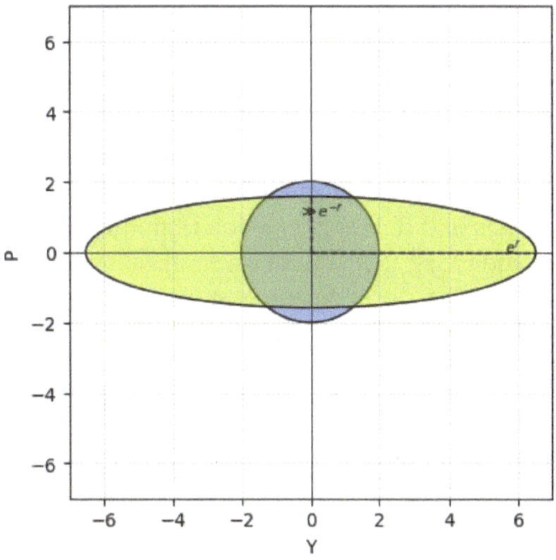

Fig. 6 Due to decoherence, the ellipse (in green) becomes broader in p-direction, while the major axis remains approximately unchanged and aligns with the y-axis

A related question is about the value of the entropy, S, for the fluctuations. If correlations were fully absent (in contradiction to observations), the entropy per mode would take for large r the value $S \approx 2r$. In our situation of pure entanglement, Eq. (16), and remaining correlation (17), the entropy per mode turns out to be limited by the value $S \approx r$, that is, only *half* of the maximum entropy $2r$ [32, 33]. This corresponds to having the field-amplitude basis as the pointer basis under decoherence (as opposed to e.g. the particle-number basis).

The calculation of the entropy is based on the von Neumann formula

$$S = -\mathrm{tr}(\rho_\xi \ln \rho_\xi);$$

see [33] for explicit calculations and expressions. If one plots the entropy in dependence of the dimensionless parameter $\chi =: \xi/\Omega_{\mathrm{R}}$ controlling the strength of decoherence, one recognizes that the asymptotic value for $\chi \gg 1$ is readily obtained; loss of one bit of information, which corresponds to $S \approx \ln 2$, is obtained already for $\chi \approx 1.5$.

The above considerations concern decoherence during inflation. Of interest is also the question about the decohering behaviour at later stages, in particular for the stage after the second Hubble-crossing during the radiation- or matter-dominated phase, see Fig. 2. This was investigated in our paper [34], where we could show that the Wigner ellipse rotates slowly and that the correlation between amplitude and momentum holds for a sufficiently long time.

As a side remark, I want to mention that similar investigations can be applied to the Hawking radiation for black holes [35]. A quantum field on the background of an object collapsing to a black hole is also turned into a (two-mode) squeezed state. Interaction with other degrees of freedom again leads to decoherence, but with a notable difference to cosmology: here, the pointer basis is the particle-number basis, leading to the maximal entropy associated with a thermal state – the thermal state usually assumed for Hawking radiation. In view of this, there is no reason to impose an information-loss "paradox", at least not at the semiclassical level.[6]

The above considerations can be put on a solid formal basis by using established methods from the quantum theory of open systems. Instead of making a case-by case analysis of different interaction Hamiltonians, it is more convenient to study master equations and their solutions. In the simplest case of an ideal interaction similar to (16), one has for a quantum-mechanical system (e.g. a dust particle) in interaction with its environment (e.g. photons or air molecules) the following master equation for the reduced density matrix [21]:

$$\frac{\mathrm{d}\hat{\rho}}{\mathrm{d}t} = -\frac{\mathrm{i}}{2m\hbar}[\hat{p}^2, \hat{\rho}] - \Lambda[\hat{x}, [\hat{x}, \hat{\rho}(t)]], \tag{18}$$

where Λ encodes the details of the interaction (e.g. Thomson cross section). Asymptotically, for large times, the density matrix becomes diagonal in position space. An interesting property for the Wigner function was proven in [36]: it becomes strictly *positive* after a certain finite time $t_\mathrm{d} = \sqrt{m/2\hbar\Lambda}$, independent of the initial state. It is straightforward to add further terms to (18) in order to include back action on the system. In that case, the density matrix does not become exactly diagonal, but a finite coherence length in position space remains [21].

A general master equation encompassing many particular interactions is the Lindblad equation for a Markovian (local in time) evolution,

$$\hbar\frac{\partial\hat{\rho}}{\partial t} = -\mathrm{i}[\hat{H}, \hat{\rho}] + \hat{L}\hat{\rho}\hat{L}^\dagger - \frac{1}{2}\hat{L}^\dagger\hat{L}\hat{\rho} - \frac{1}{2}\hat{\rho}\hat{L}^\dagger\hat{L}.$$

For the primordial fluctuations, this equation was discussed in detail in our paper [33]. In accordance with our presentation above, we found that the pointer basis is given by the field-amplitude basis and that the time scale for decoherence during inflation is given by H_I^{-1}. Results were also found for the decoherence time during radiation dominance outside the Hubble schale and for the situation after the second crossing of the Hubble scale; see also [37] for a review.

The quantum-to-classical transition for primordial fluctuations (both scalar and tensorial) continues to be of interest; see, for example, [38] and the references therein. One might wonder, for example, whether one could perform Bell-type "experiments" (observations) for the CMB in order to find possible violations of Bell inequalities and thus prove directly the quantum origin of the fluctuations. For modes being in a two-modes squeezed state, such a violation indeed occurs, but because of decoherence it seems impossible to observe such violations in the CMB [39].

---

[6] A scenario for the full quantum evolution of black holes must, of course, come from a fundamental theory of quantum gravity.

In Sect. 4 above, we have mentioned the feature of quantum Darwinism by which information about the pointer states can be obtained from observations of the environmental degrees of freedom [25]. It would be of particular interest to study quantum Darwinism in cosmology, for example through the observation of primordial gravitational waves by the LISA mission. Whether this is possible, only the future will show.

5 Arrow of Time

From the fundamental point of view of a full quantum theory of gravity, all degrees of freedom in Nature are described by quantum theory. Classical behaviour is an emergent feature only and can be understood by the process of decoherence. There is, in fact, a whole hierarchy of classicality [16]. Starting from a full quantum state of gravity and matter, one can understand how *first* the classical appearance of spacetime emerges and *then* the classical appearance of other 'background fields' such as the homogeneous part of the inflaton field [28]. Given that, the next level in the hierarchy of classicality comes from the decoherence of primordial fluctuations discussed above. In this sense, all the structure that we observe in our Universe (such as the one seen in Fig. 1) can be understood as emerging from a quantum origin. The picture is similar to the picture of a hypothetical inside view within a nucleus as described in Sect. 4—but for the Universe, we can only have an inside view.

Decoherence is also important for an understanding of the arrow of time, that is, for the origin of irreversibility in our Universe. Alexei has for a long time entertained the idea [40] that the arrow of time should emerge from a breaking of the primordial symmetry encoded in the hypothetical de Sitter state assumed in his pioneering paper [1]. In [41], I have made the proposal that the origin of irreversibility can be traced back to a quantum version of Roger Penrose's Weyl curvature hypothesis. In this, a crucial role is played by assuming the initial state (5) for the primordial fluctuations, in particular for the tensor perturbations describing primordial gravitational waves. With such an initial condition, an initial unentangled state between degrees of freedom will evolve into an entangled state, with an increase of entanglement entropy [42]. This increasing entropy may eventually lead to the Second Law and to the origin of probabilities according to the Born rule [43], although the final word on this can only be spoken after a fundamental quantum theory of all interactions will be established (if this will ever be the case [44]).

In connection with Alexei's suggestion of a symmetric initial state in Sect. 2, I have mentioned the pre-Socratic philosophers Anaximander and Anaxagoras. At least according to Aristotle in his *Eudemian Ethics*, Anaxagoras has answered to the question whether there is a reason why someone should decide to be borne instead of remaining unborne, that the reason is "to observe the celestial edifice and the order in the cosmos." Few people in our time have contributed so much to our understanding of the cosmos than Alexei Starobinsky. We greatly miss him.

Competing Interests The author has no conflicts of interest to declare that are relevant to the content of this chapter.

Acknowledgements I am most grateful to the late Alexei Starobinsky for collaboration, discussions, and many wonderful personal meetings that I will never forget. I also thank Andrei Barvinsky, Alexander Kamenshchik, and David Polarski for our enjoying collaboration on decoherence and other fascinating topics.

References

1. A.A. Starobinsky, Spectrum of relict gravitational radiation and the early state of the universe. JETP Lett. **30**, 682–685 (1979)
2. H.D. Conradi, H.D. Zeh, Quantum cosmology as an initial value problem. Phys. Lett. A **154**, 321–326 (1991)
3. K.R. Popper, *The World of Parmenides-Essays on the Presocratic Enlightenment* (Routledge, London and New York, 1998)
4. A.A. Starobinsky, A new type of isotropic cosmological models without singularity. Phys. Lett. B **91**, 99–102 (1980)
5. J. Martin, C. Ringeval, U Louvain, V. Vennin, Encyclopaedia Inflationaris: Opiparous Edition. Phys. Dark Univ. **46**, 101653 (2024)
6. D. Baumann, TASI lectures on inflation (2012). arXiv:0907.5424v2 [hep-th]
7. A.A. Starobinsky, Dynamics of phase transition in the new inflationary universe scenario and generation of perturbations. Phys. Lett. B **117**, 175–178 (1982)
8. C. Kiefer, *Quantum Gravity*, 4th edn. (Oxford University Press, Oxford, 2025)
9. C. Kiefer, Quantum geometrodynamics: Whence, whither? Gen. Relativ. Gravit. **41**, 877–901 (2009)
10. L. Chataignier, *Timeless Quantum Mechanics and the Early Universe* (Springer Nature Switzerland, 2022)
11. C. Kiefer, T.P. Singh, Quantum gravitational corrections to the functional Schrödinger equation. Phys. Rev. D **44**, 1067–1076 (1991)
12. L. Chataignier, C. Kiefer, P. Moniz, Observations in quantum cosmology. Class. Quantum Grav. **40**, 223001 (2023)
13. L.P. Grishchuk, Y.V. Sidorov, Squeezed quantum states of relic gravitons and primordial density fluctuations. Phys. Rev. D **42**, 3413–3421 (1990)
14. D. Polarski, A.A. Starobinsky, Semiclassicality and decoherence of cosmological perturbations. Class. Quantum Grav. **13**, 377–392 (1996)
15. C. Kiefer, D. Polarski, Emergence of classicality for primordial fluctuations: Concepts and analogies. Annalen Phys. **7**, 137–158 (1998)
16. E. Joos, H.D. Zeh, C. Kiefer, D. Giulini, J. Kupsch, I.-O. Stamatescu, *Decoherence and the Appearance of a Classical World in Quantum Theory*, 2nd edn. (Springer, Berlin, 2003)
17. C. Kiefer, D. Polarski, A.A. Starobinsky, Quantum to classical transition for fluctuations in the early universe. Int. J. Mod. Phys. D **7**, 455–462 (1998)
18. N.F. Mott, The wave mechanics of α-ray tracks. Proc. Roy. Soc. Lond. **A126**, 79–84 (1929). (Reprinted in: [19], 129–134)
19. J.A. Wheeler, W.H. Zurek, *Quantum Theory and Measurement* (Princeton University Press, Princeton, 1983)
20. H.D. Zeh, On the interpretation of measurement in quantum theory. Found. Phys. **1**, 69–76 (1970). (Reprinted in [19], 342–349)
21. E. Joos, H.D. Zeh, The emergence of classical properties through interaction with the environment. Z. Phys. B **59**, 223–243 (1985)

22. H.D. Zeh, Roots and fruits of decoherence (2006). arXiv:quant-ph/0512078v2
23. C. Kiefer, A. Einstein, B. Podolsky, N. Rosen (eds.), *Can Quantum-Mechanical Description of Physical Reality Be Considered Complete? Classic Texts in the Sciences* (Springer Nature Switzerland, 2022)
24. C. Kiefer (ed.), *From Quantum to Classical* (Essays in Honour of H.-Dieter Zeh, Springer Nature Switzerland, 2022)
25. W.H. Zurek, Emergence of the classical world from within our quantum Universe, in [24], 23–44
26. E. Joos, Why do we observe a classical space-time? Phys. Lett. A **116**, 6–8 (1986)
27. H.D. Zeh, Emergence of classical time from a universal wave function. Phys. Lett. A **116**, 9–12 (1986)
28. C. Kiefer, Continuous measurement of minisuperspace variables by higher multipoles. Class. Quantum Grav. **4**, 1369–1382 (1987)
29. C. Kiefer, Decoherence in quantum electrodynamics and quantum gravity. Phys. Rev. D **46**, 1658–1670 (1992)
30. A.O. Barvinsky, A.Y. Kamenshchik, C. Kiefer, I.V. Mishakov, Decoherence in quantum cosmology at the onset of inflation. Nucl. Phys. B **551**, 374–396 (1999)
31. A.O. Barvinsky, A.Yu. Kamenshchik, Preferred Basis, Decoherence and the Quantum State, in [24], 97–120
32. C. Kiefer, D. Polarski, A.A. Starobinsky, Entropy of gravitons produced in the early universe. Phys. Rev. D **62**, 043518 (2000)
33. C. Kiefer, I. Lohmar, D. Polarski, A.A. Starobinsky, Pointer states for primordial fluctuations in inflationary cosmology. Class. Quantum Grav. **24**, 1699–1718 (2007)
34. C. Kiefer, J. Lesgourgues, D. Polarski, A.A. Starobinsky, The coherence of primordial fluctuations produced during inflation. Class. Quantum Grav. **15**, L67–L72 (1998)
35. C. Kiefer, Hawking radiation from decoherence. Class. Quantum Grav. **18**, L151–L154 (2001)
36. L. Diósi, C. Kiefer, Exact positivity of the Wigner and P-functions of a Markovian open system. J. Phys. A: Math. Gen. **35**, 2675–2683 (2002)
37. C. Kiefer, D. Polarski, Why do cosmological perturbations look classical to us? Adv. Sci. Lett. **2**, 164–173 (2009). arXiv:0810.0087v2 [astro-ph]
38. C.P. Burgess, R. Holman, G. Kaplanek, J. Martin, V. Vennin, Minimal decoherence from inflation. JCAP **07**, 022 (2023)
39. J. Martin, V. Vennin, Obstructions to Bell CMB experiments. Phys. Rev. D **96**, 063501 (2017)
40. A.A. Starobinsky, Personal communication
41. C. Kiefer, On a quantum Weyl curvature hypothesis. AVS Quantum Sci. **4**, 015607 (2022). arXiv:2111.02137v2 [gr-qc]
42. C. Kiefer, L. Chataignier, M. Tyagi, Time and its arrow from quantum geometrodynamics? J. Phys: Conf. Ser. **2883**, 012008 (2024)
43. L. Chataignier, C. Kiefer, M. Tyagi, Origin of time and probability in quantum cosmology (2025). arXiv:2503.10752 [gr-qc]
44. C. Kiefer, Gödel's undecidability theorems and the search for a theory of everything. Int. J. Theor. Phys. **63**, 52 (2024)

Origin of Cosmological Density Perturbations from Quantum Fluctuations of Vacuum in General Relativity

V. N. Lukash and E. V. Mikheeva

Abstract After 45 years since the discovery of quantum-gravitational birth of the cosmological density perturbations we can try to answer the main question of cosmology what is the origin of the Universe. This has become possible because the observational data are precise enough to build a physical model of the cascade relaxation of the gravitating vacuum as a generator of the evolving Universe. We present a model of the early Universe, which is inspired by the observational data in the spatial wavenumbers $k \in (2 \cdot 10^{-4}, 10)\ \mathrm{Mpc}^{-1}$ and does not require a solution of the Friedman equations. Based on observations we present a solution in the form of a vacuum attractor of General Relativity, which provides an additional power in the form of a "bump" and a blue spectrum of perturbations at $k > 10\ \mathrm{Mpc}^{-1}$. Extending the power spectrum to the scale of hundreds of kiloparsecs and less can solve the problems of the observed cosmology—the appearance of the early star formation and supermassive black holes at $z > 10$, the birth of primordial black holes, Λ CDM model, and others.

1 Introduction

This paper is dedicated to the memory of Alexei Alexandrovich Starobinsky, whose ideas and papers formed important part of modern cosmology. The origin of the Universe was one of problems he was interested in.

For many years Alexei adhered to the idea that the initial state of the Universe corresponds to the de Sitter world with its inherent high degree of symmetry [1]. The symmetry breaking transforms the world to the Friedmannian one which is observable now on large cosmological scales. Alexei associated the transition from

V. N. Lukash · E. V. Mikheeva (✉)
ASC FIAN, Profsoyuznaya 84/32, 117997, Moscow, Russia
e-mail: helen@asc.rssi.ru

V. N. Lukash
e-mail: lukash@asc.rssi.ru

A. Barvinsky and A. Kamenshchik (eds.), *Open Issues in Gravitation and Cosmology*, Fundamental Theories of Physics 225, https://doi.org/10.1007/978-3-032-15703-4_10

the de Sitter model to the Friedmann model with the extension of GR by introducing the R^2 term in the gravitational action (see [2] and references therein).

In this paper we consider another concept of the evolving Universe, based on the idea of many fields in the vacuum state, where their initial expectation values are zero, but the total energy density is larger than zero. We call this state the polarized vacuum in GR. The exit from this state and "rolling down" successively in directions determined by different fields is responsible for the evolution of the Universe and can be tested through investigation of the power spectra of cosmological perturbations.

The equation of state of the matter ($E + P \simeq 0$), responsible for the accelerated expansion of the Universe, is close to both the vacuum ($E + P = 0$) and the scalar field equation of state with a dominant potential term ($|E + P| \ll E$). Interpretating the nature of the Λ-term as the Einstein's classical cosmological constant has led to an impasse, since the matter density cannot be stable. Another way leads to the concept of many fields, which is related to the idea of a large number of degrees of freedom (fundamental fields of matter) responsible for the polarization of vacuum in the Universe [3–8]. Since the first observational data were insufficiently precise their interpretation allowed for a large ambiguity in field interaction potentials. Different models based on those data led to different observational consequences. After 2021 (see [9]), the improved data precision is sufficient to build an observationally motivated model of the early Universe without solving the Friedman equations and without information about field potential (see [10]).

We understand by vacuum its polarization in the external gravitational field with its subsequent relaxation in time serving as a generator of the evolving Universe. In our model the solution of the vacuum attractor (VA) is suggested by the observational data obtained in the recent years [10]. The polarization fields initially were in zero states ($\varphi^{(i)} \simeq 0$). At some moment one of them (having the largest mass among others) appears as a dominating field and begins to evolve from zero to some non-zero value, ending its motion in a new vacuum state (with energy density value less than the previous one). This process could repeat itself with the help of other fields in other times. It is the ongoing gravitational process of cascade relaxation of vacuum that creates all observable cosmology. The conditions for the appearance of the supermassive black holes (SMBH) [11] and early galaxies before the formation of the large-scale structure of the Universe [12] are fulfilled when a non-power-law spectrum of cosmological density perturbations is considered. As high peaks (bumps) appear in the power spectra of density perturbations, primordial black holes (PBHs) can be borne in the Universe, as indicated by the LIGO data. The spectrum at small scales ($k > 10\,\mathrm{Mpc}^{-1}$) can be reconstructed if the observations of the J. Webb Space Telescope (JWST) are confirmed (see [13–15] and references therein).

2 The Cascade Relaxation of Vacuum

In GR the Lagrangian density has two terms: the first one consists of the metric tensor and its derivatives, $\frac{R}{16\pi G}$ (where $R = R^{\mu}_{\mu}$, $R_{\mu\nu}$ is the Ricci tensor, $G = const$), and the second one includes the fundamental degrees of freedom of the matter fields and their derivatives,

$$\mathcal{L} = \mathcal{L}\left(w^{(1)}, w^{(2)}, ..., \varphi^{(1)}, \varphi^{(2)}, ...\right), \tag{1}$$

where $w^{(1)} = \varphi^{(1)}_{,\mu}\varphi^{(1),\mu}$, etc.[1] The gravitational field is described by the metric tensor $g^{\mu\nu}$ and is included in the function (1), but its derivatives are contained only in the scalar R. The separation of this kinetic scalar is the isolation of the gravitational field, which means the appearance of GR. A further decomposition of the Lagrangian with separation of different kinetic terms of the fields occurs evolutionarily, when energy decreases.

The energy density includes entropy, particles and vacuum polarization. The temperature and particle density decrease during the gravitational expansion of the Universe, resulting in domination of the vacuum polarization. The decrease of vacuum density occurs through a sequence of relaxation epochs, where one field dominates while the others remain gravitationally frozen.

At the first stage the scalar field φ dominates with simultaneous separation of its kinetic term from other fields:

$$\mathcal{L} \to \mathcal{L}\left(w, \varphi, ...\right) \to \frac{w^2}{2} - V\left(\varphi, ...\right), \tag{2}$$

where $w^2 = \varphi_{,\mu}\varphi^{,\mu}$ is the kinetic scalar of field $\varphi = \varphi^{(1)}$, $V = V(\varphi, ...)$ is a potential of all fields. In the initial (symmetric or basic) state all fields are zero. The first field starting the motion from zero dominates the potential (has the maximal value of mass). The kinetic scalars of other fields are zero, since they do not evolve due to the equations of motion.

The process of Cascade Vacuum Relaxation (CVR) process is considered as a transition of fields in time from symmetric state $V_0 = V(0, 0, ...)$ to energetically lower states – firstly to $V_1 = V(\varphi_1, 0, ...) < V_0$ with field φ (where not otherwise specified, we assume $\varphi > 0$), then from V_1 to $V_2 = V(\varphi_1, \varphi_2, 0, ...) < V_1$ with field $\varphi^{(2)}$, etc.:

$$V_0 = V\left(0, 0, ...\right) \xrightarrow{\varphi} V_1\left(\varphi_1, 0, ...\right) \xrightarrow{\varphi^{(2)}} V_2\left(\varphi_1, \varphi_2, 0, ...\right) \longrightarrow ...$$

CRV means the sequential epochs of domination of fields, including the present dark energy epoch. All states of the potential are positive since they are larger than the cosmological Λ-term, and it is positive from observations:

---

[1] This includes all fields of all spins. To simplify the notation we do not use the space-time indices in the matter fields and their kinetic terms.

$$V_0 > V_1 > V_2 > \ldots > \Lambda > 0. \tag{3}$$

We assume that the state V_0 is a polarization of vacuum of all fields. It results in the process of accelerated expansion of the Universe along all three spatial directions, that brings any initial space-time to the Friedman symmetry. The observational model of the first stage of the CRV creates all power spectra of cosmological perturbations.

3 The Universe Model

At the first stage of CVR, the field $\varphi = \varphi(x^\mu)$ emerged from the initial state V_0 ($|\varphi| \ll \varphi_1$), being the gravitational source of the evolving spatially flat Friedman model with linear fluctuations of the metric, which are described by the density perturbation field $q = q(x^\mu)$ and gravitational waves $\tilde{q}_{ij} = \tilde{q}_{ij}(x^\mu)$ — functions of all coordinates $x^\mu = (t, x^i)$. The solution is written as a power series in the small quantities q and $\tilde{q}_{ij}$:

$$\varphi = \varphi(N) + \alpha\Delta + O\left(\Delta^2\right), \tag{4}$$

$$ds^2 = (1 + 2\Phi)\, dt^2 - a^2 (1 - 2\Phi)\left(\delta_{ij} + 2\tilde{q}_{ij}\right) dx^i dx^j,$$

where $\Delta = q - \Phi$ is matter velocity potential, $\Phi = \frac{H}{a}\int \gamma q a\, dt$ — gravitational potential, $\tilde{q}^i_i = \tilde{q}^j_{i,j} = 0$; $N = N(t) \equiv \ln(a)$, $\alpha = \alpha(N) \equiv \varphi_{,N}$, $H \equiv \dot{N}$ and $\gamma \equiv -\frac{\dot{H}}{H^2}$ depends on time, dot means the time t derivative, $X_{,x} = \frac{\dot{X}}{\dot{x}}$ (for functions depending on time), raising and lowering of space indices i and j is performed by unit tensor $\delta_{ij} = \mathrm{diag}(1, 1, 1)$.

The field $\varphi(N)$ is a source of a homogeneous vacuum background satisfying the Friedman equations. The solution (4) arises spontaneously during the gravitational expansion of a space-time region from the initial homogeneous scale $\sim 1/H_i$ with the subsequent accelerated expansion of the boundaries $\sim a/(a_i H_i)$, which constrains the function γ: $\ddot{a} > 0 \rightarrow \gamma < 1$.

All fluctuations of the backgound can be described by two gravitational fields q and $\tilde{q}_{ij}$ — the scalar (S) and tensor (T) modes of linear perturbations of the metric, respectively. These are test massless fields of the Friedman model with actions obtained by direct expansion of the general action to second order in q and $\tilde{q}_{ij}$ [6, 16, 17]:

$$\delta^{(2)} S = \int \left(L + \tilde{L}\right)\sqrt{-g}\, d^4x, \tag{5}$$

$$L = \frac{\gamma q_{,\mu} q^{,\mu}}{8\pi G} = \frac{\alpha^2 q_{,\mu} q^{,\mu}}{2}, \quad \tilde{L} = \frac{\tilde{q}_{ij,\mu}\tilde{q}^{ij,\mu}}{16\pi G} = \frac{m_P^2 \tilde{q}_{ij,\mu}\tilde{q}^{ij,\mu}}{4},$$

where $m_P = \frac{M_P}{2\sqrt{\pi}}$, $M_P = \frac{1}{\sqrt{G}}$. Let us remark that the field q is included in the velocity potential of matter and has a Newtonian limit, the field $\tilde{q}_{ij}$ has two polarizations, the speed of the S-mode tends to the speed of light according to the condition (2), and the functions α and γ are related as $\gamma = 4\pi G\alpha^2$. In addition, it is convenient to introduce the following notation for the derivatives of H and establish useful connections between them in the future:

$$\gamma = \beta^2 = -\frac{H_{,N}}{H}, \quad \beta = \frac{\alpha}{m_P} = \phi_{,N} = -\frac{H_{,\phi}}{H}, \quad \varepsilon = \beta_{,\phi}, \tag{6}$$

where $\phi = \frac{\varphi}{m_P}$ is a dimensionless field.

The expectations of the fields q and $\tilde{q}_{ij}$ are equal to zero, but there are non-zero dispersions:

$$\langle q^2 \rangle = \int_0^\infty q_k^2 \frac{dk}{k}, \quad \langle \tilde{q}_{ij}\tilde{q}^{ij} \rangle = \int_0^\infty \tilde{q}_k^2 \frac{dk}{k}, \tag{7}$$

where the brackets $\langle ... \rangle$ mean averaging over vacuum states of the fields, the functions q_k and $\tilde{q}_k$ are spectra of cosmological perturbations, k is the wavenumber. When the scale of perturbations is larger than the cosmological horizon ($k < aH$), the fields q and $\tilde{q}_{ij}$ are frozen and the spectra do not depend on time. The slopes (indices) of the spectra, their dependence on the scale and the ratio of the power spectra are determined as follows:

$$n_k = \frac{d \ln q_k}{d \ln k}, \quad \tilde{n}_k = \frac{d \ln \tilde{q}_k}{d \ln k}, \quad A_k = \frac{d \ln |n_k|}{d \ln k}, \quad r_k = \frac{\tilde{q}_k^2}{q_k^2},$$

where r_k is (by definition) four times less than the corresponding definition in [18].

The observational data [9, 19] give the following values in the range of wavenumbers in $k \in (2 \cdot 10^{-4}, 10)$ Mpc^{-1}:

$$q_k \simeq 10^{-5} \left(\frac{k_c}{k}\right)^{n_c}, \quad n_c = 0.0175 \pm 0.0025, \quad r_c < 10^{-2}, \tag{8}$$

where the boundaries of this range correspond to the present event horizon and the scale of dwarf galaxies, respectively, c refers to the "central" wavenumber $k_c = 0.05$, the wavenumber k is measured in units of Mpc^{-1} for the Hubble constant 67 km sec^{-1} Mpc^{-1}.

The cosmological scales corresponded to the event horizons $k = aH$ at the epoch of generation of S and T modes of perturbations due to their causal evolution. The background model of the early Universe was described at that time by the dominant field ϕ, which determined the function $H = H(\phi)$ and its derivatives γ, ε, and $\varepsilon_{,N}$. Assuming the last three functions to be small ($\ll 1$, which is confirmed by observations) and leaving only the main terms on γ, we obtain the following perturbation spectra, indices, and ratio of power spectra at the scale $k = He^N$ (see (17)):

$$q_k = \frac{H}{2\pi|\alpha|}, \quad \tilde{q}_k = \frac{H}{\pi m_P}, \tag{9}$$

$$n_k = -\frac{\varepsilon + \gamma}{1 - \gamma} \simeq -\varepsilon - \gamma, \quad r_k = 4\gamma \simeq -4\tilde{n}_k, \tag{10}$$

where $(\ln k)_{,N} = 1 - \gamma \simeq 1$, the index k is omitted for the model functions. A comparison of (9) and (10) with (8) shows that the functions γ, ε, and $\varepsilon_{,N}$ were indeed small, and the value of γ did not exceed the error bars of the observations:

$$\gamma < 0.0025, \quad \beta < 0.05, \quad \varepsilon + \gamma = 0.0175 \pm 0.0025, \tag{11}$$

where functions are evaluated at k_c.

The present observational data provide a small ratio of the power spectra ($r_k < 0.01$) and a constant index of the density perturbation spectrum in a certain wavenumber range ($n_k \simeq -0.02$). This allows us to consider the functions γ and ε as independent from each other and relate them directly to the observational data. By assuming the function $H = H(\phi)$ to be smooth,[2] we obtain the observational model of the early Universe without solving the Friedman equation and without information about the dominant field potential.

4 The Observational Model of Vacuum Relaxation

Starting from values close to zero, the field ϕ moves towards higher values, which allows us to present the Observational Model of vacuum relaxation (OM) as a series in the variable $\xi \equiv n_0\phi^2 \ll 1$:

$$H = H_0\left(1 - \frac{\xi}{2} + O\left(\xi^2\right)\right), \quad N = \ln\left(\frac{k}{H}\right) = \ln\left(\frac{k}{k_c}\right) - N_c, \tag{12}$$

$$\gamma = n_0\xi\left(1 + O(\xi)\right), \quad \varepsilon = n_0\left(1 + O(\xi)\right), \quad \varepsilon_{,N} = n_0 O(\gamma),$$

$$q_k = \frac{\mathrm{H}_0\left(1 + O(\xi)\right)}{2\pi n_0\phi} \simeq \frac{\mathrm{H}_0}{2\pi\beta_c}\left(\frac{k_c}{k}\right)^{n_0}, \quad r_k = 4n_0^2\phi^2\left(1 + O(\xi)\right),$$

$$n_k = -\epsilon\left(1 + \xi\left(1 + n_0\right) + O\left(\xi^2\right)\right) = -n_0\left(1 + O(\xi)\right), \quad \tilde{n}_k \simeq -n_0^2\phi^2,$$

$$\phi \simeq \phi_c e^{n_0(N+N_c)} \simeq \phi_c\left(\frac{k}{k_c}\right)^{n_0}, \quad N_c \simeq \ln\left(\frac{H_0}{k_c}\right), \quad A_k = O(\gamma),$$

[2] To obtain the perturbation spectra, we need the three first derivatives of the H-function (γ, ε, and $\varepsilon_{,N}$) and do not need the higher ones (see (16)). By recovering the initial part of the $H(\phi)$ function (from zero with $\phi > 0$) from observational data, the derivatives β, ε, and the higher terms are still within the observational error bars.

where H_0 and n_0 are constants, $\mathrm{H}_{(0)} \equiv \frac{H_{(0)}}{m_P}$, $\beta_c = n_0\phi_c$. This model provides a red power-law spectrum at scales $k \in (2 \cdot 10^{-4}, 10)$ and is built on the current observational data:

$$\mathrm{H}_0 \simeq 10^{-6}\phi_c, \quad N_c \simeq 120 + \ln\phi_c, \quad n_0 \simeq 0.017, \tag{13}$$

$$\gamma_c < 0.0025, \quad \beta_c < 0.05, \quad \phi_c < 3.$$

The OM has no free parameters, contains two known constants (H_0 and n_0) and relates to the time period with $\varphi < 0.8\, M_P$, $\alpha < 0.014\, M_P$ and $\xi \simeq \frac{\gamma}{\varepsilon} < 0.17$. It describes the beginning of the first stage of vacuum relaxation, where the field was close to zero and moved toward larger values (remaining less than the Planck mass), which ensures the constancy of the power spectrum index and the smallness of the γ-function in the wavenumber range $k = (10^{-4}, 10)$ Mpc^{-1}. Other field trajectories (e.g., models with the field moving from large values to small values) relate the spectrum index to the γ-value and are not compartible to the observational data. The OM has been obtained from the present data and does not require further verification. To extend the OM to small scales ($k > 10$), new data and information on the potential are needed (the counter terms associated with the third constant of OM are within the observational error bars).

5 Quantum-Gravitational Creation of *S* and *T* Modes of Metric Perturbations in the Friedman Universe

The decomposition of the action $S = \int(-\frac{m_P^2}{4}R + \mathcal{L})\sqrt{-g}d^4x$ on linear perturbations q and $\tilde{q}_{ij}$ leads to $S = S^{(0)} + \delta^{(1)}S + \delta^{(2)}S$, where $S^{(0)} = \int(-\frac{3}{2}m_P^2H^2 + \mathcal{L}^{(0)})a^3dtd\mathbf{x}$, $\delta^{(1)}S$ nullifies on the Friedman equations, $\delta^{(2)}S = \int(L + \tilde{L})a^3dtd\mathbf{x}$, the Lagrangian density of S-mode depends on two functions $\alpha^2 = m_P^2\gamma$ and $c_S^{-2} = \frac{w\mathcal{L}_{,w,w}}{\mathcal{L}_{,w}}$, $L = \frac{1}{2}\alpha^2(c_S^{-2}\dot{q}^2 - q_{,i}q^{,i})$ (see [6, 16, 17]). If the condition (2) is satisfied, then $c_S = 1$.

Quantum fields of 4-coordinates $x^\mu = (t, x^i)$, q and $\tilde{q}_{ij}$, and 4-momenta $k_\mu = (\frac{k}{a}, k_i)$, $q_\mathbf{k}$ and $q_{\mathbf{k}\sigma}$, in Eucledian 3-spaces of Friedmannian model $\mathbf{x} = (x^i)$ and $\mathbf{k} = (k_i)$, respectively, are related by the Fourier's transformation:

$$q = \int_{-\infty}^{\infty} q_\mathbf{k} e^{i\mathbf{kx}} \frac{d\mathbf{k}}{(2\pi)^{3/2}}, \quad q_\mathbf{k} = \nu_k a_\mathbf{k} + \nu_k^* a_{-\mathbf{k}}^\dagger, \tag{14}$$

$$\tilde{q}_{ij} = \sum_\sigma \int_{-\infty}^{\infty} p_{ij\sigma} q_{\mathbf{k}\sigma} e^{i\mathbf{kx}} \frac{d\mathbf{k}}{(2\pi)^{3/2}}, \quad q_{\mathbf{k}\sigma} = \tilde{\nu}_k a_{\mathbf{k}\sigma} + \tilde{\nu}_k^* a_{-\mathbf{k}\sigma}^\dagger,$$

where $p_{ij\sigma} = p_{ij\sigma}(\mathbf{k})$ are constant normalized polarization tensors of $\mathbf{k}$-waves ($p^i_{i\sigma} = p^j_{i\sigma} k_j = 0$, $p^*_{ij\sigma} p^{ij}_{\sigma'} = \delta_{\sigma\sigma'} = \mathrm{diag}(1, 1)$, $\sigma = \oplus, \otimes$). The commutation relations of canonically conjugated scalars q and $p = \frac{\partial L}{\partial \dot{q}} = \alpha^2 \dot{q}$ in the position space determine the commutators of constant operators of annihilation, $a_{\mathbf{k}}$, and creation, $a^\dagger_{\mathbf{k}}$, of particles (phonons) in $\mathbf{k}$-space:

$$[q\,(t, \mathbf{x})\, p\,(t, \mathbf{x}')] = qp - pq = i\frac{\delta\left(\mathbf{x} - \mathbf{x}'\right)}{a^3}, \quad \left[a_{\mathbf{k}} a^\dagger_{\mathbf{k}'}\right] = \delta\left(\mathbf{k} - \mathbf{k}'\right),$$

where $\delta = \delta(\mathbf{x})$ is Dirac function. The evolution of $q_{\mathbf{k}}$-oscillators is described by the classical normalized function $\nu_k = \bar{\nu}_k/(\alpha a)$, which depends on conformal time $\eta = \int \frac{dt}{a} = \int \frac{dN}{aH}$ and wavenumber $k = |\mathbf{k}|$:

$$\bar{\nu}''_k + \left(k^2 - U\right)\bar{\nu}_k = 0, \quad U = \frac{(\alpha a)''}{\alpha a}, \quad \bar{\nu}_k \bar{\nu}^{*\prime}_k - \bar{\nu}^*_k \bar{\nu}'_k = i.$$

Similar relations are valid for operators of annihilation ($a_{\mathbf{k}\sigma}$) and creation ($a^\dagger_{\mathbf{k}\sigma}$) of gravitons and evolutionary functions of $q_{\mathbf{k}\sigma}$-oscillators, $\tilde{\nu}_k = \sqrt{2}\bar{\tilde{\nu}}_k/(m_P a)$:

$$[a_{\mathbf{k}\sigma} a^\dagger_{\mathbf{k}'\sigma'}] = \delta\left(\mathbf{k} - \mathbf{k}'\right)\delta_{\sigma\sigma'}, \quad [a_{\mathbf{k}\sigma} a_{\mathbf{k}'\sigma'}] = 0,$$

$$\bar{\tilde{\nu}}''_k + \left(k^2 - \tilde{U}\right)\bar{\tilde{\nu}}_k = 0, \quad \tilde{U} = \frac{a''}{a}, \quad \bar{\tilde{\nu}}_k \bar{\tilde{\nu}}^{*\prime}_k - \bar{\tilde{\nu}}^*_k \bar{\tilde{\nu}}'_k = i.$$

The cosmological S and T modes of zero oscillations of the Friedmannian model are determined by the vacuum $|0\rangle$-state of all linear fields:

$$a_{\mathbf{k}(\sigma)}|0\rangle = 0, \quad \langle a_{\mathbf{k}} a^\dagger_{\mathbf{k}'}\rangle = \delta\left(\mathbf{k} - \mathbf{k}'\right), \quad \langle a_{\mathbf{k}\sigma} a^\dagger_{\mathbf{k}'\sigma'}\rangle = \delta(\mathbf{k} - \mathbf{k}')\delta_{\sigma\sigma'}.$$

The bilinear forms (7) are as follows (see (14)):

$$q_k = \frac{k^{3/2}|\nu_k|}{\sqrt{2}\pi}, \quad \tilde{q}_k = \frac{k^{3/2}|\tilde{\nu}_k|}{\pi}, \quad r_k = \frac{\tilde{q}^2_k}{q^2_k} = \frac{2|\tilde{\nu}_k|^2}{|\nu_k|^2}. \tag{15}$$

The parametric potentials of elementary oscillators are following:

$$U = a^2 H^2 \left((2 - \gamma + \varepsilon)(1 + \varepsilon) + \varepsilon_{,N}\right), \quad \tilde{U} = a^2 H^2 (2 - \gamma). \tag{16}$$

They coincide when the following inequalities are satisfied,

$$\gamma \ll 1, \ |\varepsilon| \ll 1, \ |\varepsilon_{,N}| \ll 1: \qquad U = \tilde{U} = \frac{2}{\eta^2}, \quad \eta = -\frac{1}{aH},$$

which solves the problem of cosmological perturbations in Friedmann model:

$$\bar{\nu}_k = \bar{\tilde{\nu}}_k = \frac{e^{-ik\eta}}{\sqrt{2k}}\left(1+\frac{1}{ik\eta}\right), \quad r_k = 4\gamma, \tag{17}$$

$$q_k = \frac{\sqrt{H^2+\frac{k^2}{a^2}}}{2\pi|\alpha|} \longrightarrow \frac{\mathrm{H}}{2\pi|\beta|}, \quad \tilde{q}_k = \frac{\sqrt{H^2+\frac{k^2}{a^2}}}{\pi m_P} \longrightarrow \frac{\mathrm{H}}{\pi}.$$

The last limit means time-frozen power spectra in wavelengths $k \ll aH$, where the functions $H = m_P\mathrm{H}$ and $\alpha = m_P\beta$ are calculated at the moment of time $\eta = -1/k$.

6 A Vacuum Attractor

Since the fields evolve from zero values, the potential can be decomposed in powers of the fields. Let us assume that the first three terms are essential and the rest is small (see [10]):

$$V = V_0 - \frac{m^2\varphi^2}{2} + \frac{\lambda\varphi^4}{4} + \ldots \longrightarrow V_1 + m^2\left(\frac{\varphi^2-\varphi_1^2}{2\varphi_1}\right)^2, \tag{18}$$

where $V_{0,1}$ and $\varphi_1 = \frac{m}{\sqrt{\lambda}}$ are positive constants (the coefficients of the series are determined by all fields). Owing to symmetry of V relative to the sign of φ, the field evolves from the symmetric state $V_0 = V(0)$ to a positive (or negative) φ with a lower-energy state $V_1 = V(\pm\varphi_1)$ (see Fig. 1):

$$V_0 = V_1 + \frac{m^4}{4\lambda} > V_1 > 0. \tag{19}$$

During the rolling down of the field along the potential the height of the vacuum step $V_{01} = V_0 - V_1$ depends on two fundamental constants, $V_{01} = \frac{m^4}{4\lambda} = \frac{m^2\varphi_1^2}{4}$. Since

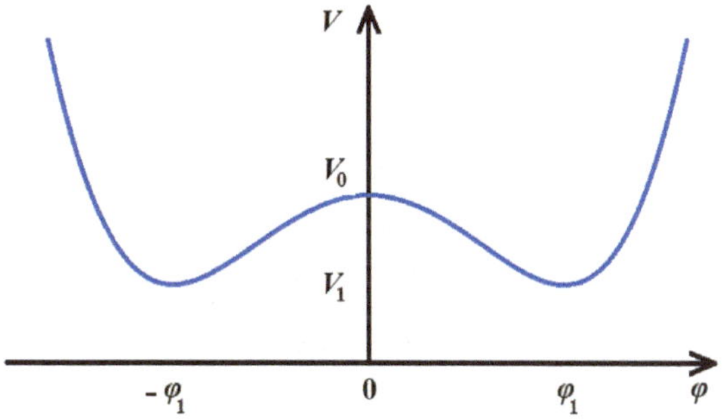

Fig. 1 The potential V as a function of φ. A minimal value V_1 corresponds to $\varphi = \pm\varphi_1$

there were three constants in (18), we take as the third (independent) one the value of the residual vacuum in units of the step height, $\upsilon_1 = \frac{V_1}{V_{01}}$, or the amplitude of the initial vacuum, $\mathrm{v}_1 = \frac{V_1}{V_0} \in (0, 1)$:

$$\upsilon = \upsilon_1 + y^2, \quad \mathrm{v} = \mathrm{v}_1 + \frac{y^2}{\upsilon_0} = 1 - \frac{\mathrm{n}_0 (1 + y) \phi^2}{2}, \tag{20}$$

where $\upsilon = \frac{V}{V_{01}}$, $\mathrm{v} = \frac{\upsilon}{\upsilon_0}$, $y = 1 - x^2$ and $x = \frac{\varphi}{\varphi_1} = \frac{\phi}{\phi_1}$ are dimensionless, $\upsilon_0 = 1 + \upsilon_1 = (1 - \mathrm{v}_1)^{-1}$, $\mathrm{n}_0 = \frac{m_P^2 m^2}{2\mathrm{V}_0}$, $\phi_1 = \frac{\varphi_1}{m_P} = \sqrt{\frac{2}{\upsilon_0 \mathrm{n}_0}}$. At $\upsilon_1 > 1$, $\mathrm{v} \simeq 1$ in the interval $x \in (0, 1)$. When $\upsilon_1 < 1$, the function v varies from $\mathrm{v} \simeq 1$ at $y \in (1, \frac{1}{\sqrt{2}})$ to $\mathrm{v} \simeq \mathrm{v}_1$ at $|y| < \sqrt{\frac{\mathrm{v}_1}{1+\mathrm{v}_1}}$, decreasing at $y^2 \in (\frac{1}{2}, \frac{\mathrm{v}_1}{1+\mathrm{v}_1})$.

Equations of motion can be derived by varying the action with respect to $g^{\mu\nu}$ and φ:

$$\phi_{,\mu}\phi^{,\nu} + \left(\mathrm{V} - \frac{\mathrm{w}^2}{2}\right)\delta^{\nu}_{\mu} = \frac{1}{2}G^{\nu}_{\mu}, \quad x^{;\mu}_{;\mu} + m^2 x \left(x^2 - 1\right) = 0,$$

where $\mathrm{V}_{(0,1)} = \frac{V_{(0,1)}}{m_P^2}$, $\mathrm{w}^2 = \phi_{,\mu}\phi^{,\mu}$, $G^{\nu}_{\mu} = R^{\nu}_{\mu} - \frac{1}{2}R\delta^{\nu}_{\mu}$, the equation for φ is included in gravity equations due to the Bianchi identities ($G^{\nu}_{\mu;\nu} = 0$).

For the case of homogeneous field the Friedman equations are as follows:

$$H^2 = \frac{\dot{\phi}^2 + 2\mathrm{V}}{3} = \frac{2\mathrm{V}}{3 - \gamma}, \quad \dot{H} = -\dot{\phi}^2, \tag{21}$$

resulting in the equation of motion of $\phi = \phi_1 x$ and/or $\dot{\phi} = H\beta$:

$$\ddot{x} + 3H\dot{x} + m^2 x \left(x^2 - 1\right) = 0, \quad \frac{\beta\beta_{,\phi}}{3 - \beta^2} + \beta = \frac{\mathrm{n}_0 y \phi}{\mathrm{v}}, \tag{22}$$

$$H = H_0\sqrt{\frac{\mathrm{v}}{1 - \frac{\gamma}{3}}} = \frac{2\mathrm{m}}{3}\sqrt{\frac{\mathbf{v}}{1 - \frac{\gamma}{3}}}, \quad \gamma = \frac{6\dot{x}^2}{2\dot{x}^2 + m^2\upsilon} \in (0, 3),$$

where $H_0 = \sqrt{\frac{2\mathrm{V}_0}{3}} = \frac{m}{\sqrt{3\mathrm{n}_0}}$, $\mathrm{m} = \sqrt{2}m$, $\mathbf{v}_{(1)} = \frac{3\mathrm{v}_{(1)}}{8\mathrm{n}_0}$, $H_1 = \sqrt{\frac{2\mathrm{V}_1}{3}} = H_0\sqrt{\mathrm{v}_1} = \frac{2\mathrm{m}}{3}\sqrt{\mathbf{v}_1}$. The equation for x can be rewriten in terms of $\mathrm{x} = 1 - x$, which is useful to describe the solutions of the residual vacuum:

$$\ddot{\mathrm{x}} + 3H\dot{\mathrm{x}} + m^2\mathrm{x}\,(1 - \mathrm{x})\,(2 - \mathrm{x}) = 0. \tag{23}$$

Let's represent this equation in terms of $X = xe^{\frac{3}{2}N}$ and $\mathrm{X} = \mathrm{x}e^{\frac{3}{2}N}$:

$$\ddot{X} - \mathrm{m}^2\mathbf{v}_+ X = 0, \quad \ddot{\mathrm{X}} + \mathrm{m}^2\mathbf{v}_-\mathrm{X} = 0, \tag{24}$$

where $\mathbf{v}_+ = \frac{y}{2} + \bar{\mathbf{v}}$, $\mathbf{v}_- = \frac{x(1+x)}{2} - \bar{\mathbf{v}}$, $\bar{\mathbf{v}} = \mathbf{v}(\frac{3-2\gamma}{3-\gamma})$. As we change from t to mt, the mass parameter drops out of the equations (m determines time/energy), and the evolution depends on the two constants n_0 and v_1.

The solutions of the equation for ϕ are trajectories on the phase plane $(\phi, \dot{\phi})$, among which there are three special points (poles): the central $\phi = \dot{\phi} = 0$ and the side ones $\phi \pm \phi_1 = \dot{\phi} = 0$, where $\gamma = \beta = \varepsilon_{,N} = 0$, $H = H_{0,1}$ and $\varepsilon = \varepsilon_{0,1}$ respectively. At the center pole, two parameters H_0 and n_0 are given, while the third v_1 is not included in the derivatives (6). All three parameters H_1, v_1 and $\mathrm{n}_1 = \frac{2\mathrm{n}_0}{\mathrm{v}_1} = \frac{3}{4\mathbf{v}_1}$ are necessary at the side poles. The constants $\varepsilon_{0,1}$ are related to the vacuum constants $\mathrm{n}_{0,1}$ by means of two independent binomials (see Eqs. (6) and (22)):

$$\varepsilon_0^2 + 3\varepsilon_0 - 3\mathrm{n}_0 = 0, \quad \varepsilon_1^2 + 3\varepsilon_1 + 3\mathrm{n}_1 = 0. \tag{25}$$

If $\varepsilon_0 > 0$ the equation has the only solution

$$\varepsilon_{0+} \equiv n_0 = \frac{3\mathrm{n}_0}{|\varepsilon_{0-}|} = \frac{2\mathrm{n}_0}{1 + \sqrt{1 + \frac{4\mathrm{n}_0}{3}}},$$

where $\varepsilon_{0\pm} = \frac{3}{2}(-1 \pm \sqrt{1 + \frac{4\mathrm{n}_0}{3}})$. The second root of equation, ε_{0-}, is negative, since the binomial is invariant under the transformation $\varepsilon_0 \to -3 - \varepsilon_0$. At $\mathbf{v}_1 > 1$ ($\mathrm{n}_1 < \frac{3}{4}$) both roots are negative, $\varepsilon_{1\pm} = \frac{3}{2}(-1 \pm \sqrt{1 - \mathbf{v}_1^{-1}})$, what is needed to build the solution of the vacuum attractor.[3]

The solution of the equations can be found in the integral form:

$$\phi = C \exp\left(\int n dN\right), \quad \beta = n\phi, \quad \varepsilon = (nx)_{,x}, \tag{26}$$

where $C = const$. Equation for $n = n(x)$ is as follows

$$n_{,N} = nn_{,x}x = (3-\gamma)(\mathrm{n} - n) - n^2 = (n_+ - n)(n - n_-),$$

where $\mathrm{n} = \frac{\mathrm{n}_0 y}{\mathrm{v}}$, $n_\pm = \frac{3-\gamma}{2}\left(-1 \pm \sqrt{1 + \frac{4\mathrm{n}}{3-\gamma}}\right)$. The functions n and n are related algebraically via the function ϵ,

$$n = \frac{\mathrm{n}}{1 + \frac{\epsilon}{3}}, \qquad \epsilon = \frac{\varepsilon}{1 - \frac{\gamma}{3}}, \tag{27}$$

[3] Two trajectories enter/exit to/from the central pole along the axes $\dot{\phi} = -(3 + n_0)H_0\phi$ and $\dot{\phi} = n_0 H_0 \phi$, respectively. Other trajectories, approaching the point $\dot{x} = x = 0$, turn without entering the pole. All trajectories, except for the central point, enter the side (stable) poles.

which satisfies the equation ($\epsilon_{,N} = n\epsilon_{,x}x$):

$$\epsilon_{,N} = 3\mathbf{n} + (\gamma - 3)\left(\epsilon + \frac{\epsilon^2}{3}\right) = \left(1 - \frac{\gamma}{3}\right)(\epsilon_+ - \epsilon)(\epsilon - \epsilon_-),$$

where $\mathbf{n} = (\mathrm{n}x)_{,x} = \frac{\mathrm{n}_0\mathbf{y}}{\mathrm{v}}$, $\epsilon_\pm = \frac{\varepsilon_\pm}{1-\frac{\gamma}{3}} = \frac{3}{2}(-1 \pm \sqrt{1 + \frac{4\mathbf{n}}{3-\gamma}})$, $\mathbf{y} = \mathrm{v}(\frac{xy}{\mathrm{v}})_{,x} = 1 + \frac{x^2(\mathrm{v}-4\mathrm{v}_1)}{\mathrm{v}}$.

In the interval $x \in (0, 1)$ the function $\mathbf{y}$ varies from 1 to –2, crossing 0 ($\mathbf{y}_* = 0$) at $x_*^2 = \frac{ly_*}{2} = \frac{l}{l+2} \in (\frac{1}{3}, 1)$:

$$\mathbf{y} = \frac{\hat{y}\left(c_0c_1 + (l-1)\tilde{y} - \tilde{y}^2\right)}{(l+2)\left(c_1 + 2\tilde{y} + \tilde{y}^2\right)} = c\hat{y} + O\left(\hat{y}^2\right),$$

where $\tilde{y} = \frac{y-y_*}{y_*}$, $\hat{y} = 2\tilde{y}$, $c = \frac{c_0}{l+2} \in (0.7, 1)$. The constants are determined by the parameter $l > 1$,

$$c_0 = l + 1 + l^{-1}, \quad c_1 = \frac{2l}{l-1}, \quad \mathrm{v}_1 = \frac{\mathrm{v}_*(l+1)}{2l} = \frac{4(l+1)}{l\ell},$$

$$v_1 = \frac{y_*^2(l+1)}{l-1}, \quad \ell = l^2 + 3l + 4 = (l_+ + 2)(l_- + 2), \quad l_\pm = l \pm \sqrt{l}.$$

The general solution has the following approximation:

$$n = \frac{n_0 y}{u\mathrm{v}}, \quad \beta = \frac{\beta_1 xy}{u\mathrm{v}}, \quad \epsilon = 3(u-1) + n_0 u = \frac{n_0\left(\mathbf{y} - \frac{u_{,x}xy}{u}\right)}{u\mathrm{v}\left(1 - \frac{\gamma}{3}\right)}, \tag{28}$$

where $u = \frac{3+\epsilon}{3+n_0}$, $\beta_1 = \sqrt{\gamma_1} = n_0\phi_1 = \sqrt{\frac{2\mathbf{n_0}}{v_0}}$, $\mathbf{n}_0 = \frac{3n_0}{3+n_0}$.

A vacuum attractor (VA) is a partial solution of the Friedman equations consisting of a sequence of eras: the motion goes from the quantum boundary to the central pole, approaches it along the axis $\dot{\phi} = -(3 + n_0)H_0\phi$ and exits along $\dot{\phi} = n_0H_0\phi$, where $n = \epsilon = \varepsilon = n_\pm = \epsilon_\pm = (n_0, -3 - n_0)$ near $\phi = \dot{\phi} = 0$.

The exit of VA from the center to the region $\phi > 0$ corresponds to the OM (12) with $\xi = \frac{6x^2}{v_0(3+n_0)}$, where the constants H_0 and n_0 are identical to those described in (13). Using the smallness of n_0, we represent the functions of $x \in (0, 1)$ as a series in $\zeta = x\bar{\zeta} = \frac{\mathbf{n}_0x^2}{\mathrm{v}}$[4] with $\gamma < 1$:

[4] The constants n_0, n_0, and $\mathbf{n}_0$ are arranged from observations, $\sqrt{\mathrm{n}_0\mathbf{n}_0} = n_0$, $\sqrt{\mathrm{n}_0/\mathbf{n}_0} = 1 + \frac{n_0}{3}$. The squares of the variables ζ and $\bar{\zeta}$ are as follows: $\zeta^2 < \zeta\bar{\zeta} < \bar{\zeta}^2 < \zeta < \bar{\zeta} < 1$.

$$u = 1 + \mathrm{u}\zeta + \frac{\mathbf{u}\bar{\zeta}^2}{3} + O\left(\frac{n_0^3 x^2}{\mathrm{v}^3}\right), \quad \gamma = \frac{2\zeta y^2}{u^2\left(y^2 + v_1\right)}, \tag{29}$$

$$\epsilon = \frac{n_0}{\mathrm{v}}\left(\mathbf{y} + \mathbf{u}\zeta + O\left(\bar{\zeta}^2\right)\right), \quad \phi \simeq \bar{\phi}_0 y^{\mathrm{v}_1/2}\left(\frac{k\, e^{\phi^2/2}}{k_0\sqrt{\mathrm{v}}}\right)^{n_0},$$

$$N + N_0 \simeq \frac{1}{n_0}\left(\ln|x| - \frac{\mathrm{v}_1}{2}\ln|y| + \frac{2y - 1 + \ln 2}{2v_0}\right) \simeq \ln\left(\frac{k}{k_0\sqrt{\mathrm{v}}}\right),$$

where functions u and $\mathbf{u}$ are constrained,[5] $|\mathrm{u}| < 1$, $|\mathbf{u}| < 1$, the condition $k = \frac{k_0\sqrt{1+3\mathrm{v}_1}}{2}$ ($N = -N_0$) is valid at the point $y = x^2 = \frac{1}{2}$ ($\phi = \phi_0$), the constants are related: $\bar{\phi}_0 = \phi_1(\sqrt{2e})^{\mathrm{v}_1 - 1}$, $N_0 = \ln(\frac{H_0}{k_0})$, $\phi_0 = \frac{\phi_1}{\sqrt{2}} = \frac{1}{\sqrt{v_0 n_0}} \simeq 7\sqrt{1 - \mathrm{v}_1}$, $\phi_1 \simeq 10\sqrt{1 - \mathrm{v}_1}$. The series of VA are valid for $x < \frac{8}{9}$, $y > \frac{2}{9}$. For smaller y, when the solution of the VA is close to the pole $x - 1 = \dot{x} = 0$, we need to consider Eq. (23).

The potential of all fields responsible for the vacuum polarization is obtained as a series in the fields with their values close to zero. At the first stage of the CVR, the first three terms of the series of the dominant field are taken without increasing the number of constants of the potential (the higher terms were assumed negligible).[6] There are three coefficients of the three terms of the series (18) which give three constants of the VA: H_0, n_0 and v_1. The first two are determined from observations, and the third is free—it does not affect the onset of VA and appears later in time, at $k > 10$, when the solution changes. The functions n and ε are close initially ($\phi < 3$) and then, approaching ϕ_1, they diverge according to the ratio of the parameters v_1 and $\mathrm{v}_{cr} = \frac{8\mathrm{n}_0}{3} \simeq 0.05$, $\mathbf{v}_1 = \frac{\mathrm{v}_1}{\mathrm{v}_{cr}} \simeq 20\mathrm{v}_1$ (see Fig. 2).

---

[5] The formulas for these functions are as follows:

$$u = 1 + \frac{\epsilon - n_0}{3 + n_0}, \quad \mathrm{u} = \frac{\mathbf{y} - \mathrm{v}}{3x^2} = 1 - \frac{2\mathrm{v}_1\,(\mathrm{v} + 2)}{3\mathrm{v}} - \frac{\mathrm{n}_0\phi^2}{6},$$

$$\mathbf{u} = \mathbf{y}\bar{\mathrm{u}} + \frac{2\mathbf{y}_{,y}y}{3}, \quad \bar{\mathrm{u}} = \frac{1}{3}\left(-5 + \frac{2\mathrm{v}_1\,(\mathrm{v} + 3)}{\mathrm{v}} + \frac{\mathrm{n}_0\phi^2}{2}\right).$$

[6] The subject of our model is not the way of reduction of the cosmological constant by means of introduction of a "specific" potential of interactions. Thus, the potential of the [20] model contains more constants and higher derivatives of the field (in the form of the field cosine), and the movement from large to small fields does not correspond to our approach. By vacuum we understand its polarization in the external gravitational field with the subsequent relaxation of vacuum in time as a generator of the evolving Universe.

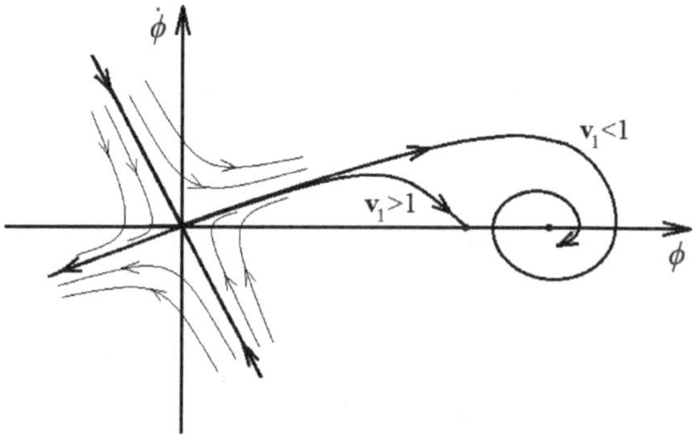

Fig. 2 The trajectories of the solution of Eq. (22) on the phase plane $(\phi, \dot{\phi})$. Bold lines correspond to the vacuum attractor, thin lines to the general solution, and arrows indicate the time direction (see [10])

7 The Case of a Small Residual Vacuum ($\mathrm{v}_1 < 1$)

The function γ increases monotonically in time, approaching $\gamma \sim 1$ at $t \simeq \mathrm{m}^{-1}$ ($x \simeq 0.87$, $y \simeq \sqrt{\mathrm{v}_{cr}} \simeq 0.2$, $H \simeq \frac{2\mathrm{m}}{3}$). If $t > \mathrm{m}^{-1}$ ($y < 0.2$), the VA trajectory circles the pole clockwise, crossing the $x = 1$ axis at time t_1, and the $\dot{x} = 0$ axis at $t_b > t_1$, moving from the top to the bottom of the phase plane. Then the field continues to oscillate around the point $y = \dot{y} = 0$, approaching it: $y \simeq 0.2\,\frac{\cos(\mathrm{m}t)}{\mathrm{m}t}$ ($H \simeq \frac{2}{3t}$) at $t \in (\mathrm{m}^{-1}, t_{\mathrm{v}})$, and $y \sim e^{-t/t_{\mathrm{v}}} \cos(\mathrm{m}_{\mathrm{v}} t)$ ($H \simeq \frac{2}{3t_{\mathrm{v}}}$) when $t > t_{\mathrm{v}}$, where $t_{\mathrm{v}} = \frac{1}{\mathrm{m}\sqrt{\mathrm{v}_1}}$, $\mathrm{m}_{\mathrm{v}} = \mathrm{m}\sqrt{1 - \mathrm{v}_1}$, we omit the ~ 1 phases of the cosines. Here the first stage of the cascade terminates and the problem is reduced to the previous one–the second stage of the CRV with a new vacuum $V = V_1$ and a new dominating field appears.

8 The Case of a Higher Residual Vacuum ($\mathrm{v}_1 > 1$)

Here $\mathrm{v}_1 = \frac{4(l+1)}{l\ell} \in (\frac{1}{20}, 1), l \in (1, 8), v_1 = \frac{4(l+1)}{(l+2)^2(l-1)} > \frac{1}{20}, \sqrt{\mathrm{v}_1} \in (\frac{2}{9}, 1)$. The function γ increases with the field and reaches the maximum value at $\gamma_r = \beta_r^2 \simeq \frac{n_0(l-1)(l+2)}{9} \simeq \frac{0{,}007}{v_1}$ at $x_r \simeq \sqrt{\frac{l}{l+2}} \in (0.6, 0.9)$ and goes back to zero at $x \to 1$, remaining small at $x \in (0, 1)$. At $x \in (0, \frac{8}{9})$, $y \in (1, \frac{2}{9})$ there is $\beta \simeq \frac{2xy}{\phi_1 v} \leq \beta_r \simeq \frac{0.08}{\sqrt{v_1}} \in (0, 0.4), \dot{x} \simeq \frac{\mathrm{m}xy}{\sqrt{3v}}$. In the region $\mathrm{x} \in (\frac{1}{9}, \frac{1}{2}\sqrt{\mathrm{v}_1})$ there is $\beta \simeq \frac{4\mathrm{x}}{\phi_1 v_1}$ and $\dot{x} \simeq \frac{2\mathrm{mx}}{\sqrt{3v_1}}$. For $|\mathrm{x}| < \frac{1}{2}\sqrt{\mathrm{v}_1}$ the solution (23) is as follows:

$$\mathrm{x} = \mathrm{x}_\times \left(\mathrm{c}_+ e^{-\bar{t}_+} + \mathrm{c}_- e^{-\bar{t}_-}\right) = \mathrm{x}_\times e^{-\bar{t}} \left(\mathrm{ch}\left(\omega \bar{t}\right) - \frac{\mathrm{c} \cdot \mathrm{sh}\left(\omega \bar{t}\right)}{\omega}\right), \tag{30}$$

$$\frac{\dot{x}}{\mathrm{m}\sqrt{\mathbf{v}_1}\mathrm{x}} = 1 + \omega + \frac{2\omega}{\mathbf{c}e^{-2\omega\bar{t}} - 1} = 1 + \mathrm{c} + \frac{\left(\mathrm{c}^2 - \omega^2\right) \mathrm{th}\left(\omega \bar{t}\right)}{\omega - \mathrm{c} \cdot \mathrm{th}\left(\omega \bar{t}\right)},$$

where $\bar{t} = \mathrm{m}\sqrt{\mathbf{v}_1}(t - t_\times) \geq 0$ and $\bar{t}_\pm = (1 \pm \omega)\bar{t}$ are the variables, $\mathrm{x}_\times, \omega = \sqrt{1 - \mathbf{v}_1^{-1}}$, $\mathrm{c}_\pm = \frac{\omega \pm \mathrm{c}}{2\omega}$, $\mathbf{c} = \frac{\mathrm{c}+\omega}{\mathrm{c}-\omega}$ and c are constants.The quantities $\mathrm{x}_\times \in (\frac{1}{9}, \frac{1}{2}\sqrt{\mathrm{v}_1})$ and $\mathrm{c} \simeq \frac{2}{\sqrt{3 v_1 \mathbf{v}_1}} - 1 \in (4.2, -1)$ can be derived from the cross-linking with VA at $\bar{t} = 0$ ($\mathrm{x} = \mathrm{x}_\times$, $\dot{\mathrm{x}} \simeq \frac{2\mathrm{m}}{\sqrt{3 v_1}}\mathrm{x}_\times$).

Let us consider two cases.

For $t > t_\times$ and $\mathrm{c} > \omega$ ($v_1 < \frac{1}{7}, \mathrm{v}_1 < \frac{1}{8}, \mathbf{v}_1 \in (1, 2.7), \omega \in (0, 0.8), l \in (4.5, 8)$),[7] a part of trajectories of VA circles the pole $\mathrm{x} = \dot{\mathrm{x}} = 0$ clockwise, at $\bar{t}_1 = \frac{\ln \mathbf{c}}{2\omega}$ it crosses the axis $x = 1$ and moves in the upper halfplane. At $\bar{t}_b$ the trajectory crosses the axis $\dot{x} = 0$ moving from the upper area to the lower area of the plane, and then at $\bar{t} > \bar{t}_2$ it goes to the pole from the right ($\phi > \phi_1$) along the axis $\dot{\phi} = \frac{\mathrm{m}(\phi_1 - \phi)}{\sqrt{\mathbf{v}_1} + \sqrt{\mathbf{v}_1 - 1}}$, where $\bar{t}_{1,2} = \bar{t}_b \pm \Delta_b$, $\Delta_b = \frac{\ln(\frac{1+\omega}{1-\omega})}{2\omega} = \frac{\ln(\sqrt{\mathbf{v}_1} + \sqrt{\mathbf{v}_1 - 1})}{\omega}$. In this interval the function n approaches zero, and ε is finite: $\varepsilon \simeq (\ln |\dot{x}|)_{,N}$. It is constant at $\bar{t} < \bar{t}_1$ ($\varepsilon \simeq -n_1$) and $\bar{t} > \bar{t}_2$ ($\varepsilon \simeq -n_2$, where $n_2 = \frac{2n_1}{1+\omega}$),[8] but in the period $\bar{t} \in (\bar{t}_1, \bar{t}_2)$ diverges, being the manifestation of the $H_1^2/\dot{\phi}$ function in the form of a high peak (δ-function): $\varepsilon \simeq (\ln |\dot{x}|)_{,N}$, which is a local feature (bump) in the spectrum of q_k at $k_b = H_1 a(t_b)$.

When $\mathrm{c} < \omega$ ($v_1 > \frac{1}{7}$, $\mathbf{v}_1 > e$, $l \in (1, 4.5)$), the trajectories coming from the axis $\dot{\phi} = 0$ (including VA) move to the pole $x = 1$ from the left ($\phi < \phi_1$), approaching it radially along $\dot{\phi} \simeq \frac{\mathrm{m}(\phi_1 - \phi)}{\sqrt{\mathbf{v}_1} + \sqrt{\mathbf{v}_1 - 1}}$ which formally takes infinite time.

9 Power Spectra of Perturbations

At $k \ll k_0$, the spectrum has a power-law (red) form in agreement with the observations (see (29)):

$$q_k = \bar{q}_0 \left(\frac{k_0}{k}\right)^{n_0}, \quad r_k = \bar{r}_0 \left(\frac{k}{k_0}\right)^{2n_0}, \quad \phi = \bar{\phi}_0 \left(\frac{k}{k_0}\right)^{n_0}, \tag{31}$$

where $\bar{q}_0 = \frac{\mathrm{H}_0}{2\pi n_0 \bar{\phi}_0} \simeq 10^{-5}(\frac{k_c}{k_0})^{n_0}$, $\bar{r}_0 = 4n_0^2\bar{\phi}_0^2$, $\bar{\phi}_0 = \phi_c(\frac{k_0}{k_c})^{n_0}$. The extension of the spectrum to the scales of hundreds of kiloparsecs or less depends on a value of the third constant $\mathbf{v}_1$, leading to an additional power in the form of a "bump" and/or a blue density perturbation spectrum.

---

[7] The condition $c^2 = \omega^2$ leads to the equation $\frac{4}{3v_1} + 1 = \phi_1 \simeq \frac{10.7}{\sqrt{v_0}}$, whence we obtain $v_1 \simeq \frac{1}{7}$, $\omega \simeq \frac{4}{5}$ (the second root of equation $v_1 \simeq 100$ is dismissed).

[8] From (30) one can derive

$$\mathrm{x} \propto \left(\mathbf{c}e^{-\tilde{t}} - 1\right) e^{-\tilde{t}_-}, \quad \dot{x} \propto \left(e^{\tilde{t}_b - \tilde{t}} - 1\right) e^{-\tilde{t}_-}, \quad \ddot{x} \propto \left(1 - e^{\tilde{t}_2 - \tilde{t}}\right) e^{-\tilde{t}_-},$$

where $\tilde{t} = 2\omega\bar{t}$, $\tilde{t}_b = \ln \mathbf{c} + 2\ln(\sqrt{\mathbf{v}_1} + \sqrt{\mathbf{v}_1 - 1})$, $\tilde{t}_{1,2} = \tilde{t}_b \pm 2\ln(\sqrt{\mathbf{v}_1} + \sqrt{\mathbf{v}_1 - 1})$.

At $\mathbf{v}_1 < 1$, the ratio of spectra r_k grows with increasing k and becomes large (~ 1) at $k = k_r \sim k_1$. When $k = k_b > k_1$, a high peak appears in q_k, potentially leading to the birth of the PBH. At higher wavenumbers $k > k_b$, the spectra fall off due to oscillations of the ϕ field near the side pole. At $\mathbf{v}_1 \ll 1$ the oscillations lead to instabilities: the birth of field particles with mass m and their further decay, which may result in the hydrodynamical phase of the radiation-dominated plasma. This, of course, refers to the decay of the oscillatory part of the field φ, but all fields included in the vacuum polarizations remain stable (in particular, the state V_1 is conserved).

For $\mathbf{v}_1 > 1$ and $x \in (0, 0.9)$ the perturbation spectra are double-fold power law:

$$q_k = \frac{q_0 \mathrm{v}^{\frac{3}{2}}}{xy}, \quad r_k = r_0 \left(\frac{xy}{\mathrm{v}}\right)^2, \quad \hat{x}^{\frac{1}{n_0}} \hat{y}^{-\frac{1}{n_1}} = \frac{k\, e^{\frac{\phi^2 - \phi_{01}^2}{2}}}{k_0 \sqrt{\mathrm{v}}}, \tag{32}$$

$$n_k \simeq -\frac{n_0}{\mathrm{v}} \left(1 + 3x^2 \left(1 - 2\frac{\mathrm{v}_1}{\mathrm{v}}\right)\right) \in (-n_0, n_1), \quad n_1 = \frac{2n_0}{\mathrm{v}_1},$$

where $q_0 = \bar{q_0}(\sqrt{2e})^{\mathrm{v}_1 - 1}, r_0 = \bar{r_0}(2e)^{1-\mathrm{v}_1}, \hat{x} = \frac{\phi}{\phi_{01}} = \sqrt{2}x, \hat{y} = 2y, n_1 \in (0.04, 0.7)$. For $k < k_0$ ($x < 0.6$), Eqs. (32) correspond to the asymptotics (31). For $k > k_0 > k_1$ ($y \in (0.2, \sqrt{\frac{\mathrm{v}_1}{1+\mathrm{v}_1}})$), the spectra are blue:

$$q_k = q_1 \left(\frac{k}{k_1}\right)^{n_1}, \quad r_k = r_1 \left(\frac{k_1}{k}\right)^{2n_1}, \quad x^2 = 1 - \left(\frac{k_1}{k}\right)^{n_1}, \tag{33}$$

where $q_1 = q_0 \mathrm{v}_1^{3/2}, r_1 = \frac{r_0}{\mathrm{v}_1^2}, k_1 = k_0 \sqrt{\mathrm{v}_1} (\sqrt{\frac{2}{e}})^{\phi_{01}^2}$.

A further growth of q_k depends on the value of the residual vacuum. At $|\mathrm{x}| < \frac{1}{2}$ the equation has the form

$$\ddot{\delta} + \mathrm{m}\sqrt{4\mathbf{v}_1 + 3\delta^2}\dot{\delta} + \mathrm{m}^2\delta = 0, \tag{34}$$

where $\delta = \phi - \phi_1 = -\phi_1 \mathrm{x}$. At $\mathbf{v}_1 \sim 1$, a bump appears in the spectrum (33) at wavenumber $k_b > k_1$, which has the form of a single peak with its amplitude $q_b \lesssim 1$.[9] At $\mathbf{v}_1 > 1$ the bump in VA spectrum is absent and it has a three-fold power law form with indices $n_k = (-n_0, n_1, n_2)$. Finishing the first stage of the CVR, the field freezes at the pole V_1 during formally an infinite time, $\phi \to \phi_1$. After the state V_1 the second stage of CVR with a new field begins (and so on in the cascade of stages of the potential V).

These changes in the spectrum at $k > 10$ Mpc^{-1} may lead to both the early star formation (as indicated by data from the JWST) and the birth of PBHs and early SMBHs (as indicated by LIGO and [11], respectively). The extension of the

[9] The peak can be approximated by a Gaussian (cf. [13]), at $k > k_b$ the spectrum has again the power-law form $q_k \sim k^{n_2}$. The bump at $k \simeq k_b$ belongs to the growing part of the three-power-law spectrum q_k with the indices $n_k = (-n_0, n_1, n_2)$.

spectrum to smaller scales can be restored if the observations of the JWST as well as the fate of the halo mass functions and evolution of subhalos in massive galaxies will be confirmed (see [13–15] and references therein). The subsequent more accurate observational data can fix the characteristics of the power spectrum at the small scales.

10 The Entrance and Exit from the Vacuum Attractor

In the neighborhood of the central pole $|\phi| < \phi_0$ and $|\dot{\phi}| < H_0$, the solution of Eq. (22) is the sum of increasing and decreasing exponentials in time:

$$\phi = C_0 a^{n_0} + \bar{C}_0 a^{-\bar{n}_0}, \quad a = \frac{k}{H_0} \propto e^{H_0 t}, \tag{35}$$

where $\bar{n}_0 = 3 + n_0$. The constants C_0 and $\bar{C}_0$ are proportional to C, and their ratio is in the scale when the exponential terms are equal $k_i = H_0 a_i$ (beginning/boundary/entrance in the VA):

$$\phi = \phi_i \kappa^{n_0} \left(1 \pm \vartheta^2 \kappa^{-n_*}\right), \quad n = n_0 \frac{1 \mp \vartheta \kappa^{-n_*}}{1 \pm \vartheta^2 \kappa^{-n_*}}, \quad \varepsilon = n_0 \frac{1 \pm \kappa^{-n_*}}{1 \mp \vartheta \kappa^{-n_*}},$$

where $\kappa = \frac{k}{k_i} = e^{H_0(t-t_i)}$, $n_* = n_0 + \bar{n}_0 = \sqrt{3(3+4\mathrm{n}_0)}$, $\vartheta = \frac{n_0}{\bar{n}_0} \simeq 0.005$. The upper sign refers to the right ($\phi_i > 0$) or left ($\phi_i < 0$) quadrants of the phase plane bounded by VA, and the lower sign—to the upper or lower quadrants.

In the vicinity of the central point, the field trajectories are under the gravitational influence of the giant swirl of the pole, which draws the trajectories to the partial automodel solution of the VA at $\kappa \sim 1$. Coming to the pole from the $|\beta| \sim 1$ side, the trajectories move along the VA, approaching the VA at $\kappa > 1$, and turn near the center, moving away from it together with the VA towards $|\phi| \sim \phi_1/2$:

$$n = n_0 \left(1 \mp \frac{\vartheta\,(1+\vartheta)}{\kappa^{n_*}} + O\left(\frac{\vartheta^3}{\kappa^{2n_*}}\right)\right), \quad \varepsilon = n_0 \left(1 \pm \frac{1+\vartheta}{\kappa^{n_*}} + O\left(\frac{\vartheta}{\kappa^{2n_*}}\right)\right). \tag{36}$$

In the evolutionary trajectory of the solution, only the growing exponent remains, since the decaying exponent quickly becomes insignificant. This means that the solution enters the VA near V_0 initially with $\phi \in (\phi_i, \phi_0)$, $\kappa \in (1, e^{N_{0i}})$ and the constant indices $n \simeq \varepsilon \simeq n_0$ (see Eqs. (28), (29)):

$$H = H_0 \left(1 - \frac{n_0 \phi^2}{2}\right), \quad \phi = \phi_i \kappa^{n_0} \simeq \phi_0 e^{n_0 (N + N_0)}, \tag{37}$$

where $N_{0i} \simeq 57 \ln(\frac{\phi_0}{\phi_i})$. The power spectrum index is weakly dependent on ϕ ($q_k \propto H^2/\dot{\phi}$, $n_k \simeq -n_0$) and agrees well with the observational model (12) with $\phi_c \in (\phi_i, \phi_0)$, $k_0 \simeq 0.05 \exp(N_{0c})$ Mpc^{-1}, $N_{0c} \simeq 57 \ln(\frac{5\sqrt{1-v_1}}{\phi_c})$. The extension of VA into the neighborhood V_1 depends on a single free parameter v_1, which is not yet constrained by the present data.

Near the side pole, when $|x| < 0.5$, $|\dot{\phi}| < H_1$ and $\mathbf{v}_1 > 1$, the solution of Eq. (23) is the sum of two decaying exponentials:

$$\phi = C_1 a^{-\bar{n}_1} + C_2 a^{-\bar{n}_2}, \quad a = \frac{k}{H_1} \propto e^{H_1 t}, \tag{38}$$

where $\bar{n}_{1,2} = -\varepsilon_{1\pm} = \frac{3}{2}(1 \mp \sqrt{1 - \mathbf{v}_1^{-1}})$. As we see, the first stage of VA leads the field to one of the side poles with $\phi = \pm\phi_1$ and $\dot{\phi} = 0$, where the field appears in a thermal bath formed by quantum fluctuations of all fields, i.e. in a thermostat with the temperature $T = \frac{H_1}{2\pi}$, during formally an unlimited time [21–23]. This situation differs from De Sitter space-time, since the landscape of V is not a constant (the state V_1 is stable for ϕ, but not for all fields). Therefore, there will always be a field (not ϕ) that will continue the CVR from V_1 to a lower state $V_2 < V_1$. The CVR itself is the VA: it can be entered once, as in the state with V_0 for our trajectory of the Universe, the CVR (as well as VA) finishes when the landscape ends.

11 Conclusions

The model of CVR is suggested by the observational data of recent years and can be called "natural" since it unties the three known parameters independent of each other (spectrum index, power spectra ratio, and number of N-epochs of evolution). The epoch of the growth of ϕ from zero is characterized by a small ratio of the power spectra ($r_k < 0.01$) and a constant density perturbation spectrum index ($n_k \simeq -0.02$). As long as r_c is unknown, the observational scales refer to any moment in time with $\varphi_c < 0.8\, M_P$. The OM is built on the current data and provides a red power-law spectrum at $k < 10$ without requiring information about a potential. The theoretical VA model has two known constants, and a third ($\mathbf{v}_1$) is necessary to extend the power spectrum at $k > 10$. The comparison of VA with OM leads to the concept of CVR as a generator of the evolving Universe, which solves the problems of the observable cosmology.

The CVR in the expanding Universe is a sequence of evolutionary epochs of polarized vacuum density decrease determining by the dominating fields (each in its own time) going from initial zero states to non-zero values, from the dominating scalar field in the early Universe to the subsequent ones, including the Λ-term in the modern Universe. The CVR can create the entire observable cosmology—from the

Friedmann model with small perturbations of the metric with non-power-law spectra, from which PBHs could have arisen, and gravitational waves in a wide wavenumber range, to the formation of dark matter and dark energy, early galaxies and SMBHs, the large-scale structure of the Universe.

Acknowledgements Authors cordially thanks P.B. Ivanov for the assistance in preparing the paper.

References

1. A.A. Starobinsky, A new type of isotropic cosmological models without singularity. Phys. Lett. B **91**, 99–102 (1980)
2. S. Gottlöber, V. Müller, A.A. Starobinsky, Analysis of inflation driven by a scalar field and a curvature-squared term. Phys. Rev. D **43**, 2510–2520 (1991)
3. V.N. Lukash, E.V. Mikheeva, Λ-inflation and CMB anisotropy. Int. J. Mod. Phys. A **15**, 3783 (2000)
4. V.N. Lukash, E.V. Mikheeva, V. Müler, A.M. Malinovsky, Generalized inflation with a gravitational wave background. MNRAS **317**, 795 (2000)
5. A.O. Barvinsky, A.Yu. Kamenshchik, A.A. Starobinsky, Inflation scenario via the Standard Model Higgs boson and LHC. JCAP **11**, 021 (2008). arXiv: 0809.2104 [hep-ph]
6. V.N. Lukash, E.V. Mikheeva, *Physical Cosmology* (FIZMATLIT, Moscow, 2010). ((in Russian))
7. D.S. Gorbunov, V.A. Rubakov, *Introduction to the very early Universe* (KRASAND, Moscow, 2010). ((in Russian))
8. V.A. Filippov, R.M. Lakhibbaev, D.I. Kazakov, D.M. Tolkachev, Dark energy due to quantum corrections to effective potential. Class. Quantum Gravity. **42**, 145005 (2025). arXiv:2405.18818 [hep-th]
9. BICEP/Keck Collaboration: P.A.R. Ade, Z. Ahmed, M. Amiri, D. Barkats, et al., Improved constraints on primordial gravitational waves using planck, WMAP, and BICEP/Keck observations through the 2018 observing season. Phys. Rev. Lett. **127**, 151301 (2021)
10. V.N. Lukash, E.V. Mikheeva, The cascade relaxation of a gravity-driven vacuum as the generator of the expanding universe. JETP Lett. **121**, 6 (2025)
11. R.L. Larson, et al., A CEERS discovery of an accreting supermassive black hole 570 Myr after the Big Bang: identifying a progenitor of massive z > 6 quasars. Astrophys. J. Lett. **953**, L29 (2023). arXiv: 2303.08918 [astro-ph.GA]
12. Y. Harikane, M. Ouchi, M. Oguri, Y. Ono, et al., A comprehensive study of galaxies at $z \simeq 9-16$ found in the early JWST data: ultraviolet luminosity functions and cosmic star formation history at the pre-reionization epoch. Astrophys. J. Supp. **265**, 5 (2023). arXiv: 2208.01612 [astro-ph]
13. M.V. Tkachev, S.V. Pilipenko, E.V. Mikheeva, V.N. Lukash, Excess of high-z galaxies as a test for bumpy power spectrum of density perturbations. MNRAS **527**, 1381–1388 (2024)
14. Y.N. Eroshenko, V.N. Lukash, E.V. Mikheeva, S.V. Pilipenko, M.V. Tkachev, Properties of central regions of the dark matter halos in the model with a bump in the power spectrum of density perturbations. JETP Lett. **120**, 83 (2024)
15. M.V. Tkachev, S.V. Pilipenko, E.V. Mikheeva, V.N. Lukash, Inner structure of dark matter halos at high z in cosmological models with non-power-law primordial spectra. Phys. Rev. D **110**, 083530 (2024)
16. V.N. Lukash, Production of sound waves in the early universe. JETP Lett. **31**, 596 (1980)
17. V.N. Lukash, Production of phonons in an isotropic universe. JETP **52**, 807 (1980)

18. Planck Collaboration: P.A.R. Ade, N. Aghanim, C. Armitage-Caplan, M. Arnaud, et al., Planck 2013 results. XXII. Constraints on inflation. Astron. Astrophys. **571**, A22 (2014). arXiv:1303.5082 [astro-ph.CO]
19. Planck Collaboration: N. Aghanim, Y. Akrami, M. Ashdown, J. Aumont, et al., Planck 2018 results. VI. Cosmological parameters. Astron. Astroph. **641**, A6 (2020). arXiv: 1807.06209 [astro-ph.CO]
20. L.F. Abbott, A mechanism for reducing the value of the cosmological constant. Phys. Lett. B **150**, 427–430 (1985)
21. G.W. Gibbons, S.W. Hawking, Cosmological event horizons, thermodynamics, and particle creation. Phys. Rev. D **15**, 2738 (1977)
22. A.A. Starobinsky, J. Yokoyama, Equilibrium state of a self-interacting scalar field in the de Sitter background. Phys. Rev. D **50**, 6357 (1994)
23. G.E. Volovik, De Sitter local thermodynamics in f(R) gravity. JETP Lett. **31**, 564 (2024)

Modified Gravity Models

Towards $\mathcal{N}=2$ Higher-Spin Supergravity

Ioseph Buchbinder, Evgeny Ivanov, and Nikita Zaigraev

Abstract We review the superfield formulation of $\mathcal{N}=2$ higher-spin supergravity theory in harmonic superspace. The analysis of both the hypermultiplet higher-spin supersymmetries and conformal supersymmetries is performed. The analytic superspace gauging of these symmetries gives rise to a set of unconstrained analytical prepotentials describing $\mathcal{N}=2$ higher-spin off-shell supermultiplets. This procedure naturally yields cubic interaction vertices of $\mathcal{N}=2$ higher spins with the hypermultiplet. Based on these results, the consistent interaction of an infinite tower of $\mathcal{N}=2$ superconformal higher spins with hypermultiplet is presented. Proceeding from this model, a method to construct a consistent interacting theory of $\mathcal{N}=2$ higher-spin supergravity is proposed by making use of the conformal compensators.

1 Introduction

This paper is dedicated to the blessed memory of Professor A.A. Starobinsky, outstanding scientist and remarkable person. His main scientific interests were focused on the problems associated with the development of modern gravity theory and cosmology, the areas to which he made a fundamental contribution. He also considered it useful to study the features of the cosmological evolution associated with a possible phenomenon of supersymmetry and/or the consequences following from generalized gravitational theories (see e.g., [1–4] and reference therein). So our work is a tribute

I. Buchbinder (✉) · E. Ivanov · N. Zaigraev
Bogoliubov Laboratory of Theoretical Physics, JINR, Dubna, Russia
e-mail: buchbinder@theor.jinr.ru

E. Ivanov
e-mail: eivanov@theor.jinr.ru

N. Zaigraev
e-mail: nikita.zaigraev@phystech.edu

N. Zaigraev
Moscow Institute of Physics and Technology, Dolgoprudny, Russia

A. Barvinsky and A. Kamenshchik (eds.), *Open Issues in Gravitation and Cosmology*,
Fundamental Theories of Physics 225, https://doi.org/10.1007/978-3-032-15703-4_11

to his endeavours of unification of precision cosmology with recent achievements in fundamental quantum gravity theory.

One of the contemporary generalizations and extensions of the theory of gravity is related to the synthesis of the concepts of supersymmetry and higher spin fields. It is expected that such a synthesis can form the basis for a consistent quantum gravity and the description of the Universe at Planck scales, where all local and global symmetries are unbroken. This paper is devoted to a brief review of one of the trends in supersymmetric higher spin field theory.

The promising modification of gravity theory is a higher-spin field theory (see reviews [5–7]), which contains, in addition to conventional fields with spins ≤ 2, also the fields with spins > 2. The presence of at least one spin higher than two in a consistent higher-spin theory immediately leads to the necessity of including all higher spins, requires an infinite dimensional algebra of gauge symmetries and also entails a modification of Riemannian geometry, which in itself proves to be not appropriate for the higher spin field theory (see, e.g., [8, 9] and references therein). Despite a significant progress in this theory (see e.g. [10–12]), there still remain many open problems. One of the most important problems is to explore the structure of higher-spin field interactions (see, e.g., recent papers [13, 14] and references therein).

A possible starting point of such studies is the Vasiliev nonlinear equations for higher-spin fields, whoes Lagrangian formulation is still unknown. Nevertheless, it can be expected that such a formulation exists. Then one of the ways to get the insight into the form of admissible Lagrangians is to further limit the set of feasible field structures. We will consider the limitations imposed by supersymmetry, which requires the introduction of superspace for its manifest realization (see, e.g., [15, 16]). Superspace formulations automatically impose restrictions on the structure of such theories and thereby significantly limit the possible interactions and forms of the Lagrangian description.

The most severe supersymmetric restrictions are naturally due to extended supersymmetries realized in extended superspaces. To date, the best formulation of theories with extended supersymmetry is achieved within the framework of $\mathcal{N} = 2$ harmonic superspace, discovered in 1984 [17, 18]. Subsequent developments have demonstrated the fundamental role of the concept of Grassmann analyticity and the harmonic condition of zero curvature in $\mathcal{N} = 2$ theories. These concepts arose due to the introduction of auxiliary harmonic coordinates and have no analogues in $\mathcal{N} = 0$ and $\mathcal{N} = 1$ theories. It was natural to expect that these fundamental concepts play the fundamental role in $\mathcal{N} = 2$ higher-spin supergravity theories as well.

In this paper we give an overview of the results obtained in the harmonic superspace formulation of $\mathcal{N} = 2$ higher spins theories[19–22]. Based upon this, we make a few assumption concerning the structure of the Lagrangians describing $\mathcal{N} = 2$ higher-spin supergravity.

The structure of the paper is as follows. Section 2 contains the introductory information about harmonic superspace. In Sect. 3 we describe the structure of global higher-spin symmetries of the free hypermultiplet action. Section 4 is devoted to the discussion of the analytic superspace gauging of $\mathcal{N} = 2$ supersymmetry and $\mathcal{N} = 2$ conformal supersymmetry, which leads to analytic prepotentials of $\mathcal{N} = 2$ Einstein supergravity and $\mathcal{N} = 2$ conformal supergravity. We compare the supermultiplets obtained, discuss relevant supersymmetric actions and the method of conformal compensators, which allows one to build Einstein supergravities from the conformal one. In Sect. 5 we show that the analytical prepotentials of supergravity naturally generalize to higher spins. In Sect. 6 we discuss a consistent model of the hypermultiplet interacting with an infinite tower of $\mathcal{N} = 2$ superconformal higher-spins and outline a number of the related open problems and possible applications. The concluding Sect. 7 involves discussions of the possible structure of $\mathcal{N} = 2$ higher-spin supergravity, as well as poses some open problems related to the construction of such a theory. For convenience of the reader, in the appendix we collect the notations used throughout the paper.

We apologize for inevitable omissions in the reference list caused by the limited size of this Contribution to the memorial volume of A.Starobinsky. A more complete list can be found in our original papers on the subject.

2 Harmonic Superspace

The supersymmetric field theories, in addition to the standard Poincarè invariance, respect a supersymmetry, the invariance under transformations with fermionic parameters mixing bosonic and fermionic fields. To construct theories with manifestly realized supersymmetry, the Minkowski space should be extended by anticommuting fermionic coordinates $\theta^\alpha_i, \bar{\theta}^i_{\dot\alpha}$ [15, 16]. This is the $\mathcal{N}$-extended Minkowski superspace

$$\mathbb{R}^{4|4\mathcal{N}} = \{x^{\alpha\dot\alpha}, \theta^\alpha_i, \bar{\theta}^i_{\dot\alpha}\}, \qquad i = 1, \dots \mathcal{N}. \tag{1}$$

In superspace, supersymmetry is geometrically implemented by the coordinate transformations

$$\delta_\epsilon x^{\alpha\dot\alpha} = 2i\left(\epsilon^{\alpha i}\bar{\theta}^{\dot\alpha}_i - \theta^{\alpha i}\bar{\epsilon}^{\dot\alpha}_i\right), \qquad \delta\theta^\alpha_i = \epsilon^\alpha_i, \qquad \delta\bar{\theta}^i_{\dot\alpha} = \bar{\epsilon}^i_{\dot\alpha}. \tag{2}$$

The analogs of relativistic fields in superspace are superfields $\Phi(x, \theta, \bar{\theta})$. When being expanded over Grassmann coordinates θ and $\bar{\theta}$, these amount to finite sets of ordinary fields. The action principle, gauge transformations, symmetries - everything can be described in terms of superfields [15, 16]. The superspace formulations of supersymmetric theories have the structure different from that of non-supersymmetric theories. Moreover, it turns out that the superspace $\mathbb{R}^{4|4\mathcal{N}}$ suits only for describing theories with the minimal $\mathcal{N} = 1$ supersymmetry. For the full-fledged formulations of theories with extended $\mathcal{N} \geq 2$ supersymmetry, some modifications of the underlying superspace are required.

The most developed version of the superspaces for extended supersymmetry is $\mathcal{N} = 2$ harmonic superspace [17, 18]. The harmonic superspace involves additional bosonic coordinates, harmonics $u_i^{\pm}$ which parametrize the auxiliary sphere $S^2 \sim SU(2)/U(1)$. The harmonics satisfy the relation $u^{+i}u_i^- = 1$ and allow one, like vielbeins in the general relativity, to convert $SU(2)$ indices into the $U(1)$ ones. In the harmonic superspace

$$\mathbb{HR}^{4+2|8} = \{x^{\alpha\dot{\alpha}}, \theta^{\alpha i}, \bar{\theta}^i_{\dot{\alpha}}, u_i^{\pm}\}, \qquad i = 1, 2, \tag{3}$$

superfields $\Phi^{(n)}(x, \theta, \bar{\theta}, u)$ with a fixed $U(1)$ charge "n" are considered. Unlike ordinary superfields, the harmonic superfields contain infinitely many component fields which appear from the harmonic expansions.

The introduction of harmonics makes it possible to distinguish a new superspace closed under supersymmetry – the *analytic superspace* containing half of the original fermionic coordinates:

$$\mathbb{HA}^{4+2|4} = \{\zeta\} = \{x_A^{\alpha\dot{\alpha}}, \theta_A^{+\alpha}, \bar{\theta}_A^{+\dot{\alpha}}, u_i^{\pm}\}. \tag{4}$$

Here we have introduced the analytic basis coordinates:

$$x_A^{\alpha\dot{\alpha}} = x^{\alpha\dot{\alpha}} - 4i\theta^{\alpha(i}\bar{\theta}^{\dot{\alpha}j)}u_i^+u_j^-, \quad \theta_A^{\pm\alpha} = \theta^{\alpha i}u_i^{\pm}, \quad \bar{\theta}_A^{\pm\dot{\alpha}} = \bar{\theta}^{\dot{\alpha}i}u_i^{\pm}. \tag{5}$$

Under $\mathcal{N} = 2$ supersymmetry (2) the analytic coordinates transform as

$$\delta_\epsilon x_A^{\alpha\dot{\alpha}} = -4i\left(\epsilon^{-\alpha}\bar{\theta}^{+\dot{\alpha}} + \theta^{+\alpha}\bar{\epsilon}^{-\dot{\alpha}}\right), \; \delta_\epsilon\theta_A^{\pm\alpha} = \epsilon^{\pm\alpha}, \; \delta_\epsilon\bar{\theta}_A^{\pm\dot{\alpha}} = \bar{\epsilon}^{\pm\dot{\alpha}}, \; \delta_\epsilon u_i^{\pm} = 0, \tag{6}$$

so the analytic superspace (4) is invariant under $\mathcal{N} = 2$ supersymmetry. Also, the analytic superspace is real with respect to the tilde conjugation (Eq. (94) in what follows), which generalizes the usual complex conjugation.

To describe a massive hypermultiplet (with the mass equal to central charge) it is also useful to introduce the auxiliary central charge coordinate x^5 with the $\mathcal{N} = 2$ supersymmetry transformations law $\delta_\epsilon x_5 = -i(\epsilon^{\alpha i}\theta_{\alpha i} - \bar{\theta}^i_{\dot{\alpha}}\bar{\theta}_i^{\dot{\alpha}})$ [17, 18]. This is an example of the Scherk-Schwartz dimension reduction method [23]. In the analytic basis we introduce $x_A^5 = x^5 + i(\theta^{+\alpha}\theta^-_\alpha - \bar{\theta}^+_{\dot{\alpha}}\bar{\theta}^{-\dot{\alpha}})$ with the analyticity-preserving transformation law under supersymmetry:

$$\delta_\epsilon x_A^5 = 2i(\epsilon^{-\alpha}\theta^+_\alpha - \bar{\epsilon}^-_{\dot{\alpha}}\bar{\theta}^{+\dot{\alpha}}). \tag{7}$$

Hereafter, unless specified, we consider x^5-independent superfields, work in the analytic basis and omit the subscript A.

The presence of harmonics allows one to introduce *harmonic derivatives* $\partial^{\pm\pm} = u^{\pm i}\frac{\partial}{\partial u^{\mp i}}$, $\partial^0 = u^{+i}\frac{\partial}{\partial u^{+i}} - u^{-i}\frac{\partial}{\partial u^{-i}}$, which are consistent with the harmonic defining relation $u^{+i}u_i^- = 1$. In the analytic basis, the harmonic derivatives take the form:

$$\mathcal{D}^{\pm\pm} = \partial^{\pm\pm} - 4i\theta^{\pm\alpha}\bar{\theta}^{\pm\dot{\alpha}}\partial_{\alpha\dot{\alpha}} + \theta^{\pm\hat{\alpha}}\partial^{\pm}_{\hat{\alpha}} + i\left[(\theta^+)^2 - (\bar{\theta}^+)^2\right]\partial_5, \tag{8a}$$

$$\mathcal{D}^0 = \partial^0 + \theta^{+\hat{\alpha}}\partial^-_{\hat{\alpha}} - \theta^{-\hat{\alpha}}\partial^+_{\hat{\alpha}}. \tag{8b}$$

The harmonic derivatives are invariant under rigid $\mathcal{N}=2$ supersymmetry (6), (7).

The analytic superspace and harmonic derivatives play the key role in the superspace formulation of $\mathcal{N}=2$ theories [17, 18]. In the next sections, we will demonstrate that the analyticity concept in conjunction with the harmonic derivatives severely constrain the structure of $\mathcal{N}=2$ supergravity, as well as of $\mathcal{N}=2$ higher-spin theories. We will pay a special attention to the ways how it is accomplished.

3 Hypermultiplet and Higher-Spin Symmetries in $\mathcal{N}=2$ Harmonic Superspace

The fundamental $\mathcal{N}=2$ matter supermultiplet, the hypermultiplet, is described in harmonic superspace by an unconstrained analytic superfield $q^+(\zeta)$ with the free action

$$S_{hyp} = -\int d\zeta^{(-4)} \tilde{q}^+ \mathcal{D}^{++} q^+ = -\frac{1}{2}\int d\zeta^{(-4)} q^{+a} \mathcal{D}^{++} q^+_a . \tag{9}$$

Here we used the notation $q^+_a = (q^+, -\tilde{q}^+)$, $q^{+a} = \epsilon^{ab} q^+_b$. In its second form, the action (9) is manifestly invariant under $SU(2)_{PG}$ group that acts on the doublet indices a.

The mass of the hypermultiplet can be introduced by allowing for a dependence on the central charge coordinate x^5

$$q^+(\zeta) \quad \rightarrow \quad q^+(\zeta, x^5) = e^{imx^5} q^+(\zeta). \tag{10}$$

The hypermultiplet mass is equal to the central charge, otherwise a massive hypermultiplet cannot be defined. Note that, due to the form of x^5-dependence, the Lagrangian density in (9) is x^5-independent.

In accordance with the general features of harmonic superfields, the hypermultiplet superfield contains an infinite number of component fields in its θ and u expansions. However, almost all the component fields are auxiliary and become zero on the free equations of motion $\mathcal{D}^{++}q^+ = 0$, leaving in q^+ only the physical ones:

$$q^+(\zeta) = f^i u^+_i + \theta^{+\alpha}\psi_\alpha + \bar{\theta}_{\dot\alpha}\bar{\chi}^{\dot\alpha} + m\left[(\theta^+)^2 - (\bar\theta^+)^2\right] f^i u^-_i + 4i\theta^{+\alpha}\bar\theta^{+\dot\alpha}\partial_{\alpha\dot\alpha} f^i u^-_i . \tag{11}$$

One of the central topics in the theory of higher spins is the higher-spin (HS) (super)symmetries. In a certain sense, the symmetry of higher spins is the maximally possible relativistic symmetry and the low-spin symmetries can be considered as low-energy symmetries inherent to the spontaneously broken phase. It turns out that the higher-spin symmetries of similar kind appear in the simplest relativistic equations. The higher-spin symmetries of the free massless scalar field were discovered by Shaynkman and Vasiliev in 3d [24] and by Eastwood for any d [25] (see also [8]). Therefore, it is quite natural to pose the problem of investigating the higher-spin (super)symmetries of the free hypermultiplet action (9), which is just $\mathcal{N}=2$ analogue of the free Klein-Gordon action.

Further in this section we will present some details related to the higher-spin $\mathcal{N}=2$ supersymmetries of the massive hypermultiplet and to $\mathcal{N}=2$ conformal supersymmetries of the massless hypermultiplet. Then, in Sects. 4 and 5, we will demonstrate that the analytic gauging of such symmetries naturally leads to the $\mathcal{N}=2$ higher spin gauge superfields.

Rigid $\mathcal{N}=2$ Poincaré supersymmetry

The free hypermultiplet action (9) is manifestly invariant under $\mathcal{N}=2$ supersymmetry (6), (7):

$$\delta^{\star}_{\epsilon} q^{+} := q'^{+}(\zeta') - q^{+}(\zeta) = 0. \tag{12}$$

For the higher-spin generalization, it is useful to consider the active form of this supersymmetry variation

$$\delta_{\epsilon} q^{+} := q'^{+}(\zeta) - q^{+}(\zeta) = -\hat{\Lambda}_{susy} q^{+}. \tag{13}$$

Here we introduced the first-order differential operator

$$\begin{aligned}
\hat{\Lambda}_{susy} =& \Big\{ a^{\alpha\dot{\alpha}} + l^{(\alpha}_{\ \beta)} x^{\beta\dot{\alpha}} + \bar{l}^{(\dot{\alpha}}_{\ \dot{\beta})} x^{\alpha\dot{\beta}} - 4i\left(\epsilon^{\alpha i} u^{-}_{i} \bar{\theta}^{+\dot{\alpha}} + \theta^{+\alpha} \bar{\epsilon}^{\dot{\alpha} i} u^{-}_{i}\right) \Big\} \partial_{\alpha\dot{\alpha}} \\
&+ \Big\{ \epsilon^{\alpha i} u^{+}_{i} + l^{(\alpha}_{\ \beta)} \theta^{+\beta} \Big\} \partial^{-}_{\alpha} + \Big\{ \bar{\epsilon}^{\dot{\alpha} i} u^{+}_{i} + \bar{l}^{(\dot{\alpha}}_{\ \dot{\beta})} \bar{\theta}^{+\dot{\beta}} \Big\} \partial^{-}_{\dot{\alpha}} \\
&+ \Big\{ \epsilon^{\alpha i} u^{-}_{i} + l^{(\alpha}_{\ \beta)} \theta^{-\beta} \Big\} \partial^{+}_{\alpha} + \Big\{ \bar{\epsilon}^{\dot{\alpha} i} u^{-}_{i} + \bar{l}^{(\dot{\alpha}}_{\ \dot{\beta})} \bar{\theta}^{-\dot{\beta}} \Big\} \partial^{+}_{\dot{\alpha}} \\
&+ \left\{ c + 2i(\epsilon^{\alpha i} u^{-}_{i} \theta^{+}_{\alpha} - \bar{\epsilon}^{i}_{\dot{\alpha}} u^{-}_{i} \bar{\theta}^{+\dot{\alpha}}) \right\} \partial_{5}.
\end{aligned} \tag{14}$$

The parameters $a^{\alpha\dot{\alpha}}$ correspond to rigid translations, $l_{(\alpha\beta)}$ and $l_{(\dot{\alpha}\dot{\beta})}$ to Lorenz transformations, $\epsilon^{\alpha\pm} = \epsilon^{i} u^{\pm}_{i}$, $\bar{\epsilon}^{\dot{\alpha}\pm} = \bar{\epsilon}^{\dot{\alpha} i} u^{\pm}_{i}$ are the supersymmetry parameters, the parameter c corresponds to $U(1) \subset SU(2)_{PG}$ symmetry. One can easily check that Lie brackets of these transformations form $\mathcal{N}=2$ super-Poincaré algebra.

The operator $\hat{\Lambda}_{susy}$ is the unique first-order superspace differential operator that satisfies the relation

$$[\mathcal{D}^{++}, \hat{\Lambda}_{susy}] = 0. \tag{15}$$

This relation amounts to the invariance of the harmonic derivative under $\mathcal{N}=2$ supersymmetry and ensures the invariance of the massive hypermultiplet action (9) under $\mathcal{N}=2$ supersymmetry transformations in the active form (13).

Rigid $\mathcal{N}=2$ superconformal symmetry

In the massless case, one can expect that the symmetry group is larger and should include conformal invariance. Accordingly, one may consider a more general condition for the hypermultiplet rigid symmetries:

$$[\mathcal{D}^{++}, \hat{\Lambda}_{sc}] = \lambda_{sc}^{++}\mathcal{D}^0. \tag{16}$$

Since $q^{+a}\mathcal{D}^0 q_a^+ = 0$, this condition implies the invariance of the free hypermultiplet action. The transformations of q_a^+ which lead to the variation

$$\delta_{sc} S_{hyp} = \frac{1}{2}\int d\zeta^{(-4)} q^{+a}[\mathcal{D}^{++}, \hat{\Lambda}_{sc}]q_a^+ = 0\,, \tag{17}$$

are found to have the form:

$$\delta_{sc} q_a^+ = -\hat{\Lambda}_{sc} q_a^+ - \frac{1}{2}\Omega_{sc} q_a^+, \tag{18}$$

where we have introduced the differential operator

$$\hat{\Lambda}_{sc} = \lambda_{sc}^{\alpha\dot{\alpha}}\partial_{\alpha\dot{\alpha}} + \lambda_{sc}^{+\alpha}\partial_\alpha^- + \lambda_{sc}^{+\dot{\alpha}}\partial_{\dot{\alpha}}^- + \lambda_{sc}^{-\alpha}\partial_\alpha^+ + \lambda_{sc}^{-\dot{\alpha}}\partial_{\dot{\alpha}}^+ + \lambda^{++}\partial^{--}, \tag{19}$$

and the weight factor

$$\Omega_{sc} = \partial_{\alpha\dot{\alpha}}\lambda_{sc}^{\alpha\dot{\alpha}} - \partial_\alpha^-\lambda_{sc}^{+\alpha} - \partial_{\dot{\alpha}}^-\lambda_{sc}^{+\dot{\alpha}} + \partial^{--}\lambda^{++}. \tag{20}$$

The operator $\hat{\Lambda}_{sc}$, in contrast to $\hat{\Lambda}_{susy}$, contains the harmonic derivative ∂^{--} but lacks the central charge derivative ∂_5. It is straightforward to verify that a non-trivial ∂_5 term that corresponds to the hypermultiplet mass does not allow solutions with non-vanishing λ^{++} parameter [22].

As a solution to the condition (16) we obtain the parameters λ_{sc}^M (to which one should add the rigid super-Poincaré parameters from (14) without the last line, corresponding to the case of trivial central charge, $\partial_5 q^+ = 0$):

$$\begin{aligned}
\lambda_{sc}^{\alpha\dot{\alpha}} &= x^{\dot{\alpha}\rho}k_{\rho\dot{\rho}}x^{\dot{\rho}\alpha} + dx^{\alpha\dot{\alpha}} - 4i\theta^{+\alpha}\bar{\theta}^{+\dot{\alpha}}\lambda^{(ij)}u_i^-u_j^- - 4i\left(x^{\alpha\dot{\rho}}\bar{\eta}^i_{\dot{\rho}}\bar{\theta}^{+\dot{\alpha}} + \theta^{+\alpha}\eta^i_\rho x^{\rho\dot{\alpha}}\right)u_i^-,\\
\lambda_{sc}^{+\alpha} &= \frac{1}{2}\theta^{+\alpha}(d+ir) + x^{\alpha\dot{\beta}}k_{\beta\dot{\beta}}\theta^{+\beta} + x^{\alpha\dot{\alpha}}\eta^i_{\dot{\alpha}}u_i^+ + \theta^{+\alpha}\left(\lambda^{(ij)}u_i^+u_j^- + 4i\theta^{+\rho}\eta^i_\rho u_i^-\right),\\
\bar{\lambda}_{sc}^{+\dot{\alpha}} &= \frac{1}{2}\bar{\theta}^{+\dot{\alpha}}(d-ir) + x^{\dot{\alpha}\beta}k_{\beta\dot{\beta}}\bar{\theta}^{+\dot{\beta}} + x^{\alpha\dot{\alpha}}\eta^i_\alpha u_i^+ + \bar{\theta}^{+\dot{\alpha}}\left(\lambda^{(ij)}u_i^+u_j^- - 4i\bar{\theta}^{+\dot{\rho}}\bar{\eta}^i_{\dot{\rho}}u_i^-\right),\\
\lambda_{sc}^{-\alpha} &= \frac{1}{2}\theta^{-\alpha}(d+ir) + x^{\alpha\dot{\beta}}k_{\beta\dot{\beta}}\theta^{-\beta} - 2i(\theta^-)^2\bar{\theta}^+_{\dot{\beta}}k^{\dot{\beta}\alpha} + \left(x^{\alpha\dot{\alpha}} + 4i\theta^{-\alpha}\bar{\theta}^{+\dot{\alpha}}\right)\bar{\eta}^i_{\dot{\alpha}}u_i^-\\
&\quad + 4i\eta^i_\beta\theta^{-\beta}\left(\theta^{-\alpha}u_i^+ - \theta^{+\alpha}u_i^-\right) + \lambda^{ij}u_i^-\left(u_j^-\theta^{+\alpha} - u_j^+\theta^{-\alpha}\right),\\
\lambda_{sc}^{-\dot{\alpha}} &= \frac{1}{2}\bar{\theta}^{-\dot{\alpha}}(d-ir) + x^{\dot{\alpha}\beta}k_{\beta\dot{\beta}}\bar{\theta}^{-\dot{\beta}} - 2i(\bar{\theta}^-)^2\theta^+_\beta k^{\beta\dot{\alpha}} + \left(x^{\alpha\dot{\alpha}} + 4i\theta^{+\alpha}\bar{\theta}^{-\dot{\alpha}}\right)\eta^i_\alpha u_i^-\\
&\quad - 4i\bar{\eta}^i_{\dot{\beta}}\bar{\theta}^{-\dot{\beta}}\left(\bar{\theta}^{-\dot{\alpha}}u_i^+ - \bar{\theta}^{+\dot{\alpha}}u_i^-\right) + \lambda^{ij}u_i^-\left(u_j^-\bar{\theta}^{+\dot{\alpha}} - u_j^+\bar{\theta}^{-\dot{\alpha}}\right),\\
\lambda_{sc}^{++} &= \lambda^{ij}u_i^+u_j^+ + 4i\theta^{+\alpha}\bar{\theta}^{+\dot{\alpha}}k_{\alpha\dot{\alpha}} + 4i\left(\theta^{+\alpha}\eta^i_\alpha + \bar{\eta}^i_{\dot{\alpha}}\bar{\theta}^{+\dot{\alpha}}\right)u_i^+.
\end{aligned} \tag{21}$$

Here the parameters $k_{\alpha\dot{\alpha}}$ d , r, $\lambda^{(ij)}$ and η^i_α, $\bar{\eta}^i_{\dot{\alpha}}$ are associated with the special conformal transformations, dilatations, $U(1)_R$ symmetry, $SU(2)_R$ symmetry and the conformal supersymmetry, respectively. The transformations (14) (without the last line) and (21) form $\mathcal{N}=2$ superconformal algebra $su(2,2|2)$.

Rigid $\mathcal{N}=2$ higher-spin supersymmetry

The active form of $\mathcal{N}=2$ supersymmetry transformations (13) of the massive hypermultiplet is convenient, when seeking for generalizations to higher spins. We introduce the differential operator involving the higher-order space-time derivatives:

$$\hat{\Lambda}^{(s)}_{susy} = \lambda^{\alpha(s-2)\dot{\alpha}(s-2)M}_{susy}\partial_M\partial^{s-2}_{\alpha(s-2)\dot{\alpha}(s-2)}. \tag{22}$$

Here $M=\{\alpha\dot{\alpha},\alpha+,\dot{\alpha}+,\alpha-,\dot{\alpha}-,5\}$ and we assume that all the Lorentz indices of the same type are symmetrized.

As in the case of $\mathcal{N}=2$ supersymmetry (15), we impose the condition

$$[\mathcal{D}^{++},\hat{\Lambda}^{(s)}_{susy}]=0. \tag{23}$$

As its general solution, we obtain:

$$\begin{aligned}
\lambda^{\alpha(s-1)\dot{\alpha}(s-1)}_{susy} &= a^{\alpha(s-1)\dot{\alpha}(s-1)} + l^{(\dot{\alpha}(s-2)(\alpha(s-1)}{}_{\beta)}x^{\beta\dot{\alpha})} + \bar{l}^{(\alpha(s-2)(\dot{\alpha}(s-1)}{}_{\dot{\beta})}x^{\alpha)\dot{\beta}}\\
&\quad - 4i\left(\epsilon^{\alpha(s-1)\dot{\alpha}(s-2)i}u^-_i\bar{\theta}^{+\dot{\alpha}} + \theta^{+(\alpha}\bar{\epsilon}^{\alpha(s-2))\dot{\alpha}(s-1)i}u^-_i\right),\\
\lambda^{\pm\alpha(s-1)\dot{\alpha}(s-2)}_{susy} &= \epsilon^{\alpha(s-1)\dot{\alpha}(s-2)i}u^\pm_i + l^{\dot{\alpha}(s-2)(\alpha(s-1)}{}_{\beta)}\theta^{\pm\beta},\\
\lambda^{\pm\alpha(s-2)\dot{\alpha}(s-1)}_{susy} &= \epsilon^{\alpha(s-2)\dot{\alpha}(s-1)i}u^\pm_i + \bar{l}^{\alpha(s-2)(\dot{\alpha}(s-1)}{}_{\dot{\beta})}\bar{\theta}^{\pm\dot{\beta}},\\
\lambda^{\alpha(s-2)\dot{\alpha}(s-2)5}_{susy} &= c^{\alpha(s-2)\dot{\alpha}(s-2)} + 2i\epsilon^{\alpha(s-2)\dot{\alpha}(s-2)\hat{\alpha}i}u^-_i\theta^+_{\hat{\alpha}}.
\end{aligned} \tag{24}$$

For $s=2$ these parameters are reduced just to the rigid $\mathcal{N}=2$ supersymmetry (13). So the relevant transformations provide the natural higher-spin generalization of $\mathcal{N}=2$ supersymmetry. The parameters $a^{\alpha(s-1)\dot{\alpha}(s-1)}$ and $c^{\alpha(s-2)\dot{\alpha}(s-2)}$ are the higher-spin generalizations of translation and $U(1)$ parameters, $\epsilon^{\alpha(s-2)\dot{\alpha}(s-2)i}$ are the parameters of $\mathcal{N}=2$ higher-spin supersymmetry.

It is interesting to consider the algebraic structure of these transformations:

$$[\hat{\Lambda}^{(s_1)}_{susy},\hat{\Lambda}^{(s_2)}_{susy}] \sim \hat{\Lambda}^{(s_1+s_2-2)}_{susy}. \tag{25}$$

For the particular case of $s_1=s_2=2$ we reproduce the standard closed $\mathcal{N}=2$ Super-Poincaré algebra. However, if there is at least one spin $s>2$, generalized $\mathcal{N}=2$ supersymmetry transformations of arbitrary spin appear.

The operators defined above generate the higher-spin supersymmetries of the hypermultiplet. Actually, the hypermultiplet transformations that lead to the vanishing variation

$$\delta^{(s)}_{susy}S_{hyp} = \frac{1}{2}\int d\zeta^{(-4)}q^{+a}[\mathcal{D}^{++},\hat{\Lambda}^{(s)}_{susy}]q^+_a = 0 \tag{26}$$

are given by

$$\delta^{(s)}_{susy}q^+_a = -\hat{\Lambda}^{(s)}_{susy}(J)^{P(s)}q^+_a, \tag{27}$$

where we have introduced the $U(1)$ generator J and the sign operator $P(s)$:

$$Jq^{+} := iq^{+}, \qquad J\tilde{q}^{+} := -i\tilde{q}^{+}, \qquad P(s) := \begin{bmatrix} 0 & \text{for even } s; \\ 1 & \text{for odd } s. \end{bmatrix} \tag{28}$$

The term $(J)^{P(s)}$ in (27) distinguishes between the charges of the hypermultiplet q^{+} and its tilde-conjugate $\tilde{q}^{+}$, when considering the odd higher spin transformations. This is the higher-spin reflection of the well-known fact that the minimal interaction of an electromagnetic field is possible only for a complex scalar field, the field and its complex conjugate having opposite electric charges.

Rigid $\mathcal{N}=2$ higher-spin superconformal symmetry

Like in the case of $\mathcal{N}=2$ superconformal symmetry, the number of higher-spin symmetries for a massless hypermultiplet increases. As in $\mathcal{N}=2$ superconformal transformations, here we also add harmonic derivative ∂^{--} and drop terms with the central charge operator ∂_5. The general ansatz for transformations has the form

$$\hat{\Lambda}^{(s)}_{sc} = \lambda^{M_1 \dots M_{s-1}} \partial_{M_{s-1}} \dots \partial_{M_1} + \text{lower derivative contributions.} \tag{29}$$

Instead of performing a complete analysis of these transformations and the relevant symmetry conditions, we will confine our attention to a non-trivial example of the higher-spin dilatation transformations:

$$\begin{aligned} \delta_{dil} q^{+}_a =& - d^{\alpha(s-2)\dot{\alpha}(s-2)} \left[x^{\alpha\dot{\alpha}} \partial_{\alpha\dot{\alpha}} + \frac{1}{2} \theta^{+\hat{\alpha}} \partial^{+}_{\hat{\alpha}} \right] \partial^{s-2}_{\alpha(s-2)\dot{\alpha}(s-2)} (J)^{P(s)} q^{+}_a \\ & + d^{\alpha(s-2)\dot{\alpha}(s-2)} \partial^{s-2}_{\alpha(s-2)\dot{\alpha}(s-2)} (J)^{P(s)} q^{+}_a . \end{aligned} \tag{30}$$

The special conformal transformations and the conformal supersymmetry are generalized in a similar way.

Now, having described the structure of global higher-spin $\mathcal{N}=2$ supersymmetries, we can move on to their analytic gauging and the definition of the corresponding gauge $\mathcal{N}=2$ superfields.

4 $\mathcal{N}=2$ Supergravity

Before proceeding to $\mathcal{N}=2$ higher spin supergravity, we consider an instructive example of $\mathcal{N}=2$ supergravity [18, 30–32]. This is a simple example of the consistent gauge theory with finite set of fields, which is a useful prototype of the higher-spin supergravities. The harmonic formulation of $\mathcal{N}=2$ supergravity suggests some conjectures concerning the structure of the complete $\mathcal{N}=2$ higher-spin theory.

$\mathcal{N}=2$ Einstein supergravity

To obtain a superfield description of the $\mathcal{N}=2$ supergravity multiplet, we consider the gauged version of rigid $\mathcal{N}=2$ supersymmetry transformations (14). Actually, since the gauge transformations contain local translations, the invariance of the action with respect to such transformations requires the introduction of gravitational field. $\mathcal{N}=2$ supersymmetry further requires that the gravitational field is accompanied by other fields from $\mathcal{N}=2$ supergravity multiplet.

We demand that the gauged transformations preserve the analytical superspace (4). So, the analytic gauging of rigid transformations takes the form:

$$\hat{\Lambda}_{susy} \quad \to \quad \hat{\Lambda}^{loc}_{susy} = \lambda^{\alpha\dot{\alpha}}(\zeta)\partial_{\alpha\dot{\alpha}} + \lambda^{\hat{\alpha}+}(\zeta)\partial^-_{\hat{\alpha}} + \lambda^{\hat{\alpha}-}(\zeta,\theta^-)\partial^+_{\hat{\alpha}} + \lambda^5(\zeta)\partial_5. \tag{31}$$

These transformations form a closed algebra, $[\hat{\Lambda}^{loc}_{susy}, \hat{\Lambda}^{loc}_{susy}] \sim \hat{\Lambda}^{loc}_{susy}$.

The harmonic derivative $\mathcal{D}^{++}$ is not invariant with respect to such transformations, $[\mathcal{D}^{++}, \hat{\Lambda}^{loc}_{susy}] \neq 0$ (compare with Eq. (15)). To restore the invariance, we covariantize the harmonic derivative as

$$\mathcal{D}^{++} \quad \to \quad \mathfrak{D}^{++} = \mathcal{D}^{++} + \kappa_2 \hat{\mathcal{H}}^{++}_{(s=2)}, \tag{32}$$

thereby introducing the vielbeins appearing in the first-rank differential operator:

$$\hat{\mathcal{H}}^{++}_{(s=2)} = h^{++\alpha\dot{\alpha}}(\zeta)\partial_{\alpha\dot{\alpha}} + h^{++\hat{\alpha}+}(\zeta)\partial^-_{\hat{\alpha}} + h^{++\hat{\alpha}-}(\zeta,\theta^-)\partial^+_{\hat{\alpha}} + h^{++5}(\zeta)\partial_5 \tag{33}$$

(in Eq. (32) κ_2 is the coupling constant, see Eq. (36) below). The requirement of invariance of the harmonic derivative $\mathfrak{D}^{++}$ fixes the transformation laws of the vielbeins

$$\begin{aligned} \delta_\lambda \mathfrak{D}^{++} &= \kappa_2[\hat{\Lambda}^{loc}_{susy}, \mathfrak{D}^{++}] + \kappa_2 \delta_\lambda \hat{\mathcal{H}}^{++}_{(s=2)} = 0 \\ \Rightarrow \quad \delta_\lambda \hat{\mathcal{H}}^{++}_{(s=2)} &= [\mathcal{D}^{++} + \kappa_2 \hat{\mathcal{H}}_{(s=2)}, \hat{\Lambda}^{loc}_{susy}]. \end{aligned} \tag{34}$$

All the vielbeins are analytic, except for $h^{++\hat{\alpha}-}(\zeta,\theta^-)$; fortunately, the latter can be entirely gauged away. Indeed, under the gauge transformations (34) the non-analytical vielbein transforms as

$$\delta_\lambda h^{++\hat{\alpha}-} = \left(\mathcal{D}^{++} + \kappa_2 \hat{\mathcal{H}}^{++}_{(s=2)}\right)\lambda^{-\hat{\alpha}} - \lambda^{+\hat{\alpha}} - \kappa_2 \hat{\Lambda}^{loc}_{susy} h^{++\hat{\alpha}-}, \tag{35}$$

and, since the non-analytical gauge parameter $\lambda^{-\hat{\alpha}}$ is unconstrained, one can impose the *analytic gauge* $h^{++\hat{\alpha}-}=0$. In this gauge the covariant harmonic derivative $\mathfrak{D}^{++}$ preserves the analyticity, $[\mathcal{D}^+_{\hat{\alpha}}, \mathfrak{D}^{++}]=0$, and the parameter $\lambda^{-\hat{\alpha}}$ is expressed in terms of the analytic $\lambda^{+\hat{\alpha}}$.

Using the operator (33) one can construct the gauge invariant coupling to the hypermultiplet:

$$S^{Ein}_{hyp} = -\frac{1}{2} \int d\zeta^{(-4)} q^{+a} \left(\mathcal{D}^{++} + \kappa_2 \hat{\mathcal{H}}^{++}_{(s=2)} \right) q^+_a . \tag{36}$$

This coupling is invariant under super-diffeomorphisms (34) accompanied by the hypermultiplet transformations

$$\delta_\lambda q^+_a = -\kappa_2 \hat{\mathcal{U}}_{susy} q^+_a = -\kappa_2 \left(\hat{\Lambda}^{loc}_{susy} + \frac{1}{2} \Omega^{loc}_{susy} \right) q^+_a . \tag{37}$$

Here we introduced the weight factor $\Omega^{loc}_{susy} := (-1)^{P(M)} \partial_M \lambda^M$, which is necessary for invariance of the hypermultiplet action [20]. Thus, by introducing the analytic prepotentials h^{++M}, it became possible to covariantize the free hypermultiplet action with respect to the super-diffeomorphisms (31).

To find out the physical field contents of the prepotentials introduced, we impose the Wess-Zumino type gauge:

$$\begin{aligned}
h^{++\alpha\dot{\alpha}}_{WZ} &= -4i\theta^{+\beta}\bar{\theta}^{+\dot{\beta}} \Phi^{\alpha\dot{\alpha}}_{\beta\dot{\beta}} + (\bar{\theta}^+)^2 \theta^{+\beta} \psi^{\alpha\dot{\alpha}i}_{\beta} u^-_i - (\theta^+)^2 \bar{\theta}^{+\dot{\beta}} \bar{\psi}^{\alpha\dot{\alpha}i}_{\dot{\beta}} u^-_i + (\theta^+)^4 V^{\alpha\dot{\alpha}(ij)} u^-_i u^-_j \\
h^{++5}_{WZ} &= -4i\theta^{+\beta}\bar{\theta}^{+\dot{\beta}} C_{\beta\dot{\beta}} + (\bar{\theta}^+)^2 \theta^{+\beta} \rho^i_\beta u^-_i - (\theta^+)^2 \bar{\theta}^{+\dot{\beta}} \bar{\rho}^i_{\dot{\beta}} u^-_i + (\theta^+)^4 S^{(ij)} u^-_i u^-_j \\
h^{++\alpha+}_{WZ} &= (\bar{\theta}^+)^2 \theta^+_\beta T^{(\alpha\beta)} + (\bar{\theta}^+)^2 \theta^{+\alpha} T + (\theta^+)^2 \bar{\theta}^+_{\dot{\beta}} P^{\alpha\dot{\beta}} + (\theta^+)^4 \chi^{\alpha i} u^-_i , \\
h^{++\dot{\alpha}+}_{WZ} &= (\theta^+)^2 \bar{\theta}^+_{\dot{\beta}} \bar{T}^{(\dot{\alpha}\dot{\beta})} + (\theta^+)^2 \bar{\theta}^{+\dot{\alpha}} \bar{T} - (\bar{\theta}^+)^2 \theta^+_\beta \bar{P}^{\beta\dot{\alpha}} + (\theta^+)^4 \bar{\chi}^{\dot{\alpha}i} u^-_i .
\end{aligned} \tag{38}$$

In the linearized limit (the leading order in κ_2) the fields in WZ gauge transform under the residual gauge freedom as [26]:

$$\begin{aligned}
&\delta_\lambda \Phi^{\alpha\dot{\alpha}}_{\beta\dot{\beta}} \sim \partial_{\beta\dot{\beta}} a^{\alpha\dot{\alpha}} - l_{(\beta}^{\ \alpha)} \delta^{\dot{\alpha}}_{\dot{\beta}} - \delta^\alpha_\beta l_{(\dot{\beta}}^{\ \dot{\alpha})} , \\
&\delta_\lambda P^{\alpha\dot{\alpha}} \sim i\partial^{\dot{\alpha}\beta} l_{(\beta}^{\ \alpha)} , \quad \delta_\lambda C_{\alpha\dot{\alpha}} \sim \partial_{\alpha\dot{\alpha}} c , \\
&\delta_\lambda \psi^{\alpha\dot{\alpha}i}_\beta \sim \partial^{\dot{\alpha}}_\beta \epsilon^{\alpha i} , \quad \delta_\lambda \chi^{\alpha i} \sim \Box \epsilon^{\alpha i} , \quad \delta_\lambda \rho^i_\alpha \sim \partial_{\alpha\dot{\alpha}} \bar{\epsilon}^{\dot{\alpha}i} .
\end{aligned} \tag{39}$$

So one can identify $\Phi^{\alpha\dot{\alpha}}_{\beta\dot{\beta}}$ with the gravitational vielbein, $\psi^{\alpha\dot{\alpha}i}_\beta$ with the doublet of gravitino fields, $C_{\alpha\dot{\alpha}}$ with the graviphoton. Using local Lorentz transformation,one can impose the symmetric gauge $\Phi_{\alpha\beta\dot{\alpha}\dot{\beta}} = \Phi_{(\alpha\beta)(\dot{\alpha}\dot{\beta})} + \epsilon_{\alpha\beta}\epsilon_{\dot{\alpha}\dot{\beta}}\Phi$. All other gauge fields $(P^{\alpha\dot{\alpha}}, \chi^i_\alpha, \rho^i_\alpha)$ can be redefined in terms of physical fields so as to become invariant under the residual gauge freedom, see [26] for details. The ultimate $\mathcal{N} = 2$ supermultiplet contains $\mathbf{40_B} + \mathbf{40_F}$ off-shell degrees of freedom:

$$\begin{aligned}
\text{Physical fields:} \quad & \left\{ \Phi_{(\alpha\beta)(\dot{\alpha}\dot{\beta})}, \Phi \right\} , \ \psi^i_{\alpha\beta\dot{\alpha}} , \ C_{\alpha\dot{\alpha}} , \\
\text{Auxiliary fields:} \quad & T, T^{(\alpha\beta)}, P^{\alpha\dot{\alpha}}, S^{(ij)}, V^{(ij)}_{\alpha\dot{\alpha}}, \rho^i_\alpha, \chi^i_\alpha .
\end{aligned} \tag{40}$$

The off-shell component content of this version of $\mathcal{N} = 2$ Einstein supergravity was firstly found by Fradkin and Vasiliev [27, 28] and, independently, by de Wit and van Holten [29]. Thus, the harmonic superspace approach allows one to obtain a $\mathcal{N} = 2$ Einstein supergravity multiplet from the analytic gauging of supersymmetry.

In order to construct the superfield action it is useful to introduce superfields H^{++M} and supersymmetry-invariant superfields G^{++M} as vielbeins in the new expansion of $\mathfrak{D}^{++}$:

$$\mathfrak{D}^{++} = \partial^{++} + \kappa_2 H^{++M}\partial_M = \mathcal{D}^{++} + \kappa_2 G^{++M}\mathcal{D}_M. \tag{41}$$

The supersymmetry-invariant derivatives $\mathcal{D}_M$ are defined in Eq. (97) of Appendix. The invariant $\mathcal{N} = 2$ Einstein supergravity action was constructed in [30]:

$$S^{full}_{sugra} = -\int d^4x d^8\theta du\, E H^{++5} H^{--5}. \tag{42}$$

Here E is $\mathcal{N} = 2$ superspace measure constructed out of h^{++M}, see [18, 30, 32]. It has a complicated structure and will not be given here. The vielbein coefficient H^{--5} is a solution of the zero curvature equation $\mathfrak{D}^{++}H^{--5} = \mathfrak{D}^{--}H^{++5}$, with the derivative $\mathcal{D}^{--}$ being defined from the curved zero-curvature condition $[\mathfrak{D}^{++}, \mathfrak{D}^{--}] = \mathcal{D}^0$.

Since in the next sections we will be interested in the free (linearized) actions for $\mathcal{N} = 2$ higher spins, it is instructive to give here the linearized action of $\mathcal{N} = 2$ Einstein supergravity [19, 33]:

$$S^{lin}_{sugra} = -\int d^4x d^8\theta du\, \left\{G^{++\alpha\dot\alpha} G^{--}_{\alpha\dot\alpha} + 4G^{++5}G^{--5}\right\}, \tag{43}$$

where we used the supersymmetry-invariant superfields $G^{\pm\pm M}$. The negatively charged potentials can be found from the flat zero-curvature equations $\mathcal{D}^{++}G^{--M} = \mathcal{D}^{--}G^{++M}$. At the component level, the action (43) is reduced to the sum of the Pauli-Fiertz, Rarita-Schwinger and Maxwell actions, for details see [26].

$\mathcal{N} = 2$ conformal supergravity

Although $\mathcal{N} = 2$ Einstein supergravity can be formulated as described above, a more convenient and universal way of its construction consists in using the multiplet of conformal supergravity in junction with the appropriate compensator multiplets, see e.g. [34]. The idea of this approach is to start from the interactions of superconformal gravity with superconformal matter multiplets and then to fix the conformal gauge freedom by choosing non-trivial vacuum values for some matter fields. Here is the simplest illustration of the conformal compensator method:

$$S[\phi, g_{mn}] = \int d^4x\sqrt{-g}\,\phi\left(\Box + \frac{1}{6}R\right)\phi \quad \xrightarrow{\phi=const} \quad S[g_{mn}] = \frac{1}{6}\int d^4x\sqrt{-g}\,R. \tag{44}$$

The most important motivation for us is that such a method of reproducing the gravity theory allows a generalization to $\mathcal{N}=2$ higher-spin supergravity.

The prepotentials of $\mathcal{N}=2$ conformal supergravity can be obtained in a way similar to that for $\mathcal{N}=2$ Einstein supergravity. In the conformal case one should gauge the conformal transformations (19):

$$\hat{\Lambda}_{sc} \quad \to \quad \hat{\Lambda}_{sc}^{loc} = \lambda^{\alpha\dot{\alpha}}(\zeta)\partial_{\alpha\dot{\alpha}} + \lambda^{\hat{\alpha}+}(\zeta)\partial^{-}_{\hat{\alpha}} + \lambda^{\hat{\alpha}-}(\zeta,\theta^{-})\partial^{+}_{\hat{\alpha}} + \lambda^{++}(\zeta)\partial^{--}. \quad (45)$$

Since in this case $[\mathcal{D}^{++}, \hat{\Lambda}_{sc}^{loc}] \neq \lambda^{++}\mathcal{D}^{0}$ (compare with (16)), the appropriate covariantization of the harmonic derivative takes the form:

$$\mathcal{D}^{++} \quad \to \quad \mathfrak{D}^{++} = \mathcal{D}^{++} + \kappa_2 \hat{\mathbb{H}}^{++}_{(s=2)}, \quad (46)$$

where the differential operator

$$\hat{\mathbb{H}}^{++}_{(s=2)} = h^{++\alpha\dot{\alpha}}(\zeta)\partial_{\alpha\dot{\alpha}} + h^{++\hat{\alpha}+}(\zeta)\partial^{-}_{\hat{\alpha}} + h^{++\hat{\alpha}-}(\zeta,\theta^{-})\partial^{+}_{\hat{\alpha}} + h^{(+4)}(\zeta)\partial^{--}. \quad (47)$$

exhibits the transformation law

$$\delta_{\lambda}\hat{\mathbb{H}}^{++}_{(s=2)} = [\mathcal{D}^{++} + \kappa_2\hat{\mathbb{H}}^{++}_{(s=2)}, \hat{\Lambda}_{sc}^{loc}] - \lambda^{++}\mathcal{D}^{0}. \quad (48)$$

It provides $\delta_{\lambda}\mathfrak{D}^{++} = \kappa_2[\hat{\Lambda}^{sc}_{susy}, \mathfrak{D}^{++}] + \kappa_2\delta_{\lambda}\hat{\mathbb{H}}^{++}_{(s=2)} = -\kappa_2\lambda^{++}\mathcal{D}^{0}$. Note that in (47) there is no derivative with respect to the auxiliary central charge coordinate x^5 (which breaks conformal symmetry), but an additional harmonic derivative ∂^{--} with the new vielbein coefficient $h^{(+4)}$ appear.

The non-analytic prepotential $h^{++\hat{\alpha}-}$ transforms according to

$$\delta h^{++\hat{\alpha}-} = \left(\mathcal{D}^{++} + \kappa_2\hat{\mathbb{H}}^{++}_{(s=2)}\right)\lambda^{-\hat{\alpha}} - \kappa_2\hat{\Lambda}_{sc}^{loc}h^{++\hat{\alpha}-} - \lambda^{+\hat{\alpha}} + \lambda^{++}\theta^{-\hat{\alpha}}, \quad (49)$$

where $\lambda^{-\hat{\alpha}}$ is non-analytic. So like in $\mathcal{N}=2$ Einstein supergravity, one can impose the analytic gauge $h^{++\hat{\alpha}-}=0$.

Using the harmonic covariant derivative (46), we specify the interaction with the massless hypermultiplet:

$$S^{sc}_{hyp} = -\frac{1}{2}\int d\zeta^{(-4)} q^{+a}\left(\mathcal{D}^{++} + \kappa_2\hat{\mathbb{H}}^{++}_{(s=2)}\right)q^{+}_{a}. \quad (50)$$

This action is invariant under the gauge transformation (48) accompanied by the hypermultiplet transformation:

$$\delta_{\lambda}q^{+}_{a} = -\kappa_2\hat{\mathcal{U}}_{sc}q^{+}_{a} = -\kappa_2\hat{\Lambda}^{loc}_{sc}q^{+}_{a} - \frac{\kappa_2}{2}\Omega^{loc}_{sc}q^{+}_{a}, \qquad \Omega^{loc}_{sc} = (-1)^{P(M)}\partial_{M}\lambda^{M}. \quad (51)$$

In order to be convinced that the introduced gauge superfields describe the multiplet of $\mathcal{N}=2$ conformal supergravity, one should impose Wess-Zumino type gauge:

$$\begin{cases} h^{++\alpha\dot{\alpha}} &= -4i\theta^{+\beta}\bar{\theta}^{+\dot{\beta}}\Phi^{\alpha\dot{\alpha}}_{\beta\dot{\beta}} + (\bar{\theta}^{+})^{2}\theta^{+\beta}\psi^{\alpha\dot{\alpha}i}_{\beta}u^{-}_{i} + (\theta^{+})^{2}\bar{\theta}^{+\dot{\rho}}\bar{\psi}^{\alpha\dot{\alpha}i}_{\dot{\rho}}u^{-}_{i} \\ &\qquad + (\theta^{+})^{4}V^{\alpha\dot{\alpha}(ij)}u^{-}_{i}u^{-}_{j}\,, \\ h^{++\alpha+} &= (\theta^{+})^{2}\bar{\theta}^{+}_{\dot{\alpha}}P^{\alpha\dot{\alpha}} + \left(\bar{\theta}^{+}\right)^{2}\theta^{+}_{\beta}T^{(\alpha\beta)} + (\theta^{+})^{4}\chi^{\alpha i}u^{-}_{i}\,, \\ h^{++\dot{\alpha}+} &= \widetilde{h^{++\alpha+}}\,, \\ h^{(+4)} &= (\theta^{+})^{4}D\,. \end{cases} \tag{52}$$

The remaining fields are determined up to the residual gauge transformations which differ from (39):

$$\begin{aligned} &\delta_{\lambda}\Phi^{\alpha\dot{\alpha}}_{\beta\dot{\beta}} \sim \partial_{\beta\dot{\beta}}a^{\alpha\dot{\alpha}} - l_{(\beta}^{\ \alpha)}\delta^{\dot{\alpha}}_{\dot{\beta}} - \delta^{\alpha}_{\beta}l_{(\dot{\beta}}^{\ \dot{\alpha})} + d\delta^{\alpha}_{\beta}\delta^{\dot{\alpha}}_{\dot{\beta}}, \\ &\delta_{\lambda}\psi^{\alpha\dot{\alpha}i}_{\beta} \sim \partial^{\dot{\alpha}}_{\beta}\epsilon^{\alpha i} + \delta^{\alpha}_{\beta}\bar{\eta}^{\dot{\alpha}i}, \quad \delta_{\lambda}\chi^{\alpha i} \sim \partial^{\alpha\dot{\alpha}}\bar{\eta}^{i}_{\dot{\alpha}} + \Box\epsilon^{\alpha i}, \\ &\delta_{\lambda}V^{(ij)}_{\alpha\dot{\alpha}} \sim \partial_{\alpha\dot{\alpha}}v^{(ij)}, \quad \delta_{\lambda}P_{\alpha\dot{\alpha}} \sim \partial_{\alpha\dot{\alpha}}(r+id) - i\partial^{\dot{\alpha}\beta}l_{(\beta}^{\ \alpha)} - ik_{\alpha\dot{\alpha}}, \\ &\delta_{\lambda}D \sim \partial^{\alpha\dot{\alpha}}k_{\alpha\dot{\alpha}}. \end{aligned} \tag{53}$$

As compared to the transformations (39), here one finds additional gauge parameters corresponding to the local dilatations (d), local special conformal transformations ($k_{\alpha\dot{\alpha}}$), local conformal supersymmetry (η^{i}_{α}, $\bar{\eta}^{i}_{\dot{\alpha}}$), local $U(1)_R$ (r) and local $SU(2)_R$ ($v^{(ij)}$) transformations. Using this freedom, one can gauge away the trace of graviton, the trace of the spin $\frac{3}{2}$ field and the imaginary part of $P_{\alpha\dot{\alpha}}$. The resulting field representation is constituted by

$$\begin{aligned} \text{Gauge fields:} \quad & \Phi_{(\alpha\beta)(\dot{\alpha}\dot{\beta})},\ \psi^{i}_{(\alpha\beta)\dot{\alpha}},\ V^{(ij)}_{\alpha\dot{\alpha}},\ R_{\alpha\dot{\alpha}} := Re(P_{\alpha\dot{\alpha}}), \\ \text{Non-gauge fields:} \quad & \chi^{i}_{\alpha},\ T^{(\alpha\beta)},\ D, \end{aligned} \tag{54}$$

that is total of $\mathbf{24}_B + \mathbf{24}_F$ off-shell degrees of freedom inherent to $\mathcal{N}=2$ Weyl multiplet [35–37]. Thus, we have shown that the analytic superfields h^{++M} can indeed be identified with the unconstrained prepotentials of $\mathcal{N}=2$ conformal supergravity.

The linearized action of conformal $\mathcal{N}=2$ supergravity can be constructed by generalizing the approach to the linearized higher-derivative invariants described in [26],

$$S_{Weyl} = \int d^{4}xd^{4}\theta\, \mathcal{W}^{(\alpha\beta)}\mathcal{W}_{(\alpha\beta)} + c.c. \tag{55}$$

Here we have introduced the linearized $\mathcal{N}=2$ super-Weyl tensor[1]

$$\mathcal{W}_{(\alpha\beta)} = (\bar{\mathcal{D}}^{+})^2 \left(\partial^{\dot\beta}_{(\alpha} G^{--}_{\beta)\dot\beta} - \mathcal{D}^{-}_{(\alpha} G^{--+}_{\beta)} + \mathcal{D}^{+}_{(\alpha} G^{---}_{\beta)} \right), \tag{56}$$

which satisfy the chirality and the harmonic independence conditions,

$$\bar{\mathcal{D}}^{\pm}_{\dot\rho} \mathcal{W}_{(\alpha\beta)} = 0, \qquad \mathcal{D}^{\pm\pm} \mathcal{W}_{(\alpha\beta)} = 0\,, \tag{57}$$

and is invariant under gauge transformations to the leading order.

In components, the $\mathcal{N}=2$ super Weyl tensor is represented as

$$\begin{aligned}\mathcal{W}_{(\alpha\beta)} \sim\ & \theta^{+\rho} \hat{C}^{i}_{(\alpha\beta\rho)} u^{-}_{i} - \theta^{-\rho} \hat{C}^{i}_{(\alpha\beta\rho)} u^{+}_{i} + \theta^{+(\rho}\theta^{-\sigma)} C_{(\alpha\beta\rho\sigma)} + \theta^{+}_{(\alpha}\theta^{-\rho)} R_{\rho\beta} \\ & + (\theta^{+})^2 V^{(ij)}_{(\alpha\beta)} u^{-}_{i} u^{-}_{j} - 2(\theta^{+}\theta^{-}) V^{(ij)}_{(\alpha\beta)} u^{+}_{i} u^{-}_{j} + (\theta^{-})^2 V^{(ij)}_{(\alpha\beta)} u^{+}_{i} u^{+}_{j} \\ & + (\theta^{+})^2 \theta^{-\rho} \check{\bar{C}}^{i}_{(\alpha\beta\rho)} u^{-}_{i} - (\theta^{-})^2 \theta^{+\rho} \check{\bar{C}}^{i}_{(\alpha\beta\rho)} u^{+}_{i} + \ldots,\end{aligned} \tag{58}$$

where we have introduced the linearized gauge-invariant tensors for conformal graviton, for doublet of conformal gravitino and for R-symmetry gauge fields:

$$C_{(\alpha\beta\rho\sigma)} = \partial^{\dot\rho}_{(\alpha} \partial^{\dot\sigma}_{\beta} \Phi_{\rho\sigma)(\dot\rho\dot\sigma)}, \quad V^{(ij)}_{(\alpha\beta)} = \partial^{\dot\alpha}_{(\alpha} V^{(ij)}_{\alpha\dot\alpha}, \quad R_{(\alpha\beta)} = \partial^{\dot\rho}_{(\alpha} R_{\beta)\dot\rho}\,, \tag{59a}$$

$$\hat{C}^{i}_{(\alpha\beta\rho)} = \partial^{\dot\rho}_{(\rho} \psi^{i}_{\alpha\beta)\dot\rho}, \qquad \check{\bar{C}}^{i}_{(\alpha\beta\rho)} = \partial^{\dot\rho}_{(\alpha} \partial^{\dot\sigma}_{\beta} \bar\psi^{i}_{\rho)(\dot\rho\dot\sigma)}. \tag{59b}$$

So the component linearized form of Weyl action (55) is written as

$$S_{Weyl} \sim \int d^4x \left\{ C^{\alpha(4)} C_{\alpha(4)} + \hat{C}^{\alpha(3)}_{i} \check{\bar{C}}^{i}_{\alpha(3)} + V^{\alpha(2)}_{(ij)} V^{(ij)}_{\alpha(2)} + R^{(\alpha\beta)} R_{(\alpha\beta)} \right\}. \tag{60}$$

This action contains higher derivatives.

The complete nonlinear action of $\mathcal{N}=2$ conformal supergravity in harmonic superspace is as yet unknown, although it has been constructed within other approaches, see [36, 38, 39].

$\mathcal{N}=2$ superconformal compensators

The comparison of the multiplets (40), (54) with the relevant gauge transformations (39), (53) shows that, nullifying the gauge "conformal" parameters d, η^{i}_{α}, $k_{\alpha\dot\alpha}$, $v^{(ij)}$ in $\mathcal{N}=2$ conformal supergravity gauge group, we can restore the physical fields of $\mathcal{N}=2$ Einstein supergravity. The natural way to do this is to consider the interaction of $\mathcal{N}=2$ conformal supergravity with matter conformal supermultiplets, for example with the hypermultiplet as in (50). Then the "conformal" gauge parameters can be

[1] The supersymmetry-invariant potentials are introduced as $\hat{\mathbb{H}}_{(s=2)} = G^{++M}\mathcal{D}_M$, $\mathcal{D}^{--}G^{++M} = \mathcal{D}^{++}G^{--M}$, $\mathcal{D}^{++}G^{---}_{\hat\beta} = G^{---}_{\hat\beta}$.

fixed by imposing the suitable set of gauge conditions corresponding to the choice of vacuums that violate these symmetries (recall the example (44)).

For $\mathcal{N} = 2$ supergravity one needs two compensators [37]: $\mathcal{N} = 2$ vector multiplet and a matter multiplet. For the latter there exist four possible options:

- Hypermultiplet
- Non-linear multiplet
- Improved tensor multiplet
- Hypermultiplet with non-trivial central charge.

A detailed discussion of these versions of $\mathcal{N} = 2$ supergravity and additional references can be found in [18, 31, 32, 40]. In particular, $\mathcal{N} = 2$ Einstein supergravity action (42) corresponds to the choice of non-linear $\mathcal{N} = 2$ multiplet as a compensator.

Schematically, the structure of the Lagrangians of such theories can be represented as:

$$S_{sg} \sim -S^{conf}_{vec} - S^{conf}_{matter}, \tag{61}$$

where the corresponding actions are covariantized with respect to the $\mathcal{N} = 2$ conformal supergravity gauge group and are taken with the wrong signs (which is typical for compensators).

Note that in the linearized approximation we considered the action (43) corresponding to the choice of non-linear compensator. It will be interesting to construct harmonic superspace actions for other versions of $\mathcal{N} = 2$ linearized supergravity.

5 $\mathcal{N} = 2$ Higher Spin Supermultiplets

The basic difference between the harmonic approach [32] and other superfield approaches to $\mathcal{N} = 2$ supergravity [39] is that the harmonic formulation naturally yields unconstrained prepotentials, the feature which is absent in other approaches. It is highly non-trivial that the harmonic formulation allows a straightforward generalization of $\mathcal{N} = 2$ supergravity prepotentials to $\mathcal{N} = 2$ higher spins.

In this section, we show how $\mathcal{N} = 2$ supermultiplets of higher spins appear in the harmonic approach, construct their cubic vertices of interaction with the hypermultiplet (which is consistent to the leading order) and present the linearized $\mathcal{N} = 2$ gauge-invariant actions.

$\mathcal{N} = 2$ Einstein-Fronsdal higher spins

Similar to the case of $\mathcal{N} = 2$ supergravity, we gauge the higher-spin supersymmetry transformations (22):

$$\hat{\Lambda}^{(s)}_{susy} \quad \rightarrow \quad \hat{\Lambda}^{(s)} = \hat{\Lambda}^{\alpha(s-2)\dot{\alpha}} \partial^{s-2}_{\alpha(s-2)\dot{\alpha}(s-2)}, \tag{62}$$

where

$$\hat{\Lambda}^{\alpha(s-2)\dot{\alpha}(s-2)} := \lambda^{\alpha(s-2)\dot{\alpha}(s-2)M}\partial_M, \qquad M = (\alpha\dot{\alpha}, \alpha+, \dot{\alpha}+, 5). \tag{63}$$

Since $[\mathcal{D}^{++}, \hat{\Lambda}^{(s)}] \neq 0$, we define the covariant harmonic derivative

$$\mathcal{D}^{++} \quad \rightarrow \quad \mathfrak{D}^{++}_{(s)} = \mathcal{D}^{++} + \kappa_s \hat{\mathcal{H}}^{++}_{(s)}, \tag{64}$$

where the spin **s** analytic differential operator is defined as

$$\hat{\mathcal{H}}^{++}_{(s)} := h^{++\alpha(s-2)\dot{\alpha}(s-2)M}\partial_M \partial^{s-2}_{\alpha(s-2)\dot{\alpha}(s-2)}, \qquad \partial_M = \{\partial_{\alpha\dot{\alpha}}, \partial^-_{\alpha}, \partial^-_{\dot{\alpha}}, \partial_5\}. \tag{65}$$

The invariance of the harmonic derivative $\mathfrak{D}^{++}_{(s)}$, in contrast to $\mathcal{N}=2$ supergravity, can be achieved only in the leading order in κ_s:

$$\delta_\lambda \mathfrak{D}^{++}_{(s)} = \kappa_s[\hat{\Lambda}^{(s)}, \mathcal{D}^{++} + \kappa_s \hat{\mathcal{H}}^{++}_{(s)}] + \kappa_s \delta_\lambda \hat{\mathcal{H}}^{++}_{(s)} = 0 + O(\kappa_s^2). \tag{66}$$

This defines the linearized transformation law of the operator (65):

$$\delta_\lambda \hat{\mathcal{H}}^{++}_{(s)} = [\mathcal{D}^{++}, \hat{\Lambda}^{\alpha(s-2)\dot{\alpha}}]\partial^{s-2}_{\alpha(s-2)\dot{\alpha}(s-2)}. \tag{67}$$

Invariance in the following orders requires the introduction of a new type of prepotentials, which are not reduced to $\hat{\mathcal{H}}^{++}_{(s)}$. This is also because of the property that the relevant transformations ((70) below) do not form a closed algebra. Thus we meet obstacles to constructing the consistent nonlinear$\mathcal{N}=2$ higher-spin actions.

Using the covariant derivative (64), one can construct the coupling consistent to the leading order:

$$S^{(s)}_{hyp} = -\frac{1}{2}\int d\zeta^{(-4)}\, q^{+a}\left(\mathcal{D}^{++} + \kappa_s \hat{\mathcal{H}}^{++}_{(s)}(J)^{P(s)}\right) q^+_a\,. \tag{68}$$

The corresponding hypermultiplet transformation has the form [20]:

$$\delta^{(s)}_\lambda q^{+a} = -\kappa_s \hat{\mathcal{U}}^{(s)}_{susy} q^{+a}, \tag{69}$$

where

$$\begin{aligned}\hat{\mathcal{U}}^{(s)}_{susy} = &-\frac{1}{2}\left\{\hat{\Lambda}^{\alpha(s-2)\dot{\alpha}(s-2)}, \partial^{s-2}_{\alpha(s-2)\dot{\alpha}(s-2)}\right\}(J)^{P(s)} \\ &-\frac{1}{2}\partial^{s-2}_{\alpha(s-2)\dot{\alpha}(s-2)}\Omega^{\alpha(s-2)\dot{\alpha}(s-2)}(J)^{P(s)},\end{aligned} \tag{70}$$

$$\Omega^{\alpha(s-2)\dot{\alpha}(s-2)} := (-1)^{P(M)}\partial_M \lambda^{M\alpha(s-2)\dot{\alpha}(s-2)}. \tag{71}$$

In order to exhibit the physical contents of the superfields introduced, we can make use of the gauge freedom (67) to impose the Wess-Zumino type gauge:

$$
\begin{aligned}
h_{WZ}^{++\alpha(s-1)\dot\alpha(s-1)} &= -4i\theta^{+\beta}\bar\theta^{+\dot\beta}\Phi^{\alpha(s-1)\dot\alpha(s-1)}_{\beta\dot\beta} \\
&+ (\bar\theta^{+})^2\theta^{+\beta}\psi^{\alpha(s-1)\dot\alpha(s-1)i}_{\beta}u^-_i + (\theta^{+})^2\bar\theta^{+\dot\beta}\bar\psi^{\alpha(s-1)\dot\alpha(s-1)i}_{\dot\beta}u^-_i \\
&+ (\theta^{+})^2(\bar\theta^{+})^2 V^{\alpha(s-1)\dot\alpha(s-1)(ij)}u^-_i u^-_j\,, \\
h_{WZ}^{++\alpha(s-2)\dot\alpha(s-2)} &= -4i\theta^{+\beta}\bar\theta^{+\dot\beta}C^{\alpha(s-2)\dot\alpha(s-2)}_{\beta\dot\beta} \\
&+ (\bar\theta^{+})^2\theta^{+\beta}\rho^{\alpha(s-2)\dot\alpha(s-2)i}_{\beta}u^-_i + (\theta^{+})^2\bar\theta^{+\dot\beta}\bar\rho^{\alpha(s-2)\dot\alpha(s-2)i}_{\dot\beta}u^-_i \\
&+ (\theta^{+})^2(\bar\theta^{+})^2 S^{\alpha(s-2)\dot\alpha(s-2)(ij)}u^-_i u^-_j\,, \\
h_{WZ}^{++\alpha(s-1)\dot\alpha(s-2)+} &= (\theta^{+})^2\bar\theta^{+}_{\dot\beta}P^{\alpha(s-1)\dot\alpha(s-2)\dot\beta} + \left(\bar\theta^{+}\right)^2\theta^{+}_{\beta}T^{\dot\alpha(s-2)\alpha(s-1)\beta} \\
&+ (\theta^{+})^2(\bar\theta^{+})^2\chi^{\alpha(s-1)\dot\alpha(s-2)i}u^-_i\,, \\
h_{WZ}^{++\dot\alpha(s-1)\alpha(s-2)+} &= \left(\widetilde{h_{WZ}^{++\alpha(s-1)\dot\alpha(s-2)+}}\right).
\end{aligned}
\tag{72}
$$

The WZ fields are defined up to the residual gauge freedom

$$
\delta_\lambda\Phi^{\alpha(s-1)\dot\alpha(s-1)}_{\beta\dot\beta} \sim \partial_{\beta\dot\beta}a^{\alpha(s-1)\dot\alpha(s-1)} - l^{\alpha(s-1))(\dot\alpha(s-2)}_{(\beta}\delta^{\dot\alpha)}_{\dot\beta} - \bar l^{\dot\alpha(s-1))(\alpha(s-2)}_{(\dot\beta}\delta^{\alpha)}_{\beta}, \tag{73a}
$$

$$
\delta_\lambda C^{\alpha(s-2)\dot\alpha(s-2)}_{\beta\dot\beta} \sim \partial_{\beta\dot\beta}a^{\alpha(s-2)\dot\alpha(s-2)} - n^{\alpha(s-2))(\dot\alpha(s-3)}_{(\beta}\delta^{\dot\alpha)}_{\dot\beta} - \bar n^{\dot\alpha(s-2))(\alpha(s-3)}_{(\dot\beta}\delta^{\alpha)}_{\beta}, \tag{73b}
$$

$$
\delta\psi^{\alpha(s-1)\dot\alpha(s-1)i}_{\beta} \sim \partial^{(\dot\alpha}_{\beta}\xi^{\alpha(s-1)\dot\alpha(s-1))i}, \qquad \delta\bar\rho^{\alpha(s-2)(\dot\alpha(s-3)\dot\beta)i}_{\dot\beta} \sim \partial_{\beta\dot\beta}\xi^{(\alpha(s-2)\beta)(\dot\alpha(s-3)\dot\beta)i}. \tag{73c}
$$

After fixing the local Lorentz invariance we come to the set of physical fields

$$
\begin{aligned}
&\left\{\Phi_{\alpha(s)\dot\alpha(s)},\,\Phi_{\alpha(s-2)\dot\alpha(s-2)}\right\},\ \left\{\psi^{\alpha(s-1)\dot\alpha(s-1)i}_{\beta},\,\bar\rho^{\alpha(s-2)(\dot\alpha(s-3)\dot\beta)i}_{\dot\beta}\right\}, \\
&\left\{C_{\alpha(s-1)\dot\alpha(s-1)},\,C_{\alpha(s-3)\dot\alpha(s-3)}\right\}.
\end{aligned}
\tag{74}
$$

They correspond to the Fronsdal spin s gauge fields, the doublet of Fang-Fronsdal spin $s-\frac{1}{2}$ fermionic gauge fields and the Fronsdal spin $s-1$ gauge fields. Other fields can be shown to be auxiliary after proper redefinitions which make them gauge-invariant (see details in [19]). Finally, we encounter total of $\mathbf{8[2s^2-2s+1]_B}+\mathbf{8[2s^2-2s+1]_F}$ off-shell degrees of freedom.

The linearized gauge invariant actions have the universal form for all superspins:

$$
\begin{aligned}
S_{(s)} = (-1)^{s+1}\int d^4x d^8\theta du\,\Big\{&G^{++\alpha(s-1)\dot\alpha(s-1)}G^{--}_{\alpha(s-1)\dot\alpha(s-1)} \\
&+ 4G^{++\alpha(s-2)\dot\alpha(s-2)}G^{--}_{\alpha(s-2)\dot\alpha(s-2)}\Big\},
\end{aligned}
\tag{75}
$$

where the supersymmetry-invariant superfields G^{++M} are defined as vielbeins in

$$\hat{\mathcal{H}}^{++}_{(s)} := G^{++\alpha(s-2)\dot{\alpha}(s-2)M}\mathcal{D}_M \partial^{s-2}_{\alpha(s-2)\dot{\alpha}(s-2)}, \tag{76}$$

and G^{--M} as solutions of the harmonic zero-curvature equations:

$$\mathcal{D}^{++}G^{--M} = \mathcal{D}^{--}G^{++M}. \tag{77}$$

In conclusion of this section we note that it is possible to construct a more general class of $\mathcal{N}=2$ higher-spin cubic vertices, using the conserved $\mathcal{N}=2$ higher-spin supercurrents [41]. We will not touch this issue here.

$\mathcal{N}=2$ superconformal higher spins

In order to construct $\mathcal{N}=2$ superconformal multiplets of higher spins, we will follow a slightly different strategy, which is more suitable for the superconformal case. We will start from the higher-spin hypermultiplet vertex of general type and, requiring invariance with respect to the superconformal transformations, determine the set of analytical potentials which are necessary to secure this invariance.

The invariance under $\mathcal{N}=2$ superconformal transformations (18) requires including all possible types of the superspace derivatives $s-1, s-3, \ldots$:

$$S^{(s)}_{hyp-conf} = -\frac{1}{2}\int d\zeta^{(-4)}\, q^{+a}\left(\mathcal{D}^{++} + \kappa_s \hat{\mathbb{H}}^{++}_{(s)}(J)^{P(s)}\right) q^{+}_{a}, \tag{78}$$

where we introduced the analytic differential operator of degree $s-1$:

$$\hat{\mathbb{H}}^{++}_{(s)} := h^{++M_1\ldots M_{s-1}}\partial_{M_{s-1}}\ldots\partial_{M_1} + h^{++M_1\ldots M_{s-3}}\partial_{M_{s-3}}\ldots\partial_{M_1} + \ldots, \tag{79}$$

Then, defining $\mathcal{N}=2$ superconformal transformations of the prepotentials as

$$\delta_{sc}\hat{\mathbb{H}}^{++}_{(s)} = [\hat{\mathbb{H}}^{++}_{(s)}, \hat{\Lambda}_{sc}] + \frac{1}{2}[\hat{\mathbb{H}}^{++}_{(s)}, \Omega], \tag{80}$$

it is possible to ensure the $\mathcal{N}=2$ superconformal invariance of the vertex (78). It is worth noting that simultaneously one covariantizes the vertex (78) under the generic conformal supergravity transformations.

The vertex (78) is invariant under the huge set of gauge transformations

$$\delta^{(s,k)}_{\lambda}\hat{\mathbb{H}}^{++}_{(s)} = \frac{1}{2}\left[\mathcal{D}^{++}, \left\{\hat{\Lambda}^{M_1\ldots M_{k-2}}, \partial_{M_{k-2}}\ldots\partial_{M_1}\right\}_{AGB}\right], \quad k = s, s-2, \ldots, \tag{81}$$

$$\begin{aligned}\delta^{(k)}_{\lambda} q^{+}_{a} &= -\kappa_s\, \hat{\mathcal{U}}^{(k)}_{s} q^{+}_{a}\\ &= -\frac{\kappa_s}{2}\left\{\hat{\Lambda}^{M_1\ldots M_{k-2}}, \partial_{M_{k-2}}\ldots\partial_{M_1}\right\}_{AGB}(J)^{P(s)}\, q^{+}_{a}\\ &\quad -\frac{\kappa_s}{4}\left\{\Omega^{M_1\ldots M_{k-2}}, \partial_{M_{k-2}}\ldots\partial_{M_1}\right\}_{AGB}(J)^{P(s)}\, q^{+}_{a}.\end{aligned} \tag{82}$$

Different values of k correspond to different spin contributions to the operator $\hat{\mathbb{H}}^{++}_{(s)}$. The algebraic structure of these transformations will be sketched in the next section. Requiring $\delta^{(s,k)}_{\lambda}\hat{\mathbb{H}}^{++}_{(s)} = 0$ gives rise to the equations which yield rigid higher-spin $\mathcal{N} = 2$ conformal supersymmetries.

The superfield gauge transformations (81) lead to the gauge transformations of the component fields (for simplicity, we consider the spin s and spin $s - \frac{1}{2}$ fields and present the transformations in a schematic form):

$$\begin{aligned}\delta_\lambda \Phi^{\alpha(s-1)\dot\alpha(s-1)}_{\beta\dot\beta} \sim\ &\partial_{\beta\dot\beta}a^{\alpha(s-1)\dot\alpha(s-1)} \\ &- l^{\alpha(s-1))(\dot\alpha(s-2)}_{(\beta}\delta^{\dot\alpha)}_{\dot\beta} - \bar{l}^{\dot\alpha(s-1))(\alpha(s-2)}_{(\dot\beta}\delta^{\alpha)}_{\beta} \\ &+ \delta^{(\alpha}_{\beta}\delta^{(\dot\alpha}_{\dot\beta} d^{\alpha(s-2))\dot\alpha(s-2))},\end{aligned} \tag{83a}$$

$$\delta_\lambda \psi^{\alpha(s-1)\dot\alpha(s-1)i}_{\beta} \sim \partial^{(\dot\alpha}_{\beta}\epsilon^{\alpha(s-1)\dot\alpha(s-2))i} + \delta^{(\alpha}_{\beta}\bar\eta^{\alpha(s-2))\dot\alpha(s-1)i}. \tag{83b}$$

It is instructive to compare these transformations with Eq. (73). As in the supergravity case, when breaking local higher-spin dilation and local conformal supersymmetry, we restore the gauge freedom of Fronsdal and Fang-Fronsdal fields. This indicates a possible connection with theories of the Fronsdal type.

Using the higher-spin Lorentz, dilatation, and conformal supersymmetry parameters to gauge away traces of fields in (83), we are finally left with the spin s and $s - \frac{1}{2}$ Fradkin-Tseytlin conformal fields [42]:

$$\begin{aligned}\delta_\lambda \Phi_{\alpha(s)\dot\alpha(s)} &\sim \partial_{(\alpha(\dot\alpha}a_{\alpha(s-1))\dot\alpha(s-1))}, \\ \delta_\lambda \psi^i_{\alpha(s)\dot\alpha(s-1)} &\sim \partial_{(\alpha(\dot\alpha}\epsilon^i_{\alpha(s-1))\dot\alpha(s-2))}.\end{aligned} \tag{84}$$

The rest of the component fields can be tackled in a similar way (see the detailed analysis of the $s = 3$ case in [22]). Everything is reduced to the Fradkin-Tseytlin type higher-spin fields:

$$\begin{aligned}h^{++\alpha(s-1)\dot\alpha(s-1)}_{WZ} &= -4i\theta^+_\rho\bar\theta^+_{\dot\rho}\Phi^{(\rho\alpha(s-1))(\dot\rho\dot\alpha(s-1))} - (\bar\theta^+)^2\theta^+_\rho\psi^{(\rho\alpha(s-1))\dot\alpha(s-1)i}u^-_i \\ &\quad - (\theta^+)^2\bar\theta^+_{\dot\rho}\bar\psi^{\alpha(s-1)(\dot\alpha(s-1)\dot\rho)i}u^-_i + (\theta^+)^4 V^{\alpha(s-1)\dot\alpha(s-1)ij}u^-_i u^-_j, \\ h^{++\alpha(s-1)\dot\alpha(s-2)+}_{WZ} &= (\theta^+)^2\bar\theta^+_{\dot\nu}P^{\alpha(s-1)(\dot\alpha(s-2)\dot\nu)} + (\bar\theta^+)^2\theta^+_\nu T^{(\alpha(s-1)\nu)\dot\alpha(s-2)} \\ &\quad + (\theta^+)^4\chi^{\alpha(s-1)\dot\alpha(s-2)i}u^-_i, \\ h^{++\alpha(s-2)\dot\alpha(s-1)+}_{WZ} &= \widetilde{h^{++\alpha(s-1)\dot\alpha(s-2)+}_{WZ}}, \\ h^{(+4)\alpha(s-2)\dot\alpha(s-2)}_{WZ} &= (\theta^+)^4 D^{\alpha(s-2)\dot\alpha(s-2)}.\end{aligned} \tag{85}$$

All other analytic prepotentials in (79) play an auxiliary role: it is possible to impose a gauge in which they are vanishing (just this gauge was chosen in Eq. (85)).

It should be emphasized that all fields in (85) are gauge fields, and this is a crucial difference of the resulting supermultiplet from the multiplet of $\mathcal{N} = 2$ conformal supergravity (**s** = **2**) and non-conformal $\mathcal{N} = 2$ higher-spin multiplets. This means that all fields appear in the action in a dynamical way, through terms containing derivatives. The component decomposition of the spin **s** superconformal multiplet can be conveniently represented as:

$$s \quad \leftrightarrow \quad 2 \times (s - 1/2) \quad \leftrightarrow \quad 4 \times (s - 1) \quad \leftrightarrow \quad 2 \times (s - 3/2) \quad \leftrightarrow \quad (s - 2)$$

plus the "hook" or conformal pseudo-spin $s - 1$ gauge field $T^{\alpha(s)\dot{\alpha}(s-2)}$. We have total of $\mathbf{8(2s-1)_B} + \mathbf{8(2s-1)_F}$ off-shell degrees of freedom.

The natural conjecture is that the free actions for $\mathcal{N} = 2$ superconformal higher spins are constructed quite analogously to the action (55) of $\mathcal{N} = 2$ conformal supergravity:

$$S^{(s)}_{Weyl} = \int d^4x d^4\theta \, \mathcal{W}^{\alpha(2s-2)} \mathcal{W}_{\alpha(2s-2)} + c.c. \tag{86}$$

Here, the higher-spin $\mathcal{N} = 2$ Weyl supertensors are defined as in [41]. At the component level, this action is reduced to the sum of actions with higher derivatives on the Fradkin-Tseytlin fields. The structure of this action needs the further analysis the results of which will be presented elsewhere.

The approaches described in this section are limited to $\mathcal{N} = 2$ multiplets with *integer* highest spins. For constructing $\mathcal{N} = 2$ multiplets with the highest *half-integer* spin, it is necessary to consider symmetries of another type. As candidates for such symmetries, the higher-spin generalizations of "hidden" supersymmetry and R-symmetry of $N = 4$ vector multiplet theory can be considered. In support of this proposal, it was shown in ref. [44] that gauging of hidden supersymmetry and R-symmetry of the hypermultiplet-vector multiplet system [45] leads to $\mathcal{N} = 2$ superconformal gravitino multiplet.

6 Consistent Superconformal Model

An important feature of the superconformal theory is the special structure of operators which generate gauge transformations. Since the highest term in the operator $\hat{\mathcal{U}}_s$ defined in (82) is the most general superfield operator of degree s, the commutation relations have the form:

$$[\hat{\mathcal{U}}_{s_1}, \hat{\mathcal{U}}_{s_2}] \sim \hat{\mathcal{U}}_{s_1+s_2-2} + \text{lower derivative } \hat{\mathcal{U}}_s \text{ terms.} \tag{87}$$

This means that the gauge algebra is closed only provided we involve into game the whole infinite sequence of operators $\hat{\mathcal{U}}_s$.

Thus, in order to build a consistent theory in all orders, it is necessary to consider an infinite tower of $\mathcal{N}=2$ superconformal higher spins:

$$\hat{\mathbb{H}}^{++} := \sum_{s=1}^{\infty} \kappa_s \hat{\mathbb{H}}^{++}_{(s)} (J)^{P(s)}. \tag{88}$$

As a result, we come to the hypermultiplet action interacting with an infinite tower of $\mathcal{N}=2$ superconformal higher-spin superfields:

$$S_{full} = -\frac{1}{2} \int d\zeta^{(-4)}\, q^{+a} \left(\mathcal{D}^{++} + \hat{\mathbb{H}}^{++}\right) q^{+}_a. \tag{89}$$

This action is invariant under the hypermultiplet transformations

$$\delta_\lambda q^{+a} := -\hat{\mathcal{U}} q^{+a} = -\sum_{s=1}^{\infty} \kappa_s \hat{\mathcal{U}}_s q^{+a}, \tag{90}$$

accompanied by non-Abelian transformations of the gauge superfields:

$$\delta_\lambda \hat{\mathbb{H}}^{++} = -\left[\mathcal{D}^{++} + \hat{\mathbb{H}}^{++}, \hat{\mathcal{U}}\right]. \tag{91}$$

Due to the generic form of the differential operator $\hat{\mathbb{H}}^{++}$, one can ensure the gauge and $\mathcal{N}=2$ superconformal invariances to all orders, see details in [22].

The action (89) is a solution for the hypermultiplet theory consistently coupled to an infinite collection of $\mathcal{N}=2$ superconformal gauge higher-spins. Emphasize that this result is completely given in terms of unconstrained off-shell harmonic superfields and generalizes the known partial cases for conformal interactions of $\mathcal{N}=0$ and $\mathcal{N}=1$ conformal higher spin fields with matter [43, 46, 47]. [2]

$\mathcal{N}=2$ higher-spin superconformal gravity
Using the consistent action (89) one can build the quantum effective action

$$e^{i\Gamma[\hat{\mathbb{H}}^{++}]} = \int \mathcal{D}q^{+} \mathcal{D}\tilde{q}^{+}\, e^{-\frac{i}{2}\int d\zeta^{(-4)}\, q^{+a}\left(\mathcal{D}^{++}+\hat{\mathbb{H}}^{++}\right)q^{+}_a}. \tag{92}$$

Since the logarithmically divergent part of such an effective action is gauge invariant by construction, it can be identified with $\mathcal{N}=2$ higher-spin superconformal gravity. In order to explicitly construct the corresponding theory, it is necessary to develop methods of calculating the effective action for the hypermultiplet in the background

[2] After appearance of our work [22] the superconformal Lagrangian formulation of the hypermultiplet coupled to $\mathcal{N}=2$ gauge higher-spin superfields was also considered in ref. [48] (as a continuation of [49]) in another approach based on the use of $\mathcal{N}=2$ projective superspace.

of $\mathcal{N} = 2$ higher spins. We expect that at the linearized level such a theory reduces to the sum of higher-spin generalizations of Weyl action and should inevitably contain higher derivatives.

7 Outlook

As another interesting application, we can use the theory (89) for a possible construction of the consistent $\mathcal{N} = 2$ higher-spin non-conformal supergravity. As such, we propose the theory of an infinite tower of $\mathcal{N} = 2$ superconformal higher spins interacting with a set of compensating superfields. The compensator set should include supermultiplets of matter, $\mathcal{N} = 2$ vector multiplet and, presumably, multiplets of half-integer $\mathcal{N} = 2$ higher spins. Such a theory can provide the appropriate arena for realization of the infinite-dimensional superconformal gauge symmetry (91) of higher spins.

As a solution to the equations of motion of this theory, we expect to gain a vacuum which can be identified with AdS background (see also a recent paper [50] on discussion of the higher-spin symmetry breaking). In this vacuum, $\mathcal{N} = 2$ higher-spin superconformal symmetries could be spontaneously broken to some non-Abelian extension of $\mathcal{N} = 2$ higher-spin AdS superalgebra and the theory be reduced to the consistent theory of interacting AdS $\mathcal{N} = 2$ supermultiplets.

To conclude, there remain many interesting open problems on the way to the completely consistent $\mathcal{N} = 2$ higher spin supergravity. First of all, it is the construction of various interactions between an infinite tower of the higher spin superfields and a vector multiplet and the study of the gauge freedom of such an extended system. Among other open problems we would highlight the construction of $\mathcal{N} = 2$ multiplets of half-integer spins (see ref. [44]), as well as of $\mathcal{N} = 2$ AdS higher spins and their interactions. It is worth pointing out that the existence of consistent conformal off-shell interaction of hypermultiplet to $\mathcal{N} = 2$ superconformal higher spin gauge superfields opens a possibility to develop an approach to study various aspects of the induced quantum effective action depending on superconformal higher-spin superfields. All the results attained in these directions would become essential ingredients of the complete eventual theory.

Acknowledgements The authors are indebted to Mikhail Vasiliev for numerous interesting discussions. Work of N.Z. was partially supported by the Foundation for the Advancement of Theoretical Physics and Mathematics "BASIS".

Competing Interests The authors have no conflicts of interest to declare that are relevant to the content of this chapter.

Appendix

In this appendix, we collect the **basic** definitions used in the text.

Spinorial notation for the space-time coordinates and derivatives:

$$x^{\alpha\dot{\alpha}} = \sigma_m^{\alpha\dot{\alpha}} x^m, \qquad \partial_{\alpha\dot{\alpha}} = \frac{1}{2}\sigma^m_{\alpha\dot{\alpha}}\partial_m, \qquad \partial_{\alpha\dot{\alpha}} x^{\beta\dot{\beta}} = \delta_\alpha^\beta \delta_{\dot{\alpha}}^{\dot{\beta}}. \tag{93}$$

The rules of tilde conjugation:

$$\widetilde{x^{\alpha\dot{\alpha}}} = x^{\alpha\dot{\alpha}}, \quad \widetilde{\theta^\pm_\alpha} = \bar{\theta}^\pm_{\dot{\alpha}}, \quad \widetilde{\bar{\theta}^\pm_{\dot{\alpha}}} = -\theta^\pm_\alpha, \quad \widetilde{u^{\pm i}} = -u^\pm_i, \quad \widetilde{u^\pm_i} = u^{\pm i}, \quad \widetilde{x^5} = x^5. \tag{94}$$

The partial spinor derivatives:

$$\partial^\pm_{\hat{\alpha}} = \frac{\partial}{\partial\theta^{\mp\hat{\alpha}}}, \qquad \hat{\alpha} = \{\alpha, \dot{\alpha}\}. \tag{95}$$

The multi-index M:

$$M = \begin{bmatrix} \alpha\dot{\alpha}, \alpha+, \dot{\alpha}+, 5 & \text{for non-conformal theories;} \\ \alpha\dot{\alpha}, \alpha+, \dot{\alpha}+, ++ & \text{for conformal theories.} \end{bmatrix} \tag{96}$$

The supersymmetry-covariant spinor derivatives:

$$\begin{aligned} \mathcal{D}^+_{\hat{\alpha}} &:= \partial^+_{\hat{\alpha}}, \\ \mathcal{D}^-_\alpha &:= -\partial^-_\alpha + 4i\bar{\theta}^{-\dot{\alpha}}\partial_{\alpha\dot{\alpha}} - 2i\theta^-_\alpha\partial_5, \\ \bar{\mathcal{D}}^-_{\dot{\alpha}} &:= -\partial^-_{\dot{\alpha}} - 4i\theta^{-\alpha}\partial_{\alpha\dot{\alpha}} - 2i\bar{\theta}^-_{\dot{\alpha}}\partial_5. \end{aligned} \tag{97}$$

The Grassmann-parity operator $P(M)$:

$$P(\alpha\dot{\alpha}) = P(++) = 0, \quad P(\alpha\pm) = P(\dot{\alpha}\pm) = 1\,. \tag{98}$$

The short-hand notation for the symmetrized spinor indices:

$$\alpha(s) = (\alpha_1\alpha_2\ldots\alpha_s), \qquad \dot{\alpha}(s) = (\dot{\alpha}_1\dot{\alpha}_2\ldots\dot{\alpha}_s). \tag{99}$$

References

1. S. V. Ketov, A. A. Starobinsky, Embedding (R + R2)-Inflation into Supergravity. Phys. Rev. D **83**, 063512 (2011), https://doi.org/10.1103/PhysRevD.83.063512
2. S.V. Ketov, A.A. Starobinsky, Inflation and non-minimal scalar-curvature coupling in gravity and supergravity. JCAP **08**, 022 (2012)

3. Alexei A. Starobinsky, Sergey V. Sushkov, Mikhail S. Volkov, Anisotropy screening in Horndeski cosmologies. Phys. Rev. D **101**, 6 (2020), https://doi.org/10.1103/PhysRevD.101.064039
4. Alexey S. Koshelev, K. Sravan Kumar, Anupam Mazumdar, Alexei A. Starobinsky, Non-Gaussianities and tensor-to-scalar ratio in non-local R^2-like inflation. JHEP **06**, 152 (2020), https://doi.org/10.1007/JHEP06(2020)152
5. M. A. Vasiliev, Consistent equation for interacting gauge fields of all spins in (3+1)-dimensions. Phys. Lett. B **243**, 378-382 (1990), https://doi.org/10.1016/0370-2693(90)91400-6
6. M. A. Vasiliev, More on equations of motion for interacting massless fields of all spins in (3+1)-dimensions. Phys. Lett. B **285**, 225-234 (1992), https://doi.org/10.1016/0370-2693(92)91457-K
7. M. A. Vasiliev, Higher spin gauge theories: Star product and AdS space. The Many Faces of the Superworld, 533-610 (2000), https://doi.org/10.1142/9789812793850_0030
8. M. A. Vasiliev, Higher-Spin Theory and Space-Time Metamorphoses. Lect. Notes Phys. **892**, 227-264 (2015), https://doi.org/10.1007/978-3-319-10070-8_9
9. A. Tomasiello, Higher spins and Finsler geometry. JHEP **10**, 047 (2024), https://doi.org/10.1007/JHEP10(2024)047
10. X. Bekaert, S. Cnockaert, C. Iazeolla, M. A. Vasiliev, Nonlinear higher spin theories in various dimensions. arXiv:hep-th/0503128 (2005)
11. X. Bekaert, N. Boulanger, P. Sundell, How higher-spin gravity surpasses the spin two barrier: no-go theorems versus yes-go examples. Rev. Mod. Phys. **84**, 987-1009 (2012), https://doi.org/10.1103/RevModPhys.84.987
12. X. Bekaert, N. Boulanger, A. Campoleoni, M. Chiodaroli, D. Francia, M. Grigoriev, E. Sezgin, E. Skvortsov, Snowmass White Paper: Higher Spin Gravity and Higher Spin Symmetry. arXiv:2205.01567 (2022)
13. Y. A. Tatarenko, M. A. Vasiliev, Bilinear Fronsdal currents in the AdS_4 higher-spin theory. JHEP **07**, 246 (2024), https://doi.org/10.1007/JHEP07(2024)246
14. V. E. Didenko, M. A. Povarnin, All vertices for unconstrained symmetric gauge fields. Phys. Rev. D **110**, 126012 (2024), https://doi.org/10.1103/PhysRevD.110.126012
15. J. Wess, J. Bagger, Supersymmetry and supergravity. (Princeton University Press, 1992)
16. I.L. Buchbinder, S.M. Kuzenko, *Ideas and Methods of Supersymmetry and Supergravity or a Walk Through Superspace* (IOP Publishing, Bristol U.K., 1998)
17. A. Galperin, E. Ivanov, S. Kalitzin, V. Ogievetsky, E. Sokatchev, Unconstrained $\mathcal{N} = 2$ Matter, Yang-Mills and Supergravity Theories in Harmonic Superspace. Class. Quant. Grav. **1**, 469-498 (1984), https://doi.org/10.1088/0264-9381/1/5/004
18. A. S. Galperin, E. A. Ivanov, V. I. Ogievetsky, E. S. Sokatchev, Harmonic superspace. (Cambridge University Press, 2001), 306 p, https://doi.org/10.1017/CBO9780511535109
19. I. Buchbinder, E. Ivanov, N. Zaigraev, Unconstrained off-shell superfield formulation of $4D, \mathcal{N} = 2$ supersymmetric higher spins. JHEP **12**, 016 (2021), https://doi.org/10.1007/JHEP12(2021)016
20. I. Buchbinder, E. Ivanov, N. Zaigraev, Off-shell cubic hypermultiplet couplings to $\mathcal{N} = 2$ higher spin gauge superfields. JHEP **05**, 104 (2022), https://doi.org/10.1007/JHEP05(2022)104
21. I. Buchbinder, E. Ivanov, N. Zaigraev, $\mathcal{N} = 2$ higher spins: superfield equations of motion, the hypermultiplet supercurrents, and the component structure. JHEP **03**, 036 (2023), https://doi.org/10.1007/JHEP03(2023)036
22. I. Buchbinder, E. Ivanov, N. Zaigraev, $\mathcal{N} = 2$ superconformal higher-spin multiplets and their hypermultiplet couplings. JHEP **08**, 120 (2024), https://doi.org/10.1007/JHEP08(2024)120
23. J. Scherk, J. H. Schwarz, How to Get Masses from Extra Dimensions. Nucl. Phys. B **153**, 61-88 (1979), https://doi.org/10.1016/0550-3213(79)90592-3
24. O. V. Shaynkman, M. A. Vasiliev, Higher spin conformal symmetry for matter fields in (2+1)-dimensions. Theor. Math. Phys. **128**, 1155-1168 (2001), https://doi.org/10.1023/A:1012399417069
25. M. G. Eastwood, Higher symmetries of the Laplacian. Annals Math. **161**, 1645-1665 (2005), https://doi.org/10.4007/annals.2005.161.1645

26. E. Ivanov, N. Zaigraev, Off-shell invariants of linearized $4D$, $\mathcal{N} = 2$ supergravity in the harmonic approach. Phys. Rev. D **110**, 066020 (2024), https://doi.org/10.1103/PhysRevD.110.066020
27. E. S. Fradkin, M. A. Vasiliev, Minimal set of auxiliary fields in $SO(2)$ extended supergravity. Phys. Lett. B **85**, 47-51 (1979), https://doi.org/10.1016/0370-2693(79)90774-3
28. E. S. Fradkin, M. A. Vasiliev, Minimal set of auxiliary fields and S-matrix for extended supergravity. Lett. Nuovo Cim. **25**, 79-90 (1979), https://doi.org/10.1007/BF02776267
29. B. de Wit, J. W. van Holten, Multiplets of Linearized $SO(2)$ Supergravity. Nucl. Phys. B **155**, 530-542 (1979), https://doi.org/10.1016/0550-3213(79)90285-2
30. A. S. Galperin, N. A. Ky, E. Sokatchev, $\mathcal{N} = 2$ Supergravity in Superspace: Solution to the Constraints. Class. Quant. Grav. **4**, 1235 (1987), https://doi.org/10.1088/0264-9381/4/5/022
31. A. S. Galperin, E. A. Ivanov, V. I. Ogievetsky, E. Sokatchev, $\mathcal{N} = 2$ Supergravity in Superspace: Different Versions and Matter Couplings. Class. Quant. Grav. **4**, 1255 (1987), https://doi.org/10.1088/0264-9381/4/5/023
32. E. Ivanov, $\mathcal{N} = 2$ Supergravities in Harmonic Superspace. Handbook of Quantum Gravity, 1-50 (Springer Nature Singapore, 2023), https://doi.org/10.1007/978-981-19-3079-9_43-1
33. B. M. Zupnik, Background harmonic superfields in $\mathcal{N} = 2$ supergravity. Theor. Math. Phys. **116**, 964-977 (1998), https://doi.org/10.1007/BF02557138
34. D. Z. Freedman, A. Van Proeyen, Supergravity. (Cambridge University Press, 2012), 607p
35. B. de Wit, J. W. van Holten, A. Van Proeyen, Transformation Rules of $\mathcal{N} = 2$ Supergravity Multiplets. Nucl. Phys. B **167**, 186 (1980), https://doi.org/10.1016/0550-3213(80)90125-X
36. E. Bergshoeff, M. de Roo, B. de Wit, Extended Conformal Supergravity. Nucl. Phys. B **182**, 173-204 (1981), https://doi.org/10.1016/0550-3213(81)90465-X
37. B. de Wit, J. W. van Holten, A. Van Proeyen, Structure of $\mathcal{N} = 2$ Supergravity. Nucl. Phys. B **184**, 77 (1981), https://doi.org/10.1016/0550-3213(81)90211-X
38. S. M. Kuzenko, J. Novak, On curvature squared terms in $\mathcal{N} = 2$ supergravity. Phys. Rev. D **92**, 085033 (2015), https://doi.org/10.1103/PhysRevD.92.085033
39. S. M. Kuzenko, E. S. N. Raptakis, G. Tartaglino-Mazzucchelli, Covariant Superspace Approaches to $\mathscr{N} = 2$ Supergravity. Handbook of Quantum Gravity, 1-61 (Springer Nature Singapore, 2023), https://doi.org/10.1007/978-981-19-3079-9_44-1
40. A. Galperin, E. Ivanov, V. Ogievetsky, Superspace Actions and Duality Transformations for $\mathcal{N} = 2$ Tensor Multiplets. Sov. J. Nucl. Phys. **45**, 157 (1987), https://doi.org/10.1088/0031-8949/1987/T15/025
41. N. Zaigraev, $\mathcal{N} = 2$ higher-spin supercurrents. Phys. Lett. B **858**, 139056 (2024), https://doi.org/10.1016/j.physletb.2024.139056
42. E. S. Fradkin, A. A. Tseytlin, Conformal supergravity. Phys. Rept. **119**, 233-362 (1985), https://doi.org/10.1016/0370-1573(85)90138-3
43. A. Y. Segal, Conformal higher spin theory. Nucl. Phys. B **664**, 59-130 (2003), https://doi.org/10.1016/S0550-3213(03)00368-7
44. E. Ivanov, N. Zaigraev, N=2 superconformal gravitino in harmonic superspace. Phys. Lett. B **862**, 139333 (2025), https://doi.org/10.1016/j.physletb.2025.139333
45. I. L. Buchbinder, E. A. Ivanov, V. A. Ivanovskiy, Superfield realization of hidden R-symmetry in extended supersymmetric gauge theories and its applications. JHEP **04**, 126 (2020), https://doi.org/10.1007/JHEP04(2020)126
46. X. Bekaert, E. Joung, J. Mourad, Effective action in a higher-spin background. JHEP **02**, 048 (2011), https://doi.org/10.1007/JHEP02(2011)048
47. S. M. Kuzenko, M. Ponds, E. S. N. Raptakis, Conformal Interactions Between Matter and Higher-Spin (Super)Fields. Fortsch. Phys. **71**, 1 (2023), https://doi.org/10.1002/prop.202200157
48. S. M. Kuzenko, E. S. N. Raptakis, Towards $\mathcal{N} = 2$ superconformal higher-spin theory. JHEP **11**, 013 (2024), https://doi.org/10.1007/JHEP11(2024)013
49. S. M. Kuzenko, E. S. N. Raptakis, Extended superconformal higher-spin gauge theories in four dimensions. JHEP **12**, 210 (2021), https://doi.org/10.1007/JHEP12(2021)210
50. V. E. Didenko, A. V. Korybut, Toward higher-spin symmetry breaking in the bulk. Phys. Rev. D **110**, 026007 (2024), https://doi.org/10.1103/PhysRevD.110.026007

Gauss-Bonnet Mimetic Gravity

Ali H. Chamseddine and Viatcheslav Mukhanov

Abstract We investigate the coupling of mimetic dark matter to the Gauss-Bonnet topological term in addition to the Einstein action. We show that such interactions can lead to a slight modification to the power law expansion of $2/3$ in the dust dominated era for small coupling constant. Alternatively, and for very large coupling constant such interaction leads to the generation of matter in the form of radiation with a slight modification to the power law of $1/2$.

1 Introduction

The modest modification of gravity by a constrained field related to the scale factor of the metric tensor [1] has proved to have far reaching consequences [2]. In particular it provided a free generation of mimetic dark matter and managed to produce non-singular cosmological [3] and black hole solutions [4, 5, 7]. This was achieved by exploiting the possibility of adding higher curvature terms, dependent only on the Laplacian of the mimetic field, without spoiling unitarity as the graviton propagator does not get contributions from higher time derivatives. In this paper we will explore the possibility of coupling the mimetic field to the Gauss-Bonnet combination of the square of curvatures which, by itself, is topological, and thus will not spoil the graviton propagator or the unitarity of the theory. The aim of this exercise is to determine whether such interactions can act as enhancing the production of dark matter and whether it can modify the expansion rate of the radius in the matter or radiation dominated era.

Dedicated to the memory of Alexei Starobinsky

A. H. Chamseddine
Physics Department, American University of Beirut, Beirut, Lebanon
e-mail: chams@aub.edu.lb

V. Mukhanov (✉)
Ludwig Maxmillian University, Munich, Germany
e-mail: Viatcheslav.Mukhanov@physik.lmu.de

A. Barvinsky and A. Kamenshchik (eds.), *Open Issues in Gravitation and Cosmology*, Fundamental Theories of Physics 225, https://doi.org/10.1007/978-3-032-15703-4_12

As a reminder, the mimetic field ϕ is introduced in order to isolate the scale factor in the metric tensor $g_{\mu\nu}$ by constraining it to have a unit kinetic term

$$g^{\mu\nu}\partial_\mu\phi\partial_\nu\phi = 1 \tag{1}$$

where we have assumed that the signature of the metric is $(+, -, -, -)$. The constraint could be solved by writing

$$g_{\mu\nu} = \widetilde{g}_{\mu\nu} S \tag{2}$$

where $\widetilde{g}_{\mu\nu}$ is some background metric and S is a scaling factor to be determined. Inverting this definition and substituting into the constraint we get

$$\frac{1}{S}\widetilde{g}^{\mu\nu}\partial_\mu\phi\partial_\nu\phi = 1, \tag{3}$$

showing that the metric $g_{\mu\nu}$ could be expressed in terms of the background metric $\widetilde{g}_{\mu\nu}$ and derivatives of the scalar field ϕ

$$g_{\mu\nu} = \widetilde{g}_{\mu\nu}\widetilde{g}^{\alpha\beta}\partial_\alpha\phi\partial_\beta\phi. \tag{4}$$

It can be immediately verified that the metric $g_{\mu\nu}$ is invariant under the scale transformation $\widetilde{g}_{\mu\nu} \to e^{\lambda}\, \widetilde{g}_{\mu\nu}$ from which we deduce that the scale factor is expressible in terms of the derivatives of ϕ. We will add to the Einstein action, the Gauss-Bonnet term interacting with the field ϕ in the form

$$\int d^4x\sqrt{-g}\, f(\phi)\left(R_{\mu\nu}{}^{\alpha\beta}R_{\alpha\beta}{}^{\mu\nu} - 4R_{\mu\nu}R^{\mu\nu} + R^2\right). \tag{5}$$

In the synchronous gauge, a solution of the constraint on ϕ implies that $\phi = t$ is the synchronous time, and making use of the fact that the combination of the curvature terms is topological and the derivative of the Chern-Simons three-form, implies, after integrating by parts, that only the projection of the three-form on the three dimensional space interacts. The plan of this paper is to study the effects of this term on the Friedmann model and to determine its influence on the expansion of the universe for both matter and radiation. We shall show that for a function $f(\phi)$ quadratic in ϕ given by $f(\phi) = -\frac{\beta}{16}\phi^2$ produces mimetic matter in the form of dust with an expansion rate slightly modified from $\frac{2}{3}$ for very small β, or mimetic matter in the form of radiation with an expansion rate slightly modified from $\frac{1}{2}$ for very large β. We mention that other interactions of the mimetic field are possible without spoiling the unitarity of the graviton propagator, by replacing the function $f(\phi)$ by $f(\Box\phi)$ where $\Box\phi$ is the trace of the intrinsic curvature in the synchronous gauge. The resultant equations, however, become too complicated to solve and will not be considered in this paper. We attempt to make this work accessible to beginning students by including step by step calculations in appendices A and B.

2 Gauss-Bonnet Mimetic Interactions

We consider the simplest model of gravity with mimetic matter coupled to the Gauss-Bonnet term, given by

$$S = \int d^4x\sqrt{-g}\left(-\frac{1}{2}R + \lambda\left(g^{\mu\nu}\partial_\mu\phi\partial_\nu\phi - 1\right) + f(\phi)\left(R_{\mu\nu}{}^{\alpha\beta}R_{\alpha\beta}{}^{\mu\nu} - 4R_{\mu\nu}R^{\mu\nu} + R^2\right)\right). \quad (6)$$

As shown in Appendix A, variation with respect to the metric $\delta g_{\mu\nu}$ gives Einstein equations

$$G^{\mu\nu} = 2\lambda\partial_\alpha\phi\partial_\beta\phi g^{\mu\alpha}g^{\nu\beta} - \frac{2}{(-g)}\epsilon^{\mu\beta\kappa\lambda}\epsilon^{\nu\alpha\gamma\delta}R_{\kappa\lambda\gamma\delta}\nabla_\beta\nabla_\alpha f(\phi), \quad (7)$$

while variation $\delta\phi$ gives the equation

$$\frac{1}{\sqrt{-g}}\partial_\mu\left(2\sqrt{-g}g^{\mu\nu}\partial_\nu\phi\lambda\right) = f'(\phi)\left(R_{\mu\nu}{}^{\alpha\beta}R_{\alpha\beta}{}^{\mu\nu} - 4R_{\mu\nu}R^{\mu\nu} + R^2\right). \quad (8)$$

Varying with respect to the Lagrange multiplier λ gives the constraint

$$g^{\mu\nu}\partial_\mu\phi\partial_\nu\phi = 1. \quad (9)$$

We will work in the synchronous gauge

$$ds^2 = dt^2 - \gamma_{ij}\left(x^k, t\right)dx^i dx^j, \qquad i, j = 1, 2, 3 \quad (10)$$

where we choose $g_{00} = 1$, $g_{0i} = 0$. A solution of the constraint (9) in the synchronous gauge is

$$\phi = t + A, \quad (11)$$

where A is constant and ϕ is synchronous time. We now specialize to the Friedmann metric

$$ds^2 = dt^2 - a^2(t)\frac{dx^i dx^j\delta_{ij}}{\left(1 + \frac{k}{4}r^2\right)^2}$$

and proceed to list all components of the Riemann tensors and its contractions. We first have

$$R_{0i0j} = \frac{\ddot{a}}{a}\gamma_{ij}, \quad (12)$$

$$R_{ijkl} = -\frac{\dot{a}^2 + k}{a^2}\left(\gamma_{ik}\gamma_{jl} - \gamma_{il}\gamma_{jk}\right), \quad (13)$$

$$R_{00} = R_{0i0}{}^{i} \quad = -3\frac{\ddot{a}}{a} \tag{14}$$

$$R_{ij} = R_{i0j}{}^{0} + R_{ikj}{}^{k}$$
$$= \gamma_{ij}\left(\frac{\ddot{a}}{a} + 2\frac{\dot{a}^2 + k}{a^2}\right), \tag{15}$$

$$R = R_{00} - \gamma^{ij} R_{ij} \tag{16}$$
$$= -6\left(\frac{\ddot{a}}{a} + \frac{\dot{a}^2 + k}{a^2}\right). \tag{17}$$

Starting with the G^{00} Eq. (62) derived in appendix B we have

$$3\frac{\dot{a}^2 + k}{a^2} = 2\lambda - 24f'(\phi)\frac{\dot{a}}{a}\frac{\dot{a}^2 + k}{a^2}. \tag{18}$$

Using the ϕ equation we get

$$\partial_0\left(a^3 2\lambda\right) = 8f'(\phi)\,\partial_0\left(\dot{a}^3 + 3k\dot{a}\right), \tag{19}$$

implying that

$$2\lambda = \frac{8}{a^3}\int dt f'(t)\,\partial_0\left(\dot{a}^3 + 3k\dot{a}\right) \tag{20}$$
$$= \frac{8\left(\dot{a}^3 + 3k\dot{a}\right)}{a^3}f'(t) - \frac{8}{a^3}\int dt f''(t)\left(\dot{a}^3 + 3k\dot{a}\right). \tag{21}$$

The ϕ equation of motion is given by

$$\frac{1}{\sqrt{-g}}\partial_\mu\left(2\sqrt{-g}g^{\mu\nu}\partial_\nu\phi\lambda\right) = f'(\phi)\left(R_{\mu\nu}{}^{\alpha\beta}R_{\alpha\beta}{}^{\mu\nu} - 4R_{\mu\nu}R^{\mu\nu} + R^2\right). \tag{22}$$

Thus, after eliminating the Lagrange multiplier λ, we have

$$3\frac{\dot{a}^2 + k}{a^2} = \frac{8\dot{a}\left(\dot{a}^2 + 3k\right)}{a^3}f'(t) - \frac{8}{a^3}\int dt f''(t)\left(\dot{a}^3 + 3k\dot{a}\right) - 24f'(t)\frac{\dot{a}}{a}\frac{\dot{a}^2 + k}{a^2}$$
$$= -\frac{16\dot{a}^3}{a^3}f'(t) - \frac{8}{a^3}\int dt f''(t)\left(\dot{a}^3 + 3k\dot{a}\right) + \epsilon, \tag{23}$$

where ϵ is the contribution of ordinary matter. Notice that the indefinite integral in Eq. (23) always implies the existence of mimetic dark matter. Here we will consider only dust with $p = 0$, for which $\epsilon \propto \frac{1}{a^3}$. In this case from the G^{ij} Eq. (64), it follows

$$2\frac{\ddot{a}}{a}+\frac{\dot{a}^{2}+k}{a^{2}}=-16\frac{\dot{a}}{a}\frac{\ddot{a}}{a}f'-8f''\frac{\dot{a}^{2}+k}{a^{2}}, \tag{24}$$

and this could also be obtained from Eq. (23) by multiplying it with a^3 and then differentiating it with respect to t.

3 Cosmological Models: A Quadratic Function

We notice from Eqs. (23) and (24) that the right hand side depends on the first and second derivatives of the function $f(\phi)$. Thus a constant term in $f(\phi)$ will have no effect on the system which is to be expected as such a term is topological. A linear term in $f(\phi)$ will effect solution for the time dependent radius $a(t)$, but it turns out that the modifications are not significant. We therefore proceed to consider the quadratic function $f(\phi)$ given by

$$f(\phi)=-\frac{\beta}{16}\phi^{2}, \tag{25}$$

specializing when $k=0$, the Eq. (64) becomes

$$2\frac{\ddot{a}}{a}+\frac{\dot{a}^{2}}{a^{2}}=2\beta t\frac{\dot{a}}{a}\frac{\ddot{a}}{a}+\beta\frac{\dot{a}^{2}}{a^{2}}. \tag{26}$$

This equation can be rewritten as a first order differential equation for the Hubble constant $H=\frac{\dot{a}}{a}$. In addition Eq. (23) gives

$$\dot{H}=\frac{(\beta-3)/2+\beta tH}{1-\beta tH}. \tag{27}$$

To get rid of the explicit time dependent term βt it is convenient to introduce the new variable

$$y=\frac{1}{tH} \tag{28}$$

instead of H. In terms of the new variable Eq. (27) simplifies to

$$\frac{dy}{d(\ln t)}=\frac{1}{\beta-y}\left(y^{2}-\frac{1}{2}(3+\beta)y+\beta\right). \tag{29}$$

First note that the numerator of this equation has roots

$$y_{\pm}=\frac{1}{4}(\beta+3\pm D), \tag{30}$$

where $D = \sqrt{(\beta - 1)(\beta - 9)}$ so that the general solutions of Eq. (29) can be either attractors or repulsors. To integrate Eq. (29) we rewrite it as

$$d\left(\ln \frac{1}{t}\right) = dy\left(\frac{n_+}{y - y_+} + \frac{n_-}{y - y_-}\right), \tag{31}$$

where

$$n_\pm = \frac{1}{2}\left(1 \mp 3\frac{\beta - 1}{D}\right). \tag{32}$$

In this form Eq. (31) integrates easily to give

$$|y - y_+|^{n_+} \, |y - y_-|^{n_-} = \frac{t_0}{t}, \tag{33}$$

where t_0 is a constant of integration. First we note that we exclude the range $1 < \beta < 9$ where D becomes imaginary. We now consider two limiting cases, namely $\beta \ll 1$ and $\beta \gg 1$.

- For $\beta \ll 1$, when the corrections due to the GB term are small compared to the leading Einstein term, Eq. (33) expanded to linear order in β simplifies to

$$\left|y - \frac{3}{2} + \frac{\beta}{6}\right|^{1-\frac{2\beta}{9}} \left|y - \frac{2\beta}{3}\right|^{\frac{2\beta}{9}} = \frac{t_0}{t}. \tag{34}$$

For very small β the second term on the left hand side goes to unity and that for $t \gg t_0$ a general solution goes to an attractor

$$y \to \frac{3}{2} - \frac{\beta}{6}. \tag{35}$$

Recall that $H = \frac{1}{yt}$ so that the Hubble constant in this case is

$$H = \left(1 + \frac{\beta}{9}\right)\frac{2}{3t}, \tag{36}$$

which integrates to give

$$a \propto t^{\frac{2}{3}\left(1+\frac{\beta}{9}\right)}. \tag{37}$$

The constant of integration in the G_{00} Einstein equation is determined by substituting this solution in Eq. (23). Thus we see that for GB mimetic dark matter, can be different at a given moment of time by few percents as compared to the case of pure dust.

- Now we consider the case when in the presence of dust, the Gauss-Bonnet term is dominant, that is $\beta \gg 1$. Then next to the subleading order in $\frac{1}{\beta}$ Eq. (33) becomes

$$\left|y - \frac{1}{2}\beta + \frac{1}{2} + \frac{2}{\beta}\right|^{-1-\frac{\beta}{6}} \left|y - 2 - \frac{2}{\beta}\right|^{2+\frac{\beta}{6}} = \frac{t_0}{t}. \tag{38}$$

It is clear that at $t \gg t_0$ the only attractor in this case is at

$$y \to 2 + \frac{2}{\beta}, \tag{39}$$

for which the Hubble constant is

$$H = \left(1 - \frac{1}{\beta}\right)\frac{1}{2t}, \tag{40}$$

and correspondingly,

$$a \propto t^{\frac{1}{2}\left(1-\frac{1}{\beta}\right)}. \tag{41}$$

Thus when β is very large the rate of expansion will deviate slightly from $\frac{1}{2}$ mimicking the radiation dominated era.

4 Conclusions

It is now well established that the constraint (1) on a scalar field ϕ replaces the scale factor of the metric tensor $g_{\mu\nu}$ with a function dependent on $\partial_\mu\phi$ and in the process yields contributions to the energy-momentum tensor that mimics pressureless dust. In the synchronous coordinate system, the scalar field ϕ represents synchronous time. Our very modest modification to General Relativity, provided a simple explanation for dark matter without the need to introduce new particles and interactions. In addition this allowed the introduction of higher derivatives interactions without introduce ghost or tachyonic modes to the graviton propagator. In particular, we can add potential terms to the action of the form $V(\phi)$ which in the synchronous gauge is just a time dependent potential, with the remarkable advantage that such term does not break diffeomorphism invariance [2]. Another possibility is to add terms of the form $f(\Box\phi)$ and that allows us to have curvatures of odd powers in time derivatives, because in the synchronous gauge $\Box\phi = \kappa$, the trace of the extrinsic curvature [3, 4]. We have shown that for certain choices of the function $f(\Box\phi)$ we can remove the singularity for Friedmann, Kasner or black hole solutions [5, 7]. We have also shown that it is possible to construct Horava type actions in mimetic gravity without breaking diffeomorphism invariance [6]. The fact that we can construct curvature invariants with odd number of derivatives suggests that mimetic gravity could also be named as the $\frac{3}{2}$ formulation of gravity. In addition, the mimetic modification has been used successfully to construct realistic cosmological models [8]. In this paper we have considered adding to the Einstein action, with the mimetic constraint, a term representing the interaction of a function $f(\phi)$ with the topological Gauss-Bonnet curvatures. The reason for this choice is not to disturb the graviton propagator by non-unitary interactions. For simplicity we have limited our study to functions of the form $f(\phi)$ and not considered instead functions of the form $f(\Box\phi)$ because in this case the resultant equations become too complicated to solve. We have shown that for an optimal function where $f(\phi)$ is quadratic in ϕ of the form $f(\phi) = -\frac{\beta}{16}\phi^2$ and for small β the mimetic field ϕ modifies the power law dependence in the expansion for $a(t) \sim t^n$ from $n = \frac{2}{3}$ to an n slightly modified by Eq. (37) and thus mimicking the dust dominated era. Alternatively, for very large value of β the mimetic field produces a power low expansion for $a(t) \sim t^n$ where n is given by Eq. (41) which deviates slightly from $n = \frac{1}{2}$ and thus mimicking the radiation dominated era.

The mimetic GB interaction does have important modifications to the Friedmann equations, thus opening the way for further studies with other functions $f(\phi)$ or $f(\Box\phi)$ in the hope that one of the models will help to solve some of the long-standing problems in cosmology.

Appendices

Appendix A: Equations of Motion

To vary the action, we first note that the Gauss-Bonnet term is topological and could be written in the form

$$\begin{aligned}
\int \epsilon_{abcd} R^{ab} \wedge R^{cd} &= \frac{1}{4}\int \epsilon_{abcd} R_{\mu\nu}{}^{ab} R_{\kappa\lambda}{}^{cd} dx^{\mu} \wedge dx^{\nu} \wedge dx^{\kappa} \wedge dx^{\lambda} \\
&= \frac{1}{4}\int \epsilon_{abcd} R_{\mu\nu}{}^{ab} R_{\kappa\lambda}{}^{cd} \epsilon^{\mu\nu\kappa\lambda} d^4x \\
&= \frac{1}{4}\int (\det e)\, \epsilon^{\mu\nu\kappa\lambda} \epsilon_{\alpha\beta\gamma\delta} R_{\mu\nu}{}^{\alpha\beta} R_{\kappa\lambda}{}^{\gamma\delta} d^4x \\
&= -\int d^4x \frac{1}{4\sqrt{-g}} g_{\mu\rho} g_{\kappa\sigma} \epsilon^{\mu\nu\kappa\lambda} \epsilon^{\alpha\beta\gamma\delta} f(\phi)\, R^{\rho}_{\ \nu\alpha\beta} R^{\sigma}_{\ \lambda\gamma\delta}, \qquad (42)
\end{aligned}$$

the reason for the minus sign is because $\epsilon^{0123} = 1$ while $\epsilon_{0123} = -1$. We use the variations

$$\frac{1}{\sqrt{-g}} \delta\left(\frac{1}{\sqrt{-g}}\right) = -\frac{1}{2} g^{\mu\nu} \delta g_{\mu\nu} \frac{1}{(-g)}$$

and

$$
\begin{aligned}
&- \int d^4x \frac{1}{2\sqrt{-g}} \delta g_{\mu\rho} g_{\kappa\sigma} \epsilon^{\mu\nu\kappa\lambda} \epsilon^{\alpha\beta\gamma\delta} f(\phi) R^{\rho}_{\ \nu\alpha\beta} R^{\sigma}_{\ \lambda\gamma\delta} \\
&= -2 \int d^4x \sqrt{-g} \delta g_{\mu\nu} f(\phi) \left(R^{\nu\mu} R - 2R^{\nu\rho\mu\beta} R_{\rho\beta} + R^{\nu\rho\kappa\lambda} R^{\mu}_{\ \rho\kappa\lambda} - 2R^{\nu\beta} R^{\mu}_{\beta} \right).
\end{aligned} \tag{43}
$$

However, the terms proportional to $f(\phi)$ vanish as can be seen from the identity

$$
0 = R_{\eta\tau\alpha\beta} R^{\kappa\lambda\gamma\delta} \delta^{\mu\eta\tau\alpha\beta}_{\nu\kappa\lambda\gamma\delta} . \tag{44}
$$

Note that $\epsilon^{\mu\nu\kappa\lambda}\epsilon^{\alpha\beta\gamma\delta}$ does not vary (it is a number) leaving only the variation of the curvature tensor square terms

$$
- \int d^4x \frac{1}{2\sqrt{-g}} g_{\mu\rho} g_{\kappa\sigma} \epsilon^{\mu\nu\kappa\lambda} \epsilon^{\alpha\beta\gamma\delta} f(\phi) \delta R^{\rho}_{\ \nu\alpha\beta} R^{\sigma}_{\ \lambda\gamma\delta}, \tag{45}
$$

where we have doubled the contribution because of symmetry. Using

$$
\delta R^{\rho}_{\ \nu\alpha\beta} = \nabla_\alpha \delta\Gamma^{\rho}_{\nu\beta} - \nabla_\beta \delta\Gamma^{\rho}_{\nu\alpha}, \tag{46}
$$

and after integrating by parts, we get

$$
\begin{aligned}
&\int d^4x \frac{1}{\sqrt{-g}} g_{\mu\rho} g_{\kappa\sigma} \epsilon^{\mu\nu\kappa\lambda} \epsilon^{\alpha\beta\gamma\delta} \delta\Gamma^{\rho}_{\nu\beta} \nabla_\alpha \left(f(\phi) R^{\sigma}_{\ \lambda\gamma\delta} \right) \\
&= \int d^4x \frac{1}{\sqrt{-g}} g_{\mu\rho} g_{\kappa\sigma} \epsilon^{\mu\nu\kappa\lambda} \epsilon^{\alpha\beta\gamma\delta} \delta\Gamma^{\rho}_{\nu\beta} R^{\sigma}_{\ \lambda\gamma\delta} \nabla_\alpha f(\phi),
\end{aligned} \tag{47}
$$

using the Bianchi identity. Writing

$$
g_{\mu\rho} \delta\Gamma^{\rho}_{\nu\beta} = \frac{1}{2} \left(\nabla_\nu \delta g_{\mu\beta} + \nabla_\beta \delta g_{\mu\nu} - \nabla_\mu \delta g_{\nu\beta} \right), \tag{48}
$$

we get, after integrating by parts, and dropping the $\delta g_{\mu\nu}$ contribution because of antisymmetry,

$$
- \int d^4x \frac{1}{2\sqrt{-g}} \epsilon^{\mu\nu\kappa\lambda} \epsilon^{\alpha\beta\gamma\delta} \left(\nabla_\nu \left(R_{\kappa\lambda\gamma\delta} \nabla_\alpha f(\phi) \right) \delta g_{\mu\beta} - \nabla_\mu \left(R_{\kappa\lambda\gamma\delta} \nabla_\alpha f(\phi) \right) \delta g_{\nu\beta} \right). \tag{49}
$$

The covariant derivative acting on the $R_{\kappa\lambda\gamma\delta}$ vanishes because of the Bianchi identity, leaving us with

$$
\int d^4x \frac{1}{\sqrt{-g}} \epsilon^{\mu\nu\kappa\lambda} \epsilon^{\alpha\beta\gamma\delta} R_{\kappa\lambda\gamma\delta} \nabla_\nu \nabla_\alpha f(\phi) \delta g_{\mu\beta} \tag{50}
$$

$$= -\int d^4x \frac{1}{\sqrt{-g}} \epsilon^{\nu\alpha\kappa\lambda} \epsilon^{\mu\beta\kappa\lambda} R_{\kappa\lambda\gamma\delta} \nabla_\beta \nabla_\alpha f(\phi) \, \delta g_{\mu\nu}. \tag{51}$$

The variation of the Einstein action and the mimetic constraint is

$$\frac{1}{2}\int d^4x \sqrt{-g} G^{\mu\nu} \delta g_{\mu\nu} - \int d^4x \sqrt{-g} \lambda \partial_\alpha \phi \partial_\beta \phi g^{\mu\alpha} g^{\nu\beta} \delta g_{\mu\nu}. \tag{52}$$

Now setting the variation $\delta g_{\mu\nu}$ to zero we get

$$G^{\mu\nu} = 2\lambda \partial_\alpha \phi \partial_\beta \phi g^{\mu\alpha} g^{\nu\beta} - \frac{2}{(-g)} \epsilon^{\mu\beta\kappa\lambda} \epsilon^{\nu\alpha\gamma\delta} R_{\kappa\lambda\gamma\delta} \nabla_\beta \nabla_\alpha f(\phi). \tag{53}$$

We are now ready to study some realistic models of mimetic dark matter interacting with the Gauss-Bonnet term by appropriate choice of the function $f(\phi)$.

Appendix B: Gauss-Bonnet Combination

To find the Gauss-Bonnet combination we first have

$$\begin{aligned} R_{\mu\nu\rho\sigma} R^{\mu\nu\rho\sigma} &= 4R_{0i0j} R^{0i0j} + R_{ijkl} R^{ijkl} \\ &= 12\left(\frac{\ddot{a}}{a}\right)^2 + \left(\frac{\dot{a}^2 + k}{a^2}\right)^2 \left(\gamma_{ik}\gamma_{jl} - \gamma_{il}\gamma_{jk}\right)\left(\gamma^{ik}\gamma^{jl} - \gamma^{il}\gamma^{jk}\right) \\ &= 12\left(\left(\frac{\ddot{a}}{a}\right)^2 + \left(\frac{\dot{a}^2 + k}{a^2}\right)^2\right). \end{aligned} \tag{54}$$

Next we have

$$\begin{aligned} R_{\mu\nu} R^{\mu\nu} &= \left((R_{00})^2 + R_{ij} R^{ij}\right) \\ &= \left(9\left(\frac{\ddot{a}}{a}\right)^2 + 3\left(\frac{\ddot{a}}{a} + 2\frac{\dot{a}^2 + k}{a^2}\right)^2\right), \end{aligned} \tag{55}$$

and finally

$$R^2 = 36\left(\frac{\ddot{a}}{a} + \frac{\dot{a}^2 + k}{a^2}\right)^2. \tag{56}$$

The GB combination is then

$$12\left(\left(\frac{\ddot{a}}{a}\right)^2+\left(\frac{\dot{a}^2+k}{a^2}\right)^2+3\left(\frac{\ddot{a}}{a}+\frac{\dot{a}^2+k}{a^2}\right)^2-3\left(\frac{\ddot{a}}{a}\right)^2-\left(\frac{\ddot{a}}{a}+2\frac{\dot{a}^2+k}{a^2}\right)^2\right)$$
$$=24\frac{\ddot{a}\left(\dot{a}^2+k\right)}{a^3}. \tag{57}$$

Notice that

$$\begin{aligned}\sqrt{-g}GB &= 24\ddot{a}\left(\dot{a}^2+k\right)=8\partial_0\left(\dot{a}^3+3k\dot{a}\right)\\ &=\partial_0\left(CS\right)\end{aligned} \tag{58}$$

so that the Chern-Simons term is

$$CS=8\left(\dot{a}^3+3k\dot{a}\right). \tag{59}$$

It is also useful to list the components of the Einstein tensor $G_{\mu\nu}$

$$\begin{aligned}G_{00} &= R_{00}-\frac{1}{2}R\\ &=3\frac{\dot{a}^2+k}{a^2}\end{aligned} \tag{60}$$

and

$$\begin{aligned}G_{ij} &= R_{ij}+\frac{1}{2}\gamma_{ij}R\\ &=\gamma_{ij}\left(\frac{\ddot{a}}{a}+2\frac{\dot{a}^2+k}{a^2}-3\left(\frac{\ddot{a}}{a}+\frac{\dot{a}^2+k}{a^2}\right)\right)\\ &=\gamma_{ij}\left(-2\frac{\ddot{a}}{a}-\frac{\dot{a}^2+k}{a^2}\right).\end{aligned} \tag{61}$$

To work out the equations of motion, we first have

$$\begin{aligned}G^{00} &= 2\lambda-2R_{klcd}\epsilon^{0jkl}\epsilon^{0icd}\left(f''(\phi)\,\partial_i\phi\partial_j\phi+f'(\phi)\,\nabla_i\nabla_j\phi\right)\\ &=2\lambda-2R_{klcd}f'(\phi)\left(-\Gamma^0_{ij}\right)\epsilon^{0jkl}\epsilon^{0icd}\\ &=2\lambda-4f'(\phi)\,\frac{\dot{a}^2+k}{a^2}\left(\gamma_{ck}\gamma_{ld}\right)\epsilon^{0jkl}\epsilon^{0icd}\Gamma^0_{ij}\\ &=2\lambda-8f'(\phi)\,\frac{\dot{a}^2+k}{a^2}\gamma^{ij}\Gamma^0_{ij}\\ &=2\lambda-24f'(\phi)\,\frac{\dot{a}}{a}\frac{\dot{a}^2+k}{a^2},\end{aligned} \tag{62}$$

where we have used $\gamma^{ij}\Gamma^0_{ij}=3\frac{\dot{a}}{a}$.

Next we look at the G_{ij} equation

$$\begin{aligned} G^{ij} &= -2R_{\kappa\lambda\rho\sigma}\epsilon^{i\nu\kappa\lambda}\epsilon^{j\mu\rho\sigma}\left(f^{''}\partial_\mu\phi\partial_\nu\phi + f'\nabla_\mu\nabla_\nu\phi\right) \\ &= -2R_{klrs}f^{''}\epsilon^{i0kl}\epsilon^{j0rs} - 2R_{\kappa\lambda\rho\sigma}\left(-\Gamma^0_{\mu\nu}\right)f'\epsilon^{i\mu\rho\sigma}\epsilon^{j\nu\kappa\lambda} \\ &= 4\frac{\dot{a}^2+k}{a^2}f^{''}\left(2\gamma^{ij}\right) + 8R_{0k0l}f'\epsilon^{im0k}\epsilon^{jn0l}\Gamma^0_{mn} \\ &= 8\gamma^{ij}\frac{\dot{a}^2+k}{a^2}f^{''} + 8f'\frac{\ddot{a}}{a}\left(\gamma^{ij}\gamma^{mn} - \gamma^{im}\gamma^{jn}\right)\Gamma^0_{mn} \\ &= \left(8f''\frac{\dot{a}^2+k}{a^2} + 16\frac{\dot{a}}{a}\frac{\ddot{a}}{a}f'\right)\gamma^{ij}. \end{aligned} \tag{63}$$

Thus the G^{ij} equation implies that

$$2\frac{\ddot{a}}{a} + \frac{\dot{a}^2+k}{a^2} = -16\frac{\dot{a}}{a}\frac{\ddot{a}}{a}f' - 8f''\frac{\dot{a}^2+k}{a^2}. \tag{64}$$

We can check the consistency of G_{00} and G_{ij} equations by multiplying the G_{00} equation by a^3 and differentiating it to get

$$\begin{aligned} 3\partial_0\left(a\left(\dot{a}^2+k\right)\right) &= 24f'(\phi)\ddot{a}\left(\dot{a}^2+k\right) - 24\partial_0\left(f'\dot{a}\left(\dot{a}^2+k\right)\right) \\ &= -48f'(\phi)\ddot{a}\dot{a}^2 - 24f''\dot{a}\left(\dot{a}^2+k\right) \end{aligned} \tag{65}$$

to obtain

$$3\dot{a}^3 + 6a\dot{a}\ddot{a} + 3k\dot{a} = -48f'(\phi)\ddot{a}\dot{a}^2 - 24f''\dot{a}\left(\dot{a}^2+k\right). \tag{66}$$

This can be rewritten as

$$2\frac{\ddot{a}}{a} + \frac{\dot{a}^2+k}{a^2} = -16\frac{\ddot{a}\dot{a}}{a^2}f' - 8f''\frac{\left(\dot{a}^2+k\right)}{a^2} \tag{67}$$

which is the same as the G^{ij} equation. It is also sometimes useful to look at the equivalent equations. To look at the trace equation $G^A_A = G_{00} - G_{aa}$ we get

$$G = 3\frac{\dot{a}^2+k}{a^2} + 3\left(2\frac{\ddot{a}}{a} + \frac{\dot{a}^2+k}{a^2}\right) = 6\left(\frac{\ddot{a}}{a} + \frac{\dot{a}^2+k}{a^2}\right) \tag{68}$$

so that

$$6\left(\frac{\ddot{a}}{a} + \frac{\dot{a}^2+k}{a^2}\right) = -\frac{16\dot{a}^3}{a^3}f'(t) - \frac{8}{a^3}\int dt f''(t)\left(\dot{a}^3 + 3k\dot{a}\right) \tag{69}$$

$$-3\left(8f''\frac{\dot{a}^2+k}{a^2}+16\frac{\dot{a}}{a}\frac{\ddot{a}}{a}f'\right)+T^m, \tag{70}$$

where

$$T^m=\sum_w (1-3w)\,\epsilon_w, \tag{71}$$

which has the advantage to cancel the contributions of radiation matter, and eliminate the dust matter by tuning the $\frac{1}{a^3}$ term resulting from the indefinite integral.

Acknowledgements The work of A. H. C is supported in part by the National Science Foundation Grant No. Phys-2207663.

References

1. A. H. Chamseddine, A. H., Mukhanov, V.F., Mimetic Dark Matter. JHEP **11**, 135 (2013)
2. A.H. Chamseddine, V. Mukhanov, A. Vikman, Cosmology with Mimetic Matter. JCAP **1406**, 017 (2014)
3. A.H. Chamseddine, V. Mukhanov, Resolving Cosmological Singularities. JCAP **1703**, 009 (2017a)
4. A.H. Chamseddine, V. Mukhanov, Nonsingular Black Hole. Eur. Phys. J. **C77**, 183 (2017b)
5. A.H. Chamseddine, V. Mukhanov, T.B. Russ, Asymptotically Free Mimetic Gravity. Eur. Phys. J. C **79**, 558 (2019a)
6. A.H. Chamseddine, V. Mukhanov, T.B. Russ, Mimetic Hořava gravity. Phys. Lett. B **798**, 134939 (2019b)
7. A.H. Chamseddine, V. Mukhanov, T.B. Russ, Black Hole Remnants. JHEP **1910**, 104 (2019c)
8. A.H. Chamseddine, M. Khaldieh, V. Mukhanov, Mimetic Inflation and Self-Reproduction. JCAP **11**, 011 (2023)

Initial Conditions for Starobinsky Inflation in General Quadratic Gravity

Daniel Müller and Alexey Toporensky

Abstract We consider initial conditions leading to Starobinsky inflation in the general quadratic gravity, where the action of the theory contains one more curvature square invariant in addition to R^2. We have chosen the corresponding coefficients in such a way that the inflationary solution is stable. Using the numerical simulations we have investigated the differences between this model and the traditional model with R^2 term in the action.

1 Introduction

Starobinsky inflation [1], being one of the oldest inflationary models is still one of the most successful models, regarding ever increasing accuracy of modern observations. In this theory there is no physical scalar field, and the inflationary behavior is achieved due to the presence of the curvature square correction R^2 to the Hilbert-Einstein action of General Relativity. Its predictions match with the observed value of primordial perturbations spectral tilt. Besides, the predicted amplitude of the primordial gravitational waves is well within the observational limits. The amplitude of the power spectrum fixes the only free constant of the theory, the coupling to the curvature square term in the action. To match with the data this constant should be chosen to be $\beta = 1.305 \times 10^9 m_{\rm P}^{-2}$ [2], where $m_{\rm P} = 1/\sqrt{8\pi G}$ is the reduced Planck mass.

On the other hand, the R^2 gravity itself is a particular case of a more general quadratic gravity, where other quadratic curvature invariants contribute to the action. It is well-known that there are three independent invariants—namely $R^{ijkl}R_{ijkl}$, $R^{ij}R_{ij}$ and R^2. However in $3+1$ dimensions it is enough to add only one more invariant, since the 3-d one can be expressed through 1-st, 2-nd and the Gauss-

D. Müller
Instituto de Física, Universidade de Brasília, Brasília, Brazil
e-mail: dmuller@unb.br

A. Toporensky (✉)
Sternberg Astronomical Institute, Moscow University, Moscow, Russian Federation
e-mail: atopor@rambler.ru

A. Barvinsky and A. Kamenshchik (eds.), *Open Issues in Gravitation and Cosmology*, Fundamental Theories of Physics 225, https://doi.org/10.1007/978-3-032-15703-4_13

Bonnet combination, which is a full derivative in $3+1$ dimensions and, thus, does not contribute to the equations of motion.

Quadratic gravity began with Weyl in 1918 [3]. After many years it was studied by Buchdahl [4]. Starobinsky inflation is the specific case when there's only the Einsten term and the R^2 term in the action. This case was studied also by Ruzmaikina and Ruzmaikin [5]. The quadratic gravity was investigated by many authors (see, for example, [6–17] and for a historical review see [18]).

We have previously investigated the initial conditions for Starobinsky inflation in $R+R^2$ theory for a spatially flat spacetime[19] and for a closed spacetime [20]. As the next step, the question of Starobinsky inflation in the framework of general quadratic gravity arises naturally. What can be shown easily, is that adding $R^{ij}R_{ij}$ term does not change properties of isotropic cosmological evolution, since such a theory can be reformulated as a theory containing the sum of Einstein and Weyl square terms, and Weyl tensor is known to vanish for isotropic metric. However, there are no *a priory* reasons why cosmological shear should vanish (or, at least, be very small) before inflation. Thus, investigations of possible impacts of other quadratic terms to the possibility of entering inflationary regimes are of interest for initially anisotropic metrics.

It is already known through the study of the behaviour of the linearized degrees of freedom that general quadratic gravity has a scalar degree of freedom of mass $m_0 = 1/\sqrt{6\beta}$, a massless spin 2 field and a spin 2 field with mass $m_2 = 1/\sqrt{-\alpha}$ [21], see also the appendix in [22]. While choosing $\beta > 0$ and $\alpha < 0$ eliminates the tachyons, it does not eliminate the ghost. The ghost comes with a different sign in the energy of the massive spin 2 field in the linear regime. For sufficiently small perturbations the coupling between these fields is negligible and these fields behave as free fields. On the other hand, by increasing the initial shear at some threshold, the coupling becomes relevant and energy can be transferred from the other fields to the spin 2 massive ghost field which have opposite signs. Linearized energy is conserved and amplitudes of the perturbations can increase. Eventually this mechanism brings the system out of the linear regime and to some singularity.

In this study we consider a Bianchi-I universe.

It is known that in the theory with both $R^{ij}R_{ij}$ and R^2 terms in the action isotropic solutions can become unstable with respect to any nonzero shear. However, for inflationary regime this happens only when a simple relation between the coefficients before these terms is not satisfied (see below). In the opposite case Starobinsky inflation is a stable regime, and a natural question is to specify initial conditions leading to it. In particular, it is important to understand if this regime appears naturally or requires some sort of fine-tunning of initial conditions. To answer this question we provide a numerical analysis of corresponding cosmological dynamics for a case when α is as big as β but satisfies the inflation stability condition.

2 The Field Equations

In this work we consider the following Lagrangian density:

$$\mathcal{L} = \frac{1}{16\pi G}\left[R + \left(\beta - \frac{1}{3}\alpha\right)R^2 + \alpha R_{ab}R^{ab}\right].$$

The equations of motion are

$$G_{ab} + \left(\beta - \frac{1}{3}\alpha\right)H^{(1)}_{ab} + \alpha H^{(2)}_{ab} = 0, \tag{1}$$

where

$$\begin{aligned}
G_{ab} &= R_{ab} - \frac{1}{2}Rg_{ab} \\
H^{(1)}_{ab} &= 2g_{ab}\Box R - 2R_{;ab} + 2RR_{ab} - \frac{1}{2}g_{ab}R^2 \\
H^{(2)}_{ab} &= \Box R_{ab} - R_{;ab} + \frac{1}{2}\Box R g_{ab} + 2R^{cd}R_{cbda} - \frac{1}{2}g_{ab}R_{cd}R^{cd}.
\end{aligned}$$

The stability of the Minkowski vacuum requires in this case that $\beta > 0$ and $\alpha < 0$, [21, 22]. Using tetrads, the metric is

$$\tilde{g}_{\alpha\beta}e^{\alpha}_{a}e^{\beta}_{b} = g_{ab} = \mathrm{diag}[-1, 1, 1, 1].$$

As is well known, in this case the connection follows from the covariant conservation of the above metric and from the zero torsion condition [23]:

$$\nabla_a g_{bc} = 0 \implies \Gamma_{bca} + \Gamma_{cba} = g_{ab|c} \tag{2}$$

$$\nabla_a e_b - \nabla_b e_a = [e_a, e_b] = D^{c}_{ab}e_c, \tag{3}$$

where $\nabla_b e_a = \Gamma^{c}_{ab}e_c$ is the connection for the tetrad base, D^{c}_{ab} is the structure function and $|$ is the directional derivative, $f_{|c} = e^{\alpha}_{c}\partial_\alpha f$. The temporal component e_0, also denoted as $n^a = (1, 0, 0, 0)$ is geodesic with zero vorticity and the decomposition of it's covariant derivative in a trace-free and a trace parts is

$$\nabla_a n_b = \sigma_{ab} + H\delta_{ab}.$$

For the diagonal shear,

$$\sigma_{ab} = \mathrm{diag}[0, -2\sigma_+, \sigma_+ + \sqrt{3}\sigma_-, \sigma_+ - \sqrt{3}\sigma_-],$$

together with (2) and (3) it gives the following non-vanishing connection coefficients:

$$\Gamma^0_{11} = \Gamma^1_{01} = -2\sigma_+ + H \quad \Gamma^0_{22} = \Gamma^2_{02} = \sigma_+ + \sqrt{3}\sigma_- \quad \Gamma^0_{33} = \Gamma^3_{03} = \sigma_+ - \sqrt{3}\sigma_-.$$

Remind that in this setting, the connection is not symmetric, for example $\Gamma^0_{10} = \Gamma^0_{20} = \Gamma^0_{30} = 0$. The Riemann tensor is given by

$$R^a_{bcd} = \Gamma^a_{bd|c} - \Gamma^a_{bc|d} + \Gamma^a_{fc}\Gamma^f_{bd} - \Gamma^a_{fd}\Gamma^f_{bc} - D^f_{cd}\Gamma^a_{bf},$$

and the Riemann scalar reads

$$R = 6\left(\sigma_-^2 + \sigma_+^2 + 2H^2 + \dot{H}\right). \tag{4}$$

In this case the Jacobi identity does not impose additional conditions and is identically satisfied by this connection, namely $R_{abcd} + R_{acdb} + R_{adbc} = 0$.

As is well known Starobinsky-Ruzmaikina-Ruzmaikin inflaton [5] is an asymptotic solution if the shear goes to zero. In this case the field equations do not depend on α. This is the advantage of the action we use, since the simple sum $\beta R + \alpha R^{ij} R_{ij}$ in the action would lead to additional term proportional to α even in the absence of shear making the comparison with pure $R + R^2$ gravity more difficult. Now for zero shear the 00 field equation gives

$$3H^2 + 36\beta H\ddot{H} + 108\beta\dot{H}H^2 - 18\beta\dot{H}^2 = 0,$$

which when $H >> \dot{H} >> \ddot{H}$ becomes

$$3H^2(1 + 36\beta\dot{H}) \approx 0 \implies H = const. - \frac{t}{36\beta}.$$

The last expression is the inflationary Ruzmaikina-Ruzmaikin-Starobinsky solution. Linearization around this solution reveals that the frequencies are

$$\lambda_1, \lambda_2 = \left(-\frac{3}{2} + \frac{1}{2}\sqrt{1 + \frac{96\beta}{\alpha}}\right)H \quad \lambda_3, \lambda_4 = \left(-\frac{3}{2} - \frac{1}{2}\sqrt{1 + \frac{96\beta}{\alpha}}\right)H \quad \lambda_5 \text{ to } \lambda_8 = -3H.$$

Stability requires $H > 0$ and that the real parts of the eigenvalues are negative. We see easily that the conditions $\beta > 0$ and $\alpha < 0$ mentioned in the Introduction are sufficient to ensure stability of the inflationary solution. In this article we choose $\alpha = -10\beta$, which is rather big, but still finds itself within the stability domain.

3 Numerical Results

In this section we present our results of numerical integration for the cosmological evolution of our model in the Bianchi-I space-time. We fix the shear so that $\sigma_- = 0$, σ_+ is specified on plots and set a grid of initial H and R given by (4). Since the

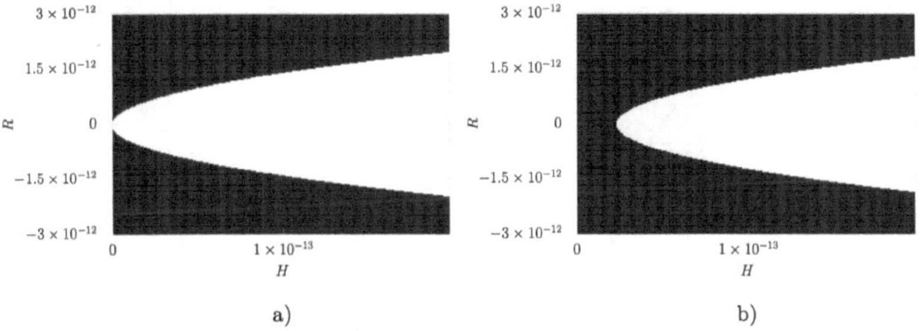

Fig. 1 **a** Basin for a very small shear $\sigma_+ = 1.0 \times 10^{-10}$. This picture looks almost the same as that for exactly zero shear presented in Fig. 6a of [19]. **b** Basin for an initially bigger shear, namely $\sigma_+ = 1.0 \times 10^{-8}$. Black points mark initial conditions which give a sufficient inflation while white points refer to an insufficient inflation

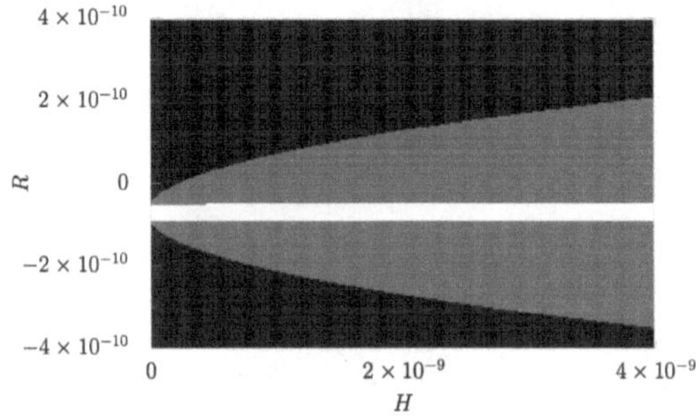

Fig. 2 For this plot the shear is increased to $\sigma_+ = 2.07 \times 10^{-6}$. As before, black points mark initial conditions with sufficient inflation, gray points refer to insufficient inflation and white points indicate initial conditions that give an evolution ended in the Big Crunch singularity

shear is fixed, the choice of a value for R corresponds to a choice of a value for $\dot{H}$. After that $\ddot{H}$ is determined by the constraint equation (5). We choose the initial conditions for the inflation asking to have a number of e-folds bigger than 60. These initial conditions are marked as black points in the coordinate plane H and R (Figs. 1, 2 and 3).

We remind the reader that in the absence of shear the condition for Starobinsky inflation to occur is roughly $R > 0$ with the exception of a thin zone near $R = 0$ (where inflation is insufficient) and the addition of a small region with $R < 0$ leading to enough inflation. Introduction of a non-zero shear changes the situation qualitatively. Very small shear, though, only slightly changes the situation (see Fig. 1). However, for the shear about $2 * 10^{-6}$ a new feature appears, and some trajectories instead of scalaron oscillations fall into a singularity (possibly, with finer grid they can be remarked for even smaller shear). This means that H changes its sign, becomes negative, and Universe ends in the Big Crunch singularity. With increasing shear scalaron oscillations regime disappears completely, moreover, initial conditions leading to inflation in the absence of shear can lead to a singularity for non-zero shear.

Describing the situation with large shear we consider only $0 < R < 1, -1 < H < 1$ region of initial conditions since going beyond 1 in Planck units is doubtful for obvious physical reasons. We remark that apart from turning points in cosmological

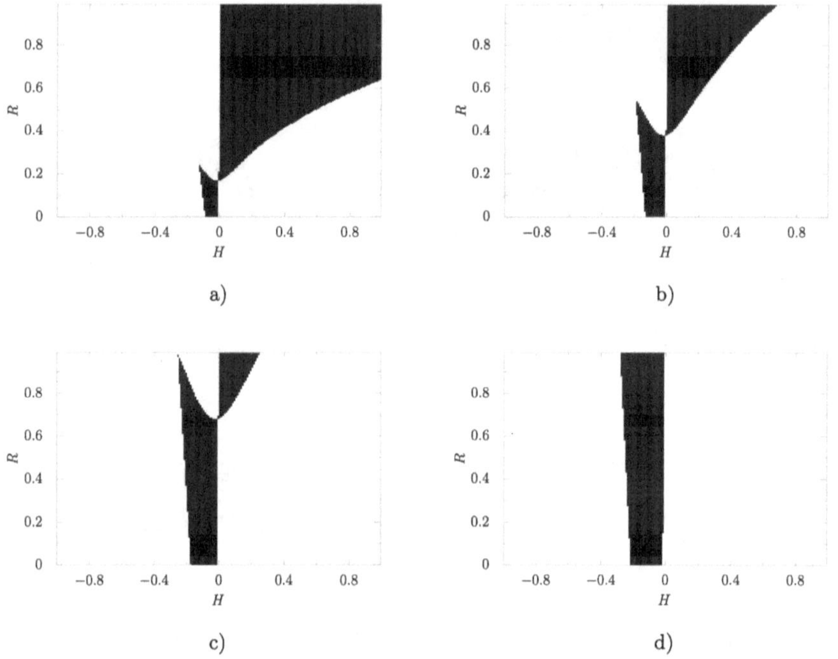

Fig. 3 We present several basin plots for different values of initial shear. Each black point marks an initial condition with sufficient inflation, while white points give the evolution towards the singularity, Big Crunch. Panel **a** $\sigma_+ = 0.1$; panel **b** $\sigma_+ = 0.15$, panel **c** $\sigma_+ 0.2$ and in panel **d** $\sigma_+ = 0.25$

evolution, transitions from contraction to expansion become possible as well, and some initial conditions with $H < 0$ appear to be good for Starobinsky inflation (see Fig. 3). As can be seen from the plots, these two effects work in the opposite directions, always leaving a room for Starobinsky inflation. While an initial shear increases, bigger and bigger parts of initial conditions for $H > 0$ leading to inflation without shear tend to singularity instead. For shear about 0.2 the zone of successful inflation moves beyond $H = 1$ limit and its physical significance can not be trusted. However, instead a zone of good initial conditions for initially contracting Universe $H < 0$ appears and grows with growing shear. For big shear all good initial conditions are those for a contracting Universe which reach Starobinsky inflation going through a bounce.

In any case, the realization of Starobinsky inflation does not need a fine-tuning of initial conditions, though an actual shape of corresponding zone in initial conditions space depends on the value of shear sharply.

4 Conclusions

We have considered initial conditions leading to Starobinsky inflation in general quadratic gravity. If the coefficient α before $R^{ij}R_{ij}$ term in the action belongs to the stability domain, the Starobinsky inflation remains stable with respect to growth of shear variables, however, zone of initial conditions leading to it can change considerably. We have fixed $\alpha = -10\beta$, so that α is big, but is located within the stability domain.

As it was anticipated in the Introduction the influence of $\alpha R^{ij}R_{ij}$ term increases with the increasing shear. Qualitative difference from the zero shear case manifests itself in the appearance of turning points of cosmological evolution, the points where $\dot{H} = 0$. For our set-up we first remarked the presence of turning points at very small initial shear of the order of 10^{-6} where Big Crunch singularities appear as a final outcome of the cosmological evolution. However, such a small shear affects only a small portion of the initial conditions space near $(0, 0)$ point in coordinates (H, R) and this change is indistinguishable in general plots.

With bigger shear turning points change the picture in two ways. First, the recollapse replaces some parts of initial conditions leading to inflation in the absence of shear. However, points of bounce appears as well, and some initial conditions with $H < 0$ now lead to inflation. This makes the shape of inflation basin more complicated for an intermediate shear.

Finally, the zone of initial conditions leading to inflation with initial $H > 0$ goes beyond $H < 1$, $R < 1$ rectangle for a big shear. However, there is always a band of good initial conditions in the initial $H < 0$ region. Trajectories from this basin experience a bounce and reach Starobinsky inflation after it.

It is reasonable to expect that the actual form of inflation basin would depend significantly on the fixed value of α and details of initial shear. However, a qualitative result of our study indicates that there is no need to fix $|\alpha|$ to be much smaller than β in the quadratic gravity (provided α is within the stability domain) in order to have Starobinsky inflation as a natural outcome of cosmological evolution without need to fine-tune initial conditions.

Appendix: Dynamical System

As has been already mentioned, we work in the diagonal Bianchi-I spacetime. The 00 field equation (1) is a constraint that must be initially satisfied

$$\begin{aligned} E = \big(&12\dot{\sigma}_+\sigma_+H + 12\dot{\sigma}_-\sigma_-H - 3\dot{\sigma}_+^2 + 6\sigma_+\ddot{\sigma}_+ - 36{\sigma_+}^4 + 9{\sigma_+}^2H^2 + 9{\sigma_-}^2H^2 - 3\dot{\sigma}_-^2 + 6\dot{H}{\sigma_+}^2 - 36{\sigma_-}^4 \\ &-72{\sigma_-}^2{\sigma_+}^2 + 6\sigma_-\ddot{\sigma}_- + 6{\sigma_-}^2\dot{H}\big)\alpha + \big(-18\dot{H}^2 + 72\dot{\sigma}_+\sigma_+H + 108\dot{H}H^2 + 72\dot{\sigma}_-\sigma_-H + 36H\ddot{H} - 54{\sigma_-}^4 \\ &-108{\sigma_-}^2H^2 - 54{\sigma_+}^4 - 108{\sigma_-}^2{\sigma_+}^2 - 72\dot{H}{\sigma_+}^2 - 72{\sigma_-}^2\dot{H} - 108{\sigma_+}^2H^2\big)\beta + 3H^2 - 3{\sigma_+}^2 - 3{\sigma_-}^2. \end{aligned} \quad (5)$$

On the other hand, the spatial field equations (1) 11, 22 and 33 can be combined to obtain

$$\dddot{H} = \frac{1}{12\beta}\Big[-48\beta\dot{\sigma}_+\sigma_+ H + 4\alpha\dot{\sigma}_+\sigma_+ H - 48\beta\dot{\sigma}_-\sigma_- H + 4\alpha\dot{\sigma}_-\sigma_- H - 3H^2 - 3{\sigma_+}^2 - 3{\sigma_-}^2 - 2\dot{H} - 54\beta\dot{H}^2$$
$$-\alpha\dot{\sigma}_+^2 - \alpha\dot{\sigma}_-^2 - 18\beta{\sigma_+}^4 - 12\alpha{\sigma_+}^4 - 18\beta{\sigma_-}^4 - 12\alpha{\sigma_-}^4 - 108\beta\dot{H}H^2 - 72\beta H\ddot{H} + 3\alpha{\sigma_+}^2H^2 - 36\beta{\sigma_+}^2H^2$$
$$+2\alpha\sigma_+\ddot{\sigma}_+ + 2\alpha\dot{H}{\sigma_+}^2 - 24\beta\dot{H}{\sigma_+}^2 + 3\alpha{\sigma_-}^2H^2 - 36\beta{\sigma_-}^2H^2 - 36\beta{\sigma_-}^2{\sigma_+}^2 - 24\alpha{\sigma_-}^2{\sigma_+}^2 + 2\alpha\sigma_-\ddot{\sigma}_-$$
$$-24\beta{\sigma_-}^2\dot{H} + 2\alpha{\sigma_-}^2\dot{H} - 24\beta\dot{\sigma}_+^2 - 24\beta\dot{\sigma}_-^2 - 24\beta\sigma_-\ddot{\sigma}_- - 24\beta\sigma_+\ddot{\sigma}_+\Big] \tag{6}$$

$$\dddot{\sigma}_+ = -\frac{1}{\alpha}\Big[+6\alpha H\ddot{\sigma}_+ - 3\sigma_+H - 12\beta\sigma_+\ddot{H} + 4\alpha\dot{\sigma}_+\dot{H} - 24\alpha\dot{\sigma}_+{\sigma_+}^2 + 11\alpha\dot{\sigma}_+H^2 - 24\alpha{\sigma_+}^3H + 6\alpha\sigma_+H^3$$
$$-24\alpha{\sigma_-}^2\sigma_+H - 8\alpha{\sigma_-}^2\dot{\sigma}_+ - 12\beta\dot{\sigma}_+\dot{H} - 36\beta\dot{\sigma}_+{\sigma_+}^2 - 24\beta\dot{\sigma}_+H^2 - 36\beta{\sigma_+}^3H - 72\beta\sigma_+H^3 - 12\beta{\sigma_-}^2\dot{\sigma}_+$$
$$-\dot{\sigma}_+ - 16\alpha\dot{\sigma}_-\sigma_+\sigma_- - 84\beta\dot{H}\sigma_+H - 36\beta{\sigma_-}^2\sigma_+H - 24\beta\dot{\sigma}_-\sigma_+\sigma_- + \alpha\sigma_+\ddot{H} + 7\alpha\dot{H}\sigma_+H\Big] \tag{7}$$

$$\dddot{\sigma}_- = -\frac{1}{\alpha}\Big[-\dot{\sigma}_- - 3\sigma_-H + 4\alpha\dot{\sigma}_-\dot{H} - 8\alpha\dot{\sigma}_-{\sigma_+}^2 + 11\alpha\dot{\sigma}_-H^2 - 24\alpha{\sigma_-}^2\dot{\sigma}_- - 24\alpha{\sigma_-}^3H + 6\alpha\sigma_-H^3$$
$$+\alpha\sigma_-\ddot{H} - 12\beta\dot{\sigma}_-\dot{H} - 12\beta\dot{\sigma}_-{\sigma_+}^2 - 24\beta\dot{\sigma}_-H^2 - 36\beta{\sigma_-}^2\dot{\sigma}_- - 36\beta{\sigma_-}^3H - 72\beta\sigma_-H^3 - 16\alpha\dot{\sigma}_+\sigma_+\sigma_-$$
$$+7\alpha\dot{H}\sigma_-H - 24\alpha{\sigma_+}^2\sigma_-H - 24\beta\dot{\sigma}_+\sigma_+\sigma_- - 84\beta\dot{H}\sigma_-H - 36\beta{\sigma_+}^2\sigma_-H + 6\alpha H\ddot{\sigma}_- - 12\beta\sigma_-\ddot{H}\Big] \tag{8}$$

References

1. A. A. Starobinsky, A New Type of Isotropic Cosmological Models Without Singularity. Phys. Lett. B **91**, 99 (1980), https://doi.org/10.1016/0370-2693(80)90670-X
2. P. A. R. Ade et al. (Planck collaboration), Planck 2015 results. XX. Constraints on inflation. Astron. Astrophys. **594**, A20 (2016), https://doi.org/10.1051/0004-6361/201525898
3. H. Weyl, Gravitation and electricity. Sitzungsber. Königl. Preuss. Akad. Wiss. **26**, 465 (1918)
4. H. Buchdahl, On the gravitational field equations arising from the square of the gaussian curvature. Il Nuovo Cimento Series **10**(23), 141 (1962)
5. T. Ruzmaikina, A. Ruzmaikin, Quadratic corrections to the lagrangian density of the gravitational field and the singularity. Sov. Phys. JETP **30**, 372 (1970)
6. K. Tomita, T. Azuma, H. Nariai, On anisotropic and homogeneous cosmological models in the renormalized theory of gravitation. Progress Theoret. Phys. **60**, 403 (1978)
7. A. L. Berkin, Contribution of the Weyl tensor to R**2 inflation. Phys. Rev. D **44**, 1020 (1991), https://doi.org/10.1103/PhysRevD.44.1020
8. S. Cotsakis, J. Miritzis, Proof of the cosmic no hair conjecture for quadratic homogeneous cosmologies. Class. Quant. Grav. **15**, 2795 (1998), https://doi.org/10.1088/0264-9381/15/9/024
9. J. Miritzis, Dynamical system approach to FRW models in higher order gravity theories. J. Math. Phys. **44**, 3900 (2003), https://doi.org/10.1063/1.1602161
10. J. D. Barrow, S. Hervik, Anisotropically inflating universes. Phys. Rev. D **73**, 023007 (2006), https://doi.org/10.1103/PhysRevD.73.023007
11. J. D. Barrow, S. Hervik, On the evolution of universes in quadratic theories of gravity. Phys. Rev. D **74**, 124017 (2006), https://doi.org/10.1103/PhysRevD.74.124017

12. T. Clifton, J.D. Barrow, Further exact cosmological solutions to higher-order gravity theories. Class. Quantum Gravity **23**, 2951 (2006)
13. J. Miritzis, Oscillatory behavior of closed isotropic models in second order gravity theory. Gen. Rel. Grav. **41**, 49 (2009), https://doi.org/10.1007/s10714-008-0651-3
14. S. Cotsakis, A. Tsokaros, Asymptotics of flat, radiation universes in quadratic gravity. Physics Letters B **651**, 341 (2007), https://doi.org/10.1016/j.physletb.2007.06.038
15. S. Cotsakis, Slice energy in higher-order gravity theories and conformal transformations. Gravitation Cosmol. **14**, 176 (2008)
16. S. Carloni, A. Troisi, P. K. S. Dunsby, Some remarks on the dynamical systems approach to fourth order gravity. Gen. Rel. Grav. **41**, 1757 (2009), https://doi.org/10.1007/s10714-008-0747-9
17. J. D. Barrow, S. Hervik, Simple Types of Anisotropic Inflation. Phys. Rev. D **81**, 023513 (2010), https://doi.org/10.1103/PhysRevD.81.023513
18. H.-J. Schmidt, Fourth order gravity: Equations, history, and applications to cosmology. International Journal of Geometric Methods in Modern Physics **4**, 209 (2007), https://doi.org/10.1142/S0219887807001977
19. S. S. Mishra, D. Müller, A. V. Toporensky, Generality of Starobinsky and Higgs inflation in the Jordan frame. Phys. Rev. D **102**, 063523 (2020), https://doi.org/10.1103/PhysRevD.102.063523
20. D. Müller, A. Toporensky, Initial conditions for Starobinsky inflation with a positive spatial curvature. Eur. Phys. J. Plus **139**, 358 (2024), https://doi.org/10.1140/epjp/s13360-024-05153-8
21. K. S. Stelle, Classical Gravity with Higher Derivatives. Gen. Rel. Grav. **9**, 353 (1978), https://doi.org/10.1007/BF00760427
22. H. van Dam, M. Veltman, Massive and massless Yang-Mills and gravitational fields. Nucl. Phys. B **22**, 397 (1970), https://doi.org/10.1016/0550-3213(70)90416-5
23. H. Stephani, D. Kramer, M. MacCallum, C. Hoenselaers, E. Herlt, *Exact solutions of Einstein's field equations*, 2nd edn. (Cambridge University Press, Cambridge, 2003)

On Stability and Isotropization of Kasner Solution in R^2 Gravity

Dmitri Pogosyan and Akash Kav

Abstract The fate of the Universe that initially expands anisotropically in the theory with R^2 quantum-gravitational term in the Lagrangian is investigated. The stability of Kasner-like expansion, specifically in the class of Kantowski-Sachs spacetimes, is analyzed. Kasner solutions are found to be unstable, with the bifurcation line between the initial conditions that lead to collapsing universes and the ones that set the universe for continuing expansion that becomes isotropic, established analytically. Under suitable conditions, the isotropized spacetime enters the intermediate slow-rolling inflationary stage similar to Starobinsky inflation.

1 Introduction

Forty five years ago, A. A. Starobinsky published a seminal paper [11] in which he suggested the novel cosmological scenario with homogeneous and isotropic inflationary expansion of the Universe based on quantum-gravitational corrections to the vacuum energy-momentum tensor of the matter fields [3]. Arguably the first complete and physically motivated model of inflation, Starobinsky scenario put what became known as R^2 modified gravity at the forefront of cosmological research into the very early Universe for decades that follow.

R^2 gravity also admits anisotropic solutions, in particular the exact Kasner [7] solutions, that play a central role in the Belinsky-Khalatnikov-Lifshitz (BKL) theory of generic anisotropic behaviour of cosmological models near singularity[2] where vacuum effects are dominant.

Shortly thereafter, Starobinsky tasked one of the authors (DP), then an undergraduate student, with the question whether there is a possible transition from anisotropic expansion of the Universe to isotropic behaviour if quantum-gravitational effects are

D. Pogosyan (✉) · A. Kav
University of Alberta, Edmonton, AB, Canada
e-mail: pogosyan@ualberta.ca

A. Kav
e-mail: kav@ualberta.ca

A. Barvinsky and A. Kamenshchik (eds.), *Open Issues in Gravitation and Cosmology*, Fundamental Theories of Physics 225, https://doi.org/10.1007/978-3-032-15703-4_14

taken into account. With the help of numerical coding and punch cards, such possibility was demonstrated, the term project successfully defended, but the result was never published. Now it is a turn for a new generation of undergraduate students to make their contribution.

The fundamental equations of cosmology are the Einstein field equations (EFEs), proposed by Albert Einstein in 1916. These equations can be derived from the principle of least action, where a Lagrangian density linear in the Ricci scalar, R is considered.

$$\delta S = \delta \int \left(\frac{R}{16\pi G} + \mathcal{L}_m \right) \sqrt{-g} d^4 x = 0 \tag{1}$$

and take the following form

$$R_{\mu\nu} - \frac{1}{2} g_{\mu\nu} R = -8\pi G T_{\mu\nu} \tag{2}$$

with $R_{\mu\nu}$ being the Ricci tensor, $g_{\mu\nu}$ the metric tensor with g its determinant, and $T_{\mu\nu}$ the energy-momentum tensor arising from the matter Lagrangian $\mathcal{L}_m$.[1]

In the early universe, quantum fluctuations in the matter fields in vacuum lead to the contribution to the energy-momentum tensor that is quadratic in R [3]

$$R_{\mu\nu} - \frac{1}{2} g_{\mu\nu} R = -\kappa T_{\mu\nu}(R^{\alpha\beta} R_{\alpha\beta}, R^2, ...). \tag{3}$$

It was found by Ginzburg et al. [5] that in four dimensions in a general space-time, only two combinations of quadratic curvature terms conserve the energy-momentum tensor. These combinations can be obtained from varying R^2 and $R_{\mu\nu} R^{\mu\nu}$ terms in the Lagrangian density

$$\mathcal{L} = \sqrt{-g} f(R_{\mu\nu} R^{\mu\nu}, R^2). \tag{4}$$

Although by now theories of ghost-free gravity have been developed in the Einstein frame [9], in order to avoid complications with ghost degrees of freedom, exclude the problematic $R_{\mu\nu} R^{\mu\nu}$ from consideration. Thus, we elect to consider only the case of a Lagrangian density with terms linear and quadratic in R.

$$\mathcal{L} = \frac{1}{16\pi G} \left(R + \frac{R^2}{m^2} \right). \tag{5}$$

Here m is a constant parameter, which defines a curvature scale upon which quantum corrections begin to become relevant.

There can be also another, model dependent contribution to the renormalized energy momentum tensor, that cannot be put as a variation of any Lagrangian density. This term is known as the conformal anomaly or trace anomaly. The effect of this

[1] $c = \hbar = 1$ system of units is used hereafter. In these units, the Newton's constant is inverse square of the Planck mass $G = 1/M_{pl}^2$.

term was explored in homogeneous and isotropic geometries first by Fischetti et al. [4] and its main role is in creating an exact de-Sitter solution for such space-times. This solution motivated Starobinsky's original paper, but its the subsequent inflation during slow roll stage after de-Sitter state decay is what provides a more realistic scenario for the Universe. This slow roll inflation including its smooth transition to Friedmann stage with dust-like equation of state is governed by R^2 Lagrangian contribution. It serves as an attractor to the majority of trajectories in the system's phase space and therefore describes the generic behaviour of isotropic universes. The complete picture was given a detailed treatment by Kofman, Mukhanov and Pogosyan in [8].

In this paper we study spatially flat homogeneous but initially anisotropic models of the Universe with the R^2 Lagrangian density given by Eq. 5 and no conformal anomaly. The metric interval that includes anisotropy can be written with three scale factors, $a(t), b(t), c(t)$,

$$ds^2 = dt^2 - a(t)^2dx^2 - b(t)^2dy^2 - d(t)^2dz^2. \tag{6}$$

In the absence of quantum effects, there are no isotropic $a = b = c$ vacuum solutions, and anisotropic Kasner power law solutions [7] are generic. With quantum correction present, homogeneous solutions are possible, and while Kasner solutions still remain the exact ones, how stable and general they are is the subject of this paper's investigation.

We limit our study to the class of Kantowski-Sachs spacetimes [6], which can be thought of as a special case of the Kasner metric, where two scale factors out of three are equal. After brief review of isotropic case in Sect. 2, we derive equations for R^2 gravity with an anisotropic metric in Sect. 3, and explore the solutions with parameters describing the anisotropy and volume for an initially expanding universe.

Noting that the Kantowski-Sachs metric is an exact solution to the EFEs in $f(R)$ gravity, as was shown, for example, by Shamir [10], we recognize that a solution which begins on the Kasner solutions exactly will remain there and not tend away from it. Thus, similar to the work of Toporensky and Müller [12], as well as Barrow and Hevrik [1], we analyze the stability of these Kasner solutions by considering a perturbation off these solutions. The work of Müller and Toporensky considered the case tending toward the BKL singularity [2], approaching which the universe oscillates between different directions that collapse or expand. We consider the case tending away from the BKL singularity, when the comoving volume of space initially grows, and look to find a bifurcation line between the regimes which develop, as well an effect of isotropization, by evaluating the equations numerically in Sect. 4.

2 R^2 Gravity Vacuum Equations

2.1 General Equations

The general Euler-Lagrange equations that arise from the Lagrangian in Eq. (5) have the following well-known form

$$R_{\mu\nu} - \frac{1}{2}g_{\mu\nu}R + \frac{2}{m^2}\left(\left(R_{\mu\nu} - \frac{1}{4}g_{\mu\nu}R\right)R + \nabla_\mu\nabla_\nu R - g_{\mu\nu}g^{\rho\sigma}\nabla_\rho\nabla_\sigma R\right) = 0, \tag{7}$$

where ∇_μ is the covariant derivative and the Ricci tensor is defined as the contraction on the first and fourth indices of the Riemann tensor[2]:

$$R_{\mu\nu} = R^\lambda_{\mu\nu\lambda} = \partial_\nu\Gamma^\lambda_{\mu\lambda} - \partial_\lambda\Gamma^\lambda_{\mu\nu} + \Gamma^\rho_{\mu\lambda}\Gamma^\lambda_{\rho\nu} - \Gamma^\rho_{\mu\nu}\Gamma^\lambda_{\rho\lambda}. \tag{8}$$

The contracted or trace equation will also be useful later:

$$g^{\mu\nu}\nabla_\mu\nabla_\nu R + \frac{m^2}{6}R = 0. \tag{9}$$

Overall, equations of R^2 gravity involve second derivatives of the Ricci tensor and, therefore, fourth derivatives of the metric coefficients. It is also clearly seen that $R_{\mu\nu} = 0$ is a solution to these equations, meaning that any Ricci-flat space-time is an exact solution for R^2 gravity.

2.2 R^2 Gravity with Homogeneous and Isotropic Flat Metric

Let us briefly review how R^2 gravity works in case of homegeneous and isotropic space-times with flat spatial sections. This will serve as a benchmark and introduction the notations that we will use for anisotropic space-times in the next section.

The Friedmann-Robertson-Walker (FRW) homogeneous and isotropic metric is given by:

$$ds^2 = dt^2 - a^2(t)[dx^2 + dy^2 + dz^2], \tag{10}$$

where $a(t)$ is the scale factor.

[2] In comparison with [8], our definition of the Ricci tensor is opposite in sign, as well as our $m^2 = 6M^2$ of [8].

The non-zero Ricci tensor components and the Ricci scalar can be expressed concisely via the Hubble parameter $H = \frac{\dot{a}}{a} = (\ln a)\dot{}$ as

$$R_0^0 = 3(\dot{H} + H^2), \quad R_j^i = (\dot{H} + 3H^2)\delta_j^i, \quad R = 6(\dot{H} + 2H^2), \quad \dot{R} = 6(\ddot{H} + 4H\dot{H}), \tag{11}$$

where Latin indices denote the 3-dimensional spatial coordinates, and Greek indices the 4-dimensional spacetime.

As the equations for the sotropic universe we write down the 00 component of Eq. 7 that reduces to the quadratic equation for the Hubble parameter

$$m^2 H^2 - 6\dot{H}^2 + 36H^2\dot{H} + 12\ddot{H}H = 0. \tag{12}$$

This equation allows for a phase space analysis in $(H, \dot{H})$ plane. Such analysis was performed (as part of a more general setup with conformal anomaly) for the relevant to cosmology expanding solutions in [8]. The most striking property is the existence of the intermediate attractor along the separatrix that follows $\ddot{H} \approx 0$, $\dot{H} \approx -\frac{m^2}{36}$ line for $|\dot{H}| \ll H^2$ and then winds up in an oscialltory spiral in the vicinity of $H = 0$, $\dot{H} = 0$ singular point (see Fig. 2 of [8]) This is, of course, a famous Starobinsky [11] inflationary solution

$$H \approx H_0 - \frac{1}{36} m^2 t\,, \tag{13}$$

that gracefully exits into the dust-like "scalaron" oscillations when the slow-roll condition $|\dot{H}| < H^2$ breaks down.

However, the Eq. 12 does not allow to invesitgate the system in all the details. Indeed, the only finite singular point, $H = 0$, $\dot{H} = 0$ in the phase diagram, is *not* a stationary point. The system can pass through it, following multiple possible trajectories that intersect at $H = 0$, $\dot{H} = 0$, including the spiralling part of the attracting separatrix (associated with a stationary point at infinity). In particular, this equation cannot be used for numerical studies. The reason is obvious. (00) Einstein equation that was used is a contraint equation that describes the integral of motion and has less degrees of freedom (three), that what the whole system does (four) (one additional silent degree of freedom is in going from Hubble parameter as a variable to the scale factor).

For numerical studies we need to use the equation with the full range of freedom, for instance the trace of Einstein equations Eq. 9 that becomes

$$\ddot{R} + 3H\dot{R} + \frac{m^2}{6} R = 0. \tag{14}$$

Let us introduce the following set of variables that will also be used for anisotropic studies

$$y = 3H = \frac{d}{dt} \log(a^3), \quad z = \dot{y}, \quad \omega = \dot{R}, \tag{15}$$

which describes the change in the comoving volume of the space through y, and the rate of change of 4-curvature R. The scale factor and Ricci curvature R can be reconstructed from these parameters as

$$a = a_0 e^{\frac{1}{3}\int y dt}, \quad R = 2z + \frac{4}{3}y^2. \tag{16}$$

Using the expression for $\dot{R}$ from Eq. 11 we can reduce the trace equation Eq. 14 to three equations of the first order

$$\begin{aligned}
\dot{w} &= -yw - \frac{m^2}{9}\left(2y^2 + 3z\right), \\
\dot{y} &= z, \\
\dot{z} &= \frac{1}{2}w - \frac{4}{3}yz,
\end{aligned} \tag{17}$$

that describe the full set of dynamical degrees of freedom. These equations will be useful for comparison with the anisotropic case. We also have the constraint equation, where Eq. 12 becomes:

$$m^2y^2 - 6z^2 - 4y^2z + 6yw = 0, \tag{18}$$

that must be satisfied by the initial conditions.

3 Derivation of R^2 Anisotropic Gravity Equations

We now turn our attention to the anisotropic Kantowski-Sachs space [6]. The Kantowski-Sachs metric is given by

$$ds^2 = dt^2 - a^2(t)dx^2 - b^2(t)[dy^2 + dz^2] \tag{19}$$

where $a(t)$ and $b(t)$ are two unequal scale factors.

Computing the Ricci tensor components from the definition of the Ricci tensor given in Eq. 8 one obtains the following non-zero components

$$R_0^0 = \frac{\ddot{a}}{a} + 2\frac{\ddot{b}}{b}, \quad R_1^1 = \frac{\ddot{a}}{a} + 2\frac{\dot{a}\dot{b}}{ab}, \quad R_i^j = \left(\frac{\ddot{b}}{b} + \frac{\dot{b}^2}{b^2} + \frac{\dot{a}\dot{b}}{ab}\right)\delta_i^j \tag{20}$$

and the Ricci scalar

$$R = 2\frac{\ddot{a}}{a} + 4\frac{\ddot{b}}{b} + 2\frac{\dot{b}^2}{b^2} + 4\frac{\dot{a}\dot{b}}{ab} \tag{21}$$

In this section Latin indices i, j vary only between 2, 3, while the first spatial direction is treated individually.

3.1 X, Y, Z, W Equations

Analogously to the R^2 isotropic gravity case, it is helpful to define two new variables $x = \frac{d}{dt}\log(\frac{b}{a})$ and $y = \frac{d}{dt}\log(ab^2)$. The variable y is again a measure of the comoving volume rate of change, but a new variable x is a measure of the anisotropy. Expressing the non-zero components of Ricci tensor in terms of these variables

$$R_0^0 = \frac{2}{3}x^2 + \frac{1}{3}y^2 + \dot{y}, \; R_1^1 = -\frac{1}{3}\left(y^2 - xy + 2\dot{x} - \dot{y}\right), \; R_i^j = \frac{1}{3}\left(xy + y^2 + \dot{x} + \dot{y}\right)\delta_i^j, \tag{22}$$

one obtains the Ricci scalar and its derivate as

$$R = \frac{2}{3}x^2 + \frac{4}{3}y^2 + 2\dot{y}, \quad \dot{R} = \frac{4}{3}\dot{x}x + \frac{8}{3}\dot{y}y + 2\ddot{y}. \tag{23}$$

Retaining the definitions $z = \dot{y}$ and $\omega = \dot{R}$, after lengthy algebraic manipulations with the help of symbolic algebra software, we obtain a set of four equations of motion

$$\begin{aligned}
\dot{w} &= -yw - \frac{m^2}{9}(x^2 + 2y^2 + 3z), \\
\dot{x} &= -xy - \frac{6xw}{3m^2 + 4x^2 + 8y^2 + 12z}, \\
\dot{y} &= z, \\
\dot{z} &= \frac{1}{2}w - \frac{4}{3}yz + \frac{2}{3}x^2 y + \frac{4x^2 w}{3m^2 + 4x^2 + 8y^2 + 12z},
\end{aligned} \tag{24}$$

subject to the constraint from the (0, 0) component of Eq. 7

$$m^2(y^2 - x^2) + 6y\omega - 6z^2 - 4z(y^2 + 2x^2) - 2x^2(x^2 + 2y^2) = 0. \tag{25}$$

It can be seen that if we set $x = 0$, we recover the isotropic case, Eqs. 17 and 18. Moreover, if one starts from the state with no initial anisotropy, the anisotropy will not be generated during the evolution, since $\dot{x} = 0$. In general, Kantowski-Sachs metric has two additional degrees of freedom in R^2 gravity as compared to the isotropic case, to the total of six. One extra degree of freedom is in the parameter x while the second one is in recovering of two scale factors instead of one from x and y.

3.2 Kasner Solution

At the focus of this paper is Kasner [7] solution in the class of Kantowski-Sachs metrics [6]. Kasner solution describes anisotropic expansion of a spatially flat space-time and is a vacumm solution of Einstein equations. As such, it is Ricci-flat, $R_{\alpha\beta} = 0$, and remains the exact solution with the addition of the R^2 term in the Lagrangian as well. In fact, this extends to general $f(R)$ gravity [10]. In case of the metric in Eq. 19, Kasner solution gives

$$a(t) = a_0(t/t_0)^{-1/3}, \quad b(t) = b_0(t/t_0)^{2/3}, \tag{26}$$

which describes the expansion of the space-time in two directions and contraction in the remaining one (assuming the increasing time variable). The comoving volume in such space is changing as $ab^2 \propto t/t_0$.

In the language of (x, y, z, ω) variables, Kasner solution corresponds to

$$x = \frac{1}{t}, \quad y = x = \frac{1}{t}, \quad z = -y^2 = -\frac{1}{t^2}, \quad \omega = 0. \tag{27}$$

It can be easily seen that it satisfies the system of Eq. 24 as well as the constraint Eq. 25 and has $R = 0$.

4 Numerical Solutions and Stability of Kasner Expansion

4.1 Problem Setup and Initial Conditions

In Einstein gravity, there are no non-trivial homogenous and isotropic flat solutions without the matter. In particular, Kasner solution with equal exponents along all three axes is not allowed. Thus, initially anisotropic space-times will remain anisotropic. With R^2 term in the Lagrangian this changes. Isotropic solutions in vacuum are possible and we can ask whether initially anisotropic spaces will isotropize as they evolve. We investigate numerically whether this happens for expanding spaces that are nearly Kasner in the beginning, or whether the solutions remain close to Kasner-like behaviour.

We specify the initial conditions at time t_0 to be nearly Kasner, subject to small deviation. We choose this perturbation to occur in the highest order derivative quantities, z and ω, to have maximally smooth deviation from Kasner law.

$$x_0 = \frac{1}{t_0}, \quad y_0 = \frac{1}{t_0}, \quad z_0 = -\frac{1}{{t_0}^2}(1 + p), \quad w_0 = \frac{p^2}{{t_0}^3}. \tag{28}$$

We parameterize the deviation by the dimensionless parameter p, such that $p \geq 0$ corresponds to solutions where the volume expansion slows down faster than in Kasner regime, while $p \leq 0$ corresponds to solutions where the volume expansion is less decelerated. Notice that the initial value of ω is determined by the $(0, 0)$ constraint, Eq. 25 and is not free. This initial configuration is not Ricci-flat, $R_0 = -\frac{2p}{{t_0}^2}$, with the sign that is determined by p, while the initial curvature derivative w_0 is always positive. Thus, for positive p initial evolution will tend to compensate the deviations of Kasner $R = 0$, while the negative p will accelerate this deviation.

4.2 Bifurcation in Initial Conditions

Even a cursory examination of Eq. 24 shows that there is a singular situation when the inital conditions are chosen to satisfy $3m^2 + 4x^2 + 8y^2 + 12z = 0$. With our starting state given by Eq. 28, this corresponds to

$$p = \frac{1}{4}m^2 t_0^2 \tag{29}$$

curve in the (t_0, p) parameter space. As we shall show numerically next, this is a bifurcation line that separates two vastly different outcomes for the model. Started below the bifurcation line, $p < \frac{1}{4}m^2 t_0^2$, including all negative p, (except $p = 0$), the anisotropic near-Kasner solution will reverse the contraction of one axis into its expansion of all three axes, eventually leading to isotropic expansion of space. While for $p > \frac{1}{4}m^2 t_0^2$, the collapsing first axis will accelerate its collapse, eventually reversing the initially growing comoving volume into a collapsing one that reaches singularity in the finite time.

Another way to formulate the bifurcation condition, is as a condition on initial scalar corvature, $R_0 = -\frac{1}{2}m^2$. The universe collapses, if the initial curvature value (zero on Kasner solution) is more negative than that.

4.3 Results

Now we numerically solve our equations. We can absorb the parameter m into definition of time $mt \rightarrow t$, by setting $m = 1$. This assumes that quantum corrections are always present, and significant for $t < 1$, while having diminishing contribution for $t > 1$. Lastly, we choose $a_0 = -\frac{1}{3}\log t_0$ and $b_0 = \frac{2}{3}\log t_0$ for the scale factors, which for purely Kasner solution gives $a = b = 1$ for $t = 1$.

As stated earlier, we will consider a universe with an initially expanding volume, which corresponds to $y > 0$, or iterating forward in time from t_0. We shall consider a set of models with parameters summarized in Fig. 1.

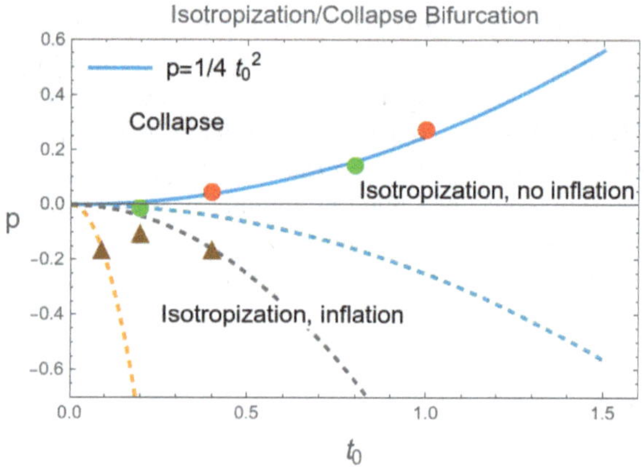

Fig. 1 The behaviour of the universe is determined by the values of t_0 and p that are chosen. We found two clear situations, an isotropizing expanding universe and an anisotropically collapsing universe. The solid blue line indicates the bifurcation line between the regimes, where any point above that line would lead to a collapsing universe. Any point below that line would lead to an isotropizing universe, except along the line $p = 0$ which gives the exact Kasner solution. The solid circles denote models that straddle the bifurcation line in our numerical tests. The red circles are just above it, and the green are just below. The dashed lines indicates the more quantitative assessment of isotropization roughly specifying the lines along each the amount of inflation is the same. There is no discernable stage of exponential behaviour above $p = -\frac{1}{4}t^2$ blue dashed line. To the left of the orange dotted line, the amount of inflation exceeds 65-efolds, the nominal value necessary for successful cosmological models. Overplotted are markers which indicate the initial conditions of other figures in this paper

As the first step, in Fig. 2 we demonstrate that the singular line $p = \frac{1}{4}t_0^2$ in initial conditions is indeed the bifurcation line. Four models plotted in Fig. 2 were chosen to straddle this line, two just above and two just below. The outcomes are discontinously different. $p > \frac{1}{4}t_0$ models collapse, first collapsing axis of Kasner solution accelerates its decay and eventually leads to the volume to shrink to a singularity as well. $p < \frac{1}{4}t_0$ models have a dramatically different outcome. The collapse of the first axis is reversed, and the models enter the epoch of asymptotic growth of equal rate along all axes, following the isotropic solutions of R^2 gravity. Thus, initially anisotropic, such spaces isotropize as $t \to \infty$. For two models of this type shown in Fig. 2, the isotropization happens shortly after $t = 1$, at $t \approx 2$. However, for p closer to zero, the transition moment shifts to ever later times, and, of course, for a special case of $p = 0$ the solution remains strictly Kasner and anisotropic forever.

The models close to the bifurcation line, as in the bottom row of Fig. 2, shows isotropic transition practically straight into the "dust-like" expansion of the isotropic Universe with $a(t), b(t) \propto t^{2/3}$, with oscillations on top, which corresponds to winding down spiral behaviour in $(H, \dot{H})$ phase space of Eq. 12 [8]. Indeed, further numerical studies show that this direct transition to "dust-like" limit is the outcome of all models in the sector $-\frac{1}{4}t_0^2 \lesssim p < \frac{1}{4}t_0^2$. One caveat to note is that while the upper

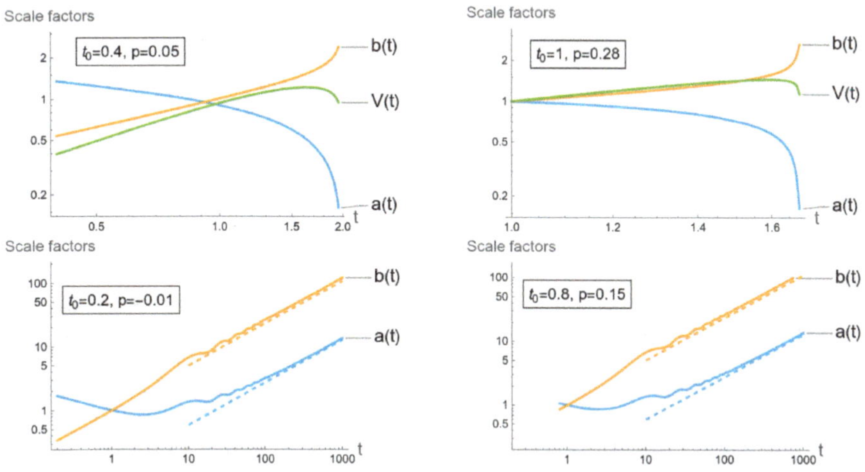

Fig. 2 A set of solutions with initial conditions that straddle the bifurcation line $p = \frac{1}{4}t_0^2$ corresponding to closed dots in Fig. 1. Above the bifurcation line (top row) the space experience fast accelerated anistropic collapse, with reversal of the initial volume expansion to construction. Below the birfucation line (bottom row) initial anistropic Kasner-like expansion transfers to isotropic one, following the "dust-like" behaviour $a, b \propto t^{2/3}$ with superimposed decaying oscillations. The dashed lines show $t^{2/3}$ asymptotics

boundary of this range is a sharp, theoretical motivated, bifurcation line, the lower boundary comes from a more qualitative interpretation of numerical outcomes.

Next, Fig. 3 shows that for $p < -\frac{1}{4}t_0^2$, the isotropization occurs before entering the "dust-like" regime, and such model experiences a transition period of near-exponential isotropic expansion that is equivalent to Starobinsky inflation in R^2 gravity.

Such stage requires $|\dot{H}| \ll H^2$ and is described by Eq. 13 which in our variables is $y(t) = y_0 - \frac{1}{12}t$. The right panel of Fig. 2 shows the presence of such slow-rolling change of the Hubble parameter from $t = 2$ to $t \approx 30$. Thus, the case investigated in this Figure has an intermediate inflationary asymptotics, which leads to approximately 8 e-fold (e^8) expansion of the scale factors. Note that dimensional value of the Hubble expansion rate in our units at is $H = \frac{1}{3}y \approx 0.7$ at the beginning of the isotropic regime, which is transitioned to at $t \approx 2$ again.

In Fig. 4 we look at the amount of exponential expansion that one obtains for different starting parameters. We find that the models lying along the lines $p = -At_0^2$, $A > 1/4$, give approximately the same number of e-folds during the inflationary stage. Left panel of Fig. 4 gives an example of a model with $A = 1$ which has a noticeable, but very short inflationary stage. Let as take the nominal value of 65 efolds as the required amount of inflation for a viable cosmological model. Such amount of expansion is achieved along $p = -20t_0^2$ line with the example case shown in the right panel of Fig. 4. Note that all models which have inflationary stage, have

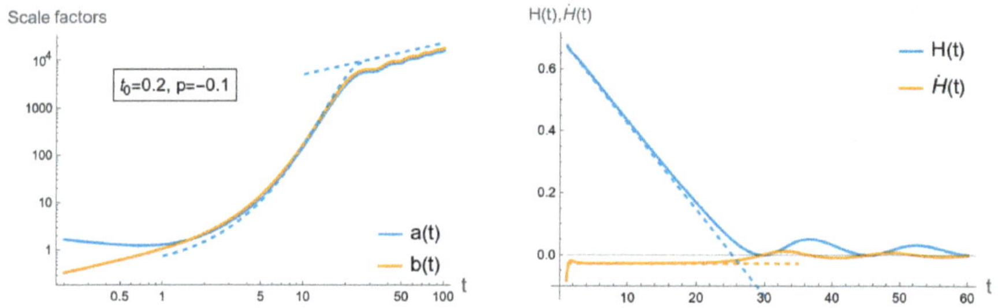

Fig. 3 A solution with the intermediate stage of near-exponential inflationary expansion. Left panel shows the scale factors $a(t)$ and $b(t)$ which approach isotropy $a(t) \propto b(t)$ at $t > 2$. The straight dashed lines corresponds to $\propto t^{2/3}$ "dust"-like asymptotics at $t \to \infty$. The rising dashed line is slow-roll Starobinsky inflationary regime of R^2 gravity $a(t) \sim e^{H_0 t - \frac{1}{72}m^2 t^2}$ that ends in this model at $mt \approx 30$. Right panel shows the behaviour of the Hubble parameter and its derivative, $H(t) = \frac{1}{3}y(t)$, $\dot{H}(t) = \frac{1}{3}z(t)$. Inflationary stage is clearly seen, corresponding to $\dot{H}(t) \approx -\frac{1}{36}m^2 = const$ and $H(t) = const - \frac{1}{36}m^2 t$, the behaviours shown by dashed lines. On subsequent "dust-like" stage at $mt > 30$, H and $\dot{H}$ oscillate, passing through (0, 0) point at every oscillation [8]

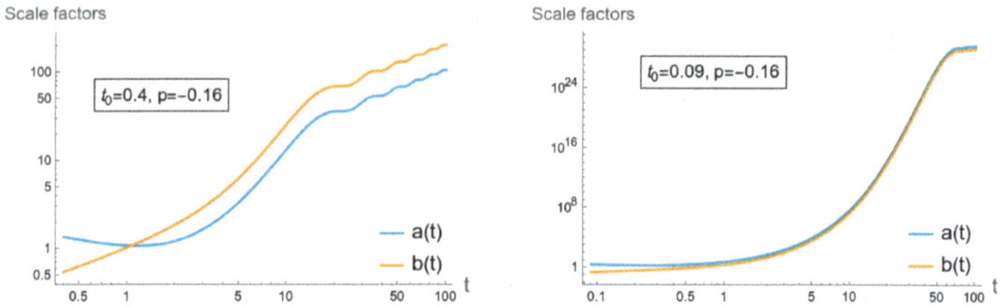

Fig. 4 Development of the inflationary stage for the parameters below $p < -\frac{1}{4}t_0^2$ line. Left panel shows the example of a model starting on $p = -t_0^2$ line (grey dashed line). These models show already a discernible, but very short inflationary stage, giving only 3–4 efold expansion. The right panel shows example of the models that achieve cosmologically significant benchmark expansion of 65 e-folds. Such models lie, approximately, along the $p \approx -20t_0^2$ line shown in Fig. 1

transitioned to isotropic stage at $t \approx 2$. What is different between the model and what governs the duration of inflation, is the value of the Hubble parameter $H = \frac{1}{3}y$ that transpires after isotropization. One need $H/m \approx 1.9$ to reach 65 e-folds.

5 Conclusion

We found that in R^2 gravity without matter fields, the anisotropic Kasner solutions, are unstable and can evolve either towards isotropization of the expansion or collapse of the Universe. What matters is how strong and how early is the deviation from the Kasner solution. In fact, there is a clear bifurcation line in $p - t_0$ space, $p = \frac{1}{4}t_0^2$, as shown in Fig. 1, that separates these two outcomes. Above this bifurcation line, for initial parameter t_0 and perturbation off Kasner solution p, the universe will collapse anisotropically. Below the bifurcation line, the universe will isotropize.

It is interesting to consider the work of Toporensky and Müller once again [12]. As they moved toward the BKL singularity, they found a bifurcation line between Kasner-type solutions and other solutions, but indicated that the Kasner-type solutions were much more stable than in our case of starting with the Universe with the expanding spatial volume, moving away from the BKL singularity. Forward-in-time Kasner solution is shown to be unstable in this work, resulting either in quick reversal of the volume expansion and collapse, or transition to isotropic expansion, but no solutions that continue to expand anisotropically except for the exact Kasner solution.

As we know, our universe today is expanding isotropically, and it is commonly accepted that there was an inflationary epoch shortly after the Big Bang. We found that in R^2 gravity if the Universe was expanding anisotropically early on, a subsequent isotropization and transition to Starobinsky isotropic model is a generic outcome. For a sufficient early and significant perturbation from Kasner behaviour there will be a period of a near exponential expansion, before continuing to power law expansion with $a, b \propto t^{\frac{2}{3}}$.

References

1. J.D. Barrow, S. Hervik, Evolution of universes in quadratic theories of gravity. Phys. Rev. D **74**, 124017 (2006). (Dec)
2. V.A. Belinskij, I.M. Khalatnikov, E.M. Lifshits, Oscillatory approach to a singular point in the relativistic cosmology. Adv. Phys. **19**, 525–573 (1970). (January)
3. P.C.W. Davies, S.A. Fulling, S.M. Christensen, T.S. Bunch, Energy-momentum tensor of a massive scalar quantum field in a Robertson-Walker universe. Ann. Phys. **109**, 108–142 (1977). (January)
4. M.V. Fischetti, J.B. Hartle, B.L. Hu, Quantum effects in the early universe. I. Influence of trace anomalies on homogeneous, isotropic, classical geometries. Phys. Rev. D **20**(8), 1757–1771 (1979)
5. V.L. Ginzburg, D.A. Kirzhnits, A.A. Lyubushin, The role of quantum fluctuations of the gravitational field in general relativity theory and cosmology. Soviet J. Exper. Theor. Phys. **33**, 242 (1971). (January)
6. R. Kantowski, R.K. Sachs, Some spatially homogeneous anisotropic relativistic cosmological models. J. Math. Phys. **7**, 443–446 (1966). (March)
7. E. Kasner, Geometrical theorems on Einstein's cosmological equations. Am. J. Math. **43**(4), 217–221 (1921). (October)
8. L.A. Kofman, V.F. Mukhanov, D.I. Pogosian, The evolution of inhomogeneities in inflationary models in the theory of gravitation with higher derivatives. Zhurnal Eksperimentalnoi i Teoreticheskoi Fiziki **93**, 769–783 (1987)
9. S. Nojiri, S.D. Odintsov, V.K. Oikonomou, A.A. Popov, Ghost-free f(r,g) gravity. Nuclear Phys. B **973**, 115617 (2021)

10. M.F. Shamir, Some Bianchi type cosmological models in f(R) gravity. Astrophys. Space Sci. **330**(1), 183–189 (2010)
11. A.A. Starobinsky, A new type of isotropic cosmological models without singularity. Phys. Lett. B **91**(1), 99–102 (1980). (March)
12. A. Toporensky, D. Müller, On stability of the Kasner solution in quadratic gravity. Gen. Relativ. Gravit. **49**(1), 8 (2017). (January)

Progress in Einstein–Cartan Gravity

Mikhail Shaposhnikov

Abstract It is well-known that the gravitational force can be obtained by gauging the Lorentz group, which puts gravity on the same footing as the Standard Model fields. The resulting theory—Einstein–Cartan gravity—has several crucial advantages. I will overview the construction of the Weyl-invariant version of this theory and discuss its applications in particle physics and cosmology, in particular for inflation and strong CP-problem.

1 Introduction

There are several approaches to the theory of gravity. Historically the first one is that of Einstein, which tells that gravity is not an ordinary force, but rather a property of spacetime geometry. Thus, the theory of gravity is geometrodynamics (see e.g. the textbook [1]). Unifying it with the other forces of nature would then require geometrizing the photon and other fields of the Standard Model (SM).

Yet another point of view (often attributed to Feynman [2]) is that gravity is just like electrodynamics, but associated with a spin-2 massless particle—graviton, instead of the spin-1 massless photon. Then geometrodynamics is simply an effective low-energy theory. The Standard Model uses the gauging of the global symmetry group SU(3)xSU(2)xU(1), leading to the existence of the photon, intermediate vector bosons, and gluons. In full analogy with this construction, the gauging of the global Lorentz group [3–5] results in the existence of the graviton and thus gravitational interactions. The corresponding gauge fields are[1] the tetrad e_μ^A and the spin connection ω_μ^{AB}, leading to the Einstein–Cartan (EC) formulation of General Relativity (GR) [6–9].

[1] General conventions: Greek and capital Latin letters are used for spacetime and Lorentz indices, respectively. The signature of both the spacetime $g_{\alpha\beta}$ and Minkowski η_{AB} metrics is mostly plus. We set $c = \hbar = 1$.

M. Shaposhnikov (✉)
Institute of Physics, École Polytechnique Fédérale de Lausanne (EPFL), Lausanne, Switzerland
e-mail: Mikhail.Shaposhnikov@epfl.ch

A. Barvinsky and A. Kamenshchik (eds.), *Open Issues in Gravitation and Cosmology*, Fundamental Theories of Physics 225, https://doi.org/10.1007/978-3-032-15703-4_15

Apriori, the EC formulation of gravity contains more degrees of freedom than the Einstein metric gravity (where the only dynamical variable is the metric $g_{\mu\nu}$), hidden in the 16 components of the tetrad field and the 24 components of the spin connection. However, many of these components happen to be non-dynamical, and different variants of the theory have just propagating graviton and at most two scalar fields. The presence of these scalar fields may lead to interesting consequences, such as cosmological inflation or a solution to the strong CP problem. The theory can be constrained further by imposing a Weyl symmetry, acting as pointwise rescalings of the metric (or of the tetrad), removing one scalar degree of freedom.

In this paper, we will provide a short overview of our works [10–16] of the construction of EC gravity and its Weyl-invariant version.[2] One of our inspirations came from the Starobinsky theory of inflation, based on the purely geometric action in metric gravity, containing, in addition to the Einstein–Hilbert term, a term quadratic in the Ricci scalar.

2 The Starobinsky Theory

Pure gravity in the metric formulation is based on the Einstein–Hilbert action

$$S_{\text{EH}} = \frac{M_P^2}{2} \int \mathrm{d}^4x \sqrt{g}(L_0 + L_2)\,, \quad L_0 = \Lambda\,, \quad L_2 = R\,, \tag{1}$$

where R is the Ricci scalar, Λ the cosmological constant, M_P the Planck scale, and g the determinant of the metric. It contains two physical degrees of freedom, those of the graviton. The action (1) includes two leading terms, L_0 and L_2, in $1/M_P^2$ expansion of a generic Lagrangian L invariant under the full group of diffeomorphisms (Diffs). The next-order terms represent operators of mass dimension four. These are:

$$L_4 = \frac{1}{4f_R^2} R^2 + C_1 W_{\mu\nu\lambda\rho} W^{\mu\nu\lambda\rho} + C_2 E_4 + C_3 \Box R\,. \tag{2}$$

Here $\Box = g^{\mu\nu}\nabla_\mu\nabla_\nu$ is the covariant D'Alembertian, and f_R^2, $C_{1,2,3}$ are dimensionless constants, while

$$E_4 = W_{\mu\nu\lambda\rho} W^{\mu\nu\lambda\rho} + \frac{2}{3} R^2 - 2R_{\mu\nu}R^{\mu\nu}\,, \tag{3}$$

is the Euler density operator, and $W_{\mu\nu\lambda\rho}$ the Weyl tensor. The terms E_4 and $\Box R$ are total derivatives. Thus, they do not contribute to the equations of motion. The square of the Weyl tensor leads to the presence of extra degrees of freedom, some of them

[2] These articles were heavily used for writing this contribution. Some new formulas and discussions are presented in Sects. 6 and 7.

with the "wrong" sign of kinetic terms, resulting in instabilities and the presence of "ghost" particles, arguably leading to an inconsistent theory.[3] The unique theory including operators of mass dimension four which is free of ghosts is that with the action

$$S_S = S_{\rm EH} + \frac{1}{4f_R^2}\int d^4x\sqrt{g}R^2 \,. \tag{4}$$

In addition to the graviton, it contains just one extra spin zero degree of freedom—the "scalaron". The simplest way to see that is to introduce an auxiliary field χ (with the mass dimension one) that allows us to make the action linear in the scalar curvature (this trick will be used several times in what follows):

$$S_S = S_{\rm EH} + \frac{1}{2}\int d^4x\sqrt{g}\left(R\chi^2 - \frac{1}{2}f_R^2\chi^4\right) \,. \tag{5}$$

The equations of motion for χ lead to $\chi^2 = R/f_R^2$, and inserting this solution to (5) gives (4). Transformation of the action (5) to the Einstein frame, with the help of conformal transformation,

$$g_{\mu\nu} \to \left(1 + \frac{\chi^2}{M_P^2}\right)^{-1} g_{\mu\nu} \tag{6}$$

reveals that (4) is nothing but

$$\int d^4x\sqrt{g}\left[\frac{M_P^2}{2}R - \frac{1}{2}(\partial_\mu\phi)^2 - \frac{f_R^2 M_P^4}{4}\left(1 - \exp\left(-\sqrt{\frac{2}{3}}\frac{\phi}{M_P}\right)\right)^2\right] \tag{7}$$

with

$$\phi = \sqrt{\frac{3}{2}}M_P \log\left(1 + \frac{\chi^2}{M_P^2}\right) \,. \tag{8}$$

The theory (4) is very interesting. It is written entirely in terms of geometrical quantities. It represents the very first theory of successful inflation—that of Starobinsky [18], consistent with all current cosmological observations.

3 An Overview of Einstein–Cartan Gravitational Theories

The Langrangian (4) is based on the metric formulation of gravity, where the role of the dynamical variable is played by $g_{\mu\nu}$. This formulation is not unique and is not even a natural one when we need to describe fermions, which are an essential

---

[3] The presence of new degrees of freedom can make the theory renormalisable [17].

part of the Standard Model. A curved space action for fermions, in addition to Diff, is invariant under local Lorentz transformations. Thus, the fermion Lagrangian contains new (in comparison with the metric) geometric objects: the tetrad—e_μ^A and the gauge field of the local Lorentz group—spin connection ω_μ^{AB} (for a review, see [19]). These variables are a priori independent (the tetrad and spin connection totalise 40 components, 16 and 24 respectively) and serve as a basis of the Einstein–Cartan (EC) formulation of gravity [6–9]. The EC formulation fits well with the gauge principle used in the construction of the SM: to get the SM vector bosons, the global internal SU(3)xSU(2)xU(1) symmetry is gauged, whereas to get gravity the global Lorentz symmetry is gauged [3–5].

To build the EC gravity Lagrangian, the following structures are essential: the curvature

$$F_{\mu\nu}^{AB} = \partial_\mu \omega_\nu^{AB} - \partial_\nu \omega_\mu^{AB} + \omega_{\mu C}^{A} \omega_\nu^{CB} - \omega_{\nu C}^{A} \omega_\mu^{CB} , \tag{9}$$

and the torsion

$$T_{\mu\nu\rho} = e_{\mu A} T_{\nu\rho}^{A} , \quad T_{\mu\nu}^{A} = \partial_\mu e_\nu^A - \partial_\nu e_\mu^A + \omega_{\mu B}^{A} e_\nu^B - \omega_{\nu B}^{A} e_\mu^B . \tag{10}$$

The torsion can be decomposed into irreducible vector (v_μ, a_μ) and tensor ($\tau_{\mu\nu\rho}$) components as follows:

$$v_\mu = T^{\nu}_{\ \mu\nu} , \quad a^\mu = E^{\mu\nu\rho\sigma} T_{\nu\rho\sigma} , \quad \tau_{\mu\nu\rho} = \frac{2}{3}\left(T_{\mu\nu\rho} - v_{[\nu} g_{\rho]\mu} - T_{[\nu\rho]\mu}\right) , \tag{11}$$

where the square brackets mean antisymmetrisation with respect to a pair of indexes, and $E^{\mu\nu\rho\sigma} = \varepsilon^{\mu\nu\rho\sigma}/\sqrt{g}$. There are two scalar curvature invariants, that read

$$F \equiv \frac{1}{8}\epsilon_{ABCD} E^{\mu\nu\rho\sigma} F_{\mu\nu}^{AB} e_\rho^C e_\sigma^D , \quad \text{and} \quad \tilde{F} \equiv E^{\mu\nu\rho\sigma} e_{\rho C} e_{\sigma D} F_{\mu\nu}^{CD} . \tag{12}$$

The second structure is called the Holst term.

To find a relation between the EC formulation of gravity and the metric one the following formulas are helpful:

$$F = \frac{R}{2} + \frac{1}{\sqrt{g}}\partial_\mu\left(\sqrt{g} v^\mu\right) - \frac{1}{3} v_\mu v^\mu + \frac{1}{48} a_\mu a^\mu + \frac{1}{4}\tau_{\mu\nu\rho}\tau^{\mu\nu\rho} , \tag{13}$$

$$\tilde{F} = -\frac{1}{\sqrt{g}}\partial_\mu\left(\sqrt{g} a^\mu\right) + \frac{2}{3} a_\mu v^\mu - \frac{1}{2}\epsilon^{\mu\nu\rho\sigma}\tau_{\lambda\mu\nu}\tau^{\lambda}_{\ \rho\sigma} , \tag{14}$$

where the torsion-free Riemannian curvature (and the Ricci scalar R) is defined as usually in the metric gravity.

The analogue of (1) in EC gravity, involving operators of dimension up to two, is

$$S_{\mathrm{EC},2} = M_P^2 \int \mathrm{d}^4x \det e[F + a\tilde{F} + bv_\mu v^\mu + ca_\mu a^\mu + da_\mu v^\mu + e\tau^{\mu\nu\rho}\tau_{\mu\nu\rho} + fE^{\mu\nu\rho\sigma}\tau^\lambda_{\mu\nu}\tau_{\lambda\rho\sigma} + \Lambda], \tag{15}$$

where a, b, ..., f are dimensionless constants. Despite the presence of many dynamical variables, this action is equivalent to (1) and contains just a graviton. Indeed, with the use of (13), (14), one can see that the equations of motion for the torsion give $v_\mu = a_\mu = \tau_{\mu\nu\rho} = 0$, leaving us with the Einstein–Hilbert action.

The theories (1) and (15), linear in curvature, can be augmented by the fields of the Standard Model. In both cases, the non-minimal coupling of the Higgs field to the curvature invariants may give rise to cosmological inflation, driven by the Higgs field [20]. In the first (metric) case, the predictions of Higgs inflation are very close to those of Starobinsky theory, whereas, for the EC gravity, they may substantially differ from them, in particular for the tensor-to-scalar ratio [11]. The torsion remains non-dynamical, though the vectors v_μ and a_μ are not equal to zero and are expressed through the SM fields. With a specific choice of non-minimal couplings, the EC gravity can be converted [11] to the metric gravity or the Palatini gravity [21], and lead to Palatini Higgs inflation [22].

The analogue of (4) in EC gravity is

$$S_{\mathrm{EC}} = S_{\mathrm{EC},2} + S_{\mathrm{EC},4}, \tag{16}$$

$$S_{\mathrm{EC},4} = \int \mathrm{d}^4x \det e \left[\frac{1}{f^2}F^2 + \frac{1}{\tilde{f}^2}\tilde{F}^2 + \frac{2}{f_m^2}F\tilde{F}\right]. \tag{17}$$

It contains the squares of the scalar curvatures (12) and is free from ghosts.[4] The dimensionless couplings f, $\tilde{f}$ and f_m can be considered as the gauge couplings of the local Lorentz group.

It is easy to see that this theory contains two extra scalar degrees of freedom in addition to the graviton. Indeed, making the same trick as for Starobinsky's theory reveals that the quadratic in curvatures part of (17) is equivalent to

$$\frac{1}{2f^2}F^2 + \frac{1}{2\tilde{f}^2}\tilde{F}^2 + \frac{1}{f_m^2}F\tilde{F} \to F\chi^2 + \tilde{F}a^2 - \frac{1}{2}\alpha\chi^4 - \frac{1}{2}\beta a^4 + \gamma\chi^2 a^2 \tag{18}$$

[4] In general, there are 10 quadratic invariants which can be constructed from the curvature. They are listed in [23]. In addition, the dimension (mass)^4 operators include the fourth power of torsion, cross terms of curvature and torsion, as well as derivatives of torsion and curvature. If taken with arbitrary coefficients, they lead to the existence of ghosts and tachyons. Some specific choices of the coefficients lead to ghost-free theories at the level of linearised excitations above the flat background, see, e.g. [24, 25]. However, non-linear effects may jeopardise this conclusion, see, e.g. [26].

with
$$\frac{1}{\alpha} = 2Df^2,\ \frac{1}{\beta} = 2D\tilde{f}^2,\ \frac{1}{\gamma} = 2Df_m^2,\ D = \frac{1}{f^2\tilde{f}^2} - \frac{1}{f_m^4}, \tag{19}$$

where χ and a are scalar auxiliary fields. As previously stated, the torsion appears algebraically, whereas the "integrating out" of it and the transformation of the theory to the Einstein frame shows that the fields χ and a acquire the non-trivial kinetic terms. As well as the Starobinsky theory, the model (17) with a certain choice of parameters can accommodate inflation. In addition, one of the scalar particles can be used as an axion, solving the strong CP problem (see below and [16]).

4 Scale Invariant Metric Gravity

The theories discussed above contain an explicit mass scale—the Planck mass. An attractive class of theories are scale-free gravities, enjoying global scale invariance. They only include dimensionless coupling constants (giving hope of renormalisability), whereas the mass scale appears because of the spontaneous breakdown of the scale symmetry.

The scale transformation acts on the metric, scalar (ϕ), vector (A_μ) and fermion (ψ) fields as

$$g_{\mu\nu} \to q^{-2} g_{\mu\nu},\ \phi \to q\phi,\ A_\mu \to A_\mu,\ \psi \to q^{3/2}\psi\,, \tag{20}$$

respectively. Here q is an arbitrary constant.

The Lagrangian of scale-invariant metric gravity is given by

$$S_{SI} = \frac{1}{4f_R^2}\int \mathrm{d}^4x\sqrt{g}R^2\,. \tag{21}$$

As previously, the square of the Ricci scalar can be removed with the help of an auxiliary field χ. After the transformation to the Einstein frame, the action is

$$S_{SI} \to \frac{1}{2}\int \mathrm{d}^4x\sqrt{g}\left(M_P^2 R - \frac{1}{2}f_R^2 M_P^4 - (\partial_\mu\tilde{\chi})^2\right)\,. \tag{22}$$

where $\tilde{\chi}$ is the canonically normalised scalar field, $\chi = M_P \exp[\tilde{\chi}/(\sqrt{6}M_P)]$. This is Einstein's gravity with a non-zero cosmological constant and a massless scalar field—dilaton, related to the spontaneous breaking of the scale invariance. If $f_R^2 > 0$ ($f_R^2 < 0$), the theory admits a de Sitter (anti-de Sitter) background. Different aspects of quadratic gravity were considered in [27–29]. Around the flat Minkowski background, it represents a strongly coupled theory with unknown excitations [29].

The theory (21) can be augmented by the fields of the Standard Model, most notably by the Higgs SU(2) doublet H. In the unitary gauge $H = (0, h/\sqrt{2})$, the relevant Lagrangian has the form

$$L_H = \frac{1}{2}\xi_h h^2 R - \frac{1}{2}(\partial_\mu h)^2 - \frac{\lambda h^4}{4}, \tag{23}$$

where λ is the scalar self-coupling. After transformation to the Einstein frame the Lagrangian, besides the Einstein–Hilbert term, contains the scalar part

$$-\frac{1}{2}K_{ij}\partial_\mu\phi_i\partial_\mu\phi_j - V, \tag{24}$$

where $\phi_1 = h,\ \ \phi_2 = \chi$, and the kinetic matrix is

$$K_{ij} = \frac{1}{2}\Omega\delta_{i1}\delta_{j1} + \frac{3}{4}M_P^2\frac{\partial\log\Omega}{\partial\phi_i}\frac{\partial\log\Omega}{\partial\phi_j}, \tag{25}$$

where $\Omega = M_P^2/(\chi^2 + \xi_h h^2)$. This is a metric of a two-sphere with scalar curvature $2/(3M_P^2)$. The potential is given by

$$V = \frac{1}{4}\Omega^2\left(f_R^2\chi^4 + h^4\lambda\right). \tag{26}$$

The ground state of this theory automatically has a flat direction

$$h = f_R\sqrt{\frac{\xi_h}{\lambda}}\chi, \tag{27}$$

corresponding to a Goldstone massless degree of freedom.[5] The vacuum energy is given by

$$V_0 = M_P^4\frac{f_R^2\lambda}{4(\lambda + f_R^2\xi_h^2)}, \tag{28}$$

and the mass of the Higgs scalar excitation is

$$m_H^2 = M_P^2\frac{f_R^2\xi_h(6\lambda + f_R^2\xi_h(1 + 6\xi_h))}{3(\lambda + f_R^2\xi_h^2)}. \tag{29}$$

In the classical theory, both the vacuum energy and the Higgs mass go to zero simultaneously if $f_R \to 0$, implying that the smallness of the Higgs mass and of

[5] Note that the massless dilaton does not lead to the fifth force and thus is harmless from the experimental point of view [30–33].

the cosmological constant may be related [16].[6] Taken at face value, this contradicts observations—the Higgs boson mass is too small if f_R^2 is chosen to fit the observed value of the cosmological constant, whereas the cosmological constant is too large if f_R^2 is fixed by comparing (29) with the observed Higgs mass. However, the classical action is not the whole story. The smallness or absence of the Higgs mass persists in scale-invariant quantum perturbation theory, as the zero values of these parameters correspond to fixed points of their renormalization group evolution [36, 37]. This makes it in principle computable in terms of the parameters of the underlying theory, once nonperturbative effects are taken into account [38–43]. A possibility certainly worth exploring is that the hierarchy between the Fermi and Planck scales be attributed to the semiclassical suppression of gravitational-Higgs instanton amplitudes, as proposed in [41, 43] (see also [42, 44]).

5 Scale Invariant EC Gravity

Let us see now what happens in the scale-invariant pure EC gravity. The scale transformations of the gravitational degrees of freedom are

$$e_\mu^A \mapsto q^{-1} e_\mu^A , \quad \omega_\mu^{AB} \mapsto \omega_\mu^{AB} . \tag{30}$$

This leads to the following transformation laws for the Ricci scalar and the irreducible components of torsion

$$R \mapsto q^2 R , \quad v_\mu \mapsto v_\mu , \quad a_\mu \mapsto a_\mu , \quad \tau_{\mu\nu\rho} \mapsto q^{-2} \tau_{\mu\nu\rho} . \tag{31}$$

The transformations of the matter fields are given by (20).

A ghost-free scale-invariant EC action is given by $S_{\mathrm{EC},4}$ defined in Eq. (17). At first sight, the resulting theory contains two extra degrees of freedom, associated with the auxiliary fields χ and a. This happens to not be the case. A peculiar feature of the action $S_{\mathrm{EC},4}$ is that in addition to the global scale symmetry, it is invariant under a wider group, namely local Weyl transformations. This symmetry acts on the gravity variables as follows: Eq. (30) remains in force, but now with q being an arbitrary function of space-time coordinates. Also, the transformations of a_μ and $\tau_{\mu\nu\rho}$ in (31) are the same, whereas the scalar curvature and the torsion vector v_μ transform as[7]

$$R \mapsto q^2 R + 6q\Box q - 12 \left(\partial_\mu q\right)^2 , \tag{32}$$

$$v_\mu \mapsto v_\mu + 3q^{-1} \partial_\mu q . \tag{33}$$

[6] In the scale-invariant theory based on the linear in curvature action the dilaton must be introduced "by hand" [32, 34, 35]. Its coupling to the Higgs field is not related to the cosmological constant.

[7] The fact that the torsion vector v_μ may be used as the Weyl gauge field was pointed out in [45–47].

As a result, the theory contains just one extra degree of freedom: one can make a Weyl transformation removing one of the scalar fields (say, χ), or, equivalently choose the gauge $\chi = M_P$. The action has the form

$$S_{\text{EC},4} \rightarrow \int d^4x\sqrt{g}\left(\frac{1}{2}M_P^2 R - \frac{1}{2}(\partial_\mu \tilde{a})^2 - V(\tilde{a})\right), \tag{34}$$

with the potential given by [48]

$$\frac{1}{32}M_P^4\left(16\alpha + \beta \sinh^2\left(\sqrt{2/3}\tilde{a}/M_P\right) - 8\gamma \sinh\left(\sqrt{2/3}\tilde{a}/M_P\right)\right), \tag{35}$$

where $\tilde{a} = \sqrt{\frac{3}{2}}M_P \operatorname{arctanh}\left(\frac{4a^2}{\sqrt{16a^4+\text{MP}^4}}\right)$ is the canonically normalised scalar field. The choice $\alpha = \gamma^2/\beta$ nullifies the cosmological constant in the minimum of the potential, while for $\gamma \gg \beta$ the theory leads to inflation with predictions close to Starobinski and Higgs inflations [48].

One can add the Higgs field h to the pure gravity theory. The scale-invariant action (in the unitary gauge) has the form

$$S_{ECSI} = S_{ECWI} + S_{ECWB}, \tag{36}$$

where S_{ECWI} is the Weyl invariant action for gravity and scalar field,

$$S_{ECWI} = S_{\text{EC},4} + S_{\text{HWI}}, \tag{37}$$

$$S_{\text{HWI}} = -\frac{1}{2}(D_\mu h)^2 - \frac{\lambda}{4}h^4 + h^2\left(\xi_h F + \zeta_h \tilde{F} + c_a a_\mu^2\right) \tag{38}$$

with

$$D_\mu = \partial_\mu + \frac{1}{3}v_\mu \tag{39}$$

being the Weyl covariant derivative. The Weyl invariance is broken explicitly by the scale-invariant term S_{ECWB}, given by

$$S_{ECWB} = h^2\left(c_v v_\mu^2 + c_{av}v_\mu a_\mu\right) + \partial_\mu(h^2)\left(\zeta_{vh}v_\mu + \zeta_{ah}a_\mu\right). \tag{40}$$

In addition to the graviton and the Higgs boson, this theory has two extra degrees of freedom, related to the auxiliary fields χ and a. The theory (36) is unlikely to have an acceptable low-energy phenomenology. Indeed, the Weyl invariance is restored in the limit of the vanishing Higgs field, leading to $S_{ECWB} \rightarrow 0$. This means that the kinetic matrix for the three fields h, a and χ (analogue of K_{ij} in (24)) has zero eigenvalue when $h \rightarrow 0$ (it can be shown that the eigenstate associated with it is $\chi + a$). Therefore, the Higgs boson interacts strongly with the massless dilaton—a

Goldstone boson of the spontaneously broken scale invariance, potentially contradicting experimental constraints on the decay rate of the Higgs boson into invisible particles.

6 Weyl Invariant Gravities

A natural generalisation of the global scale invariance is the local, Weyl symmetry. In the metric formulation of gravity, it acts on the fields as in Eq. (20), but now with q being an arbitrary function of a space-time point. A set of transformation rules of different geometric quantities can be found in [49], see Eq. (32) for the transformation of the scalar curvature. In the pure gravity case, the unique Weyl-invariant action is described by a square of the Weyl tensor $W_{\mu\nu\lambda\rho}$. The theory contains ghosts at the classical level, and whether it makes sense quantum mechanically is a subject of debate (see, e.g. [50, 51]). It also suffers from the Weyl anomaly [52], leading to its breakdown to global scale symmetry [53, 54].

Going to the EC formulation of gravity changes the situation.[8] Contrary to the metric case, the curvature given by Eq. (9) is Weyl covariant. The torsion tensor $\tau_{\mu\nu\rho}$ and tetrad e^A_μ transform in a uniform way, axial torsion vector a_μ does not change, whereas the vector torsion transforms as a gauge field corresponding to the local Weyl symmetry, Eq. (33). The latter fact allows adding matter fields in a Weyl-invariant way, by constructing a covariant derivative, as was done in Eq. (39).

Thus, a pure EC gravity theory with the action (17) is a healthy Weyl-invariant theory. The Standard Model can be added to it in a Weyl-invariant way. The Weyl-invariant Higgs action is that of Eq. (38), where the Weyl covariant derivative should be augmented by the usual SU(2)xU(1) terms. The Standard SU(3)xSU(2)xU(1) kinetic terms for the gauge fields are Weyl invariant. The Weyl-invariant fermionic action S_f (written for one generation) is

$$S_\mathrm{f} = \int \mathrm{d}^4x\sqrt{g}\left[\frac{i}{2}\overline{\Psi}\gamma^\mu D_\mu\Psi + \text{h.c.} + \left(\zeta^a_V V_\mu + z^a_A A_\mu\right)a^\mu\right], \tag{41}$$

where ζ^a_V, z^a_A are real constants, and

$$V_\mu = \bar{\Psi}\gamma_\mu\Psi\ , \quad A_\mu = \bar{\Psi}\gamma_5\gamma_\mu\Psi\ , \tag{42}$$

are the vector and axial fermionic currents, respectively. The fermionic covariant derivative D_μ is defined as

$$D_\mu = \mathcal{D}_\mu + \frac{1}{8}\omega_\mu{}^{AB}\left(\gamma_A\gamma_B - \gamma_B\gamma_A\right)\ , \tag{43}$$

---

[8] Yet another attractive possibility for construction of Weyl-invariant gravities is associated with Weyl geometry [55]. It has several similarities with the EC gravity discussed here, but there are also important differences.

with $\mathcal{D}_\mu$ the appropriate (flat spacetime) SM covariant derivative, the exact form of which depends on the specific nature of the fermion (left/right handed, lepton/quark). Yet another admitted Weyl invariant term is the coupling of the Higgs hypercharge current with the axial torsion vector,

$$\zeta_Z Im[H^\dagger \mathcal{D}_\mu H] a_\mu \,, \tag{44}$$

where ζ_Z is some real constant.

In addition to the SM fields and the graviton, this theory has just one extra degree of freedom. In the gauge $\chi = M_P$ and after transformation to the Einstein frame it can be identified with the auxiliary field a. Since The Weyl symmetry is local, there is no massless Goldstone scalar. The spectrum is massive with the exception of the graviton and the photon.

The combination of the Weyl-invariant gravity and the SM has an interesting low-energy phenomenology, allowing to unify the strong CP and hierarchy puzzles [16]. If the Lorentz gauge couplings f, $\tilde{f}$ and f_m (or, what is the same, α, β and γ) are tiny and of the same order of magnitude, the tree, classical masses of the Higgs boson, the field a, and the cosmological constant are also small, as all these quantities are proportional to them. The minimum of the potential is at

$$h^2 = 2\xi_h M_P^2 \frac{\alpha\beta - \gamma^2}{\lambda\beta}, \quad a^2 = M_P^2 \frac{\gamma}{\beta}, \tag{45}$$

leading to the vacuum energy

$$\epsilon_{vac} = M_P^4 \frac{\alpha\beta - \gamma^2}{2\beta}, \tag{46}$$

and the masses of the excitations are

$$m_h^2 = 4\xi_h M_P^2 \frac{\alpha\beta - \gamma^2}{\beta}, \quad m_a^2 = M_P^2 \frac{\beta^2 + 16\gamma^2}{24\beta}. \tag{47}$$

The discussion after Eq. (29) is fully applicable to this case, leading to a potential computability of the Higgs boson mass.

The field a—an axion-like particle (ALP) has a non-trivial coupling to the axial and vector fermionic currents, and has all the requisites to solve the strong CP-problem, provided its "gravitational" mass is small compared with the QCD induced mass [16].

In addition, this theory is fully compatible with the neutrino Minimal Standard Model (νMSM) [56, 57], which is a minimal extension of the SM in the neutrino sector capable of addressing simultaneously the experimental problems of the latter: neutrino masses and oscillations, dark matter, baryon asymmetry of the Universe. Weyl symmetry forces the tree-level Majorana masses of the heavy neutral leptons (HNLs) of the νMSM to be zero, which would be incompatible with phenomenology since successful baryogenesis cannot take place. In full analogy to the situation with the Higgs, one can speculate that non-perturbative effects generate masses for the HNLs. This is in line with the common lore that gravity breaks all global symmetries (see e.g. [58]). In this case, the classical action would have a global $B - L$ symmetry in the absence of HNL masses (B and L are the baryon and lepton numbers, respectively), and the breaking of $B - L$ à la Nambu-Jona-Lasinio [59, 60] can potentially lead to Majorana masses for the HNLs, even though the order of magnitude of this effect remains obscure and has never been computed. As an additional bonus, the EC formulation of GR provides a mechanism for generating the HNLs in the early Universe so that the lightest of them can provide the observed abundance of dark matter in a wide range of masses [10].

7 From Classical to Quantum Theory

The transition from classical to quantum Weyl-invariant theories is not straightforward for several reasons.

First, all realistic *renormalisable* conformally invariant at the classical level field theories,[9] suffer from the scale anomaly: the divergence of the dilatational current is non-zero due to quantum effects and is proportional to the β-functions of dimensionless couplings, governing their renormalisation group running [61].

Second, *in the metric formulation of gravity* the Weyl anomaly (for a review see [52]) forbids keeping the classical Weyl symmetry in a quantum theory in a non-trivial gravitational background metric.

The solution to the first problem was suggested almost 50 years ago in [62] (see also [34]). The reason for the presence of quantum scale anomalies is connected to the fact that any regularisation of divergent Feynman graphs of renormalisable field theories contains an explicit mass scale. It can be an ultraviolet (UV) cutoff Λ or mass M_{PV} in Pauli-Villars regularisation, or the scale μ in dimensional regularisation (DimReg), eliminating a mismatch between coupling constants in different dimensions. These scales break conformal invariance explicitly, resulting in the conformal anomaly. The idea of [62], who used DimReg, consists in replacing μ by a dynamical field—dilaton χ. This makes the theory conformally invariant in $D = 4 - 2\epsilon$ dimensions and allows the subtraction of divergencies in a conformally invariant way. The price to pay is the renormalisability of the theory: the Lagrangian in D dimensions contains fractional powers of the dilaton field, leading to the proliferation of different evanescent operators needed to remove the divergencies [34, 63]. The spontaneous breaking of the Weyl invariance—the non-zero dilaton vev—is automatically embedded in the formalism. Of course, the use of DimReg for the construction of Weyl invariant theories is not unique, everything works with any type of regularisation: simply replace the cutoff [30, 31], lattice spacing [64] or the Pauli-Villars mass

[9] Weyl invariance implies conformal invariance on the top of the flat Minkowski metric.

with the dynamical dilaton field. As usual, the DimReg is more suited to practical computations, which in this context can run up to several loops [65].

The solution to the second problem is the use of another gravity formulation, such as Einstein–Cartan gravity, or the Weyl geometry instead of the Riemann one [55].

In the metric formulation of gravity, the reason for the Weyl anomaly [52] can be traced to the impossibility [53, 54, 66, 67] to have local[10] Weyl and Diff invariant generalisation of the following conformally invariant action in *four dimensions* $(D = 4)$[11]

$$S_\tau = \int d^4x(\Box\tau)^2\,, \tag{48}$$

where $\tau = \log(\phi/\mu)$ and ϕ is the scalar field with mass dimension one.[12] The term $\tau\Box^2\tau$ will inevitably appear in an effective theory for the dilaton [66] in spontaneously broken conformally invariant theory in flat space-time.

If $D \neq 4$, such a generalisation exists [53], namely

$$S_\tau \rightarrow \lim_{D\rightarrow 4}\int d^Dx\sqrt{-g}\left[\tau\Delta_4\tau + 2\tau\left(-\frac{1}{6}\Box R + \frac{1}{4}E_4\right) + \frac{R^2}{36} + L_{anom}\right], \tag{49}$$

where Δ_4 is the Riegert operator [70–72],

$$\Delta_4 = \Box^2 + 2R^{\mu\nu}\nabla_\mu\nabla_\nu - \frac{2}{3}R\Box + \frac{1}{3}(\nabla^\mu R)\nabla_\mu \tag{50}$$

and

$$L_{anom} = \frac{E_4}{2(D-4)}\,. \tag{51}$$

The Euler density E_4 is a total derivative only in $D = 4$ so it cannot be dropped if $D \neq 4$, leading to an ill defined expression at $D = 4$, i.e. to Weyl anomaly.

In the EC gravity, where the vector torsion plays the role of the Weyl gauge field, every derivative can be promoted to the Weyl covariant one as in Eq. (39), removing the obstacle for Weyl invariance existing in the metric formulation of gravity (a similar argument applies to the Weyl geometry [55]).

For quantum computations, the EC theory can be formally continued into $D = 4 - 2\epsilon$ space-time, to use dimensional regularisation. The curvature invariants, transforming uniformly under Weyl transformations as $F \rightarrow q^2F,\ \tilde{F} \rightarrow q^2\tilde{F}$ are written in D-dimensions as

---

[10] However, *non-local* operators providing the Weyl invariant action can be constructed [68, 69].

[11] We define the "conformally" invariant action as that written in flat space-time and invariant under conformal group including the Poincare transformations, scale transformation, and special conformal transformations.

[12] For the scalar field τ with the mass dimension zero the action $\int d^4x(\Box\tau)^2$ allows a Weyl and Diff invariant generalisation with the help of the Riegert operator [70–72].

$$F \to \frac{R}{2} + \frac{1}{\sqrt{g}}\partial_\mu\left(\sqrt{g}v^\mu\right) - \frac{D-2}{2(D-1)}v_\mu v^\mu + \frac{1}{48}a_\mu a^\mu\,, \tag{52}$$

$$\tilde{F} \to -\frac{1}{\sqrt{g}}\partial_\mu\left(\sqrt{g}a^\mu\right) + \frac{D-2}{D-1}a_\mu v^\mu\,, \tag{53}$$

where we omitted the contribution of the tensor torsion since equations of motion nullify it (this allows us to ignore the well-known problems with the definition of the Levi-Civita tensor in DimReg). The transformation of the tetrad determinant is $det[e] \to q^{-D}det[e]$, and that of the vector torsion is

$$v_\mu \;\mapsto\; v_\mu + (D-1)q^{-1}\partial_\mu q\,. \tag{54}$$

The covariant derivative of the Higgs field (we keep its mass dimension equal to one independently of D) is given by

$$D_\mu = \partial_\mu + \frac{1}{D-1}v_\mu\,. \tag{55}$$

A construction of the Weyl-invariant action in D dimension is not unique. Namely, any Weyl covariant term in the pure gravity action (18) (e.g. $F\chi^2$), in the Higgs action (38) (e.g. Fh^2) , fermion action (41), etc. can be multiplied by $\chi^{D-4}\rho(a/\chi)$, where ρ is an arbitrary function with the property $\rho \to 1$ when $D \to 4$. For instance, following [34], the pure gravity action can be written as

$$S^D_{\rm EC} = \frac{1}{2}\int \mathrm{d}^D x\, \det e\left[F\chi^{D-2} + \tilde{F}a^{D-2} - \frac{1}{2}\alpha\chi^D - \frac{1}{2}\beta a^D + \gamma\chi^{D/2}a^{D/2}\right], \tag{56}$$

where α, β and γ are now the functions of the ratio a/χ and D with the property $\alpha, \beta, \gamma \to const$ when $D \to 4$.

All these actions give the same four-dimensional action. They lead, however, to different quantum Weyl-invariant physics because of evanescent terms related to UV divergences producing poles in ϵ. In particular, the different prescriptions on how to handle the logarithmic corrections to the effective action originate from this arbitrariness [73, 74].

8 Conclusions

The use of the Einstein–Cartan formulation of gravity may provide several benefits:

- It allows the construction of anomaly-free quantum Weyl-invariant theories of particle physics and gravity.

- The combination of the Standard Model with the Weyl-invariant Einstein–Cartan gravity automatically contains just one extra scalar degree of freedom (ALP) with all properties to solve the strong CP problem.
- The smallness of the cosmological constant, the ALP, Higgs boson, and heavy neutral lepton masses in the νMSM results from tiny values of the dimensionless gauge couplings in Einstein–Cartan gravity, which opens the possibility of potentially computing these parameters from non-perturbative effects.

Of course, there are many open problems to be solved. From the theoretical side, the quantum theory of the Weyl-invariant Einstein–Cartan gravity remains to be developed, with an understanding of its high-energy limit and non-perturbative effects, which may lead to the computation of the Higgs boson and heavy neutral lepton masses. This theory is not perturbatively renormalizable. We may think about its ultraviolet completion along the lines of asymptotic safety [75–77], classicalization [78–80], or non-renormalizable resummation of amplitudes proposed in [81]. From the phenomenological side, the most interesting questions are related to cosmology. It would be important to understand whether inflation can take place in this theory and what would be the predictions of observables. The presence of a new light scalar field poses the question of whether this can be a suitable dark matter candidate. These questions are now under investigation [82–85].

Acknowledgements I thank my collaborators Georgios Karananas, Andrey Shkerin, Inar Timiryasov, Anna Tokareva, Sebastian Zell and Daniel Zenhausern for developing together the ideas described in this contribution. The comments of Georgios Karananas and Sebastian Zell on the manuscript are greatly appreciated. This work was supported in part by the Generalitat Valenciana grant PROMETEO/2021/083.

References

1. C.W. Misner, K.S. Thorne, J.A. Wheeler, *Gravitation* (W. H. Freeman, San Francisco, 1973)
2. R.P. Feynman, Quantum theory of gravitation. Acta Phys. Polon. **24**, 697–722 (1963)
3. R. Utiyama, Invariant theoretical interpretation of interaction. Phys. Rev. **101**, 1597–1607 (1956). https://doi.org/10.1103/PhysRev.101.1597
4. T.W.B. Kibble, Lorentz invariance and the gravitational field. J. Math. Phys. **2**, 212–221 (1961). https://doi.org/10.1063/1.1703702
5. D.W. Sciama, *On the analogy between charge and spin in general relativity, in Recent Developments in General Relativity* (Pergamon Press, Oxford, 1962), p.415
6. É. Cartan, Sur une généralisation de la notion de courbure de Riemann et les espaces à torsion. Comptes Rendus. Ac. Sc. Paris **174**, 593–595 (1922)
7. É. Cartan, Sur les variétés à connexion affine et la théorie de la relativité généralisée (première partie), in Annales scientifiques de l'École normale supérieure, vol. 40 (Gauthier-Villars, 1923), pp. 325–412
8. É. Cartan, Sur les variétés à connexion affine, et la théorie de la relativité généralisée (première partie)(suite), in Annales scientifiques de l'École Normale Supérieure, vol. 41 (Gauthier-Villars, 1924), pp. 1–25

9. É. Cartan, Sur les variétés à connexion affine, et la théorie de la relativité généralisée (deuxième partie), in Annales scientifiques de l'École normale supérieure, vol. 42 (Gauthier-Villars, 1925), pp. 17–88
10. M. Shaposhnikov, A. Shkerin, I. Timiryasov, S. Zell, Einstein-Cartan portal to dark matter. Phys. Rev. Lett. 126(16), 161301 (2021). doi: 10.1103/PhysRevLett.127.169901. arXiv:2008.11686 [hep-ph]. [Erratum: Phys. Rev. Lett. 127, 169901 (2021)]
11. M. Shaposhnikov, A. Shkerin, I. Timiryasov, S. Zell, Higgs inflation in Einstein-Cartan gravity. JCAP **02**, 008 (2021). https://doi.org/10.1088/1475-7516/2021/10/E01. (arXiv:2007.14978 [hep-ph]. [Erratum: JCAP 10, E01 (2021)])
12. M. Shaposhnikov, A. Shkerin, I. Timiryasov, S. Zell, Einstein-Cartan gravity, matter, and scale-invariant generalization. JHEP **10**, 177 (2020). https://doi.org/10.1007/JHEP08(2021)162. (arXiv:2007.16158 [hep-th])
13. G.K. Karananas, M. Shaposhnikov, A. Shkerin, S. Zell, Matter matters in Einstein-Cartan gravity. Phys. Rev. D **104**(6), 064036 (2021). https://doi.org/10.1103/PhysRevD.104.064036. (arXiv:2106.13811 [hep-th])
14. G.K. Karananas, M. Shaposhnikov, A. Shkerin, S. Zell, Scale and Weyl invariance in Einstein-Cartan gravity. Phys. Rev. D **104**(12), 124014 (2021). https://doi.org/10.1103/PhysRevD.104.124014. (arXiv:2108.05897 [hep-th])
15. G.K. Karananas, M. Shaposhnikov, S. Zell, Scale invariant Einstein-Cartan gravity and flat space conformal symmetry. JHEP **11**, 171 (2023). https://doi.org/10.1007/JHEP11(2023)171. (arXiv:2307.11151 [hep-th])
16. G.K. Karananas, M. Shaposhnikov, S. Zell, Weyl-invariant Einstein-Cartan gravity: unifying the strong CP and hierarchy puzzles. JHEP **11**, 146 (2024). https://doi.org/10.1007/JHEP11(2024)146. (arXiv:2406.11956 [hep-th])
17. K.S. Stelle, Renormalization of higher derivative quantum gravity. Phys. Rev. D **16**, 953–969 (1977). https://doi.org/10.1103/PhysRevD.16.953
18. A.A. Starobinsky, A new type of isotropic cosmological models without singularity. Phys. Lett. B **91**, 99–102 (1980). https://doi.org/10.1016/0370-2693(80)90670-X
19. F.W. Hehl, P. Von Der Heyde, G.D. Kerlick, J.M. Nester, General relativity with spin and torsion: foundations and prospects. Rev. Mod. Phys. **48**, 393–416 (1976). https://doi.org/10.1103/RevModPhys.48.393
20. F.L. Bezrukov, M. Shaposhnikov, The standard model Higgs boson as the inflaton. Phys. Lett. B 659, 703–706 (2008). https://doi.org/10.1016/j.physletb.2007.11.072. arXiv:0710.3755 [hep-th]
21. A. Palatini, Deduzione invariantiva delle equazioni gravitazionali dal principio di Hamilton. Rend. Circ. Mat. Palermo **43**(1), 203–212 (1919). https://doi.org/10.1007/BF03014670
22. F. Bauer, D.A. Demir, Inflation with non-minimal coupling: metric versus Palatini formulations. Phys. Lett. B 665, 222–226 (2008). https://doi.org/10.1016/j.physletb.2008.06.014. arXiv:0803.2664 [hep-ph]
23. D. Diakonov, A.G. Tumanov, A.A. Vladimirov, Low-energy general relativity with torsion: a systematic derivative expansion. Phys. Rev. D 84, 124042 (2011). https://doi.org/10.1103/PhysRevD.84.124042. arXiv:1104.2432 [hep-th]
24. G.K. Karananas, The particle spectrum of parity-violating Poincaré gravitational theory. Class. Quant. Grav. 32(5), 055012 (2015). https://doi.org/10.1088/0264-9381/32/5/055012. arXiv:1411.5613 [gr-qc]
25. W. Barker, C. Marzo, C. Rigouzzo, *PSALTer: particle spectrum for any tensor Lagrangian* (hep-th)
26. W. Barker, S. Zell, Einstein-Proca theory from the Einstein-Cartan formulation. Phys. Rev. D 109(2), 024007 (2024). https://doi.org/10.1103/PhysRevD.109.024007. arXiv:2306.14953 [hep-th]
27. L. Alvarez-Gaume, A. Kehagias, C. Kounnas, D. Lüst, A. Riotto, Aspects of quadratic gravity. Fortsch. Phys. 64(2–3), 176–189 (2016). https://doi.org/10.1002/prop.201500100. arXiv:1505.07657 [hep-th]

28. A. Hell, D. Lust, G. Zoupanos, On the degrees of freedom of R^2 gravity in flat spacetime. JHEP 02, 039 (2024). https://doi.org/10.1007/JHEP02(2024)039. arXiv:2311.08216 [hep-th]
29. G.K. Karananas, *The particle content of R^2 gravity revisited* (hep-th)
30. C. Wetterich, Cosmology and the fate of dilatation symmetry. Nucl. Phys. B 302, 668–696 (1988). https://doi.org/10.1016/0550-3213(88)90193-9. arXiv:1711.03844 [hep-th]
31. C. Wetterich, Cosmologies with variable Newton's constant. Nucl. Phys. B **302**, 645–667 (1988). https://doi.org/10.1016/0550-3213(88)90192-7
32. M. Shaposhnikov, D. Zenhausern, Scale invariance, unimodular gravity and dark energy. Phys. Lett. B 671, 187–192 (2009). https://doi.org/10.1016/j.physletb.2008.11.054. arXiv:0809.3395 [hep-th]
33. P.G. Ferreira, C.T. Hill, G.G. Ross, No fifth force in a scale invariant universe. Phys. Rev. D 95(6), 064038 (2017). https://doi.org/10.1103/PhysRevD.95.064038. arXiv:1612.03157 [gr-qc]
34. M. Shaposhnikov, D. Zenhausern, Quantum scale invariance, cosmological constant and hierarchy problem. Phys. Lett. B 671, 162–166 (2009). https://doi.org/10.1016/j.physletb.2008.11.041. arXiv:0809.3406 [hep-th]
35. J. Garcia-Bellido, J. Rubio, M. Shaposhnikov, D. Zenhausern, Higgs-dilaton cosmology: from the early to the late universe. Phys. Rev. D 84, 123504 (2011). https://doi.org/10.1103/PhysRevD.84.123504. arXiv:1107.2163 [hep-ph]
36. G. 't Hooft, Dimensional regularization and the renormalization group. Nucl. Phys. B 61, 455–468 (1973). https://doi.org/10.1016/0550-3213(73)90376-3
37. C. Wetterich, Fine tuning problem and the renormalization group. Phys. Lett. B **140**, 215–222 (1984). https://doi.org/10.1016/0370-2693(84)90923-7
38. S.R. Coleman, E.J. Weinberg, Radiative corrections as the origin of spontaneous symmetry breaking. Phys. Rev. D **7**, 1888–1910 (1973). https://doi.org/10.1103/PhysRevD.7.1888
39. S. Weinberg, Mass of the Higgs Boson. Phys. Rev. Lett. 36, 294–296 (1976). https://doi.org/10.1103/PhysRevLett.36.294
40. A.D. Linde, On the vacuum instability and the Higgs Meson mass. Phys. Lett. B **70**, 306–308 (1977). https://doi.org/10.1016/0370-2693(77)90664-5
41. M. Shaposhnikov, A. Shkerin, Conformal symmetry: towards the link between the Fermi and the Planck scales. Phys. Lett. B 783, 253–262 (2018). https://doi.org/10.1016/j.physletb.2018.06.068. arXiv:1803.08907 [hep-th]
42. M. Shaposhnikov, A. Shkerin, Gravity, Scale Invariance and the Hierarchy Problem. JHEP 10, 024 (2018). https://doi.org/10.1007/JHEP10(2018)024. arXiv:1804.06376 [hep-th]
43. M. Shaposhnikov, A. Shkerin, S. Zell, Standard model meets gravity: electroweak symmetry breaking and inflation. Phys. Rev. D 103(3), 033006 (2021). https://doi.org/10.1103/PhysRevD.103.033006. arXiv:2001.09088 [hep-th]
44. G.K. Karananas, M. Michel, J. Rubio, One residue to rule them all: electroweak symmetry breaking, inflation and field-space geometry. Phys. Lett. B 811, 135876 (2020). https://doi.org/10.1016/j.physletb.2020.135876. arXiv:2006.11290 [hep-th]
45. H.T. Nieh, M.L. Yan, Quantized dirac field in curved Riemann-cartan background. 1. Symmetry properties, Green's Function. Ann. Phys. 138, 237 (1982). https://doi.org/10.1016/0003-4916(82)90186-5
46. T. Dereli, R.W. Tucker, Weyl scalings and spinor matter interactions in scalar - tensor theories of gravitation. Phys. Lett. B 110, 206–210 (1982). https://doi.org/10.1016/0370-2693(82)91237-0
47. Y.N. Obukhov, Conformal invariance and space-time torsion. Phys. Lett. A **90**, 13–16 (1982). https://doi.org/10.1016/0375-9601(82)90037-8
48. G.K. Karananas, Geometrical origin of inflation in Weyl-invariant Einstein-Cartan gravity. Phys. Lett. B 862, 139343 (2025). https://doi.org/10.1016/j.physletb.2025.139343. arXiv:2501.16416 [gr-qc]
49. Y. Nakayama, Scale invariance vs conformal invariance. Phys. Rept. 569, 1–93 (2015). https://doi.org/10.1016/j.physrep.2014.12.003. arXiv:1302.0884 [hep-th]

50. B. Holdom, Making sense of ghosts. Nucl. Phys. B 1008, 116696 (2024). https://doi.org/10.1016/j.nuclphysb.2024.116696. arXiv:2408.04089 [hep-th]
51. A. Salvio, A. Strumia, Agravity. JHEP 06, 080 (2014). https://doi.org/10.1007/JHEP06(2014)080. arXiv:1403.4226 [hep-ph]
52. M.J. Duff, Twenty years of the Weyl anomaly. Class. Quant. Grav 11, 1387–1404 (1994). https://doi.org/10.1088/0264-9381/11/6/004. arXiv:hep-th/9308075
53. M. Shaposhnikov, A. Tokareva, Exact quantum conformal symmetry, its spontaneous breakdown, and gravitational Weyl anomaly. Phys. Rev. D 107(6), 065015 (2023). https://doi.org/10.1103/PhysRevD.107.065015. arXiv:2212.09770 [hep-th]
54. M. Shaposhnikov, A. Tokareva, Anomaly-free scale symmetry and gravity. Phys. Lett. B 840, 137898 (2023). https://doi.org/10.1016/j.physletb.2023.137898. arXiv:2201.09232 [hep-th]
55. D.M. Ghilencea, Weyl conformal geometry vs Weyl anomaly. JHEP 10, 113 (2023). https://doi.org/10.1007/JHEP10(2023)113. arXiv:2309.11372 [hep-th]
56. T. Asaka, S. Blanchet, M. Shaposhnikov, The nuMSM, dark matter and neutrino masses. Phys. Lett. B 631, 151–156 (2005). https://doi.org/10.1016/j.physletb.2005.09.070. arXiv:hep-ph/0503065
57. T. Asaka, M. Shaposhnikov, The νMSM, dark matter and baryon asymmetry of the universe. Phys. Lett. B 620, 17–26 (2005). https://doi.org/10.1016/j.physletb.2005.06.020. arXiv:hep-ph/0505013
58. R. Kallosh, A.D. Linde, D.A. Linde, L. Susskind, Gravity and global symmetries. Phys. Rev. D 52, 912–935 (1995). https://doi.org/10.1103/PhysRevD.52.912. arXiv:hep-th/9502069
59. Y. Nambu, G. Jona-Lasinio, Dynamical Model of Elementary Particles Based on an Analogy with Superconductivity. 1. Phys. Rev. 122, 345–358 (1961). https://doi.org/10.1103/PhysRev.122.345
60. Y. Nambu, G. Jona-Lasinio, Dynamical model of elementary particles based on an analogy with superconductivity. II. Phys. Rev. 124, 246–254 (1961). https://doi.org/10.1103/PhysRev.124.246
61. S.R. Coleman, R. Jackiw, Why dilatation generators do not generate dilatations? Ann. Phys. 67, 552–598 (1971). https://doi.org/10.1016/0003-4916(71)90153-9
62. F. Englert, C. Truffin, R. Gastmans, Conformal invariance in quantum gravity. Nucl. Phys. B 117, 407–432 (1976). https://doi.org/10.1016/0550-3213(76)90406-5
63. M.E. Shaposhnikov, F.V. Tkachov, *Quantum scale-invariant models as effective field theories* (hep-th)
64. M.E. Shaposhnikov, I.I. Tkachev, Quantum scale invariance on the lattice. Phys. Lett. B 675, 403–406 (2009). https://doi.org/10.1016/j.physletb.2009.04.040. arXiv:0811.1967 [hep-th]
65. D.M. Ghilencea, Z. Lalak, P. Olszewski, Two-loop scale-invariant scalar potential and quantum effective operators. Eur. Phys. J. C 76(12), 656 (2016). https://doi.org/10.1140/epjc/s10052-016-4475-0. arXiv:1608.05336 [hep-th]
66. Z. Komargodski, A. Schwimmer, On renormalization group flows in four dimensions. JHEP 12, 099 (2011). https://doi.org/10.1007/JHEP12(2011)099. arXiv:1107.3987 [hep-th]
67. M.A. Luty, J. Polchinski, R. Rattazzi, The a-theorem and the asymptotics of 4D quantum field theory. JHEP 01, 152 (2013). https://doi.org/10.1007/JHEP01(2013)152. arXiv:1204.5221 [hep-th]
68. E.S. Fradkin, A.A. Tseytlin, Conformal anomaly in Weyl theory and anomaly free superconformal theories. Phys. Lett. B 134, 187 (1984). https://doi.org/10.1016/0370-2693(84)90668-3
69. I. Antoniadis, E. Mottola, 4-D quantum gravity in the conformal sector. Phys. Rev. D 45, 2013–2025 (1992). https://doi.org/10.1103/PhysRevD.45.2013
70. E.S. Fradkin, A.A. Tseytlin, One loop beta function in conformal supergravities. Nucl. Phys. B 203, 157–178 (1982). https://doi.org/10.1016/0550-3213(82)90481-3
71. S.M. Paneitz, *A quartic conformally covariant differential operator for arbitrary Pseudo-Riemannian manifolds (Summary)* (math.DG)
72. R.J. Riegert, A nonlocal action for the trace anomaly. Phys. Lett. B 134, 56–60 (1984). https://doi.org/10.1016/0370-2693(84)90983-3

73. F. Bezrukov, M. Shaposhnikov, Standard model Higgs boson mass from inflation: two loop analysis. JHEP 07, 089 (2009). https://doi.org/10.1088/1126-6708/2009/07/089. arXiv:0904.1537 [hep-ph]
74. M. Shaposhnikov, K. Shimada, Asymptotic scale invariance and its consequences. Phys. Rev. D 99(10), 103528 (2019). https://doi.org/10.1103/PhysRevD.99.103528. arXiv:1812.08706 [hep-ph]
75. S. Weinberg, *Ultraviolet divergences in quantum theories of gravitation, in General Relativity: An Einstein Centenary Survey* (University Press, Cambridge, 1980), pp.790–831
76. M. Reuter, Nonperturbative evolution equation for quantum gravity. Phys. Rev. D 57, 971–985 (1998). https://doi.org/10.1103/PhysRevD.57.971. arXiv:hep-th/9605030
77. J. Berges, N. Tetradis, C. Wetterich, Nonperturbative renormalization flow in quantum field theory and statistical physics. Phys. Rept 363, 223–386 (2002). https://doi.org/10.1016/S0370-1573(01)00098-9. arXiv:hep-ph/0005122
78. G. Dvali, C. Gomez, *Self-completeness of Einstein gravity* (hep-th)
79. G. Dvali, G.F. Giudice, C. Gomez, A. Kehagias, UV-completion by classicalization. JHEP 08, 108 (2011). https://doi.org/10.1007/JHEP08(2011)108. arXiv:1010.1415 [hep-ph]
80. G. Dvali, C. Gomez, A. Kehagias, Classicalization of gravitons and goldstones. JHEP 11, 070 (2011). https://doi.org/10.1007/JHEP11(2011)070. arXiv:1103.5963 [hep-th]
81. M. Shaposhnikov, S. Zell, Non-polynomial interactions as a path towards a non-renormalizable UV-completion. Phys. Lett. B **855**, 138804 (2024). (arXiv:2312.13359 [hep-th])
82. G.K. Karananas, M. Shaposhnikov, S. Zell, Gravitational origin of the QCD axion. Phys. Rev. Lett. **135**(24), 241001 (2025). (arXiv:2506.11836 [hep-th])
83. G.K. Karananas, M. Shaposhnikov, S. Zell, Weyl-invariant Einstein-Cartan gravity with a heavy ALP: higgs inflation and α-attractors. Phys. Lett. B **873**, 140151 (2026). (arXiv:2507.15927 [hep-ph])
84. G.K. Karananas, M. Shaposhnikov, Higgs inflation in Weyl-invariant Einstein-Cartan gravity. Phys. Lett. B **876**, 140376. arXiv:2511.04732 [hep-ph]
85. G.K. Karananas, M. Shaposhnikov, S. Zell, *A non-compact QCD axion*. arXiv:2512.20290 [hep-ph]

Isotropic and Homogeneous Cosmological Models in the Theory of Gravity with Non-minimal Derivative Coupling

Sergey V. Sushkov

Abstract We consider the scalar-tensor theory of gravity with the non-minimal derivative coupling of a scalar field to the curvature given by the term $(\zeta/H_0^2)G^{\mu\nu}\nabla_\mu\phi\nabla_\nu\phi$ in the Lagrangian. This theory—which is a special sub-class of Horndeski theories of gravity—has very interesting and important consequences, both astrophysical and cosmological, and in this review we focus on isotropic and homogeneous cosmological models. We show that the non-minimal derivative coupling *crucially* changes scenarios of the Universe evolution at early times. In particular, the ζ-terms turn out to be dominating at early times and totally screen the influence of the cosmological constant and the material filling. At the same time, at late times the influence of non-minimal derivative coupling on the Universe evolution completely disappears. The most important consequences of the non-minimal derivative coupling are a primary *kinetic* inflation which needs no fine-tuned potential, and the existence of a bounce scenario. Here it is also worth noting deep interest of Alexei Starobinsky in the study of cosmological models with a non-minimal derivative coupling. He was inspiring and discussing new ideas and studies in this area, and directly participated in publishing a number of papers.

1 Introduction

In recent decades the observational cosmology has been going through the period of a rapid growth. Precise measurements of the Cosmic Microwave Background (CMB) radiation [1, 2], systematic observations of nearby and distant Type Ia supernovae (SNe Ia) [3–10], the study of baryon acoustic oscillations [11–17], mapping the large-scale structure of the Universe, microlensing observations, and many other remarkable results (see, for example, the review [18]) have essentially expanded our knowledge about the Universe. Amazing discoveries, such as the accelerating expansion of the Universe and the dark matter evidence, have set new serious

S. V. Sushkov (✉)
Institute of Physics, Kazan Federal University, Kazan, Russia
e-mail: sergey_sushkov@mail.ru

A. Barvinsky and A. Kamenshchik (eds.), *Open Issues in Gravitation and Cosmology*, Fundamental Theories of Physics 225, https://doi.org/10.1007/978-3-032-15703-4_16

challenges before theoretical cosmology facing the necessity of radical modification of the standard model that has successfully been exploited for a long time. Now, any viable cosmological model has to be able to describe several qualitatively different epoches of the Universe evolution, including the primary inflation, the matter-dominated stage, and the present acceleration (or secondary inflation). Moreover, it should also explain a mechanism providing an epoch change. These challenges have prompted many speculations mostly based on phenomenological ideas which involve new dynamical sources of gravity that act as dark energy, and/or various modifications of general relativity. To date, many different versions of modified or extended theories of gravity have been proposed (see surveys [20–25] and references therein).[1]

1.1 Horndeski Theory of Gravity

One of the modified theories of gravity intensively studied today is the Horndeski theory [26] derived in the 1970s as an attempt to obtain the most general action for a scalar-tensor theory with a single scalar degree of freedom and second-order field equations. In 2011 Horndeski gravity has been rediscovered in the context of generalized Galileon theories [27], and since then the interest in this model has only been growing.[2] The action of Horndeski theory given in [27] reads

$$S^{(H)} = \int d^4x\sqrt{-g}\,(\mathcal{L}_2 + \mathcal{L}_3 + \mathcal{L}_4 + \mathcal{L}_5), \tag{1}$$

where

$$\mathcal{L}_2 = G_2(\phi, X)\,, \tag{2}$$

$$\mathcal{L}_3 = G_3(\phi, X)\,\Box\phi\,, \tag{3}$$

$$\mathcal{L}_4 = G_4(\phi, X)R - 2G_{4,X}(\phi, X)(\Box\phi^2 - \nabla^\mu\nabla^\nu\phi\,\nabla_\mu\nabla_\nu\phi)\,, \tag{4}$$

$$\begin{aligned}\mathcal{L}_5 = {} & G_5(\phi, X)G^{\mu\nu}\nabla_\mu\nabla_\nu\phi \\ & +\frac{1}{3}G_{5,X}(\phi, X)\big(\Box\phi^3 - 3\,\Box\phi\,\nabla^\mu\nabla^\nu\phi\,\nabla_\mu\nabla_\nu\phi \\ & +2\,\nabla_\mu\nabla_\nu\phi\,\nabla^\mu\nabla^\sigma\phi\,\nabla_\sigma\nabla^\nu\phi\big).\end{aligned} \tag{5}$$

[1] This plethora of models reflects a deep crisis of phenomenological approach in the modern theoretical cosmology. To date, there are no unique criteria to prefer one or another phenomenological model.

[2] The literature dedicated to various aspects of Horndeski gravity is very vast, and its survey lays out the scope of this work. The reader interested in this topic can find some references in the already mentioned surveys [21, 23].

Here R is the scalar curvature, $G_{\mu\nu}$ is the Einstein tensor, ϕ is a scalar field, $X = -\frac{1}{2}g^{\mu\nu}\nabla_\mu\phi\nabla_\nu\phi$ is its kinetic energy, and $G_a(\phi, X)$ $(a = 2, 3, 4, 5)$ are four arbitrary functions. It is worth emphasizing that all terms in Lagrangians (2)–(5) can be divided into two groups. One group contains terms consisting of the scalar field and its covariant derivatives, $\Box\phi, \Box\phi^2, \nabla^\mu\nabla^\nu\phi\,\nabla_\mu\nabla_\nu\phi$, etc. The other contains terms which are responsible for direct coupling between scalar and tensor (gravitational) degrees of freedom. More specifically, one has two qualitatively different terms:

$$\text{(I)} \quad G_4(\phi, X)R, \tag{6}$$

$$\text{(II)} \quad G_5(\phi, X)G^{\mu\nu}\nabla_\mu\nabla_\nu\phi. \tag{7}$$

The term I is of the Brans–Dicke type, i.e. it is a modified Hilbert-Einstein Lagrangian with a variable gravitational constant, $G^{-1} \sim G_4(\phi, X)$. The second term II provides direct non-minimal coupling between derivatives of the scalar field and the spacetime curvature expressed in terms of the Einstein tensor $G_{\mu\nu}$.

Since the Horndeski model is the most general theory, it contains as a particular case all known scalar-tensor theories of gravity with a single scalar field and equations of motion at most of second order. In particular,

- General Relativity (GR): $G_4(\phi, X) = \frac{1}{2}M_{Pl}^2$
- Brans–Dicke type theories: $G_4(\phi, X) = f(\phi)$
- GR with an ordinary scalar field: $G_2(\phi, X) = X - V(\phi)$
- k-essence: $G_2 = K(\phi, X)$
- Kinetic gravity braiding (KGB): $G_3 = B(\phi, X)$
- Fab Four, Gallileons, etc.

1.2 Theory of Gravity with a Non-minimal Derivative Coupling

The important subclass of the Horndeski theory of gravity is represented by models with the non-minimal derivative coupling of the scalar field ϕ with the curvature. This coupling is expressed by the term II, i.e. $G_5(\phi, X)G^{\mu\nu}\nabla_\mu\nabla_\nu\phi$, in the Horndeski lagrangian (1). The simplest choice for the function $G_5(\phi, X)$ is

$$G_5(\phi, X) = \frac{1}{2}\eta\phi, \tag{8}$$

where $\eta = const$ is a coupling parameter with dimension of $(length)^2$, and the simplest form of the action (1) with the non-minimal derivative coupling is the following[3]

[3] Here we take into account that, by means of integrating by part, the term (II) can be rewritten as $-\frac{1}{2}\eta G^{\mu\nu}\nabla_\mu\phi\nabla_\nu\phi$.

$$S^{(\eta)} = \frac{1}{2}\int d^4x\sqrt{-g}\left[\frac{1}{8\pi}(R-2\Lambda) - (\varepsilon g^{\mu\nu} + \eta G^{\mu\nu})\nabla_\mu\phi\nabla_\nu\phi - 2V(\phi)\right], \quad (9)$$

where Λ is the cosmological constant, $V(\phi)$ is a scalar potential, $\varepsilon = +1$ for an ordinary scalar field with the positive kinetic energy, and $\varepsilon = -1$ for a phantom scalar field with the negative kinetic energy. It is worth also noting that one can take $\varepsilon = 0$. In this case the standard kinetic term $X = -\frac{1}{2}g^{\mu\nu}\nabla_\mu\phi\nabla_\nu\phi$ in the action (9) just disappears, and the scalar field dynamics is provided by the term $X_\eta = -\frac{1}{2}\eta G^{\mu\nu}\nabla_\mu\phi\nabla_\nu\phi$.

The total action of the theory of gravity with non-minimal derivative coupling can be taken as

$$S = S^{(\eta)} + S^{(m)}, \quad (10)$$

where $S^{(\eta)}$ is given by Eq. (9), and $S^{(m)}$ is the action for ordinary matter fields which are supposed to be minimally coupled to gravity in the usual way.

Varying the action (10) with respect to $g_{\mu\nu}$ and ϕ gives the field equations, respectively:

$$G_{\mu\nu} = -g_{\mu\nu}\Lambda + 8\pi\big[T^{(m)}_{\mu\nu} + T^{(\phi)}_{\mu\nu} + \eta\Theta_{\mu\nu}\big], \quad (11a)$$

$$[\varepsilon g^{\mu\nu} + \eta G^{\mu\nu}]\nabla_\mu\nabla_\nu\phi = -V_\phi, \quad (11b)$$

where $V_\phi \equiv dV(\phi)/d\phi$, $T^{(m)}_{\mu\nu}$ is a stress-energy tensor of ordinary matter, and

$$T^{(\phi)}_{\mu\nu} = \varepsilon[\nabla_\mu\phi\nabla_\nu\phi - \tfrac{1}{2}g_{\mu\nu}(\nabla\phi)^2] - g_{\mu\nu}V(\phi), \quad (12)$$

$$\begin{aligned}\Theta_{\mu\nu} = &-\tfrac{1}{2}\nabla_\mu\phi\,\nabla_\nu\phi\,R + 2\nabla_\alpha\phi\,\nabla_{(\mu}\phi R^\alpha_{\nu)}\\ &+\nabla^\alpha\phi\,\nabla^\beta\phi\,R_{\mu\alpha\nu\beta} + \nabla_\mu\nabla^\alpha\phi\,\nabla_\nu\nabla_\alpha\phi\\ &-\nabla_\mu\nabla_\nu\phi\,\Box\phi - \tfrac{1}{2}(\nabla\phi)^2 G_{\mu\nu}\\ &+g_{\mu\nu}\big[-\tfrac{1}{2}\nabla^\alpha\nabla^\beta\phi\,\nabla_\alpha\nabla_\beta\phi + \tfrac{1}{2}(\Box\phi)^2 - \nabla_\alpha\phi\,\nabla_\beta\phi\,R^{\alpha\beta}\big]. \end{aligned} \quad (13)$$

Due to Bianchi identity $\nabla^\mu G_{\mu\nu} = 0$ and the conservation law $\nabla^\mu T^{(m)}_{\mu\nu} = 0$, Eq. (11a) leads to the differential consequence

$$\nabla^\mu\big[T^{(\phi)}_{\mu\nu} + \eta\Theta_{\mu\nu}\big] = 0. \quad (14)$$

Substituting Eqs. (12) and (13) into (14), one can check straightforwardly that the differential consequence (14) is equivalent to (11b). In other words, Eq. (11b) is a differential consequence of Eq. (11a).

The additional non-minimal derivative coupling term in the action (9) leads to very interesting and important consequences, both astrophysical and cosmological. In particular, black holes [28–33], wormholes [34, 35], and neutron stars [36–42] have been widely explored within this theory. Cosmological models with the non-minimal derivative coupling have been intensively studied in our recent works

[43–51]. Especially, I would like to note deep interest of Alexei Starobinsky in the study of such models, who inspired the investigations presented in Refs. [48–50].

In this review we will discuss in detail isotropic and homogeneous cosmological models with a non-minimal derivative coupling.

2 Isotropic and Homogeneous Cosmological Models with Non-minimal Derivative Coupling

Let us consider the theory of gravity with the action (10). Further, we will assume that $\varepsilon = +1$, i.e. the scalar field has the ordinary kinetic energy term, and $V(\phi) \equiv 0$, i.e. the potential is absent. Now the action reads as follows

$$S = \frac{1}{2}\int d^4x\sqrt{-g}\left[\frac{1}{8\pi}(R - 2\Lambda) - (g^{\mu\nu} + \eta G^{\mu\nu})\,\nabla_\mu\phi\nabla_\nu\phi\right] + S^{(m)}. \quad (15)$$

Then, we consider the Friedmann–Lemaître–Robertson–Walker isotropic and homogeneous cosmological models with the metric

$$ds^2 = -dt^2 + \mathrm{a}^2(t)\left[\frac{dr^2}{1 - kr^2} + r^2(d\theta^2 + \sin^2\theta d\varphi^2)\right], \quad (16)$$

where $k = 0, \pm 1$, $\mathrm{a}(t)$ is the scale factor, and $H(t) = \dot{\mathrm{a}}(t)/\mathrm{a}(t)$ is the Hubble parameter. Hereafter, the overdot denotes derivatives with respect to time, i.e. $\dot{\mathrm{a}}(t) = d\mathrm{a}/dt$, etc. Denoting the present moment of time as t_0, we have $\mathrm{a}_0 = \mathrm{a}(t_0)$ and $H_0 = H(t_0)$. Supposing homogeneity and isotropy, we also get $\phi = \phi(t)$ and $T_\nu^{\mu(m)} = \mathrm{diag}(-\rho, p, p, p)$, where $\rho = \rho(t)$ is the energy density and $p = p(t)$ is the pressure of matter.

The general field equations (11) written for the metric (16) give the following nontrivial equations:

$$3\left(H^2 + \frac{k}{\mathrm{a}^2}\right) = \Lambda + 8\pi\rho + 4\pi\dot{\phi}^2\left(1 - 9\eta\left(H^2 + \frac{k}{3\mathrm{a}^2}\right)\right), \quad (17a)$$

$$2\dot{H} + 3H^2 + \frac{k}{\mathrm{a}^2} = \Lambda - 8\pi p - 4\pi\dot{\phi}^2\left(1 + \eta\left(2\dot{H} + 3H^2 - \frac{2k}{\mathrm{a}^2} + 4H\frac{\ddot{\phi}}{\dot{\phi}}\right)\right), \quad (17b)$$

$$\frac{1}{\mathrm{a}^3}\frac{d}{dt}\left[\mathrm{a}^3\dot{\phi}\left(1 - 3\eta\left(H^2 + \frac{k}{\mathrm{a}^2}\right)\right)\right] = 0, \quad (17c)$$

Here Eq. (17a) is the modified Friedmann equation, i.e. the tt-component of (11a), Eq. (17b) is the ii-component of (11a) (i = 1, 2, 3), while Eq. (18) is the scalar field equation (11b). Note that Eq. (17c) is a differential consequence of Eqs. (17a) and (17b), therefore only two equations of the system (17) are independent. Integrating Eq. (17c) gives the first integral

$$\dot{\phi}\left(1-3\eta\left(H^2+\frac{k}{\mathrm{a}^2}\right)\right)=\frac{Q}{\mathrm{a}^3}, \tag{18}$$

where Q is a constant of integration. In fact, Q is the Noether or scalar charge associated with the shift symmetry $\phi \to \phi + \phi_0$.

Let us first consider the specific case when the scalar charge is equal to zero, i.e. $Q = 0$. In this case Eq. (18) gives two solutions, dubbed in Ref. [48] as the *GR branch* and the *screening branch*. Correspondingly,

$$\dot{\phi}=0, \tag{19}$$

$$3\left(H^2+\frac{k}{\mathrm{a}^2}\right)=\frac{1}{\eta}. \tag{20}$$

It is obviously that the first solution (19) gives $\phi = const$. Therefore, taking into account the shift symmetry $\phi \to \phi + \phi_0$, one has $\phi = 0$, thus the scalar field is completely absent in this case, and Eqs. (17a), (17b) take the standard form of the ΛCDM model. This fact justifies the term "GR branch". Note also that Eq. (17a) now reduces to the usual Friedmann equation

$$3\left(H^2+\frac{k}{\mathrm{a}^2}\right)=\Lambda+8\pi\rho. \tag{21}$$

Now, comparing Eqs. (20) and (21), one can see that the value $1/\eta$ plays the role of the cosmological constant in the second solution (20). Moreover, the Λ- and ρ-terms are screened and make no contribution to the universe acceleration. For this reason this solution was named as "screening branch". The behavior of the scalar field can be obtained from Eq. (17a), which can be now written as follows

$$\dot{\phi}^2=\frac{\eta(\Lambda+8\pi\rho)-1}{8\pi\eta(1-3k\eta/\mathrm{a}^2)}. \tag{22}$$

In order to study the general case $Q \neq 0$, we assume that the matter filling the universe is a mixture of a radiation and a non-relativistic component:

$$\rho=\rho_m+\rho_r=\rho_{m0}\left(\frac{\mathrm{a}_0}{\mathrm{a}}\right)^3+\rho_{r0}\left(\frac{\mathrm{a}_0}{\mathrm{a}}\right)^4. \tag{23}$$

Now let us introduce the dimensionless time τ, scale factor a, Hubble parameter h, and coupling parameter ζ as follows:

$$\tau=H_0t,\quad a=\frac{\mathrm{a}}{\mathrm{a}_0},\quad h=\frac{H}{H_0},\quad \zeta=\eta H_0^2, \tag{24}$$

and the following dimensionless density parameters:

$$\Omega_0 = \frac{\Lambda}{3H_0^2}, \quad \Omega_2 = \frac{k}{a_0^2 H_0^2}, \quad \Omega_3 = \frac{\rho_{m0}}{\rho_{cr}},$$

$$\Omega_4 = \frac{\rho_{r0}}{\rho_{cr}}, \quad \Omega_6 = \frac{4\pi Q^2}{3a_0^6 H_0^2}, \tag{25}$$

where $\rho_{cr} = 3H_0^2/8\pi$ is the critical density. We will assume in this work that $\Lambda \geq 0$, hence Ω_0 is always not negative, i.e. $\Omega_0 \geq 0$. The parameter Ω_2 is defined such that its sign is the same as that of k. Note that the dimensionless curvature density parameter in ΛCDM cosmology is usually defined as $\Omega_k = -\frac{k}{a_0^2 H_0^2}$, which has the opposite sign compared to Ω_2. Here it is also worth to emphasize the physical meaning of the dimensionless coupling parameter ζ. The parameter η has the dimension $(length)^2$, and so it will be convenient to use the notation $\eta = \epsilon\ell^2$, where ϵ is the sign of η, i.e. $\epsilon = \pm 1$, and ℓ is a characteristic length which characterizes the nonminimal derivative coupling between the scalar field and the curvature. The value H_0 determines the size of the Hubble horizon as $\ell_H = 1/H_0$. Hence, ζ is proportional to the square of ratio of two characteristic scales:

$$\zeta = \epsilon \left(\frac{\ell}{\ell_H} \right)^2. \tag{26}$$

Further, we will consider only positive values of ζ, so that $\epsilon = +1$ and $\zeta > 0$.

In the dimensionless terms, the Eq. (18) reads

$$\phi' \left(1 - 3\zeta \left(h^2 + \frac{\Omega_2}{a^2} \right) \right) = \frac{q}{a^3}, \tag{27}$$

where the prime means the derivative with respect to τ. Now, substituting $\dot{\phi} = H_0\phi'$ from Eq. (27) into (17a), we can rewrite the modified Friedmann equation in terms of dimensionless quantities:

$$h^2 = \Omega_0 - \frac{\Omega_2}{a^2} + \frac{\Omega_3}{a^3} + \frac{\Omega_4}{a^4} + \frac{\Omega_6\big(1 - 3\zeta(3h^2 + \frac{\Omega_2}{a^2})\big)}{a^6\big(1 - 3\zeta(h^2 + \frac{\Omega_2}{a^2})\big)^2}. \tag{28}$$

For given model parameters ζ and Ω_i Eq. (28) completely determines the scale factor $a(\tau)$ and hence the whole cosmological evolution of the Universe.

Generally, the parameters ζ and Ω_i are not independent. Really, at $\tau = \tau_0$ one has $a_0 = 1$ and $h_0 = 1$, then Eq. (28) reduces to

$$1 = \Omega_0 - \Omega_2 + \Omega_3 + \Omega_4 + \frac{\Omega_6\big(1 - 3\zeta(3 + \Omega_2)\big)}{\big(1 - 3\zeta(1 + \Omega_2)\big)^2}. \tag{29}$$

The latter represents a constraint relating the values of the parameters ζ, Ω_0, Ω_2, Ω_3, Ω_4 and Ω_6 at the present time. For practical purposes it will be convenient to rewrite the constraint (29) as follows

$$\Omega_6 = \frac{\left(1 - 3\zeta(1+\Omega_2)\right)^2}{1 - 3\zeta(3+\Omega_2)}\,(1 - \Omega_0 + \Omega_2 - \Omega_3 - \Omega_4). \tag{30}$$

Thus, one has five independent parameters ζ, Ω_0, Ω_2, Ω_3, Ω_4 with additional requirements: $\zeta > 0$, $\Omega_0 \geq 0$, $\Omega_3 \geq 0$, $\Omega_4 \geq 0$, and $\Omega_6 \geq 0$.

Denoting $y = h^2$ and bringing all terms in (28) to the common denominator yields

$$\frac{P(a,y)}{\left(1 - 3\zeta(y + \frac{\Omega_2}{a^2})\right)^2} = 0, \tag{31}$$

where

$$P(a,y) = 9\zeta^2 y^3 - 3\zeta c_2(a) y^2 + c_1(a) y - c_0(a) \tag{32}$$

is the cubic in y polynomial with the coefficients

$$\begin{aligned}
c_2(a) &= 2\left(1 - \frac{3\zeta\Omega_2}{a^2}\right) + 3\zeta\left(\Omega_0 - \frac{\Omega_2}{a^2} + \frac{\Omega_3}{a^3} + \frac{\Omega_4}{a^4}\right),\\
c_1(a) &= \left(1 - \frac{3\zeta\Omega_2}{a^2}\right)^2\\
&\quad + 6\zeta\left(1 - \frac{3\zeta\Omega_2}{a^2}\right)\left(\Omega_0 - \frac{\Omega_2}{a^2} + \frac{\Omega_3}{a^3} + \frac{\Omega_4}{a^4}\right) + \frac{9\zeta\Omega_6}{a^6},\\
c_0(a) &= \left(1 - \frac{3\zeta\Omega_2}{a^2}\right)\times\\
&\quad \times\left[\left(1 - \frac{3\zeta\Omega_2}{a^2}\right)\left(\Omega_0 - \frac{\Omega_2}{a^2} + \frac{\Omega_3}{a^3} + \frac{\Omega_4}{a^4}\right) + \frac{\Omega_6}{a^6}\right],
\end{aligned} \tag{33}$$

Equation (31) will be fulfilled if $P(a,y) = 0$ and $1 - 3\zeta(y + \frac{\Omega_2}{a^2}) \neq 0$, hence the problem reduces to studying the roots of the cubic polynomial. Finding a particular root y_i, one determines the algebraic dependence of the Hubble parameter h on the scale factor a, i.e. $h = h(a)$. The relation to the physical time is then determined by the quadrature

$$\int_{a=1}^{a} \frac{d\tilde{a}}{\tilde{a}h(\tilde{a})} = H_0(t - t_0). \tag{34}$$

2.1 Cosmological Scenarios

Various cosmological scenarios determined by the field equations (17) have been explored in detail in our works Refs. [43–52]. Here we discuss several cosmological models with different sets of parameters.

2.1.1 The Case $\zeta = 0$ and $\Omega_0 = \Omega_3 = \Omega_4 = 0$

First of all, for the sake of completeness, let us briefly discuss the simplest case with $\zeta = 0$, i.e. when the non-minimal coupling is absent, and $\Omega_0 = \Omega_3 = \Omega_4 = 0$, i.e. when the cosmological constant, radiation, and non-relativistic matter are absent. In this case Eq. (28) reduces to the simple form

$$h^2 = -\frac{\Omega_2}{a^2} + \frac{\Omega_6}{a^6}, \tag{35}$$

with the constraint $\Omega_6 = 1 + \Omega_2$, which describes a cosmological evolution of an ordinary massless scalar field in the Friedmann universe. It is obvious that at early times, when $a \to 0$, one has $h^2 \approx \Omega_6/a^6 \to \infty$, that is an initial cosmological singularity. The later evolution essentially depends on the sign of Ω_2, i.e. on the spatial curvature of the universe. As usual, in the case of zero spatial curvature, when $k = 0$ and $\Omega_2 = 0$, one has an open model with $h^2 = \Omega_6/a^6 \to 0$ at $a \to \infty$. In the case of a negative spatial curvature, when $k = -1$ and $\Omega_2 < 0$, one has an open model with $h^2 \approx |\Omega_2|/a^2 \to 0$ as $a \to \infty$. In the case when the spatial curvature is positive, i.e. $k = +1$ and $\Omega_2 > 0$, the scale factor a achieves its maximum value $a_{max} = \max(a(t))$ at $t = t_{turn}$. The moment $t = t_{turn}$ is a turning point in the universe evolution, when the expansion stage is changing to the contraction one. The value of a_{max} can be determined from the condition $h^2_{turn} = -\frac{\Omega_2}{a^2_{max}} + \frac{\Omega_6}{a^6_{max}} = 0$, so that

$$a^2_{max} = \left(\frac{\Omega_6}{\Omega_2}\right)^{1/2} = \left(1 + \frac{1}{\Omega_2}\right)^{1/2}. \tag{36}$$

Taking into account that $\Omega_2 \ll 1$, we obtain the following estimate: $a^2_{max} \approx \Omega_2^{-1/2} \gg 1$, or $\mathrm{a}^2_{max} \approx \mathrm{a}_0^2\Omega_2^{-1/2} \gg \mathrm{a}_0^2$.

The graphical illustration of the properties discussed above is given in Fig. 1.

2.1.2 The Case $\zeta \neq 0$ and $\Omega_0 = \Omega_3 = \Omega_4 = 0$

Now, let us consider the model with a non-minimal derivative coupling $\zeta \neq 0$, while $\Omega_0 = \Omega_3 = \Omega_4 = 0$, i.e. when the cosmological constant, radiation and non-relativistic matter are still absent. Hereafter, it will be convenient to

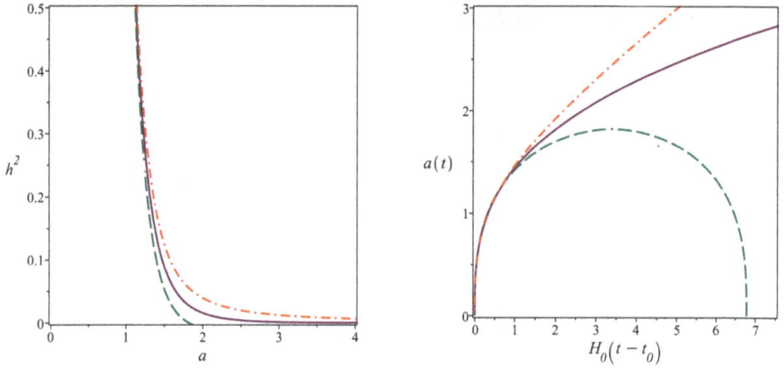

Fig. 1 The case $\zeta = 0$ and $\Omega_0 = \Omega_3 = \Omega_4 = 0$. *Left panel*: Plots of h^2 versus a. *Right panel*: Plots of $a(t)$. Here solid lines correspond to $k = \Omega_2 = 0$; dot-dash lines: $k = -1$ and $\Omega_2 = -0.1$; and dash lines: $k = +1$ and $\Omega_2 = 0.1$

consider separately cosmological models with various signs of the spatial curvature, $k = 0, -1, +1$.

Zero spatial curvature: $k = 0$ **and** $\Omega_2 = 0$
In this case Eq. (28) reads

$$h^2 = \frac{\Omega_6(1 - 9\zeta h^2)}{a^6(1 - 3\zeta h^2)^2}, \tag{37}$$

and the constraint (30) yields

$$\Omega_6 = \frac{(1 - 3\zeta)^2}{1 - 9\zeta}, \tag{38}$$

hence one has the only free parameter ζ in this case. The equation (37) has been already studied in great details in the literature (see, for example, Refs. [43–45, 48, 49]). As is well known, the non-minimal derivative coupling essentially changes the character of the cosmological evolution at early stages. Namely, in the limit $a \to 0$, the Hubble parameter h has the following asymptotic behavior:

$$h^2 = \frac{1}{9\zeta} + O(a^6). \tag{39}$$

Therefore, at early cosmological times, $t \to -\infty$, one has 'eternal' inflation with the quasi-De Sitter behavior of the scale factor: $a(t) \propto e^{H_\eta t}$, where $H_\eta = 1/\sqrt{9\eta}$. It is important to notice that the primary inflationary epoch is only driven by the non-minimal derivative or kinetic coupling between the scalar field and the curvature without introducing any fine-tuned potential, and so one can call this epoch as a

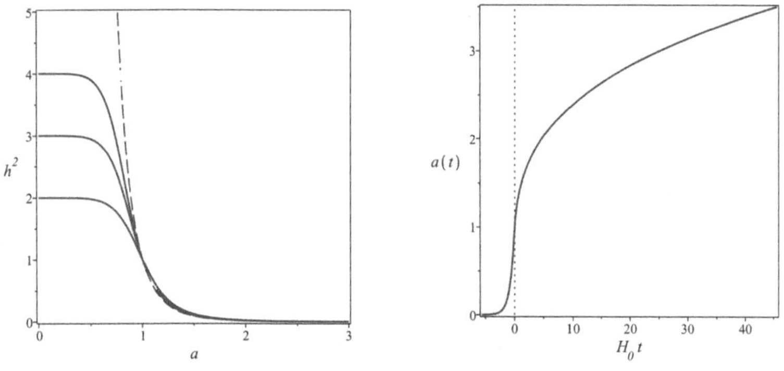

Fig. 2 The case of zero spatial curvature $k = \Omega_2 = 0$, and also $\zeta \neq 0$ (non-zero derivative coupling), and $\Omega_0 = \Omega_3 = \Omega_4 = 0$ (cosmological constant, radiation, and non-relativistic matter are absent). *Left panel*: Plots of h^2 versus a (solid curves) are given for $\zeta = 1/18, 1/27, 1/36$ from bottom to top. The dashed line corresponds to $\zeta = 0$. *Right panel*: The plot of $a(t)$ is given for $\zeta = 1/18$. The vertical dotted straight line separates two cosmological epochs: on the left side—eternal kinetic inflation era, on the right side—scalar field era

kinetic inflation. At late times, the ζ-terms in Eq. (37) become negligibly small, and one has the asymptotics

$$h^2 = \frac{\Omega_6}{a^6} + O(a^{-9}), \tag{40}$$

the same as in the case of $\zeta = 0$, when the universe evolution is driven only by the scalar field with the following behavior of the scale factor $a(t) \propto t^{1/3}$.

It is worth noting again that one needs no fine-tuning potential to provide the epochs change. The epoch of kinetic inflation changes to the scalar field epoch once the ζ-terms in Eq. (37) become negligibly small. Mathematically, it means that the second derivative $\ddot{a}$ changes its sign. Therefore, one can define a moment t_* of epoch change as $\ddot{a}(t_*) = 0$. In Fig. 2 we represent graphs of h^2 versus a for different values of ζ and show a typical dependence of the scale factor $a(t)$ on time t.

Negative spatial curvature: $k = -1$ **and** $\Omega_2 < 0$
In this case Eq. (28) reads

$$h^2 = \frac{|\Omega_2|}{a^2} + \frac{\Omega_6\left(1 - 3\zeta(3h^2 - \frac{|\Omega_2|}{a^2})\right)}{a^6\left(1 - 3\zeta(h^2 - \frac{|\Omega_2|}{a^2})\right)^2}, \tag{41}$$

where the constraint (30) yields

$$\Omega_6 = \frac{\left(1 - 3\zeta(1 - |\Omega_2|)\right)^2}{1 - 3\zeta(3 - |\Omega_2|)} (1 - |\Omega_2|). \tag{42}$$

The relation (42) means that we have two free parameters ζ and $|\Omega_2|$.

At early times, in the limit $a \to 0$, the asymptotic solution of Eq. (41) is as follows:

$$h^2 = \frac{|\Omega_2|}{3a^2} + \left(\frac{1}{9\zeta} + \frac{8\zeta|\Omega_2|^3}{27\Omega_6}\right) + \frac{4|\Omega_2|^2(3\Omega_6 - 8\zeta^2|\Omega_2|^3)}{81\Omega_6^2} a^2 + O(a^3). \quad (43)$$

One can see that in distinction to the case of $k = 0\,(\Omega_2 = 0)$ with the asymptotics (39), the Hubble parameter h has a *singular* behavior at $a \to 0$, so that $h^2 \approx |\Omega_2|/3a^2 \to \infty$. As a increases, the first term in the asymptotic (43) decreases and at $a^2 \geq a_{pse}^2 = \left(\frac{1}{3\zeta|\Omega_2|} + \frac{8\zeta|\Omega_2|}{9\Omega_6}\right)^{-1}$ it becomes negligible compared to the second term, where a_{pse} can be found from the relation $\frac{|\Omega_2|}{3a_{pse}^2} = \frac{1}{9\zeta} + \frac{8\zeta|\Omega_2|^3}{27\Omega_6}$. One can call the stage with $0 < a < a_{pse}$ as a *post-singularity era*. As the scale factor a grows further, the behavior of Hubble parameter is determined by the second term in (43), so that $h^2 \approx h_{dS}^2 = \frac{1}{9\zeta} + \frac{8\zeta|\Omega_2|^3}{27\Omega_6}$. This stage can be called as a *quasi-de Sitter era* or an epoch of kinetic inflation with the de Sitter parameter h_{dS}. As was mentioned in the previous subsection, the epoch of kinetic inflation changes to the scalar field epoch once the ζ-terms in Eq. (41) become negligibly small. Mathematically, it means that the second derivative $\ddot{a}$ changes its sign. Therefore, one can define a moment t_* of epoch change as $\ddot{a}(t_*) = 0$. At the end of quasi-de Sitter era the universe enters the last era of the late evolution, which does not depend on ζ and is determined by the following late-time asymptotics at $a \to \infty$ [48]:

$$h^2 = \frac{|\Omega_2|}{a^2} + O\left(a^{-6}\right). \quad (44)$$

In Fig. 3, where the graphical representation for h^2 versus the scale factor a is shown, one can see that plots $h^2(a)$ have a plateau at $h^2 \approx h_{dS}^2$. This plateau is the more flat the less is the value of $|\Omega_2|$, and in the limit $|\Omega_2| \to 0$ the plot of $h^2(a)$ coincides with that given in Fig. 2 for the case $\Omega_2 = 0$.

The dependence of the scale factor a on the cosmic time t can be found from the quadrature (34). In particular, the corresponding behavior of $a(t)$ near the singularity, where $h^2 \approx |\Omega_2|/3a^2$, is the following:

$$a(t) \approx \sqrt{\frac{|\Omega_2|}{3}} H_0(t - t_s), \quad (45)$$

where t_s is the moment of singularity. The example of $a(t)$ behavior is shown in Fig. 3 (right panel).

Positive spatial curvature: $k = +1$ **and** $\Omega_2 > 0$ In this case Eq. (28) reads

$$h^2 = -\frac{\Omega_2}{a^2} + \frac{\Omega_6\left(1 - 3\zeta(3h^2 + \frac{\Omega_2}{a^2})\right)}{a^6\left(1 - 3\zeta(h^2 + \frac{\Omega_2}{a^2})\right)^2}, \quad (46)$$

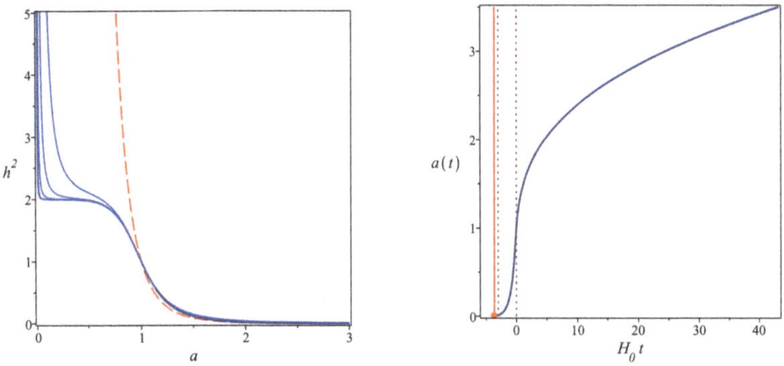

Fig. 3 The case of negative spatial curvature $k = -1$ and $\Omega_2 < 0$, and also $\zeta \neq 0$ (non-zero derivative coupling), and $\Omega_0 = \Omega_3 = \Omega_4 = 0$ (cosmological constant, radiation, and non-relativistic matter are absent). *Left panel*: Plots of h^2 versus a (solid curves) are given for $\zeta = 1/18$ and $\Omega_2 = -0.0001, -0.001, -0.01, -0.1$ from bottom to top. The dashed line corresponds to $\zeta = 0$ and $\Omega_2 = -0.0001$. *Right panel*: The plot of $a(t)$ is given for $\zeta = 1/18$ and $\Omega_2 = -0.001$. The vertical straight lines mark different epochs of the universe evolution: the solid line marks a moment of the initial singularity, the interval between two vertical lines corresponds to the kinetic inflation era, and on the right side of the dotted line one has the era of power-law expansion

and the constraint (30) yields

$$\Omega_6 = \frac{\left(1 - 3\zeta(1+\Omega_2)\right)^2}{1 - 3\zeta(3+\Omega_2)}(1+\Omega_2). \tag{47}$$

At early times, in the limit $a \to 0$, the asymptotic solution of Eq. (41) is as follows:

$$h^2 = -\frac{\Omega_2}{3a^2} + \left(\frac{1}{9\zeta} - \frac{8\zeta\Omega_2^3}{27\Omega_6}\right) + \frac{4\Omega_2^2(3\Omega_6 + 8\zeta^2\Omega_2^3)}{81\Omega_6^2}a^2 + O(a^3). \tag{48}$$

One can see that the asymptotic behavior of h^2 given by (43) essentially differs from those given by (39) and (43). Namely, since the first term in (43) is negative, at some small minimal value of $a = a_{min}$ the value of h^2 becomes zero. In the next Sect. 3 we show that

$$a_{min}^2 = 3\zeta\Omega_2. \tag{49}$$

Supposing that $\zeta \ll 1$ and $\Omega_2 \ll 1$, we have $a_{min}^2 = 3\zeta\Omega_2 \ll 1$. It must be recalled that the moment t_* when the Hubble parameter h, or $\dot{a}$, equals zero is a turning point in the universe evolution. Moreover, since at $t = t_*$ the scale factor a achieves its minimal value, $a_{min} = \min(a(t)) = a(t_*)$, the moment t_* is a *bounce*, when the stage of contraction changes to the expansion one. It is interesting that we can estimate the minimal size of the universe. Actually, returning in the relation $a_{min}^2 = 3\zeta\Omega_2$ to the dimensional quantities $\zeta = \eta H_0^2 = \ell^2 H_0^2$, $\Omega_2 = 1/(\mathrm{a}_0^2 H_0^2)$ and $a = \mathrm{a}/\mathrm{a}_0$, we obtain

$$\mathrm{a}_{min} = \sqrt{3}\,\ell, \tag{50}$$

where ℓ is the characteristic scale of the nonminimal derivative coupling. Thus, the minimal size of the universe is of order ℓ.

In the next Sect. 3, we will study in detail the turning points and bounces in the theory of gravity with a non-minimal derivative coupling. In particular, we will show that near the bounce found above the scalar field has a singular behavior. Therefore the bounce has a singular nature.

Analogously to the case of negative spatial curvature, the first term in the asymptotics (43) decreases as a increases and becomes negligible compared to the second term. As long as the second term in (43) is dominating, the Hubble parameter is approximately constant, so that $h^2 \approx h_{dS}^2 = \frac{1}{9\zeta} - \frac{8\zeta\Omega_2^3}{27\Omega_6}$, and the universe goes through the *quasi-de Sitter phase* with the Hubble parameter h_{dS}. In Fig. 4 one can see that plots of $h^2(a)$ have a plateau at $h^2 \approx h_{dS}^2$. This plateau is the more flat the less is the value of Ω_2, and in the limit $\Omega_2 \to 0$ the plot of $h^2(a)$ coincides with that given in Fig. 2 for the case $\Omega_2 = 0$.

At the end of quasi-de Sitter era the universe enters the last era of the late evolution. Characterizing this era, it is necessary to stress that h^2 turns out to be zero at some value of $a_{max} = \max(a)$. Substituting $h^2 = 0$ into Eq. (41) yields:

$$0 = -\Omega_2 + \frac{\Omega_6}{a_{max}^4}\left(1 - \frac{3\zeta\Omega_2}{a_{max}^2}\right)^{-1}. \tag{51}$$

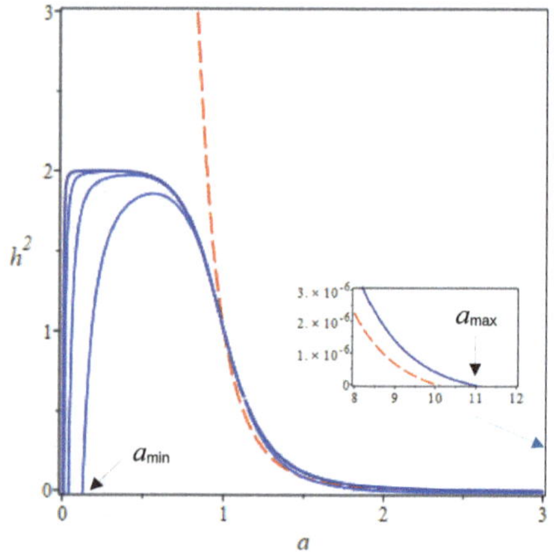

Fig. 4 The case of a positive spatial curvature $k = +1$ and $\Omega_2 > 0$, and also $\zeta \neq 0$ (non-zero derivative coupling) with $\Omega_0 = \Omega_3 = \Omega_4 = 0$ (cosmological constant, radiation and non-relativistic matter are absent). Plots of h^2 versus a (solid curves) are given for $\zeta = 1/18$ and $\Omega_2 = 0.0001$, 0.001, 0.01, 0.1 from top to bottom. The dashed line corresponds to $\zeta = 0$ and $\Omega_2 = 0.0001$

Taking into account that $3\zeta\Omega_2 \ll 0$, one can obtain

$$a_{max}^2 \approx \left(\frac{\Omega_6}{\Omega_2}\right)^{1/2}\left(1+\frac{3\zeta\Omega_2^{3/2}}{2\Omega_6^{1/2}}\right). \tag{52}$$

Comparing with the value of a_{max} obtained for $\zeta = 0$ (see (36)), one can conclude that the maximal value of the scale factor a_{max} is slightly greater in the case of $\zeta \neq 0$.

The dependence of the scale factor a on the cosmic time t follows from the quadrature (34). In particular, taking into account that near the bounce

$$h^2 \approx -\frac{\Omega_2}{3a^2} + \frac{1}{9\zeta}, \tag{53}$$

where we suppose $\zeta\Omega_2 \ll 1$, one obtains the explicit behavior of $a(t)$:

$$a^2(t) \approx 3\zeta\Omega_2 \cosh^2 \frac{H_0(t-t_B)}{\sqrt{9\zeta}}, \tag{54}$$

where t_B is a moment of bounce. The example of $a(t)$ is shown in Fig. 5. One can see that the scale factor $a(t)$ has a cyclic behavior. Each cycle begins at a bounce moment when $a(t)$ achieves its minimal value a_{min}. Then the universe comes to a quasi-de Sitter stage with $a(t) \propto e^{h_{ast}}$. After the end of quasi-de Sitter era the universe enters a stage of slow power-law expansion, which stops when the scale factor achieves its maximal value a_{max}. Further, the universe begins contracting, and its evolution goes in reverse order up to the next bounce moment. Therefore, we have a *cyclic scenario* of cosmological evolution.

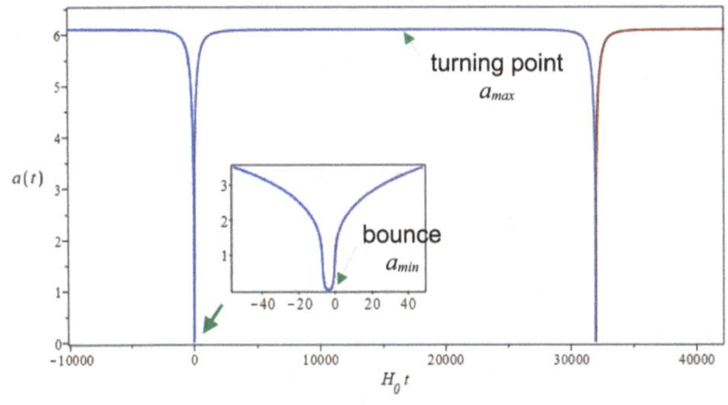

Fig. 5 The case of a positive spatial curvature $k = +1$ and $\Omega_2 > 0$, and also $\zeta \neq 0$ (non-zero derivative coupling) with $\Omega_0 = \Omega_3 = \Omega_4 = 0$ (cosmological constant, radiation, and non-relativistic matter are absent). The plot of $a(t)$ is given for $\zeta = 1/18$ and $\Omega_2 = 0.001$. On this plot one entire cycle of cyclic cosmological evolution is presented

2.2 The Case $\Omega_0 \neq 0$ and $\Omega_3 = \Omega_4 = 0$

Now let us disscus the role of the cosmological constant supposing $\Omega_0 \neq 0$, while, as before, we will assume that $\Omega_3 = \Omega_4 = 0$, i.e. when radiation and non-relativistic matter are absent,

$$h^2 = \Omega_0 - \frac{\Omega_2}{a^2} + \frac{\Omega_6\big(1 - 3\zeta(3h^2 + \frac{\Omega_2}{a^2})\big)}{a^6\big(1 - 3\zeta(h^2 + \frac{\Omega_2}{a^2})\big)^2}. \tag{55}$$

The constraint (30) now yields

$$\Omega_6 = \frac{\big(1 - 3\zeta(1 + \Omega_2)\big)^2}{1 - 3\zeta(3 + \Omega_2)}(1 - \Omega_0 + \Omega_2), \tag{56}$$

thus one has three free parameters ζ, Ω_0 and Ω_2.

At early times, in the limit $a \to 0$, the asymptotic solution of Eq. (55) reads:

$$h^2 = -\frac{\Omega_2}{3a^2} + \left(\frac{1}{9\zeta} - \frac{8\zeta\Omega_2^3}{27\Omega_6}\right) + \frac{4\Omega_2^2(3\Omega_6 + 8\zeta^2\Omega_2^3 + 9\zeta\Omega_0\Omega_6)}{81\Omega_6^2}a^2 + O(a^3). \tag{57}$$

It is important to stress here that the first two major terms in the asymptotics (57) do not contain the cosmological constant Ω_0 and coincide with those given by the asymptotics (37), (41), and (41) ($k = 0, -1, +1$, respectively). Following Ref. [48], we may say that the cosmological constant is screened at the early stage and makes no contribution to the universe evolution which, therefore, is the same as the one described in Sect. 2.1.2 for the case of $\Omega_0 = 0$. Briefly, possible scenarios of the early time universe evolution are the following:

(i) in the case $\Omega_2 = 0$ ($k = 0$) at early cosmological times, $t \to -\infty$, one has an *eternal kinetic inflation* with the quasi-de Sitter behavior of the scale factor: $a(t) \propto e^{h_{dS}(H_0 t)}$, where $h_{dS}^2 = 1/9\zeta$;

(ii) in the case $\Omega_2 < 0$ ($k = -1$) one has an *initial singularity* at $a \to 0$, so that $h^2 \approx |\Omega_2|/3a^{-2} \to \infty$. Then, after a short post-singularity era the universe enters a *primary quasi-de Sitter epoch* with the de Sitter parameter $h_{dS}^2 = \frac{1}{9\zeta} + \frac{8\zeta|\Omega_2|^3}{27\Omega_6}$;

(iii) in the case $\Omega_2 > 0$ ($k = +1$) one has a *bounce* at $t = t_B$, when the Hubble parameter turns to zero at some small minimal value of $a = a_{min}$, where $a_{min}^2 \approx 3\zeta\Omega_2$. Shortly after the bounce the universe enters a *primary quasi-de Sitter epoch* with the de Sitter parameter $h_{dS}^2 = \frac{1}{9\zeta} - \frac{8\zeta\Omega_2^3}{27\Omega_6}$.

The illustration of these scenarios is given in Fig. 6.

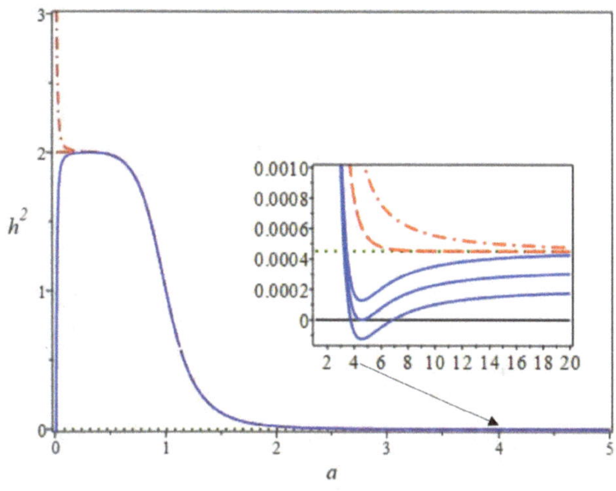

Fig. 6 The case $\Omega_0 \neq 0$ (non-zero cosmological constant), $\zeta \neq 0$ (non-zero derivative coupling), and $\Omega_3 = \Omega_4 = 0$ (radiation and non-relativistic matter are absent). *Main panel*: Plots of h^2 versus a are given for $\zeta = 1/18$, $\Omega_0 = 4.5 \times 10^{-4}$, and $\Omega_2 = 0$ (red dashed line), $\Omega_2 = -0.01$ (red dot-dashed line), $\Omega_2 = 0.01$ (blue solid line). *Auxiliary panel*: An illustration of qualitatively different behavior of h^2 depending on the value of Ω_0. Blue solid lines are plots of h^2 versus a given for $\Omega_0 = 4.5; 3.26; 2 \times 10^{-4}$ from top to bottom. The dotted line shows the asymptotic $h^2 \approx \Omega_0 = 4.5 \times 10^{-4}$

The asymptotic solution of Eq. (55) at large values of a is as follows

$$h^2 = \Omega_0 - \frac{\Omega_2}{a^2} + \frac{\Omega_6(1 - 9\zeta\Omega_0)}{(1 - 3\zeta\Omega_0)^2}\frac{1}{a^6} + O(a^{-8}). \tag{58}$$

In the case $\Omega_2 \leq 0$, i.e. when $k = 0$ or $k = -1$, it is obvious that the value of h^2 given by (58) is monotonically decreasing to Ω_0, i.e. $h^2 \approx \Omega_0$ at $a \to \infty$ (see Fig. 6). In the case $\Omega_2 > 0$ ($k = +1$) a possible scenario is more complicated. Since the second term in (58) is negative when $\Omega_2 > 0$, the behavior of h^2 is now not monotonic, so that h^2 has a minimum

$$h^2_{min} = \Omega_0 - \frac{\Omega_2}{a_*^2} + \frac{\Omega_6(1 - 9\zeta\Omega_0)}{(1 - 3\zeta\Omega_0)^2}\frac{1}{a_*^6}, \tag{59}$$

where a_* can be found from the extremum condition $d(h^2)/da = 0$ as

$$a_*^4 = \frac{3\Omega_6(1 - 9\zeta\Omega_0)}{\Omega_2(1 - 3\zeta\Omega_0)^2}. \tag{60}$$

Note that, depending on a relation between the parameters ζ, Ω_0, and Ω_2, one has $h^2_{min} > 0$ or $h^2_{min} \leq 0$. In the case of $h^2_{min} > 0$ the Hubble parameter h achieves its minimal value h_{min} at $a = a_*$ and then starts growing, so that $h^2 \to \Omega_0$ at $a \to \infty$. In the case of $h^2_{min} \leq 0$ the square of Hubble parameter h becomes equal to zero at some value of the scale factor $a = a_{max}$ at $t = t_{turn}$. The moment t_{turn} is a turning point in the universe evolution, when the expansion stage is changing to the contraction one.

Summarizing, we obtain two possible scenarios of late-time evolution of the universe:

(i) In the case of $\Omega_2 \leq 0$, at the late stage of evolution the universe enters a *secondary inflation epoch* with $h^2 = \Omega_0$, i.e. $H = H_\Lambda = \sqrt{\Lambda/3}$. In the case of $\Omega_2 > 0$ one has the same asymptotic if the value of h^2_{min} given by (59) is positive.

(ii) In the case of $\Omega_2 > 0$ and $h^2_{min} \leq 0$, there is a turning point in the universe evolution, when the expansion stage is changing to the contraction one. In this case one has a *cyclic scenario* of the universe evolution.

All possible scenarios of the cosmological evolution in the case of $\zeta \neq 0$ and $\Omega_0 \neq 0$ are shown in Fig. 7.

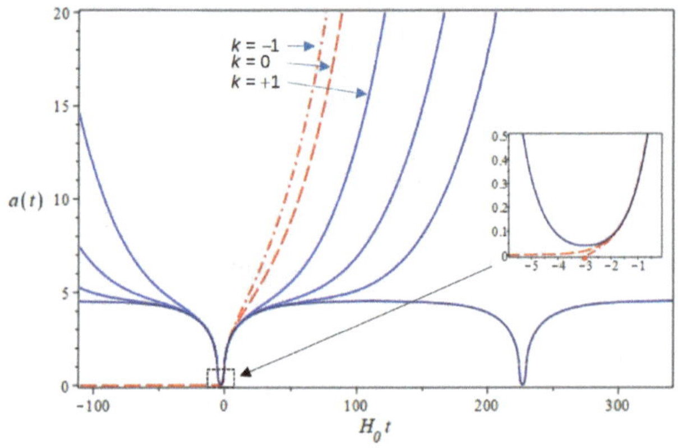

Fig. 7 The case of $\Omega_0 \neq 0$ (non-zero cosmological constant), $\zeta \neq 0$ (non-zero derivative coupling), and $\Omega_3 = \Omega_4 = 0$ (radiation and non-relativistic matter are absent). Plots of $a(t)$ are given for $\zeta = 1/18$. *Main panel*: The red dashed line corresponds to $\Omega_2 = 0$ (zero spatial curvature, $k = 0$), and $\Omega_0 = 4.5 \times 10^{-4}$. The red dot-dashed line corresponds to $\Omega_2 = -0.01$ (negative spatial curvature, $k = -1$), and $\Omega_0 = 4.5 \times 10^{-4}$. Blue solid lines correspond to $\Omega_2 = 0.01$ (positive spatial curvature, $k = +1$), and $\Omega_0 = 4.5; 3.5; 3.3; 3.26 \times 10^{-4}$ from top to bottom. *Auxiliary panel*: An illustration of qualitatively different behavior at small values of $a(t)$ depending on the value of the spatial curvature. One has (i) eternal kinetic inflation if $k = 0$ (red dashed line); (ii) initial singularity if $k = -1$ (red dot-dash line); (iii) a bounce if $k = +1$ (blue solid line)

3 Turning Points and Bounces in the Universe Evolution

In this section we study cosmological scenarios with turning points and bounces in the theory of gravity with a non-minimal derivative coupling.

A turning point in the Universe evolution may occur at a moment $t = t_*$, when the scale factor $a(t)$ reaches either its maximal or minimal value. In the first case, the turning point means that the expansion of the Universe stops and is replaced by the contraction. Such a scenario is natural, for example, for the closed Friedmann cosmological model, when the Universe is too massive and gravity is able to stop the expansion. In the second case, when $a(t)$ reaches its minimal value, the situation is reversed, that is at the turning point the contraction is replaced by the expansion. Such a scenario, if possible, is usually called a cosmological *bounce*. It is worth noting that the Hawking singularity theorem in fact forbids bounces in the standard Big Bang cosmology. To construct a cosmological model with a bounce, one needs either to suppose existing and dominating at the early stage of the evolution of some exotic anti-gravitating matter violating standard energy conditions, or to modify somehow general relativity.

Let us consider if turning points exist in the theory of gravity with a non-minimal derivative coupling. The necessary condition for the existence of an extremum (maximum or minimum) of $a(t)$ at a moment $t = t_*$ is $da(t_*)/dt = 0$, or, equivalently, $H(t_*) = h(\tau_*) = y(a_*) = 0$, where $a_* = a(\tau_*)$. Substituting $y(a_*) = 0$ into (31), we get $c_0(a_*) = 0$. Thus, a 'bounce' solution $y = 0$ of Eq. (31) can exist *if and only if* there exists a_* such that $c_0(a_*) = 0$. Using Eq. (33), we obtain two separate algebraic conditions for a_*:

$$\left(1 - \frac{3\zeta\Omega_2}{a_*^2}\right)\left(\Omega_0 - \frac{\Omega_2}{a_*^2} + \frac{\Omega_3}{a_*^3} + \frac{\Omega_4}{a_*^4}\right) + \frac{\Omega_6}{a_*^6} = 0, \tag{61}$$

and

$$\left(1 - \frac{3\zeta\Omega_2}{a_*^2}\right) = 0. \tag{62}$$

First of all, it is necessary to notice that both algebraic equations (61) and (62) have *no* solutions in the case of $\Omega_2 \leq 0$. Therefore, in cosmological models with a negative or zero spatial curvature there are no turning points. On the other hand, in the case of a positive spatial curvature, when $\Omega_2 > 0$, both equations (61), (62) have solutions, and therefore turning points can exist.

Let us first analyze the condition (61) assuming that $\Omega_2 > 0$. In the simplest case of $\zeta = 0$, $\Omega_0 = \Omega_3 = \Omega_4 = 0$, the relation (30) reads $\Omega_6 = 1 + \Omega_2$, and one easily obtains from Eq. (61)

$$a_*^2 = \left(\frac{\Omega_6}{\Omega_2}\right)^{1/2} = \left(1 + \frac{1}{\Omega_2}\right)^{1/2}. \tag{63}$$

Supposing $\Omega_2 \ll 1$, one has $a_*^2 \approx 1/\Omega_2^{1/2} \gg 1$. In general, the transcendental equation (61) can be qualitatively analyzed graphically. In particular, in the case when $\zeta \neq 0, \Omega_3 \neq 0, \Omega_4 \neq 0$, but $\Omega_0 = 0$, the condition (61) *always* has a root a_* such that $a_*^2 > 1/\Omega_2^{1/2} \gg 1$. In the case of $\Omega_0 \neq 0$, the condition (61) could still have a root $a_* \gg 1$ depending on particular values of all parameters ζ and Ω_i. Thus, the condition (61) provides the value $a_* \gg 1$ corresponding to the turning point of the Universe evolution which turns out to be at some distant future $\tau = \tau_* > \tau_0$, when the scale factor $a(\tau)$ achieves its maximal value $a_{max} = a_*$ and the Universe expansion is replaced by the contraction. It is worth emphasizing once more that the value a_{max} depends, in general, on all parameters ζ and Ω_i, so that $a_{max} = a_{max}(\zeta, \Omega_0, \Omega_2, \Omega_3, \Omega_4)$.

Now let us consider the second condition (62), which easily gives the following relation:

$$a_*^2 = 3\zeta\Omega_2. \tag{64}$$

Since $\zeta \ll 1$ and $\Omega_2 \ll 1$, one can conclude that $a_* = (3\zeta\Omega_2)^{1/2} \ll 1$, and therefore, the turning point determined by Eq. (62) turns out to be at early stages of the Universe evolution, when the scale factor $a(\tau)$ achieves its minimal value $a_{min} \equiv a_* = (3\zeta\Omega_2)^{1/2}$. Moreover, it is very important to stress that the value a_{min} depends *only* on ζ and Ω_2, and does *not* depend on Ω_0, Ω_3 and Ω_4. Following Ref. [48], one can say that the cosmological constant encoded in Ω_0 and the material filling encoded in Ω_3 and Ω_4 are screened at early stages of the Universe evolution.

To describe more accurately the behavior of $a(\tau)$ and $h(\tau)$ in the vicinity of the turning point τ_*, we will obtain a solution of Eq. (28) in the following form (for details see Ref. [52]):

$$a(\tau) = a_{min}\left(1 + \frac{\Delta\tau^2}{18\zeta}\right) + O(\Delta\tau^4), \tag{65}$$

$$h(\tau) = \frac{\Delta\tau}{9\zeta} + O(\Delta\tau^3). \tag{66}$$

where $\Delta\tau = \tau - \tau_*$. Thus, $a(\tau) \to a_{min}$ and $h(\tau) \to 0$ as $\Delta\tau \to 0$, i.e. $\tau \to \tau_*$. The spacetime metric (16) near the turning point takes the following form

$$ds^2 \approx -dt^2 + \mathrm{a}_{min}^2\left(1 + \frac{\Delta t^2}{9\eta}\right) ds_{(3)}^2, \tag{67}$$

where $\mathrm{a}_{min} = \mathrm{a}_0 a_{min}$, $\Delta t = t - t_*$, and $ds_{(3)}^2 = dr^2/(1-r^2) + r^2(d\theta^2 + \sin^2\theta d\varphi^2)$ is the metric of a unit 3d sphere. The corresponding invariant of a scalar curvature,

$$R = \frac{6(\mathrm{a}_{min}^2 + 9\eta)}{\mathrm{a}_{min}^2(\Delta t^2 + 9\eta)}, \tag{68}$$

is regular at the turning point $t = t_*$ and equal to $R_* = 6\left(\frac{1}{a_{min}^2} + \frac{1}{9\eta}\right)$.[4] Ultimately, one can conclude that the moment τ_* in the Universe evolution is a bounce such that the scale factor reaches its minimal value $a_{min} = (3\zeta\Omega_2)^{1/2}$, the Hubble parameter turns out to be zero, and the spacetime geometry remains regular.

However, the above analysis is still incomplete, because it is necessary to consider as well a behavior of the scalar field near the bounce. Substituting the asymptotics (65) and (66) into (27), we obtain

$$\phi' \approx \frac{27\zeta q}{2a_{min}^3 \Delta\tau^2}. \tag{69}$$

Thus, one has $\phi' \propto 1/\Delta\tau^2 \to \infty$ as $\Delta\tau \to 0$, i.e. $\tau \to \tau_*$. Therefore, despite the fact that the spacetime geometry is regular, the scalar field has a singular behavior near the bounce.

4 Concluding Remarks

In this paper we have explored in details homogeneous and isotropic cosmological models in the theory of gravity with the non-minimal derivative coupling given by the action (15). In general, the model depends on six dimensionless parameters: the coupling parameter ζ, density parameters Ω_0, Ω_2, Ω_3, Ω_4, Ω_6 (see Eqs. (24), (25)), and a cosmological evolution is described by the modified Friedmann equation (28). In the case of $\zeta = 0$ (no non-minimal derivative coupling) and $\Omega_6 = 0$ (no scalar field) one has the standard ΛCDM-model, while if $\Omega_6 \neq 0$—the ΛCDM-model with an ordinary scalar field. As is well-known, this model has initial singularity, the same for all k ($k = 0, \pm 1$), while its global behavior depends on k. The universe expands eternally if $k = 0$ (zero spatial curvature) or $k = -1$ (negative spatial curvature), while in the case of $k = +1$ (positive spatial curvature) the universe expansion is changed to the contraction, which terminates by a final singularity.

The situation is *crucially* changed when the scalar field possesses the non-minimal derivative coupling to the curvature, i.e. when $\zeta \neq 0$. More specifically, the ζ-terms turn out to be dominating in the modified Friedmann equation (28) and crucially changing the Universe evolution at early times. In Ref. [48] we called this effect as the screening of the cosmological constant encoded in Ω_0 with the material filling encoded in Ω_3 and Ω_4. At the same time, at late times the influence of the non-minimal derivative coupling on the Universe evolution completely disappears. The possible scenarios of the early time universe evolution are the following:

[4] Note that in the limit $\eta \to \infty$, corresponding to the transition to general relativity, one obtains the standard expression $R_* = 6/a_{min}^2$.

(i) In the case of $\Omega_2 = 0$ ($k = 0$) at early cosmological times, $t \to -\infty$, one has *eternal kinetic inflation* with the quasi-de Sitter behavior of the scale factor: $a(t) \propto e^{h_{dS}(H_0 t)}$, where $h_{dS}^2 = 1/9\zeta$;
(ii) In the case of $\Omega_2 < 0$ ($k = -1$) one has *initial singularity* at $a \to 0$, so that $h^2 \approx |\Omega_2|/3a^2 \to \infty$. Then, after a short post-singularity era the universe enters a *primary quasi-de Sitter epoch* with the de Sitter parameter $h_{dS}^2 = \frac{1}{9\zeta} + \frac{8\zeta|\Omega_2|^3}{27\Omega_6}$.
(iii) In the case of $\Omega_2 > 0$ ($k = +1$) one has a *bounce* at $t = t_*$, when the Hubble parameter becomes zero at some small minimal value of $a = a_{min}$, where $a_{min}^2 = 3\zeta\Omega_2$. Shortly after the bounce the universe enters a *primary quasi-de Sitter epoch* with the de Sitter parameter $h_{dS}^2 = \frac{1}{9\zeta} - \frac{8\zeta\Omega_2^3}{27\Omega_6}$.

It is worth emphasizing that all three scenarios possess the primary quasi-de Sitter stage with $a(t) \propto e^{h_{dS}(H_0 t)}$, where $h_{dS}^2 \approx 1/9\zeta$. Since this stage appears due to non-minimal derivative coupling term in the action (15), modifying the standard kinetic term of the scalar field, it can be dubbed as the *kinetic* inflation. Note that in the standard scenario of slow-roll inflation one usually supposes that ordinary matter is absent at this stage and appears only at the end of inflation due to the reheating process when the inflaton is transforming into ordinary matter. However, since the kinetic inflation discussion in this paper is based on the mechanism which differs from the slow-roll inflation, one has no reasons to assume *a priory* that ordinary matter is absent during the kinetic inflationary stage.

Another important feature of cosmological models in the theory of gravity with a non-minimal derivative coupling is the existence of the bounce scenario. In this paper we have explored bounce scenarios in the framework of homogeneous and isotropic cosmological models with an arbitrary spatial curvature. First of all, as expected, we found that there are no turning points and/or bounces in cosmological models with negative or zero spatial curvature. At the same time, both a turning point and a bounce can exist in the model with a positive spatial curvature. The turning point of the Universe evolution—when the expansion is replaced by the contraction—is determined by the condition (61). In the simplest case when $\zeta = 0$, $\Omega_0 = \Omega_3 = \Omega_4 = 0$, the turning point corresponds to the maximal value of the scale factor, $a_{max} = (1 + 1/\Omega_2)^{1/4}$. Generally, the value a_{max} depends on all parameters ζ and Ω_i, so that $a_{max} = a_{max}(\zeta, \Omega_0, \Omega_2, \Omega_3, \Omega_4)$.

The bounce—when the contraction is replaced by the expansion—is determined by the condition (62), which easily gives $a_{min} = (3\zeta\Omega_2)^{1/2}$. Here it is necessary once more to stress a very important fact that the value a_{min} depends *only* on ζ and Ω_2, and does *not* depend on Ω_0, Ω_3 and Ω_4. This expresses the above mentioned fact of the *screening* of the cosmological constant and matter at early stages of the Universe evolution [48]. We found that near the bounce $a(\tau) \approx a_{min}(1 + \Delta\tau^2/18\zeta)$ and $h(\tau) \approx \Delta\tau/9\zeta$, where $\Delta\tau = \tau - \tau_*$, and τ_* is a moment when the bounce happens. Thus, the scale factor $a(\tau)$, the Hubble parameter $h(\tau)$, and all corresponding geometrical invariants have a regular behavior near the bounce. In particular, $a(\tau) \to a_{min}$, $h(\tau) \to 0$ as $\tau \to \tau_*$. As well, the values characterizing matter energy densities, such as $\rho_m \sim a^{-3}$ and $\rho_r \sim a^{-4}$, are regular near the bounce.

Though the spacetime geometry and energy densities remain to be regular near the bounce, the scalar field has a singular behavior there. Really, we obtained that the time derivative of the scalar field is given by Eq. (69) and diverges near the bounce as follows: $\phi' \propto 1/\Delta\tau^2 \to \infty$ as $\Delta\tau \to 0$.

As a result, we must conclude that the complete dynamical system represented by equations (27), (28), (29) is singular near the bounce. To the best of our knowledge, such a scenario, when the spacetime geometry remains regular at approaching the moment of bounce, while the behavior of a scalar field becomes singular, was unknown before. For this reason, we can call this scenario as a *'singular' bounce*. Note that a singular behavior of the scalar field near the bounce is an inherent property, which is expressed as an exact solution of the field equations of the theory with a non-minimal derivative coupling to the curvature. This scalar singularity has a mathematical rather than a physical meaning. Actually, the scalar field is not observed directly, while all physical characteristics, such as spacetime curvature invariants and energy densities of matter, remain regular near the bounce. In order to further clarify the nature of the scalar singularity, it is at least necessary to consider anisotropic cosmological models which go beyond our review.

Competing Interests The work was supported by the Russian Science Foundation grant No. 25-22-00163.

The author has no conflicts of interest to declare that are relevant to the content of this chapter.

References

1. G.F. Smoot et al., Structure in the COBE differential microwave radiometer first year maps. Astrophys. J. **396**, L1 (1992)
2. E. Komatsu et al., [WMAP Collaboration], Seven-year Wilkinson microwave anisotropy probe (WMAP) observations: cosmological interpretation. Astrophys. J. Suppl. **192**, 18 (2011)
3. A.G. Riess et al., [Supernova Search Team Collaboration], Astron. J. **116**, 1009 (1998)
4. S. Perlmutter et al., [Supernova Cosmology Project Collaboration], Astrophys. J. **517**, 565 (1999)
5. C.L. Bennett et al., Astrophys. J. Suppl. **148**, 1 (2003)
6. M. Tegmark et al., [SDSS Collaboration], Phys. Rev. D **69**, 103501 (2004)
7. S.W. Allen et al., Mon. Not. Roy. Astron. Soc. **353**, 457 (2004)
8. R. Kessler et al., First-year sloan digital sky survey-II supernova results: hubble diagram and cosmological parameters. Astrophys. J. Suppl. **185**, 32 (2009)
9. R. Amanullah et al., [The Supernova Cosmology Project], Spectra and light curves of six type Ia supernovae at $0.511<z<1.12$ and the Union2 compilation. Astrophys. J. **716**, 712 (2010)
10. N. Susuki et al., [The Supernova Cosmology Project], Astrophys. J. **746**, 85 (2012)
11. D.J. Eisenstein et al., Detection of the Baryon acoustic peak in the large-scale correlation function of SDSS luminous red galaxies. Astrophys. J. **633**, 560 (2005)
12. N. Padmanabhan et al., The clustering of luminous red galaxies in the sloan digital sky survey imaging data. Mon. Not. R. Astron. Soc. **378**, 852 (2007)
13. E. Gaztanaga, A. Cabre, L. Hui, Clustering of luminous red galaxies IV: Baryon acoustic peak in the line-of-sight direction and a direct measurement of $H(z)$. Mon. Not. R. Astron. Soc. **399**, 1663 (2009)

14. E. Kazin et al., The Baryonic acoustic feature and large-scale clustering in the SDSS LRG sample. Astrophys. J. **710**, 1444 (2010)
15. W.J. Percival et al., Baryon acoustic oscillations in the sloan digital sky survey data release 7 galaxy sample. Mon. Not. R. Astron. Soc. **401**, 2148 (2010)
16. C. Blake et al., The WiggleZ Dark Energy Survey: mapping the distance-redshift relation with baryon acoustic oscillations (2011). arXiv:1108.2635
17. F. Beutler et al., The 6dF galaxy survey: Baryon acoustic oscillations and the local hubble constant (2011). arXiv:1106.3366
18. D.H. Weinberg et al., Observational probes of cosmic acceleration (2012). arXiv:1201.2434
19. C.M. Will, The confrontation between general relativity and experiment. Living Rev. Rel. **17**, 4 (2014)
20. S. Capozziello, M. De Laurentis, Extended theories of gravity. Phys. Rep. **509**(4–5), 167–321 (2011)
21. T. Clifton, P.G. Ferreira, A. Padilla, C. Skordis, Modified gravity and cosmology. Phys. Rept. **513**, 1–189 (2012)
22. R. Myrzakulov, L. Sebastiani, S. Zerbini, Some aspects of generalized modified gravity models. Int. J. Mod. Phys. D **22**(8), 1330017 (2013)
23. E. Berti et al., Testing general relativity with present and future astrophysical observations. Class. Quant. Grav. **32**, 243001 (2015)
24. S. Nojiri, S.D. Odintsov, V.K. Oikonomou, Modified gravity theories on a nutshell: inflation, bounce and late-time evolution. Phys. Rep. **692**, 1–104 (2017)
25. D. Langlois, Dark energy and modified gravity in degenerate higher-order scalar-tensor (DHOST) theories: A review. Int. J. Mod. Phys. D **28**(05), 1942006 (2019)
26. G.W. Horndeski, Second-order scalar-tensor field equations in a four-dimensional space. Int. J. Theor. Phys. **10**, 363–384 (1974)
27. T. Kobayashi, M. Yamaguchi, J. Yokoyama, Prog. Theor. Phys. **126**, 511 (2011)
28. M. Rinaldi, Black holes with non-minimal derivative coupling. Phys. Rev. D **86**, 084048 (2012)
29. M. Minamitsuji, Solutions in the scalar-tensor theory with nonminimal derivative coupling. Phys. Rev. D **89**, 064017 (2014)
30. A. Anabalon, A. Cisterna, J. Oliva, Asymptotically locally AdS and flat black holes in Horndeski theory. Phys. Rev. D **89**, 084050 (2014)
31. E. Babichev, C. Charmousis, Dressing a black hole with a time-dependent Galileon. JHEP **08**, 106 (2014)
32. T. Kobayashi, N. Tanahashi, Exact black hole solutions in shift symmetric scalar–tensor theories. PTEP **2014**, 073E02 (2014)
33. E. Babichev, C. Charmousis, M. Hassaine, Charged Galileon black holes. JCAP **05**, 031 (2015)
34. S.V. Sushkov, R. Korolev, Scalar wormholes with nonminimal derivative coupling. Class. Quant. Grav. **29**, 085008 (2012)
35. R.V. Korolev, S.V. Sushkov, Exact wormhole solutions with nonminimal kinetic coupling. Phys. Rev. D **90**, 124025 (2014)
36. A. Cisterna, T. Delsate, M. Rinaldi, Neutron stars in general second order scalar-tensor theory: the case of nonminimal derivative coupling. Phys. Rev. D **92**, 044050 (2015)
37. A. Cisterna, T. Delsate, L. Ducobu, M. Rinaldi, Slowly rotating neutron stars in the nonminimal derivative coupling sector of Horndeski gravity. Phys. Rev. D **93**, 084046 (2016)
38. A. Maselli, H.O. Silva, M. Minamitsuji, E. Berti, Neutron stars in Horndeski gravity. Phys. Rev. D **93**, 124056 (2016)
39. J.L. Blázquez-Salcedo, K. Eickhoff, Axial quasinormal modes of static neutron stars in the nonminimal derivative coupling sector of Horndeski gravity: spectrum and universal relations for realistic equations of state. Phys. Rev. D **97**, 104002 (2018)
40. H.O. Silva, A. Maselli, M. Minamitsuji, E. Berti, Compact objects in Horndeski gravity. Int. J. Mod. Phys. D **25**(09), 1641006 (2016)
41. P.E. Kashargin, S.V. Sushkov, Anti-de Sitter neutron stars in the theory of gravity with nonminimal derivative coupling. JCAP **01**, 005 (2023)

42. P.E. Kashargin, A.A. Lebedev, S.V. Sushkov, Neutron stars in the theory of gravity with non-minimal derivative coupling and realistic equations of state. Symmetry **17**(6), 910 (2025)
43. S.V. Sushkov, Exact cosmological solutions with nonminimal derivative coupling. Phys. Rev. D **80**, 103505 (2009)
44. E.N. Saridakis, S.V. Sushkov, Quintessence and phantom cosmology with non-minimal derivative coupling. Phys. Rev. D **81**, 083510 (2010)
45. S.V. Sushkov, Realistic cosmological scenario with nonminimal kinetic coupling. Phys. Rev. D **85**, 123520 (2012)
46. M.A. Skugoreva, S.V. Sushkov, A.V. Toporensky, Phys. Rev. D **88**, 083539 (2013)
47. J. Matsumoto, S.V. Sushkov, JCAP **1511**(11), 047 (2015)
48. A.A. Starobinsky, S.V. Sushkov, M.S. Volkov, The screening Horndeski cosmologies. JCAP **1606**(06), 007 (2016)
49. A.A. Starobinsky, S.V. Sushkov, M.S. Volkov, Anisotropy screening in Horndeski cosmologies. Phys. Rev. D **101**(6), 064039 (2020)
50. R. Galeev, R. Muharlyamov, A.A. Starobinsky, S.V. Sushkov, M.S. Volkov, Anisotropic cosmological models in Horndeski gravity. Phys. Rev. D **103**, 104015 (2021)
51. S.V. Sushkov, R. Galeev, Cosmological models with arbitrary spatial curvature in the theory of gravity with nonminimal derivative coupling. Phys. Rev. D **108**, 044028 (2023)
52. S.V. Sushkov, R.G. Galeev, Singular bounce in the theory of gravity with nonminimal derivative coupling. Phys. Rev. D **111**(8), 083554 (2025)

Dirac Singleton as a Relativistic Field Beyond Standard Model

M. A. Vasiliev

Abstract A new interpretation of Dirac singletons (Dirac in J Math Phys 4:901–909, 1963 [1]), i.e., free conformal fields in d dimensions, as relativistic fields in a $d+1$-dimensional space-time with cosmological constant, that differs from the Flato-Fronsdal dipole construction in AdS_{d+1} (Flato and Fronsdal in Commun Math Phys 108, 469, 1987 [2]), is proposed. The $d+1$-dimensional field is described at the level of both equations and Lagrangian. It forms an infinite-dimensional representation of the $d+1$-dimensional Lorentz group that relates fields at different space-time points. The associated well-known fact is that singleton cannot be localized at a point in $d+1$ dimensions, hence being unobservable via local scattering/radiation phenomena in the Standard Model ($d=3$). On the other hand, that singleton respects $d+1$ dimensional relativistic symmetries makes it possible to introduce its interactions with gravity and other relativistic fields in $d+1$ dimensions. It is speculated that the presence of singleton in a four-dimensional field theory with non-zero cosmological constant (dark energy) can be relevant to the dark matter phenomenon and baryon asymmetry generation.

1 Introduction

Aleksei Starobinsky was one of the world leaders in the field of quantum gravity and cosmology (see, e.g., [3]). He passed away tragically in a few days because of the Covid 19 decease. I had a privilege to know Aleksei personally though our scientific interests were different enough. Alexei was a deep knowledgeable person whose interests spread far beyond the scientific research. In this paper, dedicated to the memory of Aleksei, I wish to make a comment that may indicate some convergence between our seemingly different fields.

The aim of this letter is to point out that there exists an object (field) very different from the fields usually used in the Standard Model (SM), that may, in principle,

M. A. Vasiliev (✉)
I.E. Tamm Department of Theoretical Physics, Lebedev Physical Institute, Moscow, Russia
e-mail: vasiliev@lpi.ru

A. Barvinsky and A. Kamenshchik (eds.), *Open Issues in Gravitation and Cosmology*, Fundamental Theories of Physics 225, https://doi.org/10.1007/978-3-032-15703-4_17

affect general picture of the origin of some of phenomena in high-energy physics, cosmology and astrophysics. The main actor of this letter is the so-called singleton discovered by Dirac in 1963 in [1][1] as a specific branch of the solutions of a certain wave equation, that survives at infinity of AdS_4. There are two singleton fields of boson and fermion types often called *Rac* and *Di*, respectively [5] (see [6, 7] for reviews). Later on it was realized that *Rac* and *Di* and their supermultiplets considered in the context of supergravity S^7 compactifications, supermembrane and higher-spin theory, for instance, in [8–15] are nothing else as, respectively, free massless scalar and spinor fields in the $2 + 1$ dimensional space presently identified with the boundary of AdS_4 within the paradigm of holographic correspondence [16–18], which interpretation is particularly relevant in the context of higher-spin holography [19, 20].

In this letter we focus on the two issues.

Firstly, we interpret singletons as relativistic fields in $(A)dS_{d+1}$ providing for them Lorentz covariant field equations and actions which, to the best of our knowledge, were not available before. This is achieved within the unfolded dynamics approach (see [21] for an elementary review and more references). The proposed formulation differs from the Flato-Fronsdal dipole one [2] given in terms of a certain fourth-order field equation in AdS_{d+1}. In particular, it is free from the huge gauge symmetry of [2] eliminating local degrees of freedom in AdS_{d+1}. On the other hand, kinematically our formulation has some similarities with the interpretation of holography proposed in the framework of the nonlinear realisation approach in [22, 23] (though not at the Lagrangian level).

Secondly, we speculate on the possible implications of the presence of the singleton field in the context of SM and Gravity. Let us stress that our proposal differs from seemingly analogous ideas of Flato and Fronsdal engineering usual fields, starting from the electromagnetic one [24–27] (and references therein), as composites of singletons. Instead we treat singletons democratically with all other fields of SM and Gravity as independent fields.

In particular, we speculate that singleton, that exists in presence of dark energy, can be relevant to the dark matter problem as well as some other physical phenomena including the baryon asymmetry. It is important that though the genuine singleton is defined in AdS_4 our construction is applicable to dS_4 as well. Since singleton makes sense in presence of a nonzero cosmological constant (=dark energy), this idea somehow unifies dark matter with dark energy.

To put it short, our goal is to draw attention to the fact that, apart from usual local relativistic fields, there may exist exotic relativistic objects that are not local in the usual sense. Their characteristic feature is that they form an infinite-dimensional representation of the Lorentz group. A related phenomenon that singleton cannot be localised at a point in the 3*d* space of 4*d* space-time is known from the very first singleton papers [2, 7]. As a result, from the 4*d* space-time perspective it is nowhere (=everywhere). This means in particular that singleton does not affect the

[1] The name was given even earlier in the math literature [4] to emphasize its simplicity from the representation theory perspective.

local scattering and radiation SM processes. On the other hand, that singletons form representations of the Lorentz algebra makes it possible to introduce their gravitational interaction within Cartan formalism. As argued in the paper, singleton can contribute to collective physical phenomena beyond the standard collider physics.

The paper is organized as follows. In Sect. 2 we sketch main ideas of the unfolded dynamics approach underlying our construction. The mechanism allowing to formulate the same dynamics in spaces of different dimensions is recollected in Sect. 3. Unfolded formulation of the conformal scalar and spinor in d dimensions is presented in Sects. 4 and 5, respectively, at the both off-shell and on-shell levels, including conformal invariant Lagrangians. Extension of the singletons to $(A)dS_{d+1}$ is presented in Sect. 6. Spinor formulation of the on-shell $3d$ singletons and their $4d$ extensions is presented in Sect. 7. In Sect. 8 we speculate on the potential role of singletons for the resolution of some problems in SM and related aspects of cosmology and astrophysics. Main results and further research directions are summarised in Sect. 9.

2 Unfolded Dynamics

The unfolded form of dynamical equations is a generalization of the first-order form of ordinary differential equations

$$\dot{q}^i(t) = \varphi^i(q(t)) \tag{1}$$

resulting from the replacement of the time derivative by the de Rham derivation,

$$\frac{\partial}{\partial t} \to \mathrm{d} = \theta^n \partial_n, \qquad \theta^n \theta^m = -\theta^m \theta^n, \qquad \mathrm{d}x^n \equiv \theta^n,$$

and the dynamical variables q^i by a set of differential forms $W^\Omega(\theta, x)$

$$q^i(t) \to W^\Omega(\theta, x) = \sum_{p=0} \theta^{n_1} \dots \theta^{n_p} W^\Omega_{n_1 \dots n_p}(x),$$

that allows one to reformulate a system of partial differential equations in the first-order form

$$\mathrm{d}W^\Omega(\theta, x) = G^\Omega(W(\theta, x)). \tag{2}$$

Here $G^\Omega(W)$ are some functions of the "supercoordinates" W^Ω,

$$G^\Omega(W) = \sum_n f^\Omega{}_{\Lambda_1 \dots \Lambda_n} W^{\Lambda_1} \dots W^{\Lambda_n}.$$

Since $\mathrm{d}^2 = 0$, at $d > 1$ the functions $G^\Lambda(W)$ have to obey the compatibility conditions

$$G^{\Lambda}(W)\frac{\partial G^{\Omega}(W)}{\partial W^{\Lambda}} \equiv 0. \tag{3}$$

Let us stress that these are conditions on the functions $G^{\Lambda}(W)$ rather than on W^{Ω}.

The idea of the unfolded formulation was put forward in [28] where it was realized that the full system of nonlinear equations for massless higher-spin gauge fields can be searched in the form (2) as a deformation of the free unfolded equations.

As a consequence of the compatibility conditions (3) the system (2) is invariant under the gauge transformation

$$\delta W^{\Omega} = \mathrm{d}\varepsilon^{\Omega} + \varepsilon^{\Lambda}\frac{\partial G^{\Omega}(W)}{\partial W^{\Lambda}}, \tag{4}$$

where the gauge parameter $\varepsilon^{\Omega}(x)$ is a $(p_{\Omega} - 1)$-form if W^{Ω} was a p_{Ω}-form. Strictly speaking, this is true for the class of *universal* unfolded systems in which the compatibility conditions (3) hold independently of the dimension d of space-time, i.e., (3) should be true disregarding the fact that any $(d+1)$-form is zero. This is the case in all unfolded systems considered in the higher-spin literature including this paper.

As shown in [29], the variety of invariant functionals associated with the unfolded equations (2) is described by the cohomology of the operator

$$Q = G^{\Omega}\frac{\partial}{\partial W^{\Omega}}, \tag{5}$$

that obeys

$$Q^2 = 0 \tag{6}$$

as a consequence of (3). As a result, the unfolded equations can be written in the Hamiltonian-like form

$$\mathrm{d}F(W) = Q(F(W)), \quad \forall F(W). \tag{7}$$

By virtue of (7), Q-closed p-forms $L_p(W)$ are d-closed, giving rise to the gauge invariant functionals

$$S = \int_{\Sigma^p} L_p. \tag{8}$$

(See [29] for more detail and examples.)

A particular example of an unfolded system is provided by the zero-curvature (Maurer–Cartan) equations. Namely, let h be a Lie algebra with the basis $\{T_\alpha\}$. Let $\omega = \omega^\alpha T_\alpha$ be a one-form valued in h. For $G(\omega) = -\omega\omega := -\frac{1}{2}\omega^\alpha\omega^\beta[T_\alpha, T_\beta]$ Eq. (2) with $W = \omega$ reads as[2]

$$\mathrm{d}\omega + \omega\omega = 0. \tag{9}$$

[2] The exterior product wedge symbol is omitted in this paper since all products are automatically exterior due to the presence of anticommuting θ.

Relations (3) and (4) amount, respectively, to the usual Jacobi identity for h and gauge transformation for ω.

If the set W^α contains some p-forms C^i (e.g. zero-forms) and the functions G^i are linear in ω and C,

$$G^i = -\omega^\alpha (T_\alpha)^i{}_j C^j, \tag{10}$$

then (3) implies that the matrices $(T_\alpha)^i{}_j$ form some representation T of h, acting in a space V where C^i is valued. The associated equation (2) is a covariant constancy condition

$$D_\omega C = 0 \tag{11}$$

with $D_\omega \equiv \mathrm{d} + \omega$ being a covariant derivative in the h-module V.

The zero-curvature equations (9) usually describe background geometry in a coordinate independent fashion. For instance, let h be the Poincaré algebra with the gauge fields

$$\omega(x) = e^n(x) P_n + \frac{1}{2}\omega^{nm}(x) M_{nm}, \tag{12}$$

where P_n and L_{nm} are generators of translations and Lorentz transformations with $e^n(x)$ and $\omega^{nm}(x)$ identified with the frame one-form and Lorentz connection, respectively (fiber Lorentz vector indices $m, n \ldots$ run from 0 to $d-1$ and are raised and lowered by the flat Minkowski metric). It is well known that the zero-curvature condition (9) for the Poincaré algebra describes Minkowski geometry in a coordinate-independent way.

By choosing a different Lie algebra h one can describe a different background like, e.g., anti-de Sitter for $h = o(d-1, 2)$ or conformally flat for $h = o(d, 2)$. The covariant constancy equation (11) then describes h-invariant linear equations in a chosen background.

The unfolded formulation has a number of remarkable properties starting from its general applicability: every system of partial differential equations can be reformulated in the unfolded form. Due to the exterior algebra formalism, the system is coordinate independent.

Local degrees of freedom are represented by the subset of zero-forms $C^I(x_0) \in \{W^\Omega(x_0)\}$ at any $x = x_0$. This is analogous to the fact that $q^i(t_0)$ describe degrees of freedom in the first-order form of ordinary differential equations. In field-theoretic models, the zero-forms $C^I(x_0)$ realize an infinite-dimensional module dual to the space of single-particle states of the system in question [30]. This space is analogous to the phase space in the Hamiltonian dynamics.

3 Space-Time Metamorphoses and Holography

Unfolded dynamics exhibits independence of the "world-volume" space-time with coordinates x. Instead, geometry is encoded by the functions $G^{\Omega}(W)$ in the "target space" with fields W^{Ω} as local coordinates. Since the universal unfolded equations make sense in any space-time independently of a particular realization of the de Rham derivation d, one is free to extend space-time by additional coordinates z^{u},

$$\mathrm{d}W^{\Omega}(x) = G^{\Omega}(W(x)), \quad x \to X = (x, z), \quad \mathrm{d}_x \to \mathrm{d}_X = \mathrm{d}_x + \mathrm{d}_z, \quad \mathrm{d}_z = \mathrm{d}z^{u} \frac{\partial}{\partial z^{u}}. \tag{13}$$

Locally, unfolded equations reconstruct the X-dependence in terms of values of the fields $W^{\Omega}(X_0) = W^{\Omega}(x_0, z_0)$ at any X_0. Clearly, to take $W^{\Omega}(x_0, z_0)$ in space M_X with coordinates X_0 is the same as to take $W^{\Omega}(x_0)$ in the space $M_x \subset M_X$ with coordinates x.

Generally, unfolding can be interpreted as some sort of covariant twistor transform [21, 31]

$$\begin{array}{ccc} & C(Y|x) & \\ \eta \swarrow & & \searrow \nu \\ M(x) & & \mathbf{T}(Y)\,. \end{array}$$

Here $W(Y|x)$ are functions on the "correspondence space" C with local coordinates Y, x. The space-time M has local coordinates x. The twistor space $\mathbf{T}$ has local coordinates Y the expansion over which generates the components $W^{\Omega}(x)$ with various Ω (for examples see Sects. 4, 5 and 7).

Unfolded equations reconstruct the dependence of $W(Y|x)$ on x in terms of the function $W(Y|x_0)$ on $\mathbf{T}$ at some x_0. The restriction of $W(Y|x)$ or some its Y-derivatives to $Y = 0$ gives dynamical fields $\omega(x)$ in M which, in the on-shell case, solve their dynamical field equations. Hence, similarly to the Penrose transform (see [21] and references therein), unfolded equations map functions on $\mathbf{T}$ to solutions of the dynamical field equations in M.

In these terms, the holographic duality can be interpreted as the duality between different space-times M that can be associated with the same twistor space. The problem becomes most interesting provided that there is a nontrivial vacuum connection along the additional coordinates z. This is in particular the case of AdS/CFT correspondence where the conformal flat connection at the boundary is extended to the flat AdS connection in the bulk with z being a Poincaré coordinate. This mechanism has a number of interesting applications. In particular, in [30] it was applied to the description of all $4d$ massless fields in terms of a single scalar in the ten-dimensional space identified by Fronsdal with Lagrangian Grassmannian in [32] (see also [33, 34]).

Conventional holography [16–18] is based on the isomorphism of the boundary conformal group $O(d, 2)$ with the symmetry of AdS_{d+1}. Generators of the conformal algebra obey the relations

$$[D, P_a] = -P_a, \qquad [D, K^b] = K^b, \qquad [D, L_{ab}] = 0, \tag{14}$$

$$[P_a, K_b] = 2L_{ab} - 2\eta_{ab}D, \quad [P_a, P_b = 0], [K_a, K_b = 0], \tag{15}$$

$$[L_{ab}, P_c] = \eta_{bc}P_a - \eta_{ac}P_b, \qquad [L_{ab}, K_c] = \eta_{bc}K_a - \eta_{ac}K_b, \tag{16}$$

$$[L_{ab}, L_{cd}] = \eta_{bc}L_{ad} - \eta_{ac}L_{bd} - \eta_{bd}L_{ac} + \eta_{ad}L_{bc}. \tag{17}$$

Let M^d be a d-dimensional conformally flat space-time with local coordinates $\mathbf{x}$ and some $o(d, 2)$ connection $w_{\mathbf{x}}(\mathbf{x}) = w_{\mathbf{x}}^A T_A$ obeying the flatness conditions

$$\mathrm{d}_{\mathbf{x}} w_{\mathbf{x}}(\mathbf{x}) + w_{\mathbf{x}}(\mathbf{x}) w_{\mathbf{x}}(\mathbf{x}) = 0, \qquad \mathrm{d}_{\mathbf{x}} := \mathrm{d}\mathbf{x}^n \frac{\partial}{\partial \mathbf{x}^n}. \tag{18}$$

A particular flat connection, that corresponds to Cartesian coordinates in M^d, is

$$w_{\mathbf{x}}(\mathbf{x}) = \mathrm{d}\mathbf{x}^a P_a. \tag{19}$$

The dilatation generator D induces standard $\mathbb{Z}$ grading on $o(d, 2)$,

$$[D, T_A] = \Delta(T_A) T_A \tag{20}$$

with $\Delta(T_A)$ being the conformal dimension of T_A,

$$\Delta(L) = 0, \qquad \Delta(D) = 0, \qquad \Delta(K) = 1, \qquad \Delta(P) = -1. \tag{21}$$

Let us now introduce an additional coordinate z and differential $\mathrm{d}z$ so that $x = (\mathbf{x}, z)$ be local coordinates of AdS_{d+1}. A conformally foliated connection $W(x)$ of AdS_{d+1} can be introduced as follows. The components of the connection with differentials $\mathrm{d}\mathbf{x}$ are

$$W_{\mathbf{x}}^A(x) T_A = z^{\Delta(T_A)} w_{\mathbf{x}}^A(\mathbf{x}) T_A, \tag{22}$$

while the only nonzero $\mathrm{d}z$ component of the connection is associated with the dilatation generator D, having the form

$$W_z(x) D = -z^{-1} \mathrm{d}z D. \tag{23}$$

Clearly, so defined connection $W(x)$ is flat in (a local chart of) AdS_{d+1}. Poincaré coordinates result from this construction applied to the connection (19).

Analogously, unfolded equations

$$\mathrm{d}_{\mathbf{x}} C_i(\mathbf{x}) = 0, \qquad \mathrm{d}_{\mathbf{x}} := \mathrm{d}_{\mathbf{x}} + W_{\mathbf{x}}^A T_A \tag{24}$$

in M^d for a set of fields $C_i(\mathbf{x})$ carrying conformal weights Δ_i extend to the fields

$$C_i(x) = z^{\Delta_i} C_i(\mathbf{x}) \tag{25}$$

and equations

$$\mathcal{D}_{\mathbf{x}} C_i(x) = 0, \qquad \mathcal{D}_{\mathbf{x}} := \mathrm{d}_x + W_x^A T_A. \tag{26}$$

It is important to note that if a system was off-shell in M^d this is not so in the extended $d+1$-dimensional space. Indeed, the dependence on the additional coordinate z is determined by (26) in terms of that on $\mathbf{x}$ as is most obvious from (25). This means that the field in AdS_{d+1} obeys some differential equation, that determines its z-dependence.

To identify the $d+1$-dimensional space with (a local chart of) AdS_{d+1} it suffices to redefine $o(d,2)$ generators as

$$\begin{aligned} \mathcal{P}_\nu &= \big((P_a + \lambda^2 K_a), 2\lambda D\big), \\ \mathcal{M}_{\nu\mu} &= \left(L_{ab}, \frac{1}{2\lambda}(P_a - \lambda^2 K_a)\delta^d_\nu, -\frac{1}{2\lambda}(P_b - \lambda^2 K_b)\delta^d_\mu\right) \end{aligned} \tag{27}$$

with $a, b = (0, \ldots, d-1)$ $\nu, \mu = (0, \ldots, d)$, interpreting $\mathcal{P}_\nu$ and $\mathcal{M}_{\nu\mu}$ as AdS_{d+1} translation (transvection) and Lorentz generators, respectively. The respective connections are

$$W = h^\nu \mathcal{P}_\nu + \frac{1}{2}\omega^{\nu\mu}\mathcal{M}_{\nu\mu}. \tag{28}$$

In particular, this means that

$$E^a = \frac{1}{2}(h^a + 2\lambda\omega^{ad}), \tag{29}$$

where E^a is the d-dimensional vielbein rescaled in accordance with (22) (the index d in (29) indicates the direction along z). The dimensionful parameter λ is related to the cosmological constant Λ,

$$\Lambda = -\#\lambda^2 \tag{30}$$

with some positive number #. This implies that λ is real and pure imaginary in the AdS_{d+1} and dS_{d+1} cases, respectively. Naively, this suggests that the construction is not working in the de Sitter space. In fact, this is not necessarily true and, moreover, as discussed below, specificities of the dS case may be of relevance to the baryon asymmetry problem.

According to the Flato-Fronsdal theorem [5], boundary conformal currents are dual to fields in the bulk AdS_{d+1}. In other words, free relativistic fields in the bulk are associated with the bilinear currents on the boundary, the fact underlying the Klebanov-Polyakov HS holographic conjecture [20]. Here we consider an opposite situation: starting from the boundary conformal field we will see what is its bulk dual. Our construction differs from the other holographic treatments of singletons (see e.g. [6, 35, 36]) based on the dipole singleton description of [2]. The output is interesting both formally and, hopefully, physicswise shedding more light on what are bulk duals of the free conformal boundary fields, i.e., singletons.

4 Conformal Scalar in Any d Within Unfolded Formalism

Singleton Rac is a massless conformal scalar field in any dimension d. In the unfolded dynamics approach it is described as follows [37]. Let $C(y|\mathbf{x})$ be a zero-form, that depends on the space-time coordinates $\mathbf{x}^n$ and auxiliary variables y^n ($n = 0, \dots d-1$). Consider unfolded equations of the form

$$\mathrm{d}_{\mathbf{x}} C(y|\mathbf{x}) + \mathrm{d}\mathbf{x}^n \frac{\partial}{\partial y^n} C(y|\mathbf{x}) = 0. \tag{31}$$

Clearly, this equation relates the coefficients $C_{a_1 \dots a_n}(\mathbf{x})$ of the expansion

$$C(y|\mathbf{x}) = \sum_{n=0}^{\infty} \frac{1}{n!} C_{a_1 \dots a_n}(\mathbf{x}) y^{a_1} \dots y^{a_n} \tag{32}$$

to higher $\mathbf{x}^a$ derivatives of the ground component $C(\mathbf{x})$,

$$C_{a_1 \dots a_n}(\mathbf{x}) = (-1)^n \partial_{a_1} \dots \partial_{a_n} C(\mathbf{x}), \qquad \partial_a := \frac{\partial}{\partial \mathbf{x}^a}, \tag{33}$$

$$C(\mathbf{x}) := C(0|\mathbf{x}) \tag{34}$$

identified with the scalar field. The system (31) is off-shell, imposing no differential conditions on $C(\mathbf{x})$. To put the system on shell of a massless field it suffices to constrain $C(y|\mathbf{x})$ by the condition

$$\Box_y C(y|\mathbf{x}) = 0, \qquad \Box_y := \eta^{ab} \frac{\partial^2}{\partial y^a \partial y^b}. \tag{35}$$

The system (31) is equivalent to

$$\mathcal{D}_{\mathbf{x}} C(y|\mathbf{x}) = 0, \tag{36}$$

where

$$\mathcal{D}_{\mathbf{x}} := \mathrm{d}_{\mathbf{x}} + e^a P_a + f_a K^a + \frac{1}{2} \omega^{ab} L_{ab} + bD \tag{37}$$

is the covariant derivative of the conformal algebra $o(d, 2)$ with the generators P_a for translations, K^a for special conformal transformations, L_{ab} for Lorentz transformations and D for dilatations. The particular flat connection used in (31) is

$$e^a = \mathrm{d}\mathbf{x}^a, \qquad \omega^{ab} = 0, \qquad b = 0, \qquad f_a = 0. \tag{38}$$

In terms of y^a, the conformal generators, that obey (14)–(17), are realized as

$$P_a = \frac{\partial}{\partial y^a}, \qquad L_{ab} = y_a \frac{\partial}{\partial y^b} - y_b \frac{\partial}{\partial y^a}, \qquad D = y^a \frac{\partial}{\partial y^a} + \Delta, \tag{39}$$

$$K_a = y^2 \frac{\partial}{\partial y^a} - 2 y_a y^b \frac{\partial}{\partial y^b} - 2\Delta y_a, \tag{40}$$

where Δ is a number (conformal weight). The system (36) along with the consistency condition

$$\mathcal{D}_{\mathbf{x}}^2 = 0 \tag{41}$$

equivalent to the flatness condition (18) forms an unfolded system invariant under the gauge conformal transformations (4). Any choice of a particular flat connection e^a, f_a, ω_{ab}, b in (37) restricts the local conformal transformations (4) to the global ones. (Analogously the choice of Minkowski metric restricts diffeomorphisms to global Poincaré transformations. For more detail see [21]). This proves global conformal invariance of the system (36).

Note that the construction of this section is extendable to $(A)dS_d$ as well as to any other conformally flat background by the appropriate choice of the flat connection of the conformal group. In particular, conformal field theories in AdS_d and their holographic aspects were discussed in [30, 36, 38, 39].

Next one observes that if $C(Y|\mathbf{x})$ obeys the constraint (35), it is obeyed by $T_A C(Y|\mathbf{x})$ for all conformal algebra generators T_A provided that

$$\Delta = \frac{d}{2} - 1, \tag{42}$$

which is the canonical dimension of a massless scalar in d dimensions. This implies that the constraint (35) is preserved by the equation (36). Since the system (35), (36) is equivalent to the massless Klein-Gordon equation, this in turn proves conformal invariance of the latter.

From the representation theory perspective this implies that the module V^{tr} generated by the conformal generators from a constant in y^a only contains y–traceless polynomials $f(y)$ obeying $\Box_y f(y) = 0$. To see this it is instructive to check that $K^a K_a$ applied to a constant yields zero for K^a (40) and Δ (42). Hence, V^{tr} is a submodule of the module V of all polynomials, $V^{tr} \subset V$, while traceful polynomials in y^a are in the factor module V/V^{tr}.

Now we observe that the coefficients $C_{a_1 \dots a_n}(\mathbf{x})$ in the expansion (32) are in the dual module V^* with respect to the action of the covariant derivative $\mathcal{D}_{\mathbf{x}}$ (36), (37). This means that $(V/V^{tr})^*$ is a submodule of V^* while $(V^{tr})^*$ is a factor-module. As a result, there are two components of $C(y)$ the covariant derivative of which does not contain the gauge field f^a of special conformal transformations. One is the vacuum (lowest weight) component $C(\mathbf{x})$, while another is the singular vector (synonymous to be annihilated by K^a) associated with the trace component

$$C'(\mathbf{x}) := C^a{}_a(\mathbf{x}). \tag{43}$$

This implies that to prove conformal invariance of any functional built from $C(\mathbf{x})$ and $C'(\mathbf{x})$ it suffices to check its invariance under the action of the parabolic subalgebra generated by P_a, L_{ab} and D. (For more detail on the derivation of general conformal invariant equations from the representation theory of the conformal group see [40] and references therein.)

The conformal invariant Lagrangian for a scalar field is a d-form

$$L^{Rac} = \frac{1}{2}\epsilon_{a_1\ldots a_d} e^{a_1}(\mathbf{x})\ldots e^{a_d}(\mathbf{x})C(\mathbf{x})C'(\mathbf{x}), \tag{44}$$

where e^a is a vielbein one-form in d dimensions. It is easy to see that L^{Rac} is Q-closed with respect to Q (5). Indeed, the special conformal gauge field does not appear in QL since it is absent in de^a, $\mathrm{d}(C)$ and $\mathrm{d}(C')$. Lorentz connection ω^{ab} cancels because L is Lorentz invariant. Analogously, b cancels because the Lagrangian has proper scaling dimension due to (42). Finally, the contribution of e^b cancels because of antisymmetrization over $d+1$ indices $a = 0, 1, \ldots d-1$ due to the exterior product (i.e., θ–dependence) of the one-forms e^a.

The fields $C(y|\mathbf{x})$ still obey unfolded equations (36), that are off-shell just expressing higher components $C_{a_1\ldots a_n}(\mathbf{x})$ via derivatives of $C(\mathbf{x})$ according to (33), imposing no differential conditions on the latter. In particular, they imply that $C'(\mathbf{x}) = \Box_{\mathbf{x}} C(\mathbf{x})$. The Lagrangian field equations then imply the massless equation $\Box_{\mathbf{x}} C(x) = 0$.

5 Conformal Spinor in Any *d* Within Unfolded Formalism

Conformal spinor Di is described analogously to scalar Rac. Consider unfolded equations of the form

$$\mathrm{d}_{\mathbf{x}} C_\alpha(y|\mathbf{x}) + \mathrm{d}\mathbf{x}^a \frac{\partial}{\partial y^a} C_\alpha(y|\mathbf{x}) = 0 \tag{45}$$

with a spinor index α. This equation relates the coefficients $C_{\alpha,a_1\ldots a_n}(\mathbf{x})$ of the expansion

$$C_\alpha(y|\mathbf{x}) = \sum_{n=0}^{\infty} \frac{1}{n!} C_{\alpha,a_1\ldots a_n}(\mathbf{x}) y^{a_1}\ldots y^{a_n} \tag{46}$$

to higher derivatives in $\mathbf{x}^a$,

$$C_{\alpha,a_1\ldots a_n}(\mathbf{x}) = (-1)^n \partial_{a_1} \ldots \partial_{a_n} C_\alpha(\mathbf{x}), \tag{47}$$

where $C_\alpha(\mathbf{x})$ is the ground component of $C_\alpha(y|\mathbf{x})$,

$$C_\alpha(\mathbf{x}) := C_\alpha(0|\mathbf{x}) \tag{48}$$

identified with the genuine spinor field. As such, the system is off-shell, imposing no differential conditions on $C_\alpha(\mathbf{x})$.

To put the system on shell of a massless field it suffices to impose the constraint on $C(y|\mathbf{x})$

$$\gamma^a{}_\alpha{}^\beta \frac{\partial}{\partial y^a} C_\beta(y|\mathbf{x}) = 0, \tag{49}$$

where γ^a are gamma-matrices,

$$[\gamma^a, \gamma^b] = 2\eta^{ab} Id. \tag{50}$$

The system (45) is the particular case of

$$\mathcal{D}_\mathbf{x} C_\alpha(y|\mathbf{x}) = 0 \tag{51}$$

for $\mathcal{D}_\mathbf{x}$ of the form (37), (38) with the conformal generators

$$P_a = \frac{\partial}{\partial y^a}, \qquad L_{ab} = y_a \frac{\partial}{\partial y^b} - y_b \frac{\partial}{\partial y^a} + \frac{1}{4}[\gamma_a, \gamma_b], \qquad D = y^a \frac{\partial}{\partial y^a} + \Delta, \tag{52}$$

$$K_a = y^2 \frac{\partial}{\partial y^a} - 2 y_a y^b \frac{\partial}{\partial y^b} - 2\Delta y_a + \frac{1}{2} y^b [\gamma_b, \gamma_a]. \tag{53}$$

It is not hard to see that, for

$$\Delta = \frac{d-1}{2}, \tag{54}$$

which is a canonical dimension of a massless spinor, the action of $K^a \gamma_a$ on a y^a-independent element yields zero. This implies that γ^a-transversal polynomials form a submodule $V^{\gamma tr}$ of V^{sp} of all polynomials (46). Analogously to the scalar case, this has a consequence that the space of spinor fields (46) possesses a lowest weight vector $C_\alpha(\mathbf{x})$ identified with the dynamical spinor field and a singular vector $C'_\alpha(\mathbf{x})$ associated with the Dirac operator,

$$C_\alpha(\mathbf{x}) := C_\alpha(0|\mathbf{x}), \qquad C'_\alpha(\mathbf{x}) := \gamma^a{}_\alpha{}^\beta C_{\beta a}(0|\mathbf{x}). \tag{55}$$

The conformally invariant Lagrangian has the form analogous to (44),

$$L^{Di} = \frac{1}{2}\epsilon_{a_1\dots a_d} e^{a_1}(\mathbf{x})\dots e^{a_d}(\mathbf{x})\bar{C}^{\alpha}(\mathbf{x})C'_{\alpha}(\mathbf{x}), \tag{56}$$

where $\bar{C}^{\alpha}(\mathbf{x})$ is the Dirac conjugated spinor. Clearly, in the Cartesian coordinate system (38), Eq. (56) yields usual Dirac Lagrangian for the spinor field $C_{\alpha}(\mathbf{x})$ in d-dimensions.

6 Extension to $(A)dS_{d+1}$

Now we are in a position to extend the d-dimensional singleton systems to AdS_{d+1}. To this end we replace Eqs. (36), (51) by analogous equations

$$D_x C(x) = 0, \qquad D_x C_{\alpha}(x) = 0, \qquad D_x := \mathrm{d}_x + W \tag{57}$$

with W (28). The conformal generators have the form (39), (40) and (52), (53) in the scalar and spinor cases, respectively.

The AdS_{d+1} invariant Lagrangians still have the form (44) for scalar and (56) for spinor but now being d-forms in $(A)dS_{d+1}$ with the fields C and C' rescaled by (25) and E^a (29),

$$L^{Rac} = \frac{1}{2}\epsilon_{a_1\dots a_d} E^{a_1}(x)\dots E^{a_d}(x)C(x)C'(x), \tag{58}$$

$$L^{Di} = \frac{1}{2}\epsilon_{a_1\dots a_d} E^{a_1}(x)\dots E^{a_d}(x)\bar{C}^{\alpha}(x)C'_{\alpha}(x). \tag{59}$$

These Lagrangians are closed and, as a consequence of the general properties of the unfolded equations, invariant up to exact forms (i.e., total derivatives) under the symmetries (4) that leave invariant the background connections, that is global $(A)dS$ symmetries. However, now one has to take into account that Eq. (57) are no longer off-shell reconstructing the dependence of one of the coordinates, namely z, in terms of the others. This is the reason why the seemingly non-invariant form of (58), (59) in view of (29) still respects Lorentz covariance in AdS_{d+1}. In other words, the $d+1$-dimensional Lorentz transformations act on the singleton nonlocally relating fields $\phi(\mathbf{x}, z)$ at different z. As a result, being a local field in d dimensions, from the $d+1$ perspective it is nowhere (equivalently, everywhere).

As mentioned in Sect. 3, the parameter λ, that enters the $(A)dS$ connection by virtue of (27), (28), is real in the AdS case but pure imaginary in dS. Naively, this implies that the Lagrangians (58) and (59) are not Hermitian in the dS case. However, this problem can be resolved by introducing doublets of mutually conjugated fields $C^{\pm}$ and/or $C^{\pm}_{\alpha}$, associated with $\lambda = \pm i\lambda'$ with real λ'. This allows one to consider singletons as fields in the de Sitter space tantamount to dark energy [3]. The modes associated with the evolution along z are either increasing or decreasing that is not too surprising in the expansion regime.

7 Spinor Formulation for 3*d* Singletons

In $2 + 1$ dimensions it is convenient to describe massless fields in terms of spinor indices $\alpha, \beta = 1, 2$. In these terms, local coordinates are $\mathbf{x}^{\alpha\beta} = \mathbf{x}^{\beta\alpha}$ and massless field equations are

$$\frac{\partial}{\partial \mathbf{x}^{\alpha\beta}} \frac{\partial}{\partial \mathbf{x}_{\alpha\beta}} \phi(\mathbf{x}) = 0 \tag{60}$$

and

$$\frac{\partial}{\partial \mathbf{x}^{\alpha\beta}} \phi^{\beta}(\mathbf{x}) = 0 \tag{61}$$

for scalar $\phi(\mathbf{x})$ and spinor $\phi^{\beta}(\mathbf{x})$, respectively. The unfolded massless field equations are formulated [41] both for *Rac* and for *Di* with the aid of auxiliary spinor variables y^+_α, y^-_β obeying

$$[y^-_\alpha, y^+_\beta] = \varepsilon_{\alpha\beta}$$

in terms of the Fock module

$$|\phi(y^+|\mathbf{x})\rangle = \phi(y^+|\mathbf{x})|0\rangle, \quad y^-_\alpha|0\rangle = 0, \quad \phi(y^+|x) = \sum_{n=0}^{\infty} \frac{1}{n!} \phi_{\alpha_1\ldots\alpha_n}(x) y^{+\alpha_1} \ldots y^{+\alpha_n} \tag{62}$$

in the form

$$D_{\mathbf{x}}|\phi(y^+|\mathbf{x})\rangle = 0 \tag{63}$$

with the $o(3, 2) \sim sp(4|\mathbb{R})$ covariant derivative

$$D_{\mathbf{x}} := \mathrm{d}_{\mathbf{x}} + e^{\alpha\beta}(\mathbf{x}) y^{-\alpha} y^{-\beta} + \omega^{\alpha\beta}(\mathbf{x}) y^+_\alpha y^-_\beta + b(\mathbf{x})\{y^{+\alpha}, y^-_\alpha\} + f(\mathbf{x})_{\alpha\beta} y^{+\alpha} y^{+\beta} \tag{64}$$

for some flat connection obeying $D^2_{\mathbf{x}} = 0$. Cartesian coordinate system results from the flat connection

$$e^{\alpha\beta}(\mathbf{x}) = \mathrm{d}\mathbf{x}^{\alpha\beta}, \qquad \omega^{\alpha\beta} = 0, \qquad b = 0, \qquad f_{\alpha\beta} = 0. \tag{65}$$

To describe singleton as a 4*d* field it suffices to add an additional coordinate and extend the flat $o(3, 2)$ connection to the four-dimensional space. Identifying the additional coordinate with the Poincaré coordinate z we set

$$\mathbf{x}^{\alpha\beta} \to x^{\alpha\dot{\alpha}} = \left(\mathbf{x}^{\alpha\dot{\alpha}}, -\frac{i}{2}\epsilon^{\alpha\dot{\alpha}} z^{-1}\right).$$

The unfolded field equations in AdS_4 read as

$$D_x|\phi\rangle = 0 \tag{66}$$

with

$$D_x = \mathrm{d}_x + \frac{i}{z}\mathrm{d}\mathbf{x}^{\alpha\beta} y^-_\alpha y^-_\beta - \frac{\mathrm{d}z}{2z} y^-_\alpha y^{+\alpha}, \qquad \mathrm{d}_x := \mathrm{d}x^{\alpha\dot\beta}\frac{\partial}{\partial x^{\alpha\dot\beta}}$$

describing a flat AdS_4 connection in Poincaré coordinates with

$$e^{\alpha\dot\alpha} = \frac{1}{2z}\mathrm{d}x^{\alpha\dot\alpha}, \qquad \omega^{\alpha\beta} = -\frac{i}{4z}\mathrm{d}\mathbf{x}^{\alpha\beta}, \qquad \bar\omega^{\dot\alpha\dot\beta} = \frac{i}{4z}\mathrm{d}\mathbf{x}^{\dot\alpha\dot\beta}.$$

(For more detail see [42].)

Though unfolded equations are formulated in the $4d$ space-time, the dynamical equations are still three-dimensional

$$\frac{\partial^2}{\partial \mathbf{x}^{\alpha\beta}\partial \mathbf{x}_{\alpha\beta}}\phi(\mathbf{x}, z) = 0$$

at every z. Equation (66) then reconstructs z dependence of $\phi(\mathbf{x}, z)$ in terms of $\phi(\mathbf{x}, z_0)$ at any $z_0 \neq 0$. The system is relativistic in the $4d$ sense since the Lorentz algebra $o(3, 1)$ is a subalgebra of $o(3, 2)$ as well as of $o(4, 1)$. However, the singleton field treated as a relativistic field in the $4d$ space-time is very different from usual local space-time fields since it belongs to an infinite-dimensional $o(3, 1)$ module, namely, the Fock module (62). This is because the $4d$ Lorentz generators $L^{\alpha\beta}$ and $\bar L^{\dot\alpha\dot\beta}$ realized as

$$L_{\alpha\beta} = \frac{1}{2}(y^+_\alpha + iy^-_\alpha)(y^+_\beta + iy^-_\beta), \qquad \bar L_{\dot\alpha\dot\beta} = \frac{1}{2}(y^-_{\dot\alpha} + iy^+_{\dot\alpha})(y^-_{\dot\beta} + iy^+_{\dot\beta}) \tag{67}$$

(see [42]) contain the creation y^+y^+ parts.

To formulate an off-shell Lagrangian description of the $3d$ singleton in terms of spinors one has to introduce along the lines of the $4d$ construction of [43] an additional scalar variable p parameterising the (gamma)traceful components of the off-shell $3d$ fields $(C_\alpha(\mathbf{x}))C(\mathbf{x})$. The off-shell extension will be elaborated elsewhere.

8 Singleton as a Beyond SM Actor

That $4d$ Lorentz algebra acts on the singleton is crucially important making it possible to introduce interaction of the singleton with $4d$ gravity in the Cartan formalism. One can write a Lagrangian as a sum of the singleton Lagrangians with the usual $4d$ matter Lagrangians of the SM, GR, and further extensions,

$$L = L^{Rac} + L^{Di} + L^{SM} + L^{GR} + \cdots . \tag{68}$$

As such, it is a sum of the three-form Lagrangians L^{Rac} and L^{Di} and four-form Lagrangians for the genuine $4d$ local fields contributing to L^{SM} and L^{GR}. Let us stress that since the Lagrangians are closed forms, the respective actions are insensitive to the local variations of the three-cycles over which L^{Rac} and L^{Di} are integrated in the action,

$$S = \int_{\Sigma^3_{Rac}} L^{Rac} + \int_{\Sigma^3_{Di}} L^{Di} + \int_{M^4} (L^{SM} + L^{GR} + \cdots). \tag{69}$$

Because singletons are not local $4d$ fields, their direct scattering effects can unlikely be observable in the collider experiments. This implies that the presence of singletons should not affect the local high-energy $4d$ SM physics, thus avoiding a tension with the available experimental data. The same time, this raises a question whether the singletons may yield observable phenomena whatsoever.

There is an alternative mechanisms that may play a role, however, namely the prominent Flato-Fronsdal theorem [5] stating that the tensor product of two singleton fields (equivalently, their bilocal composits) amounts to the direct sum of $4d$ massless fields including the massless spin two field, i.e., graviton, and a massless scalar in the singlet representation with respect to all inner symmetries,

$$S \bigotimes S = \sum_{s=0}^{\infty} \phi_{s,m=0}(x) = graviton + singlet\ scalar + \cdots . \tag{70}$$

For the extesion of Flato-Fronsdal theorem to the case with inner symmetries see [44, 45] while its harmonic analysis interpretation was given in [46]. A related idea of bilocal fields was also put forward in [47].

That bilinears of singletons contain graviton and an inert massless scalar may have observable consequences via, e.g., an additionally induced gravitational field as well as invisible (dark) matter. In other words, though the $4d$ singleton fields cannot be localized themselves in some region of a galaxy, their nonlinear combinations can induce localizable fields. Of course, to evaluate such effects one has to introduce and analyse interactions. The first step towards gravitational interaction consists of the covariantization of D_x in (66) beyond the flat AdS_{d+1} connection. Note, however, that, as usual in the unfolded dynamics, this is only the first step that demands further extension to respect the formal consistency in the sense of (3) beyond the linearized approximation. The simplest way towards solution to this problem is probably via construction of the unfolded version of $3d$ conformal gravity theory with $3d$ massless matter. The $4d$ singleton theory then will result via the application of the space-time metamorphoses mechanism of Sect. 3 to the $3d$ conformal theory. For some progress in this direction within $3d$ conformal higher-spin gravity see [48, 49], while unfolded formulation of conformal geometry was considered in [50]. The construction of the nonlinear $3d$ conformal higher-spin theory interacting with singletons was proposed in the recent papers [51, 52]. The problem is to evaluate the effect of condensation of the gravitational and scalar fields in the galaxy induced by the singleton.

An intriguing group-theoretic singleton phenomenon found in [14] is that the compactification of eleven-dimensional supergravity on the seven sphere leads to singleton representations along with the usual $4d$ relativistic representations in the spectrum. Moreover, in the squashed seven sphere case singletons were argued to be involved into certain higgsing together with the usual $4d$ unitary representations. Since the field-theoretic interpretation of this phenomenon is still not clear it would be interesting to apply the developed unfolded machinery to its further analysis. The formalism developed in this paper may be most appropriate in this respect since unfolded dynamics essentially maps the representation theory to dynamical equations. Also a potentially useful for better understanding of this phenomenon is the coset space approach to AdS/CFT of [23]. It would be interesting to elaborate more on its relation to the unfolded dynamics approach used in this paper.

Note that Flato-Fronsdal theorem suggests that the tensor product of singletons contains massless fields of all spins. This is a group-theoretic fact true at the free field level that respects higher-spin symmetries. In the situation considered in this paper with singletons interacting with the fields of SM and gravity this is no-longer true and the tail of higher-spin fields is anticipated to be deformed to some kind of effective interactions. On the other hand, since lower-spin relativistic and inner symmetries remain unbroken, the graviton and scalar field contributions to (70) are anticipated to survive.

It is important to stress that, to be dynamically active, singleton should live in the $(A)dS$ space. In other words the proposed construction can only be working in presence of dark energy. More precisely, it works directly in the $sp(4) \sim o(3,2)$ invariant AdS_4 space with the negative cosmological constant Λ while dark energy associated with the dS_4 space has opposite sign of Λ. In dS_4, the singleton Lagrangians contain imaginary unit via $\lambda := \sqrt{-\Lambda}$. However, as discussed in Sect. 6, by an appropriate doubling of fields one can make the model Hermitean. The presence of exponential solutions is natural in the dS_4 regime of expanding Universe. (Note that singleton in the flat space was argued to be a constant [2, 53, 54].)

The presence of complex coefficients in the singleton equations may be related to such a persisting problem in cosmology and high-energy physics as baryon asymmetry. Indeed, these properties fit the prominent Sakharov necessary conditions [55] for baryon asymmetry (for review see, e.g., [56]). Positive cosmological constant may provide a non-equilibrium regime. Moreover, singletons endowed with appropriate inner structure may induce violation of the baryon number conservation. The presence of complex coefficients in the Lagrangian may induce CP violation. Though, because of its $4d$ nonlocality, the singleton field can hardly be seen in usual scattering phenomena, it can affect the formation of a charged matter via a nonlinear manifestation of the Flato-Fronsdal theorem.

A related comment is that being non-localisable in the $4d$ space such fields behave as background charges in the effective field theory if they have some non-zero VEVs responsible for violation of some global symmetries beyond the list of those manifest in the SM. In that case the proposed idea may have some relation to another Dirac's idea of evolution of fundamental (cosmological) constants [57] via evolution of the singleton fields.

9 Conclusion

The goal of this paper is to point out that there is an unusual type of 4*d* relativistic field, the Dirac singleton, that exists in presence of cosmological constant (dark energy). The new result of the paper is the formulation of its dynamics directly in the 4*d* space-time in a way free of the necessity of factorisation of the bulk modes as in the dipole approach of [2]. The singleton matter is unusual in the sense that it is 4*d* non-localisable. This is closely related to the fact that singleton forms an infinite-dimensional representation of the 4*d* Lorentz group, that does not mean, however, that singleton has more degrees of freedom than usual relativistic fields associated with finite-dimensional tensor-spinor Lorentz representations. Just other way around, singleton is essentially a 3*d* conformal field of usual type.

In this respect it is interesting to compare the proposed construction with an alternative recent dark matter candidate suggested by Bogomolny [58] based on another nonstandard relativistic matter proposed by Dirac [59]. The difference is that the latter equation describes infinitely many degrees of freedom in four dimension while singleton, being a 3*d* field, carries less than one usual relativistic field. It should be noted that Flato and Fronsdal (see [7] and references therein) suggested to use singleton as a kind of constituent matter with usual relativistic fields realized as its composites. In this paper singleton is described directly in four dimensions as a relativistic field coexisting with other relativistic fields of usual types.

Note that one can treat analogously other conformal fields in appropriate space-time dimensions. For instance, 4*d* spin-one massless field is conformal and hence can be uplifted to a relativistic singleton-type field in $(A)dS_5$ as discussed in [60]. More general types of singletons associated with unitary conformal fields in even dimensions can also be considered [61, 62].

Let us stress again that, since the singleton field is nonlocal from the 4*d* perspective, it can hardly be observed via a radiation process but can affect the surrounding gravitational field to be observed via the matter motion, that may induce a dark matter-like effect. Another potential application discussed in Sect. 8 is for the baryon asymmetry explanation. More generally, singletons may admit some cosmological manifestations affecting the cosmological evolution as well. This is what the scientific interests of Alexei Starobinsky were focused on.

Acknowledgements I wish to thank Konstantin Alkalaev, Mikhail Alfimov, Andrei Barvinsky, Olga Gelfond, Sergey Godunov, Carlo Iazeolla, Evgeny Ivanov, Andrei Mironov, Nikita Misuna, Roman Nevzorov, Bengt Nilsson, Dmitry Ponomarev, Aleksandr Pukhov, Arkady Tseytlin, Boris Voronov, and Mikhail Vysotsky for useful comments. I am grateful for hospitality to Ofer Aharony, Theoretical High Energy Physics Group of Weizmann Institute of Science where a part of this work has been done. This work was supported by Theoretical Physics and Mathematics Advancement Foundation BASIS Grant No 24-1-1-9-5.

References

1. P.A.M. Dirac, J. Math. Phys. **4**, 901–909 (1963)
2. M. Flato, C. Fronsdal, Commun. Math. Phys. **108**, 469 (1987)
3. V. Sahni, A.A. Starobinsky, Int. J. Mod. Phys. D **9**, 373–444 (2000). arXiv:astro-ph/9904398 [astro-ph]
4. J.B. Ehrman, Proc. Cambridge Philos. Soc. **53**, 290 (1957)
5. M. Flato, C. Fronsdal, Lett. Math. Phys. **2**, 421–426 (1978)
6. A. Starinets, Lett. Math. Phys. **50**, 283–300 (1999). arXiv:math-ph/9809014 [math-ph]
7. M. Flato, C. Fronsdal, D. Sternheimer. arXiv:hep-th/9901043 [hep-th]
8. E. Sezgin, Phys. Lett. B **138**, 57–62 (1984)
9. M.P. Blencowe, M.J. Duff, Phys. Lett. B **203**, 229–236 (1988)
10. H. Nicolai, E. Sezgin, Y. Tanii, Nucl. Phys. B **305**, 483–496 (1988)
11. E. Bergshoeff, A. Salam, E. Sezgin, Y. Tanii, Phys. Lett. B **205**, 237–244 (1988)
12. E. Bergshoeff, E. Sezgin, Y. Tanii, Int. J. Mod. Phys. A **5**, 3599–3616 (1990)
13. E. Bergshoeff, A. Salam, E. Sezgin, Y. Tanii, Nucl. Phys. B **305**, 497–515 (1988)
14. B.E.W. Nilsson, A. Padellaro, C.N. Pope, JHEP **07**, 124 (2019). arXiv:1811.06228 [hep-th]
15. M.J. Duff, B.E.W. Nilsson, C.N. Pope. arXiv:2502.07710 [hep-th]
16. J.M. Maldacena, Adv. Theor. Math. Phys. **2**, 231 (1998). Int. J. Theor. Phys. **38**, 1113 (1999). arXiv:hep-th/9711200
17. S.S. Gubser, I.R. Klebanov, A.M. Polyakov, Phys. Lett. B **428**, 105 (1998). arXiv:hep-th/9802109
18. E. Witten, Adv. Theor. Math. Phys. **2**, 253 (1998). arXiv:hep-th/9802150
19. E. Sezgin, P. Sundell, Nucl. Phys. B **644**, 303–370 (2002) [erratum: Nucl. Phys. B **660**, 403–403 (2003). arXiv:hep-th/0205131 [hep-th]
20. I.R. Klebanov, A.M. Polyakov, Phys. Lett. B **550**, 213 (2002). arxiv:hep-th/0210114
21. M.A. Vasiliev, Lect. Notes Phys. **892**, 227–264 (2015). arXiv:1404.1948 [hep-th]
22. S. Bellucci, E. Ivanov, S. Krivonos, Phys. Rev. D **66**, 086001 (2002) [erratum: Phys. Rev. D **67**, 049901 (2003). arXiv:hep-th/0206126 [hep-th]
23. E. Ivanov, Theor. Math. Phys. **139**, 513–528 (2004). arXiv:hep-th/0305255 [hep-th]
24. M. Flato, C. Fronsdal, Phys. Lett. B **172**, 412–416 (1986)
25. M. Flato, C. Fronsdal, J. Geom. Phys. **5**, 37–61 (1988)
26. M. Flato, C. Fronsdal, J. Math. Phys. **32**, 524–531 (1991)
27. M. Flato, C. Fronsdal, Lett. Math. Phys. **44**, 249–259 (1998)
28. M.A. Vasiliev, Phys. Lett. B **209**, 491–497 (1988)
29. M.A. Vasiliev, Int. J. Geom. Meth. Mod. Phys. **3**, 37–80 (2006). arXiv:hep-th/0504090 [hep-th]
30. M.A. Vasiliev, Phys. Rev. D **66**, 066006 (2002). arXiv:hep-th/0106149 [hep-th]
31. N. Misuna. arXiv:2408.13212 [hep-th]
32. C. Fronsdal, Massless particles, ortosymplectic symmetry and another type of Kaluza-Klein theory, preprint UCLA/85/TEP/10, in *Essays on Supersymmetry*. Mathematical Physics Studies, vol. 8 (Reidel, 1986)
33. I.A. Bandos, J. Lukierski, D.P. Sorokin, Phys. Rev. D **61**, 045002 (2000). arXiv:hep-th/9904109 [hep-th]
34. I. Bandos, X. Bekaert, J.A. de Azcarraga, D. Sorokin, M. Tsulaia, JHEP **05** (2005), 031. arXiv:hep-th/0501113 [hep-th]
35. I.I. Kogan, Phys. Lett. B **458**, 66–72 (1999). arXiv:hep-th/9903162 [hep-th]
36. T. Ohl, C.F. Uhlemann, JHEP **05**, 161 (2012). arXiv:1204.2054 [hep-th]
37. O.V. Shaynkman, M.A. Vasiliev, Theor. Math. Phys. **123**, 683–700 (2000). arXiv:hep-th/0003123 [hep-th]
38. O. Aharony, D. Marolf, M. Rangamani, JHEP **02**, 041 (2011). arXiv:1011.6144 [hep-th]
39. B.E.W. Nilsson. arXiv:1203.5090 [hep-th]
40. O.V. Shaynkman, I.Y. Tipunin, M.A. Vasiliev, Rev. Math. Phys. **18**, 823–886 (2006). arXiv:hep-th/0401086 [hep-th]

41. O.V. Shaynkman, M.A. Vasiliev, Theor. Math. Phys. **128**, 1155–1168 (2001). arXiv:hep-th/0103208 [hep-th]
42. M.A. Vasiliev, J. Phys. A **46**, 214013 (2013). arXiv:1203.5554 [hep-th]
43. N.G. Misuna, JHEP **12**, 172 (2021). arXiv:2012.06570 [hep-th]
44. S.E. Konshtein, M.A. Vasiliev, Nucl. Phys. B **312**, 402–418 (1989)
45. S.E. Konstein, M.A. Vasiliev, Nucl. Phys. B **331**, 475–499 (1990)
46. C. Iazeolla, P. Sundell, JHEP **10**, 022 (2008). arXiv:0806.1942 [hep-th]
47. R. de Mello Koch, A. Jevicki, K. Jin, J.P. Rodrigues, Phys. Rev. D **83**, 025006 (2011)
48. B.E.W. Nilsson, JHEP **09**, 078 (2015). arXiv:1312.5883 [hep-th]
49. B.E.W. Nilsson, JHEP **08**, 142 (2016). arXiv:1506.03328 [hep-th]
50. E. Joung, M.G. Kim, Y. Kim, JHEP **12**, 092 (2021). arXiv:2108.05535 [hep-th]
51. F. Diaz, C. Iazeolla, P. Sundell, JHEP **09**, 109 (2024). arXiv:2403.02283 [hep-th]
52. F. Diaz, C. Iazeolla, P. Sundell, JHEP **10**, 066 (2024). arXiv:2403.02301 [hep-th]
53. D. Ponomarev, JHEP **06**, 055 (2021). arXiv:2104.02770 [hep-th]
54. X. Bekaert, A. Campoleoni, S. Pekar, Phys. Lett. B **838**, 137734 (2023). arXiv:2211.16498 [hep-th]
55. A.D. Sakharov, Pisma Zh. Eksp. Teor. Fiz. **5**, 32–35 (1967)
56. V.A. Rubakov, M.E. Shaposhnikov, Usp. Fiz. Nauk **166**, 493–537 (1996). arXiv:hep-ph/9603208 [hep-ph]
57. P.A.M. Dirac, Nature **139**, 323 (1937)
58. E. Bogomolny, Universe **10**(5), 222 (2024). arXiv:2406.01654 [hep-th]
59. P.A.M. Dirac, Proc. Roy. Soc. Lond. A **322**, 435–445 (1971)
60. S. Ferrara, C. Fronsdal, Class. Quant. Grav. **15**, 2153–2164 (1998). arXiv:hep-th/9712239 [hep-th]
61. X. Bekaert, M. Grigoriev, SIGMA **6**, 038 (2010). arXiv:0907.3195 [hep-th]
62. X. Bekaert, arXiv:1111.4554 [math-ph]

Chaos in Horndeski Cosmologies

Mikhail S. Volkov

Abstract We analyze how a scalar field can affect the chaotic behaviour of homogeneous and isotropic Bianchi IX cosmologies. It is known that a massless, minimally coupled scalar field removes the chaos. However, in more general Horndeski theories, the situation is more complex. We find that in shift-symmetric K-essence theories, chaos persists if the scalar field contribution to the initial-value constraint is *subleading* compared to that of the anisotropies. In this case, solutions oscillate as they approach the singularity, just as in the vacuum case, and a similar behaviour is found when a non-minimal coupling is included. If the scalar field contribution is not subleading, then chaos is removed and the singularity is approached smoothly. An unusual and entirely new result appears when changing the sign in front of the scalar kinetic term, yielding the theory of a phantom scalar. If the scalar field is subleading, then solutions remain chaotic and oscillate when approaching the singularity, as before. However, if the scalar field is not subleading, solutions are also chaotic, but the spacetime singularity disappears, and the universe behaves as an apparently infinite sequence of anisotropic bounces. The spatial volume then oscillates within finite bounds, never reaching zero, while the amplitudes and positions of these oscillations appear completely random. To the best of our knowledge, this type of chaos has never been described.

1 Introduction

I had the remarkable opportunity to work with Alexei Starobinsky during the last few years of his life, when we were members of a joint research project devoted to cosmologies in Horndeski theory. Together with other colleagues, we published three papers, starting from the homogeneous and isotropic case [1] and then moving on to anisotropic models [2, 3]. We obtained an interesting result: if the scalar field

M. S. Volkov (✉)
Institut Denis Poisson, UMR—CNRS 7013, Université de Tours, Tours, France
e-mail: mvolkov@univ-tours.fr

A. Barvinsky and A. Kamenshchik (eds.), *Open Issues in Gravitation and Cosmology*, Fundamental Theories of Physics 225, https://doi.org/10.1007/978-3-032-15703-4_18

is non-minimally coupled, then in the Bianchi I case the anisotropies are suppressed (screened) and approach zero at the singularity instead of being amplified [2].

However, including spatial curvature removes the screening [2]. The anisotropies then grow toward the singularity, and the Bianchi IX cosmology with a Horndeski scalar exhibits an oscillatory behaviour similar to the "cosmological billiard" of Belinskii–Khalatnikov–Lifshitz (BKL) [4] or the "mixmaster universe" of Misner [5] (see [6] for a recent discussion). The billiard was originally discovered in the vacuum theory and is known to be chaotic, but similar behaviour persists when matter is added, provided that its energy density does not grow too rapidly near the singularity.

At the same time, it is known that adding a massless, minimally coupled scalar field suppresses the oscillations: the singularity is then approached smoothly, and the chaos disappears [7]. However, chaos persists for solutions with a non-minimally coupled scalar studied in [2]. Since in both cases the scalar is described by Horndeski models, one may naturally ask when the Horndeski theory admits chaotic solutions and when it does not.

I had intended to propose this as the next research project to Alexei Starobinsky, but we never had the opportunity to discuss it together. The project therefore remained only an idea, and it is only now that I am able to report some progress in this direction.

To clarify the issue, we shall consider below anisotropic cosmologies with a scalar field described by subclasses of Horndeski theory. In most cases, we examine the simplest shift-symmetric K-essence model. A natural conjecture is that chaos should persist if the scalar field contribution to the initial-value constraint is subleading compared to the anisotropy contribution, which scales as $\propto 1/\mathrm{a}^6$. For a minimally coupled massless scalar, its contribution also scales as $\propto 1/\mathrm{a}^6$, and therefore it is not subleading.

To test this conjecture, we generate numerical solutions and check whether they oscillate as they approach the singularity. Of course, this does not constitute a rigorous proof that the solutions are truly chaotic or oscillate infinitely many times, since numerical methods can only produce a finite number of oscillations before numerical errors grow rapidly. However, in some cases the solution profiles closely resemble those of the vacuum case, showing intervals during which one of the three scale factors is monotonic while the other two oscillate. In other cases, the oscillations cease quickly and all three scale factors become monotonic.

We therefore assume that a solution is chaotic if the numerical evolution displays oscillations that appear not to terminate. This is, of course, a very simple and naive definition of chaos, and we do not attempt to address numerous subtle aspects of chaotic dynamics (some of which are discussed in [8]). Nevertheless, we verify that even an arbitrarily small change in the initial conditions completely reshuffles the positions and amplitudes of the oscillation peaks. This sensitivity implies positive Lyapunov exponents, confirming chaotic behaviour.

We find that the above conjecture holds for standard K-essence with a positive kinetic energy of the scalar field. The scalar energy scales as $1/\mathrm{a}^\gamma$, where γ is a parameter of the theory. For $\gamma < 6$, the scalar contribution is subleading compared to the anisotropy effect, and the solutions oscillate as they approach the singularity. For

$\gamma \geq 6$, the scalar effect is not subleading, the oscillations cease, and the singularity is approached smoothly.

However, a completely new and unexpected feature arises when changing the sign of the scalar kinetic term. This yields the theory of a phantom scalar field whose energy scales as $-1/a^{\gamma}$. For $\gamma < 6$, the solutions oscillate as they approach the singularity and display chaos of the usual BKL type. Surprisingly, for $\gamma \geq 6$, they also oscillate and appear chaotic, but the spacetime singularity disappears.

The solution then becomes globally regular and exhibits a sequence of anisotropic bounces with chaotically distributed minimal and maximal sizes. Within each bounce, the three anisotropy amplitudes undergo a finite number of oscillations, but the number of bounces is seemingly infinite, and the universe extends indefinitely into both the past and future—it is eternal (see [9] for other examples of eternal cosmologies). Therefore, our conjecture fails in the case of a phantom scalar, but a new form of chaotic behaviour emerges that, to the best of our knowledge, has never been described.

Finally, we analyze the theory with a non-minimal coupling considered in [2]. The solutions approach the singularity and oscillate as in the BKL case, and we verify that the scalar contribution is subleading, which once again supports the conjecture.

In our analysis, we combine analytical and numerical methods. Of course, one cannot describe Bianchi IX solutions fully analytically, but they can be piecewise approximated by the "Kasner epochs" described by Bianchi I solutions. In the vacuum case, the Bianchi I solution (Kasner metric) is characterized by three exponents p_1, p_2, p_3, one of which is always negative, so that one of the three scale factors $a_k \propto t^{p_k}$ grows toward the singularity. This feature also appears in vacuum Bianchi IX solutions: the scale factors oscillate, and there is always one that grows toward the singularity.

When a massless scalar is included, one of the three exponents p_k in the Bianchi I case can be negative, but it also becomes possible for all three to be positive. It is this latter property that halts the oscillations in the Bianchi IX case. The values of p_k change after each oscillation cycle, and the oscillations continue as long as one of the three p_k is negative. However, sooner or later the system inevitably reaches a state in which all three p_k are positive, and then all three scale factors become monotonic, causing the oscillations to stop [7]. In what follows, we review this property.

We then consider the Bianchi I solutions with a phantom scalar field and find that among the three exponents p_k, *either one or two* can be negative, and all three can never be positive. This implies that the Bianchi IX solutions cannot become smooth, since all three scale factors cannot be monotonic, and either one or two of them grow toward the past. These properties are confirmed by the numerical solutions.

2 The Theory

We consider the theory

$$S = \frac{1}{2}\int \left(\mu\, R + \epsilon\, X^n - \alpha\, G_{\mu\nu}\,\partial^\mu\phi\,\partial^\nu\phi\right)\sqrt{-g}\,\mathrm{d}^4x \equiv \frac{1}{2}\int L\,\mathrm{d}^4x\,. \quad (1)$$

Here R and $G_{\mu\nu}$ are the Ricci scalar and Einstein tensor, and $X = -\partial^\mu\phi\,\partial_\mu\phi$, while μ is related to the Planck mass. The scalar field dynamics is determined by the K-essence term ϵX^n and by the term proportional to α, which provides the non-minimal coupling. We refer to n as the K-essence index.

The parameter ϵ takes the values $\epsilon = +1$, corresponding to an ordinary scalar field, or $\epsilon = -1$, corresponding to a phantom scalar. For $n = 2$, the theory (1) reduces to the one studied in [2] (up to a cosmological term, which we do not include).

We shall assume the scalar field to depend only on time, $\phi = \phi(t)$, and consider homogeneous and anisotropic metrics of the form

$$\mathrm{d}s^2 = -N^2(t)\,\mathrm{d}t^2 + \frac{1}{4}\left[\mathrm{a}_1^2(t)\,\omega^1\otimes\omega^1 + \mathrm{a}_2^2(t)\,\omega^2\otimes\omega^2 + \mathrm{a}_3^2(t)\,\omega^3\otimes\omega^3\right]. \quad (2)$$

Here $\omega^a = \omega^a_k\,\mathrm{d}x^k$ are spatial one-forms. In the simplest Bianchi I case, one has $\omega^a = \mathrm{d}x^a$, while in the Bianchi IX case ω^a are invariant forms on S^3 satisfying $\mathrm{d}\omega^a + \epsilon_{abc}\,\omega^b\wedge\omega^c = 0$. The factor 1/4 in (2) is introduced to match the convention used in [2].

Injecting this into (1) yields

$$L = \mu N\,(-T + V) - \alpha\,\frac{\dot\phi^2}{2N}\,(T + V) + \epsilon\left(\frac{\dot\phi}{N}\right)^n N\,\mathrm{a}_1\mathrm{a}_2\mathrm{a}_3\,, \quad (3)$$

where

$$T = \frac{1}{4N^2}\,(\mathrm{a}_1\,\dot{\mathrm{a}}_2\,\dot{\mathrm{a}}_3 + \mathrm{a}_2\,\dot{\mathrm{a}}_1\,\dot{\mathrm{a}}_3 + \mathrm{a}_3\,\dot{\mathrm{a}}_1\,\dot{\mathrm{a}}_2) \equiv \frac{3}{4}\,\mathrm{a}_1\mathrm{a}_2\mathrm{a}_3 \times y\,, \quad (4)$$

and in the Bianchi IX case one has

$$V = -\frac{(\mathrm{a}_1+\mathrm{a}_2+\mathrm{a}_3)(\mathrm{a}_1+\mathrm{a}_2-\mathrm{a}_3)(\mathrm{a}_1-\mathrm{a}_2+\mathrm{a}_3)(\mathrm{a}_1-\mathrm{a}_2-\mathrm{a}_3)}{4\mathrm{a}_1\mathrm{a}_2\mathrm{a}_3}$$
$$\equiv \frac{3}{4}\,(\mathrm{a}_2\mathrm{a}_2\mathrm{a}_3)^{1/3} \times \mathcal{K}, \quad (5)$$

whereas in the Bianchi I case $V = 0$. Setting

$$\mathrm{a}_1 = \mathrm{a}\,e^{\beta_+ + \sqrt{3}\beta_-}, \qquad \mathrm{a}_2 = \mathrm{a}\,e^{\beta_+ - \sqrt{3}\beta_-}, \qquad \mathrm{a}_3 = \mathrm{a}\,e^{-2\beta_+} \Rightarrow \mathrm{a} = \mathrm{a}_1\mathrm{a}_2\mathrm{a}_3\,, \quad (6)$$

the Lagrangian in (3) assumes the form

$$8L = 6\mu \mathrm{a}^3 N \left(\frac{\mathcal{K}}{\mathrm{a}^2} - y\right) - \frac{3\alpha \mathrm{a}^3}{N}\dot{\phi}^2 \left(y + \frac{\mathcal{K}}{\mathrm{a}^2}\right) + \epsilon \left(\frac{\dot{\phi}}{N}\right)^n N\mathrm{a}^3, \tag{7}$$

with

$$\begin{aligned} y &= \frac{1}{N^2}\left(\frac{\dot{\mathrm{a}}^2}{\mathrm{a}^2} - \dot{\beta}_+^2 - \dot{\beta}_-^2\right), \\ \mathcal{K} &= -\frac{1}{3}e^{-8\beta_+}\left(4e^{6\beta_+}\cosh^2(\sqrt{3}\beta_-) - 1\right)\left(4e^{6\beta_+}\sinh^2(\sqrt{3}\beta_-) - 1\right). \end{aligned} \tag{8}$$

Varying the Lagrangian then gives equations

$$3\mu\left(y + \frac{\mathcal{K}}{\mathrm{a}^2}\right) + \frac{3}{2}\alpha\psi^2\left(3y + \frac{\mathcal{K}}{\mathrm{a}^2}\right) - \epsilon\frac{n-1}{2}\psi \equiv \mathcal{C} = 0, \tag{9}$$

$$\frac{1}{\mathrm{a}^2 N}\left(\frac{\sigma_+ \mathrm{a}\dot{\mathrm{a}}}{N}\right)^{\cdot} = \frac{\sigma_+}{2N^2}\left(\frac{\dot{a}^2}{a^2} - 3\dot{\beta}_+^2 - 3\dot{\beta}_-^2\right) - \frac{\sigma_- \mathcal{K}}{2\mathrm{a}^2} - \frac{\epsilon}{2}\psi^n, \tag{10}$$

$$\frac{1}{N}\left(\frac{\sigma_+ \mathrm{a}^3}{N}\dot{\beta}_\pm\right)^{\cdot} = \frac{\sigma_- \mathrm{a}}{2}\frac{\partial \mathcal{K}}{\partial \beta_\pm}, \tag{11}$$

$$\mathrm{a}^3\left(-\frac{6\alpha}{n\epsilon}\left(y + \frac{\mathcal{K}}{\mathrm{a}^2}\right) + \psi^{n-2}\right)\psi = \sqrt{6\mu}\, C, \tag{12}$$

with

$$\psi = \frac{\dot{\phi}}{N}, \qquad \sigma_\pm = 2\mu \pm \alpha\psi^2. \tag{13}$$

Equation (9) is the constraint $\mathcal{C} = 0$, which restricts the initial values. This is a first integral whose derivative vanishes due to the remaining Eqs. (10)–(12), $\dot{\mathcal{C}} = 0$, so it is sufficient to impose this constraint only at the initial moment of time. Equation (12) is also a first integral, arising from the invariance of the theory under shifts

$$\phi \to \phi + const, \tag{14}$$

which implies conservation of the scalar charge—the integration constant C in (12).

The function N is a gauge parameter that enters the equations only through the combination

$$\frac{1}{N}\frac{\mathrm{d}}{\mathrm{d}t} \equiv \frac{\mathrm{d}}{\mathrm{d}x}, \tag{15}$$

so different choices of N correspond to different time coordinates. In our numerical analysis, we use two gauges, $N = 1$ and $N = \mathrm{a}^3$, and it turns out that functions such as $\beta_\pm(a)$ or $\psi(a)$ do not depend of the gauge choice. In some cases, the functions are shown against the physical time, defined in a gauge-invariant way as

$$t = \int_0^x \frac{dx}{N}. \tag{16}$$

To solve the equations numerically, we consider the second-order equations (10) and (11), together with the derivative of (12). Resolving with respect to the highest derivatives yields the system

$$\begin{aligned}\ddot{a} =& A(N, a, \dot{a}, \beta_\pm, \dot{\beta}_\pm, \psi),\\ \ddot{\beta}_\pm =& B_\pm(N, a, \dot{a}, \beta_\pm, \dot{\beta}_\pm, \psi),\\ \dot{\psi} =& F(N, a, \dot{a}, \beta_\pm, \dot{\beta}_\pm, \psi).\end{aligned} \tag{17}$$

Setting, for example, $N = 1$, and choosing the initial values at $t = 0$,

$$a(0), \quad \dot{a}(0), \quad \beta_\pm(0), \quad \dot{\beta}_\pm(0), \quad \psi(0), \tag{18}$$

which mush satisfy the constraint (9), Eq. (17) can be integrated numerically toward either negative of positive values of t. During the integration, the constraint and the scalar charge should be monitored to ensure that

$$\mathcal{C} = 0, \quad C = const. \tag{19}$$

The integration continues as long as the value of $\mathcal{C}$ remains small, and stops when it starts to grow rapidly.

Before integrating the equations numerically, some preliminary analysis is required. All formulas below correspond to the gauge choice $N = 1$.

3 Bianchi I, Vacuum Solution

It is consistent to set $\psi = C = 0$ in the equations. In the Bianchi I case, when $\mathcal{K} = 0$, the equations yield

$$\frac{\dot{a}^2}{a^2} = \dot{\beta}_+^2 + \dot{\beta}_-^2, \quad \dot{\beta}_\pm = \frac{\mathcal{B}_\pm}{a^3}, \tag{20}$$

with $\mathcal{B}_\pm$ being integration constants. Denoting

$$\mathcal{B} = \sqrt{\mathcal{B}_+^2 + \mathcal{B}_-^2} \tag{21}$$

and integrating, one obtains

$$a^3 = 3\mathcal{B}\,t, \qquad \beta_\pm = \frac{\mathcal{B}_\pm}{3\mathcal{B}} \ln(t), \tag{22}$$

which corresponds to the Kasner metric, for which $a_k \propto t^{p_k}$ with

$$p_1 = \frac{1}{3} + \frac{\mathcal{B}_+ + \sqrt{3}\,\mathcal{B}_-}{3\mathcal{B}}, \qquad p_2 = \frac{1}{3} + \frac{\mathcal{B}_+ - \sqrt{3}\,\mathcal{B}_-}{3\mathcal{B}}, \qquad p_3 = \frac{1}{3} - \frac{2\mathcal{B}_+}{3\mathcal{B}}. \quad (23)$$

One has

$$p_1 + p_2 + p_3 = p_1^2 + p_2^2 + p_3^2 = 1. \quad (24)$$

These relations imply that one of the three coefficients p_k is negative while the other two are positive. Therefore, one of the three metric functions $a_k(t)$ grows as $t \to 0$, while the other two decrease. All possible values of p_k can be parameterized as follows [4]:

$$p_1(u) = \frac{-u}{1+u+u^2}, \qquad p_2(u) = \frac{1+u}{1+u+u^2}, \qquad p_3(u) = \frac{u(1+u)}{1+u+u^2}, \quad (25)$$

where $0 \leq u \leq 1$.

4 Bianchi I with a Massless Scalar Field

Let us include the scalar field, assuming that it is minimally coupled (hence $\alpha = 0$) and that the K-essence index $n = 2$. The equations reduce to

$$\frac{\dot{a}^2}{a^2} = \dot{\beta}_+^2 + \dot{\beta}_-^2 + \frac{\epsilon}{6\mu}\,\psi^2, \qquad \dot{\beta}_\pm = \frac{\mathcal{B}_\pm}{a^3}, \qquad \psi = \frac{\sqrt{6\mu}\,C}{a^3}, \quad (26)$$

and therefore

$$\frac{\dot{a}^2}{a^2} - \frac{\mathcal{B}^2 + \epsilon\, C^2}{a^6} = 0. \quad (27)$$

If we interpret this as a sum of kinetic and potential terms, then it follows that the anisotropy contribution to the potential, $-\mathcal{B}^2/a^6$, provides an attraction toward singularity at $a = 0$. The scalar contribution $-\epsilon C^2/a^6$ is also attractive if $\epsilon = 1$, but becomes repulsive when $\epsilon = -1$.

Integrating, one obtains

$$a^3 = 3\sqrt{\mathcal{B}^2 + \epsilon\, C^2}\, t = \frac{\sqrt{6}C}{q}\, t, \qquad \dot{\beta}_\pm = \frac{q}{\sqrt{6}C}\, \frac{\mathcal{B}_\pm}{t}, \qquad \psi = \frac{\sqrt{\mu}\, q}{t}, \qquad a_k \propto t^{p_k}, \quad (28)$$

where the parameter q is related to the scalar charge C,

$$q = \frac{\sqrt{2/3}\, C}{\sqrt{\mathcal{B}^2 + \epsilon\, C^2}} \Leftrightarrow C = \frac{q\,\mathcal{B}}{\sqrt{2/3 - \epsilon\, q^2}}. \quad (29)$$

One has

$$p_1 = \frac{1}{3} + \frac{q}{\sqrt{6}\,C}\,(\mathcal{B}_+ + \sqrt{3}\mathcal{B}_-), \qquad p_2 = \frac{1}{3} + \frac{q}{\sqrt{6}\,C}\,(\mathcal{B}_+ - \sqrt{3}\mathcal{B}_-),$$
$$p_3 = \frac{1}{3} - \frac{2q}{\sqrt{6}\,C}\,\mathcal{B}_+, \tag{30}$$

which implies

$$p_1 + p_2 + p_3 = 1, \qquad p_1^2 + p_2^2 + p_3^2 = 1 - \epsilon\, q^2, \tag{31}$$

where the principal difference compared with (24) is the term $\epsilon\, q^2$ appearing on the right-hand side. In this case there is also a parametric representation similar to (25) [7],

$$\begin{aligned} p_1(u) &= \frac{-u}{1+u+u^2}, \\ p_2(u) &= \frac{1+u}{1+u+u^2}\left(u - \frac{u-1}{2}(1-\sqrt{1-B})\right), \\ p_3(u) &= \frac{1+u}{1+u+u^2}\left(1 + \frac{u-1}{2}(1-\sqrt{1-B})\right), \\ B &= 2\epsilon\, q^2\, \frac{(1+u+u^2)^2}{(u^2-1)^2}. \end{aligned} \tag{32}$$

One has

$$p_1\left(\frac{1}{u}\right) = p_1(u), \qquad p_2\left(\frac{1}{u}\right) = p_3(u), \qquad p_3\left(\frac{1}{u}\right) = p_2(u), \tag{33}$$

therefore one can assume that $u \in [-1, 1]$. Equation (32) reduces to (25) if the scalar charge C vanishes. Notice also that choosing a value of u in (32) determines the anisotropy parameters $\mathcal{B}_\pm$ in (30) as follows:

$$\mathcal{B}_+ = \frac{C}{\sqrt{2}q}(p_1(u) - p_2(u)), \qquad \mathcal{B}_- = \frac{C}{\sqrt{6}q}(1 - 3p_3(u)). \tag{34}$$

Suppose now that $C \neq 0$ and consider separately the two cases $\epsilon = +1$ and $\epsilon = -1$.

4.1 Bianchi I with an Ordinary Scalar Field, $\epsilon = 1$

In this case, as shown by (29), when the scalar charge varies in the interval $C \in [0, \infty)$, one has $q^2 \in [0, 2/3)$. The values of $p_k(u)$ for $q \neq 0$ are shown in Fig. 1, and one can see that all three of them exist only if the range of u is restricted to the interior

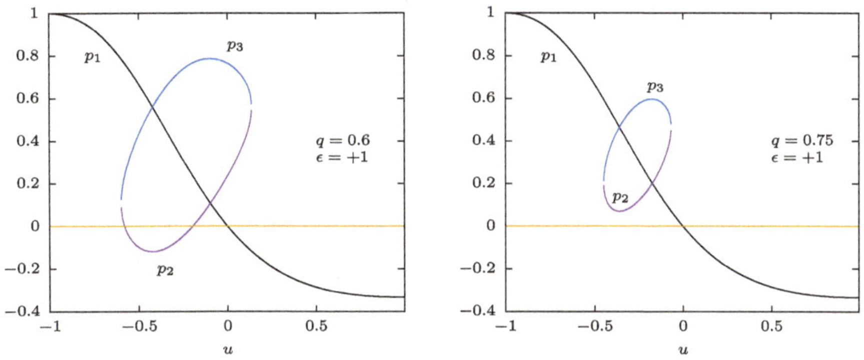

Fig. 1 Left: Coefficients $p_k(u)$ for $q = 0.6$, $\epsilon = +1$. All three $p_k(u)$'s exist for $u \in (-0.596, 0)$, and they are all positive for $u \in (-0.596, -0.578)$ and for $u \in (-0.198, 0)$. Right: Coefficients for $q = 0.75$; they exist for $u \in (-0.449, -0.065)$ and are all positive. In the limit $q \to \sqrt{2/3} \approx 0.816$, the oval formed by p_2 and p_3 shrinks to the point at $u = -2 + \sqrt{3} \approx -0.267$, and then $p_1 = p_2 = p_3 = 1/3$

of the oval region delimited by $p_2(u)$ and $p_3(u)$, hence $u \in [u_{\text{min}}(q), u_{\text{max}}(q)]$. This interval of allowed values shrinks as q grows, so that for $q = 0$ one has $u \in [-1, 1]$; if $q = 1/2$ then $u \in [-0.67, 0.25]$; and when $q \to \sqrt{2/3}$ the interval shrinks to a single point $u = -2 + \sqrt{3} \approx -0.267$.

When q is small, then for most values of u from the interval $[u_{\text{min}}(q), u_{\text{max}}(q)]$ one of the three $p_k(u)$ is negative, while the other two are positive. However, there are always values of u for which all three p_k's are positive, as seen in Fig. 1 (left panel). For $q > 1/\sqrt{2}$, all three p_k's are always positive, as shown in Fig. 1 (right panel). If all three p_k's are positive, then all three scale factors in the metric

$$\mathrm{a}_k \propto t^{p_k} \tag{35}$$

approach zero for $t \to 0$. As we shall see below, this eliminates the chaotic solutions in the Bianchi IX case. Notice also that the scale factor in (28),

$$\mathrm{a}^3 = 3\sqrt{\mathcal{B}^2 + \mathcal{C}^2}\,, \tag{36}$$

admits the isotropic limit where $\mathcal{B} \to 0$.

4.2 Bianchi I with a Phantom Scalar Field, $\epsilon = -1$

The phantom scalar provides a repulsion, and Eq. (27) assumes the form

$$\frac{\dot{\mathrm{a}}^2}{\mathrm{a}^2} = \frac{\mathcal{B}^2 - \mathcal{C}^2}{\mathrm{a}^6}, \tag{37}$$

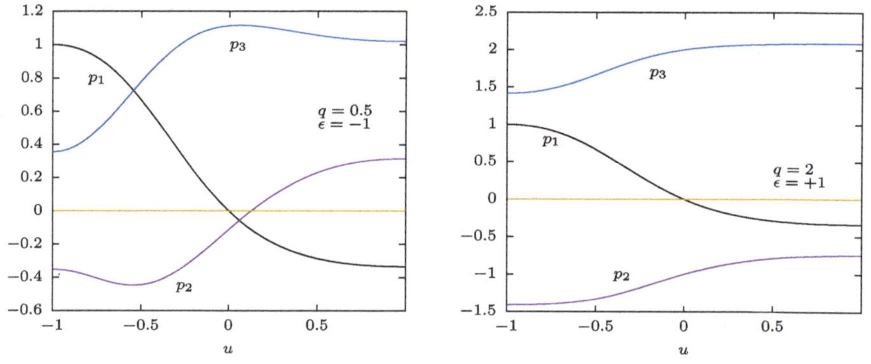

Fig. 2 Coefficients $p_k(u)$ for $\epsilon = -1$ and for either $q = 0.5$ (left) or $q = 2$ (right). They never all become positive, but two of them can be negative simultaneously

whose solution is

$$\mathrm{a}^3 = 3\sqrt{\mathcal{B}^2 - C^2}\, t\,. \tag{38}$$

The solution exists only if anisotropies do not vanish and the scalar charge does not exceed the maximal value, $|C| < \mathcal{B}$. When $|C|$ ranges within the interval $[0, \mathcal{B})$, one has $q^2 \in [0, \infty)$. Values of $p_k(u)$ for different q are shown in Fig. 2, where we see that the interval of allowed u is always $[-1, 1]$. In addition, we see that *either one or two* of the three $p_k(u)$'s are negative, and all three can never be positive. This has important consequences in the Bianchi IX case.

Summarizing, the solution is similar to the Kasner metric, with

$$\mathrm{a}_k \propto t^{p_k}, \tag{39}$$

where either one or two of the three a_k's grow toward the past.

5 Bianchi IX, Vacuum Solutions

Let us examine how the Bianchi I solutions are modified by the presence of the spatial curvature. If $\mathcal{K} \neq 0$ but $\psi = 0$, then the constraint equation (9) becomes

$$\frac{\dot{\mathrm{a}}^2}{\mathrm{a}^2} + \frac{\mathcal{K}}{\mathrm{a}^2} = \dot{\beta}_+^2 + \dot{\beta}_-^2\,. \tag{40}$$

The function $\mathcal{K}(\beta_+, \beta_-)$ defined in (8) takes values in the range

$$\mathcal{K}(\beta_+, \beta_-) \in (-\infty, 1] \quad \text{with}\ 1 = \mathcal{K}(0, 0). \tag{41}$$

We assume that at the initial moment $t = 0$ one has

$$\mathrm{a}(0) = 1, \qquad \beta_\pm(0) = 0, \qquad \dot\beta_+(0) = \mathcal{B}\sin\gamma, \qquad \dot\beta_-(0) = \mathcal{B}\cos\gamma, \tag{42}$$

where $\mathcal{B}$ and γ are constants. The constraint (40) is satisfied if

$$\dot{\mathrm{a}}(0) = 0, \qquad \mathcal{B} = 1. \tag{43}$$

By choosing γ, we use (42) and (43) as initial conditions for second-order equations (10) and (11). Their solution determines $\mathrm{a}(t)$ and $\beta_\pm(t)$, and the three amplitudes $\mathrm{a}_k(t)$.

Starting at $t = 0$, we integrate into the $t < 0$ region. The system approaches a singularity such that the three amplitudes a_k oscillate, but one of them always grows toward the past, while the product $V = \mathrm{a}_1\mathrm{a}_2\mathrm{a}_3$ decreases monotonically and approaches zero. The same patterns is observed when we integrate toward positive values of t, even if we choose in (43) $\mathcal{B} > 1$ and $\dot{\mathrm{a}}(0) \neq 0$. Therefore, the spacetime contains singularities both in the past and in the future.

We choose the value $\gamma = 3$ to illustrate the typical solution in Fig. 3. The figure shows the characteristic "billiard" behaviour: the singularity is approached through a sequence of oscillation cycles (Kasner epochs). During each cycle, one has $\partial\mathcal{K}/\partial\beta_\pm \approx 0$, and the geometry is approximately described by the Kasner metric, with $\mathrm{a}_k \propto t_k^p$ and $p_k = p_k(u)$ for some u. The term $\partial\mathcal{K}/\partial\beta_\pm$ becomes significant only at the end of each cycle, where u changes, initiating the next oscillation cycle with new values of u and $p_k(u)$.

As seen in Fig. 3 (right panel), the scale factor a is a monotonic function of t, so $\ln(\mathrm{a})$ can be used as a convenient time coordinate (left panel). Within a given interval of $\ln(\mathrm{a})$, one of the three amplitudes a_k has an exponent p_k that remains positive, making this amplitude monotonic. For example, a_2 is monotonic for $\ln(\mathrm{a}) \in$

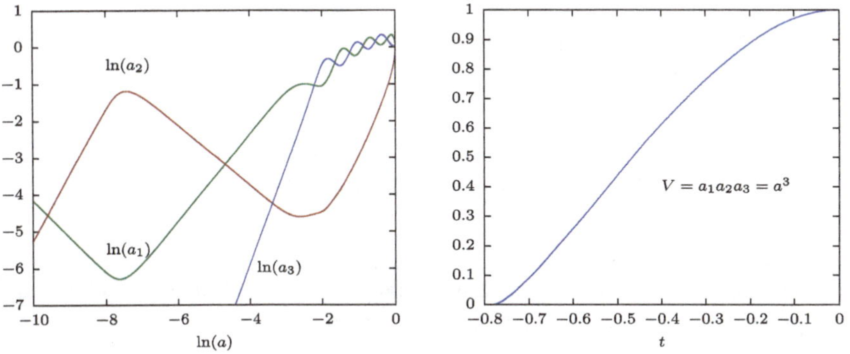

Fig. 3 The vacuum Bianchi IX solution for the initial data (42), (43) with $\gamma = 3$. The three amplitudes a_k oscillate, exhibiting the characteristic "billiard" behaviour. Their product $V = \mathrm{a}_1\mathrm{a}_2\mathrm{a}_3 = \mathrm{a}^3$ is monotonic and approaches zero at the singularity

$(-2, 0)$ in Fig. 3. The other two amplitudes exhibit several oscillations over this range, forming a "braid" pattern. After each oscillation, their exponents interchange signs, as illustrated by a_1 and a_3 for $\ln(a) \in (-2, 0)$. Outside this interval, one of the previously oscillating amplitudes becomes monotonic, e.g., $a_3(t)$ for $\ln(a) < -2$ in Fig. 3, while the remaining two begin to oscillate.

The solutions are chaotic, and the changes of u after each oscillation cycle follow a statistical pattern [4]. Although the approach to the singularity involves infinitely many oscillations, only a finite number of them can be captured in numerical simulations.

A similar oscillatory regime persists also when matter is included, provided its energy density grows more slowly than $1/a^6$ as $a \to 0$, so that its contribution to the constraint (20) remains subdominant compared to the anisotropies. This applies, for instance, to dust or a perfect fluid. However, the presence of a massless scalar field destroys the chaotic behaviour [7].

6 Bianchi IX with a Massless Scalar Field

Including the massless scalar field, the initial value constraint (9) becomes

$$\frac{\dot{a}^2}{a^2} + \frac{\mathcal{K}}{a^2} = \dot{\beta}_+^2 + \dot{\beta}_-^2 + \frac{\epsilon}{6\mu}\,\dot{\psi}^2. \tag{44}$$

This relation fixes the initial conditions, which we choose in the form analogous to (42),(43):

$$\begin{aligned} a(0) =& 1, \quad && \beta_\pm(0) = 0, \quad && \dot{\beta}_+(0) = \mathcal{B}\sin\gamma, \quad && \dot{\beta}_-(0) = \mathcal{B}\cos\gamma, \\ \dot{a}(0) =& 0, && \dot{\psi} = \sqrt{6\mu}C, && \mathcal{B} = \sqrt{1-\epsilon\, C^2}, && \end{aligned} \tag{45}$$

where C is the scalar charge. It turns out that solutions behave very differently depending on whether $\epsilon = +1$ or $\epsilon = -1$.

6.1 *Bianchi IX with an Ordinary Scalar, $\epsilon = 1$*

In this case, the scalar charge satisfies $|C| \in [0, 1]$. If $|C| = 1$, then $\mathcal{B} = 0$, and both the anisotropies and their first derivatives vanish at the initial time, so they remain zero for all times. The constraint equation (44) then reduces to

$$\frac{\dot{a}^2}{a^2} + \frac{1}{a^2} = \frac{1}{a^6}, \tag{46}$$

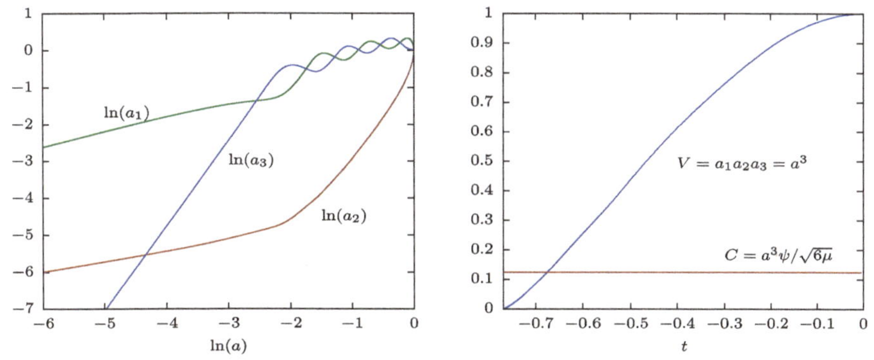

Fig. 4 The Bianchi IX solution with a massless scalar ($\epsilon = 1$), for the initial data (45) with $\gamma = 3$ and $C = 0.123$. After a few oscillations, all three functions a_k become monotonic. The volume $V = a^3$ decreases to zero, while the scalar charge C defined by (12) remains constant, implying $\dot{\psi} \propto C/a^3$

whose solution describes an isotropic universe that begins with unit size and contracts to zero volume. If instead one chooses $|C| < 1$ in (45), the solutions become anisotropic generalizations of this case: the universe undergoes oscillations while shrinking, but only a finite number of oscillation cycles occur.

The typical solution is shown in Fig. 4, corresponding to $\gamma = 3$ and $C = 0.123$. Near the initial time, for $\ln(a) \in (-2, 0)$, the solution resembles the vacuum case in Fig. 3 and exhibits oscillations. These oscillations form Kasner-type cycles, during which one of the three exponents $p_k(u)$ is negative while the other two are positive. After each cycle, the parameter u changes, yielding new values $p_k(u)$ as defined in (32). If one of the exponents remains negative, the oscillations continue.

However, as shown in [7], sooner or later the system reaches a value of u for which all $p_k(u)$ are positive, so that all three metric functions a_k become monotonic. At this point, the oscillations cease, and the singularity is approached smoothly. Indeed, Fig. 4 shows that all three a_k become monotonic for $\ln(a) < -2$, and none of them grows toward the pas.

In summary, the inclusion of a minimally coupled scalar field permits only a finite number of oscillations near the singularity, thereby eliminating chaotic behaviour [7]. By contrast, the dynamics are entirely different in the presence of a minimally coupled phantom scalar field.

6.2 Bianchi IX with a Phantom Scalar, $\epsilon = -1$

Recall that a phantom scalar field provides a repulsion from the singularity. No isotropic limit is possible in this case, since achieving it would require satisfying an equation with a sign change on the right-hand side in (46), which is not possible.

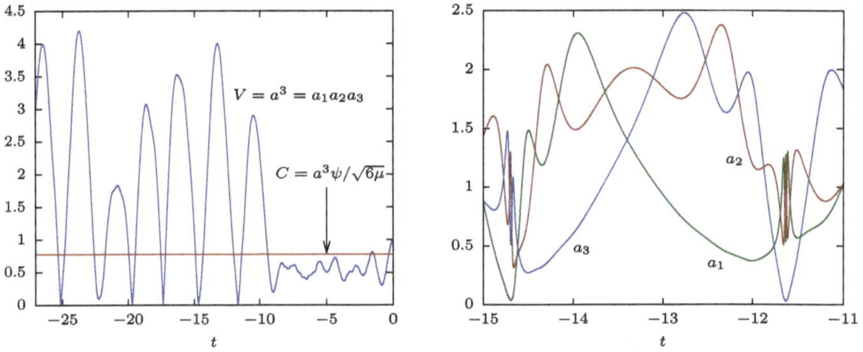

Fig. 5 Bianchi IX solution with a phantom scalar field ($\epsilon = -1$) for the initial values (45) with $\gamma = -2$ and $C = 0.774$. Left: the spatial volume $V(t) = \mathrm{a}_1\mathrm{a}_2\mathrm{a}_3$ and scalar charge $C = \mathrm{a}^3\psi/\sqrt{6\mu}$. Right: the scalar factors $\mathrm{a}_1(t)$, $\mathrm{a}_2(t)$, $\mathrm{a}_3(t)$ in the interval $t \in [-15, -11]$

For the anisotropic solutions, the scalar charge can assume any value, $|C| \in [0, \infty)$, while $\mathcal{B} = \sqrt{1 + C^2}$. Starting from the initial values (45) with $\epsilon = -1$, $C = 0.774$, and $\gamma = -2$, and integrating toward negative t, one obtains the behaviour shown in Fig. 5.

The spatial volume $V = \mathrm{a}_1\mathrm{a}_2\mathrm{a}_3 \equiv \mathrm{a}^3$ oscillates chaotically, sometimes approaching zero but never vanishing, and repeatedly bouncing back.

The function $V(t)$ in the left panel in Fig. 5 exhibits multiple minima, two of which occur in the interval $t \in [-15, -11]$. The right panel of Fig. 5 shows a zoom of this interval, revealing that all three amplitudes $\mathrm{a}_k(t)$ oscillate, with particularly dense oscillations around $t = -14.7$ and around $t = -11.6$.

Zooming further into these regions of dense oscillations, the left panel of Fig. 6 shows that $\mathrm{a}_1(t)$ passes through a non-zero minimum at $t \approx -11.69$, while $\mathrm{a}_2(t)$ and $\mathrm{a}_3(t)$ oscillate several times around each other. Similarly, the right panel of Fig. 6 shows that $\mathrm{a}_3(t)$ reaches a non-zero minimum at $t \approx -11.63$, with $\mathrm{a}_1(t)$ and $\mathrm{a}_2(t)$ oscillating around one another.

In the left panel in Fig. 6, to the right of the minimum of a_1, one of the three amplitudes grows toward the past while the two others decrease, so that among the three Kasner exponents p_k, one is negative and two are positive. Conversely, to the left of the minimum of a_1, *two* of the three amplitudes grow toward the past and the third decreases, so that among the three Kasner exponents p_k, *two* are negative and one is positive. This behaviour is consistent with the previous analysis of the allowed values of p_k in the Bianchi I case.

A similar behaviour is observed on a larger scale in the right panel of Fig. 5, where a_1 passes through a minimum and then grows toward the past, while the other two amplitudes oscillate. Subsequently, a_3 passes through a maximum and decreases toward the past, with the remaining two amplitudes continuing to oscillate.

As none of the amplitudes $\mathrm{a}_k(t)$ vanish, the curvature remains bounded and the manifold is non-singular. The solution in the negative t region can be extended to

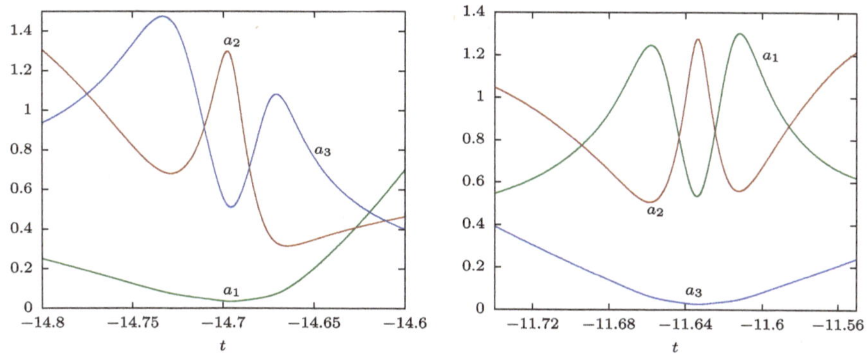

Fig. 6 Same as the right panel of Fig. 5, zoomed around the two regions where the oscillations become particularly dense

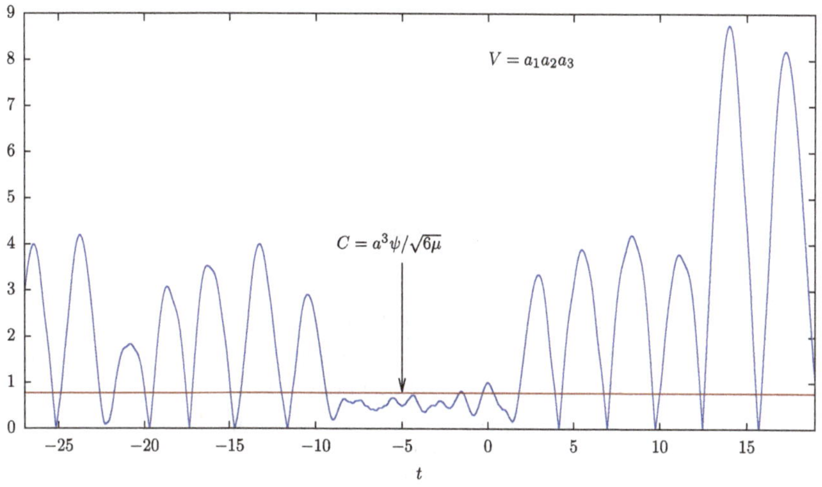

Fig. 7 Same solution as in Fig. 5, extended to positive values of t. The volume function $V(t)$ exhibits a seemingly random profile, quasi-periodically approaching zero but never reaching it

positive t, as shown in Fig. 7. Although $\dot{\mathrm{a}}(0) = 0$, the solution is not symmetric under $t \to -t$, since the initial values of the anisotropies, $\dot{\beta}_{\pm}(0)$, do not vanish. It appears that the range of t extends from $-\infty$ to ∞, although we cannot immediately prove this. In principle, it is not excluded that integrating far enough could lead to $V(t) = \mathrm{a}_1\mathrm{a}_2\mathrm{a}_3$ reaching zero at a minimum, producing a curvature singularity. However, our numerical results do not show this—the integration terminates only because numerical errors accumulate and the constraint begins to be violated.

Qualitatively, it is the repulsive character of the phantom scalar field that prevents the system from reaching a singularity.

Therefore, the solution shown in Fig. 7 appears to correspond to a completely regular, geodesically complete manifold. The distribution of peaks in Fig. 7 appears entirely random; in fact, even a very small change in the initial values parameters γ, C completely reshuffles the positions and amplitudes of the peaks. This behaviour strongly suggests that the solution is chaotic.

As a result, we obtain an anisotropic universe whose three scale functions $\mathrm{a}_k(t)$ undergo an apparently infinite sequence of oscillation cycles distributed in a seemingly random way. These functions periodically approach zero (with random periods) but never actually reach it, instead bouncing back. Consequently, the universe repeatedly enters a strong-gravity regime without developing a curvature singularity. The size of the universe is also bounded from above, resulting in an infinite sequence of bounces. To the best of our knowledge, solutions of this type have not been described previously in the literature.

Since the phantom scalar field violates the null energy condition and provides a repulsive effect, the appearance of bounce solutions is, in some sense, natural. Such bounces already arise in the isotropic limit when matter is included—for example a radiation with energy density A^2/a^4. In this case, Eq. (44) becomes

$$\frac{\dot{\mathrm{a}}^2}{\mathrm{a}^2} + \frac{1}{\mathrm{a}^2} = \frac{A^2}{\mathrm{a}^4} - \frac{C^2}{\mathrm{a}^6}, \tag{47}$$

whose solution $\mathrm{a}(t)$ oscillates between the maximal and minimal values $\mathrm{a}_\pm$ of the universe size, given by $\mathrm{a}_\pm^2 = (A^2 \pm \sqrt{A^4 - 4C^2})/2$. It is less obvious, however, that anisotropies can mimic the role of matter, leading to the anisotropic bounces described above.

In summary, the energy of a minimally coupled massless scalar field falls off as $\propto 1/\mathrm{a}^6$, just like the anisotropy contribution. As a result, the scalar is able to influence the chaotic dynamics in the Bianchi IX case: an ordinary scalar with positive energy removes the chaos, permitting only a finite number of oscillations as the singularity is approached. In contrast, a phantom scalar appears to eliminate the singularity entirely, producing a chaotic sequence of regular bounces.

7 Anisotropic Cosmologies with *K*-Essence

Let us now consider the case where the scalar field contribution is different from $\propto 1/\mathrm{a}^6$. We take a K-essence theory with index $n \neq 2$, still assuming the minimal coupling ($\alpha = 0$). Then, the initial value constraint (9) becomes

$$\frac{\dot{\mathrm{a}}^2}{\mathrm{a}^2} + \frac{\mathcal{K}}{\mathrm{a}^2} = \dot{\beta}_+^2 + \dot{\beta}_-^2 + (n-1)\frac{\epsilon}{6\mu}\,\psi^n, \tag{48}$$

while Eq. (12) yields

$$\mathrm{a}^3\psi^{n-1} = const. \Rightarrow \psi = (6\mu)^{1/n}\,\frac{C}{\mathrm{a}^{3/(n-1)}}. \tag{49}$$

The constraint then reduces to

$$\frac{\dot{a}^2}{a^2} + \frac{\mathcal{K}}{a^2} = \dot{\beta}_+^2 + \dot{\beta}_-^2 + \epsilon\,(n-1)\,\frac{C^n}{a^\gamma}, \quad \gamma = \frac{3n}{n-1}. \tag{50}$$

If $n > 2$, then $\gamma < 6$, so the scalar field contribution is $o(1/r^6)$ and subdominant compared to the anisotropy contribution during Kasner cycles, which scales as $\propto 1/a^6$. According to our conjecture, the chaotic BKL-type behaviour is therefore expected in this case. Conversely, if $n < 2$, then $\gamma > 6$, and the scalar field contribution dominates near singularity, which should modify the BKL picture.

To verify these expectations, we use the initial values similar to those in (45):

$$\begin{aligned} a(0) =&1, \quad \beta_\pm(0) = 0, \quad \dot{\beta}_+(0) = \mathcal{B}\sin\gamma, \quad \dot{\beta}_-(0) = \mathcal{B}\cos\gamma, \\ \dot{a}(0) =&0, \quad \dot{\phi} = (6\mu)^{1/n}C, \quad \mathcal{B} = \sqrt{1-\epsilon\,C^n}. \end{aligned} \tag{51}$$

Starting from these initial values, we integrate the equations assuming $\epsilon = 1$. Figure 8 shows the solutions for $\gamma = 3$ and $C = 0.123$ for two different values of the K-essence index n. For $n = 1.9 < 2$ (left panel), the scalar field dominates and the oscillations quickly stop, similarly to the $n = 2$ solution in Fig. 4. On the other hand, for $n = 2.3 > 2$ (right panel), the scalar contribution is subdominant, and the solution closely resembles the vacuum case in Fig. 3. This confirms the expectations outlined above.

Let us now consider the phantom case, $\epsilon = -1$. Figure 9 shows the solutions for $\gamma = 3$ and $C = 0.774$. For $n = 1.5 < 2$ (left panel), the scalar field dominates the dynamics and provides a repulsion from singularity, producing an apparently infinite and chaotic chain of bounces, similar to the $n = 2$ case in Fig. 5. For $n = 3 > 2$

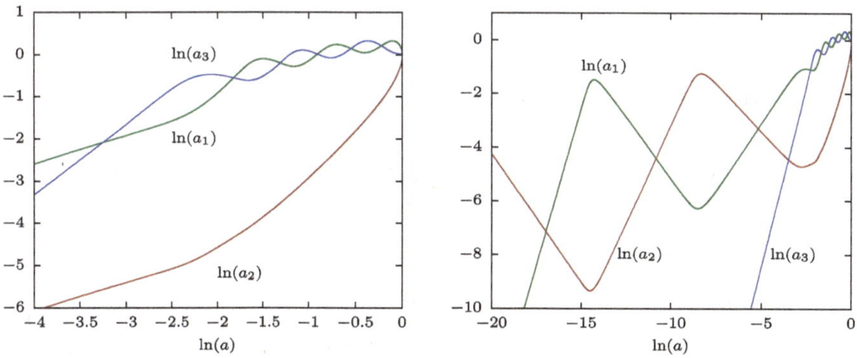

Fig. 8 Bianchi IX solutions with K-essence ($\epsilon = 1$) for the initial data (51) with $\gamma = 3$ and $C = 0.123$, and for different values of the K-essence index n. Left: $n = 1.9 < 2$; in this case, the scalar field contribution to (50) is dominant, and the oscillations quickly stop, similar to the behaviour shown in Fig. 4. Right: $n = 2.3 > 2$; here, the scalar contribution is subdominant, and the solution closely resembles the vacuum case shown in Fig. 3

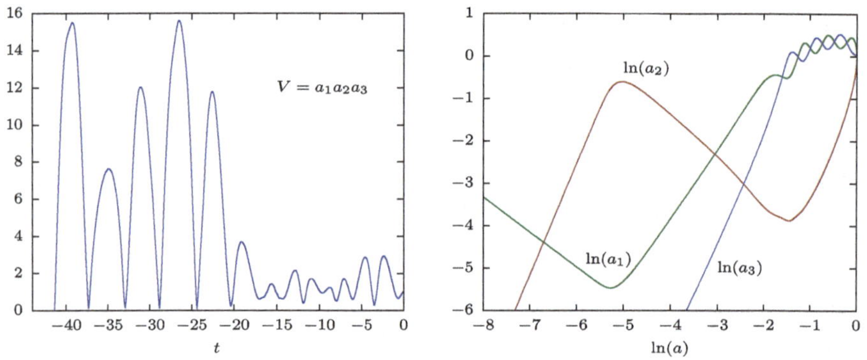

Fig. 9 Bianchi IX solutions with phantom K-essence ($\epsilon = -1$) for the parameter values $\gamma = 3$, $C = 0.774$, and for different values of the K-essence index n. Left: $n = 1.5 < 2$; the scalar field dominates and provides a repulsion from the singularity, producing an apparently infinite chain of bounces, similar to Fig. 5. Right: $n = 3$; here, the scalar contribution is subdominant, and the solution closely resembles the BKL case shown in Fig. 3

(right panel), the scalar field contribution is subdominant, and the solution closely resembles the vacuum BKL case in Fig. 3. Thus, the presence of a singularity depends on whether the scalar field is dominant or subdominant, but in all cases the dynamics remains chaotic.

Summarizing, for $\epsilon = 1$ the scalar field has positive energy. If its contribution is subdominant, $o(1/\mathrm{a}^6)$, the solutions remain chaotic, whereas if it scales as $1/\mathrm{a}^6$ or faster, chaos is eliminated, confirming our conjecture. In contrast, for $\epsilon = -1$ the scalar energy is negative, and the solutions are always chaotic: they develop a singularity when the scalar energy is subdominant ($o(1/\mathrm{a}^6)$) and remain regular otherwise.

8 Non-minimal Coupling

Let us finally switch on the non-minimal coupling by setting $\alpha \neq 0$ in Eqs. (9)–(12), while keeping the K-essence index $n = 2$. The main difference from the $\alpha = 0$ case is that, instead of $\psi = \sqrt{6\mu}C/\mathrm{a}^3$, Eq. (12) now gives

$$\psi = \frac{\sqrt{6\mu}\,C}{\mathrm{a}^3\,[1 - (y + \omega)/\zeta]}, \tag{52}$$

where

$$\zeta = \frac{\epsilon}{3\alpha}, \qquad y = \frac{\dot{\mathrm{a}}^2}{\mathrm{a}^2} - \dot{\beta}_+^2 - \dot{\beta}_-^2, \qquad \omega = \frac{\mathcal{K}}{\mathrm{a}^2}. \tag{53}$$

Inserting this to the constraint (9), the result can be represented in the form

$$a^6 = \frac{\epsilon\zeta\, C^2\,(\zeta - 3y - \omega)}{(\zeta - y - \omega)^2(y + \omega)}. \tag{54}$$

It is straightforward to verify that in the a → 0 limit this equation admits only one solution,

$$a \to 0, \qquad y \to \frac{\zeta - \omega}{3}, \tag{55}$$

so that

$$y = \frac{\epsilon}{9\alpha} - \frac{\mathcal{K}}{3a^2} + \cdots, \qquad \psi = \frac{\sqrt{6\mu}\,3\,C}{2a^3\,[1 - \mathcal{K}/(\zeta a^2)]} + \cdots, \tag{56}$$

where the dots denote subleading terms.

Let us now turn to the anisotropy equations (11). Neglecting their right-hand side parts containing $\mathcal{K}$ gives

$$\dot\beta_\pm = \frac{const.}{\sigma_+ a^3} = \frac{const.}{(2\mu + \alpha\,\psi^2)\,a^3}. \tag{57}$$

For $\alpha = 0$, this equation reduces to $\dot\beta_\pm \propto 1/a^3$, which is precisely the Kasner behaviour of the Bianchi I solutions. In the Bianchi IX case, this corresponds to the oscillatory regime, which can be approximated as a sequence of Kasner cycles. Within each cycle, the $\mathcal{K}$ term in the equations can be neglected, becoming important only at the transition between successive cycles.

If $\alpha \neq 0$, then in the Bianchi I case Eqs. (56) and (57) give

$$\mathcal{K} = 0: \quad y \approx \frac{\epsilon}{9\alpha}, \qquad \psi \propto \frac{1}{a^3} \Rightarrow \dot\beta_\pm \propto a^3. \tag{58}$$

Therefore, as a → 0, the scalar field ψ diverges while the anisotropies vanish. This effect, known as anisotropy screening, was first discovered in [2].

However, in the Bianchi IX case, one obtains from (56) and (57):

$$\mathcal{K} \neq 0: \quad y \propto \frac{\mathcal{K}}{a^2}, \qquad \psi \propto \frac{1}{\mathcal{K}a} \Rightarrow \dot\beta_\pm \propto \frac{\mathcal{K}^2}{a}, \tag{59}$$

therefore, the anisotropy screening is destroyed by the spatial curvature [2].

Using the definition of y yields

$$\frac{\dot a^2}{a^2} = \dot\beta_+^2 + \dot\beta_-^2 + y. \tag{60}$$

This is actually the constraint equation, where the right-hand side contains both the anisotropy contribution and the scalar field contribution encoded in y. Therefore, y

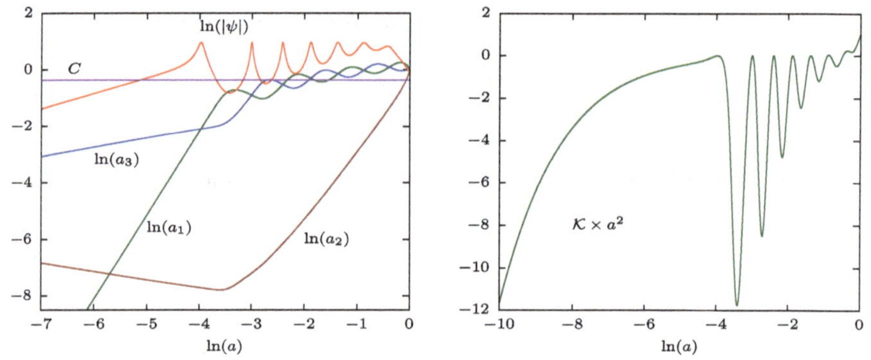

Fig. 10 Bianchi IX solutions with a non-minimally coupled scalar field for $\alpha = -0.1$, $\epsilon = 1$, n=2, and initial conditions $a(0) = 1$, $\beta_\pm(0) = 0$, $\dot\beta_+(0) = \sin(3)$, $\dot\beta_- = \cos(3)(0)$, and $\psi(0) = -0.9$. Left: the amplitudes a_k, the scalar field ψ, and the scalar charge C defined by (12). Right: The scalar curvature $\mathcal{K} \times a^2$

can be identified with the scalar field energy. The solution is expected to be chaotic if the anisotropy term dominates. We therefore consider the ratio

$$\frac{\dot\beta_+^2 + \dot\beta_-^2}{y} \propto a^6 \quad \text{if}\, \mathcal{K} = 0; \qquad \frac{\dot\beta_+^2 + \dot\beta_-^2}{y} \propto \mathcal{K}^3 \quad \text{if}\, \mathcal{K} \neq 0. \tag{61}$$

Recall that $\mathcal{K}$ takes values in the interval $(-\infty, 1]$. If $\mathcal{K}$ is close to zero, the first estimate in (8) applies. However, as shown in Fig. 10 (right panel), $\mathcal{K}$ is typically large and negative, in which case the second estimate in (8) applies. Therefore, for most of the evolution the anisotropy contribution dominates, and the solution is expected to be chaotic.

These expectations are confirmed by numerical simulations. Figure 10 (left panel) displays a solution exhibiting the characteristic BKL behaviour. The scalar curvature $\mathcal{K}$ periodically approaches zero but remains negative for most of the evolution, typically growing at least as fast as $1/a^2$.

Solutions with $\alpha \neq 0$ do not exhibit the same strong dependence on ϵ as for $\alpha = 0$. Solutions similar to that shown in Fig. 10 exist for both signs of ϵ and α. This can be attributed to the fact that, although in the non-minimally coupled case the scalar energy is not strictly positive, it is mainly positive near the singularity, where the estimate (56) applies, hence the energy is $\propto y \approx -\mathcal{K}/(3a^2) > 0$, because $\mathcal{K}$ is typically large and negative.

The theory with $\alpha \neq 0$ also admits isotropic bounces whose size oscillates between finite values $a_\pm$ (see also [10]). They can be obtained by solving Eq. (54) with $\omega = 1/a^2$, which yields $y(a)$, and this should be positive, which is indeed the case for $a \in (a_-, a_+)$. Such solutions can be extended to the anisotropic case, but no chaotic behaviour is observed.

9 Summary

We analyzed above how a scalar field affects the chaotic behaviour of homogeneous and anisotropic Bianchi IX cosmologies. Most of the time, we considered the scalar field of the simplest shift-symmetric K-essence type, defined by $X = \epsilon\,\dot{\phi}^n$, with $\epsilon = 1$ or $\epsilon = -1$. It is known that such a field removes chaos if $n = 2$ [7], provided that $\epsilon = +1$. The spacetime singularity is then approached smoothly. We have verified that chaos disappears also if $n < 2$, because in this case the field energy $\propto 1/\mathrm{a}^\gamma$ with $\gamma > 6$ dominates over the anisotropy contribution. However, for $n > 2$, the scalar field contribution is subdominant, and the solutions exhibit the usual BKL-type chaotic behaviour.

The BKL-type chaos also persists when a non-minimal coupling is included [2], and we have confirmed that in this case as well the scalar field contribution remains subdominant.

The BKL-type behaviour also persists for the phantom scalar field corresponding to $\epsilon = -1$ if $n > 2$. However, for $n \leq 2$ we obtain a completely new phenomenon: chaos remains but the spacetime singularity disappears. In this regime, the three scale factors $\mathrm{a}_k(t)$ oscillate seemingly indefinitely, never reaching zero, and the 3-volume $V = \mathrm{a}_1\mathrm{a}_2\mathrm{a}_3$ exhibits a chaotic sequence of minima and maxima. The universe thus undergoes an infinite sequence of anisotropic bounces.

It is not immediately clear whether this finding has direct physical applications, even though theories with non-positive energy have become increasingly popular in recent years. Nevertheless, these new chaotic solutions are interesting from a theoretical perspective. For example, one could study their topological entropy [8] or investigate other characteristics of their chaotic behaviour.

Acknowledgements It is a pleasure to thank Sasha Kamenshchik for valuable discussions.

References

1. A.A. Starobinsky, S.V. Sushkov, M.S. Volkov, The screening Horndeski cosmologies. JCAP **1606**(06), 007 (2016). https://doi.org/10.1088/1475-7516/2016/06/007
2. A.A. Starobinsky, S.V. Sushkov, M.S. Volkov, Anisotropy screening in Horndeski cosmologies. Phys. Rev. D **101**(6), 064039 (2020). https://doi.org/10.1103/PhysRevD.101.064039. arXiv:1912.12320
3. R. Galeev, R. Muharlyamov, A.A. Starobinsky, S.V. Sushkov, M.S. Volkov, Anisotropic cosmological models in Horndeski gravity. Phys. Rev. D **103**(10), 104015 (2021). https://doi.org/10.1103/PhysRevD.103.104015. arXiv:2102.10981
4. V. Belinskii, E. Lifshitz, I. Khalatnikov, Construction of a general cosmological solution of the Einstein equation with a time singularity. Sov. JETP **35**, 838–841 (1972)
5. C.W. Misner, Mixmaster universe. Phys. Rev. Lett. **22**, 1071–1074 (1969). https://doi.org/10.1103/PhysRevLett.22.1071
6. P.P. Goldstein, *Some Exact Results on the Belinski–Khalatnikov–Lifshitz Scenario*. arXiv:2505.15541

7. V. Belinskii, I. Khalatnikov, Effect of scalar and vector fields on the nature of the cosmological singularity. Sov. JETP **36**, 591–597 (1972)
8. A.Y. Kamenshchik, I.M. Khalatnikov, S.V. Savchenko, A.V. Toporensky, Topological entropy for some isotropic cosmological models. Phys. Rev. D **59**, 123516 (1999). https://doi.org/10.1103/PhysRevD.59.123516. arXiv:gr-qc/9809048]
9. D.A. Easson, J.E. Lesnefsky, *Eternal Universes*. https://doi.org/10.1103/5mhz-m8bg. arXiv:2404.03016
10. S.V. Sushkov, R.G. Galeev, Singular bounce in the theory of gravity with nonminimal derivative coupling. Phys. Rev. D **111**(8), 083554 (2025). https://doi.org/10.1103/PhysRevD.111.083554. arXiv:2502.05786

Effects of Gravity Theory

Alexei Starobinsky and Wormhole Physics

Kirill A. Bronnikov and Milena V. Skvortsova

Abstract Alexei Starobinsky is most famous for his great contribution to cosmology, but he has considerable achievements in other branches of gravitational physics and astrophysics, such as the theory of compact objects including black holes and wormholes. In this note, we give a brief review of Alexei's papers devoted to wormhole physics. They mostly concern such issues of common interest as the necessary conditions for wormhole existence in general relativity and its extensions as well as generic properties of some kinds of wormholes. We also extend one of the no-go theorems on thin-shell wormholes to a wider choice of their symmetry and matter content.

1 Introduction

We would like to begin our paper with some memories of Alexei Alexandrovich Starobinsky. He has been not only an outstanding scientist but a very attractive person, he paid very much attention to young scientists, students and postgraduates, when necessary, he helped them organizationally (for example, to take part in a school or conference of interest). It was a typical practice for him to suggest a problem of interest to a young colleague even if the latter worked in quite a different institute but showed a potential to solve it.

Alexei took part in a lot of conferences and was there very active, his talks did not only contain the results of new research but always had an educational function, and he was always ready to answer questions, even if quite trivial for him. When he

K. A. Bronnikov (✉)
Rostest, Moscow, Russia
e-mail: kb20@yandex.ru

K. A. Bronnikov · M. V. Skvortsova
RUDN University, Moscow, Russia
e-mail: milenas577@mail.ru

K. A. Bronnikov
National Research Nuclear University "MEPhI", Moscow, Russia

A. Barvinsky and A. Kamenshchik (eds.), *Open Issues in Gravitation and Cosmology*, Fundamental Theories of Physics 225, https://doi.org/10.1007/978-3-032-15703-4_19

listened to talks by his colleagues, he usually sat in front of his computer, probably reading something there or answering messages, but when a talk was over, he used to ask questions and make comments that were always appropriate and often deep.

His interests were not restricted to cosmology, his main area: he had also outstanding results in black hole physics (see, e.g., [1]), in quantum field theory in curved space-time [2], and he expressed interest in many others areas. Among them was wormhole physics, a subject on which we had the honor and pleasure to be his co-authors.

Wormholes are nowadays quite a popular subject of both science fiction and serious theoretical studies. These are hypothetic configurations of curved space-time having the shape of "bridges" between distant regions of the same universe or even between different universes. The possible existence of such objects in our Universe would lead to diverse remarkable effects, beginning with observation of the sky of another universe through a wormhole throat and ending with such a fantastic construct as a time machine. It is therefore not too surprising that, although nobody has ever confidently observed any wormhole, many researchers study and discuss both the conditions of wormhole emergence and existence and their possible observational signatures.

As already said, Alexei Starobinsky expressed active interest in wormhole physics. His papers in this area, written together with the present authors [3–5], are as a whole not devoted to obtaining or discussion of particular wormhole solutions of different theories of gravity (whose list is now extremely rich) but to such questions of general interest as the wormhole existence conditions and their main properties that do not depend on a particular model. The main purpose of the present paper is to recall the main results of those papers, emphasizing Alexei's contribution to this intriguing sector of gravitational physics. For one of the no-go theorems of [3–5], namely on static, spherically symmetric thin-shell wormholes, we present an extension to wider symmetry properties and arbitrary matter sources.

2 Throats and Asympotic Regions

It is quite well known that the existence of static wormhole throats (2-surfaces of minimum area in these space-time tunnels) in general relativity inevitably requires a material source of gravity whose stress-energy tensor (SET) T_μ^ν violates the weak and null energy conditions (WEC and NEC) [6, 7], and such sources have acquired the name of "exotic matter," also often called phantom or ghost matter (we use the words 'ghost' and 'phantom' as synonyms). There are strong reasons to suppose that Nature, if not completely forbids the existence of such matter, still tries to avoid its emergence. Therefore, much effort is devoted to a search for wormhole models without such matter.

One can point out two ways in attempts to achieve this goal: either consider nonstatic models in general relativity (for example, in the framework of simple cosmological models), or use theories of gravity other than general relativity. Both ways are widely used and discussed by the researchers. The second way certainly implies that the corresponding theory alternative to general relativity is sufficiently well founded and supported by observations.

Scalar-tensor theories (STT) of gravity are among the most promising alternative theories, and their numerous models sufficiently well agree with observations (see [8, 9] for recent reviews). By the year 2005, when we began our work [3], the analysis of current cosmological observations revealed some evidence that the Dark Energy (DE), responsible for the accelerated expansion of the Universe if the Einstein equations are assumed to hold, may be well described by a phantom matter equation of state parameter $w \equiv p_{\rm DE}/\rho_{\rm DE}$, slightly smaller than -1. (Let us note that this estimate remains relevant today, even though models with a cosmological constant Λ ($w = -1$) remain most popular and comprise the 'standard' (ΛCDM) cosmological paradigm.) It was, however, clear that the possibly observed $w < -1$ does not necessarily mean the existence of ghosts in Nature since there are STT models without ghosts that admit a phantom DE behavior and even a super-acceleration of the present Universe, such that $\mathrm{d}H/\mathrm{d}t > 0$, where $H(t)$ is the Hubble parameter in Friedmann–Robertson–Walker cosmological models. For example, this may take place in the framework of the Brans-Dicke STT with a nonzero scalar field potential. Thus, as was shown explicitly [10], such a model possesses sufficient freedom to account for all observations justifying the presently possible phantom-driven expansion as well as those concerning the luminosity distances and inhomogeneity growth in the Universe.

So, if a nonphantom scalar field in STT can provide a phantom behavior in cosmology, a natural question is: can such a field create local entities also usually associated with phantoms, i.e., wormholes? Our paper [3] gives a negative answer to this question, thus extending some previous results [11, 12],

To begin the discussion, it is necessary to define its subject. We here, following [3–5], consider static space-times and say that that the geometry is that of a **wormhole** if (i) the space-time is globally regular and free from horizons, (ii) its spatial sections contain a 2D surface of minimum area, called a **throat**, and (iii) the throat is a member of a family of 2D surfaces whose area is not bounded above on both sides from the throat. In the simplest case of spherical symmetry this means that a throat is a sphere of minimum spherical radius r, and that $r \to \infty$ on both sides far from this sphere. Asymptotic flatness on one or both 'ends' is not necessary though often desirable.

This definition is more narrow than those used by a number of authors but covers the cases of interest in many studies.

Now let us briefly recall the main points of our reasoning in [3]. The main feature of general (Bergman-Wagoner-Nordtvedt) STT is a nonminimal coupling between a scalar field Φ and the curvature. The Lagrangian in a (physically preferred) Jordan-frame manifold $\mathbf{M}_{\rm J}$ with the metric $g_{\mu\nu}$ may be written as[1]

---

[1] We use the following conventions: the metric signature $(+ - - -)$; the Riemann and Ricci tensors defined as $R^{\sigma}{}_{\mu\rho\nu} = \partial_\nu \Gamma^{\sigma}_{\mu\rho} - \cdots$ and $R_{\mu\nu} = R^{\sigma}{}_{\mu\sigma\nu}$, so that the Ricci scalar $R = g^{\mu\nu} R_{\mu\nu}$ is positive for both de Sitter space-time and matter-dominated cosmology; the system of units $8\pi G = c = 1$.

$$2L = f(\Phi)R + h(\Phi)g^{\mu\nu}\Phi_{,\mu}\Phi_{,\nu} - 2U(\Phi) - F^{\mu\nu}F_{\mu\nu}, \tag{1}$$

where R is the Ricci scalar and $F_{\mu\nu}$ the electromagnetic field tensor; the functions f, h and U are arbitrary. The fermion matter Lagrangian is assumed to have no coupling with Φ, hence fermion masses are constant, and atomic clocks measure the proper time t in $\mathbf{M}_\mathrm{J}$. It is, however, useful to also consider the Einstein frame, defined as a manifold $\mathbf{M}_\mathrm{E}$ with the metric

$$\overline{g}_{\mu\nu} = |f(\Phi)|g_{\mu\nu}, \tag{2}$$

in which the Lagrangian (1) takes the form

$$2L_E = (\operatorname{sign} f)\big[\,\overline{R} + (\operatorname{sign} \ell)\overline{g}^{\mu\nu}\phi_{,\mu}\phi_{,\nu}\big] - 2V(\phi) - F^{\mu\nu}F_{\mu\nu}, \tag{3}$$

where overbars mark quantities obtained from or with $\overline{g}_{\mu\nu}$, indices are raised and lowered with $\overline{g}^{\mu\nu}$ and $\overline{g}_{\mu\nu}$, and the functions involved are related by

$$\ell(\Phi) := fh + \frac{3}{2}\left(\frac{\mathrm{d}f}{\mathrm{d}\Phi}\right)^2, \qquad \frac{\mathrm{d}\phi}{\mathrm{d}\Phi} = \frac{\sqrt{|\ell(\Phi)|}}{f(\Phi)}, \qquad V(\phi) = |f|^{-2}U(\Phi). \tag{4}$$

Let us require that the theory has no ghosts, hence $f(\Phi) > 0$ (the graviton is not a ghost) and $\ell(\Phi) > 0$ (the Φ field is not a ghost). Moreover, let us assume that $f(\Phi)$ and $\ell(\phi)$ are smooth and bounded above and below by positive constants. According to [7], a static wormhole throat (defined as a minimal 2D surface in a 3D manifold) necessarily requires that the matter source in the Einstein equations violates the NEC that requires $T_{\mu\nu}\xi^\mu\xi^\nu \geq 0$, where $T_{\mu\nu}$ is the SET of matter, and ξ^μ is an arbitrary null vector. Since the matter sources in (3) satisfy the NEC, we can assert that wormholes (and even wormhole throats) are impossible in $\mathbf{M}_\mathrm{E}$.

Now we notice that under our assumptions, the conformal mapping $\overline{g}_{\mu\nu} = f(\Phi)g_{\mu\nu}$ transfers a flat spatial infinity in one frame to a flat spatial infinity in the other. Therefore, if there is a wormhole in $\mathbf{M}_\mathrm{J}$, which is asymptotically flat on each side of the throat, then its each flat infinity maps to its counterpart in $\mathbf{M}_\mathrm{E}$, the whole manifold is smooth, and we thus obtain a wormhole in $\mathbf{M}_\mathrm{E}$, contrary to the above-stated impossibility. This contradiction makes us conclude that static asymptotically flat wormholes are absent in the Jordan frame as well.

In this simple reasoning, no spatial symmetry assumption was made, hence any static wormholes are ruled out in the theory (1), if everywhere $f > 0$ and $\ell > 0$. Moreover, no assumption was made on $U(\Phi)$ (apart from smoothness and compatibility with asymptotic flatness). So the result is quite general.

The no-wormhole statement for $\mathbf{M}_\mathrm{J}$ can be further generalized in a few respects [3]:

(i) our inferences remain valid in the presence of any other ghost-free (respecting the NEC) matter;

(ii) instead of asymptotic flatness, we can simply require the existence of two spatial infinities (defined in terms of infinitely growing areas of closed 2-surfaces in the spatial sections of $\mathbf{M}_\mathrm{E}$ and $\mathbf{M}_\mathrm{J}$);
(iii) the no-wormhole statement holds not only in STT but in any metric theory of gravity (e.g., $F(R)$ or $F(R, \Phi)$ [13, 14]) admitting a conformal mapping of the physical manifold $\mathbf{M}_\mathrm{J}$ to $\mathbf{M}_\mathrm{E}$ where GR holds with ghost-free matter, provided that the conformal factor is everywhere smooth and positive;
(iv) under certain restrictions, our statement can even be extended to dynamical wormholes based on the no-wormhole results for $\mathbf{M}_\mathrm{E}$ obtained by Hochberg and Visser in [15].

The latter results rest on the definition of a dynamical wormhole throat as a marginally anti-trapped surface. There are other definitions (see, e.g., [16] for a review), some of them are less restrictive than that of [15]. Thus, a number of dynamical wormhole solutions in GR have been obtained using an intuitively clear definition of a throat referring to a physically preferable set of constant-time spacelike surfaces, see, e.g., solutions describing wormholes in a cosmological background in the framework of GR, sourced by such ghost-free kinds of matter as nonlinear electrodynamics [17, 18] and dust [19], but none of such objects can exist eternally and have initial and/or final singularities.

Returning to static systems in (not only) STT, we should stress that throats are not ruled out in $\mathbf{M}_\mathrm{J}$ (see an example below in this paper): it is a wormhole as a global entity that is ruled out. Therefore, nothing prevents one from constructing, for instance, composite wormhole models where a neighborhood of the throat is described by an appropriate STT solution, while the rest of space is either empty or has some other ghost-free matter as a source, admitting asymptotic flatness.

A further discussion in [3] concerned possible weakening of the assumptions on the function $f(\Phi)$. It is shown that if $f(\Phi)$ somewhere reaches zero, there can be exceptional twice asymptotically flat wormhole solutions in some STT, conformally mapped to solutions with extremal horizons in $\mathbf{M}_\mathrm{E}$, but such explicit examples are yet to be found. Furthermore, if $f(\Phi)$ is allowed to become negative (the effective gravitational constant is negative in such regions), then wormhole solutions are not ruled out, and a few examples of such solutions are known [20–22], with sources in the form of nonminimally coupled massless scalar fields and possible inclusion of an electromagnetic field, such that in (1) we have $f(\Phi) = 1 - \xi\Phi^2$ ($\xi = \mathrm{const}$), $h(\Phi) \equiv 1$, $U(\Phi) \equiv 0$. All such solutions involve the phenomenon of conformal continuation [23]. The point is that the surface where $f = 0$ is a singularity in $\mathbf{M}_\mathrm{E}$ but regular in $\mathbf{M}_\mathrm{J}$, and the conformal mapping (2) from the whole $\mathbf{M}_\mathrm{E}$ leads to only a part of $\mathbf{M}_\mathrm{J}$, which therefore has to be continued to a new region with another sign of f. Apart from a special nature of such constructs with respect to the whole set of solutions of a particular theory, it also happens that such wormhole solutions are generically unstable under spherically symmetric (monopole) perturbations [12, 22].

Thus we have the following **no-go theorem** [3]: in any STT (or $F(R)$, or $F(R, \Phi)$ theories of gravity), in any systems free of ghosts, no complete, infinitely extended static wormhole solutions are allowed in the physical Jordan conformal frame $\mathbf{M}_J$, even though throats, in general, can be present.

If the no-ghost conditions are weakened as described above, then special wormhole solutions can exist, but none of them can be regarded realistic.

It remains to add that if the scalar Φ is phantom, that is, if in (4) $\ell(\Phi) < 0$, wormhole solutions in STT are readily obtained, as confirmed by numerous examples beginning with [20].

3 Thin Shells and Wormholes

After the publication of [3], attempts to obtain static phantom-free wormhole models continued. In particular, it was claimed that at least in some cases, if a wormhole throat is provided by a thin shell of matter (e.g., in the Brans-Dicke STT in a certain range of the coupling constant ω), the model may satisfy the NEC and WEC [24]. Our paper [4], aimed at making clear this issue, was a kind of response to such claims. As was explicitly shown in [4], in *any* STT with a massless non-ghost scalar field, in *any* thin-shell wormhole constructed from two identical regions of vacuum static, spherically symmetric space-times, the shell that makes the throat has a negative surface energy density. We here extend this result, remaining in the framework of static spherical symmetry but canceling the requirement of vacuum ($L_m = 0$), that of $U(\Phi) = 0$, and that of identity of the two regions.

The proof in [4] again uses the mapping (4) and the fact that in the Einstein frame $\mathbf{M}_E$, trying to build a thin-shell wormhole, we inevitably obtain a negative surface density of the shell. Under the assumptions made in [4] (a massless non-ghost scalar field and no other matter), in $\mathbf{M}_E$ we are dealing with Fisher's scalar-vacuum solution [25] with the metric

$$\begin{aligned} ds^2 &= P^a dt^2 - P^{-a} dx^2 - P^{1-a} r^2 d\Omega^2, \\ P &= P(x) := 1 - \frac{2k}{x}, \qquad d\Omega^2 = d\theta^2 + \sin^2\theta d\varphi^2, \end{aligned} \tag{5}$$

where a and $k > 0$ are constants, $|a| < 1$, and x is the so-called quasiglobal (or Buchdahl) radial coordinate chosen so that $g_{tt} g_{xx} = -1$. The range of x is $2k < x < \infty$), and $x = 2k$ is a central (since the spherical radius $r = \sqrt{-g_{\theta\theta}}$ has there a zero value) naked singularity.

A thin-shell wormhole is obtained from (5) as follows: (i) consider a region (A) of the space-time (5) where $x > x_0$, and $x_0 = \text{const} > 2k$, so that $r(x_0) > 0$; (ii) consider one more copy (B) of the same region; (iii) introduce the new coordinate y such that $y = x - x_0$ in region (A) and $y = x_0 - x$ in region (B); (iv) unify the two regions by identifying their spheres $y = 0$. Then it is directly verified by a substitution to the Einstein equations that the obtained geometry corresponds to the SET in which the energy density component has the form

$$T_t^t = -\frac{4\delta(y)}{x_0^2}\left(1-\frac{2k}{x_0}\right)^{a-1}[x_0-(1+a)k] + \text{a finite quantity}, \tag{6}$$

where $\delta(y)$ is Dirac's delta function. This clearly means that the junction surface $y = 0$ has a negative surface density. The same result is obtained if we construct a thin-shell wormhole using the electrically or magnetically charged extension of Fisher's space-time obtained by Penney [26].

Suppose that we have a static, spherically symmetric thin-shell wormhole solution in an STT (1) with $U = L_m = 0$. Then, after the mapping (4), we obtain the (electro)vacuum problem in GR, on each side of the throat, hence these both regions will be desccribed in $\mathbf{M}_\mathrm{E}$ by the Fisher metric or its charged extension due to Penney. Now, it is necessary to take into consideration that the stress-energy tensor T_μ^ν of any kind of matter (other than the nonminimally coupled scalar Φ) present in $\mathbf{M}_\mathrm{J}$ transforms into $\mathbf{M}_\mathrm{E}$ by the simple law

$$T_\mu^{\nu(J)} = f^2(\Phi)T_\mu^{\nu(E)}, \tag{7}$$

where the superscripts (J) and (E) mark the SETs in the corresponding conformal frames. In particular, this law is valid for the surface SET of the thin shell.

Assuming, as before, that $f(\Phi)$ is everywhere finite and positive, we come to the conclusion that due to Eq. (6) (where $T_\mu^\nu = T_\mu^{\nu(E)}$), the surface energy density of the shell is also negative in $\mathbf{M}_\mathrm{J}$. Thus we obtain that, in fact, the no-go theorem of Sect. 2 is valid as well for thin-shell wormholes, though a proof [4] has been obtained under rather strong restrictions (static spherical symmetry, vacuum or electrovacuum, and identity of regions on different sides of the throat).

Let us now try to weaken these restrictions: assuming, as before, static spherical symmetry, let us only suppose that we are working in GR (or, equivalently, in $\mathbf{M}_\mathrm{E}$ of some STT) with matter respecting the NEC, and use the corresponding metric to obtain a thin-shell wormhole by the cut-and-paste method as described above.

To obtain the proof, let us notice that for the general static, spherically symmetric metric written in terms of the quasiglobal coordinate x,

$$ds^2 = A(x)dt^2 - \frac{dx^2}{A(x)} - r^2(x)d\Omega^2, \tag{8}$$

the difference of the $\binom{t}{t}$ and $\binom{x}{x}$ components of the Einstein equations reads

$$2\,A\frac{r''}{r} = -(\rho + p_r), \tag{9}$$

where the prime means d/dx, ρ and p_r are the energy density and radial pressure of matter, respectively, and we have $A(x) > 0$ and $r(x) > 0$ at regular points. The NEC component with the radial null vector ξ^μ requires $\rho + p_r \geq 0$, therefore, Eq. (9) leads to $r'' \leq 0$ in any regular static region with non-phantom matter.

Now suppose that there is a thin-shell wormhole with the throat at $x = 0$ (without loss of generality), where A and r are finite, and two spatial infinities ($r \to \infty$) at $x \to \pm\infty$. Then, since $r'' \le 0$, at negative x the function $r(x)$ decreases from infinity with an everywhere negative derivative $r'(x)$, while at $x > 0$ there is everywhere $r'(x) > 0$. It means that at $x = 0$, where they meet, we have inevitably a finite jump of $r'(x)$ from a negative value to a positive one, which implies that $r''(x) \propto \delta(x)$ with some positive proportionality factor. Then, according to (9), the quantity $\rho + p_r$ is a multiple of $\delta(x)$ with a negative factor. However, the whole SET of a thin shell of fixed radius r has the form

$$T_\mu^{\nu\,(\mathrm{shell})} = N\delta(x)\,\mathrm{diag}(\sigma,\ 0,\ -p_\perp,\ -p_\perp), \qquad (10)$$

where σ is the surface density, $p_\perp$ is the tangential pressure of the shell matter, and $N > 0$ is a factor depending on the particular values of $A(0)$ and $r(0)$. The radial pressure is zero since the shell is not extended in the radial direction. From $\rho + p_r < 0$ we conclude that the shell has a negative surface density, $\sigma < 0$.

At this point we can actually repeat the reasoning used in Sect. 2: if the corresponding solution of any metric theory of gravity (STT or any other) whose metric is connected with ours by an everywhere finite and positive conformal factor f, due to (7) there will be again a thin shell with negative surface density, violating the WEC and NEC. We conclude that in all such theories it is impossible to construct a static, spherically symmetric thin-shell wormhole without NEC and WEC violation.

A curious feature of our proof is that we did not apply the well-known Darmois-Israel formalism for describing thin shells dynamics: the present method is simpler and quite sufficient for static shells and is rather similar to the method applied in [27] to the case of cosmic strings described by conical singularities.

This proof did not suppose any particular form of matter filling the space-time, only assuming its non-phantom nature. Moreover, the matter and the metric can be quite different 'to the left' and 'to the right' of the throat. Thus we have significantly strengthened the no-go theorem under consideration for thin-shell wormholes.

4 More Observations on STT and $F(R)$ Theories

Our third paper [5] was also first motivated by some wrong or doubtful claims met in the literature that required making clear some subtle points of wormhole theory. As often happens in such cases, the study has led to some more general observations of interest.

Such statements mostly concerned the Brans-Dicke (BD) STT [28] that corresponds to the choice

$$f(\Phi) = \Phi, \qquad h(\Phi) = \frac{\omega}{\Phi} \qquad (11)$$

in the Lagrangian (1), where ω is a coupling constant. According to (4), the Φ field is phantom if $\omega < -3/2$, it is canonical if $\omega > -3/2$, and it has no dynamics if $\omega = -3/2$. Thus, as is clear, e.g., already from [20], vacuum static wormhole solutions of BD theory exist for all $\omega < -3/2$ and can only emerge in exceptional cases if $\omega > -3/2$. Meanwhile,

- In [29] it was asserted that vacuum wormhole solutions can be found in the BD theory at ω belonging to the non-ghost range $(-3/2, -4/3)$.
- A similar conclusion was made in [30] for a wider non-ghost range of ω, including $\omega = 0$.
- In [29], it was also asserted that vacuum BD wormholes are possible for $\omega < -2$ without mentioning the range $(-2, -3/2)$.
- In [31], wormhole solutions were indicated for some $F(R)$ theories, which are known to be equivalent to the BD theory with $\omega = 0$ and a nonzero potential $U(\Phi)$.

Apart from general objections such as that some authors use (unlike us) the term 'wormhole' for any space-time containing a throat, the above-mentioned statements of [29–31] required a closer analysis of BD vacuum solutions and a comparison of their different parametrizations. Let us reproduce the main points.

First of all, recalling [20], we can write the general vacuum static, spherically symmetric solution of any STT (1) with $U \equiv 0$ as follows:

$$ds_J^2 = \frac{1}{f(\Phi)}\left\{ e^{-2mu}dt^2 - \frac{e^{2mu}}{s^2(k,u)}\left[\frac{du^2}{s^2(k,u)} + d\Omega^2\right]\right\},$$
$$\phi = Cu, \qquad s(k,u) := \begin{cases} k^{-1}\sinh ku, & k > 0, \\ u, & k = 0, \\ k^{-1}\sin ku, & k < 0, \end{cases} \tag{12}$$

where the fields Φ (involved in the Lagrangian (1) written for $\mathbf{M}_J$) and ϕ (involved in (3) written for $\mathbf{M}_E$) are related as indicated in (4). The index J points out that it is the Jordan-frame metric. The integration constants C (scalar charge), m and k are connected by the relation

$$2k^2 \operatorname{sign} k = 2m^2 + \varepsilon C^2, \quad \varepsilon = \operatorname{sign} \ell. \tag{13}$$

The coordinate u used in (12) is the harmonic radial coordinate in $\mathbf{M}_E$, defined by the coordinate condition $\overline{g}_{uu} = \overline{g}_{tt}\overline{g}_{\theta\theta}^2$, and the expression in the curly brackets is nothing else than the metric in $\mathbf{M}_E$.

The branch $k < 0$ of the solution (existing only in the phantom case $\varepsilon = -1$) describes wormholes as long as $f(\Phi)$ is everywhere positive and finite since two adjacent zeros of the function $\sin ku$ (e.g., $u = 0$ and $u = \pi/|k|$) describe flat spatial infinities, and this fact is not spoiled by the regular conformal factor $1/f(\Phi)$. This is true, in particular, for the BD solution in which $\varepsilon = \operatorname{sign}(\omega + 3/2)$ and

$$f(\Phi) = \Phi = \Phi_0 \exp(\phi/\sqrt{|\omega + 3/2|}), \quad \Phi_0 = \text{const}, \tag{14}$$

is finite at any finite $\phi = Cu$. This removes any doubt concerning the wormhole existence at any $\omega < -3/2$. A more detailed description of static, spherically symmetric BD solutions (not only wormhole ones) with $\omega < -3/2$ may be found in [32, 33].

Let us now consider the canonical case $\varepsilon = +1$, $\omega > -3/2$. In this case, the expression in curly brackets in (12), in which $k > 0$, is Fisher's metric written using the harmonic coordinate u, as verified by the substitution

$$e^{-2ku} = 1 - \frac{2k}{x} \equiv P(x), \qquad a = \frac{m}{k} < 1, \tag{15}$$

which leads the metric in $\mathbf{M}_\mathrm{E}$ to the form (5). For the BD scalar field we then obtain

$$\Phi = P^{\xi}, \quad \xi = -C/\bigl(2k\sqrt{\omega + 3/2}\bigr), \tag{16}$$

and the relation (13) turns into

$$(2\omega + 3)\xi^2 = 1 - a^2. \tag{17}$$

The BD metric in $\mathbf{M}_\mathrm{J}$, defined and regular at $x > 2k > 0$, has the form

$$ds_J^2 = P^{-\xi}\Bigl[P^a dt^2 - P^{-a} dx^2 - P^{1-a} x^2 d\Omega^2\Bigr]. \tag{18}$$

To answer the question, whether or not this metric can describe a wormhole, one should find out (1) whether there can be a minimum of the radius $r_J = x P^{(1-a-\xi)/2}$, and (ii) if yes, is such a manifold globally regular and have an infinite radius at both ends. Since the metric (18) is asymptotically flat at $x \to \infty$, it remains to study its behavior as $x \to 2k$.

The first question has been answered as follows [4]: throats certainly do not exist in all cases, but still appear at arbitrarily large values of the parameter ω (even though the limit $\omega \to \infty$ corresponds to GR). To prove that, let us single out the subfamily of solutions given by $\xi = 2(1 - a)$. Then it is easy to obtain that $r_J(x)$ has a minimum at $x = k(3 - a) > 2k$ (because $a < 1 \Rightarrow 3 - a > 2$). Also, from (17) we find

$$2\omega + 3 = \frac{1 + a}{4(1 - a)}, \tag{19}$$

and taking a close enough to unity, the constant ω can be made arbitrarily large.

The metric behavior at $x = 2k$ is rather diverse but is generically singular in some sense, see details in [4]. The only exception, making the sphere $x = 2k$ regular, is the case $a = \xi = 1/2 \quad \Rightarrow \quad \omega = 0$. The metric (18) then has the so-called 'spatial Schwarzschild' form

$$ds_J^2 = dt^2 - \frac{dx^2}{1 - 2k/x} - x^2 d\Omega^2 \equiv dt^2 - 4(2k + y^2)dy^2 - (2k + y^2)^2 d\Omega^2, \tag{20}$$

where $y^2 = x - 2k$. The range of y is $y \in \mathbf{R}$, so that the manifold $\mathbf{M}_J$ consists of two copies ($y > 0$ and $y < 0$) of a region mapped from $\mathbf{M}_E$ defined for $x > 2k$, joined by the regular sphere $y = 0$, the throat. It is a wormhole obtained owing to the conformal continuation phenomenon [20, 23]. For the scalar field we have $\Phi = y/\sqrt{y^2 + 2k}$, it is zero at the transition sphere $y = 0$ and negative beyond it, which means that the effective gravitational constant is negative there.

Thus in BD theory we have a full confirmation of the results of [3] on wormholes in STT. As to $F(R)$ theories of gravity, the results obtained for STT are easily reformulated for them using the well-known correspondence between these two classes of theories. Specifically, if we write the Lagrangian of $F(R)$ theory as $L = F(R)/2 + L_m$, then, assuming

$$f(\Phi) = \Phi = F_R, \qquad h(\Phi) = 0, \qquad 2U(\Phi) = RF_R - F \tag{21}$$

(where $F_R \equiv dF/dR$), we obtain an equivalent representation of the same theory as the BD theory with the coupling constant $\omega = 0$, where the parametrization of the Φ field is chosen as $f(\Phi) \equiv \Phi$. Note that these both forms of the theory are presented in the same Jordan frame $\mathbf{M}_J$, with the same metric $g_{\mu\nu}$, and they have a common Einstein frame $\mathbf{M}_E$. The condition $f > 0$ in STT transforms into $F_R > 0$; furthermore, the function $\ell(\Phi)$ in (4) is nonnegative but can become zero at special values of R, where $F_{RR} \equiv d^2F/dR^2 = 0$; at such points, the parametrization $f = \Phi$ is lost. Moreover, translating the results of [3] (and Sect. 2 in the present paper) to the 'language' of $F(R)$ theory, we can assert that wormhole solutions are only possible if $F(R)$ has an extremum at some $R = R_0$ and may be obtained by a conformal continuation through such a surface, which is regular in $\mathbf{M}_J$ but singular in $\mathbf{M}_E$ [34]. Beyond such a surface there is a region where $\Phi = F_R < 0$, with a negative effective gravitational constant.

There is a subtle point [5] if at some value of R there is $F_{RR} = 0$ while $F_R \neq 0$: such cases are generically singular, with finite algebraic curvature invariants but diverging differential invariants. A similar situation in cosmological solutions is discussed in [35]. In exceptional cases such surfaces can be regular, but global wormhole solutions are still impossible by the no-go theorem of [3].

What about examples of wormhole solutions in $F(R)$ theory that can be found in the literature? In most of them, only the existence of throats is considered, and, as discussed above (see especially Eq. (19)), they are not at all excluded even in ghost-free cases. Ref. [31] discusses some examples of wormhole solutions of $F(R)$ theory in the presence of matter with anisotropic pressure, but in some of these examples there is $F_R < 0$ in certain regions, while others contain singularities on the throat [5].

5 Conclusion

In our opinion, our joint studies with Alexei Starobinsky, briefly reviewed in the present paper, are sufficiently rich in general observations of common interest. His more specific results concern subtle points of Brans-Dicke solutions ([5] and Sect. 4 here) and the existence and nature of singularities outside wormhole throats in solutions with a phantom Chaplygin gas [36]. They are also instructive and show Alexei's high level of understanding the physical picture under mathematical complexities.

As to the possible reality of wormholes in nature, our general impression is that despite the various no-go theorems, this reality is not too strongly suppressed because if throats in ghost-free theories are not ruled out, it is an easy task (at least theoretically) to take a certain neighborhood of a throat and to match it with asymptotically flat regions of vacuum or common ghost-free matter distributions. We can recall, in addition, that there can be dynamical wormholes whose lifetime is finite but comparable with the age of the Universe.

To conclude this paper, we would like to mention some developments that may be said to complement the results described above and (among many others) widen the landscape of wormhole physics. Thus, Ref. [37] presented a class of wormhole models with so-called *trapped ghosts* where a scalar field with a nonzero potential is phantom in a strong-field region near the throat but is canonical outside it. In [38, 39] it was demonstrated that stationary rotating cylindrical wormholes in GR may be in principle made of phantom-free matter though require rather special conditions for their construction. Lastly, there is a significant progress in 'wormhole astronomy', for example, Ref. [40] describes how the sky of another universe or a remote part of our own universe could be observed through the throat of a large cosmic wormhole.

Acknowledgements Our cordial thanks to Sergei Bolokhov for numerous helpful discussions and remarks. This work was supported by the RUDN University Project No. FSSF-2023-0003.

References

1. A.A. Starobinsky, Amplification of waves during reflection from a rotating ublack hole. Sov. Phys. JETP **37**, 28 (1973)
2. Y.B. Zeldovich, A.A. Starobinsky, Particle creation and vacuum polarization in an anisotropic gravitational field. Zh. Eksp. Teor. Fiz. **61**, 2161 (1971). Sov. Phys. JETP **34**, 1159 (1972)
3. K.A. Bronnikov, A.A. Starobinsky, No realistic wormholes from ghost-free scalar-tensor phantom dark energy. Pis'ma v ZhETF **85**, 1, 3–8 (2007). JETP Lett. **85**, 1, 1–5 (2007). arXiv: gr-qc/0612032
4. K.A. Bronnikov, A.A. Starobinsky, Once again on thin-shell wormholes in scalar-tensor gravity. Mod. Phys. Lett. A **24**(20), 1559–1564 (2009). arXiv: 0903.5173
5. K.A. Bronnikov, M.V. Skvortsova, A.A. Starobinsky, Notes on wormhole existence in scalar-tensor and F(R) gravity. Grav. Cosmol. **16**(3), 216–222 (2010). arXiv: 1005.3262
6. M. Morris, K.S. Thorne, Wormholes in spacetime and their use for interstellar travel: a tool for teaching general relativity. Am. J. Phys. **56**, 395 (1988)

7. D. Hochberg, M. Visser, Geometric structure of the generic static traversable wormhole throat. Phys. Rev. D **56**, 4745–4755 (1997). arXiv: gr-qc/9704082
8. S. Mironov, V. Volkova, Non-singular cosmological scenarios in scalar-tensor theories and their stability: a review. arXiv: 2409.16108
9. V. Faraoni, A. Giusti, B.H. Fahim, Spherical inhomogeneous solutions of Einstein and scalar-tensor gravity: a map of the land. arXiv: 2101.00266
10. R. Gannouji, D. Polarski, A. Ranquet, A.A. Starobinsky, JCAP **0609**, 016 (2006)
11. K.A. Bronnikov, S.B. Fadeev, A.V. Michtchenko, Scalar fields in multidimensional gravity. No-hair and other no-go theorems. Gen. Rel. Grav. **35**, 505–525 (2003)
12. K.A. Bronnikov, S.V. Grinyok, Electrically charged and neutral wormhole instability in scalar-tensor gravity. Grav. Cosmol. **11**, 75–81 (2005). arXiv: gr-qc/0509062
13. J.D. Barrow, S. Cotsakis, Inflation and the conformal structure of higher-order gravity theories. Phys. Lett. B **214**, 515 (1988)
14. K. Maeda, Towards the Einstein-Hilbert action via conformal transformation. Phys. Rev. D **39**, 3159 (1989)
15. D. Hochberg, M. Visser, Dynamic wormholes, anti-trapped surfaces, and energy conditions. Phys. Rev. D **58**, 044021 (1998). arXiv: gr-qc/9802046
16. E. Bittencourt, R. Klippert, G.B. Santos, Dynamical wormhole definitions confronted. Class. Quantum Grav. **35**, 155009 (2018). arXiv: 1707.01078
17. A.V.B. Arellano, N. Breton, R. Garcia-Salcedo, Some properties of evolving wormhole geometries within nonlinear electrodynamics. Gen. Rel. Grav. **41**, 2561–2578 (2009). arXiv: 0804.3944
18. K.A. Bronnikov, Nonlinear electrodynamics, regular black holes and wormholes. Int. J. Mod. Phys. D **27**, 1841005 (2018). arXiv: 1711.00087
19. K.A. Bronnikov, P.E. Kashargin, S.V. Sushkov, Possible wormholes in a Friedmann universe. Universe **9**, 465 (2023). arXiv: 2309.03166
20. K.A. Bronnikov, Scalar-tensor theory and scalar charge. Acta Phys. Polon. B **4**, 251 (1973)
21. C. Barceló, M. Visser, Scalar fields, energy conditions, and traversable wormholes. Class. Quantum Grav. **17**, 3843 (2000). arXiv: gr-qc/0003025
22. K.A. Bronnikov, S.V. Grinyok, Charged wormholes with non-minimally coupled scalar fields. Existence and stability, in *Festschrift in Honour of Prof. Mario Novello, Rio de Janeiro* (2002). arxiv:gr-qc/0205131
23. K.A. Bronnikov, Scalar-tensor gravity and conformal continuations. J. Math. Phys. **43**, 6096–6115 (2002). arXiv: gr-qc/0204001
24. E.F. Eiroa, M.G. Richarte, C. Simeone, Thin-shell wormholes in Brans-Dicke gravity. Phys. Lett. A **373**, 1 (2008). arXiv: 0809.1623
25. I.Z. Fisher, Scalar mesostatic field with regard for gravitational effects. Zh. Eksp. Teor. Fiz. **18**, 636 (1948). arXiv: gr-qc/9911008
26. R. Penney, Generalization of the Reissner-Nordström solution to the Einstein field equations. Phys. Rev. **182**, 1383 (1969)
27. D.D. Sokolov, A.A. Starobinsky, Structure of the curvature tensor at conical singularities. Sov. Phys. Dokl. **22**, 312–314 (1977)
28. C. Brans, R.H. Dicke, Mach's principle and a relativistic theory of gravitation. Phys. Rev. **124**, 925 (1961)
29. A. Bhattacharia, I. Nigmatzyanov, R. Izmailov, K.K. Nandi, Brans-Dicke wormhole revisited. Class. Quantum Grav. **26**, 235017 (2009). arXiv: 0910.1109
30. F.S.N. Lobo, M.A. Oliveira, General class of vacuum Brans-Dicke wormholes. Phys. Rev. D **81**, 067501 (2010). arXiv: 1001.0995
31. F.S.N. Lobo, M.A. Oliveira, Wormhole geometries in $f(R)$ modified theories of gravity. Phys. Rev. D **80**, 104012 (2009). arXiv: 0909.5539
32. K.A. Bronnikov, G. Clement, C.P. Constantinidis, J.C. Fabris, Cold scalar-tensor black holes: causal structure, geodesics, stability. Grav. Cosmol. **4**(2), 128–138 (1998)
33. K.A. Bronnikov, C.P. Constantinidis, R.L. Evangelista, J.C. Fabris, Electrically charged cold black holes in scalar-tensor theories. Int. J. Mod. Phys. D **8**, 481–505 (1999). arXiv: gr-qc/9902050

34. K.A. Bronnikov, M.S. Chernakova, F(R) theory of gravity and conformal continuations. Izv. Vuzov, Fizika 9, 46–51 (2005). Russ. Phys. J. 48, 940 (2005). arXiv: gr-qc/0503025
35. S.A. Appleby, R.A. Battye, A.A. Starobinsky, Curing singularities in cosmological evolution of F(R) gravity. JCAP 1006, 005 (2010). arXiv: 0909.1737v2
36. V. Gorini, A.Y. Kamenshchik, U. Moschella, V. Pasquier, A.A. Starobinsky, Tolman-Oppenheimer-Volkoff equations in presence of the Chaplygin gas: stars and wormhole-like solutions. Phys. Rev. D **78**, 064064 (2008). arXiv: 0807.2740
37. K.A. Bronnikov, S.V. Sushkov, Trapped ghosts: a new class of wormholes. Class. Quantum Grav. **27**, 095022 (2010). arXiv: 1001.3511
38. K.A. Bronnikov, V.G. Krechet, Potentially observable cylindrical wormholes without exotic matter in GR. Phys. Rev. D **99**, 084051 (2019). arXiv: 1807.03641
39. S.V. Bolokhov, K.A. Bronnikov, M.V. Skvortsova, Cylindrical wormholes: a search for viable phantom-free models in GR. Int. J. Mod. Phys. D **28**, 1941008 (2019). arXiv: 1903.09862
40. M. Bugaev, I. Novikov, S. Repin, P. Samorodskaya, I. Novikov Jr, Observing of background electromagnetic radiation of the real sky through the throat of a wormhole. arXiv: 2412.06872

Memories of Gravitational Shockwaves

Dmitri V. Fursaev

Abstract Gravitational shockwaves produce perturbations of field systems. We study classical scalar and electromagnetic fields and gravitational memory effects left after the action on the fields of plane-fronted gravitational shockwaves. The gravitational memory plays a key role for the choice of Cauchy data which determine the perturbations. We demonstrate that field systems'remember' only a spatial 'profile' of the shock, but not the form of the signal. Moreover the dependence on the spatial profile can be expressed in a geometric way as a transformation of fields under coordinate supertranslations defined near the shockwave front. We also discuss applications of our results to astrophysics.

1 Introduction

Gravitational shockwaves (GSW) are solutions to the Einstein equations which describe short gravitational pulses with rapid oscillations of the Riemann curvature on the shockwave fronts. Inside the shockwave the gravity is strong, as opposed to weak gravitational waves, which are ripples of the space-time ruled by the linearized Einstein equations.

Gravitational shockwaves produce perturbations of classical electromagnetic (EM), scalar, or gravitational fields, which may result in interesting physical effects. Here we study perturbations generated by plane-fronted shockwaves which propagate in flat space-times. We suppose that gravitational shockwaves are created by different sources which move with the speed of light. One may expect that this situation models gravitational waves produced by the motion of a broader class of ultra-relativistic objects. The geometries we consider here have a unique feature: their Einstein tensor is linear in the metric.

Suppose a plane-fronted shockwave is of a 'sandwich' type, that is, it is concentrated between the two fronts (null hypersurfaces): the past, $u = \delta$, and the future,

D. V. Fursaev (✉)
Bogoliubov Laboratory of Theoretical Physics, Joint Institute for Nuclear Research, Dubna, Moscow Region, Russian Federation
e-mail: fursaev@theor.jinr.ru

A. Barvinsky and A. Kamenshchik (eds.), *Open Issues in Gravitation and Cosmology*, Fundamental Theories of Physics 225, https://doi.org/10.1007/978-3-032-15703-4_20

$u = 0$, fronts, where u is a retarded time and $\delta > 0$. Consider a free field ϕ (which is a scalar or EM field). The problem for perturbations $\hat{\phi}$ of ϕ caused by the GSW is the following characteristic Cauchy problem in the causal past of the GSW:

$$P\hat{\phi} = 0, \quad u \geq \delta, \tag{1}$$

$$\hat{\phi} = \phi_0, \quad u = \delta, \tag{2}$$

where P is a corresponding relativistic second order hyperbolic differential operator. The problem can be solved once the Cauchy data ϕ_0 are known. Suppose $\bar{\phi}$ is the unperturbed field just before the shock, at $u = 0$. To solve (1), (2) one needs to determine how $\bar{\phi}$ transforms under the action of the GSW in the interval $0 < u < \delta$. In general, this may be a difficult task since the gravity in this interval is strong and field equations are essentially nonlinear in the metric.

To put it another way, we pose the question: how does ϕ_0 'remember' shock gravitational waves? We show that for a class of GSW described above field perturbations do not 'remember' the shape of the gravitational signal but depend on its spatial profile. Moreover, the gravitational memory effect appears as a special coordinate transformation near the past GSW front:

$$\phi_0 = \mathcal{L}_\zeta \bar{\phi}, \quad u = \delta. \tag{3}$$

The operation $\mathcal{L}_\zeta$ is the Lie derivative along a vector field ζ, and on the shockwave front ζ coincides with generator of well known Penrose supertranslations [1].

The rest of the work is organized as follows. In Sect. 2 we give necessary definitions and describe properties of plane-fronted shockwaves, including impulsive limit $\delta \to 0$, when the GSW geometry is treated in a sense of distributions. We start, in Sect. 3 with the memory effect for trajectories of test particles under the action of the GSW. In our approximation, test particles before and after the shockwave pulse move along geodesics in the Minkowsky space-time. After the shockwave, mutual orientations of 4-velocities $\bar{u}^\mu$ of the particles change as

$$\delta u^\mu = \mathcal{L}_\zeta \bar{u}^\mu, \quad u = \delta. \tag{4}$$

We then continue with perturbations of scalar fields and show that the effect of GSW is completely determined by the supertranslations identical to those which appear in the impulsive limit. Perturbations of EM fields by the shockwaves are analyzed in Sect. 4. Due to the gauge invariance solutions to the Maxwell equations can be uniquely determined by components of the Maxwell strength which are normal to constant u null hypersurfaces. We demonstrate that the Cauchy problem for these components has the form of (1)–(3). Concluding remarks with a review of possible applications of these results are presented in Sect. 5.

2 Plane-Fronted Gravitational Shockwaves

We start with necessary definitions. For a gravitational shockwave moving along the x direction, from the left to the right, the space-time metric is chosen to be

$$ds^2 = -dvdu - H(v, u, y)du^2 + dy_i^2. \tag{5}$$

The retarded and advanced time coordinates are $u = t - x$, $v = t + x$. With vectors $l_\mu = -\delta^u_\mu$, which are null normals to constant u hypersurfaces, the metric tensor can be written as $g_{\mu\nu} = \eta_{\mu\nu} - Hl_\mu l_\nu$, where $\eta_{\mu\nu}$ is the Minkowsky metric. The properties of the wave are described by H which we take as

$$H(v, u, y) = \chi(u)f(v, y), \tag{6}$$

where $f(v, y)$ is called the 'profile' function. Evolution of the shock in the retarded time is determined by $\chi(u)$ which can be called the 'signal' function. We also define the integrated signal function

$$\bar{\chi}(u) = \int_{-\infty}^{u} \chi(u')du'. \tag{7}$$

To normalize the profile and signal functions we assume that $\bar{\chi}(u) \to 1$ at $u \to \infty$. In most important cases listed below (which are a subclass of Kundt space-times [2]) f does not depend on v. Thus, one can call $\chi(u)$ and f the time and spatial profiles of the GSW, respectively.

In this paper we consider gravitational sandwich-like perturbations whose signal function is non-vanishing only in some time interval, say, when $0 < u < \delta$. Inside interval $\chi(u)$ can be absolutely arbitrary. At $u < 0$, $u > \delta$ metric (5) coincides with the Minkowsky metric. As has been mentioned above, the important property of (5) is that non-vanishing components of its Einstein tensor are linear in H,

$$G_{uu} = \frac{1}{2}\partial_i^2 H, \quad G_{ui} = \partial_v \partial_i H, \quad G_{ij} = 2\delta_{ij}\partial_v^2 H. \tag{8}$$

Therefore in the Einstein theory the shockwaves are produced by sources with the stress-energy tensor

$$\begin{aligned} T_{\mu\nu} &= \chi(u)\left(\sigma l_\mu l_\nu + j_i(l_\mu e^i_\nu + l_\nu e^i_\mu) + p\delta_{ij}e^i_\mu e^j_\nu\right), \\ \sigma &= f_{,ii}/\kappa, \quad j_i = -2f_{,vi}/\kappa, \quad p = 4f_{,vv}/\kappa, \quad \kappa = 16\pi G. \end{aligned} \tag{9}$$

Vectors $e^i_\mu = \delta^i_\mu$ are tangent to constant u null hypersurfaces. Examples of the profile functions for different sources are given in Table 1.

Table 1 Examples of ultra-relativistic sources with energy distribution σ

Ultrarelativistic source	Trajectory	$\sigma(y)$	$f(y)$
Massless point particle	$u=\rho=0$	$E_p\delta(\rho)$	$8GE_p \ln \rho$
Global monopole	$u=\rho=0$	E_m/ρ	$16\pi G E_m \rho$
Null straight cosmic string	$u=y_1=0$	$E_s\delta(y_1)$	$8\pi G E_s\|y_1\|$
Domain wall	$u=0$	E_w	$4\pi G E_w \rho^2$

The parameters E_p, E_m, E_s, E_w are energies of the sources, $\rho=(y_i^2)^{1/2}$

Note that in addition to our choice of H in (6) there are other types of sandwich-like gravitational plane waves. For example, one can take $H=h_{ij}(u)y^iy^j$, see [3, 4]. The source of such waves is a sequence of domain walls with the surface energies $\sigma=\frac{1}{2}h_{ii}(u)$. We do not consider these metrics here.

An immediate consequence of (8) is that $G_{\mu\nu}$ is well-defined in the limit of extremely short signals, when $\chi(u)\to\delta(u)$. The trajectory of wave front of GSW is then localized exactly on the null hypersurface $u=0$.

A singular geometry sourced by (9) with $\chi(u)=\delta(u)$ can be described by using an approach by Penrose [1] and soldering transformations of [5]. The idea is as follows. One cuts $R^{1,3}$ by the front $u=0$ into two 'halfs', $\mathcal{M}^+$ $(u>0)$ and $\mathcal{M}^-$ $(u<0)$. Then $\mathcal{M}^\pm$ are "glued" again along $u=0$ with the following identification of points:

$$v_+ = v - f(v,y), \quad y_+^i = y^i. \tag{10}$$

Supertranslations (10) leave invariant metric on $u=0$ and, thus, make an infinite dimensional group of isometries of $u=0$, also known as Carroll transformations [6, 7].

With the help of a null vector n at $u=0$, such as $(n\cdot l)=-1$, $(n\cdot e_i)=0$, one defines the transverse curvature, introduced in [8, 9],

$$C_{ab}=n_{\mu;\nu}e_a^\mu e_b^\nu, \tag{11}$$

where $e_v=l$, $e_i=e^i$. Then

$$\sigma=\frac{2}{\kappa}[C_{ii}], \quad j_i=-\frac{2}{\kappa}[C_{vi}], \quad p=\frac{2}{\kappa}[C_{vv}], \tag{12}$$

$$[C_{ab}]\equiv C_{ab}^+-C_{ab}^-, \tag{13}$$

where $C_{ab}^\pm$ correspond to $\mathcal{M}^\pm$. In the considered case

$$[C_{vv}]=2[\partial_u g_{vv}], \qquad [C_{vi}]=[\partial_u g_{vi}], \qquad [C_{ii}]=\frac{1}{2}[\partial_u g_{ii}]. \tag{14}$$

To know $\partial_u g_{\mu\nu}^+$ on $u=0$ one needs coordinate transformations of $\mathcal{M}^+$

$$x_+^\mu = x^\mu - \zeta^\mu(x), \tag{15}$$

$$\zeta^\mu(x) \simeq f\delta_v^\mu + \frac{u}{2}\eta^{\mu a} f_{,a}. \tag{16}$$

which are an "extension" of the supertranslations to a neighborhood of $u = 0$. Vector $\zeta^\mu(x)$ is completely determined [5] by the requirement that (16) do not change the metric $g_{\mu\nu}$ at $u = 0$. Since in a vicinity of $u = 0$ deviation of the metric from the flat one is nontrivial,

$$g^+_{\mu\nu} \simeq g^-_{\mu\nu} + \mathcal{L}_\zeta g^-_{\mu\nu} = \eta_{\mu\nu} + \zeta_{\mu,\nu} + \zeta_{\nu,\mu}, \tag{17}$$

one finds:

$$\partial_u g^+_{vv}|_{u=0} = f_{,vv}, \qquad \partial_u g^+_{vi}|_{u=0} = f_{,vi}, \qquad \partial_u g^+_{ij}|_{u=0} = f_{,ij}. \tag{18}$$

Substitution of (18) in (12) yields (9).

3 Test Particles and Scalar Fields After Gravitational Shocks

Consider a particle with 4-velocity $\bar{u}^\mu$ before it meets the shockwave. According to geodesic equations the velocity changes due to interaction with the shock. In the linear in H approximation

$$\frac{du^\mu}{du} = -l^\mu(H_{,\nu}\bar{u}^\nu) - \frac{1}{2}H^{,\mu} + O(H^2), \tag{19}$$

where we took into account that $(\bar{u} \cdot l) = -1$. It follows from (7), (19) that the velocity acquires two parts

$$u^\mu \simeq \bar{u}^\mu + \chi(u)u_1^\mu + \bar{\chi}(u)u_2^\mu,$$

where

$$u_1^\mu = -\frac{1}{2}l^\mu f, \qquad u_2^\mu = -\frac{1}{2}(f_{,\nu}\bar{u}^\nu) - \frac{1}{2}l^\mu f. \tag{20}$$

The term $\chi(u)u_1^\mu$ in the velocity implies that the advanced time on the particle trajectory changes as $\delta v(u) = -\bar{\chi}(u)f$, in agreement with (10). Terms proportional to $\chi(u)$ disappear after a time interval exceeding δ.

We suppose that the profile function changes slowly during the action of the shock, $|f_{,\mu}|\delta \ll |f|$. Then residual transformations of the particle coordinates and the velocity just after the shock can be written as a pure coordinate transformations

(4) generated by vector (16). To see this one should take into account that $\bar{\chi}(u) = 1$ at $u > \delta$.

Thus, the gravitational memory effect for particles moving through the shockwave has a pure geometrical form and is determined by the profile of the wave f. Particles do not 'remember' the form of the gravitational signal $\chi(u)$.

For different type of impulsive plane waves, waves given by (5) with $H = h_{ij}(u)y^i y^j$, the gravitational memory effect for particles has been studied in [10, 11].

Consider now a free scalar field ϕ with the equation:

$$(-\Box + m^2)\phi = j, \tag{21}$$

where j is a source. If one takes into into account that $g^{\mu\nu} = \eta^{\mu\nu} + Hl^\mu l^\nu$, $\det g_{\mu\nu} = -1/4$ and $l^\mu \partial_\mu = 2\partial_v$ the operator in (21) can be written as

$$\Box = \partial_\mu g^{\mu\nu}\partial_\nu = -\bar{\Box} + 4\partial_v H \partial_v.$$

$\bar{\Box} = -4\partial_u\partial_v + \partial_i^2$ is defined for the flat metric. Suppose $\bar{\phi}$ is a solution in the Minkowsky spacetime

$$(-\bar{\Box} + m^2)\bar{\phi} = \bar{j}. \tag{22}$$

We define perturbation due to the shockwave as a difference

$$\hat{\phi} \equiv \phi - \bar{\phi}, \quad u > \delta, \tag{23}$$

by assuming that

$$j = \bar{j}, \quad u > \delta. \tag{24}$$

The equation for the perturbations, which follows from (21)–(23), is

$$(-\bar{\Box} + m^2)\hat{\phi} = J + \chi(u)J_g(\phi, f), \tag{25}$$

$$J = j - \bar{j}, \qquad J_g(\phi, f) = -\partial_v(f\partial_v\phi). \tag{26}$$

Here J takes into account an action of the GSW on the current. Since we are interested in the linear in f effects one can approximate $J_g(\phi, f) \simeq J_g(\bar{\phi}, f)$. Then the term $\chi(u)J_g(\bar{\phi}, f)$ is an effective current induced by rapidly changing gravitational field on steady scalar background $\bar{\phi}$.

Now one can proceed as in the case of geodesics and look for a solution to (25) during the action of the shock in the form

$$\hat{\phi} \simeq \chi(u)\phi_1 + \bar{\chi}(u)\phi_2, \quad 0 < u < \delta, \tag{27}$$

where dependence on u of functions ϕ_k can be neglected. The same reasonigs also require that in this interval

$$J \simeq \chi(u)i_1 + \bar{\chi}(u)i_2. \tag{28}$$

However $i_2 = 0$, according to our definition of the perturbation, see (24).

If (27), (28) are substituted into (25) there appear terms, in the right hand and left hand sides of the equation, which are proportional to $\partial_u \chi(u)$, $\chi(u)$, $\bar{\chi}(u)$. Hence one comes to constraints:

$$\partial_v \phi_1 = 0, \tag{29}$$

$$4\partial_v \phi_2 - (\partial_i - m^2)\phi_1 = i_1 + 4\partial_v (f \partial_v \bar{\phi}), \tag{30}$$

which do not depend on the time profile of the GSW. Solution to (29) is trivial, $\phi_1 = 0$, since perturbation is zero in the far past, as $v \to -\infty$. Then $4\partial_v \phi_2 = i_1 + 4\partial_v \mathcal{L}_\zeta \bar{\phi}$ and just behind the wave front

$$\hat{\phi} = \mathcal{L}_\zeta \bar{\phi} + \phi_c, \quad u \simeq \delta, \tag{31}$$

where ϕ_c is related to the change of the source on the wavefront

$$\phi_c(v, y) = \frac{1}{4} \int_{-\infty}^{v} \mathrm{d}v' i_1(v', y). \tag{32}$$

One can call ϕ_c a contact term since it comes out from points where the source of the field contacts the wave front. The contact terms are not related to the gravitational memory of the field. If these terms are omitted, (31) coincides with (2), (3). For a point-like source of the field terms J and ϕ_c are related to a change of the trajectory of the source.

Effects of gravitational shockwaves on scalar fields have been studied in many publications starting with [12]. Much attention to this effect has been attracted in the context of the graviton dominance in scattering of ultra-high energy particles [13]. Scalar fields were considered there as plane waves $\bar{\phi} = e^{ikx}$, while shockwaves were taken in the impulsive limit. In this case condition (31) (without contact terms) implies transformation of particle's 4-momentum k^μ in the complete agreement with results of this Section. A comprehensive analysis of scalar field perturbations generated by gravitational shockwaves in the impulsive limit along with a discussion of how contact terms are taken into account in the Cauchy data can be found in [14].

4 Gravitational Shocks and Maxwell Fields

Now we extend this analysis to the case of EM fields under the action of the GSW. Consider Maxwell equations on shockwave spacetime (5) and in the Minkowsky spacetime

$$\partial_\mu F^{\mu\nu} = j^\nu, \qquad \partial_\mu \bar{F}^{\mu\nu} = \bar{j}^\nu, \tag{33}$$

respectively. As a result of the specific structure of (5) the both equations look similar, but indexes of $F^{\mu\nu}$ are risen with the help of the inverse metric tensor $g^{\mu\nu}$. Also conservation laws for the currents are:

$$\partial_\mu j^\mu = \partial_\mu \bar{j}^\mu = 0.$$

Let $\bar{F}^{\mu\nu}$ be the Maxwell strength of EM field before the shock. We define its perturbation as the difference

$$\hat{F}_{\mu\nu} \equiv F_{\mu\nu} - \bar{F}_{\mu\nu}, \quad u > \delta.$$

One can show that (33) lead to equations

$$\partial_\mu \hat{F}^{\mu\nu} = J^\nu + \chi(u) J_g^\nu(F, f), \tag{34}$$

where $J^\nu = j^\nu - \bar{j}^\nu$ is a variation of the current during the shock, and

$$J_g^\nu(F, f) = \partial_\mu \left[f l^\alpha (l^\nu \eta^{\mu\beta} - l^\mu \eta^{\nu\beta}) F_{\alpha\beta} \right]. \tag{35}$$

It is clear that $\partial_\mu J^\mu = \partial_\mu(\chi(u) J_g^\mu) = \partial_\mu J_g^\mu = 0$ and $l_\mu J_g^\mu = 0$. In the first approximation one can put $J_g^\nu(F, f) \simeq J_g^\nu(\bar{F}, f)$. After some algebra one can write

$$J_g^\nu(\bar{F}, f) = -2 f l_\mu \mathcal{L}_\zeta \bar{F}^{\mu\nu} - f l^\nu l_\alpha \partial_\lambda \bar{F}^{\alpha\lambda}. \tag{36}$$

One can interpret $\chi(u) J_g^\nu(\bar{F}, f)$ as an effective current, tangent to constant u null hypersurfaces in the region $0 < u < \delta$ of the GSW pulse. This current is generated by rapidly changing gravitational field on a smooth EM background $\bar{F}_{\mu\nu}$. The mechanism which yields $\hat{F}^{\mu\nu}$ is analogous to the well-known generation of transition radiation [15, 16] when a charged particle crosses the boundary of two physical media. Therefore one can also call $\hat{F}^{\mu\nu}$ transition radiation generated by gravitational shockwaves.

Our aim is to show that the transition radiation on shockwaves is determined by the gravitational memory effect and is expressed in a form of coordinate transformation (3). We suppose that the transition radiation and variation of the currents in the interval $0 \le u \le \delta$ have the form:

$$\hat{F}^{\mu\nu} \simeq \chi(u) p^{\mu\nu} + \bar{\chi}(u) q^{\mu\nu}, \tag{37}$$

$$J^\mu \simeq \chi(u) i_1^\mu + \bar{\chi}(u) i_2^\mu, \tag{38}$$

similar to variations of 4-velocities of particles. Here $p^{\mu\nu}$, $q^{\mu\nu}$, i_k^μ are some quantities which change slowly on the above interval. The gravitational memory effect is determined by $q^{\mu\nu}$ and i_2^μ.

By requiring that terms at $\partial_u \chi(u)$, $\chi(u)$, $\bar{\chi}(u)$ in the Bianchi identities and in the conservation law $\partial_\mu J^\mu = 0$ vanish one gets the restrictions:

$$p_{(\mu\nu}l_{\rho)} = 0, \quad p_{(\mu\nu,\rho)} = q_{(\mu\nu}l_{\rho)}, \tag{39}$$

$$l_\mu i_1^\mu = 0, \qquad \partial_\mu i_1^\mu = l_\nu i_2^\nu, \qquad \partial_\mu i_2^\mu = 0, \tag{40}$$

where symbol $(\mu\nu\ldots\rho)$ denotes complete symmetrization of the indices. Indices are risen and lowered with the flat metric. Substitution of (37), (38) to Maxwell equations (34) yields:

$$l_\mu p^{\mu\nu} = 0, \tag{41}$$

$$\partial_\mu p^{\mu\nu} - l_\mu q^{\mu\nu} = i_1^\nu + J_g^\nu, \tag{42}$$

$$\partial_\mu q^{\mu\nu} = i_2^\nu. \tag{43}$$

From (39), (41) we find

$$p_{\mu\nu} = \sigma_i(l_\mu e_\nu^i - l_\nu e_\mu^i), \tag{44}$$

$$q_{\mu\nu} = e_\mu^i \sigma_{i,\nu} - e_\nu^i \sigma_{i,\mu} + \rho_i(l_\mu e_\nu^i - l_\nu e_\mu^i), \tag{45}$$

where σ_i and ρ_i are some functions. Equations (44), (45) imply that

$$\partial_\mu p^{\mu\nu} + l_\mu q^{\mu\nu} = -\sigma_{i,i} l^\nu, \tag{46}$$

which together with (36), (44) yields

$$l_\mu q^{\mu\nu} = l_\mu \mathcal{L}_\zeta \bar{F}^{\mu\nu} - \mathcal{J}^\nu, \tag{47}$$

$$\mathcal{J}^\nu = -\frac{1}{2}(i_1^\nu + f l^\nu l_\alpha \bar{j}^\alpha + \sigma_{i,i} l^\nu). \tag{48}$$

Also it follows from (43), (45) that $\partial_v(\sigma_{i,i}) = i_2^v$. Thus the quantity $\mathcal{J}^\nu$ is solely determined by the current across the shockwave front. Like in the above case of scalar fields $\mathcal{J}^\nu$ indicates possible contributions to transition radiation when the source immediately contacts the shockwave.

We conclude, on quite general grounds, that normal components of the strength tensor change as

$$l_\mu \hat{F}^{\mu\nu} \simeq \bar{\chi}(u)\left(l_\mu \mathcal{L}_\zeta \bar{F}^{\mu\nu} - \mathcal{J}^\nu\right). \tag{49}$$

Outside domains, where sources contact the shockwave, the change is generated by the Lie derivative along the supertranslation vector (16).

The Cauchy data fix components $\hat{F}_{vi}$ and $\hat{F}_{vu}$. In fact, $\hat{F}_{uv}$ is not independent since it follows from the Maxwell equations that

$$\hat{F}_{uv} = -\frac{1}{2}\int_{-\infty}^{v} \partial_i \hat{F}_{vi}\, dv'. \tag{50}$$

Also it follows from the Bianchi identities that

$$\hat{F}_{ui} = \int_{-\infty}^{v} \left(\partial_u \hat{F}_{vi} + \partial_i \hat{F}_{uv}\right) dv',$$
$$\hat{F}_{ij} = \int_{-\infty}^{v} \left(\partial_i \hat{F}_{vj} - \partial_j \hat{F}_{vi}\right) dv'. \tag{51}$$

It is assumed that $\bar{F}_{\mu\nu}$ vanish as $v \to -\infty$ and so does $\hat{F}_{\mu\nu}$. Since the perturbations satisfy homogeneous Maxwell equations, we come to the characteristic problem for the two components

$$\Box \hat{F}_{vi} = 0, \quad u > \delta, \tag{52}$$

$$\hat{F}_{vi} = \mathcal{L}_\zeta \bar{F}_{vi} = \partial_v (f \bar{F}_{vi}), \quad u = \delta. \tag{53}$$

The rest components are determined by (50), (51). Equations (52), (53) prove (1)–(3) for the case of EM fields.

5 Applications of the Results

Cauchy data (31), (53) demonstrate our main statement about geometrical form of the gravitational memory in field systems after gravitational shockwaves (5), (6). Given the Cauchy data perturbations can be found by either numerical or analytical methods for shockwaves produced by different sources.

The fact that scalar fields and normal components of the Maxwell strength tensor $F_{\mu\nu}$ change in the same way as trajectories of observers crossing the shockwave front does not imply that field systems behave smoothly across the fronts. One can check that observers, after the GSW, measure jumps of $\partial_u \phi$, F_{ui} and of tangent components of the field stress-energy tensors. Thus, in general, one has two effects: conversion of non-stationary perturbations into an outgoing radiation and a plane shockwave of the field itself accompanying the initial gravitational wave. There may also appear a spherical shockwave when the gravitational wave hits the source of the field [14].

The effect of GSW on EM fields can be potentially observable for astrophysical objects, such as magnetars, which create strong magnetic fields. It has been suggested in [17] that GSW from compact ultra-relativistic objects which move near a magnetar generate short EM pulses. These pulses can be a novel engine mechanism for fast radio bursts. The fast radio bursts are experimentally observed bright pulses of radio frequencies which are not yet fully explained, see [18, 19].

Analogous effects are produced by null cosmic strings moving near magnetars [20]. Null cosmic strings are hypothetical objects, they are cosmic strings which move with the speed of light and have a finite energy per unit length. Formally the metric of a null string belongs to shockwave class (5), (6), so our analysis applies to this case as well. A number of other physical effects caused by null strings on EM fields are considered in [21]. In particular, it is shown that parts of an incoming EM wave crossing the string horizon (null hypersurface $u = 0$) from different sides of the string are refracted with respect to each other and leave behind the string a wedge-like region of interference.

We have not discussed here perturbations of gravitational fields generated by plane-fronted GSW. For the case of null cosmic strings these perturbations have been found and described in [22]. The late-time asymptotics of the perturbations have a form of outgoing gravitational waves. They have a number of interesting properties at the future null infinity. By using arguments developed above it would be interesting to extend results of [22] to GSW.

6 In the Memory of A. A. Starobinsky

I would like to say several words about Alexei Aleksandrovich Starobinsky in Dubna, a small city on the banks of the Volga river. I hope the remarks below disclose additional features of A. A.Starobinsky's personality.

Dubna is known as a location of the Joint Institute for Nuclear Research, or JINR, the famous international scientific organization. A. A. Starobinsky maintained close ties with Dubna for many years. From 2019 to 2023 he was a member of staff at the Bogoliubov Laboratory of Theoretical Physics of JINR. When the project of Dubna International Advanced School of Theoretical Physics (DIAS-TH) was launched at BLTP JINR in 2003, Alexey Aleksandrovich first became its lecturer and then the principal researcher. Here interests and main activities of Alexei Aleksandrovich have been focused on the promotion of modern ideas in cosmology and astrophysics among students and young physicsts coming to Dubna from all over the world. He was deeply involved in preparation of international schools and meetings, by working on their programmes and inviting best speakers. His talks on current trends in cosmology always attracted many people from JINR. Thanks to his efforts and authority, these schools were truly advanced.

As is known, JINR is also the host of the Russian mega-science project, a heavy ion collider called NICA. When A. A.Starobinsky also became a Chair of the Russian Gravitational Society, he, Mikhail Vasil'evich Sazhin (1951–2023) and the author of this paper had organized the Workshop "The Russian National Experiment in Field of Gravitational Physics". The Workshop took place in May 2018 at the Dubna State University, also located in Dubna. The purpose of the meeting was to discuss a Russian state-funded experimental programme in gravity, by analogy with mega-science projects in other areas of physics. Our initiative has been welcome by the leading Russian physicists and technical experts during this meeting.

Alexey Aleksandrovich strongly supported the need of the national gravitational experimental programme. I hope the work he started in Dubna may be revived in future.

References

1. R. Penrose, The geometry of impulsive gravitational waves, in *General Relativity: Papers in Honour of J. L. Synge* (L. O'Raifeartaigh, 1972), pp. 101–115
2. W. Kundt, The plane-fronted gravitational waves. Z. Phys. **163**, 77–86 (1961)
3. G.W. Gibbons, Quantized fields propagating in plane-wave spacetimes. Commun. Math. Phys. **45**, 191 (1975)
4. J. Garriga, E. Verdaguer, Scattering of quantum particles by gravitational plane waves. Phys. Rev. D **43**, 391 (1991)
5. M. Blau, M. O'Loughlin, Horizon shells and BMS-like soldering transformations. JHEP **03**, 029 (2016). e-Print: 1512.02858 [hep-th]
6. C. Duval, G.W. Gibbons, P.A. Horvathy, Conformal Carroll groups. J. Phys. A **47**(33), 335204 (2014). e-Print: 1403.4213 [hep-th]
7. C. Duval, G.W. Gibbons, P.A. Horvathy, P.M. Zhang, Carroll versus Newton and Galilei: two dual non-Einsteinian concepts of time. Class. Quantum Grav. **31**, 085016 (2014). e-Print: 1402.0657 [gr-qc]
8. C. Barrabes, W. Israel, Thin shells in general relativity and cosmology: the lightlike limit. Phys. Rev. D **43**, 1129 (1991)
9. E. Poisson, *A Reformulation of the Barrabes-Israel Null Shell Formalism*. e-Print: gr-qc/0207101 [gr-qc]
10. P.-M. Zhang, C. Duval, P.A. Horvathy, Memory effect for impulsive gravitational waves. Class. Quantum Grav. **35**(6), 065011 (2018). e-Print: 1709.02299 [gr-qc]
11. R. Steinbauer, The memory effect in impulsive plane waves: comments, corrections, clarifications. Class. Quantum Grav. **36**(9), 098001 (2019). e-Print: 1811.10940 [gr-qc]
12. T. Dray, G. 't Hooft, The gravitational shock wave of a massless particle. Nucl. Phys. B **253**, 173 (1985)
13. G. 't Hooft, Graviton dominance in ultrahigh-energy scattering. Phys. Lett. B **198**, 61 (1987)
14. D.V. Fursaev, E.A. Davydov, I.G. Pirozhenko, V.A. Tainov, Perturbations of classical fields by gravitational shockwaves. JHEP **11**, 039 (2024). e-Print: 2408.05142 [gr-qc]
15. V.L. Ginzburg, I.M. Frank, Radiation of a uniformly moving electron due to its transition from one medium into another. J. Phys. (USSR) **9**, 353 (1945)
16. V.L. Ginzburg, V.N. Tsytovich, Several problems of the theory of transition radiation and transition scattering. Phys. Rept. **49**, 1 (1979)
17. D.V. Fursaev, E.A. Davydov, V.A. Tainov, *Fast Radio Bursts and Electromagnetic Transition Radiation on Gravitational Shockwaves*. e-Print: 2503.18644 [gr-qc]
18. E. Petroff, J.W.T. Hessels, D.R. Lorimer, Fast radio bursts at the dawn of the 2020s. Astron. Astrophys. Rev. **30**(1), 2 (2022). e-Print: 2107.10113 [astro-ph.HE]
19. B. Zhang, Fast radio bursts from interacting binary neutron star systems. Astrophys. J. Lett. **890**, L24 (2020). e-Print: 2002.00335 [astro-ph.HE]
20. D.V. Fursaev, I.G. Pirozhenko, Electromagnetic waves generated by null cosmic strings passing pulsars. Phys. Rev. D **109**(2), 025012 (2024). e-Print: 2309.01272 [gr-qc]
21. D.V. Fursaev, I.G. Pirozhenko, Electrodynamics under the action of null cosmic strings. Phys. Rev. D **107**(2), 025018 (2023). e-Print: 2212.05564 [gr-qc]
22. D.V. Fursaev, E.A. Davydov, I.G. Pirozhenko, V.A. Tainov, Gravitational waves generated by null cosmic strings. Phys. Rev. D **109**(12), 125009 (2024). e-Print: 2311.01863 [gr-qc]

Constraining Non-Newtonian Gravity and Axionlike Particles in the Context of the Standard Cosmological Model

Galina L. Klimchitskaya and Vladimir M. Mostepanenko

Abstract The constraints on corrections of Yukawa type to the Newton gravitational law and on the coupling constant of axionlike particles to nucleons are considered in their relationship to the Standard Cosmological Model. It has been known that Newton's law gains the Yukawa-type correction owing to the effects of extra-dimensional physics. As to the axionlike particles, they are considered as the most probable candidates for the constituents of dark matter. After a summary of the strongest constraints on the strength of Yukawa interaction and the axion-to-nucleon coupling constant obtained from different experiments, we present the new constraints following from recent measurement of the Casimir-Polder interaction of Ar atoms with a silicon nitride nanograting. It is shown that the novel constraints on the Yukawa interaction and on the coupling constant of axionlike particles to nucleons are stronger by up to the factors 33.4 and 24.4, respectively, over the certain interaction ranges than the constraints obtained previously from measuring the Casimir force. Future progress along these lines may result not only in the stronger limits on extra dimensions and axionlike particles, but in the experimental confirmation of their existence in nature.

1 Introduction

The Yukawa- and power-type corrections to Newton's gravitational potential have long been considered in the literature [1] in connection with the so-called *fifth force* supplementary to the familiar gravitational, electromagnetic, weak and strong interactions. The power-type corrections result from an exchange of massless hypothetical particles between atoms of the closely spaced macrobodies [2–5] or from the extra-dimensional unification schemes [6, 7]. The Yukawa-type corrections to Newtonian gravity come about from an exchange of light massive scalar particles predicted in

G. L. Klimchitskaya · V. M. Mostepanenko (✉)
Central Astronomical Observatory at Pulkovo of the Russian Academy of Sciences, Saint Petersburg, Russia
e-mail: vmostepa@gmail.com

Peter the Great Saint Petersburg Polytechnic University, Saint Petersburg, Russia

A. Barvinsky and A. Kamenshchik (eds.), *Open Issues in Gravitation and Cosmology*, Fundamental Theories of Physics 225, https://doi.org/10.1007/978-3-032-15703-4_21

the supersymmetric theories and extra-dimensional models with large extra dimensions [5, 8–13]. A somewhat more complicated potential arises from an exchange of the pairs of axionlike pseudoscalar particles between hadrons [1, 14–17].

Constraints on the corrections to Newton's gravitational law have been obtained from the gravitational experiments of Eötvös- and Cavendish-type [18–22], experiments on neutron scattering [23–25], and measurements of the Casimir force [26–35]. The constraints on certain of the parameters of axionlike particles and associated potential of axion-to-nucleon interaction were obtained from the gravitational experiments [36, 37], magnitometer measurements [38], measurement of the spin-dependent interactions between two protons in the beam of molecular hydrogen [39, 40], and the Casimir force [35, 41–50]. A lot of constraints on the interaction of axions with electrons and photons have been obtained as well (see [15, 51–53] for a review).

Below we consider the constraints on axionlike particles and non-Newtonian gravity in the context of the Standard Cosmological Model. This is the most developed modern approach to understanding of our Universe. It is based on Einstein's general theory of relativity, Friedmann's solution for the homogeneous isotropic expanding spacetime, and Gamov's theory of the hot Universe. The Standard Cosmological Model assumes that the cosmological constant has a nonzero value, which explains the acceleration of the Universe expansion discovered in the end of 90th by analyzing the spectra of remote supernovae and galaxies [54, 55]. It also includes the model of invisible cold dark matter as the most plausible explanation for the nonbaryonic matter observed solely by the gravitational effects.

The Standard Cosmological Model is very successful by correctly describing the abundance of hydrogen, helium, and lithium in the Universe, the spectrum of relic radiation, and the large-scale structure in the patterns of galaxies. The Einstein equations with the cosmological term allow a reasonably good description of the discovered effect of an accelerated expansion. During the last twenty years, the picture of the evolution of the Universe, given by the Standard Cosmological Model, was confirmed by several important experiments of the observational cosmology [56–60].

The observational cosmology also confirmed that there was the cold inflationary stage in the Universe evolution when its expansion has been going on exponentially fast, as described by the de Sitter solution of Einstein's equations. The solution of this kind was obtained by substituting the vacuum expectation values of the stress-energy tensor of quantized fields in the right-hand side of Einstein's equations [61, 62] or under an impact of the classical minimally coupled inflaton field [63–65].

The great contribution to the development of inflationary cosmology was made by A.A. Starobinsky. Specifically, he demonstrated that the creation of hypothetical particles, scalarons, and their decay into usual particles transform the exponentially fast expansion of the Universe to the power-type Friedmann one [62]. Another major achievement in understanding of the early Universe made by A.A. Starobinsky is the theory of reheating after inflation [66, 67]. According to this theory, the process of reheating occurs due to the effect of exponential increase of the number of particle-antiparticle pairs of bosons created from vacuum by the oscillating in time inflaton field [66, 67] (this effect was predicted earlier for the case of the time-periodic electric field [68, 69]).

A.A. Starobinsky obtained important results in the theory of pair creation by the gravitational field of expanding Universe [70–73]. These results are cited until the present time (see, e.g., Ref. [74]). He also suggested using the Abel-Plana formula for a regularization of the Casimir energy in the closed Friedmann Universe [73]. Later on, this formula was used in hundreds of papers by different authors devoted to the Casimir effect in spaces with nontrivial topology and in restricted quantization volumes.

Below we consider first the experimental constraints on the Yukawa-type corrections to Newtonian gravity. Keeping in mind that the Newton law of gravitation follows from the general theory of relativity in the weak field limit, the obtained constraints also restrict the application region of this theory. As a consequence, these constraints place restrictions on the Standard Cosmological Model in the regions of parameters where the quantum effects of an exchange of light hypothetical particles or large extra dimensions should be taken into account. We discuss next the constraints on the coupling constant of axionlike particles to nucleons following from different experiments. The discovery of axions would be the great success of the Standard Cosmological Model because these particles are considered as the most probable constituents of cold dark matter [51, 75].

After summarizing the obtained earlier strongest constraints on the Yukawa-type corrections to Newton's law of gravitation and the axion-to-nucleon coupling constant, we present new constraints following from the recent experiment on measuring the Casimir-Polder interaction between slow Ar atoms and a silicon nitride nanograting [76]. The use of the low-density atomic beams with less than 16 m/s velocities in this experiment allowed obtaining the diffraction patterns that are mainly determined by the Casimir-Polder interaction. Based on the previous investigations employing the slow atomic beams [77, 78] and the quantum model of atomic diffraction on a nanograting [79], the strength of the Casimir-Polder interaction was measured on a metrology level with careful analysis of both the systematic and random errors [76].

According to our results [80], the new constraints on the Yukawa-type interaction are up to a factor 33.4 stronger than all the previously obtained ones from measurements of the Casimir force within the interaction range from 1 to 2 nm. In this interaction range they are second to only the constraints found from the experiments on neutron scattering [23–25]. The new constraints on the coupling constant of axionlike particles to nucleons following from the experiment [76] are stronger by up to a factor of 24.4 in the range of axion masses from 32.2 to 100 eV than that ones found previously from measuring the Casimir force.

The paper is organized as follows. Section 2 is devoted to the Yukawa-type interaction. Section 2.1 contains a summary of the constraints on the Yukawa-type corrections to Newtonian gravity obtained so far from measurements of the Casimir force and other experiments. In Sect. 2.2, we present new constraints on the Yukawa-type interaction following from recent measurements of the Casimir-Polder interaction between atoms and a nanograting. In Sect. 3, the axionlike particles are considered.

Section 3.1 is devoted to the strongest constraints on the coupling constant of axion-like particles to nucleons following from various experiments. The new constraints on the axion-to-nucleon coupling obtained from measuring the Casimir-Polder interaction are presented in Sect. 3.2. Section 4 contains our conclusions and a discussion.

We use the system of units where $\hbar = c = 1$.

2 Yukawa-Type Interaction and Constraints on Its Parameters

It is customary to present the interaction potential of Yukawa-type as a correction to Newton's gravitational potential between two pointlike particles of masses m_1 and m_2

$$V_{\text{Yu}}(r) = -\frac{Gm_1m_2}{r}\left(1 + \alpha e^{-r/\lambda}\right), \tag{1}$$

where G is the gravitational constant, r is the separation distance between particles, α is the Yukawa interaction constant, and λ is the interaction range.

If an initiation of the Yukawa-type correction to Newton's potential is caused by the exchange of scalar particles with some small mass m between the particles m_1 and m_2 (see Sect. 1) the parameter λ has the meaning of the Compton wavelength of this particle, $\lambda = 1/m$. If, however, as mentioned in Sect. 1, the Yukawa-type correction arises due to the presence of relatively large extra dimensions with the compactification scale R_*, it holds $\lambda \sim R_*$.

Thus, for the extra-dimensional models with the low compactification energy of the order of 1 TeV, the compactification scale $R_* \sim 1$ mm and 5 mm for two and three extra dimensions, respectively [11]. In this regards, it is particularly remarkable that at separations below 10 μm the available experimental data do not exclude an existence of corrections to the Newtonian potential which exceed it by many orders of magnitude [1]. Hence, by constraining the Yukawa-type corrections to Newtonian gravitational potential, one is simultaneously verifying the theoretical predictions of fundamental physics beyond the Standard Model.

2.1 Strongest Constraints on the Yukawa-Type Potential from Various Experiments

The most evident source of constraints on the Yukawa-type corrections to Newtonian gravity is the gravitational experiments of Eötvös- and Cavendish-type. If the parameter α in Eq. (1) takes a nonzero value, this results in the violation of the equivalence principle tested by the Eötvös-type experiments. The Cavendish-type experiments are sensitive to a deviation of the measured force from the inverse square power law in r. Such kind deviations arise for any nonzero α.

For different values of the interaction range λ, the strongest constraints on the strength of Yukawa potential α follow from different gravitational experiments. With the aim of constraining the extra-dimensional models, the ranges of nanometer and micrometer values of λ are the most interesting where the Newton gravitational law is lacking of sufficient experimental confirmation.

It is customary to present the constraints on the Yukawa-type interaction following from some experiment as a line in the (λ, α) plane where the region above the line is excluded by the measurement data and the region below the line is allowed. In Fig. 1, the constraints following from the Cavendish-type experiments are shown by the line 1. These constraints were obtained from the experiments of Refs. [20, 21, 81]. At larger λ up to $\lambda = 1$ cm the strongest constraints on α follow from the Cavendish-type experiments of Refs. [20–22, 82].

Within the interaction range $\lambda > 1$ cm, strong constraints on the Yukawa-type interaction are imposed by the Eötvös-type experiments [19, 83]. The constraints on the Yukawa-type corrections following from the gravitational experiments at short

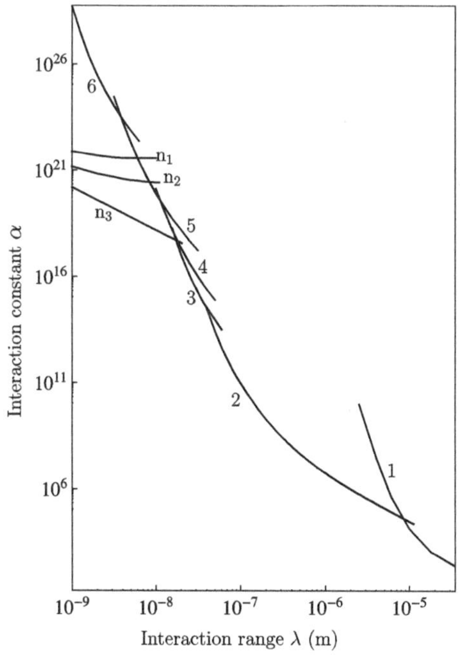

Fig. 1 Constraints on the strength of Yukawa-type interaction obtained from the Cavendish-type experiments (line 1), the Casimir-less experiment (line 2), from measuring the effective Casimir pressure (line 3), the Casimir force between corrugated surfaces (line 4), the lateral Casimir force (line 5), the Casimir force between a SiC plate and a borosilicate microsphere (line 6), and from the experiments on neutron scattering (lines labeled n_1, n_2, and n_3) are shown as the functions of the interaction range. The regions of (λ, α) plane above each line are excluded and below are allowed by the measurement data

separations and interaction ranges are very weak. For instance, from Eq. (1) and line 1 in Fig. 1 one can see that for $r = \lambda = 10\ \mu$m the correction exceeding Newtonian gravity by the factor of 10^4 is not excluded experimentally. Such a big addition to the Newtonian potential can hardly be called a "correction". However, for much larger $\lambda = 1$ cm one finds $\alpha < 4 \times 10^{-4}$ [1], i.e., the Yukawa-type correction to Newtonian gravity remains negligibly small at any separation.

As is seen in Fig. 1, line 1, the strength of constraints on the Yukawa-type potential following from the gravitational experiments quickly decreases with decreasing λ. The point is that at sufficiently small λ and r the main background force is already not the gravitational one but the Casimir force determined by the zero-point and thermal fluctuations of the electromagnetic field [84]. For this reason, with decreasing λ, the strongest constraints on the Yukawa-type interaction were obtained either from the differential experiments, where the Casimir force was nullified, or from the sufficiently precise measurements of this force. In so doing, the Newtonian gravitational force remained negligibly small.

Thus, the most precise differential force measurement [34] was performed between a sapphire sphere coated with Cr and Au layers and a plate consisting of Si and Au parts coated with Cr and Au overlayers. The strongest constraints on the Yukawa-type interaction obtained in Ref. [34] cover the wide interaction range from $\lambda = 40$ nm to $\lambda = 8\ \mu$m. They are shown by the line 2 in Fig. 1.

The constraints following from measurements of the effective Casimir pressure in the configuration of an Au-coated sphere and an Au-coated plate by means of the micromechanical torsional oscillator [31, 85] are shown by the line 3 in Fig. 1. These constraints are the strongest amongst those obtained from measurements of the Casimir force within the range 17.2 nm $< \lambda <$ 40 nm (for $\lambda <$ 17.4 nm stronger constraints follow from the experiments on neutron scattering, see below).

Line 4 in Fig. 1 shows the constraints on the Yukawa-type interaction obtained [86] from measuring the Casimir force between the Au-coated surfaces of a sphere and a plate covered with the longitudinal sinusoidal corrugations [87, 88]. This experiment was performed by means of an atomic force microscope preserving some nonzero angle between the corrugation axes. The constraints of line 4 are the strongest ones, as compared with the other constraints obtained from measuring the Casimir force within the range 11.6 nm $< \lambda <$ 17.2 nm.

By line 5 in Fig. 1, we indicate the constraints on the Yukawa-type potential obtained [32] from the experiment on measuring the lateral Casimir force between the sinusoidally corrugated Au-coated surfaces of a sphere and a plate [89, 90]. The lateral force arises when the corrugations are coaxial and possess some nonzero phase shift. These constraints are the strongest ones in comparison with the other constraints obtained from the Casimir experiments over the range of λ from 3.7 nm to 11.6 nm.

Finally, within the shortest interaction range from $\lambda = 1$ nm to $\lambda = 3.7$ nm, considered here, the strongest constraints imposed by the Casimir effect were obtained [35] from measurements of the Casimir force between a SiC plate and a borosilicate microsphere [91] performed within the separation region from 10 nm to 200 nm (i.e., at shorter separations than in all other experiments). They are shown by the line 6 in Fig. 1.

As to the strongest constraints in the nanometer separation range obtained so far by various means, they follow from measurements of the scattered intensity of neutrons on different scatterers. The scatterers used for this purpose are the xenon and helium gases [25], the gas of atomic xenon [24], and the silicon crystal [92]. The obtained constraints on Yukawa-type interaction are shown in Fig. 1 by the lines n_1, n_2, and n_3, respectively, in the region $\lambda \geqslant 1$ nm considered in this paper. It is seen that the constraints given by the line n_3 are the strongest ones up to $\lambda = 17.4$ nm, where the line n_3 intersects the line 3 representing the constraints following from measurements of the effective Casimir pressure.

2.2 *Constraints on the Yukawa-Type Potential from Measuring the Casimir-Polder Interaction Between Ar Atoms and Silicon Nitride Nanograting*

Here we demonstrate that the potentialities of the Casimir effect in constraining the hypothetical interactions are not exhausted, so that new experiments may result in major strengthening of the already obtained constraints presented in Sect. 2.1.

In Ref. [76], the Casimir-Polder interaction was measured between slow, low-density beams of Ar atoms and a silicon nitride nanograting. The measurement results were presented as a value of the coefficient C_3 where, at the nanometer separation range between an Ar atom and a silicon nitride grating, the Lifshitz formula for the Casimir-Polder interaction energy can be conveniently rewritten as [84]

$$V_{\text{C-P}}(z) = \frac{C_3 f(z)}{z^3} \tag{2}$$

with

$$C_3 \equiv \varphi(z = 3\,\text{nm}), \qquad f(z) = \begin{cases} \frac{\varphi(z)}{\varphi(z=3\,\text{nm})}, & z \geqslant 3\,\text{nm}, \\ 1, & z < 3\,\text{nm}, \end{cases}$$

$$\varphi(z) = -\frac{k_B T}{8} {\sum_{l=0}^{\infty}}{}' \alpha(i\zeta_l\omega_c) \int_{\zeta_l}^{\infty} dy\, e^{-y} \left[(2y^2 - \zeta_l^2) r_{\text{TM},l} - \zeta_l^2 r_{\text{TE},l}\right]. \tag{3}$$

Here, $\alpha(\omega)$ is the dynamic polarizability of an atom, the dimensionless Matsubara frequencies are $\zeta_l = 4\pi k_B T z l$, $l = 0, 1, 2, \ldots$, $\omega_c = 1/(2z)$, T is the temperature, k_B is the Boltzmann constant, and the prime on the sum in l divides the term with $l = 0$ by two. The coefficients $r_{\text{TM},l}$ and $r_{\text{TE},l}$ are the standard Fresnel coefficients for the transverse magnetic (TM) and transverse electric (TE) polarizations calculated at the pure imaginary Matsubara frequencies. They are expressed via the dielectric function of silicon nitride. The normalization for the quantity $\varphi(z = 3\,\text{nm})$ used in Eq. (3) leads to $f(z) \leqslant 1$ [93].

The measurement data of Ref. [76] led to the following value of the coefficient C_3 in Eq. (2):

$$C_3 = (6.9 \pm 1.2)\ \text{meV}\,\text{nm}^3. \tag{4}$$

The total error indicated in this equation at the 95% confidence level takes into account both the random and systematic errors combined in quadrature. In doing so the systematic errors include the deviations of the nanograting from the perfect rectangular geometry, the Lennard-Jones repulsion between the nanograting walls and Ar atom bypassing them in close proximity, the losses which occur when Ar atoms come into contact with the walls, the finite area of the walls and their roughness. Within the limits of measurement errors, the measured value of C_3 in Eq.(4) is in agreement with the theoretical one in Eqs. (2) and (3) [76].

The Yukawa-type interaction potential between an atom spaced at the separation z from a nanograting wall, which can be modeled with sufficient precision by the infinitely large plate of thickness $D = 97$ nm, is obtained by the integration of Eq. (1) over the wall volume. The result is [84]

$$V_{\text{Yu}}^{\text{AW}}(z) = -2\pi G m_{\text{A}} \rho_{\text{W}} \alpha \lambda^2 e^{-z/\lambda} \left(1 - e^{-D/\lambda}\right), \tag{5}$$

where, in this case, $m_{\text{A}} = 6.63 \times 10^{-23}$ g is the mass of an Ar atom and $\rho_{\text{W}} = 3.17\ \text{g/cm}^3$ is the density of a silicon nitride wall [94]. As was noted in Sect. 2.1, the Newtonian gravitational contribution is much less than (5) and can be neglected.

Now the constraints on the parameters of Yukawa-type interaction (5), which was not observed in the experiment of Ref. [76], can be obtained from the inequality

$$\left| V_{\text{Yu}}^{\text{AW}}(z) \right| < \frac{\Delta C_3}{z^3}, \tag{6}$$

where, according to Eq. (4), the total error in the measured value of the coefficient C_3 is equal to $\Delta C_3 = 1.2\ \text{meV}\,\text{nm}^3$.

The constraints on the parameters of Yukawa-type interaction α and λ following from Eq. (6) are shown by the line labeled C-P in Fig. 2 [80]. For each specific value of λ, the value of z within the experimental range of atom-wall separations was chosen for obtaining the strongest constraint. For comparison purposes, in Fig. 2 we reproduce from Fig. 1 the lines 6, 5, 4, and 3 following from different experiments on measuring the Casimir force and the lines n_1, n_2, and n_3 following from the experiments on neutron scattering.

As is seen in Fig. 2, the new constraints following from measurements of the Casimir-Polder interaction, are stronger than the constraints obtained from measuring the Casimir force between a SiC plate and a borosilicate microsphere (line 6), and the lateral and normal Casimir forces between the sinusoidally corrugated surfaces

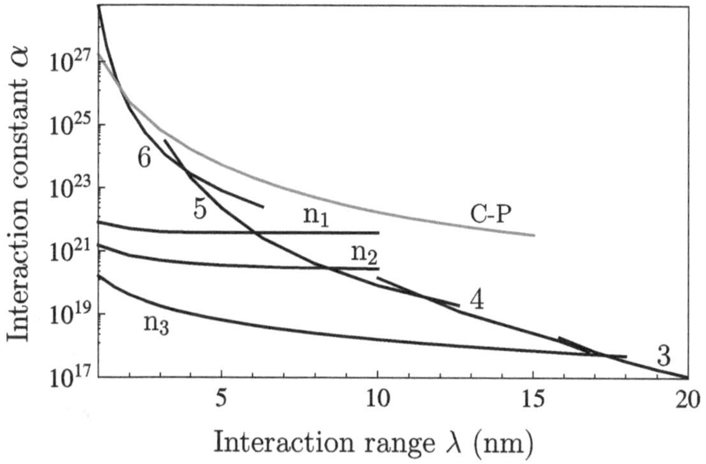

Fig. 2 Constraints on the strength of Yukawa-type interaction obtained from measurements of the Casimir-Polder interaction between Ar atoms and a silicon nitride microsphere are shown by the line labeled C-P as a function of the interaction range. For comparison purposes, the lines 3, 4, 5, and 6 are reproduced from Fig. 1. The regions of (λ, α) plane above each line are excluded and below are allowed by the measurement data

of a sphere and a plate (lines 5 and 4, respectively). The major strengthening, as compared to the line 6, is by the factor of 33.4 at $\lambda = 1$ nm.

By and large the novel constraints are stronger than those found previously from the Casimir experiments over the interaction range from $\lambda = 1$ nm to $\lambda = 2$ nm, where the line C-P intersects the line 6. The novel constraints following from measuring the Casimir-Polder interaction rank below of the constraints following from the experiments on neutron scattering [24, 25, 92] shown by the lines n_1, n_2, and n_3 in Fig. 2.

Further improvements in the experimental setup of Ref. [76] and associated decrease of the systematic errors may lead to stronger constraints on the interaction of Yukawa type than those shown by the lines n_1, n_2 and n_3.

3 Constraints on the Coupling Constant of Axionlike Particles to Nucleons

As discussed in Sect. 2, the Standard Cosmological Model includes the concept of cold dark matter for the invisible matter which manifests itself by only the gravitational interaction. An understanding of the physical nature of dark matter, which contributes up to 27% of the Universe energy, is one of the most important unresolved problems of astrophysics and cosmology. The axionlike particles predicted in many extensions of the Standard Model of elementary particles are considered as the most probable constituents of dark matter.

Originally the pseudoscalar particles, called axions, were introduced [95, 96] as the Nambu-Goldstone bosons responsible for a violation of the Peccei and Quinn symmetry [9]. An interaction of the axion field $a(x)$ with the fermion field $\psi(x)$ is described by the pseudovector Lagrangian density [14, 15]

$$\mathcal{L}_{pv}(x) = \frac{g}{2m_a}\bar{\psi}(x)\gamma_5\gamma_\mu\psi(x)\partial_\mu a(x), \tag{7}$$

where γ_5 and γ_μ are the Dirac matrices, $\mu = 0, 1, 2, 3$, the constant g is dimensionless and m_a is the axion mass. As a result, the effective coupling constant $g/(2m_a)$ is dimensional.

In the theories of the Grand Unification, the axionlike particles were also introduced which interact with the fermion field $\psi(x)$ via the pseudoscalar Lagrangian density

$$\mathcal{L}_{ps}(x) = -ig\bar{\psi}(x)\gamma_5\psi(x)a(x), \tag{8}$$

where g is the dimensionless coupling constant.

Both the pseudovector and pseudoscalar Lagrangians result in the common potential between two nucleons of mass m spaced at the distance r due to an exchange of one axion with the mass m_a [97]

$$\begin{aligned} V_{an}(r, \boldsymbol{\sigma}_1, \boldsymbol{\sigma}_2) = \frac{g^2}{16\pi m^2}\left[(\boldsymbol{\sigma}_1, \boldsymbol{n})(\boldsymbol{\sigma}_2, \boldsymbol{n})\left(\frac{m_a^2}{r} + \frac{3m_a}{r^2} + \frac{3}{r^3}\right)\right. \\ \left. - (\boldsymbol{\sigma}_1, \boldsymbol{\sigma}_2)\left(\frac{m_a}{r^2} + \frac{1}{r^3}\right)\right]. \end{aligned} \tag{9}$$

This potential depends on the nucleon spins $\boldsymbol{\sigma}_1$ and $\boldsymbol{\sigma}_2$, $\boldsymbol{n}$ is the unit vector directed from one nucleon to another. Using the potential (9), the constraints on the coupling constant g of axions and axionlike particles to nucleons can be obtained from experiments exploiting the polarized test bodies. This is, however, not the case for measurements of the Casimir force performed up to date.

The spin-independent potential arises due to the process of an exchange of two axionlike particles described by the Lagrangian density (8) between two nucleons [1, 17]

$$V_{aan}(r) = -\frac{g^4}{32\pi^3 m^2}\frac{m_a}{r^2}K_1(2m_a r), \tag{10}$$

where $K_1(z)$ is the modified Bessel function of the second kind.

Experiments on searching the additional forces caused by the potentials (9) and (10) help to constrain the parameters g and m_a and may finally lead to the discovery of axions and axionlike particles, which are also searched using the interaction of these hypothetical particles with electrons and photons [15, 51–53].

3.1 Strongest Constraints on the Interaction of Axionlike Particles with Nucleons from Various Experiments

Similar to the case of the Yukawa-type interaction, the constraints on the parameters of axionlike particles, m_a and g are presented as the lines on the (m_a, g) plane, where the regions above each line are excluded by the measurement data and below each line are allowed. It is conventional to place the constraints not on the coupling constant of axionlike particles to nucleons g but on the quantity $g^2/(4\pi)$.

In the region of the smallest axion masses from 1 neV to 1 μeV, the strongest constraints on g were obtained by measuring an interaction between the spin-polarized K and ^{3}He atoms [38]. These constraints are shown by the line m in Fig. 3. The constraints of line m are rather strong because the sought for interaction is determined by the exchange of one axion or axionlike particle [see the potential (9) which is proportional to g^2].

Within the range of axion masses from 1 μeV to 0.676 meV, the strongest constraints on g, following from the Cavendish-type experiments [20, 21], are shown by the line 1 in Fig. 3. The constraints of this line and of the lines 2-6 considered below

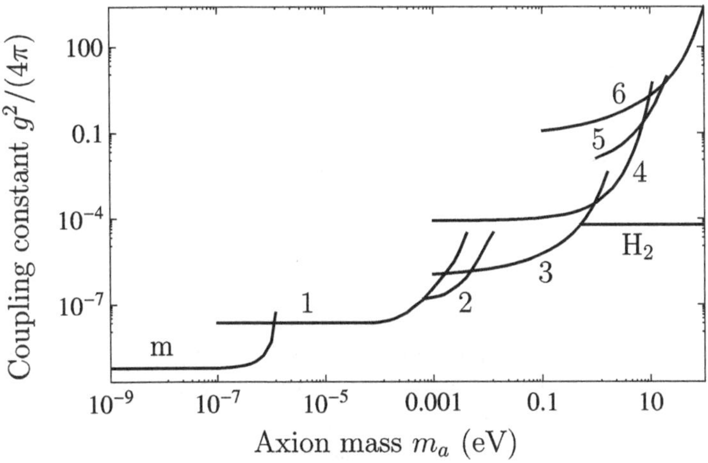

Fig. 3 Constraints on the coupling constant of axionlike particles to nucleons obtained from the magnetometer measurements (line labeled m), Cavendish-type experiments (line 1), measurements of smallest forces by means of the torsional oscillator (line 2), differential force measurements (line 3), from measuring the effective Casimir pressure (line 4), the lateral Casimir force (line 5), the Casimir force between a SiC plate and a borosilicate microsphere (line 6), and of the interaction of protons in the beam of molecular hydrogen (line labeled H_2) are shown as the functions of the axion mass. The regions of $(m_a, g^2/(4\pi))$ plane above each line are excluded and below are allowed by the measurement data

are weaker because the used test bodies were unpolarized and the interaction under study was determined by the exchange of two axionlike particles [see the potential (10) which is proportional to g^4].

The strongest constraints in the range from $m_a = 0.676$ meV to $m_a = 4.9$ meV shown by the line 2 in Fig. 3 were obtained [98] from measurements of the smallest forces of gravitational strength by means of the torsional oscillator [36, 37].

The constraints following from the differential force measurement [34], where the contribution of the Casimir force was nullified, were obtained in Ref. [45]. These constraints shown by the line 3 in Fig. 3 are the strongest ones among all the other constraints derived from the measurement data of Casimir experiments within the range of masses of axionlike particles from 4.9 meV to 0.9 eV (note that for $m_a > 0.5$ eV the stronger constraints follow from measuring the forces between protons in the beam of molecular hydrogen, see below).

Line 4 in Fig. 3 shows the constraints on the coupling constant of axionlike particles to nucleons obtained [43] from measuring the effective Casimir pressure by means of the micromechanical torsional oscillator [31, 85]. These constraints are the strongest ones comparing to other constraints found from the Casimir experiments within the range from 0.9 eV $< m_a <$ 8 eV.

By the line 5 in Fig. 3, we indicate the constraints following [44] from measurements of the lateral Casimir force between the sinusoidally corrugated surfaces [89, 90]. These are the strongest constraints imposed by the Casimir effect within the range from $m_a = 8$ eV to 17.8 eV.

Finally, the line 6 in Fig. 3 shows the constraints on the coupling constant g obtained [35] in the range 17.8 eV $< m_a <$ 100 eV from measurements of the Casimir force between a SiC plate and a borosilicate microsphere [91].

As was noted above, however, in the range of axion masses $m_a > 0.5$ eV the strongest constraints on g follow not from the Casimir physics. They were obtained [40] by comparing with theory the measurement data for the dipole-dipole forces between protons in the beam of molecular hydrogen [39] (see the line H_2 in Fig. 3). These constraints are the strongest ones up to $m_a = 200$ eV. As opposed to experiments on measuring the Casimir force, in this case the additional interaction arises due to an exchange of one axion or axionlike particle and is described by the spin-dependent potential (9) rather than (10). Even stronger constraints in the range of masses $m_a > 200$ eV were obtained in Ref. [40] by comparing with theory the measurement data for interaction of nucleons in deuterated molecular hydrogen in the nuclear magnetic resonance experiment. Here, the additional interaction is also determined by the exchange of one axion.

3.2 Constraints on the Interaction of Axionlike Particles with Nucleons from Measuring the Casimir-Polder Interaction Between Ar Atoms and Silicon Nitride Nanograting

In Sect. 2.2, we have already briefly presented the experimental results of Ref. [76] on precise measuring the Casimir-Polder interaction and obtained constraints imposed by these results on the interaction of Yukawa-type. Here, we present the constraints imposed by the same measurement results on the coupling constant of axionlike particles to nucleons [80].

Taking into account that both the Ar atoms and the nanograting are unpolarized, the interaction potential (9) due to an exchange of one axion vanishes after the averaging over the volume of the nanograting wall. Thus, the interaction in addition to the Casimir-Polder one may arise due to an exchange by two axions between the nucleons of an Ar atom and a silicon nitride wall described by the potential (10). Integrating Eq. (10) over the wall volume in the polar coordinates, one finds [41–47]

$$V_{aan}^{\mathrm{AW}}(z) = -\frac{m_a}{m^2 m_{\mathrm{H}}} C_{\mathrm{A}} C_{\mathrm{W}} \int_z^D dz_1 \int_0^\infty \rho d\rho \, \frac{K_1(2m_a\sqrt{\rho^2+z_1^2})}{\rho^2+z_1^2}, \tag{11}$$

where

$$C_{\mathrm{A}} = \frac{g^2}{4\pi}(Z_{\mathrm{A}}+N_{\mathrm{A}}), \qquad C_{\mathrm{W}} = \rho_{\mathrm{W}} \frac{g^2}{4\pi}\left(\frac{Z_{\mathrm{W}}}{\mu_{\mathrm{W}}}+\frac{N_{\mathrm{W}}}{\mu_{\mathrm{W}}}\right). \tag{12}$$

Here, $Z_{\mathrm{A,W}}$ and $N_{\mathrm{A,W}}$ are the numbers of protons and the mean numbers of neutrons in the Ar atom and in the silicon nitride molecules, respectively, ρ_{W} is the density of silicon nitride. In so doing [1]

$$Z_{\mathrm{W}} = 3Z_{\mathrm{Si}} + 4Z_{\mathrm{N}}, \qquad N_{\mathrm{W}} = 3N_{\mathrm{Si}} + 4N_{\mathrm{N}}, \tag{13}$$

and

$$\mu_{\mathrm{W}} = 3\mu_{\mathrm{Si}} + 4\mu_{\mathrm{N}}, \qquad \mu_{\mathrm{Si}} = \frac{m_{\mathrm{Si}}}{m_{\mathrm{H}}}, \qquad \mu_{\mathrm{N}} = \frac{m_{\mathrm{N}}}{m_{\mathrm{H}}}, \tag{14}$$

m_{Si} and m_{N} are the mean masses of Si and N atoms, respectively, m_{H} is the mass of atomic hydrogen.

Using the numerical data for all these parameters presented in Table 2.1 of Ref. [1], we obtain

$$C_{\mathrm{A}} = 39.985 \, \frac{g^2}{4\pi}, \qquad C_{\mathrm{W}} = 3.196 \times 10^3 \, \frac{g^2}{4\pi} \, \mathrm{kg\,m^{-3}}. \tag{15}$$

Substituting in Eq. (11) the integral representation [99]

$$K_1(x) = x \int_1^\infty du \sqrt{u^2 - 1}\, e^{-ux}, \tag{16}$$

and integrating with respect to the new variable $v = \sqrt{\rho^2 + z_1^2}$ in place of ρ, one arrives at

$$V_{aan}^{\rm AW}(z) = -\frac{1}{2m^2 m_{\rm H}} C_{\rm A} C_{\rm W} \int_1^\infty du \frac{\sqrt{u^2 - 1}}{u^2} \left(e^{-2m_a z u} - e^{-2m_a D u}\right). \tag{17}$$

Taking into account that in the experiment [76] on measuring the Casimir-Polder interaction no additional force was observed in the limits of the measurement errors, the constraints on the coupling constant of axionlike particles to nucleons follow from the inequality

$$\left| V_{aan}^{\rm AW}(z) \right| < \frac{\Delta C_3}{z^3}. \tag{18}$$

The obtained constraints on $g^2/(4\pi)$ are shown by the line labeled C-P in Fig. 4. For comparison purposes, the lines 5 and 6 in Fig. 4 reproduce the constraints already shown by these lines in Fig. 3, which were found from measuring the lateral Casimir

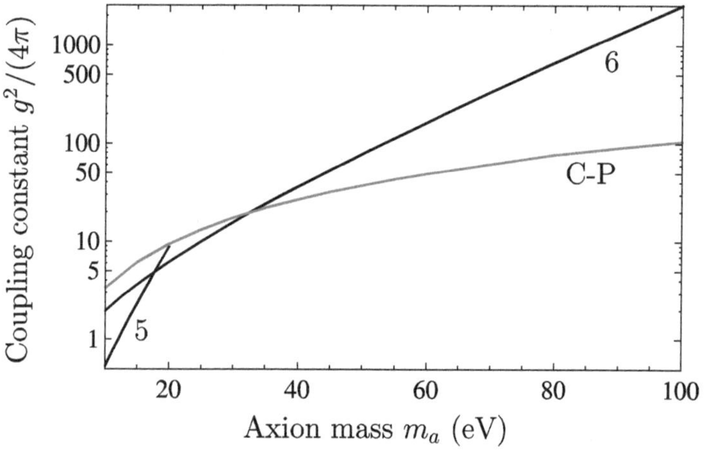

Fig. 4 Constraints on the coupling constant of axionlike particles to nucleons obtained from measurements of the Casimir-Polder interaction between Ar atoms and a silicon nitride microsphere are shown by the line labeled C-P as a function of the axion mass. For comparison purposes, the lines 5 and 6 are reproduced from Fig. 3. The regions of $(m_a,\ g^2/(4\pi))$ plane above each line are excluded and below are allowed by the measurement data

force between corrugated surfaces and the Casimir force between a SiC plate and a borosilicate microsphere, respectively.

As is seen from Fig. 4, the novel constraints are stronger than the previously known ones obtained from the Casimir physics over the range of masses of axionlike particles $m_a > 32.2$ eV. The largest strengthening by the factor of 24.4 holds for $m_a = 100$ eV. However, in this range of axion masses, the constraints on the axion-to-nucleon coupling constant following from measuring the Casimir-Polder interaction are second in strength to that ones shown by the line H_2 in Fig. 3 which were obtained from the process of an exchange of one axion between two nucleons.

4 Conclusions and Discussion

In the foregoing, we have considered the predicted corrections of Yukawa-type to Newton's law of gravitation and the interaction of axionlike particles to nucleons. Both these subjects are directly relevant to the Standard Cosmological Model which, in many respects, owes its origin to the seminal studies by A.A. Starobinsky. By constraining the Yukawa-type corrections to Newtonian gravity, we test the validity of the General Theory of Relativity in the region where the quantum effects and extra-dimensional physics may come into play. The search for additional forces determined by the exchange of axions between nucleons may finally bring not only the stronger constraints on the parameters of these hypothetical particles, but result in their discovery. This would solve one of the major problems of the Standard Cosmological Model on the nature of cold dark matter.

In this article, we have collected the current strongest constraints on the Yukawa-type corrections to Newton's gravitational law and on the coupling constant of axions to nucleons following from different experiments including measurements of the Casimir force. We have also presented new constraints following from the recent precise measurement of the Casimir-Polder interaction between Ar atoms and the silicon nitride nanograting. It was shown that the novel constraints exceed in strength all the previously known ones obtained in Casimir physics within the certain regions of parameters. For the Yukawa-type corrections to Newton's law, the obtained constraints are weaker than those found from the experiments on neutron scattering. As to the axionlike particles, they are second to only the constraints found from measuring the dipole-dipole forces between protons in the beam of molecular hydrogen. Recently it was shown [100] that under some conditions the constraints on the coupling constant of axionlike particles to nucleons obtained using the pseudoscalar Lagrangian density (8) using the process of two-axion exchange remain approximately valid for the coupling to nucleons of QCD axions described by the pseudovector Lagrangian density (7).

The experiments used above for constraining the Yukawa-type potential and the axion-to-nucleon coupling constant are also employed to constrain the hypothetical constituents of dark energy, which contributes 68% of the Universe energy (see Refs. [101, 102] for a review). An understanding of the physical nature of dark energy described by the cosmological constant is one more major problem of the Standard Cosmological Model. The expected further progress in the discussed here and other directions permits to hope that the remaining problems of the Standard Cosmological Model will be solved in not too remote future.

Acknowledgements This work was supported by the State Assignment for Basic Research (project FSEG-2026-0018).

References

1. E. Fischbach, C.L. Talmadge, *The Search for Non-Newtonian Gravity* (Springer-Verlag, New York, 1999)
2. A.A. Anselm, N.G. Uraltsev, A second massless axion? Phys. Lett. B **114**, 39–41 (1982)
3. G. Feinberg, J. Sucher, Long-Range Forces from Neutrino-Pair Exchange. Phys. Rev. **166**, 1638–1644 (1968)
4. S. Deser, B. Zumino, Broken Supersymmetry and Supergravity. Phys. Rev. Lett. **38**, 1433–1436 (1977)
5. Y. Fujii, The theoretical background of the fifth force. Int. J. Mod. Phys. A **6**, 3505–3557 (1991)
6. L. Randall, R. Sundrum, Large Mass Hierarchy from a Small Extra Dimension. Phys. Rev. Lett. **83**, 3370–3373 (1999)
7. L. Randall, R. Sundrum, An Alternative to Compactification. Phys. Rev. Lett. **83**, 4690–4693 (1999)
8. S. Dimopoulos, G.F. Guidice, Macroscopic forces from supersymmetry. Phys. Lett. B **379**, 105–114 (1996)
9. R.D. Peccei, H.R. Quinn, CP Conservation in the Presence of Pseudoparticles. Phys. Rev. Lett. **38**, 1440–1143 (1977)
10. I. Antoniadis, N. Arkani-Hamed, S. Dimopoulos, G. Dvali, New dimensions at a millimeter to a fermi and superstrings at a TeV. Phys. Lett. B **436**, 257–263 (1998)
11. N. Arkani-Hamed, S. Dimopoulos, G. Dvali, Phenomenology, astrophysics, and cosmology of theories with millimeter dimensions and TeV scale quantum gravity. Phys. Rev. D **59**, 086004 (1999)
12. E.G. Floratos, G.K. Leontaris, Low scale unification, Newton's law and extra dimensions. Phys. Lett. B **465**, 95–100 (1999)
13. A. Kehagias, K. Sfetsos, Deviations from $1/r^2$ Newton law due to extra dimensions. Phys. Lett. B **472**, 39–44 (2000)
14. J.E. Kim, Light pseudoscalars, particle physics and cosmology. Phys. Rep. **150**, 1–177 (1987)
15. G.G. Raffelt, Axions-Motivation, limits and searches. J. Phys. A: Math. Theor. **40**, 6607–6620 (2007)
16. J.E. Moody, F. Wilczek, New macroscopic forces? Phys. Rev. D **30**, 130–139 (1984)
17. F. Ferrer, M. Nowakowski, Higgs- and Goldstone-boson-mediated long range forces. Phys. Rev. D **59**, 075009 (1999)
18. J.H. Gundlach, G.L. Smith, E.G. Adelberger, B.R. Heckel, H.E. Swanson, Short-Range Test of the Equivalence Principle. Phys. Rev. Lett. **78**, 2523–2526 (1997)
19. G.L. Smith, C.D. Hoyle, J.H. Gundlach, E.G. Adelberger, B.R. Heckel, H.E. Swanson, Short-range tests of the equivalence principle. Phys. Rev. D **61**, 022001 (2000)
20. D.J. Kapner, T.S. Cook, E.G. Adelberger, J.H. Gundlach, B.R. Heckel, C.D. Hoyle, H.E. Swanson, Tests of the Gravitational Inverse-Square Law below the Dark-Energy Length Scale. Phys. Rev. Lett. **98**, 021101 (2007)

21. E.G. Adelberger, B.R. Heckel, S. Hoedl, C.D. Hoyle, D.J. Kapner, A. Upadhye, Particle-Physics Implications of a Recent Test of the Gravitational Inverse-Square Law. Phys. Rev. Lett. **98**, 131104 (2007)
22. W.-H. Tan, A.-B. Du, W.-C. Dong, S.-Q. Yang, C.-G. Shao, S.-G. Guan, Q.-L. Wang, B.-F. Zhan, P.-S. Luo, L.-C. Tu, J. Luo, Improvement for Testing the Gravitational Inverse-Square Law at the Submillimeter Range. Phys. Rev. Lett. **124**, 051301 (2020)
23. V.V. Nesvizhevsky, G. Pignol, K.V. Protasov, Neutron scattering and extra short range interactions. Phys. Rev. D **77**, 034020 (2008)
24. Y. Kamiya, K. Itagami, M. Tani, G.N. Kim, S. Komamiya, Constraints on New Gravitylike Forces in the Nanometer Range. Phys. Rev. Lett. **114**, 161101 (2015)
25. C.C. Haddock, N. Oi, K. Hirota, T. Ino, M. Kitaguchi, S. Matsumoto, K. Mishima, T. Shima, H.M. Shimizu, W.M. Snow, T. Yoshioka, Search for deviations from the inverse square law of gravity at nm range using a pulsed neutron beam. Phys. Rev. D **97**, 062002 (2018)
26. V.A. Kuzmin, I.I. Tkachev, M.E. Shaposhnikov, Restrictions imposed on light scalar particles by measurements of van der Waals forces. Pis'ma v Zh. Eksp. Teor. Fiz. **36**, 49–52 (1982). (JETP Lett. **36**, 59–62, 1982)
27. V.M. Mostepanenko, I.Y. Sokolov, The Casimir effect leads to new restrictions on long-range force constants. Phys. Lett. A **125**, 405–408 (1987)
28. M. Bordag, B. Geyer, G.L. Klimchitskaya, V.M. Mostepanenko, New constraints for non-Newtonian gravity in the nanometer range from the improved precision measurement of the Casimir force. Phys. Rev. D **62**, 011701(R) (2000)
29. V.M. Mostepanenko, M. Novello, Constraints on non-Newtonian gravity from the Casimir force measurements between two crossed cylinders. Phys. Rev. D **63**, 115003 (2001)
30. R.S. Decca, D. López, E. Fischbach, G.L. Klimchitskaya, D.E. Krause, V.M. Mostepanenkom, Precise comparison of theory and new experiment for the Casimir force leads to stronger constraints on thermal quantum effects and long-range interactions. Ann. Phys. (N.Y.) **318**, 37–80 (2005)
31. R.S. Decca, D. López, E. Fischbach, G.L. Klimchitskaya, D.E. Krause, V.M. Mostepanenko, Tests of new physics from precise measurements of the Casimir pressure between two gold-coated plates. Phys. Rev. D **75**, 077101 (2007)
32. V.B. Bezerra, G.L. Klimchitskaya, V.M. Mostepanenko, C. Romero, Advance and prospects in constraining the Yukawa-type corrections to Newtonian gravity from the Casimir effect. Phys. Rev. D **81**, 055003 (2010)
33. G.L. Klimchitskaya, U. Mohideen, V.M. Mostepanenko, Constraints on non-Newtonian gravity and light elementary particles from measurements of the Casimir force by means of a dynamic atomic microscope. Phys. Rev. D **86**, 065025 (2012)
34. Y.J. Chen, W.K. Tham, D.E. Krause, D. López, E. Fischbach, R.S. Decca, Stronger Limits on Hypothetical Yukawa Interactions in the 30–8000 Nm Range. Phys. Rev. Lett. **116**, 221102 (2016)
35. G.L. Klimchitskaya, P. Kuusk, V.M. Mostepanenko, Constraints on non-Newtonian gravity and axionlike particles from measuring the Casimir force in nanometer separation range. Phys. Rev. D **101**, 056013 (2020)
36. J.C. Long, H.W. Chan, A.B. Churnside, E.A. Gulbis, M.C.M. Varney, J.C. Price, Upper limits to submillimetre-range forces from extra space-time dimensions. Nature **421**, 922–925 (2003)
37. J.C. Long, V.A. Kostelecký, Search for Lorentz violation in short-range gravity. Phys. Rev. D **91**, 092003 (2015)
38. G. Vasilakis, J.M. Brown, T.R. Kornak, M.V. Romalis, Limits on New Long Range Nuclear Spin-Dependent Forces Set with a K-^{3}He Comagnetometer. Phys. Rev. Lett. **103**, 261801 (2009)
39. N.F. Ramsey, The tensor force between two protons at long range. Phys. A **96**, 285–289 (1979)
40. M.P. Ledbetter, M.V. Romalis, D.F. Jackson Kimball, Constraints on Short-Range Spin-Dependent Interactions from Scalar Spin-Spin Coupling in Deuterated Molecular Hydrogen. Phys. Rev. Lett. **110**, 040402 (2013)

41. V.B. Bezerra, G.L. Klimchitskaya, V.M. Mostepanenko, C. Romero, Constraints on the parameters of an axion from measurements of the thermal Casimir-Polder force. Phys. Rev. D **89**, 035010 (2014)
42. V.B. Bezerra, G.L. Klimchitskaya, V.M. Mostepanenko, C. Romero, Stronger constraints on an axion from measuring the Casimir interaction by means of a dynamic atomic force microscope. Phys. Rev. D **89**, 075002 (2014)
43. V.B. Bezerra, G.L. Klimchitskaya, V.M. Mostepanenko, C. Romero, Constraining axion-nucleon coupling constants from measurements of effective Casimir pressure by means of micromachined oscillator. Eur. Phys. J. C **74**, 2859 (2014)
44. V.B. Bezerra, G.L. Klimchitskaya, V.M. Mostepanenko, C. Romero, Constraints on axion-nucleon coupling constants from measuring the Casimir force between corrugated surfaces. Phys. Rev. D **90**, 055013 (2014)
45. G.L. Klimchitskaya, V.M. Mostepanenko, Improved constraints on the coupling constants of axion-like particles to nucleons from recent Casimir-less experiment. Eur. Phys. J. C **75**, 164 (2015)
46. G.L. Klimchitskaya, V.M. Mostepanenko, Constraints on axionlike particles and non-Newtonian gravity from measuring the difference of Casimir forces. Phys. Rev. D **95**, 123013 (2017)
47. G.L. Klimchitskaya, Recent breakthrough and outlook in constraining the non-Newtonian gravity and axion-like particles from Casimir physics. Eur. Phys. J. C **77**, 315 (2017)
48. V.M. Mostepanenko, G.L. Klimchitskaya, The State of the Art in Constraining Axion-to-Nucleon Coupling and Non-Newtonian Gravity from Laboratory Experiments. Universe **6**, 147 (2020)
49. G.L. Klimchitskaya, V.M. Mostepanenko, Dark Matter Axions, Non-Newtonian Gravity and Constraints on them from Recent Measurements of the Casimir Force in the Micrometer Separation Range. Universe **7**, 343 (2021)
50. G.L. Klimchitskaya, Constraints on Theoretical Predictions beyond the Standard Model from the Casimir Effect and Some Other Tabletop Physics. Universe **7**, 47 (2021)
51. G. Bertone, D. Hooper, J. Silk, Particle dark matter: Evidence, candidates and constraints. Phys. Rep. **405**, 279–390 (2005)
52. L.J. Rosenberg, K.A. van Bibber, Searches for invisible axions. Phys. Rep. **325**, 1–39 (2000)
53. I.G. Ivastorza, J. Redondo, New experimental approaches in the search for axion-like particles. Progr. Part. Nucl. Phys. **102**, 89–159 (2018)
54. A.G. Riess, A.V. Filippenko, P. Challis, A. Clocchiatti, A. Diercks, P.M. Garnavich, R.L. Gilliland, C.J. Hogan, S. Jha, R.P. Kirshner et al., Observational evidence from supernovae for an accelerating universe and a cosmological constant. Astron. J. **116**, 1009–1038 (1998)
55. S. Perlmutter, G. Aldering, G. Goldhaber, R.A. Knop, P. Nugent, P.G. Castro, S. Deustua, S. Fabbro, A. Goobar, D.E. Groom et al., Measurements of Ω and Λ from 42 high-redshift supernovae. Astrophys. J. **517**, 565–586 (1999)
56. M. Colless, G.B. Dalton, S.J. Maddox, W.J. Sutherland, P. Norberg, S. Cole, J. Bland-Hawthorn, T.J. Bridges, R.D. Cannon, C.A. Collins et al., The 2dF Galaxy Redshift Survey: Spectra and redshifts. Mon. Not. R. Astron. Soc. **328**, 1039–1063 (2001)
57. C.L. Bennett, D. Larson, J.L. Weiland, N. Jarosik, G. Hinshaw, N. Odegard, K.M. Smith, R.S. Hill, B. Gold, M. Halpern et al., Nine-Year Wilkinson Microwave Anisotropy Probe (WMAP) Observations: Final Maps and Results. Astrophys. J. Suppl. **208**, 20 (2013)
58. P. de Bernardis, P.A.R. Ade, J.J. Bock, J.R. Bond, J. Borrill, A. Boscaleri, K. Coble, B.P. Crill, G. De Gasperis, P.C. Farese et al., A Flat Universe from High-Resolution Maps of the Cosmic Microwave Background Radiation. Nature **404**, 955–959 (2020)
59. N. Aghanim, Y. Akrami, M. Ashdown, M. Ashdown, J. Aumont, C. Baccigalupi, M. Ballardini, A.J. Banday, R.B. Barreiro, N. Bartolo, et al., Planck Collaboration. Planck 2018 results. VI. Cosmological parameters. Astron. Astrophys. **641**, A6 (2020)
60. M. Tristram, A.J. Banday, M. Douspis, X. Garrido, K.M. Górski, S. Henrot-Versillé, L.T. Hergt, S. Iliĉ, R. Keskitalo, G. Lagache et al., Cosmological parameters derived from the final Planck data release (PR4). Astron. Astrophys. **682**, A37 (2024)

61. S.G. Mamayev, V.M. Mostepanenko, Isotropic cosmological models determined by the vacuum quantum effects. Zh. Eksp. Teor. Fiz. **78**, 20–27 (1980). (Sov. Phys. JETP **51**, 9–13, 1980)
62. A.A. Starobinsky, A new type of isotropic cosmological models without singularity. Phys. Lett. A **91**, 99–102 (1980)
63. A.H. Guth, Inflationary universe: A possible solution to the horizon and flatness problems. Phys. Rev. D **23**, 347–356 (1981)
64. A.D. Linde, A new inflationary universe scenario: A possible solution of the horizon, flatness, homogeneity, isotropy and primordial monopole problems. Phys. Lett. B **108**, 389–393 (1982)
65. A.H. Guth, E.J. Weinberg, Could the universe have recovered from a slow first-order phase transition? Nucl. Phys. B **212**, 321–364 (1983)
66. L. Kofman, A. Linde, A.A. Starobinsky, Reheating after Inflation. Phys. Rev. Lett. **73**, 3195–3198 (1994)
67. L. Kofman, A.D. Linde, A.A. Starobinsky, Towards the theory of reheating after inflation. Phys. Rev. D **56**, 3258–3295 (1997)
68. N.B. Narozhnyi, A.I. Nikishov, Pair production by a periodic electric field. Zh. Eksp. Teor. Fiz. **65**, 862–874 (1973). (Sov. Phys. JETP **38**, 427–432, 1974)
69. V.M. Mostepanenko, V.M. Frolov, Production of particles from vacuum by a uniform electric-field with periodic time-dependence. Yad. Fiz. **19**, 885–896 (1974). (Sov. J. Nucl. Phys. **19**, 451–456, 1974)
70. Ya.B. Zel'dovich, A.A. Starobinsky, Particle rpoduction and vacuum polarization in an anisotropic gravitational field. Zh. Eksp. Teor. Fiz. **61**, 2161–2175 (1971). (Sov. Phys. JETP **34**, 1159–1166, 1972)
71. V.N. Lukash, A.A. Starobinsky, The isotropization of the cosmological expansion owing to particle production. Zh. Eksp. Teor. Fiz. **66**, 1515–1527 (1974). (Sov. Phys. JETP **39**, 742–747, 1974)
72. S.G. Mamaev, V.M. Mostepanenko, A.A. Starobinsky, Energy and pressure of matter created near an isotropic cosmological singularity. Pis'ma Astron. Zh. **2** 136–139 (1976). (Soc. Astron. Lett. **2** 53–54, 1976)
73. S.G. Mamayev, V.M. Mostepanenko, A.A. Starobinsky, Particle creation from the vacuum near a homogeneous isotropic singularity. Zh. Eksp. Teor. Fiz. **70**, 1577–1591 (1976). (Sov. Phys. JETP **43**, 823–830, 1976)
74. E.W. Kolb, A.J. Long, Cosmological gravitational particle production and its implications for cosmological relics. Rev. Mod. Phys. **96**, 045005 (2024)
75. M. Kawasaki, K. Nakayama, Axions: Theory and Cosmological Role. Annu. Rev. Nucl. Part. Sci. **63**, 69–95 (2013)
76. J. Lecoffre, A. Hadi, M. Bruneau, C. Garcion, N. Fabre, É. Charron, N. Gaaloul, G. Dutier, Q. Bouton, Measurement of Casimir-Polder interaction for slow atoms through a material grating. Phys. Rev. Research **7**, 013232 (2025)
77. C. Garcion, N. Fabre, H. Bricha, F. Perales, S. Scheel, M. Ducloy, G. Dutier, Intermediate-Range Casimir-Polder Interaction Probed by High-Order Slow Atom Diffraction. Phys. Rev. Lett. **127**, 170402 (2021)
78. J. Morley, R. Flack, B.J. Hiley, P.F. Barker, Characterising a tunable, pulsed atomic beam using matter-wave interferometry. J. Phys. B: At. Mol. Opt. Phys. **54**, 155301 (2021)
79. C. Garcion, Q. Bouton, J. Lecoffre, N. Fabre, E. Charron, G. Dutier, N. Gaaloul, Quantum description of atomic diffraction by material nanostructures. Phys. Rev. Res. **6**, 023165 (2024)
80. G.L. Klimchitskaya, V.M. Mostepanenko, Constraining Yukawa-type interactions and coupling constant of axionlike particles to nucleons from recent measurement of the Casimir-Polder interaction. Eur. Phys. J. C **85**, 483 (2025); **85**, 789(E) (2025)
81. S.J. Smullin, A.A. Geraci, D.M. Weld, J. Chiaverini, S. Holmes, A. Kapitulnik, Constraints on Yukawa-type deviations from Newtonian gravity at 20 microns. Phys. Rev. D **72**, 122001 (2005)
82. J.K. Hoskins, R.D. Newman, R. Spero, J. Schultz, Experimental tests of the gravitational inverse-square law for mass separations from 2 to 105 cm. Phys. Rev. D **32**, 3084–3095 (1985)

83. S. Schlamminger, K.-J. Choi, T.A. Wagner, J.H. Gundlach, E.G. Adelberger, Test of the Equivalence Principle Using a Rotating Torsion Balance. Phys. Rev. Lett. **100**, 041101 (2008)
84. M. Bordag, G.L. Klimchitskaya, U. Mohideen, V.M. Mostepanenko, *Advances in the Casimir Effect* (Oxford University Press, Oxford, 2015)
85. R.S. Decca, D. López, E. Fischbach, G.L. Klimchitskaya, D.E. Krause, V.M. Mostepanenko, Novel constraints on light elementary particles and extra-dimensional physics from the Casimir effect. Eur. Phys. J. C **51**, 963–975 (2007)
86. G.L. Klimchitskaya, U. Mohideen, V.M. Mostepanenko, Constraints on corrections to Newtonian gravity from two recent measurements of the Casimir interaction between metallic surfaces. Phys. Rev. D **87**, 125031 (2013)
87. A.A. Banishev, J. Wagner, T. Emig, R. Zandi, U. Mohideen, Demonstration of Angle-Dependent Casimir Force between Corrugations. Phys. Rev. Lett. **110**, 250403 (2013)
88. A.A. Banishev, J. Wagner, T. Emig, R. Zandi, U. Mohideen, Experimental and theoretical investigation of the angular dependence of the Casimir force between sinusoidally corrugated surfaces. Phys. Rev. B **89**, 235436 (2014)
89. H.C. Chiu, G.L. Klimchitskaya, V.N. Marachevsky, V.M. Mostepanenko, U. Mohideen, Demonstration of the asymmetric lateral Casimir force between corrugated surfaces in the nonadditive regime. Phys. Rev. B **80**, 121402(R) (2009)
90. H.C. Chiu, G.L. Klimchitskaya, V.N. Marachevsky, V.M. Mostepanenko, U. Mohideen, Lateral Casimir force between sinusoidally corrugated surfaces: Asymmetric profiles, deviations from the proximity force approximation, and comparison with exact theory. Phys. Rev. B **81**, 115417 (2010)
91. M. Sedighi, V.B. Svetovoy, G. Palasantzas, Casimir force measurements from silicon carbide surfaces. Phys. Rev. B **93**, 085434 (2016)
92. B. Heacock, T. Fujiie, R.W. Haun, A. Henins, K. Hirota, T. Hosobata, M.G. Huber, M. Kitaguchi, D.A. Pushin, H. Shimizu et al., Pendellösung interferometry probes the neutron charge radius, lattice dynamics, and fifth forces. Science **373**, 1239–1243 (2021)
93. A.O. Caride, G.L. Klimchitskaya, V.M. Mostepanenko, S.I. Zanette, Dependences of the van der Waals atom-wall interaction on atomic and material properties. Phys. Rev. A **71**, 042901 (2005)
94. C.R.C. Handbook, of Chemistry and Physics, (eds.), *Haynes* (W.M. CRC Press, Boca Raton, 2011)
95. S. Weinberg, A New Light Boson? Phys. Rev. Lett. **40**, 223–226 (1978)
96. F. Wilczek, Problem of Strong P and T Invariance in the Presence of Instantons. Phys. Rev. Lett. **40**, 279–283 (1978)
97. E.G. Adelberger, E. Fischbach, D.E. Krause, R.D. Newman, Constraining the couplings of massive pseudoscalars using gravity and optical experiments. Phys. Rev. D **68**, 062002 (2003)
98. S. Aldaihan, D.E. Krause, J.C. Long, W.M. Snow, Calculations of the dominant long-range, spin-independent contributions to the interaction energy between two nonrelativistic Dirac fermions from double-boson exchange of spin-0 and spin-1 bosons with spin-dependent couplings. Phys. Rev. D **95**, 096005 (2017)
99. I.S. Gradshtein, I.M. Ryzhik, *Table of Integrals* (Series and Products. Academic Press, New York, 1980)
100. M. Bauer, G. Rostagni, Fifth Forces from QCD Axions Scale Differently. Phys. Rev. Lett. **132**, 101802 (2024)
101. G.L. Klimchitskaya, V.M. Mostepanenko, Testing gravity and predictions beyond the standard model at short distances: the Casimir effect, in *Modified and Quantum Gravity. From Theory to Experimental Searches on All Scales* ed. by C. Pfeifer and C. Lämmerzahl (Springer, Cham, 2023)
102. G.L. Klimchitskaya, V.M. Mostepanenko, The Nature of Dark Energy and Constraints on Its Hypothetical Constituents from Force Measurements. Universe **10**, 119 (2024)

Reconstruction of Primordial Scalar Power Spectrum from Pulsar Timing Observations

Timur Khizriev, Natalia Porayko, and Konstantin Postnov

Abstract In the summer 2023, five major pulsar timing array (PTA) collaborations reported a compelling evidence on the presence of a gravitational wave signal in their data. The prevailing hypothesis is that the detected signal is an incoherent sum of signals produced by individual merging supermassive black hole binaries embedded in the centers of galaxies. However, other scenarios associated with the physics of the early Universe, including those studied by Aleksei Starobinsky, are examined. In this contribution, the implication of the detected signature of a gravitational wave background for second-order gravitational waves produced by the primordial density perturbations is investigated. Instead of assuming a specific form of the primordial spectrum (model-dependent analysis), a novel approach has been developed based on finding a solution of the non-linear integral equation relating the primordial spectrum with amplitude of the stochastic gravitational wave background, which is a nonlinear ill-posed problem. Using the perturbation theory, the problem can be linearized to the Fredholm equation of the first kind. The algorithm has been tested on simulated PTA data and shown to be instrumental in finding a model-free reconstruction of the primordial power spectrum from the observables of the PTA experiment.

1 Introduction

In 2023, a landmark astronomical discovery was made. Major pulsar-timing arrays (PTAs) collaborations—North-American Nanohertz Observatory for Gravitational Waves (NANOGrav) [2], European and Indian PTAs (EPTA and InPTA, correspondingly) [12], Australian Parkes PTA (PPTA) [20] and Chines PTA (CPTA)

T. Khizriev · N. Porayko · K. Postnov (✉)
Sternberg Astronomical Institute, 13, Universitetsky prosp., Moscow, Russia
e-mail: kpostnov@gmail.com

T. Khizriev
e-mail: timurusff@yandex.ru

N. Porayko
Max-Planck-Institut für Radioastronomie, Bonn, Germany

A. Barvinsky and A. Kamenshchik (eds.), *Open Issues in Gravitation and Cosmology*, Fundamental Theories of Physics 225, https://doi.org/10.1007/978-3-032-15703-4_22

[24]—independently reported on registration of a stochastic signal (red noise) in the nHz frequency range showing a compelling evidence for the Hellings-Downs correlation [16] for different pulsar pairs. The Hellings-Downs correlation is a unique feature of quadrupole gravitational waves (GWs), and its detection in real PTA data indicated the beginning of GW astronomy at ultra-low frequencies. Conventionally, it is speculated that such a GW background can be produced by an ensemble of supermassive black hole binaries (SMBHBs), which are formed in the centers of galaxies. However, a number of alternative hypothesis have been put forward, which are mainly associated with the physics of the early Universe. Some of these ideas are heavily based on fundamental works of Aleksei Starobinsky, who always took seriously any observational clues about the Universe.

Already after the first tentative evidence for possible GW signal at nHZ frequency range appeared, he started thinking about possible fundamental reasons behind it, especially connected with the early Universe. In this regard, most notable are the pioneering works of Aleksei on cosmological inflation and subsequent production of tensor quantum vacuum fluctuations of the metric, which after accelerated expansion amplified to a GW background in PTA frequency band [21]. These ideas have been considered in relation to the experimental data in [1, 12].

In this paper we focus on another interesting hypothesis related to the scientific concepts of Aleksei Starobinsky. In 2023, when the EPTA/InPTA paper [12] was in preparation, we asked his opinion about the possible explanation of the signal in terms of relic stochastic GW background (SGWB). He immediately advised us to look at his earlier paper [6], where an example of SGWB produced by the scalar perturbations within the PTA frequency range due to a large peak in the primordial power spectrum was computed in the frame of two-field inflation models. While in the first order of perturbation theory of linearized Einstein equations, scalar, vector and tensor modes propagate independently, in the second order these modes intermix. In particular, second-order GWs can be sourced by the primordial scalar perturbations, which can be probed by GW observatories at modern epoch [3, 5, 7], and provide an exclusive information about the physical properties of the Universe before recombination.

While the nature of the reported PTA signal remains unclear (see [19] for more detail and references), its explanation in terms of SGWB produced by second-order scalar perturbations is of fundamental significance. In parallel to explaining the signal observed with PTA experiments, primordial perturbations can also source the production of primordial black holes (PBHs), which are the viable solution to the dark matter problem.

Previously, such implication has been limited to searches for a specific form of the primordial density perturbation power spectrum P_ζ (see, e.g., [9, 13]). However, given the diversity of existing inflation models [4, 8, 14, 15, 17], which can produce a necessary excess of power in P_ζ at wavenumbers probed by PTA measurements, the final shapes of the power spectrum can be quite different. This motivated the current investigation, in which we address the question of putting agnostic constraints on possible features in the primordial perturbations. The spectra of GWs, which is a PTA observable, is expressed as a double integral of the square of the primordial power spectrum. Therefore, the problem of P_ζ reconstruction is an example of a nonlinear ill-posed problem, which consists in solving the quadratic integral equation. Such problems can only be solved numerically, which is usually done by assuming a model parametric form of the unknown function. Here, we propose a novel method, in which the initially quadratic integral equation is converted to the Fredholm equation of the first kind to which Tikhonov regularization algorithm can be applied. To check the applicability of the algorithm, we tested it on simulated PTA data and found it instrumental for anticipated future PTA measurements with smaller errors.

The structure of the paper is as follows. In Sect. 2 we set the framework of the problem of the primordial power spectrum reconstruction. In Sect. 3 we discuss the methodology of our analysis, including several mathematical tricks to simplify the estimation of the spectrum. The results are summarized in Sect. 4. We conclude with the discussion of our findings in Sect. 5.

2 Second-order Primordial GWs

Given the power spectrum of primordial scalar perturbations $P_\zeta(k)$, one can compute the spectral density of a homogeneous Gaussian second-order SGWB $\Omega_{\rm GW}(k)$ as [3, 18]

$$\Omega_{\rm GW}(k) = \int_0^\infty dv \int_{|1-v|}^{|1+v|} du \left[K(u,v) P_\zeta(uk) P_\zeta(vk)\right], \tag{1}$$

where (u, v) are dimensionless integration parameters, k is dimensionless wavenumber in physically motivated units k_0, and $K(u, v)$ is an analytically determined function that is non-negative in the entire integration domain:

$$K(u,v) = 2A\left(\frac{4v^2-(1+v^2-u^2)^2}{4vu}\right)^2\left(\frac{3(u^2+v^2-3)}{4u^3v^3}\right)^2 \times$$
$$\times\left(\left((u^2+v^2-3)\log\left|\frac{3-(u+v)^2}{3-(u-v)^2}\right| - 4uv\right)^2 + \pi^2(u^2+v^2-3)^2\Theta(u+v-\sqrt{3})\right).$$

Here A is a coefficient derived from gravitational wave generation and propagation physics. We will also refer to the integral (1) produced by a certain power spectrum of some functional form X as $\Omega_{\rm GW}[X]$ for convenience. This integral can only be evaluated analytically for a limited variety of functions X. The numerical computation has been implemented in, e.g. SIGWfast python library,[1] which we use throughout the paper whenever such computation is required.

Pulsar timing data analysis gives $\Omega_{\rm GW}(k)$ values for a set of wavenumbers, while the subject of our research is the function P_ζ. To obtain it, one has to solve Eq. (1)—a nonlinear integral equation. This makes the reconstruction of P_ζ from real

[1] https://github.com/Lukas-T-W/SIGWfast.

data $\Omega_{\mathrm{GW}}(k)$ a non-linear ill-posed problem, which is a highly problematic task with very few solution algorithms available. To complicate the task further, our kernel function $K(u, v)$ exhibits a singularity.

3 Methodology

Even with a much higher accuracy of "Omega_GW" measurements that we have today, tackling (1) directly is not very promising. This can possibly be done by blindly searching for $P_\zeta = \mathrm{argmin}_X \|\Omega_{\mathrm{GW}} - \Omega_{\mathrm{GW}}[X]\|^2$, however, such an approach does not guarantee a physically meaningful solution.

Choosing P_ζ in a power-law form, however, will result in a GW spectrum that at least up to date fits the observed signal relatively well.[2] This suggests that reconstructing P_ζ in the form $P_\zeta(k) = P_\zeta^0(k) + \delta P_\zeta(k)$, where P_ζ^0 is some simple initial approximation (for example, a power law mentioned above), and $\delta P_\zeta(k)$ is a small correction, is a valid approach. As will become evident later, this will also simplify the integral equation, giving access to existing regularization algorithms.

3.1 Linearization

We will now substitute $P_\zeta(k) = P_\zeta^0(k) + \delta P_\zeta(k)$ into (1) in an effort to linearize it. Below we refer to the full integration domain as D:

$$\Omega_{\mathrm{GW}} = \iint_D K(P_\zeta^0 + \delta P_\zeta)(P_\zeta^0 + \delta P_\zeta) =$$
$$= \iint_D K P_\zeta^0 P_\zeta^0 + \delta\Omega_{\mathrm{GW}}^0 = \Omega_{\mathrm{GW}}^0 + \delta\Omega_{\mathrm{GW}}^0.$$

Here $\Omega_{\mathrm{GW}}^0 = \Omega_{\mathrm{GW}}[P_\zeta^0]$, while $\delta\Omega_{\mathrm{GW}}^0$ is expressed as

$$\delta\Omega_{\mathrm{GW}}^0 = \iint_D K(u, v) P_\zeta^0(uk)\delta P_\zeta(vk) du dv +$$
$$+ \iint_D K(u, v) P_\zeta^0(vk)\delta P_\zeta(uk) du dv + \Delta(\delta P_\zeta),$$

[2] Such a first approximation can at least serve as a good starting point in the frequency range of PTAs.

where Δ includes δP_ζ quadratically. Thanks to the symmetry of both function $K(u, v)$ and domain D with respect to parameters (u, v), the two linear integrals can be unified into one, leaving us with

$$\delta\Omega^0_{\rm GW} = 2\iint_D K(u, v)P^0_\zeta(uk)\delta P_\zeta(vk)dudv + \Delta(\delta P_\zeta). \tag{2}$$

We have separated a single integral that is linear in δP_ζ, but to linearize the entire equation Δ has to be removed. This term can be neglected when the linear integral constitutes a larger contribution to $\delta\Omega^0_{\rm GW}$, which can be guaranteed everywhere in D as long as $\delta P_\zeta/P^0_\zeta$ is < 1 everywhere in D. Since in this domain P^0_ζ can change by orders of magnitude, to satisfy $\delta P_\zeta/P^0_\zeta < 1$ the same may become necessary for δP_ζ, which is hard to implement and can reduce the accuracy of the solution. We instead choose to solve for $\epsilon = \delta P_\zeta/P^0_\zeta$, since for such a function, it would suffice to simply set it to be less than 1 everywhere. The resulting expression is

$$\delta\Omega^0_{\rm GW} \approx 2\iint_D \left[K(u, v)P^0_\zeta(uk)P^0_\zeta(vk)\right]\epsilon(vk)dudv,$$

Assigning $\kappa = vk$, we obtain

$$\Omega_{\rm GW}(k) = \Omega_{\rm GW}(k) - \Omega^0_{\rm GW}(k) = \int_{k_{\rm min}}^{k_{\rm max}} T^0(k, \kappa)\epsilon(\kappa)d\kappa, \tag{3}$$

$$T^0(k, \kappa) = \frac{2}{k}P^0_\zeta(\kappa)\int_{|1-\kappa/k|}^{1+\kappa/k} K(u, \kappa/k)P^0_\zeta(uk)du, \tag{4}$$

where we represent the unknown power spectrum as $P_\zeta = P^0_\zeta(1+\epsilon)$. We will also refer to Eq. (3) of some function X as $I^0[X]$ for convenience. From Eq. (1), the reader could have expected the integration limits for the second integral above to be $[0, \infty)$. However, due to the specifics of $K(u, v)$ (especially due to the singularity it has), any noticeable contribution of $\epsilon(k)$ to $\delta\Omega^0_{\rm GW}$ would come from a finite wavenumber interval $[k_{\rm min}, k_{\rm max}]$, which allows to treat ϵ as a set of values within that interval and simply 0 outside of it.

For $\epsilon < 1$ and a power-law choice of P^0_ζ this linear approximation proves to produce results close to those obtained by the calculation of the full integral. At this point, both $\Omega_{\rm GW}$ and P^0_ζ are known, which makes it possible to calculate $\Omega^0_{\rm GW}$ and T^0 right away via Eqs. (1) and (12), leaving ϵ the only unknown that we wish to approximate by a solution ϵ^0. Therefore, Eq. (3) becomes a Fredholm integral equation of the first kind, an ill-posed but well-known and extensively studied problem that can be approached with many robust and general methods, among which we choose Tikhonov regularization [22, 23].

3.2 Regularization

The main reason to consider a problem with a Fredholm equation of the first kind to be ill-posed is its instability with respect to errors in the given data. Because of this, a naive discretization of the integral operator I^0 in Eq. (3) with subsequent solution of a linear system will not yield meaningful results in most cases. Tikhonov regularization method suggests accompanying the residual norm $\left\|\delta\Omega_{\rm GW}^0 - I^0[X]\right\|^2$ by a stabilizing member $\|X\|^2$ that would not allow the solution or its derivatives to grow unconstrained, and searching for ϵ^0 as a minimum of the regularized functional $F[X]$:

$$F^0[X, \alpha] = \left\|\delta\Omega_{\rm GW}^0 - I^0[X]\right\|^2 + \alpha \|X\|^2, \tag{5}$$

$$\epsilon^0(\alpha) = \mathrm{argmin}_X(F^0[X, \alpha]), \tag{6}$$

where α is a regularization coefficient, so ϵ_α^0, while being a function of k, also depends on the choice of α. An insufficient value of α will not stabilize the solution, while an excessively large value will not give enough flexibility to approximate the data, so it should be chosen with the goal of balancing these two tendencies. In our case we select α so that for $\epsilon^0(\alpha)$ the equation residual will meet the data error:

$$\epsilon^0(\alpha) = \mathrm{argmin}_X(F^0[X, \alpha]), \tag{7}$$

$$\left\|\delta\Omega_{\rm GW}^0 - I^0[\epsilon^0(\alpha)]\right\| = \delta^2, \tag{8}$$

where δ^2 is the sum of errors of $\delta\Omega_{\rm GW}^0$ data points. Once α is chosen so that both conditions above are satisfied, we will consider ϵ^0 to be a solution.

3.3 Iterative Refinement

For $\epsilon \ll 1$ solving (3) gives an accurate solution. However, due to imperfections of the PTA data and its analysis techniques, such small corrections would not have any noticeable effect on observed $\Omega_{\rm GW}$. Meanwhile, for corrections $\epsilon > 1$ that would result in possibly observable deviations, the linear method outlined above ceases to apply.

Let us try to expand it for $\epsilon > 1$. At this stage we will have to limit the analysis to nonnegative $\epsilon(k)$. In that case, both integrals in (2) will also be nonnegative for all values of k, which means that for any k even the single linear integral (the one we used to base our Fredholm equation on) will be a better approximation of $\delta\Omega_{\rm GW}^0$ than $\epsilon \equiv 0$. Based on that, let us also assume that $P_\zeta^1 = P_\zeta^0(1 + \beta\epsilon^0)$ will too become a better estimate of P_ζ^1 than P_ζ^0 alone, with Î² being a small fixed-valued parameter that we will later illustrate to be useful. In that case, solving (3) for ϵ^0 and accepting P_ζ^1 as a new initial estimate instead of P_ζ^0, we get

$$\Omega_{\mathrm{GW}}(k) - \Omega^1_{\mathrm{GW}}(k) = \int_{k_{\min}}^{k_{\max}} T^1(k,\kappa)\epsilon^1(\kappa)d\kappa, \tag{9}$$

$$T^1(k,\kappa) = \frac{2}{k}P^1_\zeta(k)\int_{|1-\kappa/k|}^{1+\kappa/k} K(u,\kappa/k)P^1_\zeta(uk)du, \tag{10}$$

where $\Omega^1_{\mathrm{GW}} = \Omega_{\mathrm{GW}}[P^1_\zeta]$, analogous to Ω^0_{GW}. The solution of this new equation ϵ^1 will become a new correction, now for P^1_ζ, that we will use to obtain the next estimation $P^2_\zeta = P^1_\zeta(1+\beta\epsilon^1)$, and so on. We now introduce iterations by performing which one may refine the solution to extend the initial linear method to larger ϵ:

$$\Omega_{\mathrm{GW}}(k) - \Omega^n_{\mathrm{GW}}(k) = \int_{k_{\min}}^{k_{\max}} T^n(k,\kappa)\epsilon^n(\kappa)d\kappa, \tag{11}$$

$$P^n_\zeta = P^{n-1}_\zeta(1+\beta\epsilon^{n-1}), \quad T^n(k,\kappa) = \frac{2}{k}P^n_\zeta(k)\int_{|1-\kappa/k|}^{1+\kappa/k} K(u,\kappa/k)P^n_\zeta(uk)du. \tag{12}$$

We found that multiplying ϵ^n by a small value β proved to result in smoother convergence. In this work we used $\beta = 0.15$. In Fig. 1 we demonstrate how algorithm works by solving a test problem, which we will describe further.

4 Results

The developed algorithm has been tested on realistic simulations of PTA data. The simulations are comprised of timing residuals of the time of arrivals of pulses from simulate pulsars. Pulsar observations are carried out every month with a total time span of $T = 10$ years, mimicking the existing EPTA Data Release (DR) 2 New [11]. The standard deviation of timing residuals is set to 100 ns for all 10 pulsars in the model array. The simulations were carried out with the standard `libstempo` python library,[3] which is widely used software in the pulsar community. White-noise timing residuals were perturbed with GWs produced by artificially injected primordial density perturbations with the following properties. An exact function $P_\zeta = P^0_\zeta(1+\epsilon)$ (in our case a power law P^0_ζ with a Gaussian peak ϵ) with known parameters was chosen, and $\Omega_{\mathrm{GW}}[P_\zeta]$ data was generated using `SIGWfast`. The primary contribution $P^0_\zeta \sim (k/k_0)^n$ was assumed to follow a power law with the index $n = 0.9$. The excess to the power-law spectrum was described with the following functional form: $\epsilon = A_\epsilon e^{-\frac{(k-k^0)^2}{\Delta}}$: $A_\epsilon = 3.0$, $k_0 = 9$, $\Delta = 15$. The injection of such a GW signal in the simulated timing data was carried out with `createGWB` plugin of the `libstempo`. The reconstructed P_ζ was obtained after 7 iterations with $\beta = 0.15$.

To demonstrate the work of the described algorithms, we pass through the following steps. The Ω_{GW}, which is the main observable of the PTA analysis, was computed

[3] https://github.com/vallis/libstempo/tree/master.

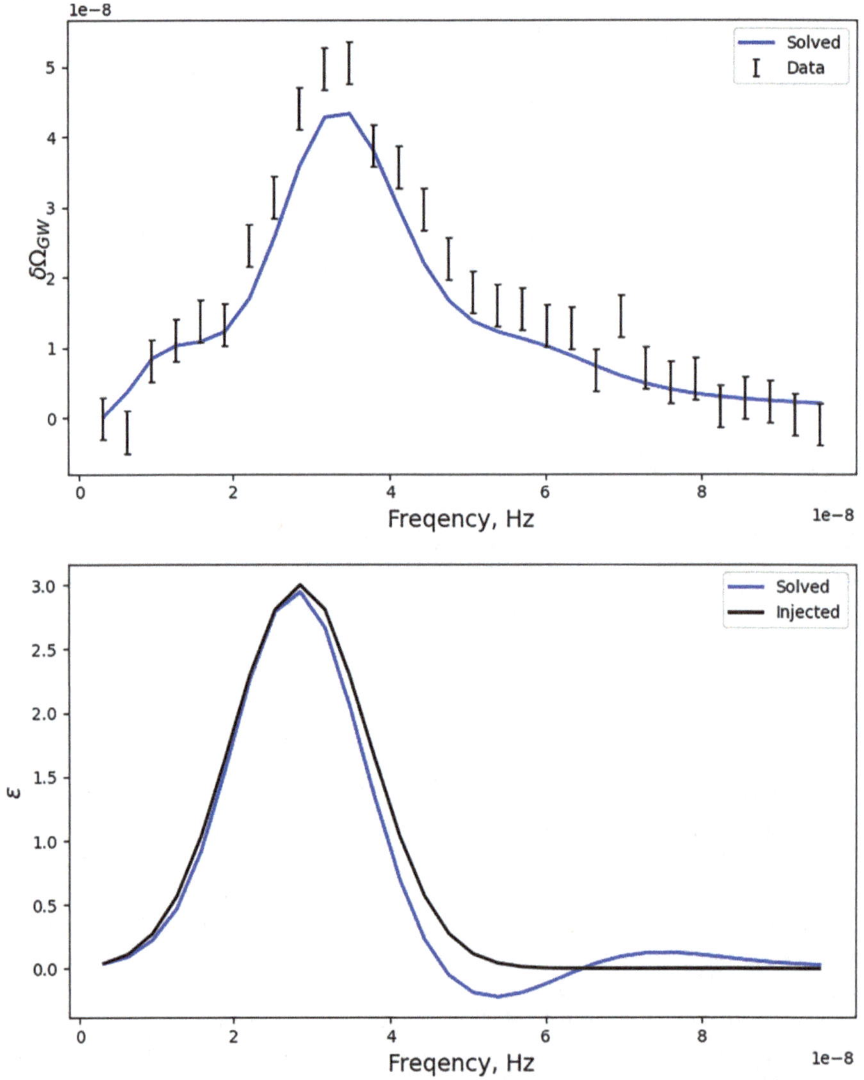

Fig. 1 Top: $\Omega_{\rm GW}[P_\zeta] - \Omega_{\rm GW}[P_\zeta^0]$ from `SIGWfast.py`. Bottom: the exact ϵ (black line) is plotted alongside the solution (blue line). In this case we used $\beta = 0.15$ and 4 iterations

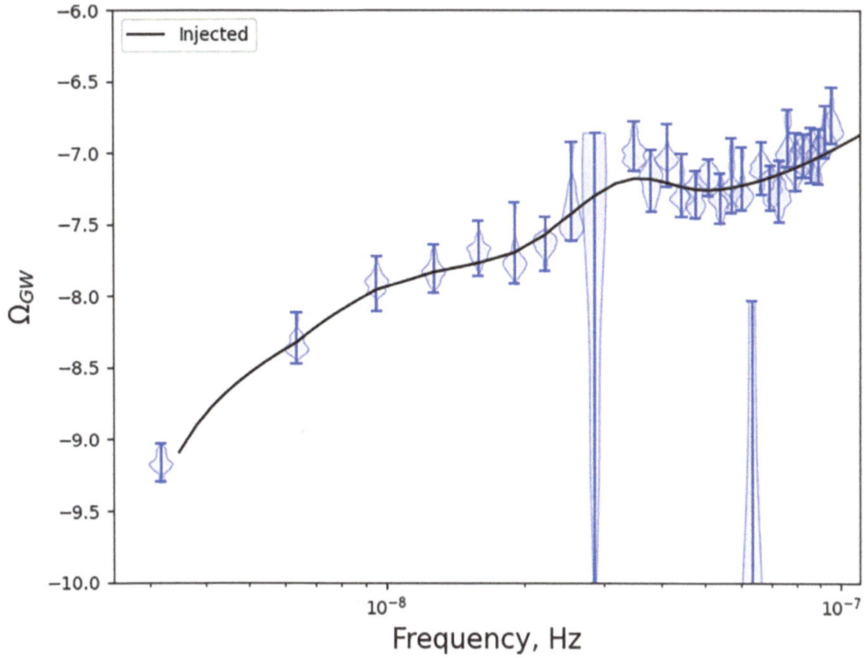

Fig. 2 Estimated power spectrum of $\Omega_{\rm GW}$ from the simulated PTA data using the free spectrum analysis (blue violins) overlaid with the injected signal (black). Higher frequencies are dominated by the white instrumental noise (only upper limits can be set)

with the well-established `enterprise` Python library [10] using the so-called free spectrum reconstruction, for which the amplitudes of the power spectrum are estimated in each frequency bin independently. The number of independent frequency components used for the spectrum decomposition is fixed to 30. As a result, the investigated wavenumber k range is [1, 30] in units of $k_0 = 2\pi/cT$. An example of the injected and obtained power spectrum $\Omega_{\rm GW}$ is depicted in Fig. 2. Starting with an initial approximation P^0_ζ we solve iterations (11) and compare the result with the originally introduced ϵ. To see how our method performs with better data, we also solved the same problem, but with $\delta\Omega_{\rm GW}$ obtained from `SIGWfast.py` (Fig. 1).

In Fig. 1 our algorithm successfully recovered the original ϵ. The reconstructed power spectrum P_ζ calculated after 4 iterations with $\beta = 0.15$ using free spectrum data shown in Fig. 2 is the main result of our analysis and is illustrated in Fig. 3. The instability of the solution at large wavenumbers is related to the increase of uncertainties at higher frequencies of the simulated PTA data, where the signal is hidden by the white instrumental noise. However, the original peak has been recovered quite accurately. Possible methods of overcoming this issue will be investigated in future works.

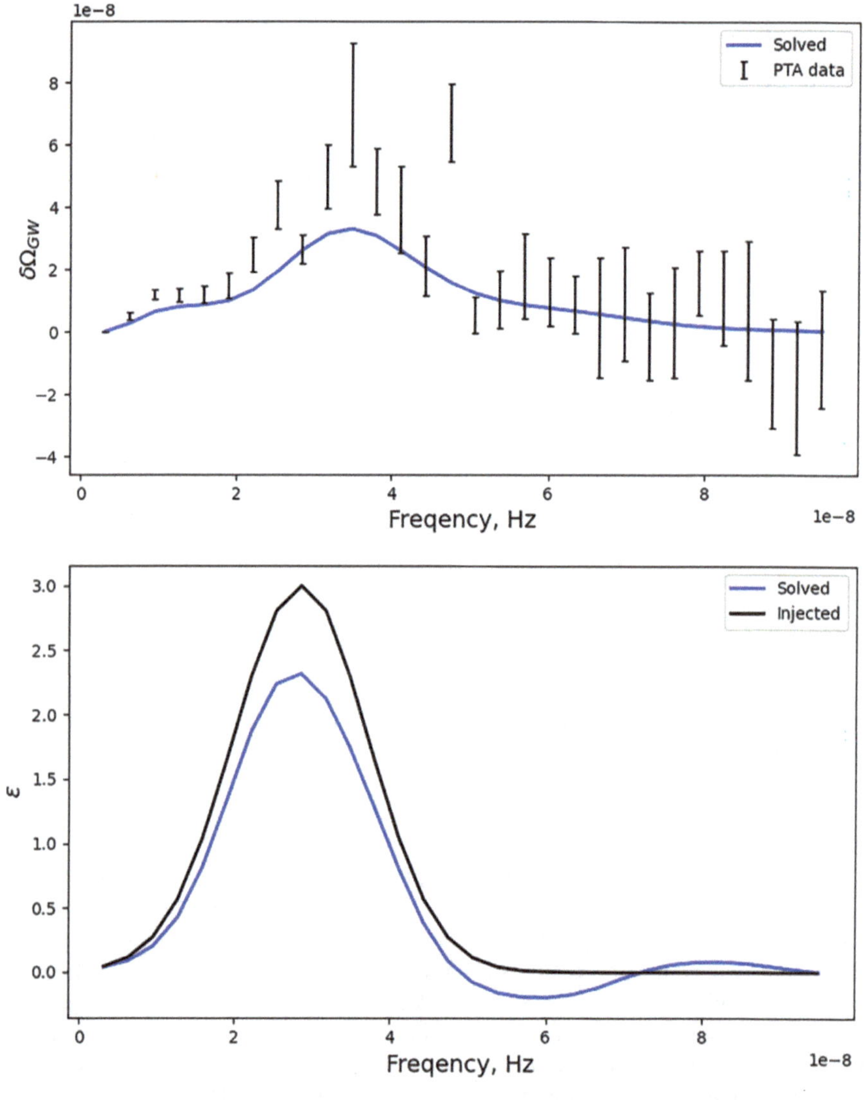

Fig. 3 Top: orange curve represents $\Omega_{\mathrm{GW}}[P_\zeta] - \Omega_{\mathrm{GW}}[P_\zeta^0]$ injected in the simulation, while the blue errorbars illustrate its free spectrum reconstruction (data from Fig. 2). Bottom: the exact injected ϵ (black line) is plotted alongside the reconstructed solution (blue line). Frequencies correspond to wavenumbers from the range [1, 30] (see text), and are the same as in Fig. 2

5 Discussion

In this contribution we have developed and tested a novel method of primordial density power spectrum reconstruction, given the GW dimensionless power spectrum as measured in PTA experiments and assuming it to be a SGWB. In order to find the exact solution of the problem stated, it is necessary to solve a non-linear integral equation, which is a challenging computational problem with poorly established mathematical framework. To simplify this task, we have expressed the sought power spectrum as a sum of the main component (which is a power law in our case) and a small addition. In the first order of perturbation theory, the problem can be narrowed down to solving a Fredholm equation of the first kind. The efficacy of the algorithm has been demonstrated with realistic PTA simulations. It has been shown that for small enough deviations from the power law spectrum the proposed algorithm provides a feasible quality of reconstruction. The accuracy can be even further boosted for larger additions to the power spectrum with the help of iterative refinement.

However, a number of challenges has been highlighted during the implementation of the developed algorithm. First and foremost, the measurement uncertainties in the dimensionless GW spectrum, from which current PTA datasets suffer, have a strong influence on the accuracy of the reconstructed primordial spectrum (see Figs. 2 and 3). Second, with a small number of frequency bins of the current PTA data (e.g., only four bins in the EPTA/InPTA data [12]), the developed scheme becomes unstable and fails to produce a valid solution. Therefore, the applicability of the described algorithm to the current PTA data (which was our initial intention) fails. Nevertheless, we emphasize that the developed framework will be of great importance to the future PTA datasets provided by the International PTA Collaboration (IPTA) and Square Kilometer Array (SKA) with a larger number of frequency bins and smaller uncertainties due to enhanced by an order of magnitude quality of the initial data (mostly, times of arrivals of pulses from pulsar arrays) .

Acknowledgements The authors thank Dr. Igor Antokhin for useful discussions. Work of KP is supported by RSF grant 23-42-00055.

References

1. A. Afzal, G. Agazie, A. Anumarlapudi, A.M. Archibald, Z. Arzoumanian, P.T. Baker, B. B'ecsy, J.J. Blanco-Pillado, L. Blecha, K.K. Boddy, A. Brazier, P.R. Brook, S. Burke-Spolaor, R. Burnette, R. Case, M. Charisi, S. Chatterjee, K. Chatziioannou, B.D. Cheeseboro, S. Chen, T. Cohen, J.M. Cordes, N.J. Cornish, F. Crawford, H.T. Cromartie, K. Crowter, C.J. Cutler, M.E. Decesar, D. Degan, P.B. Demorest, H. Deng, T. Dolch, B. Drachler, R. von Eckardstein, E.C. Ferrara, W. Fiore, E. Fonseca, G.E. Freedman, N. Garver-Daniels, P.A. Gentile, K.A. Gersbach, J. Glaser, D.C. Good, L. Guertin, K. G"ultekin, J.S. Hazboun, S. Hourihane, K. Islo, R.J. Jennings, A.D. Johnson, M.L. Jones, A.R. Kaiser, D.L. Kaplan, L.Z. Kelley, M. Kerr, J.S. Key, N. Laal, M.T. Lam, W.G. Lamb, T.J.W. Lazio, V.S.H. Lee, N. Lewandowska, R.R. Lino Dos Santos, T.B. Littenberg, T. Liu, D.R. Lorimer, J. Luo, R.S. Lynch, C.P. Ma, D.R. Madison,

A. McEwen, J.W. McKee, M.A. McLaughlin, N. McMann, B.W. Meyers, P.M. Meyers, C.M.F. Mingarelli, A. Mitridate, J. Nay, P. Natarajan, C. Ng, D.J. Nice, S.K. Ocker, K.D. Olum, T.T. Pennucci, B.B.P. Perera, P. Petrov, N.S. Pol, H.A. Radovan, S.M. Ransom, P.S. Ray, J.D. Romano, S.C. Sardesai, A. Schmiedekamp, C. Schmiedekamp, K. Schmitz, T. Schr"oder, L. Schult, B.J. Shapiro-Albert, X. Siemens, J. Simon, M.S. Siwek, I.H. Stairs, D.R. Stinebring, K. Stovall, P. Stratmann, J.P. Sun, A. Susobhanan, J.K. Swiggum, J. Taylor, S.R. Taylor, T. Trickle, J.E. Turner, C. Unal, M. Vallisneri, S. Verma, S.J. Vigeland, H.M. Wahl, Q. Wang, C.A. Witt, D. Wright, O. Young, K.M. Zurek, NANOGrav Collaboration: The NANOGrav 15 yr Data Set: Search for Signals from New Physics. ApJL **951**(1), L11 (2023). https://doi.org/10.3847/2041-8213/acdc91
2. G. Agazie, A. Anumarlapudi, A.M. Archibald, Z. Arzoumanian, P.T. Baker, B. Bécsy, L. Blecha, A. Brazier, P.R. Brook, S. Burke-Spolaor, R. Burnette, R. Case, M. Charisi, S. Chatterjee, K. Chatziioannou, B.D. Cheeseboro, S. Chen, T. Cohen, J.M. Cordes, N.J. Cornish, F. Crawford, H.T. Cromartie, K. Crowter, C.J. Cutler, M.E. Decesar, D. Degan, P.B. Demorest, H. Deng, T. Dolch, B. Drachler, J.A. Ellis, E.C. Ferrara, W. Fiore, E. Fonseca, G.E. Freedman, N. Garver-Daniels, P.A. Gentile, K.A. Gersbach, J. Glaser, D.C. Good, K. Gültekin, J.S. Hazboun, S. Hourihane, K. Islo, R.J. Jennings, A.D. Johnson, M.L. Jones, A.R. Kaiser, D.L. Kaplan, L.Z. Kelley, M. Kerr, J.S. Key, T.C. Klein, N. Laal, M.T. Lam, W.G. Lamb, T.J.W. Lazio, N. Lewandowska, T.B. Littenberg, T. Liu, A. Lommen, D.R. Lorimer, J. Luo, R.S. Lynch, C.P. Ma, D.R. Madison, M.A. Mattson, A. McEwen, J.W. McKee, M.A. McLaughlin, N. McMann, B.W. Meyers, P.M. Meyers, C.M.F. Mingarelli, A. Mitridate, P. Natarajan, C. Ng, D.J. Nice, S.K. Ocker, K.D. Olum, T.T. Pennucci, B.B.P. Perera, P. Petrov, N.S. Pol, H.A. Radovan, S.M. Ransom, P.S. Ray, J.D. Romano, S.C. Sardesai, A. Schmiedekamp, C. Schmiedekamp, K. Schmitz, L. Schult, B.J. Shapiro-Albert, X. Siemens, J. Simon, M.S. Siwek, I.H. Stairs, D.R. Stinebring, K. Stovall, J.P. Sun, A. Susobhanan, J.K. Swiggum, J. Taylor, S.R. Taylor, J.E. Turner, C. Unal, M. Vallisneri, R. van Haasteren, S.J. Vigeland, H.M. Wahl, Q. Wang, C.A. Witt, O. Young, Nanograv Collaboration: The NANOGrav 15 yr Data Set: Evidence for a Gravitational-wave Background. ApJL **951**(1), L8 (2023). https://doi.org/10.3847/2041-8213/acdac6
3. K.N. Ananda, C. Clarkson, D. Wands, Cosmological gravitational wave background from primordial density perturbations. Phys. Rev. D **75**(12), 123518 (2007). https://doi.org/10.1103/PhysRevD.75.123518
4. G. Ballesteros, J. Beltrán Jiménez, M. Pieroni, Black hole formation from a general quadratic action for inflationary primordial fluctuations. J. Cosmol. Astropart. Phys. **2019**(6), 016 (2019). https://doi.org/10.1088/1475-7516/2019/06/016
5. D. Baumann, P. Steinhardt, K. Takahashi, K. Ichiki, Gravitational wave spectrum induced by primordial scalar perturbations. Phys. Rev. D **76**(8), 084019 (2007). https://doi.org/10.1103/PhysRevD.76.084019
6. M. Braglia, D.K. Hazra, F. Finelli, G.F. Smoot, L. Sriramkumar, A.A. Starobinsky, Generating PBHs and small-scale GWs in two-field models of inflation. J. Cosmol. Astropart. Phys. **2020**(8), 001 (2020). https://doi.org/10.1088/1475-7516/2020/08/001
7. C. Carbone, S. Matarrese, Unified treatment of cosmological perturbations from superhorizon to small scales. Phys. Rev. D **71**(4), 043508 (2005). https://doi.org/10.1103/PhysRevD.71.043508
8. M. Cicoli, V.A. Diaz, F.G. Pedro, Primordial black holes from string inflation. J. Cosmol. Astropart. Phys. **2018**(6), 034 (2018). https://doi.org/10.1088/1475-7516/2018/06/034
9. Dandoy, V., Domcke, V., Rompineve, F.: Search for scalar induced gravitational waves in the international pulsar timing array data release 2 and NANOgrav 12.5 years datasets. SciPost Physics Core **6**(3), 060 (2023). https://doi.org/10.21468/SciPostPhysCore.6.3.060
10. J.A. Ellis, M. Vallisneri, S.R. Taylor, P.T. Baker, Enterprise: Enhanced numerical toolbox enabling a robust pulsar inference suite. Zenodo (2020)
11. EPTA Collaboration, J. Antoniadis, S. Babak, A.S. Bak Nielsen, C.G. Bassa, A. Berthereau, M. Bonetti, E. Bortolas, P.R. Brook, M. Burgay, R.N. Caballero, A. Chalumeau, D.J. Champion, S. Chanlaridis, S. Chen, I. Cognard, G. Desvignes, M. Falxa, R.D. Ferdman, A. Franchini,

J.R. Gair, B. Goncharov, E. Graikou, J.M. Grießmeier, L. Guillemot, Y.J. Guo, H. Hu, F. Iraci, D. Izquierdo-Villalba, J. Jang, J. Jawor, G.H. Janssen, A. Jessner, R. Karuppusamy, E.F. Keane, M.J. Keith, M. Kramer, M.A. Krishnakumar, K. Lackeos, K.J. Lee, K. Liu, Y. Liu, A.G. Lyne, J.W. McKee, R.A. Main, M.B. Mickaliger, I.C. Niţu, A. Parthasarathy, B.B.P. Perera, D. Perrodin, A. Petiteau, N.K. Porayko, A. Possenti, H. Quelquejay Leclere, A. Samajdar, S.A. Sanidas, A. Sesana, G. Shaifullah, L. Speri, R. Spiewak, B.W. Stappers, S.C. Susarla, G. Theureau, C. Tiburzi, E. van der Wateren, A. Vecchio, V. Venkatraman Krishnan, J.P.W. Verbiest, J. Wang, L. Wang, Z. Wu., The second data release from the European Pulsar Timing Array. I. The dataset and timing analysis. A&A **678**, A48 (2023). https://doi.org/10.1051/0004-6361/202346841

12. EPTA Collaboration, InPTA Collaboration, J. Antoniadis, P. Arumugam, S. Arumugam, S. Babak, M. Bagchi, A.S. Bak Nielsen, C.G. Bassa, A. Bathula, A. Berthereau, M. Bonetti, E. Bortolas, P.R. Brook, M. Burgay, R.N. Caballero, A. Chalumeau, D.J. Champion, S. Chanlaridis, S. Chen, I. Cognard, S. Dandapat, D. Deb, S. Desai, G. Desvignes, N. Dhanda-Batra, C. Dwivedi, M. Falxa, R.D. Ferdman, A. Franchini, J.R. Gair, B. Goncharov, A. Gopakumar, E. Graikou, J.M. Grießmeier, A. Gualandris, L. Guillemot, Y.J. Guo, Y. Gupta, S. Hisano, H. Hu, F. Iraci, D. Izquierdo-Villalba, J. Jang, J. Jawor, G.H. Janssen, A. Jessner, B.C. Joshi, F. Kareem, R. Karuppusamy, E.F. Keane, M.J. Keith, D. Kharbanda, T. Kikunaga, N. Kolhe, M. Kramer, M.A. Krishnakumar, K. Lackeos, K.J. Lee, K. Liu, Y. Liu, A.G. Lyne, J.W. McKee, Y. Maan, R.A. Main, M.B. Mickaliger, I.C. Niţu, K. Nobleson, A.K. Paladi, A. Parthasarathy, B.B.P. Perera, D. Perrodin, A. Petiteau, N.K. Porayko, A. Possenti, T. Prabu, H. Quelquejay Leclere, P. Rana, A. Samajdar, S.A. Sanidas, A. Sesana, G. Shaifullah, J. Singha, L. Speri, R. Spiewak, A. Srivastava, B.W. Stappers, M. Surnis, S.C. Susarla, A. Susobhanan, K. Takahashi, P. Tarafdar, G. Theureau, C. Tiburzi, E. van der Wateren, A. Vecchio, V. Venkatraman Krishnan, J.P.W. Verbiest, J. Wang, L. Wang, Z. Wu, P. Auclair, E. Barausse, C. Caprini, M. Crisostomi, F. Fastidio, T. Khizriev, H. Middleton, A. Neronov, K. Postnov, A. Roper Pol, D. Semikoz, C. Smarra, D.A. Steer, R.J. Truant, S. Valtolina, The second data release from the European Pulsar Timing Array. IV. Implications for massive black holes, dark matter, and the early Universe. A&A **685**, A94 (2024). https://doi.org/10.1051/0004-6361/202347433
13. D. Frolovsky, F. Kühnel, I. Stamou, Reconstructing primordial black hole power spectra from gravitational waves. Phys. Rev. D **111**(4), 043538 (2025). https://doi.org/10.1103/PhysRevD.111.043538
14. J. García-Bellido, E. Ruiz Morales, Primordial black holes from single field models of inflation. Physics of the Dark Universe **18**, 47–54 (2017). https://doi.org/10.1016/j.dark.2017.09.007
15. C. Germani, T. Prokopec, On primordial black holes from an inflection point. Physics of the Dark Universe **18**, 6–10 (2017). https://doi.org/10.1016/j.dark.2017.09.001
16. R.W. Hellings, G.S. Downs, Upper limits on the isotropic gravitational radiation background from pulsar timing analysis. ApJL **265**, L39–L42 (1983). https://doi.org/10.1086/183954
17. P. Ivanov, P. Naselsky, I. Novikov, Inflation and primordial black holes as dark matter. Phys. Rev. D **50**(12), 7173–7178 (1994). https://doi.org/10.1103/PhysRevD.50.7173
18. K. Kohri, T. Terada, Semianalytic calculation of gravitational wave spectrum nonlinearly induced from primordial curvature perturbations. Phys. Rev. D **97**(12), 123532 (2018). https://doi.org/10.1103/PhysRevD.97.123532
19. K.A. Postnov, N.K. Porayko, M.S. Pshirkov, Precision methods of pulsar timing and polarimetry: results and prospects. Physics Uspekhi **68**(2), 146–162 (2025). https://doi.org/10.3367/UFNe.2024.11.039812
20. D.J. Reardon, A. Zic, R.M. Shannon, G.B. Hobbs, M. Bailes, V. Di Marco, A. Kapur, A.F. Rogers, E. Thrane, J. Askew, N.D.R. Bhat, A. Cameron, M. Curyło, W.A. Coles, S. Dai, B. Goncharov, M. Kerr, A. Kulkarni, Y. Levin, M.E. Lower, R.N. Manchester, R. Mandow, M.T. Miles, R.S. Nathan, S. Osłowski, C.J. Russell, R. Spiewak, S. Zhang, X.J. Zhu, Search for an Isotropic Gravitational-wave Background with the Parkes Pulsar Timing Array. ApJL **951**(1), L6 (2023). https://doi.org/10.3847/2041-8213/acdd02
21. A.A. Starobinskii, On a nonsingular isotropic cosmological model. Sov. Astron. Lett. **4**, 82–84 (1978)

22. A.N. Tikhonov, A.S. Leonov, A.G. Yagola, Nonlinear ill-posed problems, vol. 14 (1998)
23. W.N. van Wieringen, Lecture notes on ridge regression. arXiv e-prints arXiv:1509.09169 (2015). https://doi.org/10.48550/arXiv.1509.09169
24. H. Xu, S. Chen, Y. Guo, J. Jiang, B. Wang, J. Xu, Z. Xue, R. Nicolas Caballero, J. Yuan, Y. Xu, J. Wang, L. Hao, J. Luo, K. Lee, J. Han, P. Jiang, Z. Shen, M. Wang, N. Wang, R. Xu, X. Wu, R. Manchester, L. Qian, X. Guan, M. Huang, C. Sun, Y. Zhu, Searching for the Nano-Hertz Stochastic Gravitational Wave Background with the Chinese Pulsar Timing Array Data Release I. Res. Astron. Astrophys. **23**(7), 075024 (2023). https://doi.org/10.1088/1674-4527/acdfa5

Proton Decay in de Sitter Environment

G. E. Volovik

Abstract The decay of proton in the de Sitter environment is governed by the temperature $T = H/\pi$, where H is the Hubble parameter. This temperature is twice larger than the Gibbons-Hawking temperature $T_{\rm GH} = H/2\pi$. This demonstrates the physical difference of two processes. The temperature $T = H/\pi$ determines the proton decay rate in the local process which takes well inside the cosmological horizon. While the Gibbons-Hawking temperature can be related to the processes which involve the Hawking photons or other particles radiated from the cosmological horizon. The same temperature $T = H/\pi$ determines the radiation of electron-positron pairs by positron or by other object in the de Sitter environment. Creation of matter and its thermalization in the de Sitter heat bath leads to the energy exchange between matter and quantum vacuum and finally to the decay of the de Sitter state towards the Minkowski vacuum. The lesson for the Unruh effect is also discussed. One may expect the similar separation of effects governed by two different temperatures: the effects related to the Rindler horizon with the Unruh temperature $T_{\rm U} = a/2\pi$ and the local radiation with $T = a/\pi = 2T_{\rm U}$.

1 Introduction

Hawking effect in black holes [1] and in the de Sitter spacetime [2] and also the Unruh effect (or Fulling-Davies-Unruh effect [3–5]) are the cornerstone phenomena of quantum filed theory in curved spacetimes and in non-inertial reference frames.

The Unruh effect has been applied to the problem of the decay of the accelerated proton, $p^+ \to n + e^+ + \nu$, which is a kind of the inverse β-decay, see e.g. Refs. [6–10]. Decay of the non-inertial proton was discussed in the earlier papers, see Refs. [11, 12] and references therein. Here we consider the similar proton decay in a different configuration—in the de Sitter environment. The de Sitter spacetime has high symmetry due to constant in space and time scalar Riemann curvature.

G. E. Volovik (✉)
Landau Institute for Theoretical Physics, Chernogolovka, Russia
e-mail: volovikgrigory@gmail.com

A. Barvinsky and A. Kamenshchik (eds.), *Open Issues in Gravitation and Cosmology*, Fundamental Theories of Physics 225, https://doi.org/10.1007/978-3-032-15703-4_23

This allows us to study different processes in the de Sitter state using its local thermodynamics [13]. The de Sitter thermodynamics is characterized by the local temperature $T = H/\pi$, which is twice the Gibbons-Hawking temperature, $T = 2T_{\rm GH}$. However, the holographic bulk-surface correspondence does work: the entropy integrated over the Hubble volume coincides with the Gibbons-Hawking entropy $A/4G$, where A is the area of the cosmological horizon.

Here we show that the same temperature $T = H/\pi$ governs the process of the proton decay in the de Sitter environment, and also the other processes generated by positron, such as electron-positron pair creation. It fits into the overall scheme of the particle decay in the de Sitter environment [14–19]. The processes electron-positron pair creation by proton (or by any other object) in the de Sitter environment demonstrates the instability of the de Sitter state and its power-law decay to the Minkowski vacuum.

The experience with proton in the de Sitter universe may provide some lessons for accelerated protons in Minkowski space.

2 Decay of Particles in de Sitter Spacetime

2.1 Ionization of Hydrogen Atom

In general, to consider the processes in the de Sitter environment it is convenient to use the metric in the Painlevé-Gullstrand (PG) form: [20, 21]

$$ds^2 = -dt^2 + (d\mathbf{r} - \mathbf{v}(\mathbf{r})dt)^2 \,. \tag{1}$$

Here $\mathbf{v}(\mathbf{r})$ is the shift velocity, which in the de Sitter state is $\mathbf{v}(\mathbf{r}) = H\mathbf{r}$, and the metric is:

$$ds^2 = -dt^2 + (dr - Hrdt)^2 + r^2 d\Omega^2 \,. \tag{2}$$

The PG metric is stationary, i.e. does not depend on time, and it does not have the unphysical singularity at the cosmological horizon. That is why it is appropriate for consideration of the local thermodynamics both inside and outside the horizon. This also allows us to discuss the processes related to the cosmological horizon at $r = r_h = 1/H$.

However, such processes as ionization of an atom take place well inside the cosmological horizon, at $r < r_h$, where the conventional static metric can be used:

$$ds^2 = -(1 - H^2r^2)dt^2 + \frac{dr^2}{1 - H^2r^2} + r^2 d\Omega^2 \,. \tag{3}$$

The ionization of the hydrogen atom, $H \to p^+ + e^-$, is the typical example of splitting of the composite object in the de Sitter environment. The rate of this process of ionization can be calculated using the semiclassical tunneling picture [13, 22]. The calculation of the imaginary part of the action gives the following activation exponent:

$$w \propto \exp(-2\,\mathbf{Im}\,S) = \exp\left(-\frac{\pi\epsilon}{H}\right) . \tag{4}$$

Here $\epsilon \gg H$ is ionization potential, and the semiclassical action is $S = \int dr\; p_r(r)$. The trajectory $p_r(r)$ of the radiated electron for $\epsilon \ll m_e$ is obtained from the classical non-relativistic equation

$$\frac{\mathbf{p}^2}{2m_e} - \frac{1}{2} m_e H^2 r^2 = -\epsilon\,, \tag{5}$$

which gives

$$p_r(r) = \sqrt{m_e^2 H^2 r^2 - 2m_e\epsilon}\,. \tag{6}$$

The momentum $p(r)$ is imaginary at

$$r < r_0 = \frac{1}{H}\sqrt{\frac{2\epsilon}{m_e}} \ll r_H = \frac{1}{H}\,. \tag{7}$$

Then one has

$$2\,\mathbf{Im}\,S = 2m_e H \int_0^{r_0} dr \sqrt{r_0^2 - r^2} = \frac{\pi\epsilon}{H}\,, \tag{8}$$

which gives Eq. (4).

In Eq. (4) we ignore the pre-exponential factor. In this approximation the decay rate coincides with that, which comes from the activation process in the Minkowski state in the presence of the heat bath with temperature $T = H/\pi$. This demonstrates that matter perceives the de Sitter environment as the heat bath with temperature $T = H/\pi$. The local temperature of the de Sitter heat bath governs also the other processes of the splitting of stable objects, including the decay of proton in de Sitter environment. It is important that due to the de Sitter symmetry, the local temperature is the same for all points in the de Sitter space, i.e. for all observers, that are static in the de Sitter environment. The coordinate transformation in the PG metric (1) which connects two static observers:

$$\mathbf{r} \to \mathbf{r} + \mathbf{a}\,e^{Ht}\;,\; t \to t\,. \tag{9}$$

The metric (1) is invariant under this shift of the position of the observer. Each observer considers the hydrogen atom, which is at rest in his/her reference frame.

Here we do not distinguish between composite and elementary (fundamental) particles (the so-called nuclear democracy [23, 24]). Both are considered as objects, which are stable in Minkowski vacuum and are unstable in de Sitter. In particular, we consider proton as a stable particle ignoring the possible proton decay to positron and neutral pion, which violates the conservation of baryon and lepton numbers.

2.2 General Case

Keeping in mind the inverse β-decay, $p^+ \to n + e^+ + \nu$, we consider the general case, when the initial particle with mass $M_{\rm in}$ decays into N particles with masses m_a such that the total mass $M_{\rm fin}$ of the radiated particles is larger than $M_{\rm in}$:

$$M_{\rm in} \to M_{\rm fin} = \sum_{a=1}^{N} m_a > M_{\rm in}\,. \tag{10}$$

As in the case of ionization of the hydrogen atom, such process is not allowed in the Minkowski vacuum, but is possible in the de Sitter environment or under acceleration.

It is convenient to represent the initial particle as the collection of the N virtual (off-shell) particles with energies $E_a < m_a$:

$$M_{\rm in} = \sum_a E_a\,. \tag{11}$$

Since $E_a < m_a$, the radiation of each component is classically forbidden, but in the de Sitter environment it becomes possible due to the process of quantum tunneling (see also Ref. [17]). Then the process of decay of the initial particle into N particles can be considered as N simultaneous (coherent) processes, in which each virtual (off-shell) particle with energy E_a transforms in the de Sitter environment to the on-shell particle with mass m_a.

The virtual particle becomes the real particle at some distance r_0 from the position $r = 0$ of the initial object. At this distance the gravitational potential of the de Sitter state compensates the difference between the energy E_a of the virtual particle and the mass m_a of the radiated particle, in the same way as this takes place in the process of ionization of the hydrogen atom [13, 22].

Using the static de Sitter metric in Eq. (22) one obtains the trajectory of each particle moving in the radial direction from the origin at $r = 0$ to the distance r_0:

$$p_a(r) = \frac{1}{1 - H^2 r^2}\sqrt{E_a^2 - m_a^2 + m_a^2 H^2 r^2}\,. \tag{12}$$

The momentum on this trajectory is imaginary in the classically forbidden region where the a-th particle is virtual, i.e. at $r < r_0$, where

$$r_0 = \frac{\sqrt{1 - \frac{E_a^2}{m_a^2}}}{H} < r_H = \frac{1}{H}\,. \tag{13}$$

In case of ionization of atom, one has $E_e = m_e - \epsilon$ for electron and $E_p = m_p$ for proton. For $\epsilon \ll m_e$ this gives $r_0 = \sqrt{2\epsilon/m_e}/H$ in Eq. (7).

The imaginary momentum of the a-th particle gives the imaginary contribution to its action and thus the tunneling exponent:

$$\mathbf{Im}\, S_a = \mathbf{Im} \int dr\ p_a(r) = \frac{m_a}{H} \int_0^{r_0} dr \frac{\sqrt{r_0^2 - r^2}}{r_H^2 - r^2} = \frac{\pi}{2H}(m_a - E_a)\,. \tag{14}$$

Note that since $r_0 < r_H$, the tunneling process takes place well inside the cosmological horizon and is not related to the process of Hawking radiation from the cosmological horizon.

Now we must take into account that in the quantum process of splitting of the initial object, all N virtual components tunnel simultaneously. Such coherent process of simultaneous tunneling of several component (the co-tunneling) gives the following decay rate of the initial particle into N particles (here we assume that $m_a \gg H$):

$$\Gamma(M_{\rm in} \to M_{\rm fin}) \propto \exp(-2\sum_a \mathbf{Im}\, S_a) = \exp\left(-\frac{\pi}{H}\sum_a (m_a - E_a)\right) = \exp\left(-\frac{\pi(M_{\rm fin} - M_{\rm in})}{H}\right)\,. \tag{15}$$

This demonstrates that rate of the particle decay in the de Sitter environment in the limit $m_a \gg H$ is the same as in the heat bath with temperature $T = H/\pi$ in the Minkowski spacetime:

$$\Gamma(M_{\rm in} \to M_{\rm fin}) \sim \exp\left(-\frac{\Delta M}{T}\right)\,,\ T = \frac{H}{\pi}\,. \tag{16}$$

Note that in the limit of small H and ignoring the pre-exponential factor, which depends on the coupling constants of interactions responsible for the decay, this result is universal. It does not depend on details of the process. In particular it does not depend on the distribution of energies E_a between the virtual particles in Eq. (11) and also on the number N of the virtual states (a kind of nuclear democracy [23, 24]).

3 Proton in de Sitter Environment

3.1 Decay of Proton in de Sitter Environment

Equation (16) can be applied to the decay of proton in the inverse β-decay in the de Sitter environment:

$$p^+ \to n + e^+ + \nu\,. \tag{17}$$

Equation (16) for splitting into $N = 3$ components gives the proton decay rate:

$$\Gamma(p^+ \to n + e^+ + \nu) \sim \exp\left(-\frac{\pi \Delta M}{H}\right), \tag{18}$$

$$\Delta M = M_{\rm fin} - M_{\rm in} = m_n + m_e + m_\nu - m_p \,. \tag{19}$$

Here for m_ν we can use the mass eigenstates for neutrino, where there is no distinction between inertial and gravitational masses [25].

In a similar way one can consider the following channel of decay of the hydrogen atom:

$$\Gamma(H \to n + \nu) \sim \exp\left(-\frac{\pi \Delta M}{H}\right), \tag{20}$$

$$\Delta M = M_{\rm fin} - M_{\rm in} = m_n + m_\nu - m_H \,. \tag{21}$$

3.2 Decay of Proton in the Schwarzschild-de Sitter Environment

We can use the conventional static metric for the Schwarzschild-de Sitter (SdS) black hole:

$$ds^2 = -\left(1 - \frac{2M}{r} - H^2 r^2\right) dt^2 + \frac{dr^2}{1 - \frac{2M}{r} - H^2 r^2} + r^2 d\Omega^2 \,. \tag{22}$$

Let us place the proton at the point $r = r_0$, where

$$r_0^3 = \frac{GM}{H^2} \,. \tag{23}$$

The spherical surface at $r = r_0$ is the zero-gravity surface, and thus we can find the proton decay rate, when proton is rest. Neat this point, at $|r - r_0| \ll r_0$, the metric is

$$ds^2 = -\left(1 - 3H^2(r - r_0)^2\right) dt^2 + \frac{dr^2}{1 - 3H^2(r - r_0)^2} + r^2 d\Omega^2 \,. \tag{24}$$

The decay rate in such environment corresponds to the decay rate in the de Sitter state with the effective Hubble parameter $\tilde{H} = \sqrt{3} H$. Then the temperature, which determines the decay of proton in the SdS environment, is (see also Ref. [26]):

$$T_{SdS} = \frac{\sqrt{3}}{\pi} H \,. \tag{25}$$

It is interesting that this temperature does not depend on the mass of a black hole and is fully determined by the Hubble parameter. The process of decay takes place in the region between the black hole horizon and the cosmological horizon, $r_- < r_0 < r_+$, and thus it is the local process, which has no relation to the Hawking radiation from the horizons. However, in the limit when the two horizons merge we can compare this activation temperature with the Bousso-Hawking temperature [27]. In this Nariai limit the Bousso-Hawking temperature $T_{\rm BH}$ is twice smaller than the temperature which determines the proton decay in the SdS spacetime, $T_{\rm BH} = \frac{\sqrt{3}}{2\pi} H = T_{SdS}/2$.

3.3 Proton as the Source of Electron-Positron Pairs

If the neutrino mass is neglected, the mass difference in proton decay in Eq. (19) is:

$$\Delta M \approx m_e + (m_n - m_p) \approx 0.51 MeV + 1.29 MeV \approx 3.5\, m_e > 2m_e \,. \qquad (26)$$

Since $\Delta M > 2m_e$, another process related to proton has the higher rate. Since neutron is unstable, the proton in the de Sitter environment serves as the source of the electron-positron pair creation:

$$p^+ \to n + e^+ + \nu \to p^+ + e^+ + e^- + \nu + \bar{\nu} \,. \qquad (27)$$

The rate of the direct electron-positron pair creation process generated by proton (i.e. without the intermediate step of inverse β-decay) is:

$$\Gamma(p^+ \to p^+ + e^+ + e^-) \sim \exp\left(-\frac{2\pi m_e}{H}\right) = \exp\left(-\frac{2m_e}{T}\right). \qquad (28)$$

Here we consider the process in which neutrino and antineutrino are either annihilated or are not formed. In the latter case the virtual energies in Eq. (11) are $E_{e^+} = E_{e^-} = 0$, while proton is on-shell, $E_{p^+} = m_p$.

Note that in the Gibbons-Hawking radiation from the cosmological horizon, with such rate only a single electron (or a single positron) can be created, $w \propto \exp(-m_e/T_{GH})$. For the coherent creation of pairs, the double Gibbons-Hawking temperature $T = 2T_{GH}$ is needed. The reason for such difference is that in our case the process of pair creation takes place well inside the cosmological horizon. It is generated by the positron: its presence in the de Sitter vacuum violates the de Sitter symmetry and allows for the creation of pairs from the vacuum by providing the non-zero matrix elements. This demonstrates the physical difference between the local processes governed by temperature $T = H/\pi$ and the processes generated by the cosmological horizon with the Gibbons-Hawking temperature $T_{GH} = H/2\pi$, which is by the factor 2 smaller.

The analogous process of creation of the neutrino-antineutrino pairs generated by proton in the de Sitter environment has the rate:

$$\Gamma(p^+ \to p^+ + \nu + \bar{\nu}) \sim \exp\left(-\frac{2\pi m_\nu}{H}\right). \quad (29)$$

4 Vacuum Back Reaction and de Sitter Decay

The process of creating electron-positron and neutrino-antineutrino pairs by a positron in Eqs. (28) and (29) can be repeated many times. The same positron inserted into the de Sitter vacuum may create more and more pairs. This demonstrates the instability of the de Sitter vacuum with respect to the creation of matter if a positron or other external object is present in the de Sitter Universe. This instability has no relation to the cosmological horizon. It is initiated by the external objects. These external objects serve as sources of particle radiation in the de Sitter medium, while the emitted particles in turn become sources of further radiation.

This leads to the following consequences. First, since the processes of radiation are determined by the characteristic temperature $T = H/\pi$ of the de Sitter heat bath, one may expect that the temperature of the radiated matter will tend to approach the temperature of the heat bath. Second, the creation of matter with its energy density must be followed by the back reaction of the vacuum, which looses energy.

The decay of the quantum vacuum due to activation and then the thermalization of matter can be considered using the extension of the Starobinsky approach [28–35]. Here we consider the simple phenomenological scenario [13] based on the energy exchange between the de Sitter heat bath and the thermal matter under the slow-roll condition, $|\dot{H}| \ll H^2$. Without the back reaction, the Friedmann equations give the following decay of matter density due to the Hubble friction,

$$\partial_t \rho_M = -3(1+w)H\rho_M . \quad (30)$$

This equation is modified, when we take into account the heat exchange. We consider here the case when matter is represented by the relativistic gas of massless particles, i.e. $P_M = w\rho_M$ with $w = 1/3$.

Since the relativistic gas tends to approach the temperature $T = H/\pi$ of the de Sitter environment, its energy density ρ_M tends to approach the value $\rho_M(H) = bH^4$, where the dimensionless parameter b depends on the number of massless relativistic fields. Then the natural modification of Eq. (30) is:

$$\partial_t \rho_M = -4H(\rho_M - \rho_M(H)) \ , \ \rho_M(H) = bH^4 . \quad (31)$$

The extra gain of the matter energy, $4H\rho_M(H)$, must be compensated by the corresponding loss of the vacuum energy:

$$\partial_t \rho_{\rm vac} = -4H\rho_M(H)\,. \tag{32}$$

Since the vacuum energy density is $\rho_{\rm vac} = 3H^2/8\pi G$, one obtains from Eq. (32) the following time dependence of the Hubble parameter and of the energy densities of vacuum and matter:

$$H(t) = b^{-1/3} M_{\rm Pl} \left(\frac{t_{\rm Pl}}{t+t_0}\right)^{1/3}\,, \tag{33}$$

$$\rho_{\rm vac}(t) = 6b^{-2/3} M_{\rm Pl}^4 \left(\frac{t_{\rm Pl}}{t+t_0}\right)^{2/3}\,. \tag{34}$$

Here $M_{\rm Pl}^2 = 1/16\pi G$ and $t_{\rm Pl} = 1/M_{\rm Pl}$, and we assume that $t_0 \gg t_{\rm Pl}$, and thus $|\dot{H}| \ll H^2$.

The power law $1/3$ of decay of H in Eq. (33) agrees with the results obtained using several different approaches. [36–42] This demonstrates that the phenomenological scenario of thermalization of matter by the de Sitter heat bath in Eqs. (31) and (32) is reasonable. This scenario does not require any microscopic or macroscopic model and describes the Starobinsky inflation in terms of the phenomenological parameters t_0. This parameter determines the initial value of the Hubble parameter at the beginning of inflation: $H(t=0) = b^{-1/3} M_{\rm Pl} \left(\frac{t_{\rm Pl}}{t_0}\right)^{1/3} \ll M_{\rm Pl}$, which corresponds to the scaleron mass in Starobinsky inflation.

In this approach there is no cosmological constant problem. Even if the initial vacuum energy is on the order of the Planck scale, the vacuum looses its energy and finally decays towards the perfect vacuum—the Minkowski vacuum. The energy of the stable Minkowski vacuum is always zero, [43] as it follows from the thermodynamic Gibbs-Duhem relation, $\rho = -P = 0$, which is applicable to the ground state of any macroscopic system (relativistic and non-relativistic) in the absence of external environment.

On the other hand, the pure de Sitter vacuum can be considered as the combination of dark energy with the equation of state $w = -1$ and dark matter with the equation of state $w = +1$ [44]. The dark matter component of the de Sitter vacuum has the same equation of state as the Zel'dovich stiff matter [45].

5 Discussion

We considered the decay of proton in the de Sitter environment. What can be the lessons for the decay of the accelerated proton considered in Refs. [6–10]?

One lesson is that there are two different temperatures, which govern the decay of stable particles in the de Sitter environment: the local temperature $T = H/\pi$

and the twice smaller Gibbons-Hawking temperature $T_{\rm GH} = H/2\pi$ related to the cosmological horizon. The local temperature determines the proton decay rate in the local processes which occur well inside the cosmological horizon. On the other hand, the Gibbons-Hawking temperature can be related to the processes which involve the Hawking photons, electrons and other particles radiated from the cosmological horizon.

The similar division may take place for the accelerated protons. There can be the local processes and the processes related to the Rindler horizon. For example, the doubling of the Unruh temperature takes place in the particular case of the combined Schwinger-Unruh process [46]. This is similar to the doubling of the Gibbons-Hawking temperature. This demonstrates that the FDU phenomenon is rather delicate giving rise to some controversies or requires some modifications [47–49]. In particular, it depends on the way how the observer measures the temperature, and on his/her ability to get the full information about the process.

Another lesson from de Sitter is that in the de Sitter environment the proton radiates electron-positron pairs. The rate of the emission of pairs is also determined by the local temperature $T = H/\pi$. This is distinct from the Hawking radiation of electrons and positrons caused by the cosmological horizon, which is determined by the twice smaller Gibbons-Hawking temperature. One may expect the similar doubling of the Unruh temperature in the radiation of pairs by accelerated proton. If so this would be in favour of the separation of two effects with two different temperatures: the effect related to the Rindler horizon with Unruh temperature $T_{\rm U} = a/2\pi$ and the local radiation with $T = a/\pi = 2T_{\rm U}$.

It would be interesting to consider, how these two sides of thermodynamics (the local thermodynamics and the thermodynamics related to the cosmological and Rindler horizons) are modified in the time-dependent environment. The particular case is when the de Sitter state is perturbed towards the Friedmann–Lemaitre–Robertson–Walker (FLRW) cosmology [50]. Such transformation from de Sitter to FLRW takes place due to processes of creation of electron-positron pairs initiated by proton or by any other object inserted into the de Sitter Universe. These processes lead to the power-law decay of the de Sitter state towards the Minkowski vacuum due to the energy exchange between the created matter and the de Sitter heat bath.

Acknowledgements I thank Oleg Andreev for discussions.

Competing Interests The author has no conflicts of interest to declare that are relevant to the content of this chapter.

References

1. S.W. Hawking, Black hole explosions? Nature **248**, 30–31 (1974)
2. G.W. Gibbons, S.W. Hawking, Cosmological event horizons, thermodynamics, and particle creation. Phys. Rev. D **15**, 2738 (1977)

3. S.A. Fulling, Nonuniqueness of Canonical Field Quantization in Riemannian Space-Time. Phys. Rev. D **7**, 2850–2862 (1973)
4. P.C.W. Davies, Scalar production in Schwarzschild and Rindler metrics. J. Phys. A **8**, 609–616 (1975)
5. W.G. Unruh, Notes on black-hole evaporation. Phys. Rev D **14**, 870–892 (1976)
6. Daniel A. T. Vanzella and George E. A. Matsas, Decay of Accelerated Protons and the Existence of the Fulling-Davies-Unruh Effect, Phys. Rev. Lett. **87**, 151301 (2001)
7. H. Suzuki, K. Yamada, Analytic Evaluation of the Decay Rate for an Accelerated Proton. Phys. Rev. D **67**, 065002 (2003). ((arXiv:gr-qc/0211056))
8. D. Vir Ahluwalia, L. Labun, G. Torrieri, The Unruh effect and oscillating neutrinos, J. Phys.: Conf. Ser. **706**, 042006 (2016)
9. Gabriel Cozzella, Andre G.S. Landulfo, George E.A. Matsas and Daniel A.T. Vanzella, Proposal for Observing the Unruh Effect using Classical Electrodynamics, Phys. Rev. Lett. **118**, 161102 (2017), arXiv:1701.03446 [gr-qc]
10. M. Blasone, G. Lambiase, Giuseppe Gaetano Luciano and Luciano Petruzziello, On the β-decay of the accelerated proton and neutrino oscillations: a three-flavor description with CP violation. Eur. Phys. J. C **80**, 130 (2020)
11. V.L. Ginzburg and G,F. Zharkov, Pion and Beta-ray Emission by Protons Moving in a Magnetic Field, JETP **20**, 1525 (1965)
12. V.L. Ginzburg, S.I. Syrovatsky, Cosmic magnetic bremsstrahlung (synchrotron radiation). Usp. Fiz. Nauk **87**, 65 (1965)
13. G.E. Volovik, Thermodynamics and decay of de Sitter vacuum, Symmetry **16**, 763 (2024), https://doi.org/10.3390/sym16060763
14. J. Bros, H. Epstein, U. Moschella, Lifetime of a massive particle in a de Sitter universe. JCAP **0802**, 003 (2008). ((hep-th/0612184))
15. G.E. Volovik, Particle decay in de Sitter spacetime via quantum tunneling, Pis'ma ZhETF **90**, 3–6 (2009); JETP Lett. **90**, 1–4 (2009); arXiv:0905.4639 [gr-qc]
16. J. Bros, H. Epstein, U. Moschella, Particle decays and stability on the de Sitter universe. Annales Henri Poincare **11**, 611–658 (2010). (arXiv:0812.3513 [hep-th])
17. L. Vanzo, G. Acquaviva, R. Di Criscienzo, Tunnelling Methods and Hawking's radiation: achievements and prospects. Topical Review. Class. Quantum Grav. **28**, 183001 (2011). ((arXiv:1106.4153))
18. D.P. Jatkar, L. Leblond, A. Rajaraman, On the Decay of Massive Fields in de Sitter. Phys. Rev. D **85**, 024047 (2012). (arXiv:1107.3513 [hep-th])
19. Nima Arkani-Hamed and Juan Maldacena, Cosmological Collider Physics, arXiv:1503.08043
20. P. Painlevé, La mécanique classique et la théorie de la relativité, C. R. Acad. Sci. (Paris) **173** , 677 (1921)
21. A. Gullstrand, Allgemeine Lösung des statischen Einkörper-problems in der Einsteinschen Gravitations-theorie. Arkiv. Mat. Astron. Fys. **16**, 1–15 (1922)
22. H. Maxfield, Z. Zahraee, Holographic solar systems and hydrogen atoms: non-relativistic physics in AdS and its CFT dual. JHEP **11**, 093 (2022)
23. G. Chew, N. Democracy, B. Dynamics, *in: Maurice Jacob and Chew, Strong-Interaction Physics: A Lecture Note Volume* (Benjamin, New York, 1964), pp.103–152
24. G. F. Chew, The analytic S-matrix: a basis for nuclear democracy, W.A. Benjamin, U.S.A. (1966)
25. M. Blasone, P. Jizba, G. Lambiase, L. Petruzziello, Non-relativistic neutrinos and the weak equivalence principle apparent violation. Phys. Lett. B **811**, 135883 (2020)
26. G.E. Volovik, On the global temperature of the Schwarzschild-de Sitter spacetime, Pis'ma v ZhETF **118**, 5–6 (2023), JETP Lett. **118**, 8–13 (2023), arXiv:2304.09847
27. R. Bousso, S.W. Hawking, Pair creation of black holes during inflation. Phys. Rev. D **54**, 6312 (1996)
28. Y.B. Zel'dovich, A.A. Starobinsky, Rate of particle production in gravitational fields. JETP Lett. **26**, 252 (1977)

29. A.A. Starobinsky, Dynamics of phase transition in the new inflationary universe scenario and generation of perturbations. Phys. Lett. B **117**, 175–178 (1982)
30. A.A. Starobinsky, STOCHASTIC DE SITTER (INFLATIONARY) STAGE IN THE EARLY UNIVERSE. Lect. Notes Phys. **246**, 107–126 (1986)
31. A.A. Starobinsky, J. Yokoyama, Equilibrium state of a self-interacting scalar field in the de Sitter background. Phys. Rev. D **50**, 6357 (1994)
32. D. Polarski, A.A. Starobinsky, Semiclassicality and decoherence of cosmological perturbations. Class. Quantum Grav. **13**, 377 (1996)
33. L. Kofman, A. Linde, A.A. Starobinsky, Towards the theory of reheating after inflation. Phys. Rev. D **56**, 3258 (1997)
34. A. Dobado, A.L. Maroto, Particle production from non-local gravitational effective action. Phys. Rev. D **60**, 104045 (1999). ((arXiv:gr-qc/9803076))
35. Hyun Jeong, K. Kamada, A.A. Starobinsky and J. Yokoyama, Reheating process in the R^2 inflationary model with the baryogenesis scenario, JCAP11(2023)023
36. T. Padmanabhan, Cosmological constant-the weight of the vacuum. Phys. Rep. **380**, 235–320 (2003)
37. T. Padmanabhan, Gravity and the thermodynamics of horizons. Phys. Rep. **406**, 49–125 (2005)
38. F.R. Klinkhamer, G.E. Volovik, Dynamic cancellation of a cosmological constant and approach to the Minkowski vacuum. Mod. Phys. Lett. A **31**, 1650160 (2016). (arXiv:1601.00601 [hep-th])
39. T. Markkanen, De Sitter stability and coarse graining. Eur. Phys. J. C **78**, 97 (2018)
40. M. Fairbairn, T. Markkanen and D. Rodriguez Roman, Horizon feedback inflation, Eur. Phys. J. C **78**, 347 (2018)
41. D. Rodriguez Roman, Gravitational particle creation in the early universe, Thesis: PhD King's Coll. London (2020)
42. J.-O. Gong, M.-S. Seo, Instability of de Sitter space under thermal radiation in different vacua, JCAP10(2021)042, arXiv:2011.01794
43. F.R. Klinkhamer, G.E. Volovik, Self-tuning vacuum variable and cosmological constant. Phys. Rev. D **77**, 085015 (2008). ((arXiv:0711.3170))
44. G.E. Volovik, From Landau two-fluid model to de Sitter Universe. Uspekhi Fizicheskikh Nauk **195**, 1137–1156 (2025), arXiv:2410.04392
45. Ya.B. Zel'dovich, The equation of state at ultrahigh densities and its relativistic limitations, ZhETF **41**, 1609–1615 (1961), JETP **14**, 1143–1147 (1962)
46. G.E. Volovik, Schwinger vs Unruh, Pis'ma v ZhETF **120**, 659 (2024), JETP Lett. **120**, (2024) 10.1134/S0021364024603324, arXiv:2409.03316
47. G.W. Ford, R.F. O'Connell, Is there Unruh radiation? Phys. Lett. A **350**, 17–26 (2006). ((arXiv:quant-ph/0509151))
48. G. Gaetano Luciano, M. Blasone, Nonextensive Tsallis statistics in Unruh effect for Dirac neutrinos. Eur. Phys. J. C **81**, 995 (2021)
49. G. Gregori, G. Marocco, S. Sarkar, R. Bingham, C. Wang, Measuring Unruh radiation from accelerated electrons. Eur. Phys. J. C **84**, 475 (2024)
50. S.D.Odintsov, T. Paul, S. SenGupta, Natural validation of the second law of thermodynamics in cosmology. Phys. Rev. D **111**, 043544 (2025), arXiv:2409.05009

Quantum Gravity and Quantum Field Theory in Curved

An Unfinished Collaboration with A. A. Starobinsky

Thibault Damour

Abstract The present text summarizes some of the results obtained in 2008 during the initial stages of a collaboration with Starobinsky which remained unfinished. The collaboration was an attempt to apply the stochastic approach to infrared (IR) quantum-gravity effects in an inflationary spacetime.

1 Introduction

My dear friend Alexei Alexander Starobinsky (henceforth Alyosha) visited the Institut des Hautes Etudes Scientifiques (IHES) in October 2008. During this visit, we started a collaboration on the impact of infrared quantum gravity effects on the global structure of de Sitter (or inflationary) spacetimes. I had always hoped that Alyosha would come back at IHES to allow us to continue our discussions, and to hopefully finalize the preliminary results we had reached in October 2008. Alas this did not happen. Though I had the pleasure of meeting again Alyosha in person during some later visits of mine in Moscow (notably during the 2019 conference dedicated to the 100th anniversary of Khalatnikov), and though Alyosha kindly agreed to speak (remotely) during the 2021 "Damour Fest" at IHES, we were both dragged away from our collaborative project by other scientific pursuits, and we never finalized our collaboration. As an homage to him, and as a token of my admiration for all the deep and beautiful ideas he contributed to physics, the present text will present the gist of our discussions, as they were left at the time (together with some additional considerations on Brownian motion in the space of 3-by-3 matrices). Hopefully, the approach we pursued (crucially based on Alyosha's stochastic approach to inflation) has still some value, in spite of the many developments that have since happened (and are still happening) in the study of infrared effects in de Sitter spacetimes. Before entering into details, let me mention that the stochastic approach to inflation is the idea, pioneered in Ref. [1] (see also [2, 3], as well as [4, 5]), that the infrared

T. Damour (✉)
Institut des Hautes Etudes Scientifiques, Bures sur Yvette, France
e-mail: damour@ihes.fr

A. Barvinsky and A. Kamenshchik (eds.), *Open Issues in Gravitation and Cosmology*, Fundamental Theories of Physics 225, https://doi.org/10.1007/978-3-032-15703-4_24

part of a quantum fluctuating field during inflation may be considered as a classical (i.e., c-number) space-dependent stochastic field satisfying a local (i.e., without spatial derivatives) evolution equation where the classical evolution is completed by an additional Langevin-like fluctuating term (coming from the higher-frequency quantum fluctuations). For recent reviews of the stochastic approach to inflation see Ref. [6, 7]. For entries into the still growing literature on infrared effects in de Sitter (or inflationary) spacetimes, see [8–14]. The last two references, of which I became aware while writing the present account, have close links with the aim of my unfinished collaboration with Alyosha. I leave to the future to decide whether they make it irrelevant by superseding it.

2 First Starting Point

The first starting point of our discussions was the following classic result of Alyosha. In 1983, Alyosha derived the asymptotic structure, at large times, of the general solution of Einstein's equations in presence of a (positive) cosmological constant $\Lambda = 3H^2$ [15] (in $3+1$ dimensions, and when neglecting other types of stress-energy tensor), namely (in synchronous gauge)

$$ds^2 = -dt^2 + e^{2Ht}\hat{g}_{ij}(\mathbf{x}, t)dx^i dx^j \,, \tag{1}$$

where the rescaled spatial metric $\hat{g}_{ij}(\mathbf{x}, t)$ admits an expansion in powers of e^{-Ht} of the form

$$\hat{g}_{ij}(\mathbf{x}, t) = a_{ij}(\mathbf{x}) + b_{ij}(\mathbf{x})e^{-2Ht} + c_{ij}(\mathbf{x})e^{-3Ht} + \cdots \,. \tag{2}$$

One can further impose the unimodularity condition $\det a_{ij}(\mathbf{x}) = 1$. The latter expanded solution[1] is general in that it contains four physical arbitrary functions of three variables: there are two arbitrary functions in the unimodular spatial metric $a_{ij}(\mathbf{x})$ (considered modulo 3-diffeomorphisms); the function $b_{ij}(\mathbf{x})$ is determined by the Ricci tensor of $a_{ij}(\mathbf{x})$ (namely $H^2 b_{ij}(\mathbf{x}) = R_{ij}[a] - \frac{1}{4}R[a]a_{ij}$); and the other two arbitrary functions of three variables are contained in $c_{ij}(\mathbf{x})$ (which must satisfy $c_i^i = 0$ and $\nabla^j c_{ij} = 0$). [The latter constraints use $a_{ij}(\mathbf{x})$ as background metric.] Terms beyond $c_{ij}(\mathbf{x})$ in the expansion (2) are fully determined by $a_{ij}(\mathbf{x})$ and $c_{ij}(\mathbf{x})$.

The expansion (2) is valid in many spatial coordinate systems. E.g., one recovers the global de Sitter metric with $\hat{g}_{ij}(\mathbf{x}, t)dx^i dx^j = H^{-2}e^{-2Ht}\cosh^2 Ht d\Omega_3^2$ (where $d\Omega_3^2$ is a round 3-sphere metric), which leads to $a_{ij}(\mathbf{x})dx^i dx^j = \frac{1}{4}H^{-2}d\Omega_3^2$ and $b_{ij}(\mathbf{x}) = 2a_{ij}(\mathbf{x})$. Alternatively, the Poincaré-patch case corresponds to $a_{ij}(\mathbf{x}) = \delta_{ij}$ and $b_{ij}(\mathbf{x}) = 0$.

The crucial point here is that though this general solution exhibits a phenomenon of local isotropization, dubbed the cosmic no-hair behavior (with an exponentially

[1] In an anti-de-Sitter context, it would be called a Fefferman-Graham expansion.

vanishing Weyl tensor, $C_{\mu\nu\kappa\lambda}C^{\mu\nu\kappa\lambda} \propto e^{-6Ht}$), its global spatial geometry is conformal, at large times, to an arbitrary (unimodular) 3-metric $a_{ij}(\mathbf{x}) = \lim_{t\to\infty} \hat{g}_{ij}(\mathbf{x}, t)$. The conformal spatial geometry defined by $a_{ij}(\mathbf{x})$ is globally inhomogeneous and anisotropic. It can be arbitrarily different from the round sphere obtained for a global de Sitter spacetime. On the other hand, it is time-independent.

3 Second Starting Point

As just reviewed, in a classical context the large-time limit of the conformal spatial metric exists and defines a globally inhomogeneous and anisotropic, but time-independent geometry. The second basic starting point of my collaboration with Alyosha was to extend his stochastic approach to inflation to an approximate description of the evolution of $\hat{g}_{ij}(\mathbf{x}, t)$ in a quantum context, and to argue that a suitably coarse-grained version of the quantum $\hat{g}_{ij}(\mathbf{x}, t)$ would define a stochastically evolving (unimodular) 3-metric $a_{ij}(\mathbf{x}, t)$ becoming exponentially anisotropic as $t \to +\infty$.

Let us first recall another classic result of Alyosha, concerning the production of "relict gravitational radiation" from quantum gravitational inflationary fluctuations [16]. Linearized quantum gravitational fluctuations in a de Sitter background, i.e. small fluctuations in the (conformally rescaled) spatial metric of the type,

$$\hat{g}_{ij}(\mathbf{x}, t) = \delta_{ij} + h_{ij}(\mathbf{x}, t)\,, \tag{3}$$

can be Fourier-decomposed as (with $\kappa \equiv \sqrt{32\pi G}$)

$$h_{ij}(\mathbf{x}, t) = \sum_e \kappa \int \frac{d^3k}{(2\pi)^{\frac{3}{2}}} e_{ij} a_{\mathbf{k}} e^{i\mathbf{k}\cdot\mathbf{x}} \frac{\chi_{\mathbf{k}}(\eta)}{a(\eta)} + \text{h.c.} \tag{4}$$

Here, we assumed for simplicity that we work in a Poincaré patch, and we introduced the conformal time η, varying between $-\infty$ and 0^-, and such that

$$a = e^{Ht} = -\frac{1}{H\eta}\,. \tag{5}$$

In Eq. (4), e_{ij} labels two different normalized ($e_{ij}e_{ij} = 1$) transverse-traceless polarization tensors, and the corresponding mode functions[2] $\chi_{\mathbf{k}}(\eta)$ satisfy the equation (with $' = d/d\eta$)

$$\chi''_{\mathbf{k}}(\eta) + \left(k^2 - \frac{a''}{a}\right)\chi_{\mathbf{k}}(\eta) = 0\,, \tag{6}$$

[2] For simplicity we leave implicit the extra label, say $e = +, \times$, on the mode functions, and on the annihilation operators.

whose solution (corresponding to the Bunch-Davies vacuum) is

$$\chi_{\mathbf{k}}(\eta) = \frac{e^{-ik\eta}}{\sqrt{2k}} \left(1 - \frac{i}{k\eta}\right) . \tag{7}$$

Finally, $a_{\mathbf{k}}$ is a corresponding quantum annihilation operator satisfying the canonical commutation relation

$$[a_{\mathbf{k}}\,, a^{\dagger}_{\mathbf{k}'}] = \delta(\mathbf{k} - \mathbf{k}')\,, \tag{8}$$

and defining the Bunch-Davies vacuum through $a_{\mathbf{k}}|0\rangle = 0$.

A first way to see the physics associated with the quantum gravitational field (4) is to compute the VEV of $h^2_{ij}(\mathbf{x}, t)$, namely

$$\langle 0|h^2_{ij}(\mathbf{x}, t)|0\rangle = 2\kappa^2 \left(\frac{H}{2\pi}\right)^2 \int_{k_{\min}}^{k_{\max}} \frac{dk}{k}(1 + (k\eta)^2)\,. \tag{9}$$

The latter integral is both UV and IR divergent. The UV (power-law) divergence is the usual local-Minkowski divergence of the two polarization modes of the gravitational field (with $h_{ij}(L) \sim \kappa/L$ over a small (sub-Hubble) spatial length $L \ll H^{-1}$). On the other hand, the IR (logarithmic) divergence involves the lower bound $k_{\min} = Ha(\eta_0)$ corresponding to the Hubble scale of some initial time, η_0. When integrated over the interval of momenta $k_{\min} < k < Ha(\eta)$ (where the upper bound $Ha(\eta)$ corresponds to the Hubble scale at the considered current time, η), the contribution of the IR logarithm then gives (using $a(\eta) = e^{Ht}$)

$$\langle 0|h^2_{ij}(\mathbf{x}, t)|0\rangle_{\mathrm{IR}} \approx 2\kappa^2 \left(\frac{H}{2\pi}\right)^2 \ln \frac{a(\eta)}{a(\eta_0)} = 2\kappa^2 \left(\frac{H}{2\pi}\right)^2 H\,(t - t_0)\,. \tag{10}$$

We see on the latter equation that the IR part of the quantum gravitational fluctuations leads to a linear-in-time wandering of the squared metric fluctuation. We recognize a Brownian motion of $h_{ij}(t)$ due to quantum modes coming from the UV, being redshifted by the expansion, and passing through the Hubble scale at time t. Essentially, Eq. (10) says that each one of the two polarization modes contributes a random amplitude $\pm\kappa\frac{H}{2\pi}$, during each time interval $\Delta t = H^{-1}$.

4 Towards a Stochastic Approach to Super-Hubble Quasi-classical IR Fluctuations of the De Sitter Metric

Alyosha's stochastic approach to quantum fluctuating light scalar fields (with mass $m \ll H$) during inflation consists in representing the Heisenberg operator of the quantum field ϕ as [1, 4, 5]

$$\phi(\mathbf{x},t) = \bar{\phi}(\mathbf{x},t) + \int \frac{d^3k}{(2\pi)^{\frac{3}{2}}} \theta(k - \epsilon a(t) H) \left[a_{\mathbf{k}} \phi_{\mathbf{k}}(t) e^{i\mathbf{k}\cdot\mathbf{x}} + \text{h.c.} \right] . \tag{11}$$

Here $\bar{\phi}(\mathbf{x},t)$ is a coarse-grained version of $\phi(\mathbf{x},t)$, i.e. averaged over a constant physical volume larger (given some $\epsilon < 1$) than a Hubble volume. The second term in Eq. (11), which involves a step function $\theta(x)$ and annihilation/creation operators $a_{\mathbf{k}}, a_{\mathbf{k}}^{\dagger}$, represents the short-wavelength counterpart of $\phi(\mathbf{x},t)$. The presence of the step function means that quantum modes of wave number k coming from the UV get redshifted by the expansion, $k^{\text{phys}} = \frac{k}{a(t)}$, and thereby continuously transit from the quantum high-frequency part of $\phi(\mathbf{x},t)$ to the low-frequency part $\bar{\phi}(\mathbf{x},t)$ as $k^{\text{phys}} = \frac{k}{a(t)}$ becomes smaller than ϵH. The basic idea is to approximate the low-frequency part $\bar{\phi}(\mathbf{x},t)$ by a classical stochastic field which can wander to large values in field space, contrary to the high-frequency quantum part, which is treated as staying "small" (after suitably subtracting UV divergences). When taking the time derivative of Eq. (11), the derivative of the step function generates a term in $\partial\phi(\mathbf{x},t)/\partial t$ of the form $-f(\mathbf{x},t)$, where

$$f(\mathbf{x},t) \equiv \epsilon a(t) H^2 \int \frac{d^3k}{(2\pi)^{\frac{3}{2}}} \delta(k - \epsilon a(t) H) \left[a_{\mathbf{k}} \phi_{\mathbf{k}}(t) e^{i\mathbf{k}\cdot\mathbf{x}} + \text{h.c.} \right] . \tag{12}$$

The term $+f(\mathbf{x},t)$ then must appear as an additional term in the time derivative of the quasi-classical low-frequency field $\bar{\phi}(\mathbf{x},t)$. The main technical feature of the stochastic approach is then to consider the term $+f(\mathbf{x},t)$ as a Langevin-type stochastic force driving (in addition to classical terms) the evolution of $\bar{\phi}(\mathbf{x},t)$. E.g., in the case of a scalar field with self-interacting potential $V(\phi)$, one gets the approximate first-order evolution equation

$$\frac{\partial\bar{\phi}(\mathbf{x},t)}{\partial t} = -\frac{1}{3H} V'(\bar{\phi}) + f(\mathbf{x},t) , \tag{13}$$

where $\bar{\phi}(\mathbf{x},t)$ is treated as a classical stochastic variable, and $f(\mathbf{x},t)$ as a random Gaussian noise, having the statistical properties defined by the quantum averages of $a_{\mathbf{k}}$ and $a_{\mathbf{k}}^{\dagger}$. References [5, 7] have shown that this stochastic approach does correctly describe IR effects in various field models. See also Ref. [17] for a discussion of why the low-frequency quantum fluctuations can be treated as classical stochastic variables.

The aim of our unfinished collaboration was to use this stochastic approach to derive an approximate evolution of the low-frequency part (say $\bar{g}_{ij}(\mathbf{x},t)$) of the conformal metric $\hat{g}_{ij}(\mathbf{x},t)$. We explored several possibilities for defining $\bar{g}_{ij}(\mathbf{x},t)$ and for writing down its Langevin-type evolution equation. Let me mention one of them.

If we start from the 3 + 1 decomposition of Einstein's equations (with Λ) in a synchronous (or Gauss) gauge, and rewrite them in terms of $\hat{g}_{ij}(\mathbf{x},t) \equiv a^{-2} g_{ij}(\mathbf{x},t)$, with $a(t) = e^{Ht}$, they read (with $\hat{K} \equiv \hat{K}^s_{\ s}$)

$$\hat{g}^{is}\partial_t\hat{g}_{sj} = 2\hat{K}^i{}_j\,,$$
$$\partial_t\hat{K}^i{}_j + 3\,H\hat{K}^i{}_j + H\hat{K}\delta^i{}_j + \hat{K}\hat{K}^i{}_j + a^{-2}\hat{R}^i{}_j = 0\,. \tag{14}$$

Let us start, around some spatial point, and at some time, with a local value of the metric of the type

$$\hat{g}_{ij}(\mathbf{x},t) = \bar{g}_{ij}(\mathbf{x},t) + h^{\mathrm{UV}}_{ij}(\mathbf{x},t)\,, \tag{15}$$

where the initial value of the low-frequency part is the flat Euclidean metric, $\bar{g}_{ij}(\mathbf{x}, t_0) = \delta_{ij}$, while the high-frequency part reads

$$h^{\mathrm{UV}}_{ij}(\mathbf{x},t) = \sum_e \kappa \int \frac{d^3k}{(2\pi)^{\frac{3}{2}}}\theta(k - \epsilon a(t)H)\left[e_{ij}a_{\mathbf{k}}e^{i\mathbf{k}\cdot\mathbf{x}}\frac{\chi_{\mathbf{k}}(\eta)}{a(\eta)} + \mathrm{h.c.}\right]. \tag{16}$$

Taking the time derivative of the step function in $h^{\mathrm{UV}}_{ij}(\mathbf{x},t)$ suggests that the low-frequency part will approximately evolve as

$$\frac{\partial\bar{g}_{ij}(\mathbf{x},t)}{\partial t} \approx f_{ij}(\mathbf{x},t)\,, \tag{17}$$

where the Langevin force $f_{ij}(\mathbf{x},t)$ (induced by the time derivative of $h^{\mathrm{UV}}_{ij}(\mathbf{x},t)$) reads

$$f_{ij}(\mathbf{x},t) \equiv \epsilon\kappa a(t)H^2\sum_e \int \frac{d^3k}{(2\pi)^{\frac{3}{2}}}\delta(k - \epsilon a(t)H)\left[e_{ij}a_{\mathbf{k}}e^{i\mathbf{k}\cdot\mathbf{x}}\frac{\chi_{\mathbf{k}}(\eta)}{a(\eta)} + \mathrm{h.c.}\right]. \tag{18}$$

The trace, $\delta^{ij}f_{ij}$, of this Langevin force vanishes, and its two-point correlation has (when considering for simplicity a fixed spatial point[3]) the following structure:

$$\langle f_{ij}(t)f_{kl}(t')\rangle \propto (\delta_{ik}\delta_{jl} + \delta_{il}\delta_{jk} - \frac{2}{3}\delta_{ij}\delta_{kl})\left(\frac{\kappa H}{2\pi}\right)^2 H\delta(t-t')\,. \tag{19}$$

When going beyond the approximation of linear stochastic perturbations of $\bar{g}_{ij}(\mathbf{x},t)$ around an Euclidean metric δ_{ij}, described by Eq. (17), it seems consistent to define the nonlinear stochastic evolution of $\bar{g}_{ij}(\mathbf{x},t)$ by requiring that $\bar{g}_{ij}(\mathbf{x},t)$ stays unimodular, and satisfies a *geometrically-defined* first-order stochastic evolution equation in the space, say S, of unimodular, positive-definite symmetric 3-by-3 matrices.

Let us recall that the space of unimodular, positive-definite symmetric 3-by-3 matrices $\bar{g}$ is a 5-dimensional Riemannian space, endowed with the metric

$$ds^2 = \mathrm{Tr}[(\bar{g})^{-1}d\bar{g}]^2 = \bar{g}^{ik}\bar{g}^{jl}d\bar{g}_{ij}d\bar{g}_{kl}, \tag{20}$$

---

[3] Alyosha told me that he expected the (physical) spatial correlation length R_c to be (asymptotically) of (logarithmic) order $\ln(HR_c) \sim (\kappa H)^{-1}$.

and that it is equivalent to the symmetric space $S \equiv SL(3)/SO(3)$. The equivalence between these two 5-dimensional spaces is made particularly clear by parametrizing the matrix $\bar{g} = (\bar{g}_{ij}(\mathbf{x}, t))$ by means of an Iwasawa decomposition KAN of the group $SL(3)$. Namely, a generic group element $\gamma \in SL(3)$ can be uniquely decomposed as the product kan of three 3-by-3 matrices, with: $k \in SO(3)$; a being diagonal, positive, and unimodular; and with n being upper triangular (with ones on the diagonal). The coset $S \equiv SL(3)/SO(3)$ is described by the an part, which is in one-to-one correspondence with $\bar{g} = \gamma^t \gamma = n^t a^t a n = n^t A n$ (t denoting a transpose, and $A = a^t a$ being diagonal, positive and unimodular). More explicitly, we can write at each spatial point $\mathbf{x}$

$$\bar{g}_{ij}(t) = \sum_{a=1}^{3} N^a{}_i e^{2\beta_a} N^a{}_j \,; \text{ with } \beta_1 + \beta_2 + \beta_3 = 0\,. \tag{21}$$

Here, $N^a{}_i$ denotes the components of the upper triangular matrix n, while $A = a^t a = \text{diag}[e^{2\beta_a}]$. The explicit expression of the metric ds^2 in the Iwasawa coordinates β_a, $N^a{}_i$ can be found in [18].

The infinitesimal neighbourhood of $\bar{g}_{ij} = \delta_{ij}$, say $\bar{g}_{ij} = \delta_{ij} + \epsilon f_{ij}$ (with f_{ij} symmetric and traceless), is parametrized by $\beta_a \to \epsilon \beta_a$ and $N^a{}_i = \delta_{ai} + \epsilon n_{ai}$, with n_{ai} strictly upper triangular (i.e. with n_{12}, n_{13}, n_{23} as only non-zero elements). The correspondence between f_{ij} and β_a, n_{ai} is $f_{aa} = 2\beta_a$, $f_{12} = n_{12} (= f_{21})$, $f_{13} = n_{13}$, $f_{23} = n_{23}$. The Langevin term (18) in the evolution of $\bar{g}_{ij}(t)$ near δ_{ij} then defines a corresponding Langevin term in the evolution of the coordinates, $\beta_a, n_{12}, n_{13}, n_{23}$, of the coset element an near the identity.

The issue is then to define, at the nonlinear level (when $\bar{g}$ starts wandering away from the identity), a definite stochastic evolution of $\bar{g}$ over the curved manifold S. The classic work of Itô [19] has shown how to define stochastic differential equations on a curved manifold in a geometrically covariant way, i.e. independently of the choice of coordinates on the considered manifold. In the case of the coset space S, the five numbers $\beta_1, \beta_2, n_{12}, n_{13}, n_{23}$ parametrizing $\bar{g}$ are particular coordinates. One wants to consider a first-order stochastic differential system that is equivariant under the choice of parametrization of the elements of S. Indeed, one might have used coordinates on S based, e.g., on a Gauss decomposition of the matrix $\bar{g}_{ij}$, as used, e.g., in [20].

Itô found that a geometrically covariant stochastic differential equation for the motion of the coordinates X^μ of a point on a Riemannian manifold (with metric tensor $G_{\mu\nu}(X)$ in the coordinates X^μ) must involve an additional (non-covariant) drift vector. The most natural Brownian motion equation for $X^\mu(T)$ (with a suitable scaled time variable T) reads

$$dX^\mu(T) = e^\mu_a(X)\, dw^a(T) - \frac{1}{2} G^{\alpha\beta} \Gamma^\mu_{\alpha\beta} dT\,, \tag{22}$$

where $e_a^\mu(X)$ is a local orthonormal frame ("repère mobile", satisfying $G_{\mu\nu}e_a^\mu e_b^\nu = \delta_{ab}$), $dw^a(T)$ a normalized white noise in the tangent space ($\langle dw^a(T)dw^b(T')\rangle = \delta^{ab}\delta(T-T')$), and $\Gamma^\mu_{\alpha\beta}$ the Christoffel coefficients of $G_{\mu\nu}(X)$. The diffusion (or Fokker-Planck) equation describing the evolution of the (one-point) probability distribution function (pdf) $\rho(X)$ then reads

$$\frac{\partial\rho(X)}{\partial T} = \frac{1}{2}\Delta_G\,\rho(X)\,, \tag{23}$$

where Δ_G denotes the Laplace-Beltrami operator defined by the metric $G_{\mu\nu}(x)$.

It was argued above that the IR quantum fluctuations of the spatial metric $\bar{g}_{ij}(\mathbf{x},t)$, when starting from the flat metric δ_{ij}, satisfy $\frac{\partial\bar{g}_{ij}(t)}{\partial t} \approx f_{ij}(t)$ with a white noise f_{ij} defined by Eq. (18). Viewing $\bar{g}_{ij}(\mathbf{x},t)$, at each spatial point $\mathbf{x}$, as a point X^μ in the 5-dimensional symmetric space $S \equiv SL(3)/SO(3)$ (endowed with the metric (20)), it seems natural to assume that, at the nonlinear level (when $\bar{g}$ starts wandering away from the identity), the stochastic wandering of $\bar{g}_{ij}(\mathbf{x},t)$ over the 5-dimensional symmetric space S will be described by the simple geometric Brownian motion (22), with associated diffusion equation

$$\frac{\partial\rho(\bar{g})}{\partial T} = \frac{1}{2}\Delta_S\,\rho(\bar{g})\,, \tag{24}$$

for the one-point pdf of $\bar{g}$ on $S \equiv SL(3)/SO(3)$. Here Δ_S denotes the Laplace-Beltrami operator on S, endowed with the metric (20). [We also wondered whether in Eq. (24) Δ_S should be augmented by a scalar-curvature term, say $\Delta_S + \xi R_S$ with $\xi = O(1)$.] The heat equation (24) involves a suitably rescaled (dimensionless) time variable,

$$T \sim \left(\frac{\kappa H}{2\pi}\right)^2 Ht\,. \tag{25}$$

While writing up the results of my discussions with Alyosha, I became aware of the large mathematical literature discussing Brownian motions on symmetric spaces. For entries in the literature see notably Refs. [21, 22].

A remarkable fact is that the long-time behavior of a Brownian motion on a non-compact symmetric space (such as S) exhibits special features linked to the negative sectional curvature of such spaces. Namely, when using an Iwasawa parametrization, say $\bar{g} = n^t a^t a n = n^t A n$, heat diffusion in symmetric spaces G/K of rank $\ell \geq 2$ asymptotically (for large times T) proceeds in the following way: (i) on the one hand, the off-diagonal component $n(T)$ (parametrized by $N^a{}_i$ in Eq. (21)), tends to some (random) finite limit as $T \to +\infty$; while, (ii) on the other hand, the logarithm $\alpha = \ln A$ of the diagonal component (equal to $\alpha = \mathrm{diag}(2\beta^a)$ in the notation of Eq. (21)) asymptotically proceeds in a fixed direction (Weyl vector) in the linear Cartan space, and with a fixed, finite speed.

In our case, $S = G/K$ with $G = SL(3,R)$ and $K = SO(3)$, the rank (i.e. the number of independent β's) is $\ell = 2$. Actually, in this case, one can explicitly write down the solution of the stochastic differential equation satisfied by the five Iwasawa coordinates $N^a{}_i$ and β_a (with the constraint $\beta_1+\beta_2+\beta_3=0$), in terms of six Wiener processes $w_I(T)$, $I=1,2,\cdots,6$, as was done in the recent Ref. [23]. [One must subtract from the first three Wiener processes in Sect. 3.2 of [23] their trace $\frac{1}{3}(w_1+w_2+w_3)$, i.e. replace w_a by $\bar{w}_a=w_a-\frac{1}{3}(w_1+w_2+w_3)$]. This leads to a result of the following form for the asymptotic (large T) stochastic evolution of the metric $\bar{g}_{ij}(T)$ (considered at each given spatial point $\mathbf{x}$):

$$\bar{g}_{ij}(T)=\sum_{a=1}^{3} N^a{}_i(T)e^{2\beta_a(T)}N^a{}_j(T)\,, \tag{26}$$

where the three diagonal (logarithmic) components are given by simple explicit expressions involving Wiener processes (i.e. Brownian motions), namely

$$\begin{aligned}
2\beta_1(T)&=2\beta_1(0)+\frac{1}{2}T+\bar{w}_1(T),\\
2\beta_2(T)&=2\beta_2(0)+\bar{w}_2(T),\\
2\beta_3(T)&=2\beta_3(0)-\frac{1}{2}T+\bar{w}_3(T),
\end{aligned} \tag{27}$$

while the off-diagonal components are given by stochastic integrals (involving three more independent Wiener processes $w_{12}(T)$, $w_{13}(T)$, $w_{23}(T)$) of the type

$$N^1{}_2(T)-N^1{}_2(0)\sim\int_0^T \exp(\beta_2(T')-\beta_1(T'))dw_{12}(T'). \tag{28}$$

The latter stochastic integrals converge to finite (random) limits as $T\to+\infty$. The existence of a finite limit follows from the fact that the exponents entering the differential of $N^a{}_i(T)$ are decreasing approximately linearly in T; e.g. (recalling that Brownian motions $w_a(T)$ behave as $O(T^{\frac{1}{2}})$ when $T\to+\infty$)

$$\beta_2(T)-\beta_1(T)=\beta_2(0)-\beta_1(0)-\frac{1}{4}T+\frac{1}{2}\bar{w}_2(T)-\frac{1}{2}\bar{w}_1(T)\approx-\frac{1}{4}T+O(T^{\frac{1}{2}}). \tag{29}$$

We note in passing that the asymptotic (large T) stochastic evolution of the metric $\bar{g}_{ij}(T)$ is reminiscent of the Belinsky-Khalatnikov-Lifshitz [24] dynamics of the spatial metric $g_{ij}(\tau)$ (considered at a fixed spatial point) near a cosmological singularity. Indeed, when parametrizing small cosmic proper times $t\to 0$ by $\tau=\int\frac{dt}{\sqrt{g}}\sim-\ln t\to+\infty$, and using an Iwasawa parametrization, the off-diagonal components $N^a{}_i(\tau)$ of the metric have finite limits as $\tau\to+\infty$, while the diagonal components of the metric, $a_1^2=e^{2\beta_1}$, $a_2^2=e^{2\beta_2}$, $a_3^2=e^{2\beta_3}$, have a stochastic behavior made of an infinite succession of Kasnerlike periods during which the β_a's evolve approximately linearly with τ, say $\beta_a\approx\beta_a^0+c_a\tau$, with velocities $c_a=O(1)$, see, e.g., [25, 26].

When going back from the heat-diffusion of the rescaled metric $\bar{g}_{ij}$ over S with respect to the rescaled time T, Eq. (25), to the stochastic evolution of the original spatial metric $g_{ij}(\mathbf{x},t) = e^{2Ht}\hat{g}_{ij}(\mathbf{x},t)$ in synchronous gauge, the final result is that the spatial metric at each spatial point would stochastically wander in the following approximate way

$$g_{ij}(\mathbf{x},t) \approx e^{2\,Ht}\Big(e^{\frac{1}{2}T+2\beta_1^0+\bar{w}_1(T)}N^1{}_i(T)N^1{}_j(T) + e^{2\beta_2^0+\bar{w}_2(T)}N^2{}_i(T)N^2{}_j(T) + e^{-\frac{1}{2}T+2\beta_3^0+\bar{w}_3(T)}N^3{}_i(T)N^3{}_j(T)\Big)\,, \tag{30}$$

where we recall that $T \sim \left(\frac{\kappa H}{2\pi}\right)^2 Ht$.

This quantum asymptotic behavior differs from the classical cosmic-no-hair asymptotic behavior of Eqs. (1), (2):

$$g_{ij}(\mathbf{x},t) \approx e^{2\,Ht}a_{ij}(\mathbf{x})\,, \tag{31}$$

in that the time-independent limiting classical rescaled (unimodular) metric $a_{ij}(\mathbf{x})$ is replaced by a (unimodular) metric whose diagonal elements (in an Iwasawa parametrization) keep evolving in an exponential way. This corresponds to a *quantum violation of the cosmic-no-hair behavior* (the rescaled metric becoming more and more anisotropic in the future). However, we note that the anisotropic deviations from a uniformly expanding de Sitter metric are fractionally small. Indeed, we have an anisotropic Hubble expansion for the diagonal components a_a^2 of the type $a_a \approx \exp H_a t$, where $a = 1, 2, 3$ and

$$H_a = H + c_a\left(\frac{\kappa H}{2\pi}\right)^2 H\,, \tag{32}$$

with $c_a \propto (+\frac{1}{4}, 0, -\frac{1}{4})$ (modulo an overall factor of order unity). Here the dimensionless (one-loop) factor $\left(\frac{\kappa H}{2\pi}\right)^2$ (which is proportional to G) must be assumed to be small in the present treatment. Note also that, in the approximation used here, we have no indication of the presence of a quantum back reaction slowing down the exponential inflation (as was suggested by several authors, notably by Polyakov [27], and by Tsamis and Woodard [28]). This is not unexpected as our approximation is a resummed version of one-loop effects, while the perturbative computations of Ref. [28] (see also [13, 14]) found a slowing down of inflation at the two-loop level, i.e., $H_{\text{effective}} \sim H\left[1 - c\left(\frac{\kappa H}{2\pi}\right)^4 (Ht)^2\right]$ with a positive coefficient $c = O(1)$.

I leave to future work the task[4] of comparing the results sketched above to the recent leading-logarithm quantum gravity results of Ref. [13]. This task poses the delicate problem of defining observable quantities that could be compared in both

[4] I also leave to future work the task of putting on a firmer ground the sketchy derivation given above, with precise numerical estimates of the various $O(1)$ factors that entered it.

approaches (and that are physically meaningful within a quantum cosmological spacetime).

Summarizing: a possibly too naive, and certainly preliminary, application of Starobinsky's stochastic approach to IR quantum-gravity effects in an inflationary spacetime suggests a picture where the low-frequency part of the metric (coarse-grained on Hubble scales) undergoes (modulo the overall exponential expansion factor e^{2Ht}) a Brownian motion in the 5-dimensional symmetric space $SL(3)/SO(3)$. This leads to a quantum violation of the cosmic-no-hair behavior in which the rescaled metric $\bar{g}_{ij}(\mathbf{x}, t)$ does not tend to a finite limit in the asymptotic future, but becomes more and more anisotropic in a way reminiscent of the Belinsky-Khalatnikov-Lifshitz behavior near a singularity.

Acknowledgements It is a pleasure to thank Sasha Polyakov and Richard Woodard for enlightening discussions about IR effects in de Sitter spacetimes, and Yvain Bruned and Laurence Field for informative exchanges about stochastic differential equations. The present work was partly supported by the *"2021 Balzan Prize for Gravitation: Physical and Astrophysical Aspects"*.

References

1. A.A. Starobinsky, Dynamics of phase transition in the new inflationary universe scenario and generation of perturbations. Phys. Lett. B **117**, 175–178 (1982)
2. A. Vilenkin, The birth of inflationary universes. Phys. Rev. D **27**, 2848 (1983)
3. A.D. Linde, Eternally existing selfreproducing chaotic inflationary universe. Phys. Lett. B **175**, 395–400 (1986)
4. A.A. Starobinsky, Stochastic de Sitter (inflationary) stage in the early universe. Lect. Notes Phys. **246**, 107–126 (1986)
5. A.A. Starobinsky, J. Yokoyama, Equilibrium state of a selfinteracting scalar field in the De Sitter background. Phys. Rev. D **50**, 6357–6368 (1994) [arXiv:astro-ph/9407016 [astro-ph]]
6. D. Cruces, Review on stochastic approach to inflation. Universe **8**(6), 334 (2022) [arXiv:2203.13852 [gr-qc]]
7. R. P. Woodard, Recent developments in stochastic inflation. [arXiv:2501.15843 [gr-qc]]
8. A. M. Polyakov, Infrared instability of the de Sitter space. [arXiv:1209.4135 [hep-th]]
9. P.R. Anderson, E. Mottola, D.H. Sanders, Decay of the de Sitter Vacuum. Phys. Rev. D **97**(6), 065016 (2018) [arXiv:1712.04522 [gr-qc]]
10. E.T. Akhmedov, U. Moschella, F.K. Popov, Characters of different secular effects in various patches of de Sitter space. Phys. Rev. D **99**(8), 086009 (2019) [arXiv:1901.07293 [hep-th]]
11. E.T. Akhmedov, K.V. Bazarov, D.V. Diakonov, U. Moschella, F.K. Popov, C. Schubert, Propagators and Gaussian effective actions in various patches of de Sitter space. Phys. Rev. D **100**(10), 105011 (2019) [arXiv:1905.09344 [hep-th]]
12. T. Cohen, D. Green, Soft de Sitter effective theory. JHEP **12**, 041 (2020). ([arXiv:2007.03693 [hep-th]])
13. S.P. Miao, N.C. Tsamis, R.P. Woodard, Leading Logarithm Quantum Gravity, [arXiv:2409.12003 [gr-qc]]
14. R. P. Woodard, Resummations for inflationary quantum gravity. Int. J. Mod. Phys. D **34**(10), 2542002 (2025). https://doi.org/10.1142/S0218271825420027. [arXiv:2501.05077 [gr-qc]]
15. A.A. Starobinsky, Isotropization of arbitrary cosmological expansion given an effective cosmological constant. JETP Lett. **37**, 66–69 (1983)

16. A.A. Starobinsky, Spectrum of relict gravitational radiation and the early state of the universe. JETP Lett. **30**, 682–685 (1979)
17. C. Kiefer, D. Polarski, A.A. Starobinsky, Quantum to classical transition for fluctuations in the early universe. Int. J. Mod. Phys. D **7**, 455–462 (1998) [arXiv:gr-qc/9802003 [gr-qc]]
18. M. Henneaux, M. Pilati, C. Teitelboim, Explicit solution for the zero signature (strong coupling) limit of the propagation amplitude in quantum gravity. Phys. Lett. B **110**, 123–128 (1982)
19. K. Itô, Stochastic differential equations in a differentiable manifold. Nagoya Math. J. **1**, 35–47 (1950)
20. T. Damour, P. Spindel, Quantum Einstein-Dirac Bianchi universes. Phys. Rev. D **83**, 123520 (2011). [arXiv:1103.2927 [gr-qc]]
21. M.-P. Malliavin, P. Malliavin, Factorisations et lois limites de la diffusion horizontale au-dessus d'un espace riemannien symétrique. Lect. Notes Math. **404**, 166–217 (1974)
22. J.-P. Anker, A.G. Setti, Asymptotic finite propagation speed for heat diffusion on certain Riemannian manifolds. J. Funct. Anal. **103**, 50 (1992)
23. H. Matsumoto, J. Otani, Laplacian and Brownian motion on positive definite matrices, revisited. Statist. Probab. Lett. **193**, 109696 (2022)
24. V.A. Belinsky, I.M. Khalatnikov, E.M. Lifshitz, Oscillatory approach to a singular point in the relativistic cosmology. Adv. Phys. **19**, 525–573 (1970)
25. T. Damour, M. Henneaux, H. Nicolai, Cosmological billiards. Class. Quant. Grav. **20**, R145–R200 (2003) [arXiv:hep-th/0212256 [hep-th]]
26. T. Damour, S. de Buyl, Describing general cosmological singularities in Iwasawa variables. Phys. Rev. D **77**, 043520 (2008). [arXiv:0710.5692 [gr-qc]]
27. A.M. Polyakov, Phase transitions and the Universe. Sov. Phys. Usp. **25**, 187 (1982)
28. N.C. Tsamis, R.P. Woodard, Quantum gravity slows inflation. Nucl. Phys. B **474**, 235–248 (1996) [arXiv:hep-ph/9602315 [hep-ph]]

Loop Integrals on the De Sitter Sphere and Some Summation Formulae of the Generalized Hypergeometric Series

Ugo Moschella

Abstract Among his many contributions to cosmology Alexei Starobinsky wrote several important papers about de Sitter Quantum Field Theory. He devoted his attention to this subject during all his lifetime. Here, to honor his memory, I review some recent results concerning one and two-loop de Sitter Schwinger diagrams with arbitrary masses in the dimensional regularization approach. I also present some new infinite summation formulae of higher order hypergeometric series which are deduced from the 1-loop diagram.

1 Alexei Starobinsky and the De Sitter Universe

In 1979 Alexei Starobinsky was the first to propose a model for the early universe where a period of extremely rapid exponential expansion, which is nowadays called inflation, takes place [1].

In the Starobinsky model the source of the cosmic inflation are quantum gravity effects; gravitational waves are produced that are now detectable as a background. The model also implies that primordial quantum fluctuations are stretched beyond the quantum scale eventually giving rise to the structures of the universe today [2].

Exponential expansion points towards the de Sitter (dS) geometry and (a portion of) the de Sitter spacetime in first approximation mimicks the inflationary phase of the early universe. This is why Quantum Field Theory on the de Sitter background has occupied a central place in Alexei Starobinsky's interests and researches in theoretical cosmology. His seminal studies in this subject are fundamental [3–6].

U. Moschella (✉)
Disat, Università dell'Insubria, Como, Italy
e-mail: ugo.moschella@uninsubria.it

INFN, Sezione di Milano, Milano, Italy

A. Barvinsky and A. Kamenshchik (eds.), *Open Issues in Gravitation and Cosmology*, Fundamental Theories of Physics 225, https://doi.org/10.1007/978-3-032-15703-4_25

To honor his memory I review here some recent results [7–10] about one and two-loop calculations of de Sitter Quantum Field Theory (QFT) and I present some previously unpublished new summation formulae of higher order hypergeometric series of Ramanujan type which may be computed in closed form by applying the results obtained in computing the 1-loop integral.

2 Generalities

Let M_{d+1} be the $(d+1)$-dimensional Minkowski space-time and $M^{(c)}_{d+1}$ be its complexification. In a chosen Lorentz frame the scalar product of two (complex) events is

$$z_1 \cdot z_2 = z_1^0 z_2^0 - z_1^1 z_2^1 - \ldots - z_1^d z_2^d. \tag{1}$$

The future cone V_+ and the future and past tubes $T_\pm$ of the (complex) Minkowski spacetime are defined as follows:

$$V_+ = \{x \in M_{d+1} \;:\; x \cdot x > 0, \quad x^0 > 0\}, \tag{2}$$

$$T_\pm = \{x + iy \in M^{(c)}_{d+1} \;:\; y \in \pm V_+\}\,. \tag{3}$$

The tubes $T_\pm$ are the geometrical image of the spectral condition of QFT, which is the requirement the positivity of the spectrum of the energy operator in every Lorentz frame [10, 11]. This characteristic property lies at the heart of QFT at zero temperature. The Euclidean formulation of quantum field theory and all the known renormalization schemes would not exist without the spectral condition.

The complex de Sitter universe may be represented as the complex quadric

$$dS_d^{(c)} = \left\{ z = x + iy \in M^{(c)}_{d+1} \;:\; z \cdot z = x \cdot x - y \cdot y + 2ix \cdot y = -R^2 = -1 \right\}. \tag{4}$$

The real de Sitter spacetime dS_d corresponds to $\Im z = y = 0$ i.e. $\{x \in M_{d+1} \; x \cdot x = -1\}$.

Given two complex events $z_1, z_2 \in dS_d^{(c)}$, the de Sitter invariant complex variable ζ is their scalar product in the ambient spacetime: $\zeta = z_1 \cdot z_2$. Two real events x_1 and x_2 in dS_d are timelike separated if and only if $(x_1 - x_2)^2 = -2 - 2x_1 \cdot x_2 > 0$.

As regards Quantum Field Theory, the main difficulty is that the de Sitter geometry admits no global timelike Killing vector. There are however complex *domains of analyticity* which closely corresponds to the tubes of flat spacetime: these are the de Sitter future and past tuboids $\mathcal{T}_\pm$ which are actually obtained as the intersections of the ambient tubes $T_\pm$ with the complex de Sitter manifold:

$$\mathcal{T}_\pm = \{x + iy \in X_d^{(c)} \;:\; y \in \pm V_+\}\,. \tag{5}$$

A spectral condition may then be formulated as the following requirement of *normal analyticity* [12, 13]: the two-point distributions are boundary values of functions analytic in the domain $\mathcal{T}_- \times \mathcal{T}_+$; this condition, together with de Sitter invariance and the Canonical Commutation Relations, selects a unique two-point function for any de Sitter Klein-Gordon field [13]. The most natural construction of such two-point function is done in terms of a basis of plane-wave solutions of the de Sitter Klein-Gordon equation

$$\Box\psi(z) + m^2\psi(z) = 0, \tag{6}$$

which is parameterized by the choice of a lightlike vector $\xi \in C^+ = \partial V^+$ and a complex number λ which is in turn parametrized by the spacetime dimension and another complex number ν as follows:

$$\psi_\lambda(z, \xi) = (z \cdot \xi)^\lambda, \quad \lambda = -\frac{d-1}{2} + i\nu. \tag{7}$$

The parameters λ and ν are related to the complex mass squared:

$$m^2 = -\lambda(\lambda + d - 1) = \frac{(d-1)^2}{4} + \nu^2. \tag{8}$$

Plane waves are well-defined and analytic in each of the tubes $\mathcal{T}^+$ and $\mathcal{T}^-$ [12, 13]. Of course their squared mass is real and positive only when:

(a) ν is real; this correspond in a group-theoretical language to the principal series of unitary representations of the Lorentz group;

(b) ν is purely imaginary such the $|\nu| < \frac{d-1}{2}$; this correspond to the complementary series of unitary representations of the Lorentz group.

Correspondingly, plane waves may be used to construct [12, 13] a manifestly dS covariant integral representation of the two-point Wightman function of a Klein-Gordon field (in general spacetime dimension d):

$$W_\nu^d(z_1, z_2) = \frac{\Gamma\left(\frac{d-1}{2} + i\nu\right)\Gamma\left(\frac{d-1}{2} - i\nu\right)e^{\pi\nu}}{2^{d+1}\pi^d} \int_\gamma (\xi \cdot z_1)^{-\frac{d-1}{2} - i\nu}(\xi \cdot z_2)^{-\frac{d-1}{2} + i\nu}\,\alpha(\xi) \tag{9}$$

$$= w_\nu^d(\zeta) = \frac{\Gamma\left(\frac{d-1}{2} + i\nu\right)\Gamma\left(\frac{d-1}{2} - i\nu\right)}{2(2\pi)^{d/2}}(\zeta^2 - 1)^{-\frac{d-2}{4}}\, P^{-\frac{d-2}{2}}_{-\frac{1}{2}+i\nu}(\zeta) \tag{10}$$

$$= \frac{\Gamma\left(\frac{d-1}{2} + i\nu\right)\Gamma\left(\frac{d-1}{2} - i\nu\right)}{(4\pi)^{d/2}\Gamma\left(\frac{d}{2}\right)}\, {}_2F_1\left(\frac{d-1}{2} + i\nu, \frac{d-1}{2} - i\nu; \frac{d}{2}; \frac{1-\zeta}{2}\right). \tag{11}$$

The plane wave representation (9) is holomorphic for $z_1 \in \mathcal{T}_-$ and $z_2 \in \mathcal{T}_+$: this is the *normal analyticity property*.

In standard coordinates, the $(d-1)$-form $\alpha(\xi)$ in Eq. (9) is written

$$\alpha(\xi) = (\xi^0)^{-1} \sum_{j=1}^{d} (-1)^{j+1} \xi^j \, d\xi^1 \ldots \widehat{d\xi^j} \ldots d\xi^d \,. \tag{12}$$

γ denotes any $(d-1)$-cycle in the forward light-cone C^+. The rhs of (9) does not depend on the choice of γ being the integral of a closed differential form. In particular we may choose the unit sphere S_{d-1} (equipped with its canonical orientation):

$$\gamma_0 = S_{d-1} = C^+ \cap \{\xi \ : \ \xi^0 = 1\} = \{\xi \in C^+ : {\xi^1}^2 + \ldots + {\xi^d}^2 = 1\}. \tag{13}$$

With this choice $\alpha(\xi)$ coincides with the rotation invariant measure $d\xi$ on S_{d-1} normalized as follows:

$$\omega_d = \int_{\gamma_0} d\xi = \frac{2\pi^{\frac{d}{2}}}{\Gamma\left(\frac{d}{2}\right)}. \tag{14}$$

By choosing any two points in the respective tubes such that $\zeta = z_1 \cdot z_2$ one can show the validity of Eq. (10). The Legendre function [14] of the first kind $P^{\mu}_{\nu}(\zeta)$ in that formula is holomorphic in the cut-plane $\mathbf{C} \setminus (-\infty, \ 1]$ but the reduced two-point function $w^d_\nu(\zeta)$ is holomorphic in the larger domain[1]

$$\zeta = z_1 \cdot z_2 \in \Delta = \{\mathbf{C} \setminus (-\infty, \ -1]\}, \tag{15}$$

i.e. everywhere except on the causality cut (2); this is the *maximal analyticity property*.

When ν is either real or is purely imaginary and such that $|\nu| < \frac{d-1}{2}$ the corresponding two-point function (9) is *positive-definite* and admits a direct quantum probabilistic interpretation.

This is not the end of the story for free fields. Even though it is often stated "there is no de Sitter invariant state for $m^2 < 0$" [5] this is not true: when $\frac{d-1}{2} + i\nu = -n$, where n is zero or a positive integer, a more involved but yet acceptable quantum interpretation of formula (9) is also possible; on the de Sitter universe there exist *tachyonic fields* having no counterpart in flat space [15, 16] and this is possible precisely because there is no *global* timelike Killing vector field on the de Sitter manifold.

Finally, the Schwinger function (in short: the propagator) is the restriction of the maximally analytic two-point function to the Euclidean sphere so that $z_1 \cdot z_2 = -\cos(s)$:

$$G_\nu(-\cos s) = \frac{\Gamma(\frac{d-1}{2} + i\nu)\Gamma(\frac{d-1}{2} - i\nu)}{2(2\pi)^{d/2}} (\sin s)^{-\frac{d-2}{2}} \mathbf{P}^{-\frac{d-2}{2}}_{-\frac{1}{2}+i\nu}(-\cos s) \tag{16}$$

[1] The prefactor $(\zeta^2 - 1)^{-\frac{d-2}{4}}$ exactly compensates the singularity of $P^{-\frac{d-2}{2}}_{-\frac{1}{2}+i\nu}(\zeta)$ at $\zeta = 1$, making the reduced two-point function regular there.

where $\mathbf{P}^{\mu}_{\rho}(z)$ is the so called "Legendre function on the cut" or Ferrers function of the first kind [14]. It is important to keep in mind that Ferrers functions $\mathbf{P}^{\alpha}_{\beta}(z)$ and Legendre functions $P^{\alpha}_{\beta}(z)$ are holomorphic in different cut-planes; as regards Ferrers function this is

$$\Delta_2 = \mathbf{C} \setminus \{(-\infty, -1] \cup [1, \infty)]\}. \tag{17}$$

In terms of the above Schwinger function we can write the n-loop banana integral on the sphere with $n+1$ propagators as follows:

$$I_n(\nu_1, \ldots, \nu_{n+1}, d) = \int G^d_{\nu_1}(x_0 \cdot x) G^d_{\nu_2}(x_0 \cdot x) \ldots G^d_{\nu_{n+1}}(x_0 \cdot x) \sqrt{g}\, dx; \tag{18}$$

here x varies on the Euclidean de Sitter sphere, x_0 is a fixed reference point over there and $\sqrt{g}\, dx$ is the rotation invariant measure. By integrating over the angles we get

$$I_n(\nu_1, \ldots, \nu_{n+1}, d) =$$
$$= \frac{2\pi^{\frac{d}{2}}}{\Gamma\left(\frac{d}{2}\right)} \int_0^{\pi} G^d_{\nu_1}(-\cos s) G^d_{\nu_2}(-\cos s) \ldots G^d_{\nu_{n+1}}(-\cos s)(\sin s)^{d-1} ds. \tag{19}$$

3 1-Loop: The Bubble

Let us start with the already nontrivial two-line case, i.e. the bubble:

$$I(\lambda, \nu, d) = \frac{\Gamma(\frac{d-1}{2} - i\lambda)\Gamma(\frac{d-1}{2} + i\lambda)\Gamma(\frac{d-1}{2} - i\nu)\Gamma(\frac{d-1}{2} + i\nu)}{2\left(2\sqrt{\pi}\right)^d \Gamma\left(\frac{d}{2}\right)} \int_{-1}^{1} \mathbf{P}^{-\frac{d-2}{2}}_{-\frac{1}{2}+i\lambda}(u) \mathbf{P}^{-\frac{d-2}{2}}_{-\frac{1}{2}+i\nu}(u) du \tag{20}$$

As simple as it may look, the integral at the rhs of Eq. (20) has been computed explicitly only recently [8, 9].

The simplest calculation goes as follows: let us consider the auxiliary two-line Feynman diagram where the external points do not coincide:

$$F_{\lambda\,\nu}(x, y) = \int_{S_d} G^d_{\lambda}(x \cdot z) G^d_{\nu}(y \cdot z) \sqrt{g(z)}\, dz. \tag{21}$$

The evaluation of the analogous of Eq. (21) in flat space is elementary by Fourier transform. Here the three points belong to the Euclidean de Sitter sphere: $(x \cdot x) = (y \cdot y) = (z \cdot z) = 1$ and we do not have translation invariance. A direct calculation is however possible but we prefer to proceed in another way by observing that $F_{\lambda\,\nu}(x, y)$ solves the following equations:

$$(-\nabla_x^2 + m_\lambda^2) F_{\lambda\nu}(x, y) = \int \delta(z, x) G_\nu^d(z, y) \sqrt{g(z)}\, dz = G_\nu^d(x \cdot y), \quad (22)$$

$$(-\nabla_y^2 + m_\nu^2) F_{\lambda\nu}(x, y) = \int G_\lambda^d(x \cdot z) \delta(y, z) \sqrt{g(z)}\, dz = G_\lambda^d(x \cdot y). \quad (23)$$

de Sitter invariance implies that $F_{\lambda\nu}(x, y) = F_{\lambda\nu}(x \cdot y)$ depends only on the scalar product $x \cdot y$ and therefore

$$\nabla_x^2 F_{\lambda\nu}(x, y) = \nabla_y^2 F_{\lambda\nu}(x, y), \quad (24)$$

in view of $x \cdot x = y \cdot y = 1$. Subtracting Eq. (23) from Eq. (22) it follows that

$$F_{\lambda\nu}(x, y) = -\frac{G_\lambda^d(x \cdot y) - G_\nu^d(x \cdot y)}{m_\lambda^2 - m_\nu^2}. \quad (25)$$

The above elementary derivation shows that Eq. (25) is valid for the Schwinger propagator of a Klein-Gordon field in a curved background $\mathcal{E}_d$ with Euclidean signature whenever Eq. (24) is verified:

$$F_{m_1 m_2}(x, y) = \int_{\mathcal{E}_d} G_{m_1}^{(\mathcal{E})}(x, z) G_{m_2}^{(\mathcal{E})}(z, y) \sqrt{g(z)}\, dz = -\frac{G_{m_1}^{(\mathcal{E})}(x, y) - G_{m_2}^{(\mathcal{E})}(x, y)}{m_1^2 - m_2^2}. \quad (26)$$

In the limiting case when the two masses coincide this becomes

$$F_{mm}(x, y) = -\frac{\partial}{\partial m^2} G_m^{(\mathcal{E})}(x, y). \quad (27)$$

Taking the limit $y \to x$ we deduce a general formula for the bubble in a curved background with Euclidean signature (similar remarks apply to the chronological propagator in Lorentzian signature):

$$bubble(m_1, m_2) = \int_{\mathcal{E}_d} G_{m_1}^{(\mathcal{E})}(x, z) G_{m_2}^{(\mathcal{E})}(z, x) \sqrt{g(z)}\, dz = -\lim_{y \to x} \frac{G_{m_1}^{(\mathcal{E})}(x, y) - G_{m_2}^{(\mathcal{E})}(x, y)}{m_1^2 - m_2^2}. \quad (28)$$

To apply this formula in the de Sitter case let us at first suppose $d < 2$; Eq. (11) at $\zeta = 1$ then gives

$$\lim_{\zeta \to -1} \frac{\Gamma\left(\frac{d-1}{2} + i\kappa\right) \Gamma\left(\frac{d-1}{2} - i\kappa\right)}{(4\pi)^{\frac{d}{2}} \Gamma\left(\frac{d}{2}\right)} {}_2F_1\left(\frac{d-1}{2} + i\kappa, \frac{d-1}{2} - i\kappa; \frac{d}{2}; \frac{1-\zeta}{2}\right) =$$
$$= \frac{\Gamma\left(1 - \frac{d}{2}\right) \Gamma\left(\frac{d-1}{2} - i\kappa\right) \Gamma\left(\frac{d-1}{2} + i\kappa\right)}{(4\pi)^{\frac{d}{2}} \Gamma\left(\frac{1}{2} - i\kappa\right) \Gamma\left(\frac{1}{2} + i\kappa\right)} \quad (d < 2) \quad (29)$$

so that

$$I(\lambda, \nu, d) = \frac{\Gamma\left(1-\frac{d}{2}\right)}{(4\pi)^{\frac{d}{2}}(\lambda^2-\nu^2)}\left(\frac{\Gamma\left(\frac{d-1}{2}-i\nu\right)\Gamma\left(\frac{d-1}{2}+i\nu\right)}{\Gamma\left(\frac{1}{2}-i\nu\right)\Gamma\left(\frac{1}{2}+i\nu\right)} - \frac{\Gamma\left(\frac{d-1}{2}-i\lambda\right)\Gamma\left(\frac{d-1}{2}+i\lambda\right)}{\Gamma\left(\frac{1}{2}-i\lambda\right)\Gamma\left(\frac{1}{2}+i\lambda\right)}\right). \tag{30}$$

This formula defines the bubble by analytic continuation in d as a meromorphic function of d. Here are a few consequences of the above result:

1. The bubble is regular at $d = 2$:

$$I(\lambda, \nu, 2) = \frac{\left(\psi\left(\frac{1}{2}-i\lambda\right)+\psi\left(\frac{1}{2}+i\lambda\right)-\psi\left(\frac{1}{2}-i\nu\right)-\psi\left(\frac{1}{2}+i\nu\right)\right)}{4\pi\left(\lambda^2-\nu^2\right)}. \tag{31}$$

2. In odd spacetime dimension the formula becomes very simple; for instance at $d = 3$

$$I(\lambda, \nu, 3) = \frac{\lambda\coth(\pi\lambda) - \nu\coth(\pi\nu)}{4\pi\lambda^2 - 4\pi\nu^2}. \tag{32}$$

3. At $d = 4$ we encounter the first divergence. The Laurent expansion of the formula near $d = 4$ gives

$$\begin{aligned} I(\lambda, \nu, 4) \simeq &-\frac{1}{8\pi^2(d-4)} + \frac{1-\gamma+\log(4\pi)}{16\pi^2} \\ &- \frac{\left(4\lambda^2+1\right)\left(\psi\left(\frac{3}{2}-i\lambda\right)+\psi\left(\frac{3}{2}+i\lambda\right)\right)-\left(4\nu^2+1\right)\left(\psi\left(\frac{3}{2}-i\nu\right)+\psi\left(\frac{3}{2}+i\nu\right)\right)}{64\pi^2\left(\lambda^2-\nu^2\right)} \\ &+ \mathrm{O}(d-4). \end{aligned} \tag{33}$$

4 The Effective Potential for the $O(N)$ Model

Let us consider a scalar multi-component field

$$\phi : S_d \longrightarrow R^N, \tag{34}$$

where S_d is the (Euclidean) de Sitter sphere of radius R. The radius R and the cosmological constant

$$\Lambda = \frac{(d-1)(d-2)}{2R^2} \tag{35}$$

will reappear later. For the moment we take $R = 1$.

The action for ϕ is the quartic $O(N)$-invariant action

$$S[\phi] = \int_{S_d} \Big[\Lambda_0 + \frac{1}{2}\langle \partial_\mu \phi \mid \partial^\mu \phi\rangle + \frac{m_0^2}{2}\langle \phi \mid \phi\rangle + \frac{c_0}{4}\langle \phi \mid \phi\rangle^2\Big]\sqrt{g}\, d^d x, \tag{36}$$

where m_0 and c_0 are respectively the bare mass and bare self-coupling constant; x^μ are local coordinates on the sphere $\langle \cdot \mid \cdot\rangle$ denotes the scalar product in R^N. We included also an extra cosmological constant Λ_0 for convenience when renormalizing.

We will compute the effective potential at $d = 4$ for the constant configuration

$$\bar{\phi} \equiv \varphi\, e_0, \tag{37}$$

where φ is a real constant and e_0 is a given vector of norm 1 in E. Of course a nonzero expectation value of ϕ breaks the symmetry down to $O(N-1)$. After choosing any orthonormal basis $\{e_j\}_{j=0}^{N-1}$ of E whose first element is e_0, we may write

$$\phi = (\varphi + \psi_0)e_0 + \sum_{j=1}^{N-1} \psi_j e_j, \tag{38}$$

The effective potential $\mathcal{V}(\varphi)$ is defined as follows

$$\exp(-\Omega_d \mathcal{V}(\varphi)) = \int [\prod_j D\psi_j] \exp(-S[\phi]); \tag{39}$$

here Ω_d is the volume of the Euclidean de Sitter spacetime (which is finite) and $[\prod_j D\psi_j]$ is the formal path integral measure. At 1-loop, we get

$$\mathcal{V}(\varphi) = \Lambda_0 + \frac{m_0^2}{2}\varphi^2 + \frac{c_0}{4}\varphi^4 - \frac{1}{\Omega_d}\log \int [\prod_j D\psi_j] \exp(-S_2[\phi]), \tag{40}$$

where

$$S_2[\psi;\varphi] = \frac{1}{2}\int_{S_d} \Big[\sum_{j=0}^{N-1} \partial_\mu \psi_j \partial^\mu \psi_j + (m_0^2 + 3c_0\varphi^2)\psi_0^2 + (m_0^2 + c_0\varphi^2)\sum_{j=1}^{N-1} \psi_j^2\Big]\sqrt{g}\, d^d x. \tag{41}$$

In standard form:

$$\mathcal{V}(\varphi) = \Lambda_0 + \frac{m_0^2}{2}\varphi^2 + \frac{c_0}{4}\varphi^4 + \mathcal{V}_0(\varphi) + (N-1)\mathcal{V}_1(\varphi), \tag{42}$$

where, for $a = 0, 1$,

$$\mathcal{V}_a(\varphi) = -\frac{1}{\Omega_d}\log\int[D\Phi]\exp\left(-\frac{1}{2}\int_{M_d}\sqrt{g}\Big[\partial_\mu\Phi\partial^\mu\Phi + M_a^2(\varphi)\Phi^2\Big]d^dx\right), \tag{43}$$

$$M_0^2(\varphi) = m_0^2 + 3c_0\varphi^2, \tag{44}$$

$$M_1^2(\varphi) = m_0^2 + c_0\varphi^2. \tag{45}$$

Instead of looking at $\mathcal{V}_a(\varphi)$ as a function of φ, it is useful to see it as a function of $M_a^2 \equiv z_a$. By taking the first derivative w.r.t. z we get the tadpole integral which leads to the Lee-Sciaccaluga equation for determining the effective potential [17].

At $d=4$, we set $d=4-2\epsilon$. Since $d\nu^2 = dm^2$, we have to integrate twice $-I(\nu,\nu,4)/2$ in $d\nu^2 = 2\nu d\nu$. From (33), we have

$$I(\nu,\nu,4-2\epsilon) = \frac{1}{16\pi^2\epsilon} + \frac{1-\gamma+\log(4\pi)}{16\pi^2} - \frac{1}{128\pi^2\nu}\frac{d}{d\nu}\left[\left(4\nu^2+1\right)\left(\psi\left(\frac{3}{2}-i\nu\right)+\psi\left(\frac{3}{2}+i\nu\right)\right)\right] + O(\epsilon). \tag{46}$$

The nontrivial part comes from the last term:

$$\begin{aligned}&\int\int\frac{1}{8\nu}\frac{d}{d\nu}\left[(4\nu^2+1)\left(\psi\left(\frac{3}{2}-i\nu\right)+\psi\left(\frac{3}{2}+i\nu\right)\right)\right]2\nu d\nu = \\ &= C\nu^2 + \frac{1}{2}\left(\nu^2+\frac{1}{4}\right)^2\left(\psi\left(-\frac{1}{2}-i\nu\right)+\psi\left(-\frac{1}{2}+i\nu\right)\right) - \frac{i}{2}\int d\nu\left(\nu^2+\frac{1}{4}\right)^2\left(\psi'\left(-\frac{1}{2}-i\nu\right)-\psi'\left(-\frac{1}{2}+i\nu\right)\right)\end{aligned} \tag{47}$$

where C is an integration constant. After setting $y := \nu^2 + \frac{1}{4}$, we can introduce the function

$$B(y) = \frac{y^2}{4} - \frac{i}{2}\int d\nu\left(\nu^2+\frac{1}{4}\right)^2\left(\psi'\left(-\frac{1}{2}-i\nu\right)-\psi'\left(-\frac{1}{2}+i\nu\right)\right) = -\frac{1}{2}\int dy\, y^2\left(\sum_{n=0}^{\infty}\frac{2n-1}{(n(n-1)+y)^2}-\frac{1}{y}\right); \tag{48}$$

in the last equality we have used the series expansion

$$\psi'(z) = \zeta(2,z) = \sum_{n=0}^{\infty}\frac{1}{(z+n)^2}, \tag{49}$$

By using the standard Abel-Plana's formula [18] we can write

$$\sum_{n=0}^{\infty}\frac{2n-1}{(n(n-1)+y)^2} = -\frac{1}{2y^2} - 4\int_0^\infty\frac{t}{e^{2\pi t}-1}\frac{(t^2-y)^2-y}{[(t^2-y)^2+t^2]^2}, \tag{50}$$

so that

$$B(y) = \frac{y}{3} + \int_0^\infty \frac{2t^4 + t^2 - 1 + y(4-2t^2)}{(t^2-y)^2 + t^2} \frac{t^3 dt}{e^{2\pi t} - 1} + 6\int_0^\infty \arctan\left(t - \frac{y}{t}\right) \frac{t^2 dt}{e^{2\pi t} - 1}$$
$$+ 2\int_0^\infty \left(t^2 - \frac{1}{2}\right) \log[t^2 + (t^2 - y)^2] \frac{t dt}{e^{2\pi t} - 1}, \tag{51}$$

where we have set to zero an integration constant. With these notations, the (renormalized) effective potential is

$$\mathcal{V}_R = \frac{m_R^2}{2}\varphi_R^2 + \frac{c_g}{4}\varphi_R^4 + \frac{1}{64\pi^2}(m_R^2 + 3c_g\varphi_R^2 - 2)^2 \left(\psi\left(-\frac{1}{2} - i\nu_0\right) + \psi\left(-\frac{1}{2} + i\nu_0\right)\right)$$
$$+ \frac{1}{32\pi^2} B(m_R^2 + 3c_g\varphi_R^2 - 2) + \frac{N-1}{32\pi^2} B(m_R^2 + c_g\varphi_R^2 - 2)$$
$$+ \frac{N-1}{64\pi^2}(m_R^2 + c_g\varphi_R^2 - 2)^2 \left(\psi\left(-\frac{1}{2} - i\nu_1\right) + \psi\left(-\frac{1}{2} + i\nu_1\right)\right)$$
$$- \frac{1}{64\pi^2}\left[(m_R^2 + 3c_g\varphi_R^2 - 2)^2 + (N-1)(m_R^2 + c_g\varphi_R^2 - 2)^2\right] \log \mu_R^2. \tag{52}$$

The last line is just the introduction of a reference energy scale, related to the renormalization, while

$$\nu_0 = \sqrt{m_R^2 + 3c_g\varphi_R^2 - \frac{9}{4}}, \tag{53}$$

$$\nu_1 = \sqrt{m_R^2 + c_g\varphi_R^2 - \frac{9}{4}}. \tag{54}$$

R refers to the de Sitter radius and appears through the formula $m_R = mR$, $\mu_R = \mu R$, and $\varphi_R = R\varphi$, m_R, μ_R and φ_R being adimensional quantities.

To get the flat limit of the above formula we need to compute the behaviour of $B(\nu)$ for $\nu \to \infty$. This is obtained from (51) by noticing that the integral in the first line goes to zero and using

$$\int_0^\infty \frac{t^k dt}{e^{2\pi t} - 1} = \frac{k!}{(2\pi)^{k+1}} \zeta(k+1). \tag{55}$$

Therefore,

$$B\left(\nu^2 + \frac{1}{4}\right) = \frac{1}{3}\nu^2 - \frac{1}{15}\log\left(\nu^2 + \frac{1}{4}\right) + \frac{1}{12} - \frac{3\zeta(3)}{4\pi^3} + \ldots, \tag{56}$$

where the dots stay for terms vanishing for $\nu \to \infty$.
On the other hand, for $|\arg(x)| \leq \pi - \delta$, $\delta > 0$, one has

$$\psi(x) = \log x - \frac{1}{2x} - \sum_{j=1}^{m} \frac{B_{2j}}{2j} \frac{1}{x^{2j}} + O\left(\frac{1}{x^{2m+2}}\right), \tag{57}$$

where B_j are the Bernoulli numbers. Since $B_2 = \frac{1}{6}$ and $B_4 = -\frac{1}{30}$, it follows that

$$\frac{1}{2}\left(\nu^2+\frac{1}{4}\right)^2\left(\psi\left(-\frac{1}{2}-i\nu\right)+\psi\left(-\frac{1}{2}+i\nu\right)\right)=\frac{1}{2}\left(\nu^2+\frac{1}{4}\right)^2\log\left(\nu^2+\frac{1}{4}\right)+\frac{1}{3}\nu^2+\frac{1}{20}+\ldots. \quad (58)$$

In this limit we have

$$\begin{aligned}\mathcal{V}_R=&\frac{m_R^2}{2}\varphi_R^2+\frac{c_g}{4}\varphi_R^4+\frac{1}{32\pi^2}\left[\frac{(m_R^2+3c_g\varphi_R^2-2)^2}{2}\log\frac{m_R^2+3c_g\varphi_R^2-2}{\mu_R^2}\right.\\&\left.+\frac{2}{3}(m_R^2+3c_g\varphi_R^2-2)-\frac{1}{15}\log(m_R^2+3c_g\varphi_R^2-2)-\frac{1}{30}-\frac{3\zeta(3)}{4\pi^3}+\ldots\right]\\&+\frac{N-1}{32\pi^2}\left[\frac{(m_R^2+c_g\varphi_R^2-2)^2}{2}\log\frac{m_R^2+c_g\varphi_R^2-2}{\mu_R^2}\right.\\&\left.+\frac{2}{3}(m_R^2+c_g\varphi_R^2-2)-\frac{1}{15}\log(m_R^2+c_g\varphi_R^2-2)-\frac{1}{30}-\frac{3\zeta(3)}{4\pi^3}+\ldots\right].\end{aligned} \quad (59)$$

The potential in the flat limit is defined by $\mathcal{V}=\lim_{R\to\infty}R^{-4}\mathcal{V}_R$ and gives exactly the well-known flat space result. For very large but finite R, we can write

$$\begin{aligned}\mathcal{V}_\Lambda=&\frac{m^2}{2}\varphi^2+\frac{c_g}{4}\varphi^4+\frac{1}{64\pi^2}\left[(m^2+3c_g\varphi^2)^2\log\frac{m^2+3c_g\varphi^2}{\mu^2}\right.\\&\left.+(N-1)(m^2+c_g\varphi^2)^2\log\frac{m^2+c_g\varphi^2}{\mu^2}\right]-\frac{\Lambda}{48\pi^2}\left[(m^2+3c_g\varphi^2)\left(\log\frac{m^2+3c_g\varphi^2}{\mu^2}+\frac{1}{6}\right)\right.\\&\left.+(N-1)(m^2+c_g\varphi^2)\left(\log\frac{m^2+c_g\varphi^2}{\mu^2}+\frac{1}{6}\right)\right]\\&+\frac{\Lambda^2}{144\pi^2}\left[-\frac{1}{30}\log\frac{3(m^2+3c_g\varphi^2)}{\Lambda}+\log\frac{m^2+3c_g\varphi^2}{\mu^2}+\frac{49}{60}-\frac{3\zeta(3)}{4\pi^3}\right.\\&\left.+(N-1)\left(-\frac{1}{30}\log\frac{3(m^2+c_g\varphi^2)}{\Lambda}+\log\frac{m^2+c_g\varphi^2}{\mu^2}+\frac{49}{60}-\frac{3\zeta(3)}{4\pi^3}\right)\right]+O(\Lambda^3),\end{aligned} \quad (60)$$

where $\Lambda = 3R^{-2}$ is the cosmological constant. The first two lines are the standard Coleman-Weinberg effective potential [19, 20], while the remaining terms provide the perturbative corrections in the cosmological constant, up to order two. The exact expression is given by (52), which in terms of the full dimensional quantities is

$$
\begin{aligned}
\mathcal{V}_\Lambda =& \frac{m^2}{2}\varphi^2 + \frac{c_g}{4}\varphi^4 + \frac{1}{64\pi^2}(m^2 + 3c_g\varphi^2 - \frac{2}{3}\Lambda)^2\left(\psi\left(-\frac{1}{2} - i\nu_0\right) + \psi\left(-\frac{1}{2} + i\nu_0\right)\right) \\
&+ \frac{\Lambda^2}{288\pi^2} B\left(3\frac{m^2}{\Lambda} + 9c_g\frac{\varphi^2}{\Lambda} - 2\right) + \frac{(N-1)\Lambda^2}{288\pi^2} B\left(3\frac{m^2}{\Lambda} + 3c_g\frac{\varphi^2}{\Lambda} - 2\right) \\
&+ \frac{N-1}{64\pi^2}(m^2 + c_g\varphi^2 - \frac{2}{3}\Lambda)^2\left(\psi\left(-\frac{1}{2} - i\nu_1\right) + \psi\left(-\frac{1}{2} + i\nu_1\right)\right) \\
&- \frac{1}{64\pi^2}\left[(m^2 + 3c_g\varphi^2 - \frac{2}{3}\Lambda)^2 + (N-1)(m^2 + c_g\varphi^2 - \frac{2}{3}\Lambda)^2\right]\log\frac{3\mu^2}{\Lambda},
\end{aligned} \tag{61}
$$

with

$$
\nu_0 = 3\sqrt{\frac{m^2}{3\Lambda} + c_g\frac{\varphi^2}{\Lambda} - \frac{1}{4}}, \tag{62}
$$

$$
\nu_1 = 3\sqrt{\frac{m^2}{3\Lambda} + c_g\frac{\varphi^2}{3\Lambda} - \frac{1}{4}}. \tag{63}
$$

5 The Bubble: Calculation by Linearization

Here we discuss a second way to compute the bubble which is based on the Källén-Lehmann representation of the product of two Wightman function of different masses [21]; the fact that this product of distribution exists is another consequence of the spectral condition which implies the maximal analyticity property (15):

$$
w_\lambda^d(\zeta)w_\nu^d(\zeta) = \int_{-\infty}^{\infty} \rho_d(\lambda, \nu, \kappa) w_\kappa^d(\zeta)\kappa d\kappa, \tag{64}
$$

where

$$
\rho_d(\lambda, \nu, \kappa) = \frac{1}{2^d\pi^{\frac{d-1}{2}}\kappa\Gamma\left(\frac{d-1}{2}\right)} \frac{\prod_{\epsilon,\epsilon',\epsilon''=\pm 1}\Gamma\left(\frac{d-1}{4} + \frac{i\epsilon\lambda + i\epsilon'\nu + i\epsilon''\kappa}{2}\right)}{\prod_{\epsilon=\pm 1}\Gamma\left(\frac{i\epsilon\kappa}{2}\right)\Gamma\left(\frac{1}{2} + \frac{i\epsilon\kappa}{2}\right)\Gamma\left(\frac{d-1}{4} + \frac{i\epsilon\kappa}{2}\right)\Gamma\left(\frac{d+1}{4} + \frac{i\epsilon\kappa}{2}\right)}. \tag{65}
$$

The integral of the Schwinger function over the sphere gives

$$
\int G_\kappa(x_0\cdot x)dx = \frac{\Gamma(\frac{d-1}{2} + i\kappa)\Gamma(\frac{d-1}{2} - i\kappa)}{\Gamma\left(\frac{d+1}{2} + i\kappa\right)\Gamma\left(\frac{d+1}{2} - i\kappa\right)} = \frac{1}{\kappa^2 + \frac{(d-1)^2}{4}} \tag{66}
$$

(see [14] Eq. 3.12.27). The bubble may then be computed in terms of Källén-Lehmann weight as follows:

$$
I(\lambda, \nu, d) = \int\int \rho(\lambda, \nu, \kappa)G_\kappa(x_0\cdot x)\kappa\, d\kappa\, dx = \int \rho(\lambda, \nu, \kappa)\frac{\Gamma(\frac{d-1}{2} + i\kappa)\Gamma(\frac{d-1}{2} - i\kappa)}{\Gamma\left(\frac{d+1}{2} + i\kappa\right)\Gamma\left(\frac{d+1}{2} - i\kappa\right)}\kappa\, d\kappa. \tag{67}
$$

By inserting (65) at the rhs of Eq. (67) and posing

$$s = \frac{i\kappa}{2}, \quad \delta = \frac{d-1}{4}, \quad x = \frac{i\lambda}{2}, \quad y = \frac{i\nu}{2}, \quad u = x + y \tag{68}$$

we may express the bubble as a Meijer[2] integral;

$$I(\lambda, \nu, d) = \frac{\pi^{-2\delta+1}}{2^{4\delta+1}\Gamma(2\delta)} \int_{-i\infty}^{i\infty} \frac{ds}{2\pi i} \prod_{\epsilon,\epsilon',\epsilon''=\pm} \frac{\Gamma\left(\epsilon s + \delta + i\epsilon' x + i\epsilon'' y\right)}{\Gamma(\epsilon s)\,\Gamma\left(\frac{1}{2}+\epsilon s\right)\Gamma\left(\delta+\frac{1}{2}+\epsilon s\right)\Gamma(\delta+1+\epsilon s)} \tag{71}$$

$$= \frac{1}{2^{4\delta+1}\pi^{2\delta-1}\Gamma(2\delta)} G^{4,4}_{8,8}\left(\begin{matrix} a_1, \ldots, a_8 \\ 1-a_1, \ldots, 1-a_8 \end{matrix} \,\middle|\, 1\right) \tag{72}$$

where the parameters are given by

$$\begin{matrix} a_1 = 1-\delta-x-y & a_2 = 1-\delta+x+y & a_3 = 1-\delta+x-y & a_4 = 1-\delta-x+y \\ a_5 = 0 & a_6 = 1+\delta & a_7 = \frac{1}{2}+\delta & a_8 = \frac{1}{2} \end{matrix} \tag{73}$$

and $b_j = 1 - a_j$. The two relevant sets of poles

$$s_a = -\delta \pm x \pm y - n, \qquad s_b = \delta \pm x \pm y + n \tag{74}$$

are such that the imaginary axis may be chosen as integration path.

The integral (71) is ultraviolet divergent. Power counting is the same as in flat space. By using Stirling's formula we have that

$$\left|\frac{\kappa\,\rho(\nu,\lambda,\kappa)}{\kappa^2+4\delta^2}\right| \sim \text{Const}\,|\kappa|^{2\sum_{j=1}^{4}\Re a_j - 2\sum_{j=5}^{8}\Re a_j} = \text{Const}\,|\kappa|^{4\delta-4} = \text{Const}\,|\kappa|^{d-5} \tag{75}$$

and therefore the integral converges if $d < 4$.

---

[2] For the reader's convenience we recall here the general definition of Meijer's G-function [14]

$$G^{m,n}_{p,q}\left(\begin{matrix} a_1, \ldots, a_p \\ b_1, \ldots, b_q \end{matrix} \,\middle|\, z\right) = \frac{1}{2\pi i}\int_L \frac{\prod_{j=1}^{m}\Gamma(b_j - s)\prod_{j=1}^{n}\Gamma(1-a_j+s)}{\prod_{j=m+1}^{q}\Gamma(1-b_j+s)\prod_{j=n+1}^{p}\Gamma(a_j-s)} z^s\,ds; \tag{69}$$

By using the results of [27] the above function may be written as a sum of hypergeometric functions

$$G^{m,n}_{p,q}\left(\begin{matrix} a_1, \ldots, a_p \\ b_1, \ldots, b_q \end{matrix} \,\middle|\, z\right) = \sum_{h=1}^{m} \frac{\prod_{j=1}^{m}\Gamma(b_j-b_h)^*\prod_{j=1}^{n}\Gamma(1+b_h-a_j)\,z^{b_h}}{\prod_{j=m+1}^{q}\Gamma(1+b_h-b_j)\prod_{j=n+1}^{p}\Gamma(a_j-b_h)} \times$$
$$\times\, {}_pF_{q-1}\left(\begin{matrix} 1+b_h-\mathbf{a_p} \\ (1+b_h-\mathbf{b_q})^* \end{matrix} \,\middle|\, (-1)^{p-m-n}\,z\right). \tag{70}$$

One should however be careful as the addends at rhs may be more singular than their sum at lhs.

By comparing Eqs. (30) and (72) we get a nontrivial resummation for that particular Meijer's function. The result may be in turn re-expressed as a symmetric sum of hypergeometric series as in Eq. (70): let

$$A_\delta(\lambda, \nu) = A_\delta(x, y) = -\frac{\pi^{-2\delta-1}\Gamma(-2x)\Gamma(-2y)\Gamma(2\delta+2x)\Gamma(2\delta+2y)}{16(x+y)\,\Gamma(-2\delta-2x-2y)\Gamma(4\delta+1+2x+2y)} \times$$

$$\times\, {}_7F_6\left(\begin{matrix} x+y, x+y+\frac{1}{2}, 2\delta, \delta+x+y+1, 2\delta+2x, 2\delta+2y, 2\delta+2x+2y \\ 2x+1, 2y+1, 2x+2y+1, \delta+x+y, 2\delta+x+y+\frac{1}{2}, 2\delta+x+y+1 \end{matrix}; 1\right) \tag{76}$$

where

$${}_pF_q(a_1, \ldots, a_p; b_1, \ldots, b_q; z) = \sum_{n=0}^{\infty} \frac{(a_1)_n \cdots (a_p)_n}{(b_1)_n \cdots (b_q)_n} \frac{z^n}{n!}, \tag{77}$$

and

$$(x)_n = x(x+1)\ldots(x+n).$$

We deduce the following nontrivial identity:

$$I(\lambda, \nu, d) = A_\delta(\lambda, \nu) + A_\delta(-\lambda, \nu) + A_\delta(\lambda, -\nu) + A_\delta(-\lambda, -\nu) =$$
$$= \frac{\Gamma\left(1-\frac{d}{2}\right)}{2^d \pi^{\frac{d}{2}}(\lambda^2-\nu^2)} \left(\frac{\Gamma\left(\frac{d-1}{2}-i\nu\right)\Gamma\left(\frac{d-1}{2}+i\nu\right)}{\Gamma\left(\frac{1}{2}-i\nu\right)\Gamma\left(\frac{1}{2}+i\nu\right)} - \frac{\Gamma\left(\frac{d-1}{2}-i\lambda\right)\Gamma\left(\frac{d-1}{2}+i\lambda\right)}{\Gamma\left(\frac{1}{2}-i\lambda\right)\Gamma\left(\frac{1}{2}+i\lambda\right)} \right) \tag{78}$$

6 Summing the Hypergeometric Series

The relation (78) offers unexpected possibilities to produce sums of non trivial generalized hypergeometric series. The strategy is to use the Mellin-Barnes integral representation

$$\Gamma\left[\begin{matrix}\{a_1, \ldots, a_7\} \\ \{b_1, \ldots, b_6\}\end{matrix}\right] {}_7F_6\left[\begin{matrix}\{a_1, \ldots, a_7\} \\ \{b_1, \ldots, b_6\}\end{matrix}\right] = \frac{1}{2\pi i}\int_{-\epsilon-i\infty}^{-\epsilon+i\infty} ds\, e^{\pm i\pi s} \frac{\Gamma(-s)\prod_{j=1}^{7}\Gamma(s+a_j)}{\prod_{k=1}^{6}\Gamma(s+b_k)} \tag{79}$$

in two separate ways to produce a set of identities among hypergeometric series. Let us single out the parameter a_1; after closing the path either to the left or to the right and doing the same manipulations as in [24] we get

$$
{}_7F_6\left[\begin{matrix}\{a_1,\ldots,a_7\}\\ \{b_1,\ldots,b_6\}\end{matrix}\right]\sin(\pi a_1)\frac{\prod_{j=1}^{7}\Gamma(a_j)}{\prod_{k=1}^{6}\Gamma(b_j)}+
$$

$$
+\sum_{j=2}^{7}\sin(\pi(a_j-a_1))\frac{\prod_{k\neq j}\Gamma(a_k-a_j)}{\prod_{k=1}^{6}\Gamma(b_k-a_j)}\Gamma(a_j)\,{}_7F_6\left[\begin{matrix}\{a_j,1+a_j-b_1\ldots,1+a_j-b_6\}\\ \{1+a_j-a_k\}_{k\neq j}\end{matrix}\right]=0. \tag{80}
$$

We may write seven relations of this kind involving eight different hypergeometric series but it is immediately seen that only two of them are independent.

We do not investigate here the general case but focus on our concrete case; in the hypergeometric at the rhs of Eq. (76) there are only three independent parameters x, y and δ:

$$
\begin{aligned}
&\alpha_1 = x+y, \quad \alpha_2 = \tfrac{1}{2}+x+y, \alpha_3 = 2\delta, \qquad \alpha_4 = 1+\delta+x+y,\\
&\alpha_5 = 2\delta+2x, \alpha_6 = 2\delta+2y, \quad \alpha_7 = 2\delta+2x+2y,\\
&\beta_1 = 2x+1, \quad \beta_2 = 2y+1, \qquad \beta_3 = 2x+2y+1,\\
&\beta_4 = \delta+x+y, \beta_5 = 2\delta+2+x+y+\tfrac{1}{2}, \beta_6 = 2\delta+2+x+y+1.
\end{aligned} \tag{81}
$$

It is seen that with the above choice there are only seven independent hypergeometrics (one is repeated twice):

$$
\begin{aligned}
X_{pp} &= {}_7F_6\left(\begin{matrix}x+y, x+y+\frac{1}{2}, 2\delta, \delta+x+y+1, 2\delta+2x, 2\delta+2y, 2\delta+2x+2y\\ 2x+1, 2y+1, 2x+2y+1, \delta+x+y, 2\delta+x+y+\frac{1}{2}, 2\delta+x+y+1\end{matrix};1\right)\\
X_{pm} &= {}_7F_6\left(\begin{matrix}x-y, x-y+\frac{1}{2}, 2\delta, \delta+x-y+1, 2\delta+2x, 2\delta-2y, 2\delta+2x-2y\\ 2x+1, -2y+1, 2x-2y+1, \delta+x-y, 2\delta+x-y+\frac{1}{2}, 2\delta+x-y+1\end{matrix};1\right)\\
X_{mp} &= {}_7F_6\left(\begin{matrix}-x+y, -x+y+\frac{1}{2}, 2\delta, \delta-x+y+1, 2\delta-2x, 2\delta+2y, 2\delta-2x+2y\\ -2x+1, 2y+1, -2x+2y+1, \delta-x+y, 2\delta-x+y+\frac{1}{2}, 2\delta-x+y+1\end{matrix};1\right)\\
X_{mm} &= {}_7F_6\left(\begin{matrix}-x-y, -x-y+\frac{1}{2}, 2\delta, \delta-x-y+1, 2\delta-2x, 2\delta-2y, 2\delta-2x-2y\\ -2x+1, -2y+1, -2x-2y+1, \delta-x-y, 2\delta-x-y+\frac{1}{2}, 2\delta-x-y+1\end{matrix};1\right)\\
X_0 &= {}_7F_6\left(\begin{matrix}x+y, -x+y, x-y, -x-y, 1-\delta, \frac{1}{2}-2\delta, -2\delta\\ 1-2\delta+x+y, 1-2\delta-x+y, 1-2\delta+x-y, 1-2\delta-x-y, \frac{1}{2}, -\delta\end{matrix};1\right)\\
X_1 &= {}_7F_6\left(\begin{matrix}\frac{1}{2}+x+y, \frac{1}{2}-x+y, \frac{1}{2}+x-y, \frac{1}{2}-x-y, \frac{3}{2}-\delta, 1-2\delta, \frac{1}{2}-2\delta\\ \frac{3}{2}-2\delta+x+y, \frac{3}{2}-2\delta-x+y, \frac{3}{2}-2\delta+x-y, \frac{3}{2}-2\delta-y-x, \frac{3}{2}, \frac{1}{2}-\delta,\end{matrix};1\right)\\
X_2 &= {}_7F_6\left(\begin{matrix}1+\delta+x+y, 1+\delta+y-x, 1+\delta+x-y, 1+\delta-x-y, 2, \frac{3}{2}-\delta, 1-\delta\\ 2-\delta+y+x, 2-\delta+y-x, 2-\delta+x-y, 2-\delta-y-x, 2+\delta, \frac{3}{2}+\delta,\end{matrix}\right)
\end{aligned} \tag{82}
$$

There are seven possible identities depending on the choice $a_1 = \alpha_j$, but, as we said, only two of them are independent. Furthermore, in our particular case, X_2 does not enter in any of them and we are left with two independent equations in six variables. Let us choose in particular $a_1 = \alpha_1 = x + y$ and get the first identity:

$$
\begin{aligned}
X_1 = {} & \frac{X_{pp}\,\pi\Gamma(4\delta)\Gamma(-1+2\delta)\Gamma(2x+2\delta)\Gamma(2y+2\delta)\Gamma(1+2x+2y+2\delta)\sec(\pi x-\pi y)\sin(\pi(x+y))}{(x+y)\Gamma(1+2x)\Gamma(1+2y)\Gamma\left(-\frac{1}{2}-x-y+2\delta\right)\Gamma\left(-\frac{1}{2}+x-y+2\delta\right)\Gamma\left(-\frac{1}{2}-x+y+2\delta\right)\Gamma\left(-\frac{1}{2}+x+y+2\delta\right)\Gamma(1+2x+2y+4\delta)}+ \\
& -\frac{X_{pp}\,\pi\Gamma(-2x)\Gamma(-2y)\Gamma(-2x-2y-4\delta)\Gamma(4\delta)\Gamma(1+2x+2y+2\delta)\sec(\pi x-\pi y)\sin(\pi(x+y+2\delta))}{(x+y)\Gamma(2-2\delta)\Gamma(1-2x-2\delta)\Gamma(1-2y-2\delta)\Gamma\left(-\frac{1}{2}-x-y+2\delta\right)\Gamma\left(-\frac{1}{2}+x-y+2\delta\right)\Gamma\left(-\frac{1}{2}-x+y+2\delta\right)\Gamma\left(-\frac{1}{2}+x+y+2\delta\right)}+ \\
& +\frac{X_{mm}\,\pi\Gamma(2x)\Gamma(2y)\Gamma(2x+2y-4\delta)\Gamma(4\delta)\Gamma(-1+2\delta)\sec(\pi x-\pi y)\sin(\pi(-x-y+2\delta))}{(x+y)\Gamma(1+2x-2\delta)\Gamma(1+2y-2\delta)\Gamma(2x+2y-2\delta)\Gamma\left(-\frac{1}{2}-x-y+2\delta\right)\Gamma\left(-\frac{1}{2}+x-y+2\delta\right)\Gamma\left(-\frac{1}{2}-x+y+2\delta\right)\Gamma\left(-\frac{1}{2}+x+y+2\delta\right)}+ \\
& +\frac{X_{pm}\,\pi\Gamma(-2x)\Gamma(2y)\Gamma(-2x+2y-4\delta)\Gamma(4\delta)\Gamma(2x+2\delta)\sec(\pi x-\pi y)\sin(\pi(x-y+2\delta))}{(x-y)\Gamma(2-2\delta)\Gamma(1+2y-2\delta)\Gamma(-2x+2y-2\delta)\Gamma\left(-\frac{1}{2}-x-y+2\delta\right)\Gamma\left(-\frac{1}{2}+x-y+2\delta\right)\Gamma\left(-\frac{1}{2}-x+y+2\delta\right)\Gamma\left(-\frac{1}{2}+x+y+2\delta\right)}+ \\
& -\frac{X_{mp}\pi\Gamma(2x)\Gamma(-2y)\Gamma(2x-2y-4\delta)\Gamma(4\delta)\Gamma(2y+2\delta)\sec(\pi x-\pi y)\sin(\pi(-x+y+2\delta))}{(x-y)\Gamma(2-2\delta)\Gamma(1+2x-2\delta)\Gamma(2x-2y-2\delta)\Gamma\left(-\frac{1}{2}-x-y+2\delta\right)\Gamma\left(-\frac{1}{2}+x-y+2\delta\right)\Gamma\left(-\frac{1}{2}-x+y+2\delta\right)\Gamma\left(-\frac{1}{2}+x+y+2\delta\right)}
\end{aligned} \tag{83}
$$

By choosing $a_1 = \alpha_2 = \frac{1}{2} + x + y$ we get the second identity:

$$
\begin{aligned}
X_0 = {} & \frac{X_{pp}\,2\pi(x-y)\cos(\pi(x+y))\csc(\pi(x-y))\Gamma(2\delta)\Gamma(4\delta)\Gamma(2(x+\delta))\Gamma(2(y+\delta))\Gamma(1+2x+2y+2\delta)}{\Gamma(1+2x)\Gamma(1+2y)\Gamma(-x-y+2\delta)\Gamma(x-y+2\delta)\Gamma(-x+y+2\delta)\Gamma(x+y+2\delta)\Gamma(1+2x+2y+4\delta)}+ \\
& +\frac{X_{pp}\,2\pi(-x+y)\cos(\pi(x+y+2\delta))\csc(\pi(x-y))\Gamma(-2x)\Gamma(-2y)\Gamma(4\delta)\Gamma(-2(x+y+2\delta))\Gamma(1+2x+2y+2\delta)}{\Gamma(1-2\delta)\Gamma(1-2x-2\delta)\Gamma(1-2y-2\delta)\Gamma(-x-y+2\delta)\Gamma(x-y+2\delta)\Gamma(-x+y+2\delta)\Gamma(x+y+2\delta)}+ \\
& -\frac{X_{pm}\,2\pi(x+y)\csc(\pi x-\pi y)\Gamma(-2x)\Gamma(2y)\Gamma(-2x+2y-4\delta)\Gamma(4\delta)\Gamma(2x+2\delta)\sin\left(\pi\left(-\frac{1}{2}+x-y+2\delta\right)\right)}{\Gamma(1-2\delta)\Gamma(1+2y-2\delta)\Gamma(-2x+2y-2\delta)\Gamma(-x-y+2\delta)\Gamma(x-y+2\delta)\Gamma(-x+y+2\delta)\Gamma(x+y+2\delta)}+ \\
& +\frac{X_{mp}\,2\pi(x+y)\csc(\pi x-\pi y)\Gamma(2x)\Gamma(-2y)\Gamma(2x-2y-4\delta)\Gamma(4\delta)\Gamma(2y+2\delta)\sin\left(\pi\left(-\frac{1}{2}-x+y+2\delta\right)\right)}{\Gamma(1-2\delta)\Gamma(1+2x-2\delta)\Gamma(2x-2y-2\delta)\Gamma(-x-y+2\delta)\Gamma(x-y+2\delta)\Gamma(-x+y+2\delta)\Gamma(x+y+2\delta)}+ \\
& +\frac{X_{mm}\,2\pi(x-y)\csc(\pi x-\pi y)\Gamma(2x)\Gamma(2y)\Gamma(2x+2y-4\delta)\Gamma(2\delta)\Gamma(4\delta)\sin\left(\pi\left(-\frac{1}{2}-x-y+2\delta\right)\right)}{\Gamma(1+2x-2\delta)\Gamma(1+2y-2\delta)\Gamma(2x+2y-2\delta)\Gamma(-x-y+2\delta)\Gamma(x-y+2\delta)\Gamma(-x+y+2\delta)\Gamma(x+y+2\delta)}.
\end{aligned} \tag{84}
$$

The other five relations give nothing new and are identically satisfied assuming the validity of (83) and (84). On the other hand, a third independent (inhomogeneous) equation comes from (78), which we rewrite as follows:

$$
\begin{aligned}
& \frac{X_{mm}\,2^{-3+4\delta}\Gamma(2x)\Gamma(2y)\Gamma(-2x+2\delta)\Gamma(-2y+2\delta)}{\sqrt{\pi}(x+y)\Gamma\left(\frac{1}{2}-2\delta\right)\Gamma(2(x+y-\delta))\Gamma(1-2x-2y+4\delta)} - \frac{X_{pm}\,2^{-3+4\delta}\Gamma(-2x)\Gamma(2y)\Gamma(2(x+\delta))\Gamma(-2y+2\delta)}{\sqrt{\pi}(x-y)\Gamma\left(\frac{1}{2}-2\delta\right)\Gamma(-2(x-y+\delta))\Gamma(1+2x-2y+4\delta)}+ \\
& +\frac{X_{mp}\,2^{-3+4\delta}\Gamma(2x)\Gamma(-2y)\Gamma(2(y+\delta))\Gamma(-2x+2\delta)}{\sqrt{\pi}(x-y)\Gamma\left(\frac{1}{2}-2\delta\right)\Gamma(2x-2y-2\delta)\Gamma(1-2x+2y+4\delta)} - \frac{X_{pp}\,2^{-3+4\delta}\Gamma(-2x)\Gamma(-2y)\Gamma(2(x+\delta))\Gamma(2(y+\delta))}{\sqrt{\pi}(x+y)\Gamma\left(\frac{1}{2}-2\delta\right)\Gamma(-2(x+y+\delta))\Gamma(1+2x+2y+4\delta)}+ \\
& = \frac{\dfrac{\Gamma(-2x+2\delta)\Gamma(2x+2\delta)}{\Gamma\left(\frac{1}{2}-2x\right)\Gamma\left(\frac{1}{2}+2x\right)} - \dfrac{\Gamma(-2y+2\delta)\Gamma(2y+2\delta)}{\Gamma\left(\frac{1}{2}-2y\right)\Gamma\left(\frac{1}{2}+2y\right)}}{4x^2-4y^2}.
\end{aligned} \tag{85}
$$

Starting from here we may obtain new highly nontrivial summation formulae for generalized hypergeometric series as we show below.

6.1 Case of Equal Masses $x = y$

For $x \neq y$ the linear system LS1 composed by Eqs. (83), (84) and (85) can be solved for X_{pp} as follows:

$$X_{pp}(x,y) = a_{pp}(x,y)X_0(x,y) + b_{pp}(x,y)X_{pm}(x,y) + c_{pp}(x,y)X_{mp}(x,y) + r_{pp}(x,y) \tag{86}$$

At $x = y$ three hypergeometric series become trivial: $X_{pm}(x,x) = X_{mp}(x,x) = X_0(x,x) = 1$. An explicit computation shows that at $x = y$ none of the coefficients at the rhs of Eq. (86) is singular; we may thus simply take $x = y$ and obtain - after some work - the following result:

$$X_{pp}(x,x) = {}_7F_6\left(\begin{matrix} 2x, \frac{1}{2}+2x, 2\delta, 1+2x+\delta, 2x+2\delta, 2x+2\delta, 4x+2\delta \\ 1+2x, 1+2x, 1+4x, 2x+\delta, \frac{1}{2}+2x+2\delta, 1+2x+2\delta \end{matrix} ; 1 \right) =$$
$$= \frac{2\pi\Gamma(1+2x)^2\Gamma(1+4x+4\delta)(\pi\cot(2\pi(x-\delta)) + \psi(2x+2\delta) - \psi(-2x+2\delta))}{\sin(4\pi\delta)\Gamma(1+2x-2\delta)\Gamma(-2\delta)\Gamma(2x+2\delta)\Gamma(1+4x+2\delta)\Gamma(1+4\delta)}. \tag{87}$$

This is the first summation of an infinite series ${}_7F_6(1)$ depending on two independent complex parameters I am aware of (see [25, pp. 570, 571] for a partial list of known examples). The famous Ramanujan-Dougall formula is in fact only a finite sum. Also

$$X_{mm}(x,x) = X_{pp}(-x,-x). \tag{88}$$

As regards $X_1(x,x)$ the situation is a bit more complicated as some of the coefficients diverge at $x = y$. The shortest way to sum $X_1(x,x)$ is to go back to Eq. (83): there are no singularities in the first three lines; to compute them we can simply set $x = y$ and use Eqs. (87) and (88). On the other hand the coefficients in last two lines have poles at $x = y$ having opposite residues. Since

$$X_{pm}(x,y) \simeq 1 - (y-x)\, {}_7\dot{F}_6(x,\delta), \tag{89}$$
$$X_{mp}(x,y) \simeq 1 + (y-x)\, {}_7\dot{F}_6(x,\delta), \tag{90}$$

the singular terms cancel each other and we obtain a finite result which, however, is expressed in terms of the unknown derivative of the hypergeometric series

$${}_7\dot{F}_6(x,\delta) = \frac{\partial}{\partial a}\, {}_7F_6\left(\begin{matrix} a, \frac{1}{2}, 2\delta, 2\delta, \delta+1, 2\delta-2x, 2\delta+2x \\ 1, 1-2x, 2x+1, \delta, 2\delta+\frac{1}{2}, 2\delta+1 \end{matrix} ; 1 \right)\Bigg|_{a=0}. \tag{91}$$

Luckily, ${}_7\dot{F}_6(x,\delta)$ can be obtained by taking the limit $y \to x$ of Eq. (85):

$${}_7\dot{F}_6(x,\delta) = \pi(\cot(2\pi x) - 2\cot(2\pi\delta)) + 2\psi(2x) - 2\psi(2\delta) + 2\psi(1+4\delta) + 4\pi\csc(4\pi\delta)\sin^2(2\pi x)$$
$$+ \frac{1}{2x} - \frac{1}{\delta} + \psi(-2x+2\delta)(2\csc(4\pi\delta)\sin(4\pi x) - 1) - \psi(2x+2\delta))(1 + 2\csc(4\pi\delta)\sin(4\pi x)) +$$
$$+ \frac{X_{mm}\Gamma(2x)^2\Gamma(1-2\delta)\Gamma(4\delta)\Gamma(-2x+2\delta)\sin(2\pi x)}{\pi\Gamma(4x-2\delta)\Gamma(2x+2\delta)\Gamma(1-4x+4\delta)} - \frac{X_{pp}\Gamma(-2x)^2\Gamma(1-2\delta)\Gamma(4\delta)\Gamma(2x+2\delta)\sin(2\pi x)}{\pi\Gamma(-4x-2\delta)\Gamma(-2x+2\delta)\Gamma(1+4x+4\delta)} \tag{92}$$

By inserting the expressions (87) and (88) may deduce a self-contained formula. After collecting all the terms together, a concrete amount of work provides a surprisingly simple final result:

$$X_1(x,x) = {}_7F_6\left(\begin{matrix} \frac{1}{2}+2x, \frac{1}{2}, \frac{1}{2}, \frac{1}{2}-2x, \frac{3}{2}-\delta, 1-2\delta, \frac{1}{2}-2\delta \\ \frac{3}{2}-2\delta+2x, \frac{3}{2}-2\delta, \frac{3}{2}-2\delta, \frac{3}{2}-2\delta-2x, \frac{3}{2}, \frac{1}{2}-\delta, \end{matrix}; 1\right) =$$
$$= \frac{2^{-2-4x+4\delta}\pi^{5/2}\Gamma\left(\frac{3}{2}-2\delta\right)^2 (\psi(2x+2\delta)-\psi(-2x+2\delta)+\pi\csc(2\pi(x-\delta))\csc(2\pi(x+\delta))\sin(4\pi x))}{x\Gamma(1-2\delta)\Gamma(2-2\delta)\Gamma(1-2x-2\delta)\Gamma(1+2x-2\delta)^2\Gamma\left(-\frac{1}{2}+2x+2\delta\right)\Gamma(-1-4x+4\delta)(\sin(2\pi(x-3\delta))+\sin(2\pi(3x-\delta)))} \tag{93}$$

6.2 Unequal Masses $y = x + n$, $y = x + n + \frac{1}{2}$

When $x - y$ is either integer of half integer but not zero, only two of the six hypergeometric series given in (82) are polynomial and three equations are not enough to determine the remaining unknown ones.

The key observation to obtain more relations is that the following three families of parameters

$$\hat{\alpha}_i(x,y) = \alpha_i(-x,y), \quad \hat{\beta}_i(x,y) = \beta_i(-x,y), \tag{94}$$

$$\check{\alpha}_i(x,y) = \alpha_i(x,-y), \quad \hat{\beta}_i(x,y) = \beta_i(x,-y), \tag{95}$$

$$\tilde{\alpha}_i(x,y) = \alpha_i(-x,-y), \quad \hat{\beta}_i(x,y) = \beta_i(-x,-y) \tag{96}$$

(see Eq. 81) give rise by the Hardy-Bailey trick [24] to the same family of hypergeometric series (82); we may therefore repeat the whole procedure for all the above possibilities and write 21 more relations among the hypergeometric series (82). Only two of them are independent from Eqs. (83) and (84); we chose the following ones:

$$\begin{aligned}
X_0 = {} & X_{mp}\frac{2\pi(x+y)\cos(\pi(x-y))\csc(\pi(x+y))\Gamma(2\delta)\Gamma(4\delta)\Gamma(2(y+\delta))\Gamma(-2x+2\delta)\Gamma(1-2x+2y+2\delta)}{\Gamma(1-2x)\Gamma(1+2y)\Gamma(-x-y+2\delta)\Gamma(x-y+2\delta)\Gamma(-x+y+2\delta)\Gamma(x+y+2\delta)\Gamma(1-2x+2y+4\delta)}+ \\
& - X_{mm}\frac{2\pi(x-y)\cos(\pi(1+x+y-2\delta))\csc(\pi(x+y))\Gamma(2x)\Gamma(2y)\Gamma(2(x+y-2\delta))\Gamma(4\delta)\Gamma(-2x+2\delta)}{\Gamma(1-2\delta)\Gamma(1+2y-2\delta)\Gamma(2(x+y-\delta))\Gamma(-x-y+2\delta)\Gamma(x-y+2\delta)\Gamma(-x+y+2\delta)\Gamma(x+y+2\delta)}+ \\
& - X_{pm}\frac{2\pi(x+y)\cos(\pi(x-y+2\delta))\csc(\pi(x+y))\Gamma(-2x)\Gamma(2y)\Gamma(-2x+2y-4\delta)\Gamma(2\delta)\Gamma(4\delta)}{\Gamma(1-2x-2\delta)\Gamma(1+2y-2\delta)\Gamma(-2(x-y+\delta))\Gamma(-x-y+2\delta)\Gamma(x-y+2\delta)\Gamma(-x+y+2\delta)\Gamma(x+y+2\delta)}+ \\
& + X_{mp}\frac{2\pi(x+y)\cos(\pi(1+x-y-2\delta))\csc(\pi(x+y))\Gamma(2x)\Gamma(-2y)\Gamma(2x-2y-4\delta)\Gamma(4\delta)\Gamma(1-2x+2y+2\delta)}{\Gamma(1-2\delta)\Gamma(1+2x-2\delta)\Gamma(1-2y-2\delta)\Gamma(-x-y+2\delta)\Gamma(x-y+2\delta)\Gamma(-x+y+2\delta)\Gamma(x+y+2\delta)}+ \\
& + X_{pp}\frac{2\pi(-x+y)\cos(\pi(x+y+2\delta))\csc(\pi(x+y))\Gamma(-2x)\Gamma(-2y)\Gamma(4\delta)\Gamma(2(y+\delta))\Gamma(-2(x+y+2\delta))}{\Gamma(1-2\delta)\Gamma(1-2x-2\delta)\Gamma(-2(x+y+\delta))\Gamma(-x-y+2\delta)\Gamma(x-y+2\delta)\Gamma(-x+y+2\delta)\Gamma(x+y+2\delta)}
\end{aligned} \tag{97}$$

and

$$X_0 = X_{pp}\frac{2\pi(-x+y)\cos(\pi(x+y+2\delta))\csc(\pi(x+y))\Gamma(-2x)\Gamma(-2y)\Gamma(4\delta)\Gamma(2(y+\delta))\Gamma(-2(x+y+2\delta))}{\Gamma(1-2\delta)\Gamma(1-2x-2\delta)\Gamma(-2(x+y+\delta))\Gamma(-x-y+2\delta)\Gamma(x-y+2\delta)\Gamma(-x+y+2\delta)\Gamma(x+y+2\delta)}+$$
$$+X_{mm}\frac{2\pi(x-y)\csc(\pi x+\pi y)\Gamma(2x)\Gamma(2y)\Gamma(2x+2y-4\delta)\Gamma(4\delta)\Gamma(-2y+2\delta)\sin\left(\pi\left(-\frac{1}{2}-x-y+2\delta\right)\right)}{\Gamma(1-2\delta)\Gamma(1+2x-2\delta)\Gamma(2x+2y-2\delta)\Gamma(-x-y+2\delta)\Gamma(x-y+2\delta)\Gamma(-x+y+2\delta)\Gamma(x+y+2\delta)}$$
$$+X_{pm}\frac{2\pi(x+y)\csc(\pi x+\pi y)\Gamma(-2x)\Gamma(2y)\Gamma(-2x+2y-4\delta)\Gamma(4\delta)\Gamma(1+2x-2y+2\delta)\sin\left(\pi\left(-\frac{1}{2}+x-y+2\delta\right)\right)}{\Gamma(1-2\delta)\Gamma(1-2x-2\delta)\Gamma(1+2y-2\delta)\Gamma(-x-y+2\delta)\Gamma(x-y+2\delta)\Gamma(-x+y+2\delta)\Gamma(x+y+2\delta)}+$$
$$+X_{mp}\frac{2\pi(x+y)\csc(\pi x+\pi y)\Gamma(2x)\Gamma(-2y)\Gamma(2x-2y-4\delta)\Gamma(2\delta)\Gamma(4\delta)\sin\left(\pi\left(-\frac{1}{2}-x+y+2\delta\right)\right)}{\Gamma(1+2x-2\delta)\Gamma(1-2y-2\delta)\Gamma(2x-2y-2\delta)\Gamma(-x-y+2\delta)\Gamma(x-y+2\delta)\Gamma(-x+y+2\delta)\Gamma(x+y+2\delta)}+$$
$$-X_{pp}\frac{2\pi(x-y)\csc(\pi x+\pi y)\Gamma(-2x)\Gamma(-2y)\Gamma(-2x-2y-4\delta)\Gamma(4\delta)\Gamma(2x+2\delta)\sin\left(\pi\left(-\frac{1}{2}+x+y+2\delta\right)\right)}{\Gamma(1-2\delta)\Gamma(1-2y-2\delta)\Gamma(-2x-2y-2\delta)\Gamma(-x-y+2\delta)\Gamma(x-y+2\delta)\Gamma(-x+y+2\delta)\Gamma(x+y+2\delta)}. \tag{98}$$

There is still one equation lacking to solve for the full set of hypergeometric series depending on three parameters. But we can express five of the hypergeometrics (82) in term of a sixth, say X_0; for instance we may write

$$X_{pp}(x, y, \delta) = a(x, y, \delta)X_0(x, y, \delta) + b(x, y, \delta) \tag{99}$$

and compute X_{pp} in the special cases where X_0 is explicitly known. For instance we may compute the above formula at $y = x + 1$ and discover that there are no singularities in the coefficients there. Curiously enough, one finds that for n integer

$$a(x, x+n, \delta) = 0 \tag{100}$$

and therefore the actual knowledge of the polynomial $X_0(x, x+n, \delta)$ is irrelevant in calculating $X_{pp}(x, x+n, \delta)$.

The actual calculation is however not so simple because of many apparent divergences that are to be tamed. After a lengthy and laborious procedure we obtain the following result:

$$\begin{aligned}
&X_{pp}(x, x+n) = \\
&= {}_7F_6\left(\begin{matrix}2x+n, 2x+n+\frac{1}{2}, 2x+n+1+\delta, 2x+2n+2\delta, 4x+2n+2\delta, 2x+2\delta, 2\delta\\ 2x+1, 2x+2n+1, 4x+2n+1, 2x+n+\delta, 2x+n+\frac{1}{2}+2\delta, 2x+n+1+2\delta\end{matrix}; 1\right) = \\
&= \frac{\pi^{-\frac{1}{2}}2^{-1-4\delta}\cos(2\pi n)\Gamma(-2x)\Gamma(1+2x)^2\Gamma\left(\frac{1}{2}-2\delta\right)\Gamma(-2n-4x-2\delta)\Gamma(1+2n+4x+4\delta)}{n\Gamma(-2n-2x)\Gamma(1-4x-2\delta)\Gamma(2x-2\delta)\Gamma(2x+2\delta)\Gamma(4x+2\delta)\Gamma(1-2x+2\delta)\Gamma(1-2x-2\delta)} \\
&\times(\Gamma(1-2n-2x-2\delta)\Gamma(-2x+2\delta)-\Gamma(1-2x-2\delta)\Gamma(-2n-2x+2\delta))
\end{aligned} \tag{101}$$

The above result indeed also holds for half-integer n but not for general y.

Similar formulae hold for $X_{pm}(x, x+n)$, $X_{mp}(x, x+n)$, $X_{mm}(x, x+n)$ and $X_1(x, x+n)$ for all half-integers $n \in \mathbf{Z}/2$. We do not reproduce them here but the interested reader may reconstruct them following the previous steps.

7 2-Loop: The Watermelon

In this last section, using the same idea based on the Källen-Lehmann representation (64), we may construct an integral representation of the two-loop watermelon as a superposition of one-loop diagrams:

$$I_3(\nu_1, \nu_2, \nu_3, d) = \int G_{\nu_1}(x_0 \cdot x) G_{\nu_2}(x_0 \cdot x) G_{\nu_3}(x_0 \cdot x)\sqrt{g}dx =$$
$$= \int \rho(\nu_1, \nu_2, \kappa) I_2(\kappa, \nu_3, d)\kappa d\kappa = I_3^{(1)}(\nu_1, \nu_2, \nu_3, d) - I_3^{(2)}(\nu_1, \nu_2, \nu_3, d). \quad (102)$$

where

$$I_3^{(1)}(\nu_1, \nu_2, \nu_3, d) = \frac{\left(\frac{\Gamma\left(\frac{d-1}{2}-i\nu_3\right)\Gamma\left(\frac{d-1}{2}+i\nu_3\right)}{\Gamma\left(\frac{1}{2}-i\nu_3\right)\Gamma\left(\frac{1}{2}+i\nu_3\right)}\right)}{2^d \pi^{\frac{d-1}{2}}\Gamma\left(\frac{d-1}{2}\right)} \frac{\Gamma\left(1-\frac{d}{2}\right)}{(4\pi)^{\frac{d}{2}}} \mathcal{A}(\nu_1, \nu_2, \nu_3, d) \quad (103)$$

$$\mathcal{A}(\nu_1, \nu_2, \nu_3, d) = \int \frac{\prod_{\epsilon,\epsilon',\epsilon''} \Gamma\left(\frac{d-1}{4} + \frac{i\epsilon'\nu_1 + i\epsilon'\nu_2 + i\epsilon''\kappa}{2}\right)}{\prod_\epsilon \Gamma\left(\frac{i\epsilon\kappa}{2}\right)\Gamma\left(\frac{1}{2}+\frac{i\epsilon\kappa}{2}\right)\Gamma\left(\frac{d-1}{4}+\frac{i\epsilon\kappa}{2}\right)\Gamma\left(\frac{d+1}{4}+\frac{i\epsilon\kappa}{2}\right)} \frac{d\kappa}{(\kappa^2 - \nu_3^2)} \quad (104)$$

and

$$I_3^{(2)}(\nu_1, \nu_2, \nu_3, d) = \frac{1}{2^d \pi^{\frac{d-1}{2}}\Gamma\left(\frac{d-1}{2}\right)} \frac{\Gamma\left(1-\frac{d}{2}\right)}{(4\pi)^{\frac{d}{2}}} \mathcal{B}(\nu_1, \nu_2, \nu_3) \quad (105)$$

$$\mathcal{B}(\nu_1, \nu_2, \nu_3) = \int \frac{\frac{\Gamma\left(\frac{d-1}{2}-i\kappa\right)\Gamma\left(\frac{d-1}{2}+i\kappa\right)}{\Gamma\left(\frac{1}{2}-i\kappa\right)\Gamma\left(\frac{1}{2}+i\kappa\right)} \prod_{\epsilon,\epsilon',\epsilon''} \Gamma\left(\frac{d-1}{4} + \frac{i\epsilon'\nu_1 + i\epsilon'\nu_2 + i\epsilon''\kappa}{2}\right)}{\prod_\epsilon \Gamma\left(\frac{i\epsilon\kappa}{2}\right)\Gamma\left(\frac{1}{2}+\frac{i\epsilon\kappa}{2}\right)\Gamma\left(\frac{d-1}{4}+\frac{i\epsilon\kappa}{2}\right)\Gamma\left(\frac{d+1}{4}+\frac{i\epsilon\kappa}{2}\right)} \frac{d\kappa}{(\kappa^2 - \nu_3^2)}. \quad (106)$$

$I_3^{(1)}(\nu_1, \nu_2, \nu_3, d)$ and $I_3^{(2)}(\nu_1, \nu_2, \nu_3, d)$ are symmetric only in the exchange of ν_1 and ν_2. To work with symmetric expressions we might replace them by their totally symmetric counterparts but this is not really helpful.

Let us consider the first term $I_3^{(1)}(\nu_1, \nu_2, \nu_3, d)$; the change of variables[3]

$$s = \frac{i\kappa}{2}, \quad \delta = \frac{d-1}{4}, \quad x = \frac{i\nu_1}{2}, \quad y = \frac{i\nu_2}{2}, \quad w = \frac{i\nu_3}{2}, \quad u = x + y \quad (107)$$

[3] In the following we will use either (ν_1, ν_2, ν_3) or (x, y, w) and either d or $\delta(d)$ interchangeably.

allows to rewrite the integral as follows:

$$\mathcal{A}(x, y, w, d) = \int_{-i\infty}^{i\infty} \frac{ds}{2i} \prod_{\epsilon,\epsilon',\epsilon''=\pm} \frac{\Gamma\left(\epsilon s + \delta + \epsilon' x + i\epsilon'' y\right)\Gamma\left(\epsilon s + w\right)}{\Gamma\left(\epsilon s\right)\Gamma\left(\frac{1}{2} + \epsilon s\right)\Gamma\left(\delta + \epsilon s\right)\Gamma\left(\delta + \frac{1}{2} + \epsilon s\right)\Gamma\left(\epsilon s + w + 1\right)}. \tag{108}$$

There are poles on the integration path at

$$s = -\delta + \frac{\pm i\nu_1 \pm i\nu_2}{2} - n, \qquad s = -\frac{i\nu_3}{2} - n, \tag{109}$$

$$s = -\frac{i\nu_3}{2} + n, \qquad s = \delta + \frac{\pm i\nu \pm i\lambda}{2} + n. \tag{110}$$

By integrating along the imaginary axis (with suitable indentations) the result may be expressed as a linear combination of four hypergeometric functions

$$I_3^{(1)}(x, y, z, d) = \sum_{\epsilon,\epsilon'=\pm 1} A_d(\epsilon x, \epsilon' y, w) \tag{111}$$

where

$$A_d(x, y, w) = a_d(x, y, w) \times$$
$${}_9F_8\left(\begin{matrix} 2\delta, u + \frac{1}{2}, u + 1, \delta + u + 1, 2\delta + 2x, 2\delta + 2y, 2\delta + 2u, \delta - w + u, \delta + w + u \\ 2x + 1, 2y + 1, 2u + 1, \delta + u, 2\delta + u, 2\delta + u + \frac{1}{2}, \delta - w + u + 1, \delta + w + u + 1 \end{matrix}; 1\right), \tag{112}$$

$$a_d(x, y, w) = -\frac{4^{-2\delta-3}\pi^{-4\delta-\frac{1}{2}}\Gamma\left(\frac{1}{2}-2\delta\right)\cos(2\pi w)\Gamma(2\delta-2w)\Gamma(2w+2\delta)\Gamma(2x+2\delta)\Gamma(2y+2\delta)}{\sin(2\pi x)\sin(2\pi y)\Gamma(2x+1)\Gamma(2y+1)\left(w^2-(\delta+u)^2\right)\Gamma(-2u-2\delta)\Gamma(2u+4\delta)}. \tag{113}$$

The hypergeometric series in (112) converges absolutely for $3 - 4\delta = 4 - d > 0$.

It is possible to proceed to a simplification of the above expressions by observing that in each term of the hypergeometric series (112) the following product of Pochhammers reduces nicely to a rational function:

$$\frac{(u + \delta + 1)_n (u - w + \delta)_n (u + w + \delta)_n}{(u + \delta)_n (u - w + \delta + 1)_n (u + w + \delta + 1)_n} = \frac{((\delta + u)^2 - w^2)\left(\frac{1}{\delta+n+u+w} + \frac{1}{\delta+n+u-w}\right)}{2(\delta + u)}. \tag{114}$$

We obtain in this way a formula consisting of eight terms ${}_7F_6$:

$$I_3^{(1)}(x, y, z, d) = \sum_{\epsilon,\epsilon',\epsilon''=\pm 1} A'_d(\epsilon x, \epsilon' y, \epsilon'' w) \tag{115}$$

where

$$A'_d(x,y,w) = a'_d(x,y,w)\times$$
$$ {}_7F_6\left(\begin{matrix} 2\delta, u+\frac{1}{2}, u+1, \delta+u-w, 2\delta+2u, 2\delta+2x, 2\delta+2y \\ 2u+1, 2x+1, 2y+1, \delta+u-w+1, 2\delta+u, 2\delta+u+\frac{1}{2} \end{matrix}; 1\right) \tag{116}$$

$$a'_d(x,y,w) =$$
$$= \frac{2^{-4\delta-7}\pi^{-4\delta-\frac{1}{2}}\Gamma\left(\frac{1}{2}-2\delta\right)\cos(2\pi w)\Gamma(2\delta-2w)\Gamma(2w+2\delta)\Gamma(2x+2\delta)\Gamma(2y+2\delta)}{\sin(2\pi x)\sin(2\pi y)(\delta+u)\Gamma(2x+1)\Gamma(2y+1)\Gamma(-2u-2\delta)\Gamma(2u+4\delta)(\delta+u-w)}. \tag{117}$$

This simplification has not worsened the convergence: (116) converges absolutely under the same conditions: $4\delta - 3 = -d + 4 > 0$.

As regards the second term, we have

$$\mathcal{B}(x,y,w,d) = \int \frac{ds}{2i} \prod_{\epsilon,\epsilon',\epsilon''=\pm 1} \frac{2^{d-2}\Gamma(\epsilon s+w)\Gamma\left(\epsilon s+\delta+\epsilon' x+\epsilon'' y\right)}{\Gamma(\epsilon s)\Gamma\left(\epsilon s+\frac{1}{4}\right)\Gamma\left(\epsilon s+\frac{1}{2}\right)\Gamma\left(\epsilon s+\frac{3}{4}\right)\Gamma(\epsilon s+w+1)} \tag{118}$$

Again, the result may be expressed as a linear combination of four hypergeometric functions:

$$I_3^{(2)} = \sum_{\epsilon,\epsilon'=\pm 1} B_d(\epsilon x, \epsilon' y, w) \tag{119}$$

where

$$B_d(x,y,w) = b_d(x,y,w)$$
$$\times\ {}_7F_6\left(\begin{matrix} 2\delta, \delta+u+1, \delta-w+u, \delta+w+u, 2\delta+2x, 2\delta+2y, 2\delta+2u \\ 2x+1, 2y+1, 2x+2y+1, \delta+u, \delta-w+u+1, \delta+w+u+1 \end{matrix}; 1\right) \tag{120}$$

$$b_d(x,y,w) = \frac{2^{-8\delta-4u-7}\pi^{1-4\delta}\Gamma\left(\frac{1}{2}-2\delta\right)\csc(2\pi x)\csc(2\pi y)\csc(2\pi u)\Gamma(2x+2\delta)\Gamma(2y+2\delta)}{\left((\delta+u)^2-w^2\right)\Gamma(2x+1)\Gamma(2y+1)\Gamma(2u+1)\Gamma(-4u-4\delta)\Gamma\left(2u+2\delta+\frac{1}{2}\right)}. \tag{121}$$

The hypergeometric series in (120) converges absolutely for $4 - 8\delta = 2(3-d) > 0$. Proceeding exactly as before we may find a simplification of the above formula:

$$I_3^{(2)}(x,y,z,d) = \sum_{\epsilon,\epsilon',\epsilon''=\pm 1} B'_d(\epsilon x, \epsilon' y, \epsilon'' w) \tag{122}$$

where

$$B'_d(x, y, w) = b'_d(x, y, w)\ {}_5F_4\left(\begin{matrix} 2\delta, \delta+u+w, 2\delta+2u, 2\delta+2x, 2\delta+2y \\ 2u+1, 2x+1, 2y+1, \delta+u+w+1 \end{matrix}; 1\right) \tag{123}$$

$$b'_d(x, y, w) = -\frac{2^{-6-4u-8\delta}\pi^{1-4\delta}\csc(2\pi x)\csc(2\pi y)\csc(2\pi u)\Gamma\left(\frac{1}{2}-2\delta\right)\Gamma(2x+2\delta)\Gamma(2y+2\delta)}{(w+u+\delta)\Gamma(1+2x)\Gamma(1+2y)\Gamma(1+2u)\Gamma(1-4u-4\delta)\Gamma\left(\frac{1}{2}+2u+2\delta\right)}. \tag{124}$$

Again, the simplification has not worsened the convergence: (123) converges absolutely for $4 - 8\delta = 2(3 - d) > 0$.

In conclusion, $I_3^{(1)}$ is the most regular of the two terms; it starts diverging at $d = 4$. The second term diverges already at $d = 3$. This means that no compensation among the two terms may be expected. The only exception is at $d = 2$: in that case the watermelon is finite; the two terms have to compensate each other to render the divergence in the coefficient $\Gamma\left(1 - \frac{d}{2}\right)$ harmless.

8 Conclusion

de Sitter QFT was one of Alexei Starobinsky favorite topics because of its importance to understand the physics of our universe. Here we have shown how the simple one-loop diagram with no external lines on the Euclidean de Sitter sphere already contains a great deal of mathematical and physical information. Among other things, this shows that there is still plenty of room for new discoveries in de Sitter QFT and this very possibility would have pleased Alexei.

References

1. A.A. Starobinsky, A new type of isotropic cosmological models without singularity. Phys. Lett. B **91**, 99–102 (1980). https://doi.org/10.1016/0370-2693(80)90670-X
2. V.F. Mukhanov, G.V. Chibisov, Quantum fluctuations and a nonsingular universe. J. Exp. Theor. Phys. 33(10):532–535
3. A.A. Starobinsky, The perturbation spectrum evolving from a nonsingular initially De-Sitter cosmology and the microwave background anisotropy. Sov. Astron. Lett. **9**, 302 (1983)
4. A.A. Starobinsky, Multicomponent de Sitter (inflationary) stages and the generation of perturbations. JETP Lett. **42**, 152–155 (1985)
5. A.A. Starobinsky, Stochastic deSitter (inflationary) stage in the early universe. Lect. Notes Phys. **246**, 107–126 (1986)
6. A.A. Starobinsky, J. Yokoyama, Equilibrium state of a selfinteracting scalar field in the De Sitter background. Phys. Rev. D **50**, 6357–6368 (1994)
7. S.L. Cacciatori, H. Epstein, U. Moschella, Banana integrals in configuration space. Nucl. Phys. B **995**, 116343 (2023)
8. S.L. Cacciatori, H. Epstein, U. Moschella, Loops in de Sitter space. JHEP **07**, 182 (2024)
9. S.L. Cacciatori, H. Epstein, U. Moschella, Loops in anti de Sitter space. JHEP **08**, 109 (2024)
10. U. Moschella, The spectral condition, plane waves, and harmonic analysis in de Sitter and Anti-de Sitter quantum field theories. Universe **10**(5), 199 (2024)

11. R.F. Streater, A.S. Wightman, *PCT, Spin and Statistics, and All That* (Princeton University Press , 2000), Phys. Rev. Lett. **107**, 191103 (2011)
12. J. Bros, U. Moschella, J.P. Gazeau, Quantum field theory in the de Sitter universe. Phys. Rev. Lett. **73**, 1746–1749 (1994)
13. J. Bros, U. Moschella, Two point functions and quantum fields in de Sitter universe. Rev. Math. Phys. **8**, 327–392 (1996)
14. A. Erdélyi (ed.), *The Bateman Project: Higher Transcendental Functions*, vol. I (McGraw-Hill Book Company, New York, 1953)
15. J. Bros, H. Epstein, U. Moschella, Scalar tachyons in the de Sitter universe. Lett. Math. Phys. **93**, 203–211 (2010)
16. H. Epstein, U. Moschella, de Sitter tachyons and related topics. Commun. Math. Phys. **336**(1), 381–430 (2015)
17. S.Y. Lee, A.M. Sciaccaluga, Evaluation of higher order effective potentials with dimensional regularization. Nucl. Phys. B **96**, 435–444 (1975)
18. P.L. Butzer, P.J.S.G. Ferreira, G. Schmeisser, et al., The summation formulae of Euler–Maclaurin, Abel–Plana, Poisson, and their interconnections with the approximate sampling formula of signal analysis. Results. Math. **59**, 359–400 (2011)
19. S. Coleman, E. Weinberg, Radiative corrections as the origin of spontaneous symmetry breaking. Phys. Rev. D **7**, 1888 (1973)
20. C. Ford, D.R.T. Jones, The Effective potential and the differential equations method for Feynman integrals. Phys. Lett. B **274**, 409–414 (1992) [erratum: Phys. Lett. B **285**, 399 (1992)]
21. J. Bros, H. Epstein, M. Gaudin, U. Moschella, V. Pasquier, Triangular invariants, three-point functions and particle stability on the de Sitter universe. Commun. Math. Phys. **295**, 261–288 (2010)
22. J. Bros, H. Epstein, U. Moschella, Towards a general theory of quantized fields on the anti-de Sitter space-time. Commun. Math. Phys. **231**, 481–528 (2002)
23. G. Szegö, *Orthogonal Polynomials*, 4th edn. (American Mathematical Society Colloquium Publications, Providence, 1975)
24. W.N. Bailey, *Generalized Hypergeometric Series* (Cambridge University Press, 1964)
25. A.P. Prudnikov, Y.A. Brychkov, O.I. Marichev, *Integrals and Series*, vol. 3 (Gordon and Breach, New york, 1990)
26. S.L. Cacciatori, H. Epstein, U. Moschella, Loops in Anti de Sitter space, to appear
27. L. Slater, *Generalized Hypergeometric Functions* (Cambridge University Press, Cambridge, 1966)
28. NIST Digital Library of Mathematical Functions. https://dlmf.nist.gov

The Starobinsky Model of Inflation and Renormalizability

R. Percacci and G. P. Vacca

Abstract The Starobinsky model was born in a cosmological scenario where conformally coupled matter quantum field fluctuations on the vacuum drive a non trivial semiclassical energy momentum tensor quadratic in curvature. The presence of an unstable de Sitter solution of the semiclassical Einstein equations contributed to spread the idea that the early universe could have experienced an inflationary epoch. Effective "$R + R^2$" models of gravity have later gained much attention since their predictions are in very good agreement with the measurements of CMB data and tensor to scalar ratio bounds. In this paper we observe how the Starobinsky model can be well approximated by the asymptotically free quadratic gravity on a part of a renormalization group trajectory (below some high UV scale) which is free from tachyonic instabilities, if a definition of "physical" running is employed.

1 Introduction

One cannot underestimate the enormous contributions given by Alexei A. Starobinsky to the development of many aspects of early universe Cosmology. We shall discuss here some quantum features of gravity, which are specifically related to just one aspect of Alexei's work, known nowadays simply as *Starobinsky inflation*.

This scenario originated from his work [1], which was very quantum and geometrical at the same time. Maybe picking up the path, started by A. Sakharov's idea of "induced gravity" [2], Alexei studied a model of Einstein gravity in presence of conformally coupled massless fields, whose quantum vacuum fluctuations were the source of a non trivial effective energy momentum tensor, quadratic in the curvature, associated to the trace anomaly. Solving the semiclassical Einstein equation of

R. Percacci
SISSA, Trieste, Italy
e-mail: percacci@sissa.it

G. P. Vacca (✉)
INFN - Sezione di Bologna, Bologna, Italy
e-mail: vacca@bo.infn.it

A. Barvinsky and A. Kamenshchik (eds.), *Open Issues in Gravitation and Cosmology*, Fundamental Theories of Physics 225, https://doi.org/10.1007/978-3-032-15703-4_26

motion for the metric he was finding an (unstable) de Sitter solution, describing an early universe expanding phase.

Later it was understood that this kind of quadratic gravity models were interesting in cosmology [3, 4] for the inflationary paradigm and the consequent properties of particle production.

It is natural to think about gravitation interacting with quantum fields to be itself part of the quantum world, with propagating quantum fluctuations of its own. In an interacting theory like gravity, these quantum fluctuations give rise to quantum corrections to the dynamics and this leads to the necessity of a renormalization procedure. It was shown that while standard General Relativity, possibly together with matter, is not a consistent quantum field theory at perturbative level, i.e. is not perturbatively renormalizable, quadratic gravity is, and can play the role of UV completion of General Relativity. In this case the Einstein-Hilbert term, linear in the curvature, is just a relevant operator which becomes dominant in the low energy, large distance and low curvature regime, whereas the curvature squared terms dominate at Planck energies. Therefore these are expected to play an important role in the very early universe, and possibly also during inflation.

Quadratic gravity is the theory of gravity with action (in signature $(-,+,+,+)$, natural units and omitting total derivatives)

$$S=\int d^4x\sqrt{|g|}\left[\frac{M_{\rm Pl}{}^2}{2}(R-2\Lambda)-\frac{1}{2\lambda}C^2-\frac{1}{\xi}R^2\right], \tag{1}$$

where $M_{\rm Pl}=1/\sqrt{8\pi G_N}$ is the (reduced) Planck mass, Λ is the cosmological constant, $C_{\mu\nu\rho\sigma}$ is the Weyl tensor. Quadratic gravity is renormalizable [5] but, at least in a naive perturbative analysis in flat space, leads to pathological propagation. Indeed, in the spin two sector the propagator is

$$\frac{-4i\lambda}{q^4+\frac{1}{2}\lambda M_{\rm Pl}{}^2q^2}=\frac{8i}{M_{\rm Pl}{}^2}\left[\frac{1}{-q^2}-\frac{1}{-q^2-\frac{1}{2}\lambda M_{\rm Pl}{}^2}\right],$$

showing the massless graviton and a massive state that is a ghost and, for $\lambda<0$, also tachyonic. In the spin zero sector the propagator is

$$\frac{-i\xi/3}{q^4-\frac{1}{12}\xi M_{\rm Pl}{}^2q^2}=\frac{4i}{M_{\rm Pl}{}^2}\left[-\frac{1}{-q^2}+\frac{1}{-q^2+\frac{1}{12}\xi M_{\rm Pl}{}^2}\right].$$

The massless pole (a ghost) can be eliminated by gauge transformations, as in Einstein's General Relativity, whereas the massive state is a tachyon for $\xi>0$. Thus, to avoid tachyons, we must choose $\lambda>0$ and $\xi<0$.

The theory without the square of the Weyl tensor, which is relevant to the Starobinsky model, does not have ghosts but is not renormalizable. Here we shall consider the Starobinsky model as a limit of the renormalizable quadratic gravity theory, in which the square of the Ricci scalar is in some sense dominant with respect to the square of the Weyl tensor. This point of view is motivated by the better quantum properties of the general quadratic gravity theory. There remains then the problem of the ghosts and tachyons. There has been some progress recently on both fronts. Ghosts are generally assumed to imply violation of unitarity, but several authors have suggested that massive ghosts cannot appear as asymptotic states and that the theory is unitary [6–9]. Instead, there would be a violation of causality at microscopic scales (scales comparable to the mass of the ghost) and this may be acceptable. We will not discuss ghosts further and focus instead on the problem of tachyons, whose presence depends on the signs of the quadratic curvature terms. In addition to being renormalizable, quadratic gravity had also been shown to be asymptotcally free, but only for a choice of signs that implies the existence of a massive spin 0 tachyon, signalling an intrinsically unstable vacuum [10, 11]. A recent recalculation of the beta functions [12], based on scattering amplitudes and including in the running the contribution of the infrared logs (i.e. terms related to a mass threshold like $\log(E^2/m^2)$), leads to a slighly different set of beta functions which admit asymptotically free theories lying outside the tachyonic region. The main aim of this paper is to discuss possible implications of this result for Starobinsky inflation.

We will briefly review Starobinsky inflation in Sect. 2, the new beta functions in Sect. 3, and their implications in Sect. 4. Finally, in Sect. 5 we summarize and conclude with some open questions.

One of us (G.P. V.) had the privilege to collaborate with Alexei: "We met several times when he visited our department starting from the early "2000". Our work together was in the context of his stochastic inflation techniques [13, 14] *to define renormalized correlators, extending the results to the description of renormalized correlators of gauge invariant scalar fluctuations in a quasi de Sitter inflationary phase* [15, 16]. *I remember him with great pleasure as a very nice person beyond being a great scientist who left us a profound legacy."*

2 Starobinsky Model and Inflation

Here we shall briefly review the inflationary predictions of the "$R + R^2$" Starobinsky model. This model has an action containing terms up to second order in the Ricci scalar that is usually written in the form

$$S = \frac{{M_{\rm Pl}}^2}{2} \int d^4x \sqrt{-g} \left[R + \frac{1}{6m^2} R^2 \right] . \tag{2}$$

Here m is the mass of the propagating spin-zero degree of freedom present in the R^2 term. One can appreciate in a simple way the dynamical features of this model in relation to inflation, changing description after moving from the Jordan frame to the Einstein frame. First one introduces a Lagrange multiplier field X to rewrite

$$S=\frac{{M_{\mathrm{Pl}}}^2}{2}\int d^4x\sqrt{-g}\left[\left(1+\frac{X}{3m^2}\right)R-\frac{1}{6m^2}X^2\right], \tag{3}$$

then performs a Weyl transformation $g_{\mu\nu}\to\Omega^2 g_{\mu\nu}$, $R\to\Omega^{-2}R-6\Omega^{-3}\Box\Omega$, and in order to obtain a standard Einstein term one chooses $\Omega=\left(1+\frac{X}{3m^2}\right)^{-1/2}$. The last step, after an integration by parts and neglecting total derivatives, is to define the field $\phi=\sqrt{\frac{3}{2}}M_{\mathrm{Pl}}\ln\left(1+\frac{X}{3m^2}\right)$ which has a canonical kinetic term. The rest, written in terms of ϕ defines its potential. One obtains

$$S=\int d^4x\sqrt{-g}\left[\frac{{M_{\mathrm{Pl}}}^2}{2}R-\frac{1}{2}\partial_\mu\phi\partial^\mu\phi-V(\phi)\right], \tag{4}$$

where $V(\phi)=\frac{3m^2{M_{\mathrm{Pl}}}^2}{4}\left(1-e^{-\sqrt{\frac{2}{3}}\frac{\phi}{M_{\mathrm{Pl}}}}\right)^2$. This is the Starobinsky model in the Einstein frame.

At this point a standard analysis for the study of the FLRW spacetime evolution and gauge invariant fluctuations can be performed [17, 18]. For such a flat potential the slow roll approximation is well suited and as usual one can study the evolution of the scale factor $a(t)$ with Hubble parameter $H=\dot{a}/a$ using the equations: $H^2\simeq V(\phi)/(3{M_{\mathrm{Pl}}}^2)$ and $3H\dot{\phi}+V'(\phi)\simeq 0$. Using the first equation one can also write the relation for H in terms of the field ϕ as

$$H(\phi)=\frac{m}{2}\left(1-e^{-\sqrt{\frac{2}{3}}\frac{\phi}{M_{\mathrm{Pl}}}}\right). \tag{5}$$

Clearly the inflationary expansion is of a quasi de Sitter kind, involving in the Einstein frame a potential (which is almost flat) energy density of about $3/4\,m^2{M_{\mathrm{Pl}}}^2$. Fixing the pivotal momentum scale and then the time of horizon crossing when the corresponding fluctuation freezes, one can define the number of e-folds till the end of inflation N_e, and the slow roll parameters ϵ_V and η_V (whose values, when close to one, set the end of the inflationary phase) as

$$N_e=\frac{1}{{M_{\mathrm{Pl}}}^2}\int_{\phi_e}^{\phi_*}d\phi\frac{V}{V'},\quad \epsilon_V=\frac{{M_{\mathrm{Pl}}}^2}{2}\left(\frac{V'}{V}\right)^2,\quad \eta_V={M_{\mathrm{Pl}}}^2\frac{V''}{V}. \tag{6}$$

One computes the power spectrum of the primordial gauge invariant scalar P_ζ and tensor (spin 2) P_t density perturbations which in terms of the spectral tilts around a pivot scale k_0 can be written as

$$P_\zeta=A_s\left(\frac{k}{k_0}\right)^{n_s-1},\quad P_t=A_t\left(\frac{k}{k_0}\right)^{n_t}. \tag{7}$$

In the lowest approximation the spectral indices can be assumed to be momentum independent (no running). In particular the tilt n_s of the nearly scale invariant scalar power spectrum and the ratio $r = A_t/A_s$ are the most used observables to discriminate between inflationary models. They can be computed in the slow roll regime using

$$n_s = 1 + 2\eta_V - 6\epsilon_V, \quad r = 16\epsilon_V . \tag{8}$$

Since tensor perturbations have not been observed yet, there is a strong experimental upper bound $r < 0.036$ at 95% C.L. [19]. From the form of the potential of the Starobinsky model in the Einstein frame in Eq. (4) one obtains very simple relations for such observables in terms of the number of e-folds N_e:

$$n_s \approx 1 - \frac{2}{N_e}, \quad r \approx \frac{12}{N_e^2} \quad \Rightarrow \quad r \approx 3(n_s - 1)^2 . \tag{9}$$

Moreover at horizon crossing the amplitude of the scalar power spectrum can be evaluated in the slow roll approximation to give $A_{s*} = \frac{V_*^3}{12\pi^2 M_{\text{Pl}}{}^6 V_*'^2}$. The predictions of the quadratic Starobinsky model are in very good agreement with the measurements of CMB data from Planck for n_s (with also r well below the BICEP/KecK experimental constraint) having a central value for the number of e-folds $N_e = 56$ and $m/M_{\text{Pl}} = O(10^{-5})$ to fit the observed values of scalar perturbations.

3 On the Running Couplings in Quadratic Gravity

Even though gravity is probably the most interesting context for higher derivative theories, our interest in the subject started from scalar models. In particular, a family of scalar theories with $(\phi(-\Box)^k\phi)$ kinetic term and a shift-invariant $(\partial_\mu\phi\partial^\mu\phi)^2$ interaction can define conformal field theories in dimensions $d = 4(k-1)$ [20]. A special case is the case of $k = 2$ and dimension $d = 4$, for which the field is dimensionless. When a standard kinetic term $\partial_\mu\phi\partial^\mu\phi$ is also present, it comes with a mass parameter and the theory propagates a massless particle and a massive ghost. It has two free fixed points and one can flow from the $(\Box\phi)^2$ free theory to the $\partial_\mu\phi\partial^\mu\phi$ free theory, with the dimension of the field changing continuously from zero to one along the renormalization group trajectory [21].

This was thought to be a possible toy model for gravity, with the Hilbert term containing the two-derivative kinetic term and the quadratic curvature terms containing the four-derivative kinetic term and interactions. This turned out to be only a poor analogy, for reasons that we shall mention later, but it spurred a re-examination of the beta functions of quadratic gravity.

Let us recall here that the main application of the renormalization group in particle physics is to capture and improve certain properties of scattering amplitudes or correlators. In (perturbatively) renormalizable theories, the scattering amplitude receives quantum corrections that in the high energy limit are of the same form as the classical one and therefore can be subsumed in a change of the coupling constant with energy. This energy dependence of the amplitude is logarithmic and for sufficiently high energies would invalidate perturbation theory. By considering the infinitesimal change of the coupling with the energy E and integrating it, one effectively resums certain pieces of the perturbative expansion obtaining an expression for the amplitude that is valid over a much larger range of energies. Following [22], we call the beta functions obtained in this way the "physical" beta functions, as opposed to the ones obtained by taking derivatives with respect to the parameter μ of dimensional regularization, that we shall call the "mu" beta functions.

We learn from textbooks that in the limit $E \to \infty$, these beta functions are the same. However, this is not universally true for all quantum field theories. It holds in standard renormalizable quantum field theories, because in the high energy limit the masses can be neglected and in the absence of any other dimensionful parameters, the arguments of the logarithms must necessarily involve the ratio E/μ. However, in theories with four derivatives, something different can happen. The propagator behaves typically like $1/(q^4 + m^2 q^2)$, which improves the convergence of loop integrals in the UV, but it makes it worse in the IR when we take $m \to 0$. In particular, depending on details of the interaction, a one loop integral could behave like $\int \frac{d^4 q}{q^4+m^2q^2}$, that is logarithmically divergent for $m \to 0$, and would also show up in dimensional regularization. In such cases the mass cannot be neglected and could appear in the arguments of the logs. On dimensional grounds, an amplitude could typically have the form

$$\mathcal{M}(E) = \lambda(\mu) + a\lambda^2(\mu)\log\left(\frac{m^2}{\mu^2}\right) + b\lambda^2(\mu)\log\left(\frac{E^2}{\mu^2}\right) + c\lambda^2(\mu)\log\left(\frac{E^2}{m^2}\right), \tag{10}$$

where λ is the coupling. If there are no infrared divergences in the massless limit, as is the case in four-dimensional theories with standard two-derivative kinetic terms, $c = a$. Then, the mass drops out, the amplitude is

$$\mathcal{M}(E) = \lambda(\mu) + (a+b)\lambda^2(\mu)\log\left(\frac{E^2}{\mu^2}\right), \tag{11}$$

and the "physical" beta function is identical to the "mu" beta function. In four-derivative theories with infrared divergences, $c \neq a$ and taking the derivative with respect to E gives a different result from taking the derivative with respect to μ. This is the case in quadratic gravity, where infrared divergences appear if we neglect the Hilbert term. The Planck mass acts as a IR regulator and its effect is felt even in the $E \to \infty$ limit.

The explicit calculation leads to the following result for the "mu" beta functions

$$\beta_\lambda = -\frac{1}{(4\pi)^2}\frac{133}{10}\lambda^2, \tag{12}$$

$$\beta_\xi = -\frac{1}{(4\pi)^2}\frac{5(72\lambda^2 - 36\lambda\xi + \xi^2)}{36}, \tag{13}$$

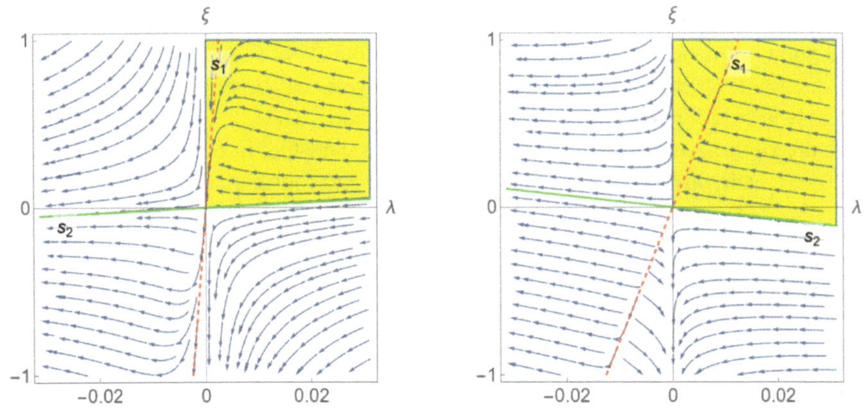

Fig. 1 Left: old "mu" flow. Right: new "physical" flow

while the "physical" beta functions are

$$\beta_\lambda = -\frac{1}{(4\pi)^2}\frac{(1617\lambda - 20\xi)\lambda}{90}\,, \tag{14}$$

$$\beta_\xi = -\frac{1}{(4\pi)^2}\frac{\xi^2 - 36\lambda\xi - 2520\lambda^2}{36}\,. \tag{15}$$

The renormalization group flows defined by these beta functions are shown in Fig. 1. Their main qualitative features are determined by the presence of a Gaussian (free) fixed point in the origin, and by various separatrix curves that delimit different basins of attraction. In this case the separatrices are loci where the flow is purely radial and correspond to fixed points for the ratio $\omega = -3\lambda/\xi$. In the case of the "mu" flow they are

$$s_1:\quad \xi = \frac{1291 + \sqrt{1637881}}{20}\lambda \approx 128.5\lambda \quad \Rightarrow \omega = -0.0233$$

$$s_2:\quad \xi = \frac{1291 - \sqrt{1637881}}{20}\lambda \approx 0.5601\lambda \quad \Rightarrow \omega = -5.3558$$

whereas for the "physical" flow they are

$$s_1:\quad \xi = \frac{569 + \sqrt{386761}}{15}\lambda \approx 79.4\lambda \quad \Rightarrow \omega = -0.03778$$

$$s_2:\quad \xi = \frac{569 - \sqrt{386761}}{15}\lambda \approx -3.53\lambda \quad \Rightarrow \omega = 0.8506$$

The main new feature to observe is that the basin of attraction of the Gaussian fixed point in the origin, shown in yellow in both panels of Fig. 1, lies entirely in the first quadrant, where the spin zero state is a tachyon, for the old beta functions but for the new beta functions also extends to a wedge in the fourth quadrant, where the signs are right to avoid tachyons. The spin 2 massive state in both cases is a ghost (and not a tachyon) for $\lambda > 0$, namely in the first and fourth quadrants.

4 Discussion

The Starobinsky inflation is doing very well as an inflationary model. Pushing the idea that this model should come from a fundamental UV description at high energies, one can ask what happens in the context of a quantum field theory description of gravity, which is perturbatively renormalizable when the lagrangian is quadratic in the curvature. Focusing on the one loop renormalized framework of quadratic quantum gravity, which was recalled in the last section, we see that the Starobinsky model, which has no tachyon states is obtained in the limit $\lambda \to \infty$ and $\xi = -\frac{12m^2}{M_{\rm Pl}{}^2}$.

Therefore only in an asymptotically free quadratic gravity theory described by couplings with the "physical" running encoded in Eqs. (14) and (15) one can find RG renormalized trajectories connecting a theory without tachionic instabilities in the "low" energy region to the UV Gaussian fixed point.

Pragmatically, one is tempted to consider the region in the theory space of a renormalizable quadratic gravity, where the Weyl square C^2 term is strongly suppressed with respect to the R^2 one, i.e. $\lambda \gg -\xi$ (or $\omega \gg 1$) at some suitable energy scale M. We could imagine for example M related to the scale of inflation, possibly defined by the potential energy density and therefore of the order $M \approx \sqrt{m M_{\rm Pl}}$. One might expect these theories to be good candidates to describe a dynamics, in particular an inflationary dynamics, which is very close to the one of the Starobinsky model. Then we observe that a further condition on the couplings must be satisfied: one would require the theory to be free from tachyons at least many order of magnitudes above the scale M which means that we require the RG time T needed for ξ to reach zero in the flow towards UV to be large enough.

Let us discuss how the RG trajectories associated to the "physical" beta functions are realized. The special trajectory given by the s_2 separatrix is characterized by the relation $\lambda \simeq -0.28\,\xi$, which is free from tachyons at any scale. Along this special trajectory one has a fixed ratio of the renormalized couplings λ/ξ, with even a slightly dominant Weyl square term with respect to the R^2 one, which is therefore far from the Starobinsky model. Any other trajectory connecting a point with the same negative ξ above this separatrix has a larger value of λ. In this case the trajectory starting at (λ_i, ξ_i), directed towards the UV Gaussian fixed point, crosses the λ-axis after some positive renormalization group time T and enters the scalar tachyonic region of positive ξ, approaches the separatrix s_1 and then stays close to it until it reaches the origin after an infinite amount of RG time. In particular in the case $-\xi_i \ll \lambda_i \ll 1$, one can find a simple relation for the RG time T needed for a negative ξ to flow to the zero value, since λ changes much more slowly than ξ, which then changes almost linearly with RG time. One finds

$$T \simeq -\frac{\xi_i}{\lambda_i} \frac{1}{\frac{35}{8\pi^2}\lambda_i + \frac{539}{480\pi^2}\xi_i} \simeq -\frac{8\pi^2}{35} \frac{\xi_i}{\lambda_i^2} \,. \tag{16}$$

It is also interesting to look at the RG flow trajectory towards the IR starting from this region of parameters. One can immediately see that an IR Landau pole is typically reached after some finite (negative) RG time T_{LP}. A good estimate of its position can be obtained on neglecting the mixed $\xi\lambda$ term in the beta function β_λ so that one obtains

$$T_{LP} \simeq -\frac{480\pi^2}{539} \frac{1}{\lambda_i} \,, \tag{17}$$

which for $\lambda \ll 1$ is very large.

Let us consider a numerical example, considering for an inflationary regime a model at the energy scale M with $\xi \simeq -10^{-9}$ and $\lambda = 10^{-5}$, so that the above mentioned suppression factor for the term C^2 w.r.t. R^2 is 10^{-4}. In this case one has the RG time $T \simeq 22.6$ for ξ to become zero towards the UV, which means that, following the RG flow, the system would enter the tachyonic region at a scale $M_T = Me^T \simeq 10^{10} M$, while on half a way along the flow since ξ changes almost linearly, one would have $\xi = -0.5 \times 10^{-9}$ at the energy scale $M_{T/2} \approx 10^5 M$. This shows that the effective theory which approximates the Starobinsky model for the given C^2 suppression is valid in a wide interval of energy scales. One can also easily check that the RG flow in the opposite direction towards the IR with the one loop "physical" beta functions indeed flows to hit a Landau pole, but located at extreme low energy scale. In the numerical example given above this would happen at an RG time $T_{LP} \simeq -8.6 \times 10^5$, which is beyond any meaningful physical scale, corresponding to lengths much larger that the extension of the known universe. Instead for a much shorter flow towards the IR, for example of $T = -20$ (corresponding to an energy scale e^T smaller, i.e. about 9 order of magnitude smaller) one finds a small change in the value of ξ, with $\xi = -1.88 \times 10^{-9}$, and λ practically unchanged, which means that there is a wide range of energy scales with slowly changing parameters of the model.

5 Conclusions

We have discussed how the Starobinsky model, which makes very successful inflationary predictions, could be thought of as an approximation to a renormalizable and UV complete quadratic gravity theory. Then, the use of the so called "physical" scheme for the running couplings leads to an extension, free of tachionic instabilities, of the domain of validity of the model over a wide range of energy scales along the RG trajectory, both towards the UV and the IR.

Nevertheless within this picture there are several aspects that deserve to be investigated in greater detail. The homogeneous scale factor evolution of the FLRW spacetime is not affected by the presence of the conformally invariant Weyl square term but certainly at the level of fluctuations we expect non trivial effects, possibly both of classical and quantum origin. Therefore one has to check that the differences from the Starobinsky model are not spoiling its predictions. In a very recent work for example the spectral tilt and amplitude of tensorial power spectrum has been computed in the presence of the spin 2 ghost states and shown to be slightly suppressed [23]. On the other hand recently a linear classical perturbative analysis of scalar metric gauge invariant perturbations in presence of the Weyl square term has shown an exponential grow and oscillating behavior [24]. This might signal an instability which should be further investigated, eventually performing a non linear classical analysis. Indeed there are known examples of dynamical higher derivative systems which one would naively expect to be unstable and which, instead, show within regions of initial condition to have a bounded evolution [25], a behavior completely unseen by a linear analysis. In a recent numerical analysis of quadratic gravity evolution [26], which focus on metric perturbations around a black hole, it has been shown that while at linear level instabilities can be encountered, the full non linear evolution is physically stable. Therefore further investigations are needed in the context of inflationary dynamics. Moreover one should later address also the quantum aspect with the treatment of the ghost states.

There are also some more theoretical aspects that require further analysis. One might wonder if the "physical" beta functions considered here are the relevant way to define a perturbative renormalized model to be used in inflationary studies. These have been derived in the context of on shell scattering amplitudes (on an almost flat background), that is are based on an analysis of in-out correlators of the theory. Normally in cosmology one deals with in-in correlators. There have been recently some analysis suggested by the so called cosmological bootstrap approach, where it has been shown that one can deduce in-in correlators from in-out ones [27]. But this equivalence has been derived in a context where several assumptions have been made: a maximally symmetric spacetime background of de Sitter type is considered, unitarity of the theory at hands was satisfied, and IR divergences were absent. It is therefore important to understand up to which point this correspondence can be further pushed.

Acknowledgements G.P.V. thanks Michele Cicoli and R.P. thanks Luca Buoninfante for useful discussions.

References

1. A.A. Starobinsky, A new type of isotropic cosmological models without singularity. Phys. Lett. B Ser. **91**, 99–102 (1980)
2. A.D. Sakharov, Vacuum quantum fluctuations in curved space and the theory of gravitation. Dokl. Akad. Nauk Ser. Fiz. **177**, 70–71 (1967)
3. A.A. Starobinsky, The perturbation spectrum evolving from a nonsingular initially De-Sitter cosmology and the microwave background anisotropy. Sov. Astron. Lett. series **9**, 302 (1983)
4. A. Vilenkin, Classical and quantum cosmology of the Starobinsky inflationary model. Phys. Rev. D Ser. **32**, 2511 (1985)
5. K.S. Stelle, Renormalization of higher derivative quantum gravity. Phys. Rev. D **16**, 953 (1977)
6. D. Anselmi, M. Piva, The ultraviolet behavior of quantum gravity. JHEP **05**, 027 (2018). [arXiv:1803.07777 [hep-th]]
7. A. Salvio, Quadratic gravity. Front. Phys. **6**, 77 (2018). [arXiv:1804.09944 [hep-th]]
8. J.F. Donoghue, G. Menezes, On quadratic gravity. Nuovo Cim. C **45**(2), 26 (2022) [arXiv:2112.01974 [hep-th]]
9. L. Buoninfante, Massless and partially massless limits in quadratic gravity. JHEP **12**, 111 (2023). [arXiv:2308.11324 [hep-th]]
10. E.S. Fradkin, A.A. Tseytlin, Renormalizable asymptotically free quantum theory of gravity. Phys. Lett. B **104**, 377 (1981); Nucl. Phys. B **201** 469 (1982)
11. I.G. Avramidi, A.O. Barvinsky, Asymptotic freedom in higher derivative quantum gravity. Phys. Lett. B **159**, 269–274 (1985)
12. D. Buccio, J.F. Donoghue, G. Menezes, R. Percacci, Physical running of couplings in quadratic gravity. Phys. Rev. Lett. **133**(2), 021604 (2024)
13. A.A. Starobinsky, in *Field Theory, Quantum Gravity and Strings*, eds. H.J. De Vega, N. Sanchez, Lect. Notes Phys. **246**, 107 (1986)
14. A.A. Starobinsky, J. Yokoyama, Equilibrium state of a selfinteracting scalar field in the De Sitter background. Phys. Rev. D **50**, 6357–6368 (1994)
15. F. Finelli, G. Marozzi, A.A. Starobinsky, G.P. Vacca, G. Venturi, Generation of fluctuations during inflation: comparison of stochastic and field-theoretic approaches. Phys. Rev. D **79**, 044007 (2009). [arXiv:0808.1786 [hep-th]]
16. F. Finelli, G. Marozzi, A.A. Starobinsky, G.P. Vacca, G. Venturi, Stochastic growth of quantum fluctuations during slow-roll inflation. Phys. Rev. D **82**, 064020 (2010)
17. V.F. Mukhanov, H.A. Feldman, R.H. Brandenberger, Theory of cosmological perturbations. Part 1. Classical perturbations. Part 2. Quantum theory of perturbations. Part 3. Extensions. Phys. Rept. **215**, 203–333 (1992)
18. J. Ellis, M.A.G. Garcia, D.V. Nanopoulos, K.A. Olive, Calculations of inflaton decays and reheating: with applications to no-scale inflation models. JCAP **07**, 050 (2015). [arXiv:1505.06986 [hep-ph]]
19. P.A.R. Ade et al. [BICEP and Keck], Improved constraints on primordial gravitational waves using Planck, WMAP, and BICEP/Keck observations through the 2018 observing season. Phys. Rev. Lett. **127**(15), 151301 (2021). [arXiv:2110.00483 [astro-ph.CO]]
20. M. Safari, A. Stergiou, G.P. Vacca, O. Zanusso, Scale and conformal invariance in higher derivative shift symmetric theories. JHEP **02**, 034 (2022)
21. D. Buccio, R. Percacci, Renormalization group flows between Gaussian fixed points. JHEP **10**, 113 (2022)
22. D. Buccio, J.F. Donoghue, R. Percacci, Amplitudes and renormalization group techniques: a case study. Phys. Rev. D **109**(4), 045008 (2024) [arXiv:2307.00055 [hep-th]]
23. J. Kubo, J. Kuntz, Primordial Gravitational Waves in Quadratic Gravity, [arXiv:2502.03543 [gr-qc]]
24. A. De Felice, R. Kawaguchi, K. Mizui, S. Tsujikawa, Starobinsky inflation with a quadratic Weyl tensor. Phys. Rev. D **108**(12), 123524 (2023), [arXiv:2309.01835 [gr-qc]]
25. C. Deffayet, A. Held, S. Mukohyama, A. Vikman, Global and local stability for ghosts coupled to positive energy degrees of freedom. JCAP **11**, 031 (2023), [arXiv:2305.09631 [gr-qc]]

26. A. Held, H. Lim, Nonlinear evolution of quadratic gravity in 3+1 dimensions. Phys. Rev. D **108**(10), 104025 (2023), [arXiv:2306.04725 [gr-qc]]
27. Y. Donath, E. Pajer, The in-out formalism for in-in correlators. JHEP **07**, 064 (2024). [arXiv:2402.05999 [hep-th]]

The Infra-Red Road to Quantum Gravity

S. Tajik, M. J. Desrochers, and P. C. E. Stamp

Abstract We review work in areas ranging from condensed matter physics to quantum gravity, with the following interconnected questions in mind: (i) what is the nature of the vacuum in condensed matter systems, in quantum field theory, and in classical and quantum gravity; (ii) how do analogies between these systems work, how well do they work, and how useful are they; (iii) what modifications can we make to quantum mechanics to deal with quantum gravity, and (iv) how and why low-energy theories of quantum gravity are, in our view, the right way to make progress in this field. We use many different examples to illustrate our arguments.

1 Introduction

This article is written to honour Starobinsky [1] whose remarkable work in cosmology has fundamentally influenced the subject. Indeed, it now appears that the Starobinsky version [2] of inflation theory—which was the very first version, and which thus started the whole field—may in fact be the only version that has survived the verdict of observational test. The inflation scenario is one of the more remarkable examples of the application—to the entire universe—of a more fundamental theory that we don't yet have, viz. a widely accepted theory of quantum gravity. Such a theory is supposed to somehow unify General Relativity (GR) with Quantum Field Theory (QFT), in a third theory which reduces to one or another in the appropriate limits.

Starobinksy himself brought to bear on these questions a very interesting background. His training followed the remarkable Soviet scientific ethos of that time—a broad understanding of the whole of physics, exemplified in the famous Landau-Lifshitz course of 10 volumes in theoretical physics. In fact Starobinsky's first research experience [1] was in work with the well-known solid-state theorist Ilya

S. Tajik · M. J. Desrochers · P. C. E. Stamp (✉)
Physics and Astronomy, University of British Columbia, Vancouver, BC, Canada
e-mail: stamp@phas.ubc.ca

P. C. E. Stamp
Pacific Institute of Theoretical Physics, University of British Columbia, Vancouver, BC, Canada

A. Barvinsky and A. Kamenshchik (eds.), *Open Issues in Gravitation and Cosmology*, Fundamental Theories of Physics 225, https://doi.org/10.1007/978-3-032-15703-4_27

M. Lifshitz (brother of Evgenii M. Lifshitz), on the theory of disordered conductors (a subject developed in the Soviet Union by Lifshitz in the early 1960s, in work following that of P. W. Anderson in 1958).

Starobinsky then fell under the influence of Ya. B. Zeldovich, a theorist who was himself largely self-taught, having left school at 16 years old. Zeldovich began as a theoretical chemist, before doing pioneering work in theoretical physics, relativistic astrophysics, and cosmology (and on the way, developing with A. D. Sakharov the theory behind Soviet nuclear weapons).

Starobinsky's early work in quantum gravity and cosmology anticipated much of the work that is commonly highlighted in the West. This included work on "acceleration radiation" by rotating objects, along with Zeldovich [3], and of course his invention of inflation theory (before the work of Guth and Linde). In recorded discussions one of us had [1] with Starobinsky and several Russian colleagues in 2016, Starobinsky expressed a long-standing interest in the physics of decoherence processes in both condensed matter and cosmology, including their role in the early universe. He planned with PCES an extended visit to UBC in 2020. Unfortunately and tragically, COVID intervened to prevent this.

This article involves the interests of all three co-authors in topics ranging from superfluids on earth, all the way to quantum gravity and cosmology—all of these topics interested Starobinsky.

The common theme running through all the topics addressed here is the application of quantum mechanics (QM) and quantum field theory (QFT) to macroscopic systems, from the lab to the universe. We begin by asking about the nature of the vacuum in physics. Our discussion of this question will range from 2d and 3d superfluids, and disordered insulators, up to systems at the cosmological scale. Along the way we decided to address questions about (a) analogies in physics between phenomena in condensed matter systems and in quantum and classical gravity; and (b) the foundations of QM. This leads finally to our original intention, viz., (c) to discuss the search for a consistent theory of quantum gravity (QGr). We argue that most theory, which has been looking at Planck-scale physics, has simply been looking in the wrong place, and that the key is to look for a low-energy replacement for quantum mechanics, in which QM breaks down at macroscopic scales, because of gravity; and we examine a consistent theory of this kind (the CWL theory).

2 Nature of the Vacuum in Physics

We begin with a simple physical question. Suppose we have a large cubical box of volume L^3, containing just a single H atom, which sits in one of the huge voids between galactic superclusters. What is the vacuum state of this system?

The answer seems obvious. If the box is at $T = 0$, one has a bound H atom in its ground state, with orbital state being in the lowest energy state of translational motion (with energy $3\hbar^2\pi^2/2ML^2$, where M is the H atom mass). Suppose however, the box is $T = 2.7\,\mathrm{K}$ (as it will be, since we assume it is in thermal equilibrium with

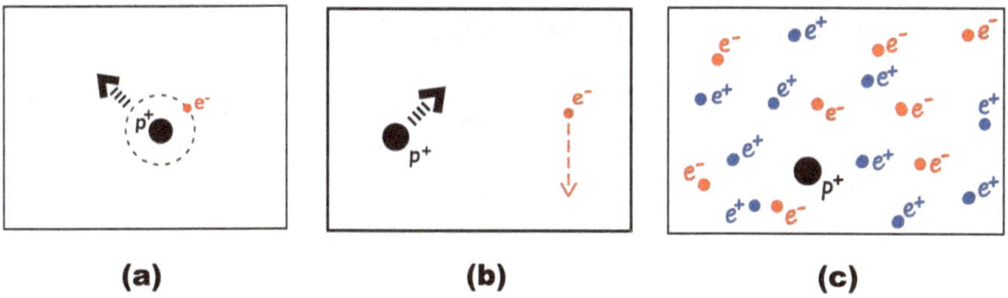

Fig. 1 Thought experiment for a H atom in a box. In **a** we show the physical vacuum at $T = 0$; one has a simple atom in its ground state. In **b** we shown the same vacuum at $T = 2.7$ K; for a large enough box the atom dissociates. In **c** we show the vacuum when $T - 10^{10}$ K; it is now full of thermally excited e^+/e^- pairs. Finally, in **d** we show the system at $T = 0$ but with a very strong electric field $E_o \sim 10^{18}$ V/m applied. The vacuum is now unstable, with e^+/e^- pairs dissociating out of the vacuum, the H atom completely dissociated, and all particle accelerating to infinity

the microwave background)? Then, apart from low-energy photons in the box, in equilibrium with the H atom, we will find that the atom in a mixture of states, in which it has partially dissociated into separate e^+ and e^- particles, according to the Saha equation. What now if the temperature is 10^{10} K? Then the QED vacuum will be full of e^+/e^- pairs, which will strongly modify its dielectric properties, along with a dense sea of very hot photons. The situation is depicted in Fig. 1.

One may object that while we are clearly describing the physical vacuum here, we are not describing the theoretical construct which we also call the "vacuum" or "ground" state, unless we are at $T = 0$. Fine; but suppose we do now go back to $T = 0$, and now apply an electric field E? Then, according to the Schwinger 1-loop formula [4], one sees e^+/e^- pairs being spontaneously produced by vacuum tunneling, so that there is no well-defined vacuum at all. So, what do we mean by the vacuum?

In what follows we discuss the vacuum in several condensed matter systems, finishing with a discussion of the vacuum in classical General Relativity (GR), and in relativistic Quantum Field Theory (QFT). We will discuss some of the key conceptual questions for each vacuum, and discuss how one can expose these vacua, and see some of its hidden properties, using various probes. These probes include external current flow (for superfluids), strong electric fields (for QED), and a combination of singularities (like black holes) and low-energy experiments (for GR). In the next section (Sect. 3), we spend some time discussing the role of analogies in physics, using the examples discussed in this section (Sect. 2).

2.1 Vacuum in Condensed Matter Systems

The above discussion shows that even a simple 1-particle problem becomes a many-body problem when one looks at it properly. So what does condensed matter theory say about all of this?

2.1(a): General Remarks

Let's make a few general remarks—these are by and large well-known to any condensed matter physicist. In condensed matter systems, we can say that:

(i) One never deals with the true N-body states: they are far too complicated, and there are far too many of them. Thus the density of states $D(E)$ for a very small grain of diameter $d \sim 0.4$ mm, and mass $M \sim M_p = 2 \times 10^{-8}$ kg (so $N \sim 10^{18}$), goes like $D(E) \sim e^{S(E)}$, where the entropy $S(E) \sim N(E/E_0)^{\frac{3}{4}}$ and $E_0 \sim 3\pi^2\hbar^2/2M_p d^2$. Even for a small molecule, the number of states is beyond astronomical, even at very low energies.

(ii): Instead, one deals with an "effective" theory and an "effective Hamiltonian" H_{eff}, written in terms of "quasiparticles" or some other kind of effective excitation. The effective Hilbert space is drastically restricted, and the effective vacuum $|0\rangle_{\text{eff}}$ is the ground state of H_{eff}.

(iii): Unfortunately, the real behaviour of the system may be quite different from that of H_{eff}. A quick glance shows that we are surrounded, not by perfect fluids or perfect crystals, but instead by highly disordered systems. For such systems any putative ground state is physically meaningless—the system would need a time $\sim O(\tau_0 \exp(N^\alpha))$ to reach it, if it exists at all (here $\alpha \simeq 2 - 3$, and τ_0 is a microscopic timescale). On any meaningful timescale, the physical behaviour is entirely controlled by excitations in a narrow band of many-body energies, usually far above the "ground state" in energy.

(iv): Only two condensed matter systems exist where one can meaningfully discuss, and do experiments on, the vacuum state. These are superfluid ^{4}He and superfluid ^{3}He, with the superfluidity discovered in 1937 and 1972 respectively. Because only ^{3}He dissolves in ^{4}He, one can flush ^{4}He to fantastic purity; it has been understood since the 1950s that it can be treated at very low T as, for all practical purposes, a perfect vacuum (the mean free path of quasiparticles, at the lowest attainable temperatures in these two superfluids, is many light years).

The example of superfluid ^{3}He is interesting because it has often been claimed that it can provide an analogy for processes in cosmology. Analogies between radically differing physical systems have been employed throughout the history of physics. Famous examples include Maxwell's analogy between the EM field and different mechanical systems; the analogies which were employed, in the early history of GR, between the spacetime metric field and various fluid, and superfluid, and solid media; and an enormous number of analogies made between various condensed matter media and the fields in QFT—perhaps the most famous being the Anderson–Higgs boson [5].

2.1(b): Key Points About Superfluids

Here we recall some key points about (uncharged) superfluids that have a bearing on the topics in this article.

(i) Definition of Superfluidity: Operationally, this is provided by the "Hess–Fairbank" effect [6], which is the analogue for a neutral superfluid of the Meissner effect. One imagines cooling a rotating container of normal fluid (a cylinder in 3D, a circular disc in 2D), which while still in the normal state is precisely equivalent to Newton's famous "rotating bucket" thought experiment. The 3D normal fluid will rotate with the bucket, and show a parabolic depression on the liquid surface. The 2 superfluid will rotate with the disc. However, as soon as we go through the superfluid transition, the fluid "expels rotation" (see Fig. 2), and thereby comes to rest in a local inertial frame, even while the containers continue to rotate (indeed their angular velocity increases!).

We note that "superflow", i.e., the flow of the superfluid without viscosity, is not part of the fundamental definition of a superfluid, but rather an ancillary concomitant of it (and when vortices are in the system, one has dissipative superflow—see below). For many years superfluidity was thought to be a consequence of "Off-Diagonal Long- Range Order" (ODLRO), along with Bose condensation (BEC), in the superfluid density matrix [7, 8]. However ODLRO and BEC are actually unnec-

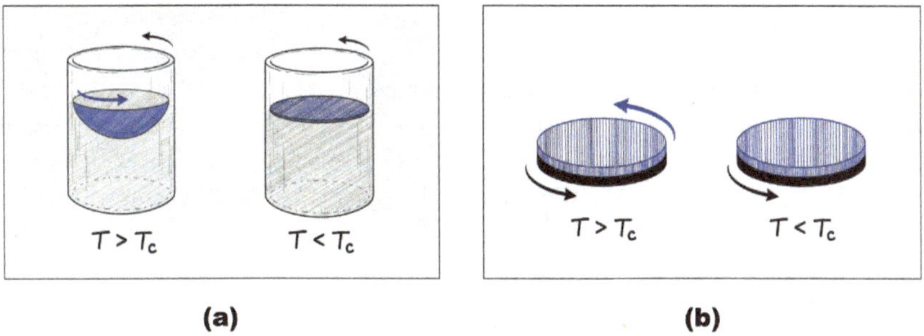

Fig. 2 Operational definition of a superfluid. In **a** we show the equilibrium state of a superfluid 3-d rotating cylindrical bucket. If $T > T_c$, the superfluid transition temperature, then the normal fluid rotated with the cylinder (and shows a parabolic depression of the fluid surface). Below T_c, if the angular velocity $\Omega < \Omega_c$, where Ω_c is a critical angular velocity, then the superfluid is stationary in a local inertial frame. When $\Omega > \Omega_c$, the equilibrium state has single quantized vortices present. In **b** we show the same situation for a 2-d superfluid on a rotating disc, where the same argument applies, **provided** no vortex/anti-vortex pairs are excited—these pairs destabilize the superfluid state, even at $T = 0$. Note that in (**b**), we do not show the parabolic depression of the film surface that occurs for a rotating disc above T_c

essary for superfluidity—indeed 2D superfluids have neither ODLRO nor BEC at any temperature. The definition of superfluidity is more subtle for a 2D superfluid—we return to this below.

So far we have not discussed vortices. The system can lower its energy, above a critical rotation velocity, by having quantized vortices. These enter the superfluid from the walls, either by thermal activation (for higher T), or by tunneling (at very low T). This is a process we will call "extrinsic vortex nucleation". The system is still superfluid, but it no longer shows superflow—the flow is now dissipative, because the superflow drives vortex motion, taking energy from the superflow, and unless one drives the superflow externally, it decays with time.

Much more difficult is the process of "intrinsic vortex nucleation", in which the vortex nucleates in the bulk superfluid, far from any containing boundary [9]. In fact it has apparently never been seen—we will have a lot to say about it in this article, as well as its analogues in particle physics and quantum gravity.

(ii) Excitation spectra: In 3D superfluids like ^{4}He or several BECs, the excitation spectrum is well understood (see Fig. 3). For weak coupling (typically this means a low-density BEC) one uses the results of Bogoliubov [10]. We assume a low-density gas of bosonic particles, each of mass M, with short range repulsive interactions; then only s-wave scattering is important, with scattering length a_s. This defines a coupling constant g, given in 3D by $g = 4\pi \hbar a_s/M$.

Then the unbounded 3D system has single-particle quasiparticle excitations with dispersion

$$\epsilon_p^2 = \hbar^2\omega_k^2 = c_s^2\hbar^2k^2 + \frac{\hbar^4k^4}{4M^2} \equiv c_s^2p^2 + (\epsilon_p^o)^2, \quad c_s = \sqrt{\frac{g\rho_0}{M}}, \quad \text{(weak coupling)} \tag{1}$$

where $\epsilon_p^o = p^2/2m$ is the free particle dispersion, and ρ_0 is the density of particles in the Bose condensate; for weak coupling $\rho_0 \sim \rho$, the total mass density. Then for momenta $p \equiv \hbar k$ such that $p \ll 2Mc_s$, we have simple phonons with energy $\epsilon_p = c_s p$ moving at the rather low sound velocity c_s, whereas for $p \gg 2Mc_s$, this crosses over to the free particle form $\epsilon_p = \epsilon_p^o + Mc_s^2$. This latter motion is simply that of a free particle in a uniform medium with a refractive index $n > 1$. We can write a very simple effective Hamiltonian for this system in two different ways. First, we define the condensate wave-function $\psi(\mathbf{x}, t)$, and write [11]

$$\hat{H}_{eff}^{BEC} = -\frac{\hbar^2}{2M}\nabla^2 + V(\mathbf{r}) + g|\psi(\mathbf{r}, t)|^2 \qquad \text{(weak coupling)} \tag{2}$$

which gives the Gross–Pitaevski equation of motion $i\hbar\partial_t\psi(t, \mathbf{r}) = \hat{H}_{eff}\,\psi(t, \mathbf{r})$. Here $V(\mathbf{x})$ is some external potential acting on the superfluid (for cold BEC gases this would be a trapping potential). Alternatively, since ϵ_p simply gives the eigenvalues of H_{eff}, we write [10]

$$H_{eff}^{BEC} = E_0 + \sum_p \epsilon_p \hat{a}_p^\dagger \hat{a}_p \qquad \text{(weak coupling)} \tag{3}$$

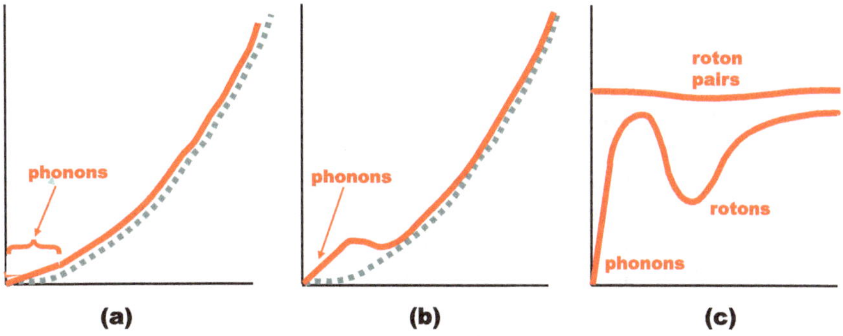

Fig. 3 The form of the quasiparticle spectrum in Bose superfluids. In **a** we show the result for weak coupling, along with the free particle spectrum. In **b** we show what happens when one increases the coupling strength for a dilute 3D Bose gas; and In **c** we show the result for strongly-coupled 3D ^{4}He superfluid. In **c** we show the main quasiparticle spectrum, and just one of the multiparticle branches (the bound roton pair states). In none of these three cases do we show the broad band of incoherent multi-quasiparticle excitations

where $\hat{a}_p^\dagger, \hat{a}_p$ are second quantized operators for the quasiparticles. The Hamiltonian $\hat{H}_{eff}$ is of course extremely simple—it is just a quadratic form describing a set of non-interacting quasiparticles.

For very strong repulsive coupling things are very different (Fig. 3c). The well-known exemplar of this is 3D superfluid ^{4}He, in which the interactions induce ϵ_p to turn over completely [14], and descend to form roton quasiparticles at a momentum $p_o \sim 1/a_o$, where a_o is the mean interparticle spacing. Phonons and rotons exist in both 3D and 2D superfluids; in 3D they can be created in many ways, by injecting everything from neutrons to picosecond laser pulses [14, 15]. In 2D, this is more difficult [16]. One also has quantized surface capillary waves (3rd sound), and rotons are strongly attracted to the surface and to each other [12–16].

The interactions between the quasiparticles are now not negligible; we must write, following Landau [17]

$$\mathcal{H}_{eff} = E_0 + \sum_p \epsilon_p \hat{a}_p^\dagger \hat{a}_p + \frac{1}{2} \sum_{p,p'} \sum_q V(p,p',q)\, \hat{a}_{p+q}^\dagger \hat{a}_p^\dagger \hat{a}_{p'-q} \hat{a}_{p'} \quad \text{(strong coupling)} \qquad (4)$$

where the effect of the interaction $V(p, p', q)$ is to strongly renormalize all the quasiparticle properties, and to cause scattering between them. In superfluid ^{4}He the effect of these interactions is very important—it makes, *inter alia*, the quasiparticle properties very strongly temperature dependent. The interactions between the quasiparticles also lead in 3D to bound roton pair states [14], and even bound triplet roton states [18].

It is extremely hard to calculate the parameters in the strongly-coupled effective Hamiltonian in any reliable way. For the weakly-coupled Bose gas it is of course straightforward, using perturbation theory: then the BEC fraction $N_0/N \sim O(1)$

at $T = 0$ (in this weak interaction limit, $1 - (N_o/N) = (8/3\pi^{1/2})(na_s^3)^{1/2}$, where $n = N/V$ is the density of particles). For the strongly-coupled system $N_o/N \ll 1$ (for ^{4}He, $N_o/N \sim 0.08$ at $T = 0$). Of course in both systems the superfluid fraction $\rho_s/\rho \to 1$ as $T \to 0$; but ρ_s/ρ has no connection to N_0/N.

In the intermediate coupling range (often called strong coupling by people working on BECs), the quasiparticle spectrum begins to turn over (see Fig. 3b). As far as we are aware there are no direct measurements of the quasiparticle spectrum in BECs; in 3D ^{4}He it has been looked at in great detail.

Well-known kinematic arguments of Landau [19] predicted that the form of the excitation spectrum would yield a critical fluid velocity v_c below which one would see superflow, and above which quasiparticles are excited from the vacuum. This has been verified in detail in 3D superfluid ^{4}He and ^{3}He, by moving ions rapidly through the superfluid at high pressure [20] (at low pressure in ^{4}He, the moving ions nucleate vortex rings [21, 22]).

All of the above remarks about weakly- and strongly coupled bosons and their excitation spectra apply to 3D boson systems. Many of them do *not* apply to 2D superfluids, which have to be treated differently.

2.2 The Vacuum in Superfluid ^{4}He Films

We now look at our first key example. Two-dimensional ^{4}He superfluid films behave very differently from bulk 3d superfluids, and the property of superfluidity in 2d has to be understood quite differently. As shown by Berezinskii [23] and Kosterlitz and Thouless [24], there exists a critical temperature T_K (the KT transition temperature), below which vortex/anti-vortex pairs bind, leaving only quasiparticle excitations in a 2d superfluid with short-range ordering, but without long-range order. However above T_K, the vortex/anti-vortex pairs unbind in profusion, destroying the superfluid state, and destroying even the short-range order.

This argument, amply confirmed by later experiments on strongly-coupled ^{4}He films [16], is a thermodynamic one: the free energy $F = U - TS$ is minimized in each phase, because both U and S depend logarithmically on the distance between the nearest vortex/anti-vortex pairs. In 2D BECs the physics is rather subtle [25], because the interparticle interactions are weaker, making quantum fluctuations even more important; however these can be suppressed somewhat by a trapping potential.

For these and other reasons, the superfluid vacuum state is quite different in the two cases. The 3d vacuum state has ODLRO, i.e., it Bose condenses, with a macroscopic occupation of zero-momentum particle states ($\mathbf{k} = 0$ modes). Neither quantum nor thermal fluctuations of the finite-$\mathbf{k}$ quasiparticle modes can destroy this order until we reach the superfluid transition temperature T_C. Thus the 3d vacuum state, although fantastically complex because of the strong interparticle interactions, is ordered in a fundamental way.

The 2d vacuum state, on the other hand, is not ordered. Right down to $T = 0$, long-wavelength quantum fluctuations of the gapless phonon modes disorder the system:

there is no ODLRO even at $T = 0$. There are instead strong quantum fluctuations of the vortex/anti-vortex modes; and all of this feeds into the final vacuum state of the 2d superfluid. Basically, the vacuum state of the 2-d superfluid is quantum disordered, and although it can show superfluidity, there is no Bose condensate (the appellation "BEC" given to cold 2D superfluid atomic gases is thus completely misleading—they are superfluid, but there is no condensate!). Instead one defines a "quasi-condensate" (just another name for short-range order); if the system size is less than the length scale beyond which order disappears, then the idea of a quasi-condensate works.

Many of the theoretical approaches that work for 3D Bose systems fail in 2D. Thus the Gross–Pitaevski equation works poorly, and "small fluctuation" expansions of density modes about a (non-existent) condensate, or even about a quasi-condensate, are clearly doubtful unless the fluctuations are "pinned" by the boundaries. So one needs to find alternative approaches.

2.2(a): Long-Wavelength Analysis

One way we can look at this theoretically **in the long wavelength limit** is by writing a long-wavelength "quantum hydrodynamic" form for the action of the system as [9, 26],

$$S = \int dt \int d^2 r \left[\frac{\hbar}{m_4} \rho(\boldsymbol{r}, t)\, \dot{\Phi}(\boldsymbol{r}, t) - \mathcal{H}(\boldsymbol{r}, t) \right] \tag{5}$$

where m_4 is the mass of the ^{4}He atom, $\rho(\boldsymbol{r}, t)$ is the areal mass density (proportional to the film thickness L_z), $\Phi(\boldsymbol{r}, t)$ is the local superfluid phase (so that the local superfluid velocity is $\boldsymbol{v}_s(\boldsymbol{r}, t) = (\hbar/m_4)\nabla\Phi(\boldsymbol{r}, t)$, and $\mathcal{H}(\boldsymbol{r}, t)$ is the Hamiltonian density

$$\mathcal{H}(\boldsymbol{r}, t) = \left[\frac{\rho(\boldsymbol{r}, t)}{2m_4^2} (\hbar\nabla\Phi(\boldsymbol{r}, t))^2 - \epsilon(\boldsymbol{r}, t) \right] \tag{6}$$

where $\epsilon(\boldsymbol{r}, t)$ parametrizes energy fluctuations around the homogeneous state (i.e., around the vacuum state if $T = 0$). For a film of uniform thickness $\rho(\boldsymbol{r}, t) \to \rho_s(T)$, the T-dependent superfluid density. Since this is a 2D system, $\rho_s(T) \to \rho$ as $T \to 0$, but the BEC fraction $N_0/N = 0$ always.

Quite generally we can write $\rho(\boldsymbol{r}, t) = \rho_s + \eta(\boldsymbol{r}, t)$, where $\eta(\boldsymbol{r}, t)$ parametrizes density fluctuations, and $\Phi(\boldsymbol{r}, t) = \Phi_S(\boldsymbol{r}) + \phi(\boldsymbol{r}, t)$, where $\phi(\boldsymbol{r}, t)$ parametrizes phase fluctuations around the time-independent background phase $\Phi_S(\boldsymbol{r})$. One assumes a background superflow $\boldsymbol{v}_S^0(\boldsymbol{r}) = (\hbar/m_4)\nabla\Phi_S(\boldsymbol{r}, t)$, created by the geometry of the system and by any external driving flow (which is often imposed by boundary conditions). As already emphasized, in a 2-d superfluid, the phase and density fluctuations are **not small**! As one lowers the dimensionality, quantum fluctuations become more important, and 2D is the critical dimension in this context. The most serious are the phase fluctuations, and these occur between multiple vacuum states, between which the system is constantly fluctuating. These states have different numbers of topological excitations in the system, these being the quantum solitonic vortices, which have quantized circulation [27, 28].

The flow field of the vortex falls off like $1/|\boldsymbol{r} - \boldsymbol{R}_V(t)|$, where $\mathbf{R}_V(t)$ is the vortex coordinate. The total flow energy of the system, and the vacuum state, must then depend on the behaviour of the system at any boundary, no matter how far away. This dependence on boundary conditions is common to any system (and the QFT that describes it) in which some effective long-range gauge field exists, i.e., for almost any macroscopic system in Nature! For this reason, no vacuum state can be defined independently of the boundaries, and indeed, the whole idea of a thermodynamic limit is quite meaningless [29]. Two well-known examples are:

(a): the 2D Fractional Quantum Hall effect (FQHE), where all of the properties (quasiparticle excitations, thermodynamics, dynamics) are carried by the boundaries, except for the bulk phonons [30].

(b): a simple ferromagnet, where the long-range dipolar fields create a demagnetization field which controls both the thermodynamics and the spin quasiparticle spectrum no matter how large the system, and which is controlled by the shape of the system; it also controls the topological excitations of the system (domain walls, vortices, etc.), in both the classical and quantum regimes [31–33]. This example applies in both 2 and 3 dimensions.

We will see below that the same dependence of the vacuum state on boundaries and topological excitations also obtains in 3-dimensional superfluids (^{4}He, ^{3}He).

The analogy between the 2-superfluid and an EM gauge theory like QED comes from the mapping [9, 26] between the hydrodynamic action above and a (2+1)-dimensional QED action of the form (with a single vortex present):

$$S_V\left[\boldsymbol{R}_V, \dot{\boldsymbol{R}}_V\right] = \int dt \left\{ \left[\frac{1}{2} M_V^0 \dot{R}_V^2 - \mathcal{T}(\boldsymbol{R}_V) \right] + \left[\dot{\boldsymbol{R}}_V \cdot \boldsymbol{A}_V(\boldsymbol{R}_V, t) - V_V(\boldsymbol{R}_V, t) \right] \right\} \tag{7}$$

Here $\boldsymbol{\kappa} = \pm (h/m_4)\,\hat{z}$ is the circulation quantum, and $\boldsymbol{R}_V$ is the vortex coordinate. The energy $\mathcal{T}(\boldsymbol{R}_V) = \frac{1}{2}\int d^2r\, \rho_S(\boldsymbol{r} - \boldsymbol{R}_V(t))\, \boldsymbol{v}_s^2(\boldsymbol{r}, t)$ is the superflow energy in the system, which creates a bare potential acting on the vortex. The gauge field components $\boldsymbol{A}_V$ and V_V are:

$$\boldsymbol{A}_V = \boldsymbol{A}_0 + \boldsymbol{A}_{QP} = \frac{1}{2}\rho_S(\boldsymbol{R}_V \times \boldsymbol{\kappa}) + \frac{\hbar}{2m_4}\int d^2r\, \boldsymbol{j}_{QP}(\boldsymbol{r}, t) \tag{8}$$

where $\boldsymbol{J}_{QP}(\boldsymbol{r}, t)$ is the quasiparticle **pair** current, given by

$$\boldsymbol{j}_{QP}(\boldsymbol{r}, t) = \frac{\hbar}{m_4}\left[\eta(\boldsymbol{r}, t)\nabla\Phi(\boldsymbol{r}, t) - \phi(\boldsymbol{r}, t)\nabla\eta(\boldsymbol{r}, t)\right] \tag{9}$$

and where M_V^0 is the "bare" hydrodynamic mass of the vortex (which also depends on the superflow throughout the system, so that $M_V^0 = M_V^0(\boldsymbol{R}_V(t))$. We note that the vortex does not couple to single quasiparticles [26], but only to pairs of quasiparticles, at lowest order in the quasiparticle variables, because the vortex is a quantum soliton (which in the QED analogy becomes an electric charge). Thus, a photon in the QED analogy is equivalent to a pair of quasiparticles. Finally, the analogue to the electric potential is

$$V_V(\boldsymbol{R}_V, t) = \frac{\hbar}{2m} \int d^2r \, \boldsymbol{j}_{QP}(\boldsymbol{r}, t) \cdot \nabla \Phi_V(\boldsymbol{r} - \boldsymbol{R}_V(t)) \tag{10}$$

in which quasiparticle pairs couple to the superflow from the vortex - this coupling is long-range, i.e., even quasiparticle pairs at an arbitrary distance from the vortex core still couple to it.

The true vacuum of the system is a superposition of states with different numbers of vortex/anti-vortex pairs, with vacuum tunneling constantly taking place between these states, and vacuum fluctuations largely controlled by this tunneling.

One can reveal these vortex/anti-vortex vacuum fluctuations by applying an external superflow $\boldsymbol{v}_S^0(\boldsymbol{r})$ to the system. Then for the vortex/anti-vortex pair, *if we ignore completely the effect of the quasiparticles*, we find an interaction energy

$$\mathcal{T}(\boldsymbol{r}_{12}) = \frac{\rho_s \kappa^2}{4\pi} \ln \left| \frac{r_{12}}{\xi_0} \right| - \frac{1}{2} \kappa \rho_s \boldsymbol{v}_s^0(\boldsymbol{r}) \cdot \boldsymbol{r}_{12} \tag{11}$$

where $|\boldsymbol{r}_{12}|$ is the distance between the virtual vortex and the virtual anti-vortex, and ξ_0 is the vortex core radius (the "healing length").

We plot (Fig. 4) the way in which the inter-pair vortex potential $\mathcal{T}(\boldsymbol{r}_{12})$ varies with $\boldsymbol{r}_{12}$ for different external superflow velocities $\boldsymbol{v}_s^0$ (with $\boldsymbol{v}_s^0$ perpendicular to $\boldsymbol{r}_{12}$, so that the Magnus forces from the superflow, acting equally and oppositely on the two vortices, pull them apart). At high T vortex pairs will unbind by thermal

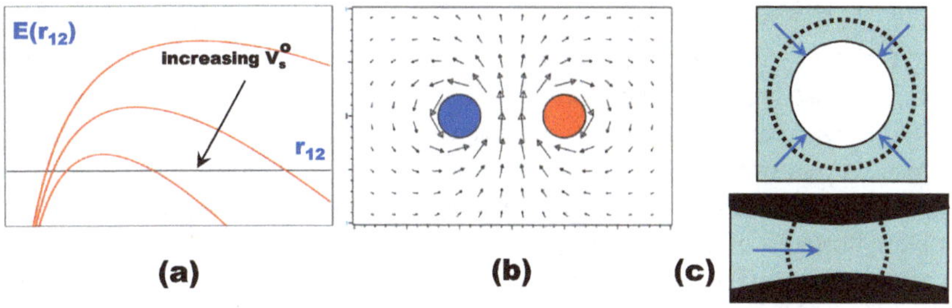

Fig. 4 The form of the tunneling potential $\mathcal{T}(\mathbf{r}_12)$ is shown here in (**a**) for different external flow velocities $\boldsymbol{v}_s^0$. In **b** we show the superflow pattern as the vortex/anti-vortex pair separates. Finally, in **c** we show two other interesting geometries in which one can imagine looking for vacuum tunneling in 2D superfluid films. In the upper figure, superfluid is shown accelerating towards a circular "drain"; in the figure below, through a constriction. The "horizons", defined by the critical crossover velocity $\bar{v}_c(\mathbf{r})$ where vacuum tunneling switches on, are shown by dotted lines; and the superflow directions in (**b**) and (**c**) are shown by arrows

activation over the energy barrier—although we note that as $|\boldsymbol{v}_s^0| \to 0$, the energy barrier blocking this unbinding diverges logarithmically with $|\mathbf{r}_{12}|$, meaning that in reality this unbinding could never occur!

From this argument we see that (i) the Kosterlitz–Thouless thermal unbinding transition is really only physically meaningful when $|\boldsymbol{v}_s^0|$ is large enough—otherwise it will never happen; and (ii) it is not really a thermodynamic phase transition at all, but rather a rapid crossover between two regimes, one in which there are no unbound pairs, and the other where they proliferate. This crossover occurs around a critical "crossover velocity" $\bar{v}_c$. Moreover (iii) traditional Kosterlitz–Thouless theory, which only takes account of thermal fluctuations out of the vacuum, clearly needs to be revised to incorporate the effect of the very large quantum fluctuations in the vacuum.

In any case we see that the crossover around $\bar{v}_c$ is from what (misleadingly) seems to be a quiescent superfluid vacuum state (for low $|\boldsymbol{v}_s^0| < \bar{v}_c$) to a state where the already rather unstable vacuum has now been completely destabilized (at high $|\boldsymbol{v}_s^0| > \bar{v}_c$). In fact, strictly speaking, for finite $|\boldsymbol{v}_s^0|$, there is no vacuum state at all! The system is condemned to remain forever in a non-equilibrium steady state, in which the superflow field $\boldsymbol{v}_s^0$ is constantly pulling virtual vortex pairs from the vacuum, and making them "real". At low T, this happens via vacuum tunneling out of the vacuum state (but we note that vacuum tunneling is constantly taking place in the vacuum state already, between different vortex states). The fact that ^{4}He superfluid can be made in a completely pure form means that this may be the best available laboratory we have for looking at vacuum tunneling between topologically different vacua.

The above considerations all assume a uniform superflow. However, as noted in Ref. [9], one can also introduce a non-uniform background flow field $\boldsymbol{v}_s^0(\boldsymbol{r})$, so that the critical crossover velocity $\bar{v}_c \to \bar{v}_c(\mathbf{r})$, a position-dependent crossover. We can now, if we fall back on analogical thinking, think of the line surface where $\boldsymbol{v}_s^0(\boldsymbol{r}) = \bar{v}_c(\mathbf{r})$ as a "horizon", where the vacuum instability switches on (note that these "tunneling horizons" are quite different from the ones discussed in Sect. 3.2 below, which arise when superflow velocities equal phonon velocities or "slow light" velocities; these are more like "Cerenkov" horizons". On one side of the tunneling horizon the vacuum is, within a broad range of experimental timescales, stable for all practical purposes. On the other side it is unstable, with vortex/anti-vortex pairs being created in profusion. In Fig. 4c we show two geometries in which one can define tunneling horizons of this kind (we note that one can also have a situation where the superfluid flows fast at a boundary, in which case one has "*extrinsic*" nucleation at the boundary [9]).

2.2(b): Vortex Avalanches in 2-d Geometry

As we've seen, one can take an apparently pristine superfluid vacuum at zero temperature, with no free vortices, and destabilize it by applying a static external superfluid flow field $\mathbf{v}_s^0(\mathbf{r}$. However, we have also emphasized that the underlying vacuum state is actually **not** pristine, but is actually brimming with incipient vortex/anti-vortex pairs—it is in fact extremely close to instability even without the external flow. It is then no surprise that the picture of vortex pairs, nucleating via vacuum tunneling independently of each other [9], is actually not a good approximation. A simple argument shows that the real critical velocity for vacuum tunneling ought to be much lower than that estimated above and in Ref. [9]. This is because the presence of vortex\anti-vortex pairs enhances the production of future pairs under the correct conditions.

In the presence of a single vortex, the effective flow velocity $\boldsymbol{v}_s^0$ is altered slightly by the presence of a single nearby vortex. Imagine then that a vortex/anti-vortex pair has just nucleated and we are now watching a new virtual pair trying to nucleate nearby. It is easy to see that the largest effect of the existing pair on the new pair will be if they all lie along a line perpendicular to $\mathbf{v}_s^0(\mathbf{r})$. One then sees that either (a) The barrier height for the new pair is increased, and the tunneling rate for the new pair is reduced, if their orientation is opposite to that of the existing pair; or (b) if they are parallel, the barrier height is reduced, and the tunneling rate increased, thereby lowering the critical velocity for the nucleation of the pair.

By the same argument, interactions between virtual vortex/anti-vortex pairs will lower the critical velocity—if the quantum fluctuations are strong enough, they will lower it to zero and the system will never become superfluid at all (as it is, thermal fluctuations prevent ODLRO from developing at any temperature). In ^{4}He films, the quantum fluctuations are not quite strong enough to destroy $T = 0$ superfluidity, but not too far from it.

This is not the only consequence of the interaction between different vortex/anti-vortex pairs. One should also in principle see "vortex avalanches". This is a kind of **quantum avalanche**, of the kind first described recently for quantum Ising systems [33]. A quantum avalanche is simply a process occurring at sufficiently low temperature that dissipation is very small, and one remains in the "quantum regime". Then, when interactions between quantum nucleation events (quantum fluctuations or vacuum tunneling) occur, an avalanche process of similar nucleations then follows, with very little heating—and so the system remains in the quantum regime. This process can then occur almost without limit, with the system still remaining in the quantum regime, so that the entire process is quantum-mechanical. In experiments on a quantum Ising system [33], quantum avalanche processes involving $\sim 10^{16}$ spins were seen—by far the largest macroscopic quantum tunneling process hitherto seen in the lab.

If the theory is correct, one should certainly be able to see such quantum avalanche processes in 2-d superfluids as well. It will be interesting to look for this experimentally. The critical velocity required to initiate these vacuum tunneling and subsequent vortex avalanche process will be much lower than the value $\sim 6\,\mathrm{ms}^{-1}$ estimated in Ref. [9], because of the interactions between vortex pairs. Detailed calculations will be given elsewhere.

2.3 The Vacuum in 3-D Superfluids

In three dimensions we have a plethora of superfluids: ^{4}He, ^{3}He, various ultracold BECs, and of course, a huge variety of charged superconductors (not to mention the interiors of neutron stars by far the largest contribution to the supply of superfluids and superconductors in the universe). However most of these are not satisfactory as test beds for ideas about quantum vacua. Experiments on superfluid ^{3}He are difficult—few labs current operating can work below 2.7 °mK. Fewer still can do experiments on nanoscopic short-lived samples of cold BEC's at nano Kelvin temperatures. Superconductors are typically dirty and have many other complications. Here we will focus on ^{4}He and BEC's, which have the great advantage of being simple (^{3}He superfluid, with its 18-component order parameter, is endlessly fascinating, and many experiments have been done to investigate analogies with cosmological phenomena. However this lies beyond the scope of this article). So in what follows we first look at bulk ^{4}He superfluid, and then at BEC's.

2.3(a): Bulk ^{4}He Superfluid

One can use the same kind of low-energy effective action for 3-d superfluid ^{4}He as we described for the 2-d film. Interactions are again strong, and the coherence/healing length is basically unchanged. However, this description is less useful here, because the 3d excitations are more complex, and the short-range physics is more important. In the quasiparticle spectrum one now includes phonons, rotons, and bound roton pairs, as well as vortex lines [34, 35]. The rotons and roton pairs have structure in the length scale range 4−20Å; and the vortex lines form "vortex rings", as well as complicated "vortex tangles", which have very complicated configurations. All this physics has been studied exhaustively over the last 85 years.

In 3D ^{4}He, quantum fluctuations are much less pronounced than in 2d; the vacuum is more "classical". We can remove the vorticity completely in experiments, by applying a heat flush (a "phonon wind") to sweep them out. This leaves thermally excited quasiparticles; at very low T, the rotons disappear, because they are gapped, but just as in 2d, we still have gapless phonons.

One can probe the superfluid ^{4}He vacuum using neutrons, or by re-injecting either vortices or quasiparticles. It is quite hard to do the letter in a controllable way in bulk ^{4}He superfluid. Vorticity nucleates at boundaries with sharp edges, and it is hard to do experiments on individual quasiparticles (although remarkable experiments by Wyatt [36] have succeeded). A new technique using femtosecond laser pulses has succeeded in creating dense "fireballs" of roton pairs [15]; and one can produce individual vortex rings by injecting electrons into the superfluid using field emission tips, and then accelerating them in an electric field [21]. Depending on the pressure, the moving ions then reach a critical velocity $\sim$50–60 ms^{-1}, where they either emit pairs of rotons, or a vortex loop is nucleated at the surface of the ion bubble. Such experiments [20, 22] give very direct evidence for the tunneling nucleation of vortex rings in 3d ^{4}He.

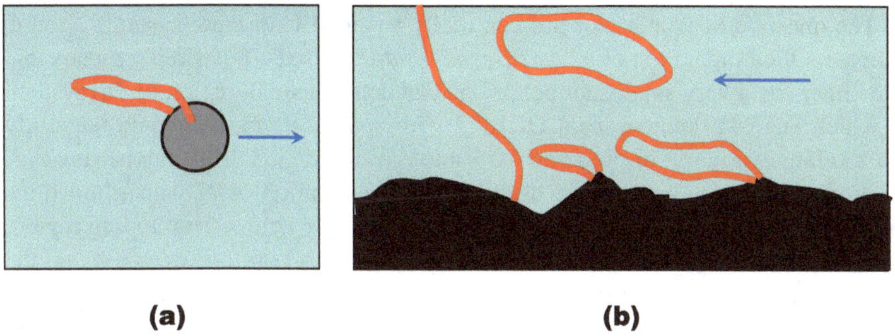

Fig. 5 Some processes that can produce vortices in 3D ^{4}He superfluid. In **a** we see a vortex loop nucleating on the surface of a moving ion (whose motion is denoted by an arrow). In **b** we show vortices nucleating at a boundary, past which the superfluid is flowing. We also show how the vortex loops can escape from the boundary by forming vortex rings. The vortex loops nucleate around boundary regions where the local boundary curvature and the neighbouring superflow velocity are high. Once formed they can migrate along the surface

But it is much harder to understand these experiments (and other vortex tunneling experiments in the bulk) than it is for the 2d films previously described, because the variety of shapes that 3d vortex lines can take makes any comparison between theory and experiment very hard (see Fig. 5).

2.3(b): Cold Gas Bose–Einstein Condensates (BECs)

Since the first real experimental realizations of Bose–Einstein condensation (BEC) in dilute alkali gases in 1995 [37, 38], ultracold atomic condensates have become an important platform for probing macroscopic quantum phenomena. The underlying system consists of alkali atoms in magnetic or optical traps, with interactions, at least in weak coupling, well described by a single s-wave scattering length a_s. Different confinement geometries (1D, 2D, 3D) can be made by optical lattices or highly anisotropic traps.

In an ideal BEC, all the particles share a common wavefunction, giving rise to a coherent matter wave with phase stiffness (i.e., a "winding energy" associated with phase gradients) across the entire sample. In a true BEC, at sufficiently low temperatures (typically in the nanokelvin regime for atomic BECs), a non-zero fraction N_o/N of atoms occupies the ground state of the external trap, representing a striking manifestation of quantum statistics at the macroscopic scale. As discussed above, for weakly interacting BECs, $N_o/N \sim O(1)$. In atomic BECs, one can vary the coupling strength across several orders of magnitude, from very weak (where $na_s^3 \ll 1$, and the healing/coherence length may be up to several microns in size) up to intermediate values, by tuning the scattering around a Feschbach resonance [39]. However at no point can one approach the very strong coupling regime existing in ^{4}He.

The question of whether or not one really has a BEC in these systems is quite delicate. Often the notion of a "*quasi-condensate*" is used—this simply means that one either has a very small sample, or one can define a small region of the system, in which one can define a local ODLRO with density $n_s(\mathbf{r})$, and some hopefully not too large variation of phase across sample or the region. This can be done for arbitrary dimension—indeed we used it in our description of 2D superfluid films above. However in a trap $n_s(\mathbf{r})$ typically varies strongly across the trapping region, and both thermal and quantum fluctuations become larger as one decreases (i) the dimensionality, and (ii) the strength of the interaction [25, 39]. The best approximation to a uniform quasi-condensate comes in 3D with sufficiently strong coupling in "box" potential wells.

The numbers are important here. In typical trap experiments, the healing length $\xi_o \sim 0.1 - 0.2\ \mu\text{m}$, whereas the trap size $L_o \sim 10 - 40\ \mu\text{m}$. Thus $\xi_o/L_o < 10^{-2}$ in these experiments. In harmonic traps, where $n_s(\mathbf{r})$ varies strongly across the sample, one is dealing with a fundamentally non-uniform system. However in box potentials, with $\xi_o/L_o \ll 1$, the quasi-condensate can be treated as uniform except near the boundaries.

In any case, in the same way as described in the previous section on 2D ^{4}He films, one can use a low-energy effective action for BEC quasi-condensates. The long-wavelength degrees of freedom are again phonons quasiparticles. As one increases the coupling strength, the quasiparticle dispersion begins to turn over [40, 41], but at no point do we get anything like the spectrum seen in strongly-coupled ^{4}He superfluid.

Under appropriate circumstances (not too strong or not too weak coupling, 3 dimensions), the transition to quasi-condensation can be captured by the Gross–Pitaevskii equation (2), which encodes both the mean-field quasi-condensate and its collective excitations [42, 43]. In these systems, quantum fluctuations are small, and finite traps can strongly suppress thermal fluctuations.

The situation is less clear for lower dimensions, or for very strong coupling (and in the very weak coupling limit, the requirement that $\xi_o \ll$ system size breaks down). In uniform 2D Bose fluids, there is no condensate. In a finite trapped 2D system, a KT transition can be seen. Below T_K one can try to argue that a quasi-condensate exists with well-defined phase coherence across the sample [25, 44]. This argument is the same as that used to argue for "local ordering" (but not long-range order) in any 2-D system without long-range forces. Note that in 2D we do **not** expect the fluctuations to be small unless the system itself is so small that the local order parameter is locked in by boundary conditions.

In later sections we will discuss how, under specific conditions, these properties of dilute BECs have made them a candidate for analogue quantum field experiments.

2.4 Quantum Mechanics and Gravity

We now move to "the real thing", i.e., the deepest problem currently facing physics. This is the problem of reconciling two apparently incompatible theories (GR vs. QM/QFT) into a new third theory. We look at this problem from several points of view in Section V. Here, we just discuss what one can say about the vacuum in GR and QM/QFT separately.

2.4(a): The Vacuum in Classical General Relativity

The metric $g^{\mu\nu}(x)$, with 10 independent components, is far more complex than anything we have looked at up until now (with the exception of superfluid ^{3}He). In spite of attempts [45] to make mappings between ^{3}He superfluid, with its 18-component order parameter, and a theory in which the standard model is added to the 10-component metric field, these two systems are vary different, and we shall see that this analogy fails. Let's first discuss key features of the vacuum state in GR.

The metric tensor configuration is sourced by the stress-energy tensor $T_{\mu\nu}$ in Einstein's field equation,

$$G_{\mu\nu}(x) = \kappa T_{\mu\nu} \tag{12}$$

where $\kappa = 8\pi G/c^3$ and $G_{\mu\nu}(x) = R_{\mu\nu}(x) - \frac{1}{2}R(x)\, g_{\mu\nu}(x)$, with $R_{\mu\nu}(x)$ the Ricci tensor. Many thousands of solutions are known to this equation [46]; it is certain that these known solutions constitute a miniscule fraction of the total set of solutions. Those solutions that are discussed in cosmology and in black hole theory are highly symmetric and extremely simple. Cosmological models look almost exclusively at homogeneous solutions (whereas a generic solution of (12) involves 6 functions of 2 variables and 4 functions of 3 variables). Black hole theory often only discusses three solutions, viz., the Schwarzchild, the Reissner–Nordstrom, and the Kerr-Newman solutions, these being the unique stationary solutions.

In spite of this, some rather general results are known. In the absence of any matter, we know that singulariteis must exist somewhere, and it is generally believed that these will take the BKL or Mixmaster form [47, 48], with metric such that

$$ds^2 = -c^2dt^2 + R^2(t) \sum_{i,j=1}^{3} e^{\beta_{ij}} \sigma_i \sigma_j \tag{13}$$

where $R(t) = R_0 a(t)$, with R_0 a constant and $a(t)$ a scale factor (so that $a^3(t)$ ~volume), and the differential forms $\{\sigma_j\}$ satisfy $d\sigma_i = \epsilon_{ijk}\sigma_j \wedge \sigma_k$ and parameterize rotations on the 3-sphere. Diagonalizing β_{ij}, we find $\sum_j \beta_{jj} = 1$, and so the dynamics can be written in terms of a "$\boldsymbol{\beta}$−particle" moving in a 2-dimensional $\boldsymbol{\beta}$-plane, spanned by orthogonal vectors $\beta_\pm$ such that $\beta_+ = -\beta_{33}$ and $\sqrt{3}\beta_- = \beta_{11} - \beta_{22}$. As Misner observed [48], the dynamics then assumes the BKL form, with an infinite sequence of chaotic oscillations accumulating as the singularity is approached (during which $a(t) \to 0$). This behaviour is most easily understood by noting that the $\boldsymbol{\beta}$-particle is confined to a triangular potential well, with very steep-sided walls, in the 2-dimensional $\boldsymbol{\beta}$-plane. It bounces off these walls, which themselves are moving outwards. This mapping of the dynamics of the shape of the 3-sphere to the motion of the $\boldsymbol{\beta}$-particle is very helpful; one sees that the 3-sphere suddenly distorts to a new shape each time the $\boldsymbol{\beta}$-particle bounces off one of the walls.

None of this behaviour can be described in any way by phenomena in superfluid ^{3}He (where singularities in the flow field, like the "boojum" in ^{3}He-A, are approached smoothly, with no oscillations at all). The point of course, is that no field theory, in condensed matter or anywhere else, has the attractive self-interaction which exists in the spacetime metric.

There is a further twist to this story, the subject of current research [49]. This is that if we add matter fields to the background spacetime, the dynamics in the $\boldsymbol{\beta}$-plane changes again. Bosonic fields simply hasten the approach to the singularity, but fermionic fields have a negative zero-point energy [50], and if one adds a combination of bosonic and fermionic fields, one finds that the formation of a singularity is prevented, and the spacetime contraction reverses to an expansion.

In any case, even without the quantized matter fields, the vacuum of GR is very peculiar. The presence of any spacetime curvature leads to a non-localizable source of stress-energy, which if it is allowed to grow, or is otherwise focussed (e.g., in the collision between two gravitational shock waves), will "blow up" into a singularity. From a physical standpoint, this is why GR, when quantized, is non-renormalizable: a single gravitational wave or gravitational pulse with energy E_1 will create an attractive potential for another one of energy E_2, and the attraction diverges as $E_1, E_2 \to \infty$, and is $\sim O(1)$ if $E_1, E_2 \sim E_p$, the Planck energy.

Thus the vacuum in GR is already unstable for reasons having nothing to do with quantum mechanics at all.

2.4(b): The Vacuum in Quantum Field Theory

Since condensed matter media are already described by QFT's, many of the features we have seen, notably those connected with vacuum fluctuations, occur in the same way in relativistic QFT's. However, there are some key differences which are crucial when we come to discuss Quantum Gravity (Sect. 5). The most important is the existence of UV divergences. In renormalizable QFT's these are regularized, and any divergent zero-point energy is eliminated. However this option is not open to us when incorporating QFT into some quantum gravity theory, or even when using QFT on a background curved spacetime—see below.

Such UV divergences do not exist in QFT's for condensed matter systems, because, as already emphasized above, condensed matter QFT's are effective field theories with built-in UV cut-offs. Typically these cut-offs arise in a natural way, at energies where the effective $\mathcal{H}$ breaks down as new structures and new physics come into play. Note that exactly the same point can be made about relativistic QFTs. Any

QFT, whether it is for a non-relativistic superfluid, for ordinary QED or for some attempt at a full quantum-gravitational theory, has some built-in UV cutoff: if this does not appear naturally, it is imposed theoretically in some way. The vacuum state, in the theory, is adjusted accordingly—this is just as true for a relativistic QFT as it is for our discussion of a H atom in a box, at the beginning of this section.

However what is not open to choice is the existence of divergent zero-point vacuum energy contributions in quantum gravity. This is because they must contribute to the cosmological constant [50]. The contribution is $\sim\Lambda_o^4$, where Λ_o is the UV cutoff. A proper discussion of this must include fermionic as well as bosonic quantum fields—crucially, the fermions give a negative contribution, and counteract the attractive contribution of gravitational and bosonic fields. In work to be published elsewhere, we show how we can use this result to give a non-divergent theory [49].

3 A Dialogue About Analogues in Physics

We have seen that in discussions of the vacuum in condensed matter systems, one can adopt rather different views about the value of analogies between phenomena in physically very different systems. In discussions between two of us (PCES and ST) it has become clear that it would be interesting to air the arguments for and against the use of such analogies. We shall do this in a somewhat Socratic form, in the form of a dialogue between the two of us. The sceptical or "pessimistic" point of view will be adopted by PCES, and the optimistic antithesis to this view by ST.

Let's begin by saying roughly what we mean by an analogy in physics. Suppose we imagine two quite distinct physical systems, S_{A} and S_{B}, and two theories T_{A} and T_{B} describing each (See Fig. 6). We also imagine that we are going to observe, in experiments, some set P_{A} of phenomena in S_{A}, and try to infer from these observations, what may be seen of some other putative set of phenomena P_{B} in system S_{B}. Usually this kind of reasoning is employed when, for some reason or another, one cannot (at least at the time), do experiments or make observations on system S_{B}. So one is either trying to guess or infer what might be seen on S_B. Alternatively, one might be trying to confirm some ideas about S_B by observations of P_A made on S_A.

We observe that this reasoning involves the existence of some mapping $\mathcal{M}_{AB}$—more or less well-defined—between a theory T_{A} describing S_{A} and another theory T_{B} describing S_{B}. Since the two theories are usually very different, this mapping will almost never be a 1-1 mapping. In fact it is often the case that we instead devise a mapping/analogy between some model t_{A}, which is an approximation of some sort to T_{A}, another model t_{B} which is a similar adaptation to T_{B}.

We also observe that the interpretation $\mathcal{I}_A$ of any experiment/observation on some phenomenon P_{A} is always very complicated. It not only will involve the theory T_A and the model t_A in an essential way, but also a large number of auxiliary theoretical assumptions about the experiment or observation—the way one interprets any experiment or observation always comes "laden with theory".

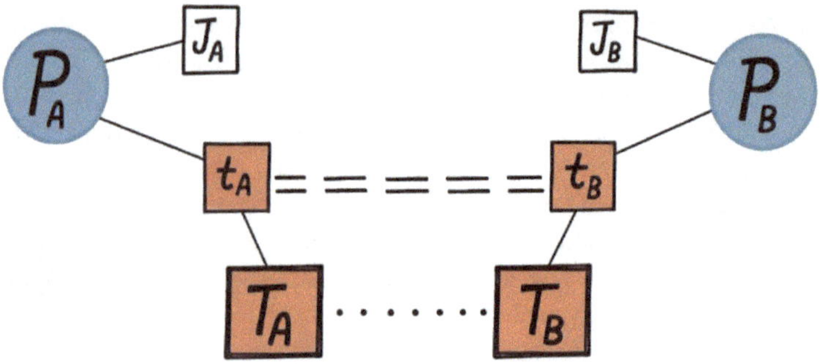

Fig. 6 Our rough model of how analogues work in physics. We imagine two quite different physical systems S_A and S_B, in which phenomena P_A and P_B are supposed to occur. We describe S_A and S_B using theories T_A and T_B; there is assumed to be some sort of (partial) mapping $\mathcal{M}_{AB}$ between the two theories. We also assume subsidiary theories t_A, t_B which are supposed to model P_A and P_B; they are typically simplifications of or approximations to T_A and T_B. Often, instead of a mapping $\mathcal{M}_{AB}$ between T_A and T_B, one makes a mapping $\Updownarrow_{AB}$ between t_A and t_B. Finally, experiments or observations on P_A and/or P_B require interpretive frameworks $\mathcal{I}_A$ and $\mathcal{I}_B$ to make the connection between what is seen and how it is to be understood. See text for further explanation

All of this is summarized in Fig. 6. The idea of the analogue is to then argue that the observation of some phenomenon P_{A} in system S_{A} provides evidence that an analogous phenomenon P_{B} will occur in system S_{B}.

In this paper we have so far discussed two systems for which analogues can be developed in some detail, viz., (i) the analogy between vortex/anti-vortex tunneling in 2-d superfluids, and vacuum tunneling in either QED or quantum gravity; and (ii) the analogy between the propagation of phonons in superfluid BECs, or of light in optical media, and the propagation of light or other excitations in curved spacetime near, e.g., black holes.

In what follows we first describe how each of these analogues work, and then give a critical discussion of the general question using these examples.

3.1 Example A: 2-D Superfluids and QFT Analogues

We begin with the analogy already mentioned, between the Schwinger effect and vortex/anti-vortex unbinding. The Schwinger effect was originally discussed before modern QFT [51], when it was found that strong electric and/or magnetic fields would induce effective photon-photon interactions in the QED Lagrangian. In 1951 Schwinger [4] wrote a remarkable paper—the first non-perturbative calculation in modern QFT—which showed how the QED vacuum would become unstable in strong fields to electron-positron pair production. In related work Dyson [52] pointed out that QED had an essential singularity, as a function of complex coupling constant α, with a branch cut extending away from the origin along the negative α axis.

It is immediately obvious what is the parallel between vortex tunneling and the Schwinger effect. In fact, in (2+1)-D QED with no topological terms in the action, the form of the electron-positron potential is precisely the same as $\mathcal{T}(\boldsymbol{r}_{12})$, if we substitute a static electric field $\boldsymbol{E}^0(\boldsymbol{r})$ in place of $\boldsymbol{v}_s^0(\boldsymbol{r})$, and the 2-D logarithmically attractive e^-/p^+ interaction between charges for the same interaction between vortex pairs. This seems like a very straightforward and successful analogy between two theories T_1 and T_2, and two different vacuum tunneling phenomena P_1 and P_2 in the two different systems. So, have we found a successful and useful analogy here?

The current theory of the Schwinger effect is confined to 1-loop and 2-loop expansions, both in $3+1$ dimensions and in $(2+1)$-D QED without topological terms. In the latter case the 1-loop result for vacuum tunneling is [53]

$$\Gamma_{2+1}^{\mathrm{QED}} = \frac{(eE)^{\frac{3}{2}}}{4\pi^2}\sum_{n=1}^{\infty}\frac{1}{n^{\frac{3}{2}}}\exp\left\{-\pi n\frac{m^2}{eE}\right\} \tag{14}$$

We see immediately that there are 2 effects contained in the superfluid film calculation that are not included in this result, viz.,

(a): One can assign an effective mass to the vortex pair dynamics as a function of the separation coordinate $\boldsymbol{r}_{12}$; this is given by $M_V^{\mathrm{pair}}(\boldsymbol{r}_{12}) = \left(\rho_s\kappa^2/4\pi a_0^2\right)\ln|r_{12}/\xi_0|$, which increases as the vortices separate. Physically, this happens because as the vortices separate, their mutual dipolar flow field extends over ever larger regions, and the resulting inertial mass increases accordingly. The analogous effect in the QED calculation would have to include radiative corrections to all orders—i.e., it would need to go to infinite order in a loop expansion. One would have to carefully preserve all the Ward identities in this calculation.

(b): When we couple the tunneling vortex/anti-vortex pair to the quasiparticles, two things happen. First, there are radiative corrections to the vortex effective mass; and second, the vortex/quasiparticle coupling, even at $T = 0$, causes dissipation, thereby reducing the tunneling rate and modifying its dependence on the background $\boldsymbol{v}_s^0$. This mechanism, which involves IR divergent processes, is not in the Schwinger calculation at all.

There are also 2 key features not included in either the Schwinger calculation [53] or our recent superfluid vortex tunneling calculation [9]. The first is simply the emission of photons/quasiparticles off to infinity caused by the appearance of the new pair excitations. The second is the effect of interactions between two different vortex/anti-vortex pairs (a 4-way interaction) or between single vortices and a vortex pair (a 3-way interaction). As noted in Sect. 2.2(b), such interactions must be included in any proper calculation, and will lead to important physical processes, including "quantum avalanche" production of a dense interacting vortex/anti-vortex system, which will show quantum turbulence (recall Sect. 2.2(b) above).

Of course if we are really to see how good the QED/superfuid analogy is, experiments will have to be done, both in superfluid films (to see the intrinsic vacuum tunneling process) and in QED vacua (to see the Schwinger effect)! It is probable that the superfluid vacuum tunneling will be seen first. Note that in real vacuum QED, the electron-positron-coupling is weak, so that some of the effects that are crucial in vortex tunneling may be quite small in any experiment on the Schwinger effect, whenever this may be done. If history is any guide, the experiments will certainly be more complicated than the theory, and will bring surprises. One must never forget that the real QED vacuum and the real 2d superfluid vacuum are **not at all the same**!

Similar remarks can be made about phenomena in superfluid films and putative analogues in the behaviour of quantum fields in curved spacetime (we noted in Ref. [9], and reiterated above, how one could simulate processes around a black hole in superfluid films—cf. Fig. 4c). In this analogue, the critical surface $\bar{v}_c$ around which vortex/anti-vortex vacuum tunneling occurs is the analogue of the black hole horizon around which Hawking radiation is supposed to be seen by an external observer. If we pursue this analogy then we can call this a "vacuum quantum tunnelling horizon". Note that this is not at all the same as the acoustic horizons discussed for BEC analogies with phenomena in curved spacetime (see below), which are entirely classical.

3.2 Example B: BECs and Analogue Gravity

We now consider possible laboratory tests of quantum field theory in curved spacetime (QFTCS), focussing on "analogue gravity", which is a framework to study analogues of general relativistic geometries within condensed matter, fluid, and optical systems. The central idea is that as far as the purely kinematic characteristics of QFTCS are concerned, one can to some extent overlook the structure of the underlying system and work with an effective metric, allowing one to apply purely geometric concepts from GR. One then maps structural properties of QFTCS such as the amplification of fluctuations, the creation of quasiparticles, and mode mixing across horizons, onto excitations in various analog systems.

Analogue models in physics have a long history; as an early example, one can refer to Walter Gordon's 1923 formulation of an effective metric for light propagating in moving dielectrics [54], and similar analogical approaches have appeared in various areas of physics since then. In the modern sense, the foundation of the field of analogue gravity dates from Unruh's seminal paper in 1981 [55], in which he shows that sound waves in an irrotational, inviscid fluid propagate according to an effective Lorentzian "acoustic metric", obtaining a "sonic horizon" if the fluid flow passes from subsonic to supersonic, and phonon emission from that horizon mimics thermal Hawking radiation.

Formally, when one linearizes the equations of motion for small perturbations in a suitable background, the quadratic action for the perturbation field ϕ takes the following universal form:

$$S^{(2)} = \frac{1}{2} \int d^4x \sqrt{-g_{\text{eff}}}\, g_{\text{eff}}^{\mu\nu} \partial_\mu \phi \partial_\nu \phi, \tag{15}$$

so that ϕ obeys a Klein–Gordon equation in an effective metric:

$$\Box_{g_{\text{eff}}}\phi = \frac{1}{\sqrt{-g_{\text{eff}}}}\partial_\mu\left(\sqrt{-g_{\text{eff}}}\,g_{\text{eff}}^{\mu\nu}\,\partial_\nu\phi\right) = 0. \tag{16}$$

Hence, phonons in a flowing fluid can be mapped to a quantum field in a curved spacetime, provided that the effective geometry of that curved spacetime can be described by the Painleve–Gullstrand–Lemaitre type metric [56].

Building on this, Jacobson in 1991 emphasized the deep connection between horizon thermodynamics and the underlying microphysics, introducing the idea that Einstein's equations themselves may be viewed as an equation of state [57], and while not experimental, his work motivated the analogue program by showing that the thermodynamic and the causal-structure aspects of horizons are central, not the full Einstein dynamics. One of the most prolific voices in the field of analogue gravity, Matt Visser, in the 1990s, derived the general form of the acoustic metric for a wide class of fluids, showing how analogue models could mimic black hole, cosmological, and rotating spacetimes [58]

Multiple research works by Schützhold and Unruh extended the scope of analogue gravity even further, showing that horizons and Hawking-like radiation can also be realized in different fluid [59]. In these works the importance of dispersion relations and mode mixing near horizons have been highlighted, and it has been clarified under which conditions analogue systems reproduce Hawking-like particle creation. From a broader perspective, as Hawking and Ellis had already explained [60] in 1973, much of the predictive power of GR derives from its causal structure: the light cones, horizons, and the associated kinematics. This insight shed some light on why analogue systems are valuable and how they imitate the causal kinematics of fields in curved spacetimes.

Comprehensive reviews (especially Barceló, Liberati, and Visser [58], and more recently by Ralf Schützhold [61]) have framed analogue gravity as a kinematics-not-dynamics correspondence, highlighting the idea that excitations in condensed matter and optical systems propagate according to effective metrics. Analogue systems are valuable precisely because they can imitate these causal and kinematical aspects of curved spacetimes, even though they do not reproduce the full dynamical content of Einstein's equations. They thereby allow laboratory studies of phenomena such as Hawking radiation, cosmological particle creation, or superradiance under controlled conditions, while leaving the problem of quantum gravity dynamics untouched. In the following we consider a few examples of these systems, noting that it would not be possible to give all of the references here.

Bose–Einstein condensate as a probe of quantum fields and gravity: Bose–Einstein condensates (BEC) are one of the laboratory analogues for relativistic quantum fields, where, their low energy collective excitations can be mapped to an effective Klein–Gordon field in curved spacetime under the right circumstances.

Using the Madelung split of the condensate wave function ψ as $\psi = \sqrt{\rho_0}\exp\{iS/\hbar\}$, it has been shown that in the long wave limit one can find the analogy between sound waves in Bose–Einstein condensates and mass-less and minimally coupled scalar fields in curved space-times [61]

In some of the recent analogue gravity tests in particular in proposals such as the interferometric Unruh detector schemes of Gooding et al. [62] it is common to employ a quasi-two-dimensional Bose–Einstein condensate rather than a full three-dimensional one. The reason is that in a 2D geometry, in the light-matter interaction the effective overlap between probe light and condensate is stronger and easier to control, and the excitations behave more like 'purely scalar' modes, with no transverse modes and no polarization structure. In these models, 2D BECs are realized by applying a strong trapping potential along one direction, creating a 'pancake' condensate [44], and to probe it with laser beams, one can send light through the plane and measure density/phase fluctuations directly (see also Ref. [63]).

In these recent proposals, it is shown how through a mapping of the Gross–Pitaevskii dynamics to an effective Klein–Gordon field, one can examine phenomena such as the Unruh and Hawking effects, vacuum decay, or entanglement harvesting. For the sample of a 2D BEC obeying the two-dimensional Gross–Pitaevskii field equation, one can write small fluctuations of this field as [64]:

$$\psi = \psi_0 + \Psi. \tag{17}$$

Then, defining the background density of the BEC as $\rho_0 = \psi_0^2$, and choosing $V + g\rho_0 = 0$, one can rewrite the corresponding action in terms of real and imaginary parts as:

$$S = 2\int \left[\Psi_R^\dagger \partial_t \Psi_I - \frac{1}{4M}\left[(\nabla\Psi_R)^2 + (\nabla\Psi_I)^2\right] - g\rho_0\Psi_R^2\right] d\sigma dt, \tag{18}$$

where considering long wave limit for the dispersion relation, one can obtain the corresponding Lagrangian mimicking Klein Gordon field:

$$L_{\text{BEC}} = \frac{1}{c_s^2}\left(\partial_t \frac{\Psi_R}{\sqrt{M}}\right)^2 - \left(\nabla \frac{\Psi_R}{\sqrt{M}}\right)^2 \tag{19}$$

where $c_s = \sqrt{g\rho_0/M}$ is the effective light velocity as before; and $\phi(t, \boldsymbol{r}) = \Psi_R(t, \boldsymbol{r})/\sqrt{m}$ is identified as the effective (2+1)dimensional Klein-Gordon field (see [63]. Through this scheme, the condensate functions as a laboratory analogue of a relativistic quantum field.

Employing a BEC as an analogue system for different spacetimes was originally proposed by Fedichev and Fischer in 2003 [65], who showed that an expanding BEC can mimic a de Sitter universe, with the excitations obeying the same equations as fields in curved spacetime, in which the effective phonon metric takes the form:

$$ds^2 = -c_s^2 dt^2 + a^2(t) d\mathbf{r}^2, \tag{20}$$

with scale factor $a(t)$ determined by trap dynamics. They predicted analogue Gibbons–Hawking radiation with temperature:

$$T_{\text{GH}} = \frac{\hbar H}{2\pi k_B}, \qquad H = \frac{\dot{a}}{a}, \tag{21}$$

emerging as thermal phonon emission from the expanding condensate.

Shortly thereafter, A. Retzker, J. I. Cirac, M. B. Plenio, and B. Reznik, in their 2008 proposal [66], modelled the embedding of accelerated detectors in BECs to probe thermal radiation. They assumed a UDW detector model, with an interaction Hamiltonian:

$$H_{\text{int}} = \lambda(\sigma^+ + \sigma^-)\phi(t, \mathbf{r}(t)), \tag{22}$$

where ϕ is the phonon field along the accelerated trajectory $x(t)$, and showed that the transition probability per unit time is thermal. Hence, they recovered the Unruh temperature by verifying Einstein's detailed balance condition between excitation and de-excitation rates of their detector, derived from the BEC correlation functions.

Gooding, Biermann, Erne, Louko, Unruh, Schmiedmayer, and Weinfurtner in their theoretical proposal extended this idea to interferometric Unruh detectors, where a continuous laser probe acts as a quasiparticle detector. They showed that a continuous quasiparticle detector in the form of a laser beam interacting with a BEC can cleanly distinguish the Unruh effect from other sources of noise [62]. Their detector's response function was expressed through the Wightman function of the BEC phonon field:

$$\mathcal{F}(\omega) = \int d\tau\, d\tau'\, e^{-i\omega(\tau-\tau')} G^+(\tau, \tau'), \tag{23}$$

with

$$G^+(\tau, \tau') = \langle \phi(\tau)\phi(\tau') \rangle. \tag{24}$$

These theoretical ideas connect with a broader analogue-gravity experiments. For instance, the Technion group (Steinhauer, 2016) observed Hawking-like correlations across an analogue horizon in a BEC [67]. Their relevant observable is the density–density correlation function:

$$G^{(2)}(x, x') = \langle \delta n(x), \delta n(x') \rangle, \tag{25}$$

showing that entangled quasiparticle pairs are consistent with analogue Hawking emission. Furthermore, Jaskula et al., in 2012, reported the first experimental realization of an acoustic analog of the dynamical Casimir effect in a Bose–Einstein condensate (BEC), where they observed quasiparticle pair creation in a time-modulated condensate. The mode occupation followed a squeezed-state spectrum:

$$n_k = \sinh^2 r_k, \tag{26}$$

where the squeezing parameter r_k is set by modulation amplitude and frequency [68]. Together, these experiments establish BECs as robust analogues of quantum fields in curved spacetime.

BECs have also been suggested as analogues for false-vacuum decay by Jenkins et al. [69]. In their theoretical proposal, they modeled bubble nucleation and expansion by tunneling through an effective potential barrier in the non-linear Schrödinger dynamics. Recently, Mann and Gooding have shown how a BEC vacuum can "harvest" entanglement between two detectors by transferring nonclassical correlations from the condensate to a pair of pulsed laser beams, which highlights the potential use of BEC not only for analogue gravity tests but also for quantum information protocols [70].

Hence, BECs have been used in multiple scenarios to investigate the structure of the vacuum. Taking the approach of mapping Gross–Pitaevskii dynamics to effective Klein–Gordon fields in curved spacetimes, one can examine phenomena such as Unruh and Hawking effects, vacuum decay, and entanglement harvesting. In this way, BECs allow us to test, within the laboratory, features of quantum field theory and quantum gravity that would otherwise be inaccessible.

3.3 Dialogues: For and Against

It will be clear that there is a divergence of opinion between two of us on the use of analogies between phenomena in condensed matter and QFT/gravity. One of us (PCES) is very suspicious of their use, and the other (ST) is an enthusiastic proponent of them. We feel that this is an important question, which only came up in discussions between the two of us while we were finishing the article. The best way we could think of to air the various issues is in the form of a "Socratic dialogue". We caution the reader that we have had very little time to prepare this, and we do not think that what is written here is anywhere near the last word on this topic. All we are doing is introducing some of the issues.

(A) The Pessimistic View (PCES)
Unfortunately any reasoning in favour of using analogies is only as good as the links in the chain between the different components in Fig. 6, extending from P_{A} to P_{B}, as well as the quality of the components themselves. It is usually the case that the "mapping" between T_{A} and T_{B} is imprecise and incomplete, and it is never 1-1. The theories T_{A} and T_{B} are themselves often incomplete, and have their own imprecisions, and the connections that one tries to establish between T_{A} and S_{A}, or T_{B} and S_{B}, usually leave a lot to be desired. A real physical system is almost always far more complex than the theory being used to describe it; T_{A} is usually a crude idealization of S_{A}. Moreover—a point often neglected in philosophical discussions—experiments are never simple, and always come "laden with theory". Thus, unless one is very careful, it is easy to draw false or misleading conclusions about phenomena

$\{P_B\}$ which one thinks might exist in S_B, based on phenomena $\{P_A\}$ which might have been observed in S_A.

Consider how things turned out for some of the analogies mentioned earlier:

(i) Maxwell's analogy [72] between the EM field and mechanical systems turned out to be helpful to him in visualizing the field and its consequences. It was actually a pretty good analogy in that the two theories matched each other rather well. However from a physical point of view it turned out to make no sense at all, because of the principle of special relativity; and it was quickly abandoned. A related example, starting even earlier, began from the apparently identical behaviour of waves in fluids and light waves in vacua (including interference phenomena and diffraction). Use of this analogy began with Huyghens in 1690, in his debate with Newton over the nature of light: he argued that empty space was filled with a fluid-like "aether", through which light propagated [73, 74]. This argument of course turned out to be completely wrong (as were Newton's 1604 counterarguments!), although Huyghens's ideas led to a large number of predictions (e.g., 2-sit interference) that were confirmed. One sees that any analogy between condensed media and spacetime is fraught with pitfalls. There are many more examples like this, some not so well known—in writing the history of physics, we tend to forget the theories that failed.

(ii) In the same way, the analogies made since at least the 1940s between curved spacetime in GR and different condensed matter media can look rather awkward. The spacetime metric $g^{\mu\nu}$ has 10 °C of freedom, and no analogy with, e.g., ideal Eulerian hydrodynamics or with simple superfluids (using, e.g., Clebsch theory) can possibly hope to capture this. Much has been made of the analogy with ^{3}He superfluid, which has an 18-component order parameter—but again, the analogy is very misleading (imagine trying to simulate the approach towards a singularity with a superfluid—all of the BKL/mixmaster complexity would be completely absent!). Current attempts to match ideas in string theory with the "landscape" picture of spin glass theory seem even more far-fetched.

Of course one can employ condensed matter/curved spacetime analogies in a much more limited way. The most commonly used analogy involves optical media and curved spacetime, employed by many since 1920. The best-known use of this was by Penrose, to understand the Raychaudhuri equation and the focusing of null vectors, and to derive his famous results on trapped surfaces and singularities [75, 76]. Here one is analyzing smooth classical spacetime far from any singularities, and using the analogy with a smoothly varying optical medium in the long-wavelength regime. In the same way, Starobinsky and Zeldovich, in their work on acceleration radiation [3] (work which then led Hawking to Hawking radiation), were inspired by an EM analogy.

(iii) One analogy that worked rather spectacularly, in a predictive way, was that leading to the Anderson–Higgs boson. By extending the 1957 BCS theory to include long-range Coulomb interactions, both Bogoliubov [77] and Anderson [78] understood in 1958 that these interactions ought to gap certain collective modes in real s-wave superconductors (this prediction was later experimentally verified). Anderson then argued [5] in 1963 that if a similar model applied in particle physics, then one should see the same gapped excitation. In 1964 Higgs took this same model, and applied it directly to make the same prediction [79] for what later became the Anderson–Higgs boson. In the much more complicated "standard model" of particle physics this boson survived. The experimental verification at CERN of the existence of this boson had to wait nearly 50 more years. This story had a happy ending, but as we've just seen, many similar stories do not.

Let us also consider the 2 examples we discussed in detail above.

(a) Vortex tunneling/QED analogy: this is the analogy between vacuum tunneling in 2D superfluids, and the same tunneling in the QED Schwinger effect. At first glance this seems like a fruitful analogy—it has already indicated that the original Schwinger calculation, and subsequent work on it, may have missed out several key effects. It will be interesting to see how the dialogue works between the 2 calculations in the future. But my expectation is that it will not work very well. We are dealing with two very different vacua, and the more one probes them, the more these differences will become obvious. Thus one should be very careful with this analogy, and try to avoid circular argument. My feeling is that while the analogy is interesting, it probably will not lead to new physics.

(b) BEC/curved spacetime analogy: In this case I would argue that if one treats the background superfluid as classical (which is what is being done here), then all that is being done is to compare two classical mathematical models. Thus, e.g., experiments on phonons or classical sound waves in flowing fluids are just testing whether well-known classical models of fluid flow match the classical theory of Hawking radiation, and they tell us nothing about how real Hawking radiation would behave around a real black hole. In fact all they do is confirm that the simple classical causal structure outside black hole horizons is the same as the causal structure in the analogue classical fluid or classical optical system. But these have been deliberately contrived to have the same causal structure! Thus all that has been done is to construct a tautology.

Having looked at all these examples, I would now like to ask a key question, viz,

Question: *Suppose we have an analogy between two theories T_A and T_B, and we then do experiments on the supposedly analogous phenomena P_A and P_B, and we find that the analogy* **does not work** *—that predictions made about P_B on the basis of the known phenomenon P_A turn out to be wrong. What then?*

What *ought* to happen, of course, is that the analogy is then rejected. Unfortunately, as history all too frequently shows, this is **not** what happens. Very often we see a quite different behaviour, in which experiments that do not confirm theory are not published. If they are published, they are either ignored, or attempts are made to dismiss them. The criterion for a good experiment then becomes whether or not it agrees with the theory (or in the present case, whether it confirms the analogy).

This is of course a very dangerous tendency; the dialogue between theories then becomes a "dialogue des sourds", in which circular arguments on both sides prevail.

Examples of this are all too frequent. Without going into detail, I just mention several of them:

– the history of experiments on the Kondo effect in the 1960s and early 1970s, in which every new version of the theory led to experiments which agreed with it. As Anderson later remarked, the agreement with each theory was just as good as the last, even though the theoretical predictions and the experiments changed each time! Anderson's recommendation was that the experimental groups simply ignore the theory. But he also remarked that the way the scientific system works made it very hard for these groups to do this—it is much easier for experimenters to get recognition and funding if one simply confirms theory.

– the lamentable history of experiments on strongly-correlated 2D and quasi-2D conducting systems. This has led repeatedly to notorious scandals (e.g., the "Batlogg-Schoen" scandal [80], and numerous more recent scandals) over experimental data purporting to show high-T_c superconductivity, or other exotic states. Common to all these scandals is the suppression of data not supporting the desired theoretical picture, or the outright fabrication of data designed to support some theory. The situation has become so bad in this field that physicists can now unmask such work without necessarily losing face—so this black cloud may have a silver lining.

– the even more lamentable story of multiple attempts to show that different quantum computation systems work in the lab, in which the very real effects of decoherence are simply ignored (or treated theoretically using inapplicable models of classical noise).

– the history of "confirmations" of the Kibble-Zurek (KZ) defect nucleation phenomenon. Initial experiments on superfluid ^{4}He (for which the predictions were made) initially confirmed the predictions, but it was then found by the same group [81] that their initial data were incorrect—the correct results showed that the predictions were quite wrong. One can only salute the honesty and courage of the experimenters here. On the theory side these experiments have been largely ignored, or dismissed using completely spurious arguments.

I have little doubt that most readers of this article will be familiar with similar stories. It is unfortunate that they are usually suppressed—for the most part only assiduous science journalists are willing to expose what is really going on in many of these cases.

The consequences can be really bad for the science, even when all parties are being honest. If physicists had spent less time in the century before 1905 trying, in all good conscience, to justify the "aether analogy", they might have come much earlier to special relativity. In my view the very large effort being made to force fit experiments on BECs and similar systems into analogies with quantum gravity, is not doing either field much good. The study of BECs—which are very interesting systems in their own right—is in no way justified or enhanced by showing that they might look vaguely like curved spacetime, and as far as I know, no results obtained on BECs have had any impact on work in classical GR. In the same way, vortex vacuum tunneling in 2D superfluids is not made more interesting by a possible analogy with vacuum tunneling in QED. 2D superfluids are spectacularly interesting systems in their own right (and very likely behave differently from QED).

To make this clear, let's return to the question above, and ask what should be the correct reaction of the physics community if experiments give different results from theory for any of the following 3 analogies (i) the 2D vortex/QED Schwinger analogy; (ii) the analogy between phonon propagation in BECs and classical GR; and (iii) the analogy between fluid experiments and black hole Hawking radiation (assuming this latter is ever observed!). It seems to me that what *ought* to happen in these three cases is that

(i) for the vortex/QED analogy, we would reject the analogy, and assume that since QED is well understood, that therefore 2D superfluids are not yet well understood.

(ii) for the BEC/GR analogy, we would not reject classical GR, which for smooth spacetimes far from singularities, is very well understood indeed. We would instead reject our description of BECs, and the analogy with GR.

(iii) for the fluid flow/Hawking radiation analogy: since our understanding of classical fluid dynamics is pretty good, especially for smooth flows without turbulence, we would reject the analogy, and assume that our understanding of genuine quantum gravity and of real black holes is just not good enough.

In other words, if any disagreement occurs, we would reject the analogy. If this is the case, then the only use of the analogy, if it works, is to confirm what we already know, or confirm prejudices we already have. If we accept this, then the analogy has no predictive value at all.

On the other hand, what analogies **are really good for** is suggesting **new physics**, or new approaches to physical questions. The examples given above, of the Anderson–Higgs boson, and the Penrose-Hawking work on black holes, illustrate this rather well. Nobody would try to push the analogy between charged superconductors and the current version of the standard model very far. But as a means of intuiting important new physics, it worked extremely well.

I therefore conclude that whereas analogies can be of enormous heuristic value in deriving new physics (as in the case of the work done at different times by Anderson, Penrose, Starobinski, Zeldovich, and Hawking) they can also lead to very serious error (as in the case of the aether). In the case of analogies between condensed media and spacetime (as in the QED/superfluid and the BEC/spacetime analogies discussed above) my feeling is that the jury is still out—but that the use of analogies in these two fields runs the clear risk of producing either more errors, or of just restating what we already know.

(B) The Optimistic View (ST)

Here, I want to first agree that many analogies in physics' history have been misleading, but I would also argue that analogue gravity has already demonstrated itself to be of a different order. I agree with the concern that analogies between condensed matter systems and general relativity must be handled carefully, since the mapping is never one-to-one and can risk overstating what is being tested. On the other hand, however, in my view, this critique can be avoided if one reinterprets the scope of analogue systems and remembers that analogue gravity is not intended to "solve quantum gravity" or to reproduce the Einstein field equations in a laboratory, but rather to probe the robustness of quantum field theory in curved spacetime (QFTCS).

One can argue that this is similar to the framework that Hawking originally used as well; i.e., he did not quantize gravity or aim to do so, and Hawking radiation was derived from treating matter fields quantum mechanically in a fixed classical spacetime. In other words, by using only minimal assumptions such as the existence of a horizon, causal field propagation, and quantum vacuum fluctuations, and without invoking any quantum gravity effects, Hawking showed that black holes shrink as they emit radiation. The lack of any input from quantum gravity means of course that short lengthscale gravitational phenomena, along with conceptual issues such as the trans-Planckian problem, are inaccessible to this long wavelength classical approach. Yet Hawking's result remains one of the most remarkable successes of theoretical physics.

What makes analogue gravity practical is not the claim that one can reproduce the full dynamics of Einstein's theory, or solve the Einstein equation, or provide a microscopic theory of quantum gravity. The point is that one can isolate and experimentally probe the kinematical structures of quantum field theory in curved spacetime, and that the strength of analogue gravity lies in its ability to test the kinematical and causal structures that underlie quantum field theory in curved spacetime, rather than in reproducing the full dynamics of general relativity.

As discussed in a recent paper by Ralf Schützhold [61], one can argue that limitations due to degrees of freedom should not be taken as a barrier; rather, as an outline of what analogue gravity can and cannot do. For example, the effective sonic or acoustic metric [55], involving degrees of freedom from the flow velocity, the sound speed, and the conformal factor does not span the full 10 independent components of a 4D Lorentzian metric, and, we cannot reproduce arbitrary spacetimes but, nevertheless it is reliable enough to simulate many QFTCS phenomena within the subset of metrics that can be realized, and to capture the essential causal structure underpinning Hawking and Unruh effects.

A recent work by Baak, Datta, and Fischer [71] uses Petrov classification to provide a systematic taxonomy of which algebraic curvature classes can arise in analogue systems. The fact that analogue models correspond to restricted Petrov types highlights that while not in any way a replacement for GR, analog systems remain entirely valid laboratory tests for universality of the physics of horizons, mode mixing, and quantum field kinematics in curved backgrounds.

As discussed in the main body, real experiments in BECs, water waves, and nonlinear optics have reproduced Hawking-like spectra, Unruh-type detector responses, and even cosmological particle creation. One can argue that these are direct observations of the same mathematics. In my view, one of the most exciting aspects of analogue models is their ability to probe the ability of entanglement transfer and partner correlations across analogue horizons, something impossible in astrophysical black holes. The fact that in BECs or optical analogues, as shown by Unruh and Schützhold, and more recently by Mann and Gooding, analogue radiation necessarily involves entangled pairs straddling the horizon, and BEC studies have proposed and demonstrated how these can be extracted via non-local density correlations.

Finally, I believe these results demonstrate why analogue gravity has matured into a genuine research field as it validates in the laboratory the universality of causal-kinematical predictions of QFT in curved spacetime, despite the fact that it leaves the dynamical questions of full quantum gravity for future theory.

4 Quantum Mechanics Under the Gun

Let us now turn to a topic on which all three of us are agreed. Up to this point we have assumed that we can use conventional QM and QFT without question. But we now wish to look at genuine quantum gravity, and this means that both conventional GR and conventional QM/QFT must be examined with a critical eye. In Sect. 5 we will argue that it is QM/QFT that must be modified when trying to find a theory of quantum gravity. Here we prepare the ground for this, by recalling key features of QM: our ultimate goal is to see how QM can be modified.

4.1 The Fundamental Objects in QM: Orthodox View

In the standard texts on QM, one imbibes almost by osmosis the orthodox view of QM, according to which [82–85]:

(i): A non-relativistic system S is described either by a state vector $|\Psi_s\rangle$ or a density matrix operator $\hat{\rho}_s$, the latter involving some average over an environment $\mathcal{E}$. These are defined using Hilbert space $\mathcal{H}_S$, spanned by some set $\{|\psi_j\rangle\}$ of orthonormal states, so that $|\Psi_S\rangle = \sum_j c_j|\psi_j\rangle$. Physical quantities are determined by "measurements"; these are defined by operators $\{\hat{O}_\alpha\}$ acting in $\mathcal{H}_S$, so that $\langle O_\alpha\rangle = \langle\Psi_S|\hat{O}_\alpha|\Psi_S\rangle$ (or by $\langle\hat{O}_\alpha\rangle = \mathrm{Tr}\,\hat{\rho}_S\hat{O}_\alpha$ when we describe $\mathcal{S}$ using $\hat{\rho}_S$). We also define a Hamiltonian operator $\hat{\mathcal{H}}_S$, so that $|\Psi_S\,(t)\rangle$ evolves in time according to $\hat{\mathcal{H}}_S\,|\Psi_S\rangle = i\hbar\partial_t\,|\Psi_S\rangle$, the Schrödinger equation.

(ii): Measurements of $\langle\hat{O}_\alpha\rangle$ involve the interaction of $\mathcal{S}$ with some measuring apparatus $\mathcal{M}_\alpha$; if we treat $\mathcal{M}_\alpha$ as a quantum system, described by states $\{\Phi_\alpha^{(k)}\}$, then the measurement process involves entanglement between $\mathcal{S}$ and $\mathcal{M}_\alpha$ in such a way that the state of $\mathcal{M}_\alpha$ after the measurement interaction is correlated with that of $\mathcal{S}$ before the interaction. The simplest example is an "ideal measurement" (called by Pauli [82], a "measurement of the 1st kind"), in which $\mathcal{M}_\alpha$ begins in a state $|\Phi_\alpha^{(k)}\rangle$, and one has the transition

$$|\chi_{\rm in}\rangle \;=\; |\Psi_S\rangle|\Phi_\alpha^{(0)}\rangle \;=\; |\Phi_\alpha^{(0)}\rangle\sum_j c_j|\psi_j\rangle \;\underset{\text{"mmt"}}{\longrightarrow}\; \sum_j c_j|\psi_j\rangle|\Phi_\alpha^{(j)}\rangle \qquad \text{(1st kind)} \quad (27)$$

in which the $|\Phi_\alpha^{(j)}\rangle$ are an orthonormal (sub)set of the states $\mathcal{M}_\alpha$ which are in 1-1 correlation with the $\{|\psi_j\rangle\}$ of $\mathcal{S}$. We note that the state $\mathcal{S}$ is left unchanged (in a non-ideal measurement of the 2nd kind, the state $\mathcal{S}$ would change; indeed $\mathcal{S}$ might even be destroyed), and that $\mathcal{S}$ and $\mathcal{M}_\alpha$ are now entangled. One also defines a "reproducible measurement" or "state preparation" if the measurement has a unique outcome, so that we know what the state of $\mathcal{S}$ is after the measurement (this is of course not the only way that state preparation can occur).

(iii): The world is divided into a classical part, in which definite results and states exist, and the quantum part, described above. At some point the final state

$|\chi_f\rangle = \sum_j c_j |\psi_j\rangle|\Phi_\alpha^{(j)}\rangle$ undergoes "wave-function collapse" to one of the states $|\chi_j\rangle = |\psi_j\rangle|\Phi_\alpha^{(j)}\rangle$, with probability $|c_j|^2 = p_j$; at the same point the density matrix collapses to the diagonal mixed state $\hat{\rho}_s = |\chi_j\rangle p_j \langle \chi_j|$. From this moment onwards, the measurement has been performed, and we have a definite result, and the measuring system $\mathcal{M}_\alpha$ is in a definite classical state in the classical world. Neither the ineffable wave-function collapse, nor the behaviour in the classical world, is describable in QM, or indeed at all in the current theory.

Since the early period of quantum mechanics (up to c.1970) a number of other key ideas have been added to the canon of orthodox QM. (although textbooks have been very slow to catch up!). Amongst these are:

(a): The possibility of high-order entanglement and coherence involving many degrees of freedom. From this vague idea of multi-particle entanglement were born more precise ideas about macroscopic quantum coherence [86] and quantum computation [87], involving multi-particle entanglement in an essential way. Although the experimental realization of genuinely macroscopic coherence, and of viable quantum computation, still need a lot of work (see Refs. [88–90]), we do have a much better theoretical understanding of multi-particle entanglement [91, 92].

(b): The key phenomenon of environmental decoherence. This is the process by which system $\mathcal{S}$ entangles with the environment $\mathcal{E}$, so that when one averages over $\mathcal{E}$ to get a reduced density operator ρ_s for $\mathcal{S}$, coherence and interference phenomenon are lost. Although the idea of decoherence has a very long history (see Appendix), it was not made precise until realistic models for $\mathcal{E}$, and the coupling between $\mathcal{S}$ and $\mathcal{E}$ became available, starting with the work of Caldeira and Leggett [93]. We now have very detailed models describing both delocalized modes [93] and localized modes [94] in the environment $\mathcal{E}$.

However, this raises an interesting question: since decoherence is a one-way process in time, and very difficult to reverse, how are state preparation and coherent states even possible? We note that state preparation and independent coherent states evolve quite naturally both on earth and in intergalactic space, quite independent of any "intelligent" intervention. The answer is illustrated in Fig. 7; one can "disentangle" two systems A and B from each other, just by allowing one of them (e.g. system B) to interact and entangle with another system C. Then B and C are entangled, leaving A disentangled from both of them. This result is an example of "quantum monogamy" and happens all the time, throughout the universe. If system B is the environment $\mathcal{E}$, then we see how we can decouple A from the environment by entangling it with system C.

From all of this we can see that in orthodox QM, the important theoretical objects are the state vector $|\Psi_s\rangle$ in Hilbert space $\mathcal{H}_s$, the set of operators $\{\hat{O}_\alpha\}$ which operate on $|\Psi_s\rangle$, and the idea of a measuring system $\mathcal{M}_\alpha$ which, through the process of wave-function collapse, leads to a projection of $|\Psi_s\rangle$ onto some specific eigenstate of $\hat{O}_\alpha$, with a probability specified by Born's rule.

No fundamental changes occur when we go to QFT; wave-functions are replaced by wave-functionals, defined again in a Hilbert space, and we still have measuring systems and wave-functional collapse. In conventional quantum gravity, things

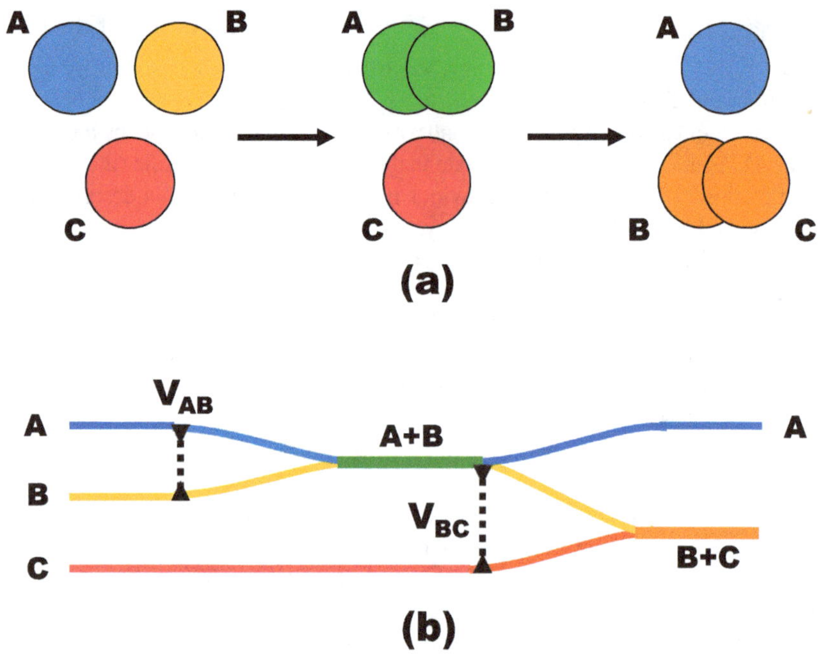

Fig. 7 Illustration of the entanglement process between 3 systems, discussed in the text. In **a** at top we show schematically the 3 different states of entanglement between systems A, B, and C. Initially they are unentangled, then A and B are entangled, and finally B and C are entangled. In **b** below, the time evolution of the entanglement is shown schematically, with time along the horizontal axis—the interaction V_{AB} initiates entanglement between A and B, and then the interaction V_{BC} decouples (and disentangles) A and B, and initiates entanglement between B and C

become a little more complicated, since it is not clear what the correct physical observables are. However, treated as a gauge theory, conventional quantum gravity is no different in its structure from, e.g., non-Abelian gauge theory.

Up to this point we have not mentioned the path integral formulation of QM. This is because we will take the view that it differs in very important ways from the orthodox view just given—we discuss it in Sect. 4.3 below.

4.2 *Key Problems and Questions in Quantum Mechanics*

Why would anyone doubt the validity of QM at the macroscopic scale (even now, most physicists believe that there is nothing wrong with quantum mechanics, although this view is certainly not shared by any of the three co-authors of this paper!). This question has of course has been controversial; some of the views on this are as follows:

(i): The "Plain Vanilla" orthodox view: This was formulated long ago [95]. It is an amalgam of Copenhagen and von Neumann arguments, and appears in well-known textbooks written in the early days of QM [82, 96]. In this view, we know how to use QM and it works, therefore we understand it and there is nothing to worry about. One simply has to learn the textbook rules involving states in (some) Hilbert space, projections and measurements, and a fundamental division between the classical and quantum worlds. (with the quantum world defined in terms of objects in the classical world).

There are many problems with this view [97]. In the first place, being able to use something, or follow instructions for its use, does not mean one understands it! Children and adults with no scientific understanding at all can use cell phones, laptops, and all manner of other sophisticated devices with no understanding of how they work. The simple fact is that the orthodox interpretation has deep and ineluctable problems which could be buried so long as the domain of QM was restricted to microscopic systems, or (as in the case of superfluids and superconductors) to systems described by product states over microscopic degrees of freedom [8].

All of this changed when Leggett pointed out [86] the possibility that under very specific circumstances—an absence of decoherence sources—one could envisage Schrodinger Cat states **in the lab**, of form $|\Psi\rangle = \left(a \prod_j^N |\varphi_j^a\rangle + b \prod_j^N |\varphi_j^b\rangle\right)$ for an N-particle system, with N taken to be macroscopic. Unfortunately, the states of this kind observed so far either have values of N no larger than $\sim O\left(10^4\right)$ (as with interference experiments with molecules), or have almost complete overlap between $\left|\varphi_j^a\right\rangle$ and $\left|\varphi_j^b\right\rangle$ (i.e., one has both $1 - |\langle\varphi_j^a|\varphi_j^b\rangle| = \delta_j \ll 1$, for all j, and also $\sum_j \delta_j \sim O(1)$). For a proper discussion, see the papers of Whaley et al. [88].

If, however, experiments do succeed in preparing systems in which genuine macroscopic coherence is found, then the orthodox interpretation (and many other interpretations) of QM will be in big trouble!

(ii) The "Information Theoretic" interpretation: According to this view, QM is no more nor less than an account of the information possessed by experimenters, based on outcomes of measurements (see, e.g., Laloe [85]). While popular in some quarters, this extraordinarily anthropocentric interpretation apparently abandons any attempt to find a description of a physical world which is independent of observers and observations—it has been ridiculed for this reason [97]. It is clearly incompatible with any extrapolation of QM to regions of spacetime that are unobserved, either because they lie behind horizons or because no measurements have ever been performed on them.

(iii) The "Multi-Time" interpretations: these include the "two-time" or "weak" measurement approaches, and began with the well-known work of Aharonov, Bergmann, and Lebowitz [98], and in the case of 2-time measurements, with the key result that for some operator $O_\alpha(x)$,

$$\chi_\alpha(2,1|x) = \frac{\left\langle \psi_2(t_2) \left| \hat{O}_\alpha(x) \right| \psi_1(t_1) \right\rangle}{\langle \psi_2(t_2) | \psi_1(t_1) \rangle} = \frac{\int_1^2 \mathcal{D}q(t)\, e^{\frac{i}{\hbar}S[q]} O_\alpha(x)}{\int_1^2 \mathcal{D}q(t)\, e^{\frac{i}{\hbar}S[q]}} \tag{28}$$

where $\int_1^2 \mathcal{D}q(t) = \langle \psi_2(t_2) | q_2(t_2) \rangle \int_{q(t_1)=q_1}^{q(t_2)=q_2} \mathcal{D}q(t) \langle q_1(t_1) | \psi_1(t_1) \rangle$. We see that $\chi_\alpha(2,1|x)$ is just the amplitude for the system (here a particle) to propagate from the state $\psi_1(t_1)$ at time t_1, to state $\psi_2(t_2)$ at time t_2, conditional on there being a measurement of $\hat{O}_\alpha$ performed at x somewhere along the trajectory between q_1 and q_2. Alternatively, we can think of it as a measurement of $\hat{O}_\alpha$ at x, conditional on the system having started out in state $|\psi_1(t_1)\rangle$ and finished at $|\psi_2(t_2)\rangle$. This latter interpretation is the one that has been adopted in recent years [99].

There has been considerable discussion, particularly in the context of delayed choice [100] and quantum eraser [101] experiments, of the extent to which one can think of ideas like prediction and retrodiction when QM is formulated in this way [102]. The same questions arise when one considers interpretations like the "transactional interpretation" of QM, in which both retarded and advanced signals propagate between (a) the QM system of interest and (b) both past and future measuring systems [103].

We note that all of these multi-time interpretations still maintain the central role of measurements. As such they are open to the same criticisms as orthodox QM, connected with the division of the world into quantum and classical parts, and the anthropocentric role assigned to measurements.

(iv) Deterministic interpretations: in all of the approaches to QM just described, one keeps the usual QM paraphernalia of measurements and the operators associated with them, and one also assumes that the element of choice on the part of observers is an essential part of the measurement process (so that QM probabilities for measurement outcomes are assigned to different choices). The idea that conscious choices are made in experiments is considered to be more or less axiomatic to the whole business of experimental science by most physicists. It was built into the QM formalism from the very beginning (without however, ever making it explicit in the form of an "axiom"). Choices are made, and as a consequence the deterministic Schodinger equation is superseded by a probabilistic "wave function collapse". As noted by several authors [104], all of the paradoxical features of QM, along with Bell's inequalities, disappear if one instead assumes an entirely deterministic theory of QM, where "choice" does not exist, and where the Schrodinger equation (or some QFT generalization) is universally valid.

The question of whether one can use a "wave function for the universe" is of course intimately bound up with this discussion; the moment one tries to describe the entire universe by QM/QFT, one can no longer invoke the existence of external observers, and wave function collapse becomes moot; we end up with a deterministic universe.

There have been two prominent attempts at a deterministic theory of QM. The first was by Bohm [105], who reprised older work of de Broglie [106] to come up with a very interesting *non-linear* version of QM, in which a "quantum potential" $Q(\mathbf{r})$ is added to the Schrodinger equation of motion; this function itself depends on the "pilot wave" function $\psi(\mathbf{r}, t)$ (hence the non-linearity). However $\psi(\mathbf{r}, t)$ now determines real trajectories for the system—the probabilistic features now come in entirely traditional fashion, from our lack of knowledge of the initial conditions (hence the description of Bohm's theory as a "hidden variable theory"). Equally interesting is Bohm's view of quantum measurements, in which the hidden variables depend on both the system $\mathcal{S}$ and the measuring system $\mathcal{M}$ in interaction with it.

This brings us to the somewhat controversial "relative state" view of QM, which since the influential work of DeWitt [107] has been known as the "many worlds" interpretation. This appellation is unfortunate, since the original formulation of Everett [108] did **not** involve a multitude of worlds - it instead simply demands that the **single** world we live in be described by a single wave-function. Crucially, the special role of measurements is completely removed from the theory, and wave-function collapse never occurs. Measurements are then just another physical process occurring in the universe, and observers like us are physical systems like any other. Clearly a relative state formulation of QM/QFT is appealing to those working in quantum gravity; hence its popularity amongst cosmologists.

It is often argued that the weakness of deterministic approaches is that they do not explain how definite results are obtained from quantum mechanical superpositions, unless one invokes non-linearity (which is what Bohm did [105]). In an Appendix we say a little more about the ideas of Bohm and Everett, and some of the related history, since some of these ideas, and the related history, are misunderstood.

(v) The Role of Decoherence and Disentanglement: one way to try and bridge the gap between quantum superpositions and definite results has gained currency in recent years. This comes when one considers how different sectors of the universe "entangle" with, and "disentangle" from, each other. Entanglement of a system $\mathcal{S}$ with its environment $\mathcal{E}$, either directly or via "3rd parties" [29] leads to environmental decoherence. Decoherence is a complex process, not well captured by toy models; but it is known that one can divide the "environments" responsible for decoherence into "oscillator bath" environments [93], describing delocalized modes (phonons, photons, etc.) and "spin bath" environments [94], describing localized modes (defects, dislocations, nuclear and paramagnetic spins, etc.).

However, systems are also constantly disentangling from their surroundings, and from each other. That this can happen is primarily because of "quantum monogamy"; If a system $\mathcal{A}$ is entangled with another system $\mathcal{B}$ (and $\mathcal{B}$ might be the environment), we can disentangle them by entangling one of them (e.g., $\mathcal{A}$) with another system C; this automatically disentangles $\mathcal{A}$ from $\mathcal{B}$. This happens quite naturally without intervention from humans or other complex systems—it is happening all over the universe, all the time.

All this is interesting, purely as physics. But there is a school of thought which argues that decoherence also "solves the measurement problem", by suppressing off-diagonal "interference" matrix elements in the basis of the system-environment interaction, in the reduced density matrix for the system. That this argument is incorrect has been repeatedly emphasized [109]; one can always in principle (and increasingly in practise) reverse the effects of decoherence, and the interference terms are always "out there", at least in the framework of orthodox QM. Thus, in the end, decoherence does not help solve the paradoxes of QM.

In the Appendix we also say a little about the history of ideas about decoherence (which has some interesting relations to the history of Bohm and Everett's work).

We see that none of these attempts to "interpret" QM is really satisfactory. The only approach that does seem to offer any way out of the QM paradoxes is to try and alter QM; for example, by making it non-linear. There are both pitfalls and promise in this approach, as we will now see.

4.3 Alternative Objects and Structures in QM

In this sub-section we will argue that the orthodox structure of QM is not inappropriate for the discussion of certain well-established physical phenomena, and that there are better ways to do things. In particular we will argue that (i) the path integral formulation of QM, which is **not the same as** orthodox QM, captures key features of QM that are not accessible to the orthodox framework; and (ii) that although non-linear theories of QM have problems, one can bypass them (and indeed solve them) in a proper theory of quantum gravity.

4.3(a): Path Integrals and QM

It was first remarked by C. Morette [112] that the path integral formulation of QM is *not* equivalent to orthodox QM. Her point was that there are non-local features of QM that exist in path integrals, but not in wave-functions.

Let's begin with three physical phenomena which make this point very clearly (and which are depicted in Fig. 8):

(a) The Aharanov–Bohm effect: In its original formulation [113], the Aharonov–Bohm effect was formulated as a scattering effect; but it is more interesting (and experimentally relevant) to formulate it in a ring geometry. As first noted by Feynman [114], it is very simple to formulate it in terms of a sum over paths, with each path accumulating an extra phase $\phi_{AB} = (q/h) \oint d\mathbf{r} \cdot \mathbf{A}(\mathbf{r})$ around a ring, where $\mathbf{A}(\mathbf{r})$ is the vector potential and q the charge. A wave-function has to be defined on an infinite number of sheets; defined as a simple function of $\mathbf{r}, t$ it is multi-valued, and nonsensical. The path integral description is extremely simple [114], and very natural.

(b) Anyon particles/quasiparticles: as first shown by Laidlaw and C. Morette [115], in 2 dimensions, particles can show fractional statistics—these are now called "anyons" and exist as, e.g., quasiparticles in Fractional Quantum Hall (FQHE) systems [30]. The phase accumulated by moving one anyon around another is fixed but

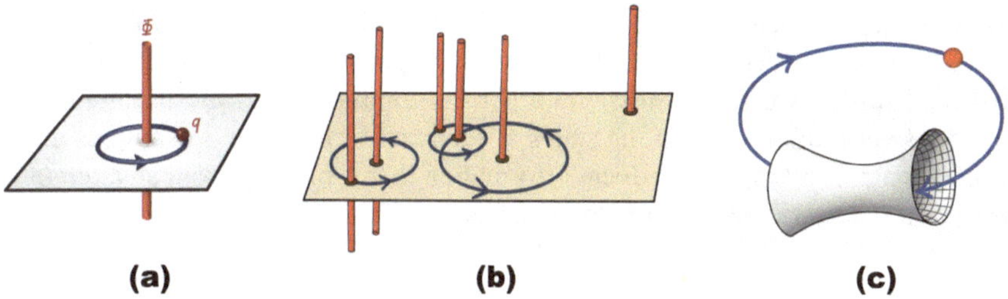

Fig. 8 Three phenomena which require path integrals for their description. In **a** we show the Aharonov–Bohm effect; motion of a charged particle around the flux tube adds a topological phase to the particle amplitude. In **b** we show how topological phases are incurred when anyons move around each other. In **c** we show a system moving through an achronal region of spacetime—here it is a wormhole, and again an extra topological phase is incurred. If the system undergoes a circuit, as shown here, then we have a closed time curve (CTC)

arbitrary; the theory is easily formulated using path integrals but the N-anyon wavefunction is multi-valued as a function of all of its arguments—again, nonsensical. The path integral description [116] is simple and illuminating.

(c) Particles in achronal spacetime: particles or fields passing through achronal regions of spacetime (e.g., through a Kerr black hole, or a wormhole) have no Hamiltonian at all—their dynamics can *only* be defined using path integrals [117], and this is the only way that we are able to discuss, e.g., closed time curves in quantum gravity.

We see that to try to use orthodox QM for the first 2 examples is misguided, and for the 3rd example, impossible. In all three cases, the non-local features of the QM system require path integrals for their description. All we will need for this paper is contained in the formulas for the "generating functional" $\mathcal{Z}_o[\mathbf{j}]$ and the propagator $K_o(2,1) \equiv K_o(\mathbf{r}_2, \mathbf{r}_1; t_2, t_1)$ for the system. The simplest example is a particle with coordinate $\mathbf{r}(t)$, for which we have [118]

$$\mathcal{Z}_o[\mathbf{j}] = \oint \mathcal{D}\mathbf{r}(t)\, e^{i(S_o[\mathbf{r},\mathbf{P}] + \int \mathbf{j}\cdot\mathbf{r})} \tag{29}$$

where the path integration $\oint \mathcal{D}\mathbf{r}(t)$ is taken over a set of closed "ring paths" and $\mathbf{j}(t)$ is an arbitrary "force" applied to the system. The propagator is then:

$$K_o(2,1) = \int_1^2 \mathcal{D}\mathbf{r}(t) e^{iS_o[\mathbf{r},\mathbf{P}]} \tag{30}$$

where the path integration $\int_1^2 \mathcal{D}\mathbf{r}(t)$ now extends from 1 to 2, i.e., from $\mathbf{r}_1(t_1)$ to $\mathbf{r}_2(t_2)$. One obtains $K_o(2,1)$ from $\mathcal{Z}_o[\mathbf{j}]$ by a process of cutting the closed ring path [118]. Note that the action in equations. (29) and (30) will in general contain topological terms like the Aharonov–Bohm contribution ϕ_{AB}.

The key point to emphasize here is that both $\mathcal{Z}_o[\mathbf{j}]$ and $K_o(2, 1)$ are *non-local objects*. They are defined throughout spacetime: the propagator has information about all of the paths between the initial state $|1\rangle$ and the final state $|2\rangle$, and the object $K_o(2, 1)$ is very different from the states at each end of it. In the 3 examples given above, it contains all of the non-local information that is missing from an attempt at a definition in terms of states in Hilbert space.

4.3(b): Non-linear QM

It is not immediately obvious what one means by a non-linear QM: a large variety have been suggested. One can define it simply as a quantum theory in which the equation of motion for the quantum states evolves non-linearly (so that superpositions are not preserved). A related definition required the Hamiltonian to be a function of the state vector. But these definitions are too restrictive—what if a Hamiltonian can't be defined, or if the theory is not written in terms of state vectors (cf. our discussion immediately above)? The problem becomes particularly acute in dealing with non-linear versions of QFT.

Here we will simply define a non-linear QM as one in which the superposition principle is violated, no matter how the theory is set up.

Examples include (i) the work of Kibble [119], who introduced non-linearity by coupling the QM state vector $|\psi\rangle$ to the spacetime metric $g^{\mu\nu}(x)$, and letting the Einstein equations of motion induce the non-linearity in the dynamics of $|\psi\rangle$; and (ii) the work of Weinberg [120], who used the standard toolbox of Hilbert space, measurements, etc., but with a built-in dependence of the Hamiltonian on the state vector. It was quickly shown that a broad class of non-linear theories, including Weinberg's, led to the very undesirable feature of superluminal propagation of information.

It is then extremely interesting to notice that the much earlier 1952 Bohm theory [105] is also a genuinely non-linear theory of QM. Yet no superluminal communication appears in Bohm's theory! Thus remarkably, Bohm's theory not only undermines von Neumann's "proof" that no consistent hidden variable theory exists [121]; it also undermines the more recent claims that non-linear QM theories must necessarily have superluminal signal propagation.

A very careful analysis of this question was given by Polchinski [122], who pointed out that the observables in a non-linear QM which would **not** lead to superluminal communication were those which could be written as a function of the density matrix of the system. Elsewhere we will discuss Bohm's theory in this light.

Notice that none of these analyses applies to a quantum field theory (although Kibble did look at a non-linear dynamics for a scalar field functional [123]). As Kibble himself remarked, the principal problem with all of these efforts was that they were too closely based on the existing structure of QM, and assumed that the usual apparatus of measurements, Hilbert space, etc., could be taken over to the non-linear theory.

— —

Let's summarize our point of view on all of this. This is that without the use of alternative structures to describe QM phenomena, it is impossible to solve the paradoxes and contradictions of the subject. In the next section we discuss how path integrals can be used to give a non-linear theory of quantum gravity, which then results in a non-linear QM in the non-relativistic domain. This leads to different predictions for low-energy experiments than the predictions of QM (and, we argue, it deals with some of the problems found in orthodox QM).

5 The Road to Quantum Gravity

The most popular approach to quantum gravity is to regard it as a high-energy problem. This makes it hard to test. In this article we adopt a different view, viz., that the key to resolving the conflict between GR and QFT is to be found at low energies. We begin by comparing the high- and low-energy approaches, with their radically different assumptions and methodologies. We then go on to discuss a theory—the CWL theory—which we believe surmounts the main problems confronting a theory of low-E gravity, and which also solves the problems confronting QM that were listed in Sect. 4. Finally, we briefly discuss experiments in quantum gravity.

5.1 *High Energy Versus Low-Energy Approaches*

In the high-energy approach, one argues that the key is to be found at the Plank energy scale. This is a very high energy scale. To get a feeling for the numbers, note that the Plank mass $M_P = (\hbar c/G)^{1/2} \simeq 2.18 \times 10^{-8}$ kg, the Planck energy $E_p = (\hbar c^5/G)^{1/2} \simeq 1.96 \times 10^9$ J $\equiv 1.22 \times 10^{19}$ GeV $\equiv 1.42 \times 10^{32}$ K, and the Planck energy density $\rho_P = E_p/\ell_p^3 \simeq 2.61 \times 10^{123}$ GeV/m^3, with the Planck length $\ell_p = (\hbar G/c^3)^{1/2} \simeq 1.616 \times 10^{-35}$ m.

We see that a very small sand grain, with diameter ~0.4 mm and having a Planck mass, would if converted to a Planck energy, heat 5 tons of H_2O from 0 to 100 °C. To experimentally test theory at the Planck scale, we would have to deposit this colossal energy into a volume ℓ_p^3 in size. The resulting energy density ρ_p is hard to imagine; it is 10^{45} times higher than the highest energy densities realized in particle collisions in the Large Hadron Collider (LHC), and only imaginable in the very early universe, at timescales $\sim t_p = 5.39 \times 10^{-44}$ after the initial singularity (if such a singularity even exists).

These numbers create a methodological problem for Planck-scale theories. The problem is illustrated in Fig. 9, which depicts the way in which theory has been tested at progressively higher energies in the last 120 years or so. One sees that if history is a guide, Planck scale theories will be untestable for another 300 years. The only way out of this impasse is if the universe itself can provide us with indirect evidence

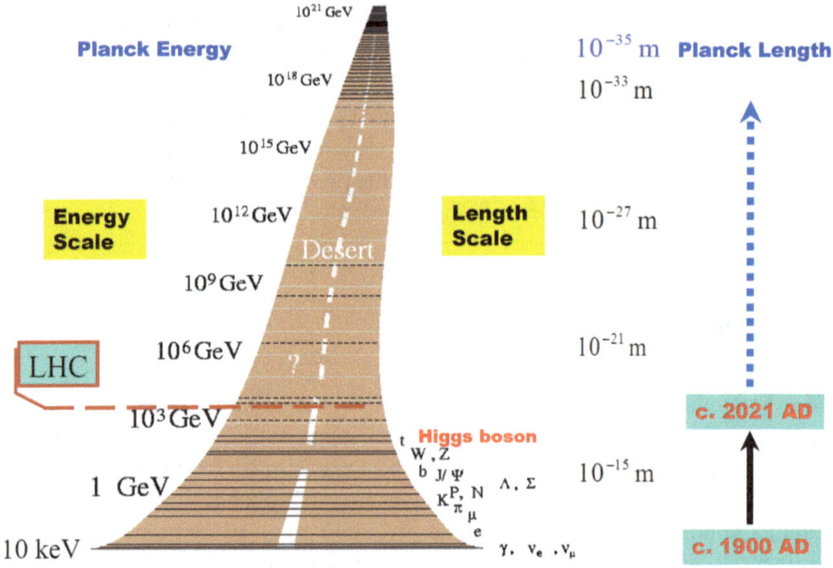

Fig. 9 The high-energy road to quantum gravity. In the last 125 yrs., experiments have gone from roughly 10 keV to centre of mass energies $\sim 1.4 \times 10^4$ GeV, increasing roughly exponentially with time. To get to the Planck scale $E_p \sim 1.22 \times 10^{19}$ GeV will take a long time

for one approach or another, in gravitational waves created at the beginning of the universe—this was in fact Starobinsky's hope.

In the last section we reviewed some of the conceptual problems that arise in QM and QFT. How does the high-energy approach deal with these? Basically it ignores them. To explain the "quantum/classical division", and the validity of GR at the macroscopic scale, one either invokes some orthodox interpretation like the Copenhagen interpretation, in which this division is simply assumed; or it is argued that QM effects are washed out (e.g., by decoherence) once one reaches macroscopic scales. But in any case, the realm of high-energy theories is the "very small", far from the problems that arise for QM at the macroscopic scale.

Rather than deal with the problems of QM, it is instead argued that some variety of superstring theory is *internally consistent*, i.e., that it has passed all the consistency tests one might think of (no unphysical sectors in the space of states; satisfaction of all Ward identities and their concomitant conservation laws, along with all the other requisite symmetries; correct behaviour as $\hbar \to 0$, consistent behaviour as coupling constants $\to 0$; and so on). Since one can't do experimental tests of string theory, and one has already dismissed the problems of QM at the macroscopic scale, these arguments of consistency, along with (highly subjective) considerations of mathematical beauty, are regarded as sufficient.

There is however another methodological hurdle facing any theory of quantum gravity, which has nothing to do with experimental tests, or with the problems of QM *per se*. This is the question of internal consistency at *low energies*. There are various ways of describing this. In a canonical formulation of the theory, one has to calculate field commutators; one such is the commutator of the metric field $C^{\mu\nu\alpha\beta}(x, x') = [g^{\mu\nu}(x), g^{\alpha\beta}(x')]$, which for spacelike intervals between x and x' must be zero. But, as Wald remarks [124], an operator identity of form $C^{\mu\nu\alpha\beta}(x, x') = 0$ cannot depend on what specific gauge one is in (i.e., on the metric), and yet the question of whether the interval is spacelike or not depends on the metric. In a quantum theory, one will in any case have some probability distribution for the different metrics, with both spacelike and timelike intervals.

Another way to think of this argument is to realize that in a path integral for quantum gravity, one must functionally integrate over different metric field configurations, with both spacelike and timelike intervals between 2 specific coordinates x, x', and different lightcones centred on any coordinate for different paths.

Yet another way to think about this argument [125] is to imagine a 2-path experiment involving a massive object (Fig. 10). In conventional QM this would be described by the superposition

$$|\Psi\rangle \;=\; a_1|\Phi_1; \tilde{\mathfrak{g}}^{\mu\nu}_{(1)}(x)\rangle + a_1|\Phi_2; \tilde{\mathfrak{g}}^{\mu\nu}_{(2)}(x)\rangle \tag{31}$$

where those gravitational degrees of freedom that are tied to matter, in the gravitational part $|\tilde{\mathfrak{g}}^{\mu\nu}(x)\rangle$ of the state vector, are quantized, and completely entangled with $|\Phi\rangle$. Clearly, once we tie the metric to matter in this way, we must quantize the metric field. But now if we try to evaluate the interference term, we must look at the inner product $\langle\tilde{\mathfrak{g}}^{\mu\nu}_{(1)}(x)|\tilde{\mathfrak{g}}^{\mu\nu}_{(2)}(x')\rangle$. How are we to interpret or evaluate this?

All of these arguments tell us that there is a fundamental problem associated with the application of QM at *low energies* to spacetime, if we quantize it—and it becomes much worse for massive bodies.

So what do we do? An idea that goes back to Feynman [126] is to guess that the gravitational field as formulated in GR is not only incompatible with QM, but that in a proper theory, gravitation will cause a breakdown of QM. We avoid discussing the lengthy history of attempts to find a theory which tries to treat this idea, and instead just go directly to the theory that we favour.

5.2 *Correlated Worldline Theory*

In what follows we focus on one particular candidate for a theory of quantum gravity which is intended to address the incompatibilities between QM/QFT and GR. This theory's basic structure is actually rather rigid, in the sense that there are no adjustable parameters in it, once one has accepted the initial assumptions.

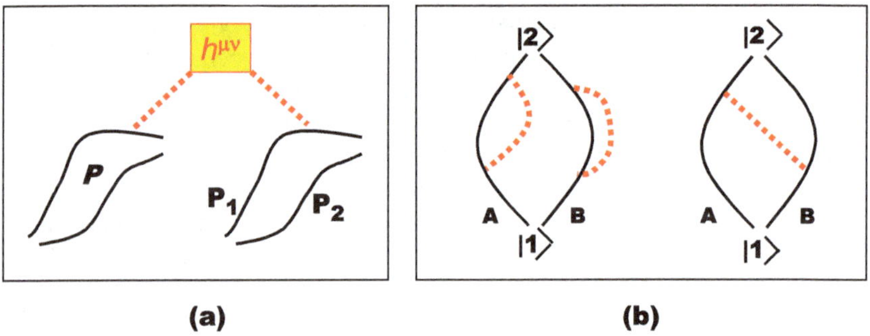

Fig. 10 Two-path thought experiments in quantum gravity. In **a** we compare a pair of paths for a single particle, with a pair of paths for 2 identical particles. The external gravitational field probe $h^{\mu\nu}(x)$ cannot distinguish them. In **b** we show, at left, a situation in standard QM where a single matter system can propagate along a pair of paths labelled A and B from an initial state $|1\rangle$ to a final state $|2\rangle$; each matter path is decorated with graviton lines, which contribute to the gravitational self-energy. In **b** at right we show the same situation according to CWL theory, in which gravitational interactions between a pair of paths for the *same matter system* are allowed

5.2(a): Basic Ideas of CWL Theory

In a series of papers (see Refs. [127, 128], finishing with Ref. [118]) the "Correlated Worldline" theory was developed formally (with some of the ideas having been described in earlier papers [125, 129]). Here the discussion will be informal: the ideal is to provide intuition rather than mathematics [130]. We begin from the view adopted in the previous section, viz., that the fundamental objects in any quantum theory are paths/histories, rather than state vectors in Hilbert space. Consider then the thought experiment in Fig. 10a, where we compare (i) a system where we have two identical particles of mass M, moving along 2 paths $q_1^{\mu}(\tau)$ and $q_2^{\mu}(\tau)$, and (ii) a single mass M moving along a superposition of these paths. In a path integral formulation of QM, can an external probe gravitational field distinguish these two cases? Since the field only sees $T_{\mu\nu}(x)$, the answer is that it cannot: in both cases it sees a sum $\left(T^{(1)}_{\mu\nu}(q_1|x) + T^{(2)}_{\mu\nu}(q_2|x)\right)$. Thus these two situations are equivalent, in the same sense as an accelerated frame and a frame in a gravitational field were held to be equivalent by Einstein (although Einstein's equivalence is actually quite misleading [131].

As a consequence of this reasoning, because we know that the pair of masses will interact via gravitation, then there must also be interactions between a pair of paths A and B for a single mass. In conventional quantum gravity, one is only allowed diagrams like that shown in Fig. 10b, in which a matter line emits and re-absorbs gravitons. However in CWL theory, diagrams like that in Fig. 10c must occur, in which we show the interaction between paths A and B. A little more thought [125] then shows that there will be higher-order interactions between n-tuples of matter lines, and a consistent treatment of this in perturbation theory [118] shows that one can set up a "diagrammar" in which all non-zero contributions can be summed.

However the mathematical development of the CWL theory is not limited to perturbative expansions in powers of the gravitational coupling G_N (Newton's constant). In fact it turns out that results for the matter propagator (as well as other quantities like the generating functional) can be obtained exactly in CWL theory, which also turns out to be a renormalizable theory. Moreover, the theory has passed every consistency test that has so far been thrown at it. These include:

1. If we take $\hbar \to 0$ in CWL theory, we immediately recover classical GR [128]
2. If we let $G_N \to 0$ in CWL theory, we immediately recover conventional QM or QFT in flat spacetime [128].
3. The theory has consistent expansions to all orders in $\hbar$ (about classical GR) and in G_N (about conventional QM/QFT) [118, 128]
4. All relevant Ward identities are satisfied, including those connected with diffeomorphisms; so all the relevant conservation laws are obeyed [128].
5. Amongst the invariances following from the Ward identities, one may also add Lorentz invariance in flat spacetime; and there is no superluminal propagation of information in CWL theory.

To understand all of this better, it is helpful to first discuss a specific object in CWL theory, and then look at how the theory can be applied to experiments.

5.2(b): Propagator in CWL Theory

In conventional QFT, one has a propagator $K_0(2,1)$ for the propagation of some matter system between states $|1\rangle$ and $|2\rangle$. As we have seen above, states in a path integral formulation of QM/QFT can be formulated as resulting from "cuts" in a ring propagator, these cuts representing the interaction of the system with some external "measuring system". It is helpful, in understanding CWL theory, to see what we get.

Suppose we have some scalar field $\phi(x)$, and we wish to determine its propagator between two configruations $\Phi_1(x)$ and $\Phi_2(x)$, on some background and fixed metric g_0. This is just the question asked since the 1960s, about the dynamics of quantum fields in a static curved spacetime. We write this (letting $\hbar = 1$):

$$K_0(2,1|g_0) = \int_{\Phi_1}^{\Phi_2} \mathcal{D}\phi \, e^{iS_\phi[\phi,g_0]} = e^{i\psi_0(2,1|g_0)} \tag{32}$$

so that ψ_0 is the total phase accumulated between the two states, which are assumed to be field configurations on spacelike surfaces Σ_1 and Σ_2.

If we now go to conventional quantum gravity, this generalizes to

$$\begin{aligned} K(2,1) &= K\left(\Phi_1\Phi_2; \mathfrak{h}_2^{ab}, \mathfrak{h}_1^{ab}\right) \\ &= \int_{h_1}^{h_2} \mathcal{D}g \, e^{iS_G[g]} \Delta(g)\, \delta(\chi^\mu)\, K_0(2,1\, g) \\ &= e^{i\Psi_0(2,1)} \end{aligned} \tag{33}$$

where we functionally integrate over $g^{\mu\nu}(x)$, using the Einstein action $S_G[g]$, between induced metrics $\mathfrak{h}_1^{ab}$ and $\mathfrak{h}_2^{ab}$ on Σ_1 and Σ_2 respectively. Here $\Delta(g)$ is the Fadeev–Popov functional determinant and $\chi^\mu(g)$ is a gauge fixing function. Finally $K_0(2, 1|g)$ is just the scalar field propagator in some fixed field configuration $g = g^{\mu\nu}(x)$.

The result for CWL theory is rather different. Let's start by defining the "conditional" stress tensor

$$\chi_{\mu\nu}^{\mathfrak{T}}(2, 1|x, g) = \frac{\langle \Phi_2 | T_{\mu\nu}(x|g) | \Phi_1 \rangle}{\langle \Phi_2 | \Phi_1 \rangle} \tag{34}$$

i.e., a stress tensor defined at x for a metric g, conditional on the system beginning at state $|\Phi_1\rangle$ and ending at state $|\Phi_2\rangle$. Alternatively, we can write it as

$$\chi_{\mu\nu}^{\mathfrak{T}}(2, 1|x, g) = \frac{\int_{\Phi_1}^{\Phi_2} \mathcal{D}\phi e^{iS_\phi[\phi,g]} T_{\mu\nu}(x|g)}{\int_{\Phi_1}^{\Phi_2} \mathcal{D}\phi T_{\mu\nu}(x|g)} \tag{35}$$

i.e., a normalized path integral conditional on $T_{\mu\nu}(x|g)$ being "measured" at x. Another way to think of this as a "weak measurement" of $T_{\mu\nu}(x|g)$, or a "2-time measurement" (compare Ref. [99]). Note that $\chi_{\mu\nu}^{\mathfrak{T}}(x|g)$ is a complex quantity (a conditional amplitude).

The rather remarkable <u>exact</u> result in CWL theory is that the propagator for the scalar field is given by [118],

$$\mathcal{K}_{\mathrm{CWL}}(2, 1) \;=\; \exp i\left[\left(S_G\left[\overline{g}_{21}\right] + \psi_0\left(2, 1|\overline{g}_{21}\right)\right)\right] \tag{36}$$

and the analogue to Einstein's field equation is (again, $\kappa = 8\pi G_N/c^4$):

$$G_{\mu\nu}\left(\overline{g}_{21}(x)\right) = \kappa\chi_{\mu\nu}^{\mathfrak{T}}(2, 1|x, g) \tag{37}$$

where $\overline{g}_{21}(x)$ is the conditional metric satisfying,

$$\frac{\delta}{\delta g}\left[S_G[g] + \psi_0(2, 1\, g)\right]|_{g=\overline{g}_{21}} = 0 \tag{38}$$

Let's understand all of this physically. We have now defined a new spacetime, a quantum spacetime, which is <u>complex</u>; the new Einstein tensor can also be written as $G^{\mu\nu}\left(\overline{g}_{21}(x)\right) = R^{\mu\nu}\left(\overline{g}_{12}\right) - \frac{1}{2}R\left(\overline{g}_{12}\right)\overline{g}_{12}^{\mu\nu}$. We see that $G_{\mu\nu}$ in CWL theory is <u>not</u> the same as what one finds in semiclassical gravity, where

$$G_{\mu\nu}[g(x)] = \kappa\left\langle T_{\mu\nu}(x|g)\right\rangle \tag{39}$$

Just how different equations (5) and (7) are from each other is seen by considering again the 2-path experiment [118]. Denoting the two paths again by A and B respectively, we note that in semiclassical gravity, a slow-moving point object sees the stress energy expectation value $\langle T_{00}(x)\rangle = M\delta\left(\boldsymbol{r} - \frac{1}{2}\left|\boldsymbol{r}_{\mathrm{A}}(t) + \boldsymbol{r}_{\mathrm{B}}(t)\right|\right)$ where $\boldsymbol{r}_{\mathrm{A}}(t)$ and $\boldsymbol{r}_{\mathrm{B}}(t)$ are the two path coordinates. The semiclassical Einstein equation then reads,

$$G_{\mu\nu}(g(\boldsymbol{r},t)) = \frac{\kappa}{2}\sum_{\alpha}^{\mathrm{A,B}}\left\langle \boldsymbol{r}_\alpha(t)\left|T_{\mu\nu}(\boldsymbol{r},t)\right|\boldsymbol{r}_\alpha(t)\right\rangle \tag{40}$$

which is, of course real on both sides. The conditional correlator $\chi^{\mathfrak{T}}_{\mu\nu}(2,1|x)$ is also real in semiclassical theory:

$$\chi^{\mathfrak{T}}_{\mu\nu}(2,1|x) = \frac{1}{2}\left(T^{(\mathrm{A})}_{\mu\nu}(x) + T^{(\mathrm{B})}_{\mu\nu}(x)\right) \tag{41}$$

and does not depend on the initial and final states.

On the other hand, if we treat everything quantum-mechanically, we have a quite different result for the stress energy expectation value: instead of a result which is peaked at the mean position of the mass, we get $\langle T_{00}(x)\rangle = M\left[\delta\left(\boldsymbol{r} - \boldsymbol{r}_{\mathrm{A}}(t)\right) + \delta\left(\boldsymbol{r} - \boldsymbol{r}_{\mathrm{B}}(t)\right)\right]/2$, with two peaks on the two different paths. The conditional correlator is now complex

$$\chi^{\mathfrak{T}}(2,1|x) = \frac{1}{2}\left[\left(T^{(\mathrm{A})}_{\mu\nu}(x) + T^{(\mathrm{B})}_{\mu\nu}(x)\right) + i\left(T^{(\mathrm{A})}_{\mu\nu}(x) - T^{(\mathrm{B})}_{\mu\nu}(x)\right)\tan\left(\Delta S^{\mathrm{AB}}_{21}\right)\right] \tag{42}$$

where $\Delta S^{\mathrm{AB}}_{12}$ is the phase difference between paths A and B, along the trajectories from $|1\rangle$ to $|2\rangle$. The resulting field equation is then complex, and is given by (37) above.

If we now ask what the propagator for this system is, we find there different results for the propagator of our massive object, depending on which theory we choose; we get $\mathcal{K}(2,1) = e^{i\Psi_0(2,1)}$ for conventional quantum gravity, $\mathcal{K}_{\mathrm{sc}}(2,1) = e^{i\chi_{\mathrm{sc}}(2,1)}$ for semiclassical gravity, and $\mathcal{K}_{\mathrm{CWL}}(2,1) = e^{i\Phi_{\mathrm{CWL}}(2,1)}$ for CWL theory. Both $\Psi_0(2,1)$ and $\chi_{\mathrm{sc}}(2,1)$ are real, but they are not the same; and $\Phi_{\mathrm{CWL}}(2,1)$ is complex.

At first glance this makes one think of a simple decoherence effect, but this is not actually what is going on—the key is that **spacetime is now complex**. This is to be expected—the metric is a quantum object, and so the amplitude for a given spacetime configuration will in general be complex. The imaginary part of $\Phi_{\mathrm{CWL}}(2,1)$ comes entirely from the interference between paths A and B (see Eq. (42) above), and its source can be pinned down to the gravitational interaction taking place between paths A and B. Thus the imaginary part comes from a quantum communication between the two paths, mediated by gravitation.

This of course is the way in which CWL theory violates the QM superposition principle, and leads to a non-linear form of QM. The inter-path communication is reminiscent of the description given by Polchinski [122] of the physics arising in attempts to generalize QM to non-linear versions of the theory, which we already discussed above (Sect. 4.3(b)). Polchinski was particularly concerned with the non-linear version of QM described by Weinberg [120], and he described this communication between different branches or elements of a superposition as an "Everett telephone". Amongst other things, Polchinski's work showed that non-linear QM would have problems with causality—typically such theories allow superluminal communication.

It is therefore of considerable interest that CWL theory does not permit superluminal communication. This actually follows from the proofs given [128] of the Ward identities for the theory, but can also be shown by simply noting that one may define a complete set of observables for CWL theory, and we find that the set of operators $\langle \hat{O} \rangle_{CWL}$ corresponding to these is in the class of operators excluded by Polchinski in his demonstration of superluminality [122, 132].

5.3 Experiments and Predictions of CWL Theory

It is interesting to continue with the technical details of CWL theory, but these have been discussed fairly exhaustively in the original papers [118, 127, 128]. What is more interesting is to ask what kinds of experimental predictions are made by CWL theory. This is not really a question about "observables" (a rather abstract notion of more interest to mathematicians, at least in the context of conventional QM).

It is instead a question about real experiments, and about how one formulates the idea of a measurement in CWL theory.

A number of different kinds of real measurement, and predictions for experiments, have already been described in CWL theory. These include (1) optomechanical measurements of mirrors in cavities [133], two-path experiments [118], the dynamics of rotating macroscopic bodies [134], and "4-path" experiments [135]. Here we will focus on the first two, which illustrate crucial features of CWL theory.

5.3(a): Extended Bodies in CWL Theory

Because any test of CWL theory has to be carried out on a massive body, we need to work out the way in which different paths for spatially extended bodies will interact gravitationally. This is a technical exercise which requires detailed knowledge of the phonons in the body (here assumed solid). The reason is that all of the gravitational mass of the body is concentrated in the nuclei, and their motion (i.e., their paths) are described by phonon collective coordinates. Individual nuclear paths are then confined to oscillate over a length scale $\xi_0 (T)$, such that

$$\xi_0^2 (T) = \frac{\hbar^2}{2m} \int \frac{dE}{E} g (E) (1 + 2n (E, T))$$

where $g (E)$ is the phonon density of states, and $n (E, T)$ is the Bose distribution function.

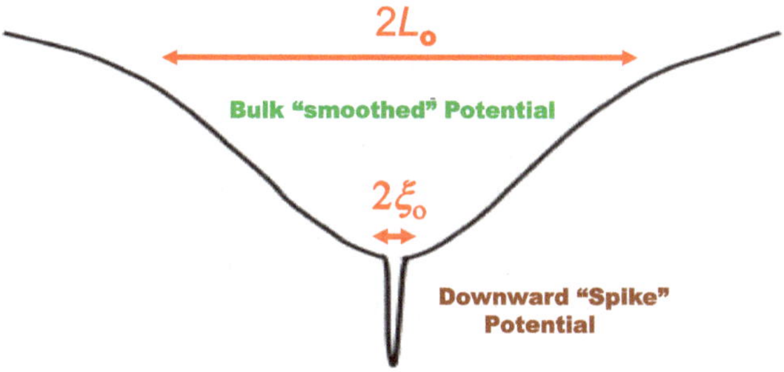

Fig. 11 A plot of the effective gravitational potential $U(\mathbf{Q})$ between paths $\boldsymbol{R}_0(t)$ and $\boldsymbol{R}'_0(t)$ of a massive extended body. Here $\mathbf{Q} \equiv (\boldsymbol{R}_0(t) - \boldsymbol{R}'_0(t))$, and $\boldsymbol{R}_0(t)$ and $\boldsymbol{R}'_0(t)$ are the centre of mass coordinates of the 2 paths. The sharp downward spike and the broad potential well have widths $2\xi_o$ and L_o respectively (see main text)

The net interaction between a pair of paths for an extended mass is then given by a function shown in Fig. 11, which we call $U(\boldsymbol{R}_0, \boldsymbol{R}'_0)$; here $\boldsymbol{R}_0(t)$ and $\boldsymbol{R}'_0(t)$ are the centre of mass coordinates of the body when following these paths. The downward "spike" comes from paths where the nuclei for one path overlap with the nuclei for the other path; their width is $\sim 2\xi_0$, which for typical solids is $\sim 10^{-12}$m. The broad potential well is produced by the interaction of the smoothed mean distribution of the entire extended mass, on one path, with it's counterpart on the other—so that this potential well is spread across a length-scale L_o equal to the size of the extended mass.

Detailed analysis shows that

1. For typical parameters which obtain in macroscopic solid objects, one must go to low temperatures for the inter-path interaction $U(\boldsymbol{R}_0, \boldsymbol{R}'_0)$ to be relevant. For temperature $T \ll 25\,°$K, the two paths will be closely bound to each other, within a length scale $\sim \xi_0$; their paths will oscillate around each other at frequencies ~ 0.4 Hz, in a good approximation to a harmonic potential (at the bottom of the spike).
2. At room temperature, the two paths will move much further apart, and the spike potential will have very little effect. The relative motion $\boldsymbol{R}_0 - \boldsymbol{R}'_0$ will then be in the broad smooth potential, with frequency $\sim 10^{-3}$ Hz.

5.3(b): Harmonic Experiments in CWL Theory

Consider first experiments on massive oscillating objects like mirrors, in which the centre of mass coordinate $\mathbf{Q}(t)$ moves in a harmonic potential. From the results just given, it follows that when $T \ll 25\,°$K, or at room temperature, the motion of the relative coordinate $z(t) = (\boldsymbol{R}_0(t) - \boldsymbol{R}'_0(t))$ between any two CWL paths $\boldsymbol{R}_0(t)$ and $\boldsymbol{R}'_0(t)$ is also harmonic. Under these circumstances, one can show [133] that the motion of the centre of mass coordinate $\mathbf{Q}(t)$ for the massive object will be indistinguishable from the corresponding motion of the oscillator in QM. This is because the CWL corrections—coming from the interactions between all of the different CWL paths—completely decouple from the motion of **Q**.

Thus tests of CWL on simple oscillators are pointless; they give the same results as QM. One can try to evade this problem by employing highly non-Gaussian states in the experiment; or by doing pulsed experiments; or by trying to make the potential anharmonic. It remains to be seen if this can be done [133].

Why is this conclusion important? Simply because any attempt to test QM at the macroscopic scale with harmonic systems will be unable to differentiate QM from CWL, and doubtless from other theories as well. We believe this point is important in view of claims [136] that experiments on the LIGO system have verified that QM works for the motion of 40 kg mirrors. All that these experiments have done is verify that a certain class of theories can describe the mirror behaviour. That multiple different theories can describe the same experiment is of course very common in physics—a well known example is the way in which the Lorentz-Fitzgerald approach gave identical results to Einstein's special relativity, for the ensemble of electrodynamic experiments known at that time.

Thus we can say pretty clearly that the LIGO experiments in no way prove that QM is valid at the macroscopic scale, and that the mirrors are in a "macroscopic quantum state". Other experiments will be necessary.

5.3(c): 2-Path Experiments in CWL Theory

One can fairly easily calculate the dynamics for a 2-path experiment in CWL theory, for a simple particle [118]. To adapt these calculations to an extended body is also straightforward. The main result is fairly simple to state. Suppose we have a transition amplitude, in ordinary QM, of form $K_o(2,1) = \langle \mathbf{r}_2, t_2 | e^{-iHt/\hbar} | \psi_1, t_1 \rangle$, for a system to pass from an 'input' state $|\psi_1\rangle$ to a final state at point $\mathbf{r}_2$ in real space. Suppose also that the system can follow one or other of 2 paths, labelled A and B.

Then the standard QM amplitude will take the form

$$K_o(2,1) \;=\; \langle \mathbf{r}_2, t_2 | e^{-iHt/\hbar} | \psi_1, t_1 \rangle \;\propto\; \sum_{\alpha}^{A,B} K_o^{(\alpha)}(2,1) \tag{43}$$

where $K_o^{(\alpha)}(2,1)$ is the amplitude to go along path α. However the amplitude in CWL theory will take the form

$$\mathcal{K}_{CWL}(2,1) \;\propto\; e^{i\Theta_{21}^{AB}} K_o(2,1) \;\;\equiv\;\; e^{iS_G[\langle \bar{g}_{21}^{AB} \rangle]} \sum_{\alpha}^{A,B} e^{iS_M[\mathbf{r}_\alpha | \langle \bar{g}_{21}^{AB} \rangle]} \tag{44}$$

where the phase Θ_{21}^{AB} is *complex*, as was the conditional correlator for this same 2-path system (see Eq. (42) above); the phase Θ_{21}^{AB} here is incorporated into the complex phase $\Phi_{CWL}(2, 1)$ defined previously. We have written the answer in the final form in (44) so that one can see that the "conditional" metric field $\langle \bar{g}_{21}^{AB} \rangle$, defined above in Eq. (38), comes in. This field is sourced by both paths A and B simultaneously. The matter (i.e., the moving massive object) is still propagating quantum-mechanically, in a superposition of paths A and B, in the presence of the same background conditional metric field $\langle \bar{g}_{21}^{AB} \rangle$.

It is very important that the effect of the CWL correlations is to multiply the conventional QM amplitude by the exponential of a global phase. This means the CWL correlations do not select out any one of the paths for special consideration. As we will show elsewhere, this leads to very interesting experimental possibilities, which there is no space to discuss here. For the 2-path system, the main effect is to alter the form (but not the position) of the light and dark fringes in the 2-path interference [CWL3].

5.3(c): Measurements in CWL Theory

Until now we have discussed specific measurement schemes in CWL theory. But can one devise a general theory of measurements for CWL theory, analogous to that in orthodox QM? This is a topic of some complexity, which has only been discussed briefly so far [125]. The situation in CWL theory is similar to that in any QM theory in which measurements are treated as just another QM process, involving entanglement between a system S and some measurement apparatus $\mathcal{M}$ (in other words, there is no special "measurement process", and there is no "collapse of the wave-function").

However CWL theory is very different from QM in that is non-linear: however this non-linearity emerges only once one begins to displace macroscopic objects. Elsewhere we have discussed how big an object is required (see Refs. [118, 133] and also [125]). But as we have just seen above, a mass displacement is not enough—a body controlled by a harmonic potential will show, in CWL theory, identical behaviour to that in QM, no matter how massive it is—one requires fundamentally non-harmonic systems (of which the 2-path system is a good example).

Under these circumstances the key feature to pay attention to is the phenomenon of "path-bunching" [118, 133], whereby different paths for a massive object collapse towards each other onto one dominant trajectory. We note that path-bunching is **not a decoherence process**, and can happen quite independently of any environmental couplings, purely because of the gravitational interaction between paths. It happens however at a rate determined both by coupling to the external environment and by the internal non-linear dynamics.

For a microscopic object the path-bunching time is astronomically long, and path-bunching is then completely obscured by conventional decoherence and dissipation processes. For a macroscopic object it can be quite short (although not for an oscillator). We now observe that if some microscopic system S then entangles with a macroscopic system $\mathcal{M}$ which is undergoing or has undergone path punching, then S will also be induced to path bunch with $\mathcal{M}$ [125]. Again we stress that this is not a process that can in principle happen without decoherence. Since $\mathcal{M}$ will very easily entangle with other objects, we see that this path-bunching chain will propagate to a succession of other massive objects, and eventually to everything within the light cone of $\mathcal{M}$.

We now see that CWL theory distinguishes between the purely quantum microscopic realm of superpositions, and the classical realm of apparently definite states, not on the basis of some bizarre (and arbitrary) idea like consciousness, or wave-function collapse, but instead on the completely objective feature of path bunching, appropriate to certain kinds of macroscopic object. The objects are distinguished by (i) a large gravitational mass, and (ii) a dynamics which is inherently non-harmonic.

At this point one can, if one likes, look at different kinds of measurement scheme, and ask which ones satisfy these two criteria. These measurements will then lead to strong departures from orthodox quantum behaviour, and will look, for all practical purposes, like wave-function collapse. However we also see that one can no longer connect measurements in general with path bunching, and that macroscopic states that look quantum-mechanical are possible in CWL theory.

A detailed mathematical treatment of measurements on CWL theory, and the comparison with QM, will appear elsewhere.

6 Conclusions

It is clear from parts of this article that there was an interesting debate going on while it was being written, between ST and PCES, about the use of analogies in physics. We chose to focus the debate somewhat on 2 key examples, viz., the analogy between 2-d superfluid films and QED, and the analogy between dilute BEC gases and curved spacetime. We note that all of these systems are interesting in their own right. Rather than hide our differences, or attempt to arrive at a synthesis (for which there was no time) we simply chose to air them. The result is very preliminary and we hope to return to this topic.

Two things we were all agreed on were (i) that QM has serious problems, which one should not try to hide; and (ii) that these cannot be solved unless we modify QM in some way. We pointed out that arguments that are commonly used to dismiss non-linear modifications of QM are not nearly as general as often thought. We also pointed out that there are good reasons to believe an argument first suggested in 1957 by Feynman, that QM may break down for macroscopic systems because of gravity, We discussed a theory (the CWL theory) that implements this idea in a form which passes all theoretical consistency tests—the big test that it now needs to pass is the comparison with experiment. In future it will also be interesting to apply this theory to cosmological phenomena—work in this direction is already underway.

It is unfortunate that we are unable to discuss the contents of this article with Starobinsky, since his work overlapped at one time or another with all of the themes discussed here, and he was interested in all of them. We hope that this article will help to sustain interest in these topics.

Acknowledgements Several parts of this article were produced by a lengthy and stimulating dialogue between two of us (PCES and ST), which we have tried to reproduce in the paper. One of us (PCES) would like to thank the other two (MJD and ST) for typing all his contributions, as well as doing the figures and references, since he is currently prevented from looking at computer screens. ST thanks W. G. Unruh, S. Weinfurtner, and members of the Weinfurtner group for discussions. MJDR thanks V. Milner and members of the Milner research group for discussions, and he thanks V. Milner for funding. Finally, PCES thanks J. Dalibard, M. O. Scully and C. DeWitt for discussions. This work was funded by NSERC (the National Science and Engineering Research Council) of Canada.
The article began with a collaboration between ST and PCES, and we were later joined by MJDR. Section 1 was written by PCES, Sect. 2 by MJDR and PCES, Sect. 3 by ST and PCES, Sects. 4 and 5 by PCES, and the appendix by ST and PCES. The article was proof-read by all of us, until we were all agreed on the contents.

Appendix: Some Historical Remarks

Two of us (PCES and ST) have spent some time looking at the historical development of (i) ideas about decoherence, along with ideas about non-linear quantum mechanics and Everett's relative state theory, and (ii) the development of the Aharonov–Bohm effect and the way it was later understood in path integral theory. The is no space here (and currently we have no time, since the deadline for submitting this article is past!), to give more than a quick sketch of some highlights—a proper discussion will have to wait.

A.1: Decoherence and Related Topics—Early History

From the very beginning of QM, it was realized that any extrapolation of the theory to the macroscopic scale would imply superpositions and interference at that scale. It is very well known that in the 1930s this led to a divergence between those like Bohr who felt that QM at the microscopic level had to be defined by classical physics at the macroscopic level (and who therefore rejected the use of QM in the classical world) and those like Schrodinger who deliberately extrapolated QM to the macroscopic scale (notably in the form of the Schrodinger Cat) to show the problems that such extrapolations led to. Von Neumann argued that the "von Neumann chain" of measurement correlations could only be truncated by the introduction of consciousness.

Many of these ideas were summarized in the well known work of London and Bauer in 1938, and later in Bohm's 1951 text, where he analyzed what he called the "destruction of interference in the process of measurement", i.e., the loss of phase relations when the system becomes correlated with a macroscopic apparatus.

In the 1950s and early 1960s interest grew in what later came to be called "decoherence", in which the loss of phase coherence came not from interaction with a measuring system but instead with some generic environment. Amongst a large number of papers we can single out the work of Ludwig [137], Green [138], and Daneri et al. [139], as well as the extensive discussions at Princeton involving, at different times, Everett and DeWitt [108, 110], and the group of people surrounding Wigner [111] (all of this after Bohm had left the USA for Brazil in Oct. 1951).

In 1970 Zeh [140] was able to gather these ideas into a general thesis, viz., that decoherence was a generic feature of quantum mechanics, able to explain why one could not see macroscopic interference. Other papers (see, e.g., Simonius [141] and Peres [142]) continued in this vein. But the situation changed fundamentally with the publications of Leggett and co-workers, starting in 1978 [143], and continuing with work in 1980 [86] and 1981–83 [93]. This work accomplished two key steps. First, it argued, contrary to all previous work, that one **should** be able to see macroscopic superpositions in certain systems, if QM worked at the macroscopic scale - indeed it advocated a search for these in superconducting SQUID devices. Second, it gave for the first time a detailed predictive theoretical framework for how to evaluate the effects of the environment—up to that point there had been no quantitative theory of any kind. This work then ushered in what one might call the "modern period" of decoherence studies, focussed on detailed predictions for experiments. It is impossible to imagine any of the subsequent work in fields ranging from quantum computation to quantum devices without this theoretical framework. Even a superficial review of this modern period would require an entire book.

It is interesting to see how this early history is bound up with Everett and DeWitt's parallel work. As far as we know, Everett never met Bohm, and Bohm did not visit Princeton again until many years later, well after he had moved to the UK (curiously, Wheeler tried to organize a boycott of Bohm's lecture at Princeton, because of their political differences). However, although he never discussed hidden variables with Bohm, Everett was apparently familiar with Bohm's 1952 theory (regarding it as "more cumbersome and artificial" than his own theory [110]).

Apparently neither Everett nor DeWitt properly appreciated the key introduction of non-linearity into QM by Bohm's quantum potential, which fundamentally differentiates Bohm's theory from QM and from Everett's work. The essential point is very simple—if we write the wave function as $\psi(\mathbf{r}, t) = R(\mathbf{r}, t)\, e^{iS/\hbar}$, and add to the Hamilton-Jacobi equation for $S[\mathbf{r}, t]$ a new "quantum potential" given by $Q(\mathbf{r}, t) = -(\hbar^2/2m)(\nabla^2 R/R)$, along with a "guiding equation" $m\mathbf{P} = \nabla S$ for the particle, then the system becomes non-linear, and the particle follows a trajectory different from that given by the Hamilton-Jacobi equation without $Q(\mathbf{r}, t)$ (this equation having been found by Schrodinger himself, in the led up to his discover of the Schrodinger wave equation). None of this was acceptable to Everett, for whom QM had to remain linear [144].

One thing that would however have united Bohm and Everett very firmly, if they had ever met, was their attitude towards the Copenhagen interpretation and the Bohr–Rosenfeld school. Everett was openly disdainful (regarding it as a "philosophical monstrosity" [110]). This sentiment was returned by Rosenfeld (and Bohr's disapproval led Wheeler to withdraw his former enthusiastic support for Everett's ideas). And Bohm's 1952 departure from Copenhagen was complete and irrevocable.

A.2: The Aharonov–Bohm effect

Only a few years after leaving Princeton, and by way of Brazil and Israel, Bohm moved to Bristol in the UK. His peregrinations owed much to M. H. L. Pryce, who along with Einstein had helped Bohm in his move to Brazil (Pryce having been on sabbatical in Princeton when Bohm's problems started there). Pryce also helped Bohm get a position in Israel, and then, having become moved from oxford to become head of the physics dept. in Bristol, Pryce brought Bohm along with his student Aharonov to Bristol. At Bristol Bohm found a lively intellectual atmosphere, which included P. K. Feyerabend; the intellectual debates amongst Pryce, Feyerabend, and Bohm, often took place in a weekly seminar led by Feyerabend, and led to a book by Bohm with a related long article by Feyerabend [145], and eventually a book in which both Bohm and Pryce were involved [146]. Much later, Pryce characterized the discussions at that time [147] as "amongst the most exciting discussions of his life".

Most famously, it was this atmosphere that produced (amongst other things) the extraordinary Aharonov–Bohm papers on the non-local aspects of QM [113]. According to Aharonov [147] the formulation of the "Aharonov–Bohm" effect suggested to him by Bohm (involving an infinitesimal flux tube) was too difficult for him to solve, so he went to Pryce, who then solved it on a blackboard. Pryce however refused to let Aharonov and Bohm put Pryce's name on the first paper, saying that what he had done was "too trivial". There was considerable scepticism at the time about the effect: according to Pryce [147], Bohr did not believe it existed at all, and a number of other authors were doubtful (including DeWitt [148]).

It is therefore interesting that one of the first people to realize the full power of the Feynman path integral method was C. Morette, who married B. DeWitt in 1951 (and who thereafter signed herself as either C. M. DeWitt, or as C. DeWitt-Morette, in her publications). Morette also overlapped with Feynman for several months during Feynman's 1951–52 sabbatical in Rio de Janeiro, by which time she had already made an important contribution to the subject of path integrals (in the form of the "Morette/Pauli/van Vleck" formula). Years later, just before she and Bryce DeWitt left Chapel Hill for Austin Texas, she made a profound contribution to quantum mechanics, by inventing [115], along with her student Michael Laidlaw, what are now called "anyons", i.e., particles with fractional statistics. Curiously, according to Laidlaw, at no time during his PhD work did Morette discuss with Laidlaw her views on how one might discuss fractional statistics in terms of path integrals [149]. The full story of how Morette discovered anyons, and the subsequent history, will have to await another article.

References

1. In writing this article we have availed ourselves of 7 hrs of recordings made by PCES in 2016 in Moscow, with A. Starobinsky
2. A.A. Starobinsky, Pis. Zh. Eksp. Teor. Fiz. **30**, 719 (1979) [J.E.T.P. Lett. **30**, 682 (1979)]
3. A.A. Starobinsky, S.M. Churilov, Zh. Eksp. Teor. Fiz. **65**, 65 (1973) [J.E.T.P. **38**, 1 (1974)]. See also Ya. B. Zel'dovich, ZhETF Pis. Red. **14**, 270 (1971) [JETP Lett. **14**, 180 (1971)]; and Zh. Eksp. Teor. Fiz. **62**, 2076 (1972) [Sov. Phys. JETP **35**, 108 (1972)]
4. J. Schwinger, Phys. Rev. **82**, 664 (1951). See also J. Schwinger, Phys. Rev. **93**, 615 (1954), and Phys. Rev. **94**, 1362 (1954)
5. P.W. Anderson, Phys. Rev. **130**, 439 (1963)
6. G.B. Hess, W.M. Fairbank, Phys. Rev. Lett. **19**, 216 (1967)
7. O. Penrose, L. Onsager, Phys. Rev. **104**, 576 (1956). See also O. Penrose, Phil. Mag. **42**, 1378 (1951)
8. C.N. Yang, Rev. Mod. Phys. **34**, 694 (1962)
9. M.J. DesRochers, D. Marchand, P.C.E. Stamp, PNAS **122**, e2421273122 (2025)
10. N.N. Bogoliubov, J. Phys USSR **11**, 23 (1947)
11. E.P. Gross, Il Nouvo Cimento **20**, 454 (1961); L.P. Pitaevskii, Sov.Phys. J.E.T.P. **13**, 451 (1961)
12. Wilks, textitThe Properties of Liqiuid and Solid Helium (Clarendon Press, Oxford, 1967)
13. D.R. Tilley, J. Tilley, *Superfluidity and Superconductivity*, 3rd edn. (CRC Press, 1990)
14. A. Griffin, *Excitations in a Bose Condensed Liquid* (Aambridge University Press, 1993)
15. A.A. Milner, P.C.E. Stamp, V. Milner, PNAS **120**, e2303231120 (2023)
16. R.B. Hallock, J. Low Temp. Phys. **205**, 160 (2021)
17. The form here for the quasiparticle effective Hamiltonian is adapted to a strongly-correlated boson system from the famous Fermi liquid theory formulation of Landau. See, eg., K. Bedell, D. Pines, I. Fomin, J. Low Temp. Phys. **48**, 417 (1982); or K. Bedell, D. Pines, A. Zawadowski, Phys. Rev. B**29**, 102 (1984)
18. D.S. Hirashima, T. Yamashita, F. Iwamoto, Phys. Rev. B **40**, 10891 (1989)
19. L. D. Landau, J. Phys. (Moscow) **5**, 71 (1941); *ibid.***11**, 91 (1947)
20. R.M. Bowley, P.V.E. McClintock, F.E. Moss, G.G. Nancolas, P.C.E. Stamp, Phil. Trans. Roy. Soc. A **307**, 201 (1982)
21. G.W. Rayfield, F. Reif, Phys. Rev. **136**, A1194 (1964)
22. P.C.E. Stamp, P.V.E. McClintock, W.M. Fairbairn, J. Phys. C **12**, L589 (1979)
23. V.L. Berezinskii, Sov. Phys. JETP **34**, 610 (1972)
24. J. M. Kosterlitz, D. J. Thouless, J. Phys. C**5**, L124 (1973), and J. Phys. C **6**, 1181 (1973); and J. M. Kosterlitz, J. Phys. C **7**, 1046 (1974)
25. Z. Hadzibabic, J. Dalibard, Riv. del N. Cimento **34**, 389 (2011)
26. L. Thompson, P.C.E. Stamp, Phys. Rev. Lett. **108**, 184501 (2012)
27. L. Onsager, Il Nuovo Cimento. **6** (Suppl 2) (2): 279–287 (1949)
28. R.P. Feynman, Prog. Low Temp. Phys. **1**, 17 (1955)
29. P.C.E. Stamp, Stud. Hist. Phil. Mod. Phys. **37**, 467 (2006)
30. See, eg., B.I. Halperin, J.K. Jain, *Fractional Quantum Hall Effects: New Developments* (World Scientific, 2020)
31. A. Aharoni, *Introduction to the Theory of Ferromagnetism*, 2nd edn. (Oxford University Press, 1996)
32. P.C.E. Stamp, Phys. Rev. Lett. **66**, 2802 (1991)
33. C. Simon, D.M. Silevitch, P.C.E. Stamp, T.F. Rosenbaum, PNAS **121**, e2315598121 (2024)
34. R.J. Donnelly, *Quantized Vortices in Helium II* (Cambridge University Press, 1991)
35. E. B. Sonin, *Dynamics of Quantized Vortices in Superfluids* (Cambridge University Press, 2016)
36. M. Brown, A.F.G. Wyatt, J. Phys. Condens. Matter **2**, 5025 (1990); see also A.F.G. Wyatt, J. Phys. Condens. Matter **8**, 9249 (1996); A.F.G. Wyatt, Phys. Scr. **T49**, 59 (1993)

37. M.H. Anderson, J.R. Ensher, M.R. Matthews, C.E. Wieman, E.A. Cornell, Science **269**, 198 (1995)
38. K.B. Davis, M.-O. Mewes, M.R. Andrews, N.J. van Druten, D.S. Durfee, Kurn, W. Ketterle, Phys. Rev. Lett. **75**, 3969 (1995)
39. I. Bloch, J. Dalibard, W. Zwerger, Rev. Mod. Phys. **80**, 885 (2008)
40. R. Lopes, C. Eigen, A. Barker, K.G.H. Viebahn, M. Robert-de-Saint-Vincent, N. Navon, Z. Hadzibabic, R.P. Smith, Phys. Rev. Lett. **118**, 210401 (2017)
41. S.V. Andreev, Phys. Rev. Lett. **134**, 166001 (2025)
42. L.P. Pitaevskii, S. Stringari, *Bose-Einstein Condensation* (Oxford University Press, 2003)
43. F. Dalfovo, S. Giorgini, L.P. Pitaevskii, S. Stringari, Rev. Mod. Phys. **71**, 463 (1999)
44. Z. Hadzibabic, P. Kruger, M. Cheneau, B. Battelier, J. Dalibard, Nature **441**, 1118 (2006)
45. G.E. Volovik, Phys. Rep. **351**, 195 (2001)
46. H. Stephani, D. Kramer, M. MacCallum, C. Hoenselaers, E. Herlt, *Exact Solutions to Einstein's Field Equations*, 2nd edn. (Cambridge University Press, 2003)
47. V.A. Belinsky, I.M. Khalatnikov, E.M. Lifshitz, Adv. Phy. **31**, 639 (1982). (and references therein)
48. C.W. Misner, Phys. Rev. Lett. **22**, 1071 (1969)
49. Q. Wang, Z. Zhao, P.C.E. Stamp, to be published
50. J. Martin, C R Phys. **13**, 566 (2012)
51. W. Heisenberg, H. Euler, Z. Physik **98**, 714 (1936)
52. F.J. Dyson, Phys. Rev. **85**, 631 (1952)
53. W. Dittrich, H. Gies, *Probing the Quantum Vacuum* (Springer, 2000)
54. W. Gordon, Ann. Phys. **72**, 421 (1923)
55. W.G. Unruh, Phys. Rev. Lett. **46**, 1351 (1981)
56. P. Painlevé, C.R. Acad, Sci. (Paris) **173**, 677 (1921)
57. T. Jacobson, Phys. Rev. Lett. **75**, 1260 (1995)
58. M. Visser, Class. Q. Grav. **15**, 1767 (1995)
59. R. Sch*ü*tzhold, W.G. Unruh, Phys. Rev. D **66**, 044019 (2002)
60. S.W. Hawking, G.F.R. Ellis, *The Large Scale Structure of Space-Time* (Cambridge University Press, 1973)
61. R. Sch*ü*tzhold, preprint (2025)
62. C. Gooding, S. Biermann, S. Erne, J. Louko, W.G. Unruh, J. Schmiedmayer, S. Weinfurtner, Phys. Rev. Lett. **125**, 213603 (2020)
63. C. Gooding, C. Bunney, R.D. Cameron, S. Tajik, S. Erne, S. Biermann, J. Schmiedmayer, J. Louko, W.G. Unruh, S. Weinfurtner (2025). arXiv:2508.01080
64. W.G. Unruh, J. Low Temp. Phys. **208**, 196 (2022)
65. P.O. Fedichev, U.W. Fischer, Phys. Rev. Lett. **91**, 240407 (2003)
66. A. Retzker, J.I. Cirac, M.B. Plenio, B. Reznik, Phys. Rev. Lett. **101**, 110402 (2008)
67. J. Steinhauer, Nat. Phys. **12**, 959 (2016)
68. J.C. Jaskula, G.B. Partridge, M. Bonneau, R. Lopes, J. Ruaudel, D. Boiron, C.I. Westbrook, Phys. Rev. Lett. **109**, 220401 (2012)
69. A. Jenkins, Phys. Rev. D **109**, 023506 (2024)
70. C. Gooding, New J. Phys. **26**, 105001 (2024)
71. M. Baak, S. Datta, U.W. Fischer, Class. Q. Grav. **40**, 215001 (2023)
72. J.C. Maxwell, *A Treatise on Electricity and Magnetism*, vols. I and II (Clarendon, Oxford, 1873)
73. C. Huygens, *Traité de la Lumière* (Pieter van der Aa, Leiden, 1690)
74. I. Newton, *it Opticks* (Royal Society, 1704)
75. Penrose discusses the optical analogy in several places - an example is R. Penrose, pp. 259–274 in *Perspectives in Geometry and Relativity: Essays in Honour of Vaclav Havaty*, ed. by B. Hoffmann (Indiana University Press, 1966). Note that this work was done in 1960–61
76. Interviews with R. Penrose, conducted between 2013 and 2019
77. N.N. Bogoloiubov, V.V. Tolmachev, D.V. Shirkov, *A New Method in the Theory of Superconductivity* (Consultants Bureau Inc., 1959)

78. P.W. Anderson, Phys. Rev. **112**, 1900 (1958)
79. P. Higgs, Phys. Rev. Lett. **13**, 508 (1964). See also F. Englert, R. Brout, Phys. Rev. Lett. **13**, 321 (1964); and G. Guralnik. C.R. Hagen, T.W.B. Kibble, Phys. Rev. Lett. **13**, 585 (1964)
80. E.S. Reich, *Plastic Fantastic: How the Biggest Fraud in Physics Shook the Scientific World* (Pargrave MacMillan, 2009)
81. M.E. Dodd, P.C. Hendry, N.S. Lawson, P.V.E. McClintock, C.D.H. Williams, Phys. Rev. Lett. **81**, 3703 (1998), and refs. therein
82. Some of the early texts which defined "orthodox QM" were: P.A.M Dirac, *The Principles of Quantum Mechanics*, 1st edn. (Oxford University Press, 1930); W. Pauli, *The General Principles of Quantum Mechanics*, 1st edn. (Springer, 1934); L.D. Landau, E.M. Lifshitz, *Quantum Mechanics: Non-Relativistic Theory*, 1st English edn. (Pergamon, 1958)
83. J. van Neumann, *Mathematical Foundations of Quantum Mechanics* (Princeton University Press, 1955), translated from *Mathematische Grundlagen der Quantenmechanik* (Springer, 1932)
84. D. Bohm, *Quantum Theory* (Prentice-Hall, 1951)
85. F. Laloe, *Do we really understand Quantum Mechanics?*, 2nd edn. (Cambridge University Press, 2019)
86. A.J. Leggett, Sup. Prog. Th. Phys (Jap) **69**, 80 (1980); see also A.J. Leggett, Ch. 6, pp. 395–506 in *Chance and Matter*, Les Houches vol. XLVI (1986), ed. J. Souletie, J. Vannimenus, R. Stora (Elsevier, 1987)
87. R.P. Feynman, Optics News, Feb 1985, pp. 11–20; see also R.P. Feynman, *Lectures on Computation* (Addison-Wesley, 1996), and P. Benioff, Phys. Rev. Lett. **48**, 1581 (1982)
88. J.I. Korsbakken, K.B. Whaley, J. Dubois, J.I. Cirac, Phys. Rev. A **75**, 042106 (2007); J. Korsbakken, F. Wilhelm, K. Whaley, Europhys. Lett. **89**, 30003 (2010); T.J. Volkoff, K.B. Whaley, Phys. Rev. A **89**, 012122 (2014).. See also G.C. Knee, K. Kakuyanagi, M.-C. Yeh, Y. Matsuzaki, H. Toida, H. Yamaguchi, S. Saito, A.J. Leggett, W.J. Munro, Nat. Commun. **7**, 13253 (2016); and A.J. Leggett arXiv:1603.03992
89. S. Takahashi, I.S. Tupitsyn, J. van Tol, C.C. Beedle, D.N. Hendrickson, P.C.E. Stamp, Nature **476**, 76 (2011)
90. M. Arndt, K. Hornberger, Nat. Phys. **10**, 271 (2014)
91. R. Horodecki, P. Horodecki, M. Horodecki, K. Horodecki, Rev. Mod. Phys. **81**, 865 (2009)
92. T. Cox, P.C.E. Stamp, Phys. Rev. A **98**, 062110 (2018)
93. A.O. Caldeira, A.J. Leggett, Phys. Rev. Lett. **46**, 211 (1981); A.O. Caldeira, A.J. Leggett, Ann. Phys. **149**, 374 (1983)
94. N.V. Prokof'ev, P.C.E. Stamp, Rep. Prog. Phys. **63**, 669 (2000)
95. M. Jammer, *The Philosophy of Quantum Mechanics* (Wiley, 1974)
96. F. London, E. Bauer, *La Theorie de l'Observation en Mecanique Quantique* (Hermann et Cie, Paris, 1938)
97. J.S. Bell, *Speakable and Unspeakable in Quantum Mechanics*, 2nd edn. (Cambridge University Press, 2004)
98. Y. Aharonov, P.G. Bergmann, J.L. Lebowitz, Phys. **134**, B1410 (1964)
99. J. Dressel, M. Malik, F.M. Miatto, A.N. Jordan, R.W. Boyd, Rev. Mod. Phys. **86**, 307 (2014); L. Qin, W. Feng, X.-Q Li, Sci. Rep. **6**, 20286 (2016), p. 19
100. X.S. Ma, J. Kofler, A. Zeilinger, Rev. Mod. Phys. **88**, 015005 (2016)
101. M.O. Scully, K. Drühl, Phys. Rev. A **25**, 2208 (1982): and K. Yoon-Ho, R. Yu, S.P. Kulik, Y.H. Shih, M.O. Scully, Phys. Rev. Lett. **84**, 1 (2000); X.S. Ma et al., PNAS **110**, 1221 (2013)
102. For a review of ideas about retrocausality in QM, see S.Friederich, P.W. Evans, *Retrocausality in Quantum Mechanics*. Stanford Encyclopedia of Philosophy (Nov 13, 2023)
103. J.G. Cramer, Rev. Mod. Phys. **58**, 647 (1986); R.E. Kastner, *The Transactional Interpretation of Quantum Mechanics: The Reality of Possibility* (Cambridge University Press, 2012)
104. Clearly any theory that assumes that QM can be entirely described using the Schrodinger equation, or some other deterministic equation, will make the idea of "choice" redundant. In recent years this idea has been renamed "super-determinism", and discussed by a number of authors - see eg., C.H. Brans, Int. J. Theor. Phys. **27** 219 (1988); or S. Hossendfelder, T. Palmer, Frontiers Phys. **8**, 139 (2020)

105. D. Bohm, Phys. Rev. **85**, 166 (1952), and *ibid***85**, 180 (1952). See also D. Bohm, Phys. Rev. **87**, 389 (1953), and *ibid***89**, 319 (1953) and **89**, 459 (1953)
106. L. De Broglie, Comptes Rendus **183**, 447 (1926); *ibid* **184**, 273 (1927); and L. de Brogie, J. de Physique et du Radium **8**, 225 (1927)
107. B. DeWitt, N. Graham (eds.), *The Many World Interpretation of Quantum Mechanics* (Princeton University Press, 1973), with reprints of several articles by DeWitt; see also B. DeWitt, Physics Today, **23**, 30 (1970)
108. H. Everett, Rev. Mod. Phys. **29**, 454 (1957); and H. Everett, *The theory of the universal wave-function* (Ph.D. thesis, Princeton, 1957), reprinted in B. DeWitt, N. Graham [107]
109. That the interaction with the environment did not solve the problem of macroscopic interference terms was of course well understood by, eg., Einstein and Schrodinger in the 1930s. Considerable debate took place in Princeton after the war about this - this is reflected in Bohm's 1951 book [84], in Everett's work [108] (see also a letter from Everett to Dewitt in 1957 [110]), and in lecture notes and articles by Wigner in the early 1960s [111]. For more discussion of this see the Appendix
110. H. Everett, letter to B. DeWitt, May 31 (1957)
111. E.P. Wigner, *Symmetries and Reflections.* (Indiana University Press, 1967), contains several chapters by Wigner on foundational questions in QM. See also E.P. Wigenr, Am. J. Phys. **31**, 6 (1963), and E.P. Wigner, The Monist **48**, 248 (1964)
112. C. Morette-DeWitt, Comm. Math. Phys. **28**, 47 (1972)
113. Y. Aharonov, D. Bohm, Phys. Rev. **115**, 485 (1959); and *ibid* **123**, 1511 (1961)
114. R.P. Feynman, R. Leighton, M. Sands, *The Feynman Lectures on Physics* (Addison-Wesley, 1965), vol. 2, Chap. 15, and vol. 3, Chap. 21
115. M.G.G. Laidlaw, C. Morette-DeWitt, Phys. Rev. D **3**, 1375 (1971); see also the later paper of J.M. Leinaas, J. Myrheim, Nuovo Cim. **37**B, 1 (1977)
116. W. Yong-Shi, Phys. Rev. Lett. **52**, 2103 (1984)
117. J.B. Hartle, Phys. Rev. D **37**, 2818 (1988); Phys. Rev. D **38**, 2985 (1988)(1994); and Phys. Rev. D **49**, 6543; G. Klinkhammer, K.S. Thorne, unpublished; and D.S. Goldwirth, M.J. Perry, T. Piran, K.S. Thorne, Phys. Rev. D **49**, 3951 (1994)
118. J. Wilson-Gerow, P.C.E. Stamp, Phys. Rev. D **105**, 084015 (2022)
119. T.W.B. Kibble, Comm. Math. Phys. **64**, 73 (1978); T.W.B. Kibble, S. Randjbar-Daemi, J. Phys A **13**, 141 (1980)
120. S. Weinberg, Phys. Rev. Lett. **62**, 485 (1989); S. Weinberg, Ann. Phys. (NY) **194**, 336 (1989)
121. See J.S. Bell, Rev. Mod. Phys. **38**, 447 (1966)
122. J. Polchinski, Phys. Rev. Lett. **66**, 397 (1991)
123. T.W.B. Kibble, Comm. Math. Phys. **65**, 189 (1979)
124. R.M. Wald, *General Relativity* (University of Chicago Press, 1984), section 14.1
125. P.C.E. Stamp, New J. Phys. **17**, 065017 (2015)
126. See R.P. Feynman et al., in *The Role of Gravitation in Physics: Report from the 1957 Chapel Hill Conference*, ed. C.M. DeWitt, D. Rickles (Max-Planck-Gesellschaft zur Förderung der Wissenschaften, Berlin, 1957), particularly session VIII, Secs. 22, 23
127. A.O. Barvinsky, D. Carney, P.C.E. Stamp, Phys. Rev. D **98**, 084052 (2018)
128. A.O. Barvinsky, J. Wilson-Gerow, P.C.E. Stamp, Phys. Rev. D **103**, 064028 (2021)
129. P.C.E. Stamp, Phil. Trans. R. Soc. A **370**, 4429 (2012)
130. D. Carney, P.C.E. Stamp, J.M. Taylor, Classical Quantum Gravity **36**, 034001 (2019)
131. J.L. Synge, *Relativity: The General Theory: Pub* (North-Holland, 1960)
132. On a historical note - it was actually J. Polchinski who first encouraged PCES to publish the CWL ideas in their preliminary form, during a visit by PCES to Santa Barbara in 2009. But at the time Polchinski never told PCES of the paper [122] published in 1991!
133. J. Wilson-Gerow, Y. Chen, P.C.E. Stamp, Phys. Rev. D **109**, 064078 (2024)
134. P. Feldmann, S. Tajik, P.C.E. Stamp, to be published
135. J. Wilson-Gerow, Y. Malicki, M.O. Scully, P.C.E. Stamp, to be published
136. H. Yu, L. McCuller, et al., Nature **583**, 43 (2020). To give some context to this work, see K.S. Thorne, in *The Future of Spacetime*, ed. by S.W. Hawking, K.S. Thorne, I. Novikov, T. Ferris, A. Lightman (WW Norton, 2002), pp. 109–152; in particular, note "prediction No. 5"

137. G. Ludwig, Zeit. Physik **135**, 483 (1953); *ibid* **150**, 486 (1958), *ibid 152*, 98 (1958)
138. H.S. Green, Nuovo Cim. **9**, 880 (1958)
139. A. Daneri, A. Loinger, G.M. Prosperi, Nucl. Phys. **33**, 297 (1962), and Nuovo Cimento **44B**, 119 (1966)
140. H.D. Zeh, Found. Phys. **1**, 69 (1970)
141. M. Simonius, Phys. Rev. Lett. **40**, 980 (1978)
142. A. Peres, Phys. Rev. D **22**(879), 28 (1980)
143. A.J. Leggett, Journal de Physique **39**, C6-1264 (1978)
144. For much more detail on Everett's work, and his life, see P. Byrne, *The Many Worlds of Hugh Everett III: Multiple Universes, Mutual Assured Destruction, and the Meltdown of a Nuclear Family* (Oxford University Press, 2010); and also *Many Worlds? Everett, Quantum Theory, and Reality*, ed. by S. Saunders, J. Barrett, A. Kemt, D. Wallace (Oxford University Press, 2010)
145. D. Bohm, *Causality and Chance in Modern Physics* (Routledge, 1957); see also the discussion of P.K. Feyerabend, Brit. J. Phil. Sci. **10**, 321 (1960)
146. *Quanta And Reality: A Symposium*, D. Bohm, N.R. Hanson, M.B. Hesse, Nicholas Kemmer, A.B. Pippard, Maurice Pryce (American Research Council, 1962)
147. Discussions between PCES, Y. Aharonov, M.H.L. Pryce at UBC (1994)
148. B. DeWitt, Phys. Rev. **125**, 2189 (1962)
149. Interviews with M. Laidlaw, 9th Nov and 14 Dec, 2020 (H. Brown also attending)

Recent Developments in Stochastic Inflation

R. P. Woodard

Abstract This article is dedicated to the memory of Alexei Starobinsky. I begin with some recollections of him and then review the generalization of his wonderful stochastic formalism from scalar potential models to theories which interact with fermions and photons, and finally to theories with derivative interactions such as nonlinear sigma models and gravity. This entails effective potentials generated by the usual field-dependent masses, as well as by field-dependent field strengths, and by field-dependent Hubble parameters. I also discuss secular loop corrections which cannot be captured by stochastic techniques.

1 Introduction

Alexei Starobinsky was a genius whose contributions to cosmology include one of the earliest models of primordial inflation [1] which is still completely viable [2]. He also made the remarkable suggestion that quantum fluctuations from the epoch of primordial inflation might be observable today [3] and can even be used to constrain models of inflation [4].

Of particular interest to me is the stochastic technique Alexei devised to explain the secular growth one finds in the correlation functions of some quantum field theories on inflationary backgrounds [5]. With Yokoyama, he even developed a way of summing up the leading secular corrections to all orders [6]. I believe this technique will eventually be recognized as equal in significance to the development of the renormalization group by the late Ken Wilson for flat space quantum field theory and statistical mechanics [7, 8]. Wilson won the 1982 Nobel Prize for this work and I feel that Alexei's contributions would have been similarly honored, had he only lived.

This article consists of 7 sections. In the next one I share some personal recollections of Alexei from the two decades during which I was privileged to have known

R. P. Woodard (✉)
Department of Physics, University of Florida, Gainesville, FL, USA
e-mail: woodard@phys.ufl.edu

A. Barvinsky and A. Kamenshchik (eds.), *Open Issues in Gravitation and Cosmology*, Fundamental Theories of Physics 225, https://doi.org/10.1007/978-3-032-15703-4_28

him. In Sect. 3 I review the large secular loop corrections which occur in scalar potential models on inflationary backgrounds, and I prove that Alexei's stochastic formalism recovers the leading logarithms produced at each order in perturbation theory. Section 4 explains how to handle charged scalars and those with Yukawa couplings. In Sect. 5 I discuss derivative interactions involving scalars and gravity. Theories with such derivative interactions can experience another sort of leading logarithm loop correction which cannot be described stochastically but is instead amenable to treatment by a variant of the renormalization group. This is described in Sect. 6. My conclusions comprise Sect. 7.

2 Remembering Alexei Starobinsky

American elites are infamous for their view that the only interesting parts of my country are portions of the coasts, extending roughly from Boston to Washington in the East, and from Seattle to Los Angeles in the West. Many dismiss the rest of the nation as "fly-over country" which they only see while traveling between the parts they consider civilized. I'm told that Russian elites take a similar view of Moscow and Saint Petersburg. Alexei was not like that. Although he had many friends and collaborators outside of Russia, and could surely have emigrated had he wished, he stood by his country during hard times, and supported physics throughout its vast expanse. Indeed, he contracted the illness which took his life while attending an event in Kazan. And I first met Alexei in 2002, at a conference in the Siberian city of Tomsk.

At that time of our first meeting I had been applying the Schwinger–Keldysh formalism [9–13], with dimensional regularization on de Sitter background, to work out loop corrections in renormalizable theories of massless, minimally coupled scalars which interact with themselves [14], or with electromagnetism [15, 16]. It had long been known that the coincidence limit of the massless, minimally coupled scalar propagator on de Sitter shows secular growth [4, 17, 18]. This observation had eventually led to a full-blown proof that there are no normalizable de Sitter invariant states in this theory [19]. What I was showing is that the secular growth persists when realistic interactions are added.

I confess to having felt a bit persecuted at the time of the Tomsk meeting because both quantum field theory experts and cosmologists were, for different reasons, dismissing my carefully derived results as nonsense. The point which bothered quantum field theory people is that the secular growth I found breaks de Sitter invariance. They were convinced that de Sitter invariance should prove as powerful an organizing principle for QFT on de Sitter background as Poincaré invariance does on flat space background. They therefore suspected that my de Sitter breaking effects must derive from a poorly chosen vacuum, or from having posed unphysical questions [20], or from outright mistakes. On the other hand, some cosmologists altogether rejected the application of QFT on inflationary backgrounds. They argued that QFT expectation values amounted to taking unphysical averages over portions of the wave function which had de-cohered and could not be simultaneously accessible to any real observer. Therefore, they concluded that my results were as unphysical as averaging over the alive-dead cat superposition in the famous paradox of Schrödinger [21].

I spoke on my work and ended by asking, with some asperity, "What is *wrong* with this?" Alexei raised his hand and said that he didn't think anything was wrong with my results, and that the secular growth they contained could be predicted by his stochastic formalism. I was dumbfounded (pleasantly) by a reaction so different from the ones I usually encountered. But I was also suspicious that there could be a trivial way of capturing the most important parts of dimensionally regulated and fully renormalized QFT calculations which required weeks of hard work. However, the possibility was too exciting to ignore, and I agreed to check Alexei's predictions against some three loop results I had not published. I screwed up the check at first, but Alexei patiently explained how to do it correctly, at which point I became convinced that his formalism must be right. Alexei was used to the skepticism of QFT experts that an ultraviolet finite, stochastic random variable could reproduce important features of an ultraviolet divergent, quantum field theory, and he must have sensed in me a willing tool to clarify for them a truth which he could see directly. In any case, he asked me to devise a proof from quantum field theory.

Alexei's friends and collaborators will confirm that he sometimes posed difficult problems! That particular one took me and Nick Tsamis three years to solve. When we finally found a proof [22] it was my great pleasure to present it to Alexei during a conference at the Lebedev Institute. Since then Alexei and I have overlapped at many sites: Dubna, Moscow, Paris, Ann Arbor, Utrecht, Cargese, Tallinn, as well as innumerable virtual conferences. I treasured these opportunities to learn from him, and I was seldom disappointed.

Like many brilliant scientists, Alexei was sometimes impatient with lesser men. It was my good fortune that he was always willing to talk with me. We twice began working together on projects, but never brought anything to completion. It saddens me that the opportunity to do so has been forever foreclosed by his untimely passing.

Let me close this section by commenting that, for someone so renowned, Alexei was quite shy. My wife, Shun-Pei, who is also a shy person, picked up on this immediately when she first met him, during the fall of 2006. She had recently discovered the first large logarithm induced by a loop of gravitons in a dimensionally regulated and fully renormalized computation [23, 24], and was attending a 3-month advanced school in Paris while still a graduate student. She and Alexei used to chat quietly in the corner during coffee breaks, while more gregarious and less accomplished physicists held forth loudly in the center. Years later, as a full professor in her own right, Shun-Pei organized the Taiwan portion of a Russia-Taiwan funding proposal with Alexei and Andrei Barvinsky. It was a good proposal and I'm sure it would have won support, however, the entire program was peremptorily terminated due to the worsening of relations that followed the Russian invasion of Ukraine, which Alexei opposed.

3 Why Starobinsky's Formalism Works

One can characterize the geometry of cosmology by its scale factor $a(t)$, Hubble parameter $H(t)$ and first slow roll parameter $\epsilon(t)$,

$$ds^2 = -dt^2 + a^2(t)d\mathbf{x}\cdot d\mathbf{x}, \qquad H(t) \equiv \frac{\dot{a}}{a}, \qquad \epsilon(t) \equiv -\frac{\dot{H}}{H^2}. \tag{1}$$

Spacetime expansion means that the first derivative of $a(t)$ is positive, which corresponds to $H(t) > 0$. Inflation means that the second derivative is also positive, which corresponds to $0 \leq \epsilon(t) < 1$. We know that this can occur because it's taking place now [25, 26]. The simplest interpretation of cosmological data [27] is that it also took place during the very early universe [28], and Alexei's model [1] offers one of the best explanations for what caused it.

For light fields which are not conformally invariant the accelerated expansion of inflation can rip virtual, long wavelength quanta out of the vacuum so that they become real particles. This occurs maximally for massless, minimally coupled scalars, and for gravitons whose linearized equation of motion is the same [29]. On de Sitter background ($a(t) = e^{Ht}$ with H constant) the occupation number for a single wave vector $\mathbf{k}$ experiences staggering growth,

$$N(t,k) = \left[\frac{a(t)H}{2k}\right]^2. \tag{2}$$

This is what caused the (still unresolved) tensor power spectrum [3] and its scalar cousin [30], which has been measured to three significant figures.

The exponentially increasing occupation numbers (2) for super-horizon modes is what makes the coincident scalar propagator grow [4, 17, 18]. The thing which got me in so much trouble was asserting that this growth must persist when interactions are present. The simplest example is a scalar potential model,

$$\mathcal{L} = -\frac{1}{2}\partial_\mu\phi\partial_\nu\phi g^{\mu\nu}\sqrt{-g} - V(\phi)\sqrt{-g}. \tag{3}$$

Few would argue that the expectation value of the stress tensor is unphysical. For any reasonable vacuum on the cosmological background (1) it must take the perfect fluid form,

$$\langle\Omega|T_{\mu\nu}(t,\mathbf{x})|\Omega\rangle = [\rho(t) + p(t)]\delta^0_{\ \mu}\delta^0_{\ \nu} + p(t)g_{\mu\nu}. \tag{4}$$

For the case of $V(\phi) = \frac{1}{4!}\lambda\phi^4$ on de Sitter background the 1-loop result can be absorbed into a renormalization of the cosmological constant but the 2-loop results (suitably renormalized and with appropriate 1-loop state corrections to absorb exponentially falling terms) show secular growth [14, 31, 32],

$$\rho(t) = \frac{\lambda H^4}{2^7\pi^4} \times \ln^2[a(t)] + O(\lambda^2), \tag{5}$$

$$p(t) = \frac{\lambda H^4}{2^7\pi^4} \times \Big\{-\ln^2[a(t)] - \frac{2}{3}\ln[a(t)]\Big\} + O(\lambda^2). \tag{6}$$

It is time for a crucial digression on notation. The $\lambda \ln^2[a(t)]$ corrections in (5)–(6) are *leading logarithm*; the $\lambda \ln[a(t)]$ contribution to (6) is a *sub-leading logarithm.* For order λ^N corrections in a quartic potential the leading logarithm contribution is $\ln^{2N}[a(t)]$, and any term with fewer factors of $\ln[a(t)]$ is sub-leading. For a general monomial potential $V \sim \lambda\phi^K$ the leading logarithm contribution at order λ^{2N} is $\ln^{K \cdot N}[a(t)]$ [22].

Vakif Onemli and I derived (5)–(6) the hard way, by explicitly computing the various diagrams in dimensional regularization, then adding counterterms and taking the unregulated limit. It took weeks. Starobinsky devised a way of trivially getting the leading logarithm result, not just for the stress tensor of a quartic potential but for any correlator in any scalar potential model (3), and quite a bit before us [5]. The technique is to replace the exact Heisenberg equation for the ultraviolet divergent quantum field $\phi(t, \mathbf{x})$,

$$-\frac{1}{\sqrt{-g}}\partial_\mu[\sqrt{-g}\, g^{\mu\nu}\partial_\nu\phi] = -V'(\phi), \tag{7}$$

with a Langevin equation for the ultraviolet finite stochastic field $\varphi(t, \mathbf{x})$,

$$3\, H[\dot{\varphi} - \dot{\varphi}_0] = -V'(\varphi). \tag{8}$$

The stochastic jitter $\dot{\varphi}_0$ is the time derivative of the infrared truncated, free field mode sum,

$$\varphi_0(t, \mathbf{x}) = \int_H^{Ha(t)} \frac{d^3k}{(2\pi)^3} \frac{He^{i\mathbf{k}\cdot\mathbf{x}}}{\sqrt{2k^3}} \Big\{\alpha(\mathbf{k}) + \alpha^\dagger(-\mathbf{k})\Big\}, \quad [\alpha(\mathbf{k}), \alpha^\dagger(\mathbf{p})] = (2\pi)^3\delta^3(\mathbf{k}-\mathbf{p}). \tag{9}$$

There are two ways to use Starobinsky's Langevin equation (8). The first is to solve it perturbatively for $\varphi(t, \mathbf{x})$ as an expansion in powers of $\varphi_0(t, \mathbf{x})$. For the quartic potential this gives,

$$\varphi(t, \mathbf{x}) = \varphi_0(t, \mathbf{x}) - \frac{\lambda}{18\, H}\int_0^t dt'\, \varphi_0^3(t', \mathbf{x}) + \frac{\lambda^2}{108\, H^2}\int_0^t dt' \varphi_0^2(t', \mathbf{x})\int_0^{t'} dt''\, \varphi_0^3(t'', \mathbf{x}) + O(\lambda^3). \tag{10}$$

Substituting in the stress tensor and taking the expectation value gives the leading logarithm expansion for the energy density (5) [22],

$$\rho(t) \longrightarrow \frac{\lambda H^4}{2^7\pi^4} \ln^2[a(t)] \times \Big\{1 - \frac{\lambda}{12\pi^2}\ln^2[a(t)] + \frac{53\lambda^2}{2^4 3^4 5\pi^4}\ln^4[a(t)] + O\Big(\lambda^3 \ln^6[a(t)]\Big)\Big\}. \tag{11}$$

Note that I have carried this to *fourth order* in the loop expansion.

The second way of using Starobinsky's Langevin equation (8) is to derive a Fokker-Planck equation for the time-dependent probability density $\varrho(t, \varphi)$ [6],

$$\dot{\varrho}(t, \varphi) = \frac{\partial}{\partial \varphi}\Big[\frac{V'(\varphi)}{3H}\varrho(t, \varphi)\Big] + \frac{H3}{8\pi^2}\frac{\partial^2 \varrho(t, \varphi)}{\partial \varphi^2}. \tag{12}$$

If the potential is bounded below, and steep enough, the system will approach an equilibrium when the inward pressure of the classical force balances the outward pressure of inflationary particle production. In that case the left hand side of equation (12) vanishes and we can solve for the late time probability density,

$$\varrho(t, \varphi) \longrightarrow N \exp\Big[-\frac{8\pi^2 V(\varphi)}{3H^4}\Big]. \tag{13}$$

This allows one to sum the series of leading logarithms. For example, the late time limit of the energy density for the quartic potential is,

$$\rho(t) \longrightarrow \frac{3H^4}{8\pi^2}\frac{\Gamma(\frac{3}{4})}{\Gamma(\frac{1}{4})}. \tag{14}$$

The physical interpretation is that inflationary particle production pushes the scalar up its quartic potential until an equilibrium is reached with the downward classical force.

Before giving the proof of Starobinsky's formalism let me comment on how very strange it seems from the perspective of quantum field theory. The original scalar potential model (3) describes a quantum field which obeys a second order equation and whose commutator $[\phi(t, \mathbf{x}), \phi(t', \mathbf{x}')]$ is nonzero for timelike and lightlike separations. The correlators of this quantum field are typically ultraviolet divergent and require both regularization and renormalization. In contrast, the infrared truncated mode sum (9) for $\varphi_0(t, \mathbf{x})$ is a stochastic random variable which commutes with itself everywhere, $[\varphi_0(t, \mathbf{x}), \varphi_0(t', \mathbf{x}')] = 0$. Because the mode sum is cut off at $k = Ha(t)$, its correlators are ultraviolet finite. These same two properties are inherited by the full field $\varphi(t, \mathbf{x})$, which obeys the first order equation (8). Yet correlators of φ agree with those of ϕ at leading logarithm order. You can see why quantum field theory experts were suspicious of Alexei's formalism!

The first step in proving Starobinsky's formalism is to act the retarded Green's function on the dimensionally regulated Heisenberg field equation (7) to derive the Yang-Feldman equation [22],

$$\phi(x) = \phi_0(x) - \int d^D x' \sqrt{-g(x')}\, i\theta(t-t')[\phi_0(x), \phi(x')] V'\Big(\phi(x')\Big). \tag{15}$$

The free field $\phi_0(x)$ is,

$$\phi_0(t, \mathbf{x}) = \int_H \frac{d^{D-1}k}{(2\pi)^{D-1}}\Big\{\alpha(\mathbf{k})u(t, k)e^{i\mathbf{k}\cdot\mathbf{x}} + \alpha^\dagger(\mathbf{k})u^*(t, k)e^{-i\mathbf{k}\cdot\mathbf{x}}\Big\}, \tag{16}$$

where $\alpha^\dagger$ and α are canonically normalized creation and annihilation operators and the free field mode function in Bunch–Davies vacuum [33–35] is,

$$u(t,k) \equiv i\sqrt{\frac{\pi}{4Ha^{D-1}}}\,H^{(1)}_{\frac{D-1}{2}}\Big(\frac{k}{aH}\Big). \tag{17}$$

Note that I have expressed the retarded Green's function as $i\theta(t-t')$ times the commutator of two free fields. Note also that the infrared limit of the mode function is,

$$k \ll Ha \qquad \Longrightarrow \qquad u(t,k) = \frac{\Gamma(\frac{D-1}{2})}{\sqrt{4\pi H}}\Big(\frac{2H}{k}\Big)^{\frac{D-1}{2}}\Big\{1 + O\Big(\frac{k^2}{a^2H^2}\Big)\Big\}. \tag{18}$$

This is why the free field mode sum (16) has been infrared cut off at $k = H$ to exclude modes which are in correlated Bunch–Davies vacuum at the beginning of inflation [17].

The Yang–Feldman equation (15) is exact; iterating it would produce the usual interaction picture expansion of the full field ϕ in powers of the free field ϕ_0. Factors of $\ln[a(t)] = Ht$ in correlators derive from two sources:

1. The integral of $u(t,k)u^*(t',k)$ from the infrared cutoff at $k = H$ to the point $k = H \times \min[a(t), a(t')]$ at which the infrared limit (18) breaks down; and
2. Vertex integrations over the retarded Green's function.

The first source is associated with a pair of free fields which combine to give a propagator; the second source is also associated with the pair of free fields in the commutator of the retarded Green's function. Now recall that reaching leading logarithm order requires that each extra pair of vertices must contribute one factor of $\ln[a]$ for each pair of fields. In the interaction picture this translates to each pair of free fields (including those in the retarded Green's function) contributing a factor of $\ln[a]$. Because these factors derive from the two sources given above, correlators will be unaffected, at leading logarithms order, if we cut off the mode sum in (16) at $k = Ha(t)$ (at which point we can take $D = 4$), and also retain only the term (18) in the small k/aH expansion of the mode function.

Because the retarded Green's function involves a commutator, we must carry the expansion for these terms out to include the first imaginary part,

$$D = 4 \qquad \Longrightarrow \qquad u(t,k) = \frac{H}{\sqrt{2k^3}}\Big\{1 + \frac{1}{2}\big(\tfrac{k}{aH}\big)^2 + \frac{i}{3}\big(\tfrac{k}{aH}\big)^3 + \dots\Big\}. \tag{19}$$

Hence the retarded Green's function becomes,

$$G(x;x') = i\theta(t-t')\int \frac{d^3k}{(2\pi)^3}\,e^{i\mathbf{k}\cdot(\mathbf{x}-\mathbf{x}')}\Big\{u(t,k)u^*(t',k) - u^*(t,k)u(t',k)\Big\}, \tag{20}$$

$$\longrightarrow \frac{\theta(t-t')}{3H} \times \Big[\frac{1}{a^3(t')} - \frac{1}{a^3(t)}\Big] \times \delta^3(\mathbf{x}-\mathbf{x}'). \tag{21}$$

Of course this allows us to perform the spatial integrations in the Yang–Feldman equation (15). A final simplification becomes apparent after multiplying (21) by the measure factor $\sqrt{-g(x')} = a^3(t') = e^{3Ht'}$ in expression (15). It follows that the factor of $1/a^3(t)$ in (21) can be neglected,

$$\int_0^t dt' a^3(t') \times [\tfrac{1}{a^3(t')} - \tfrac{1}{a^3(t)}] = t - \tfrac{1}{3H}[1 - e^{-3Ht}]. \tag{22}$$

The various truncations produce a completely different field $\varphi(t, \mathbf{x})$ whose correlators nonetheless agree with those of $\phi(t, \mathbf{x})$ at leading logarithm order. The truncated Yang–Feldman equation obeyed by this field is,

$$\varphi(t, \mathbf{x}) = \varphi_0(t, \mathbf{x}) - \tfrac{1}{3H}\int_0^t dt' \, V\Big(\varphi(t', \mathbf{x})\Big). \tag{23}$$

Here $\varphi_0(t, \mathbf{x})$ is precisely the infrared truncated free field expansion given in (9). Taking the time derivative of (23), and making some trivial rearrangements, gives Starobinsky's Langevin equation (8). Note that we have not only established its validity, we have clarified its precise relation to quantum field theory: *Starobinsky's formalism reproduces the leading logarithms of scalar potential models* (3).

4 Passive Fields

It is important to distinguish between *Active Fields*, which experience inflationary particle production (2), and *Passive Fields*, which do not. The prime examples of active fields are massless, minimally coupled scalars and gravitons; examples of passive fields include conformally coupled scalars, fermions and photons. We have seen that loops of active fields induce secular growth such as (5)–(6). Passive fields do not induces such factors, but they can mediate them, and they can modify the way that active fields interact.

Passive fields typically contribute nonzero constants to correlators. For example, one of several 1-loop electromagnetic contributions to the dimensionally regulated graviton self-energy on de Sitter background involves the coincidence limit of two field strengths [36],

$$\langle \Omega | F_{\mu\nu}(x) F_{\rho\sigma}(x) | \Omega \rangle = \frac{H^D}{(4\pi)^{\frac{D}{2}}} \frac{\Gamma(D-1)}{\Gamma(\frac{D}{2}+1)} \times [g_{\mu\rho} g_{\nu\sigma} - g_{\mu\sigma} g_{\nu\rho}]. \tag{24}$$

Such contributions come from the full range of modes, from ultraviolet to infrared. They also involve the full passive field mode function, not just its infrared truncation. This is why no stochastic formulation of a passive field can be correct. For example, stochastically truncating electromagnetism would render the field strength an ultraviolet finite, Hermitian operator whose square must be positive. That is obviously inconsistent with the exact result (24) for the case of $\mu = \rho$ and $\nu = \sigma$.

The right thing to do with passive fields is integrate them out in the presence of a constant active field background. For passive fields coupled to an active scalar field setting this scalar to a constant induces a mass for the passive field, and integrating out the passive field results in a Coleman–Weinberg potential [37]. At this stage the active field has been reduced to a scalar potential model on which Starobinsky's formalism can be applied. Two theories for which this procedure has been applied are a real scalar which is Yukawa-coupled to a fermion [38], and a complex scalar which is coupled to electromagnetism [39].

The Lagrangian for Yukawa theory is,

$$\mathcal{L} = -\tfrac{1}{2}\partial_\mu\phi\partial_\nu\phi g^{\mu\nu}\sqrt{-g} - \tfrac{1}{2}\delta\xi\phi^2 R\sqrt{-g} - V(\phi)\sqrt{-g} - f\phi\overline{\psi}\psi\sqrt{-g} + \overline{\psi}e^{\mu}{}_{a}\gamma^a\Big(i\partial_\mu - \tfrac{1}{2}A_{\mu bc}J^{bc}\Big)\psi\sqrt{-g}. \tag{25}$$

The usual notation applies: γ^a denotes the gamma matrices, the vierbein is $e^{\mu}{}_{a}(x)$, the spin connection is $A_{\mu bc}(x)$, and the spin generator is J^{bc},

$$g^{\mu\nu} = e^{\mu}{}_{a}e^{\nu}{}_{b}\eta^{ab}, \quad A_{\mu bc} \equiv e^{\nu}{}_{b}(e_{\nu c,\mu} - \Gamma^{\rho}{}_{\mu\nu}e_{\rho c}), \quad J^{bc} \equiv \tfrac{i}{4}[\gamma^b, \gamma^c]. \tag{26}$$

For constant ϕ the last term on the first line of (25) can be recognized as a fermion mass of $m = f\phi$. The massive fermion propagator on de Sitter $i[{}_iS_j](x; x')$ was derived by Candelas and Raine [40]. Integrating the fermion out of the scalar field equation (with suitable renormalizations) gives,

$$\begin{aligned}
&\partial_\mu(\sqrt{-g}\, g^{\mu\nu}\partial_\nu\phi) - \delta\xi\phi R\sqrt{-g} - V'(\phi)\sqrt{-g} - f\overline{\psi}\psi\sqrt{-g} \\
&\longrightarrow \partial_\mu(\sqrt{-g}\, g^{\mu\nu}\partial_\nu\phi) - [\delta\xi\phi R - V'(\phi)]\sqrt{-g} - fi[{}_iS_i](x; x)\sqrt{-g}, \qquad (27)\\
&\longrightarrow \partial_\mu(\sqrt{-g}\, g^{\mu\nu}\partial_\nu\phi) - [\alpha H^2\phi + \beta\phi^3]\sqrt{-g} \\
&\qquad + \tfrac{fH^3}{4\pi^2}\tfrac{f\phi}{H}\big(1 + \tfrac{f^2\phi^2}{H^2}\big)\Big\{\psi\big(1 + \tfrac{if\phi}{H}\big) + \psi\big(1 - \tfrac{if\phi}{H}\big)\Big\}\sqrt{-g}.
\end{aligned} \tag{28}$$

Here $\psi(x) \equiv \frac{d}{dx}\ln[\Gamma(x)]$ is the digamma function and the constants $\alpha \sim f^2$ and $\beta \sim f^4$ are arbitrary.

The Lagrangian of scalar quantum electrodynamics (SQED) is,

$$\mathcal{L} = -(D_\mu\phi)^* D_\nu\phi g^{\mu\nu}\sqrt{-g} - \delta\xi\phi^*\phi R\sqrt{-g} - V(\phi^*\phi)\sqrt{-g} - \tfrac{1}{4}F_{\mu\nu}F_{\rho\sigma}g^{\mu\rho}g^{\nu\sigma}\sqrt{-g}. \tag{29}$$

Here $D_\mu \equiv \partial_\mu + ieA_\mu$ is the covariant derivative operator. Setting the scalar equal to a constant endows the photon with a mass $m^2 = e^2\phi^*\phi$. The Lorentz gauge massive photon propagator on de Sitter background $i[{}_\mu\Delta_\nu](x; x')$ was derived in [41]. Using it to integrate the photon out of the scalar field equation (with suitable renormalization) gives,[1]

$$D_\mu(\sqrt{-g}\, g^{\mu\nu} D_\nu\phi) - \delta\xi\phi R\sqrt{-g} - \phi V'(\phi^*\phi)\sqrt{-g}$$
$$\longrightarrow \partial_\mu(\sqrt{-g}\, g^{\mu\nu}\partial_\nu\phi) - \phi[\delta\xi R - V']\sqrt{-g} - e^2\phi\, i[_\mu\Delta_\nu](x;x)g^{\mu\nu}\sqrt{-g}, \tag{30}$$

$$\longrightarrow \partial_\mu(\sqrt{-g}\, g^{\mu\nu}\partial_\nu\phi) - \phi[\overline{\alpha} + \overline{\beta}\phi^*\phi]\sqrt{-g} - \tfrac{3e^2 H^2\phi}{8\pi^2}(1 + \tfrac{e^2\phi^*\phi}{H^2})$$
$$\times \Big\{\psi\Big(\tfrac{3}{2} + \tfrac{1}{2}[1 - \tfrac{8e^2\phi^*\phi}{H^2}]^{\frac{1}{2}}\Big) + \psi\Big(\tfrac{3}{2} - \tfrac{1}{2}[1 - \tfrac{8e^2\phi^*\phi}{H^2}]^{\frac{1}{2}}\Big)\Big\}\sqrt{-g}. \tag{31}$$

The constants $\overline{\alpha} \sim e^2$ and $\overline{\beta} \sim e^4$ are again arbitrary.

The stochastic formulations of the scalar potential models (28) and (31) agree with explicit dimensionally regulated and fully renormalized computations at 1-loop and 2-loop orders [38, 39, 43, 44]. "Leading logarithm" in these two theories means that each additional factor of f^2 in Yukawa, or of e^2 in SQED, brings an extra factor of $\ln[a(t)]$. Note that ultraviolet regularization and renormalization are required to derive the Coleman–Weinberg potentials, however, the resulting scalar potential model is afterwards UV finite at leading logarithm order. The need for renormalization is reflected in the arbitrary finite parts associated with the constants α and β for (28), and $\overline{\alpha}$ and $\overline{\beta}$ for (31).

Finally, it is worth commenting on signs. The Yukawa effective potential (28) is negative, so this system never reaches a stable equilibrium; the universe instead decays in a Big Rip singularity [38]. In contrast, the effective potential for SQED is bounded below, which means that the system approaches a stable equilibrium. However, it turns out that the vacuum energy is slightly reduced [39], for essentially the same reason that a slab of dielectric is pulled into a charged, parallel plate capacitor. With the quartic potential result (14) this means that scalar models span the range of possibilities:

- The system may or may not approach a stable equilibrium; and
- The vacuum energy may be increased or decreased.

5 Differentiated Active Fields

Because the secular growth factors of $\ln[a(t)]$ derive from logarithmic integrations, the presence of a single derivative suppresses them. The status of differentiated active fields is very similar to that of passive fields: although they induce no secular loop corrections, they make important contributions of order one which come as much from the ultraviolet as from the infrared, and involve the full mode function. Hence it

[1] The result was first obtained by Allen as a mode sum [42].

is also wrong to infrared truncate differentiated active fields. For example, in dimensional regularization the coincidence limit of the doubly differentiated, massless, minimally coupled scalar propagator is [14, 31],

$$\lim_{x'\to x} \partial_\mu \partial'_\nu i\Delta(x;x') = \langle\Omega|\partial_\mu\phi_0(x)\partial_\nu\phi_0(x)|\Omega\rangle = -\frac{H^D}{2(4\pi)^{\frac{D}{2}}}\frac{\Gamma(D)}{\Gamma(\frac{D}{2}+1)} \times g_{\mu\nu}(x). \quad (32)$$

Were we to replace $\phi_0(x)$ by its infrared truncation $\varphi_0(x)$ the expectation value would be positive for $\mu = \nu$ because it is the square of a finite, Hermitian operator. However, the exact result (32) is *negative* for spatial $\mu = \nu$.

Rather than stochastically simplifying differentiated active fields, the correct procedure is to integrate them out of the field equations in the presence of a constant active field background. One then applies Starobinsky's formalism. This is also the same as for passive fields. The key difference is that imposing a constant active field background does not induce a mass but rather a field strength—for nonlinear sigma models—or a modification in the background geometry—for differentiated actives coupled to gravity.

Another important point is that none of these theories is renormalizable. However, all theories can be renormalized by subtracting BPHZ (Bogoliubov and Parasiuk [45], Hepp [46] and Zimmermann [47, 48]) counterterms, order-by-order in the loop expansion. When this is done, the leading secular growth factors are uniquely determined by low energy effective field theory in the sense of Donoghue [49–51].

5.1 Nonlinear Sigma Models

Quantum gravity is *hard*. The weeks it requires to perform a 2-loop scalar computation on de Sitter background [14, 31, 52] become months to perform a 1-loop graviton computation [53–55]. The fundamental $h\partial h\partial h$ interaction of quantum gravity also occurs in nonlinear sigma models such as,

$$\mathcal{L}_1 = -\tfrac{1}{2}(1+\tfrac{\lambda}{2}\phi)^2\partial_\mu\phi\partial_\nu\phi g^{\mu\nu}\sqrt{-g}, \quad (33)$$

$$\mathcal{L}_2 = -\tfrac{1}{2}\partial_\mu A\partial_\nu A g^{\mu\nu}\sqrt{-g} - \tfrac{1}{2}(1+\tfrac{\lambda}{2}A)^2\partial_\mu B\partial_\nu B g^{\mu\nu}\sqrt{-g}. \quad (34)$$

For this reason nonlinear sigma models have long been employed as a simple venue to sort out the complexities of derivative interactions, without the tensor indices, the gauge issue and the vast proliferation of interactions which make quantum gravity so difficult [22, 56–59].

The key to integrating out differentiated fields in (33)–(34) is the observation that working in a constant ϕ or A background just changes the field strengths of the ϕ and B propagators from that of the massless, minimally coupled scalar $i\Delta(x;x')$,

$$\langle\Omega|T[\phi(x)\phi(x')]|\Omega\rangle_{\phi_0} = \frac{i\Delta(x;x')}{(1+\frac{\lambda}{2}\phi_0)^2}, \quad \langle\Omega|T[B(x)B(x')]|\Omega\rangle_{A_0} = \frac{i\Delta(x;x')}{(1+\frac{\lambda}{2}A_0)^2}. \quad (35)$$

Integrating out the $\partial B \partial B$ term (for constant A_0) from the A field equation (and taking $D = 4$) gives,

$$\partial_\mu[\sqrt{-g}\, g^{\mu\nu}\partial_\nu A] - \tfrac{\lambda}{2}(1 + \tfrac{\lambda}{2}A)\partial_\mu B \partial_\nu B g^{\mu\nu}\sqrt{-g}$$
$$\longrightarrow \partial_\mu[\sqrt{-g}\, g^{\mu\nu}\partial_\nu A] - \tfrac{\lambda}{2}(1 + \tfrac{\lambda}{2}A_0) \times -\frac{\frac{3H^4}{8\pi^2}\sqrt{-g}}{(1+\frac{\lambda}{2}A_0)^2}. \quad (36)$$

Applying Starobinsky's formalism to this scalar potential model gives a Langevin equation for the stochastic field $\mathcal{A}(t, \mathbf{x})$,

$$3\, H(\dot{\mathcal{A}} - \dot{\mathcal{A}}_0) = \frac{3\lambda H^4}{16\pi^2}\frac{1}{1+\frac{\lambda}{2}\mathcal{A}}. \quad (37)$$

The effective potential for A induced in (36) is $V(A) = -\frac{3H^4}{8\pi^2}\ln|1 + \frac{\lambda}{2}A|$. This is unbounded below so no static limit is approached. However, the time evolution of the background is sedate. It consists of a "classical" contribution, obtained by ignoring the jitter $\mathcal{A}_0$, plus a series in powers of $\mathcal{A}_0$,

$$\mathcal{A}(t, \mathbf{x}) = -\tfrac{2}{\lambda} + \tfrac{2}{\lambda}\Big[1 + \frac{\lambda^2 H^2 \ln[a(t)]}{16\pi^2}\Big]^{\frac{1}{2}}$$
$$+ \mathcal{A}_0(t, \mathbf{x}) - \frac{\lambda^2 H^3}{32\pi^2}\int_0^t dt' \mathcal{A}_0(t', \mathbf{x}) + \frac{\lambda^3 H^3}{64\pi^2}\int_0^t dt' \mathcal{A}_0^2(t', \mathbf{x}) + O(\lambda^4). \quad (38)$$

Correlators of $\mathcal{A}$ agree, at leading logarithm order, with dimensionally regulated and fully BPHZ renormalized 1-loop and 2-loop computations of those of A, for example [59, 60],

$$\langle\Omega|\mathcal{A}(t, \mathbf{x})|\Omega\rangle = \frac{\lambda H^2 \ln[a(t)]}{16\pi^2} + \frac{\lambda^3 H^4 \ln^2[a(t)]}{2^{10}\pi^4} + O(\lambda^5). \quad (39)$$

Note that the basic time dependence of (39) derives from the classical part on the first line of (38), and is understandable in terms of the field rolling down its potential. The stochastic jitter accelerates the roll, essentially because it is easier to fluctuate down a potential than up.

The stochastic formulation of the single field model (33) is crucial for gravity because, in both cases, it is the *same* field whose derivatives are integrated out and which experiences stochastic fluctuations. It is straightforward to integrate out differentiated scalars from the ϕ field equation,

$$(1 + \tfrac{\lambda}{2}\phi)\partial_\mu\Big[(1 + \tfrac{\lambda}{2}\phi)\sqrt{-g}\, g^{\mu\nu}\partial_\nu\phi\Big]$$
$$\longrightarrow (1 + \tfrac{\lambda}{2}\phi_0)\partial_\mu\Big[\tfrac{\lambda}{4}\sqrt{-g}\, g^{\mu\nu}\partial_\nu\langle\Omega|\phi^2|\Omega\rangle_{\phi_0}\Big], \quad (40)$$
$$\longrightarrow (1 + \tfrac{\lambda}{2}\phi_0) \times -\frac{3\lambda H^4}{16\pi^2}\frac{\sqrt{-g}}{(1+\frac{\lambda}{2}\phi_0)^2}. \quad (41)$$

This corresponds to another logarithmic potential $V(\phi) = \frac{3H^4}{8\pi^2}\ln|1 + \frac{\lambda}{2}\phi|$. The key point is that we add it to the stochastic simplification of the same derivative terms from which it emerged,

$$3\,H(1+\tfrac{\lambda}{2}\varphi)^2[\dot{\varphi}-\dot{\varphi}_0] = -\tfrac{3\lambda H^4}{16\pi^2}\tfrac{1}{1+\frac{\lambda}{2}\varphi}. \tag{42}$$

As with the double field model (38), the solution of this Langevin equation consists of a classical part plus a series in powers of φ_0,

$$\varphi(t,\mathbf{x}) = -\tfrac{2}{\lambda} + \tfrac{2}{\lambda}\Big[1 - \tfrac{\lambda^2 H^2 \ln[a(t)]}{8\pi^2}\Big]^{\frac{1}{4}} + \varphi_0(t,\mathbf{x}) + \tfrac{3\lambda^2 H^3}{32\pi^2}\int_0^t dt' \varphi_0(t',\mathbf{x}) - \tfrac{3\lambda^3 H^3}{32\pi^2}\int_0^t dt' \varphi_0^2(t',\mathbf{x}) + O(\lambda^4). \tag{43}$$

Also like the double field model (39), correlators of φ agree, at leading logarithm order, with explicit 1-loop and 2-loop computations performed using dimensional regularization and BPHZ renormalization [59],

$$\langle\Omega|\varphi(t,\mathbf{x})|\Omega\rangle = -\tfrac{\lambda H^2 \ln[a(t)]}{16\pi^2} - \tfrac{15\lambda^3 H^4 \ln^2[a(t)]}{2^{10}\pi^4} + O(\lambda^5). \tag{44}$$

5.2 Scalar Corrections to Gravity

Most matter fields are not active, but the massless, minimally coupled scalar is an exception,

$$\mathcal{L}_{\rm MMC} = -\tfrac{1}{2}\partial_\mu\phi\partial_\nu\phi g^{\mu\nu}\sqrt{-g}. \tag{45}$$

To integrate differentiated scalar out of the gravitational field equation,

$$R_{\mu\nu} - \tfrac{1}{2}g_{\mu\nu}R + (\tfrac{D-2}{2})\Lambda g_{\mu\nu} = 8\pi G\Big\{\partial_\mu\phi\partial_\nu\phi - \tfrac{1}{2}g_{\mu\nu}g^{\rho\sigma}\partial_\rho\phi\partial_\sigma\phi\Big\}, \tag{46}$$

we must define what it means to be in a "constant graviton background". If the temporal coordinate is changed from co-moving time t to conformal time η, such that $d\eta = dt/a(t)$, the background geometry of cosmology (1) takes the form $a^2\eta_{\mu\nu}$. We define the graviton field $h_{\mu\nu}(x)$ by conformally transforming the full metric,

$$g_{\mu\nu} \equiv a^2\widetilde{g}_{\mu\nu} \equiv a^2(\eta_{\mu\nu} + \kappa h_{\mu\nu}), \qquad \kappa^2 \equiv 16\pi G. \tag{47}$$

The transverse-traceless and purely spatial components of $h_{\mu\nu}$ obey the same equation of motion as that of the massless, minimally coupled scalar [29]. Consequently, it is $h_{\mu\nu}$ which experiences inflationary particle production (2), and $h_{\mu\nu}$ which should be held constant when we integrate out differentiated scalars from equation (46).

Constant $h_{\mu\nu}$ means that $\widetilde{g}_{\mu\nu}$ is also constant. It turns out that setting $g_{\mu\nu} = a^2\widetilde{g}_{\mu\nu}$, with constant $\widetilde{g}_{\mu\nu}$ amounts to being in de Sitter with a different Hubble parameter [61, 62],

$$\partial_\rho \widetilde{g}_{\mu\nu} = 0 \qquad \Longrightarrow \qquad H^2 \longrightarrow -\widetilde{g}^{00} H^2 \equiv \widetilde{H}^2. \tag{48}$$

This makes it simple to integrate out the scalars from equation (46),

$$\tfrac{\kappa^2}{2}\Bigl\{\partial_\mu\phi\partial_\nu\phi - \tfrac{1}{2} g_{\mu\nu} g^{\rho\sigma}\partial_\rho\phi\partial_\sigma\phi\Bigr\} \longrightarrow \tfrac{\kappa^2}{2}\Bigl(\delta^\rho_{\ \mu}\delta^\sigma_{\ \nu} - \tfrac{1}{2} g_{\mu\nu} g^{\rho\sigma}\Bigr) \times \langle\Omega|\partial_\rho\phi\partial_\sigma\phi|\Omega\rangle_h, \tag{49}$$

$$\longrightarrow \tfrac{\kappa^2}{2}\Bigl(\delta^\rho_{\ \mu}\delta^\sigma_{\ \nu} - \tfrac{1}{2} g_{\mu\nu} g^{\rho\sigma}\Bigr) \times -\tfrac{3\widetilde{H}^4}{32\pi^2}\, g_{\rho\sigma} = \tfrac{3\kappa^2 \widetilde{H}^4}{64\pi^2}\, g_{\mu\nu}. \tag{50}$$

Expression (50) corresponds to a finite renormalization of the cosmological constant [63],

$$\delta\Lambda = -\tfrac{3\kappa^2 \widetilde{H}^4}{64\pi^2}. \tag{51}$$

The need for this renormalization to make the graviton self-energy conserved had been discovered previously through explicit computations [55, 64].

5.3 Pure Gravity

There is no more postponing the complexities of gravity. The vast proliferation of interactions is apparent from expressing the invariant Lagrangian in terms of the graviton field $h_{\mu\nu}$ and the conformally transformed metric $\widetilde{g}_{\mu\nu} \equiv \eta_{\mu\nu} + \kappa h_{\mu\nu}$ [65],

$$\begin{aligned}\mathcal{L}_{\rm inv} = a^{D-2}\sqrt{-\widetilde{g}}\,\widetilde{g}^{\alpha\beta}\widetilde{g}^{\rho\sigma}\widetilde{g}^{\mu\nu}\Bigl\{\tfrac{1}{2}h_{\alpha\rho,\mu}h_{\nu\sigma,\beta} - \tfrac{1}{2}h_{\alpha\beta,\rho}h_{\sigma\mu,\nu} \\ + \tfrac{1}{4}h_{\alpha\beta,\rho}h_{\mu\nu,\sigma} - \tfrac{1}{4}h_{\alpha\rho,\mu}h_{\beta\sigma,\nu}\Bigr\} + (\tfrac{D-2}{2}) a^{D-1} H\sqrt{-\widetilde{g}}\,\widetilde{g}^{\rho\sigma}\widetilde{g}^{\mu\nu} h_{\rho\sigma,\mu}h_{\nu 0}.\end{aligned} \tag{52}$$

The simplest gauge is based on adding the gauge fixing term [65, 66],

$$\mathcal{L}_{\rm GF} = -\tfrac{1}{2}a^{D-2}\eta^{\mu\nu}F_\mu F_\nu, \quad F_\mu = \eta^{\rho\sigma}[h_{\mu\rho,\sigma} - \tfrac{1}{2}h_{\rho\sigma,\mu} + (D-2)aHh_{\mu\rho}\delta^0_{\ \sigma}]. \tag{53}$$

The associated ghost Lagrangian is,

$$\mathcal{L}_{\rm gh} = -a^{D-2}\eta^{\mu\nu}\overline{c}_\mu \delta F_\nu, \tag{54}$$

where δF_ν represents the infinitesimal transformation $x^\mu \rightarrow x^\mu + \kappa\epsilon^\mu$ of F_ν with the transformation parameter replaced by the ghost field.

The gauge (53)–(54) results in ghost and graviton propagators whose tensor structure consists of constants formed from $\eta_{\mu\nu}$ and $\delta^0_{\ \mu}$,

$$i[{}_{\mu\nu}\Delta_{\rho\sigma}](x;x') = \sum_{I=A,B,C} [{}_{\mu\nu}T^I_{\rho\sigma}] \times i\Delta_I(x;x'), \tag{55}$$

$$i[{}_{\mu}\Delta_{\rho}](x;x') = \overline{\eta}_{\mu\nu} \times i\Delta_A(x;x') - \delta^0_{\ \mu}\delta^0_{\ \nu} \times i\Delta_B(x;x'). \tag{56}$$

Here and henceforth the purely spatial Minkowski metric is $\overline{\eta}_{\mu\nu} \equiv \eta_{\mu\nu} + \delta^0_{\ \mu}\delta^0_{\ \nu}$, and the other tensor factors are,

$$[{}_{\mu\nu}T^A_{\rho\sigma}] = 2\overline{\eta}_{\mu(\rho}\overline{\eta}_{\sigma)\nu} - \tfrac{2}{D-3}\overline{\eta}_{\mu\nu}\overline{\eta}_{\rho\sigma}, \qquad [{}_{\mu\nu}T^B_{\rho\sigma}] = -4\delta^0_{\ (\mu}\overline{\eta}_{\nu)(\rho}\delta^0_{\ \sigma)}, \tag{57}$$

$$[{}_{\mu\nu}T^C_{\rho\sigma}] = \tfrac{2}{(D-2)(D-3)}[(D-3)\delta^0_{\ \mu}\delta^0_{\ \nu} + \overline{\eta}_{\mu\nu}][(D-3)\delta^0_{\ \rho}\delta^0_{\ \sigma} + \overline{\eta}_{\rho\sigma}]. \tag{58}$$

Parenthesized indices are symmetrized. The three propagators $i\Delta_I(x;x')$ are those of a minimally coupled scalar with masses,

$$m_A^2 = 0, \qquad m_B^2 = (D-2)H^2, \qquad m_C^2 = 2(D-3)H^2. \tag{59}$$

Because graviton propagators possess indices, working in a constant $\widetilde{g}_{\mu\nu}$ background requires more than just replacing all the factors of H^2 by $\widetilde{H}^2$, as per equation (48), in the scalar propagators $i\Delta_I(x;x')$. One must also modify the gauge fixing function to [67],

$$\widetilde{\mathcal{L}}_{\rm GF} = -\tfrac{1}{2}a^{D-2}\sqrt{-\widetilde{g}}\,\widetilde{g}^{\mu\nu}\widetilde{F}_\mu\widetilde{F}_\nu, \quad \widetilde{F}_\mu = \widetilde{g}^{\rho\sigma}[h_{\mu\rho,\sigma} - \tfrac{1}{2}h_{\rho\sigma,\mu} + (D-2)aHh_{\mu\rho}\delta^0_{\ \sigma}]. \tag{60}$$

The ghost Lagrangian (54) suffers a similar change [67],

$$\widetilde{\mathcal{L}}_{\rm gh} = -a^{D-2}\sqrt{-\widetilde{g}}\,\widetilde{g}^{\mu\nu}\overline{c}_\mu\delta\widetilde{F}_\nu. \tag{61}$$

The tensor structures of the ghost and graviton propagators (55)–(58) are constructed from the Minkowski metric $\eta_{\mu\nu}$ and the timelike 4-vector $\delta^0_{\ \mu}$. Generalizing to a constant $\widetilde{g}_{\mu\nu}$ background amounts to stating what becomes of these two tensors. The most straightforward is $\eta_{\mu\nu}$: it generalizes to $\widetilde{g}_{\mu\nu}$. Explaining what happens to $\delta^0_{\ \mu}$ is facilitated by the ADM decomposition [68],

$$\widetilde{g}_{\mu\nu}dx^\mu dx^\nu = -N^2 d\eta^2 + \gamma_{ij}(dx^i - N^i d\eta)(dx^j - N^j d\eta). \tag{62}$$

This suggests the generalization [67],

$$\delta^0_{\ \mu} \longrightarrow u_\mu \equiv -N\delta^0_{\ \mu}, \tag{63}$$

$$\overline{\eta}_{\mu\nu} \longrightarrow \overline{\gamma}_{\mu\nu} \equiv \widetilde{g}_{\mu\nu} + u_\mu u_\nu. \tag{64}$$

So the graviton and ghost propagators in a constant $\widetilde{g}_{\mu\nu}$ background are,

$$i[{}_{\mu\nu}\widetilde{\Delta}_{\rho\sigma}](x;x') = \sum_{I=A,B,C} [{}_{\mu\nu}\widetilde{T}^{I}_{\rho\sigma}] \times i\widetilde{\Delta}_I(x;x'), \tag{65}$$

$$i[{}_{\mu}\widetilde{\Delta}_{\rho}](x;x') = \overline{\gamma}_{\mu\nu} \times i\widetilde{\Delta}_A(x;x') - u_\mu u_\nu \times i\widetilde{\Delta}_B(x;x'). \tag{66}$$

The scalar propagators $i\widetilde{\Delta}_I(x;x')$ are obtained by making the replacement (48) in $i\Delta_I(x;x')$, and the tensor factors $[{}_{\mu\nu}\widetilde{T}^{I}_{\rho\sigma}]$ are obtained by making the replacements (63)–(64) in (57)–(58).

The sum of the invariant Lagrangian (52) and the new gauge fixing term (60) can be grouped into six terms [67],

$$\mathcal{L}_{1+2+3} = a^{D-2}\sqrt{-\widetilde{g}}\,\widetilde{g}^{\alpha\beta}[-\tfrac{1}{4}\widetilde{g}^{\gamma\rho}\widetilde{g}^{\delta\sigma} + \tfrac{1}{8}\widetilde{g}^{\gamma\delta}\widetilde{g}^{\rho\sigma}]h_{\gamma\delta,\alpha}h_{\rho\sigma,\beta} \\ +(\tfrac{D-2}{2})a^{D}\sqrt{-\widetilde{g}}\,\widetilde{g}^{\gamma\alpha}\widetilde{g}^{\beta\rho}\widetilde{g}^{\sigma\delta}h_{\alpha\beta}h_{\rho\sigma}\widetilde{H}^2 u_\gamma u_\delta, \tag{67}$$

$$\mathcal{L}_{4+5} = \sqrt{-\widetilde{g}}\,\widetilde{g}^{\alpha\beta}\widetilde{g}^{\gamma\delta}\widetilde{g}^{\rho\sigma}\partial_\beta\Big[-\tfrac{1}{2}\partial_\sigma(a^{D-2}h_{\gamma\rho}h_{\delta\alpha}) + a^{D-2}h_{\gamma\rho}h_{\delta\alpha,\sigma}\Big], \tag{68}$$

$$\mathcal{L}_{6} = (\tfrac{D-2}{2})\kappa H a^{D-1}\sqrt{-\widetilde{g}}\,\widetilde{g}^{\alpha\beta}\widetilde{g}^{\gamma\delta}\widetilde{g}^{\rho\sigma}h_{\rho\sigma,\gamma}h_{\delta\alpha}h_{\beta 0}. \tag{69}$$

Completing the stochastic simplification of pure gravity entails three steps:

1. Find the contribution to the equation of motion from (61) and (67)–(69);
2. Integrate out the ghost fields and the differentiated graviton fields from each contribution; and
3. Derive the Langevin kinetic operator from each contribution.

The final step involves distinguishing between the spatial components of the graviton field, which are active, and the other components, which are passive, constrained fields whose dynamics is driven by the active components [69],

$$\kappa h_{\mu\nu} = A_{\mu\nu} + 2u_{(\mu}B_{\nu)} + [u_\mu u_\nu + \overline{\gamma}_{\mu\nu}]C, \qquad u^\rho A_{\rho\sigma} = 0 = u^\rho B_\rho. \tag{70}$$

Note that the active components $A_{\mu\nu}$ have a stochastic jitter $a_{\mu\nu}$ whereas the passive, constrained fields B_μ and C do not.

The first two steps have been implemented [67], and the third is in progress [69]. The complete result for (67) is,

$$\frac{\kappa a^{-4}}{\sqrt{-g}} \times \frac{\delta S_{1+2+3}}{\delta h_{\mu\nu}} \longrightarrow \frac{\kappa^2\widetilde{H}^4}{8\pi^2}[13\widetilde{g}^{\mu\nu} + 6u^\mu u^\nu] - [\widetilde{g}^{\mu\rho}\widetilde{g}^{\nu\sigma} - \tfrac{1}{2}\widetilde{g}^{\mu\nu}\widetilde{g}^{\rho\sigma}]a^{-1}\widetilde{H}u^\alpha\partial_\alpha(A_{\rho\sigma} - a_{\rho\sigma}) \\ + u^{(\mu}\widetilde{g}^{\nu)\rho}\widetilde{D}_B B_\rho + u^\mu u^\nu \widetilde{D}_C C + \widetilde{H}^2\Big\{\tfrac{1}{2}\widetilde{g}^{\mu\nu}(B^2 - C^2) \\ + 2u^{(\mu}A^{\nu)\rho}B_\rho - B^\mu B^\nu - 2u^{(\mu}B^{\nu)}C + u^\mu u^\nu(2B^2 - 3C^2)\Big\}. \tag{71}$$

Although the two operators agree for $D = 4$, it is useful to retain dimensional regularization in order to see the difference between $\widetilde{D}_B$ and $\widetilde{D}_C$,

$$\widetilde{D}_B = \partial_\alpha [a^{D-2}\sqrt{-\widetilde{g}}\,\widetilde{g}^{\alpha\beta}\partial_\beta] - (D-2)\widetilde{H}^2 a^D \sqrt{-\widetilde{g}}, \tag{72}$$

$$\widetilde{D}_C = \partial_\alpha [a^{D-2}\sqrt{-\widetilde{g}}\,\widetilde{g}^{\alpha\beta}\partial_\beta] - 2(D-3)\widetilde{H}^2 a^D \sqrt{-\widetilde{g}}. \tag{73}$$

In $D = 4$ both operators degenerate to the conformal d'Alembertian.

Whereas nonlinear sigma models and scalar corrections to gravity had seen extensive dimensionally regulated and fully renormalized computations prior to their stochastic realizations, the stochastic formulation of pure gravity is far ahead of the explicit calculations with which it might be compared. The 1PI (one-particle-irreducible) 1-graviton function has been computed in the original gauge (53) [70], but it must be re-done in the new gauge (60) to check the stochastic equation. Quantum gravity presumably requires the same sort of finite renormalization of the cosmological constant that was found for scalar corrections to gravity [64]. Also in the old gauge (53), the 1PI 2-point function (the "graviton self-energy") was computed, away from coincidence and without using dimensional regularization or renormalizing [53]. This needs to re-done in the new gauge (60) using dimensional regularization and BPHZ renormalization. A technique was devised for extending the old calculation to a fully renormalized result [71]. This extension was then employed to compute 1-loop corrections to the graviton mode function [72] and to the gravitational response to a point mass [73]. Both of these things need to re-done with a exact computation in the new gauge.

6 The Other Source of Large Logarithms

The first large logarithm from a loop of gravitons was discovered 20 years ago [23, 24]. It was soon shown not to follow from a naive realization of the stochastic formalism [74]. In the meantime a number of similar effects were found from graviton loop corrections to matter [75–81] and in graviton corrections to gravity [71–73]. The same kinds of derivative interactions in matter corrections to gravity also induce large logarithms [36, 55, 82, 83]. Explaining these effects, and devising techniques to resum them, has been a long and confusing struggle for two reasons:

1. The correct stochastic formulation involves integrating differentiated active fields out, which entails new sorts of effective potentials based on background-dependent field strengths and Hubble parameters; and
2. There is another way large logarithms can be induced which cannot be captured stochastically.

Sect. 5 concerned the first problem, this section concerns the second.

The second way of inducing large logarithms is the incomplete cancellation between primitive divergences and counterterms. When using dimensional regularization it turns out that the factors of a^{D-2} from vertices are canceled by inverse factors from propagators. For example, the $i\Delta_I(x;x')$ propagators which appear in the graviton and ghost propagators (55)–(56) have the same leading singularity,

$$i\Delta_I(x;x') = \frac{\Gamma(\frac{D}{2}-1)}{4\pi^{\frac{D}{2}}} \frac{1}{[aa'(x-x')^2]^{\frac{D}{2}-1}} + \dots, \tag{74}$$

where $(x-x')^2 \equiv (x-x')^\mu (x-x')^\nu \eta_{\mu\nu}$ is the conformal coordinate interval. Higher terms in the propagator can give rise to additional, integer powers of (aa'), but the D-dependence will always be $(aa')^{-\frac{D}{2}}$. On the other hand, counterterms inherit a factor of a^D from the $\sqrt{-g}$ measure. It follows that there is an incomplete cancellation between primitive divergences and counterterms,

$$\frac{(2H)^{D-4}}{D-4} - \frac{(a\mu)^{D-4}}{D-4} = -\ln(\tfrac{a\mu}{2H}) + O(D-4). \tag{75}$$

Nonlinear sigma models exhibit both stochastic logarithms such as (39) and (44), and also renormalization-induced logarithms (74). The large logarithms of most graviton loop corrections to matter fields [75–78, 81], and all known matter loop corrections to gravity [36, 54, 55, 82, 83], are induced by renormalization. The case of pure gravity [71–73] has yet to be decided.

Renormalization-induced logarithms occur even for passive fields which do not experience inflationary particle production (2), so they cannot be explained stochastically. On the other hand, the close association in expression (75) between the dimensional regularization scale μ and the scale factor $a(t)$ makes these logarithms amenable to a modified renormalization group treatment. I will explain the technique in the context of matter loop corrections to gravity [36, 54, 55, 82, 83] for which a unified treatment is possible.

What has been done is to compute matter loop contributions to the graviton self-energy $-i[^{\mu\nu}\Sigma^{\rho\sigma}](x;x')$ on de Sitter background [36, 54, 55, 82, 83]. This can be used to quantum-correct the linearized Einstein equation,

$$\mathcal{L}^{\mu\nu\rho\sigma}\kappa h_{\rho\sigma} - \int d^4x' \, [^{\mu\nu}\Sigma^{\rho\sigma}](x;x')\kappa h_{\rho\sigma}(x') = \tfrac{\kappa^2}{2} T^{\mu\nu}(x). \tag{76}$$

Here $\mathcal{L}^{\mu\nu\sigma\rho}$ is the Lichnerowicz operator on de Sitter background and $T^{\mu\nu}$ is the "graviton stress tensor," defined as minus the variation of the matter action with respect to $h_{\mu\nu}$. With $T^{\mu\nu} = 0$, one is solving for 1-loop corrections to gravitational radiation. There are no corrections in flat space background, but a loop of massless, minimally coupled scalars gives rise to a secular enhancement of the Weyl tensor [55],

$$C_{0i0j}(t,\mathbf{x}) = C^{\text{tree}}_{0i0j}(t,\mathbf{x})\Big\{1 - \tfrac{3\kappa^2 H^2}{160\pi^2} \ln[a(t)] + \dots\Big\}. \tag{77}$$

Setting $T^{\mu\nu}(t, \mathbf{x}) = -M\delta^{\mu}{}_{0}\delta^{\nu}{}_{0}a(t)\delta^{3}(\mathbf{x})$ gives the response to a point mass, which can be parameterized in terms of two scalar potentials,

$$ds^2 = -[1 - 2\Psi(t, r)]dt^2 + a^2(t)[1 - 2\Phi(t, r)]d\mathbf{x} \cdot d\mathbf{x}. \tag{78}$$

For a loop of massless, minimally coupled scalars one finds [55, 84],

$$\Psi(t, r) = \frac{GM}{a(t)r}\Big\{1 + \frac{\kappa^2}{320\pi^2 a^2(t) r^2} - \frac{3\kappa^2 H^2}{160\pi^2}\ln[a(t)] + \dots\Big\}, \tag{79}$$

$$\Psi(t, r) + \Phi(t, r) = \frac{GM}{a(t)r}\Big\{0 + \frac{\kappa^2}{240\pi^2 a^2(t) r^2} + \frac{3\kappa^2 H^2}{160\pi^2} + \dots\Big\}. \tag{80}$$

The fractional κ^2/a^2r^2 correction in (79) is the de Sitter descendant of an old flat space effect [85], whereas the $\kappa^2 H^2$ corrections are new. Corrections to the Weyl tensor and to the two potentials have also been computed for conformally invariant matter theories. For a massless, conformally coupled scalar the results are [83],

$$C_{0i0j}(t, \mathbf{x}) = C^{\text{tree}}_{0i0j}(t, \mathbf{x})\Big\{1 + \frac{\kappa^2 H^2}{480\pi^2}\ln[a(t)] + \dots\Big\}, \tag{81}$$

$$\Psi(t, r) = \frac{GM}{a(t)r}\Big\{1 + \frac{\kappa^2}{720\pi^2 a^2(t) r^2} + \frac{\kappa^2 H^2}{480\pi^2}\ln[a(t)] + \dots\Big\}, \tag{82}$$

$$\Psi(t, r) + \Phi(t, r) = \frac{GM}{a(t)r}\Big\{0 + \frac{\kappa^2}{1440\pi^2 a^2(t) r^2} - \frac{\kappa^2 H^2}{480\pi^2} + \dots\Big\}. \tag{83}$$

As with its minimally coupled cousin, the κ^2/a^2r^2 corrections to (82)–(83) are well known from flat space background [86]. All (new and old) of the 1-loop corrections from massless, Dirac fermions are a factor of 6 times those in (81)–(83) [83, 87]. For a loop of photons the factor is 12 [36, 82, 88, 89].

The Lagrangian of general relativity with a cosmological constant Λ is,

$$\mathcal{L}_{\text{GR}} = \frac{[R - (D-2)\Lambda]\sqrt{-g}}{16\pi G}. \tag{84}$$

On the other hand, all single matter loop corrections to gravity require the same two counterterms [90],

$$\Delta\mathcal{L} = c_1 R^2\sqrt{-g} + c_2 C^{\alpha\beta\gamma\delta}C_{\alpha\beta\gamma\delta}\sqrt{-g}. \tag{85}$$

The first problem in applying the renormalization group to general relativity (84) is that the counterterms (85) do not seem to represent renormalizations of the classical action. However, one can easily decompose the Eddington (R^2) counterterm into a higher derivative part and two lower derivative parts which can be viewed as part of the classical action [63],

$$R^2 = [R - D\Lambda + D\Lambda]^2 = (R - D\Lambda)^2 + 2D\Lambda[R - (D-2)\Lambda] + (D-4)D\Lambda^2. \tag{86}$$

The final term in (86) could be regarded as renormalizing the cosmological constant, but the factor of $(D-4)$ makes it irrelevant. In contrast, the middle term of (86) could be regarded as a renormalization of the graviton field strength.

Making a similar decomposition of the Weyl $(C^{\alpha\beta\gamma\delta}C_{\alpha\beta\gamma\delta})$ counterterm is more challenging. We first express it in terms of the Gauss-Bonnet scalar G,

$$C^{\mu\nu\rho\sigma}C_{\mu\nu\rho\sigma} = R^{\mu\nu\rho\sigma}R_{\mu\nu\rho\sigma} - \tfrac{4}{D-2}R^{\mu\nu}R_{\mu\nu} + \tfrac{2R^2}{(D-1)(D-2)}, \tag{87}$$

$$= G + 4(\tfrac{D-3}{D-2})R^{\mu\nu}R_{\mu\nu} - \tfrac{D(D-3)}{(D-1)(D-2)}R^2. \tag{88}$$

The Gauss-Bonnet scalar is,

$$G \equiv R^{\mu\nu\rho\sigma}R_{\mu\nu\rho\sigma} - 4R^{\mu\nu}R_{\mu\nu} + R^2. \tag{89}$$

It is significant by virtue of being a total derivative in $D=4$ dimensions, which means it cannot affect 1-loop divergences. We now decompose the square of the Ricci tensor similar to (86) into a higher derivative term and two lower derivatives,

$$R^{\mu\nu}R_{\mu\nu} = (R^{\mu\nu}-\Lambda g^{\mu\nu})(R_{\mu\nu}-\Lambda g_{\mu\nu}) + 2\Lambda[R-(D-2)\Lambda] + (D-4)\Lambda^2. \tag{90}$$

Substituting (86) and (90) in equation (88) gives,

$$\begin{aligned} C^{\mu\nu\rho\sigma}C_{\mu\nu\rho\sigma} &= 4(\tfrac{D-3}{D-2})(R^{\mu\nu}-\Lambda g^{\mu\nu})(R_{\mu\nu}-\Lambda g_{\mu\nu}) - \tfrac{D(D-3)}{(D-1)(D-2)}(R-D\Lambda)^2 \\ &\quad - \tfrac{2(D-2)(D-3)}{D-1}\Lambda[R-(D-2)\Lambda)] - \tfrac{(D-2)(D-3)(D-4)}{D-1}\Lambda^2 + G. \end{aligned} \tag{91}$$

Expressions (86) and (91) imply that counterterms (85) can be viewed as a graviton field strength renormalization, which gives rise to a gamma function,

$$\delta Z = 2\Big[D(D-1)c_1 - (D-2)(D-3)c_2\Big]\kappa^2 H^2 \qquad \Longrightarrow \qquad \gamma = \tfrac{\partial \ln(1+\delta Z)}{\partial \ln(\mu^2)}. \tag{92}$$

The linearized Weyl tensor and the Newtonian potential can be viewed as 2-point Green's functions, for which the Callan-Symanzik equation implies,

$$\Big[\tfrac{\partial}{\partial \ln(\mu)} + 2\gamma\Big]G^{(2)} = 0. \tag{93}$$

Relation (75) shows that factors of $\ln(\mu)$ always come in the form $\ln(\mu a)$, so it should be valid to replace the derivative with respect to $\ln(\mu)$ in expression (93) by the derivative with respect to $\ln(a)$. At this point one sees that it is possible to predict the large temporal logarithms from renormalization.

The results are impressive. For a loop of massless, minimally coupled scalars the constants c_1 and c_2 in (85) are [54, 55],

$$c_1 = \frac{\mu^{D-4}\Gamma(\frac{D}{2})}{2^8\pi^{\frac{D}{2}}} \frac{(D-2)}{(D-1)^2(D-3)(D-4)}, \tag{94}$$

$$c_2 = \frac{\mu^{D-4}\Gamma(\frac{D}{2})}{2^8\pi^{\frac{D}{2}}} \frac{2}{(D+1)(D-1)^2(D-3)^2(D-4)}. \tag{95}$$

Substituting these results in expression (92) give the gamma function,

$$\gamma_{\text{MMCS}} = \frac{3\kappa^2 H^2}{320\pi^2}. \tag{96}$$

Using γ_{MMCS} in the Callan-Symanzik equation (93) not only explains the leading logarithms (77)–(79), but also permits a full resummation,

$$C_{0i0j}(t, \mathbf{x}) \longrightarrow C^{\text{tree}}_{0i0j}(t, \mathbf{x}) \times \Big[a(t)\Big]^{-\frac{3\kappa^2 H^2}{160\pi^2}}, \tag{97}$$

$$\Psi(t, r) \longrightarrow \frac{GM}{a(r)r} \times \Big[a(t)rH\Big]^{-\frac{3\kappa^2 H^2}{160\pi^2}}. \tag{98}$$

For massless, conformally coupled scalars one has $c_1 = 0$ and c_2 equal to [83],

$$c_2 = \frac{\mu^{D-4}\Gamma(\frac{D}{2})}{2^8\pi^{\frac{D}{2}}} \frac{2}{(D+1)(D-1)(D-3)^2(D-4)} \qquad \Longrightarrow \qquad \gamma_{\text{MCCS}} = -\frac{\kappa^2 H^2}{960\pi^2}. \tag{99}$$

This not only explains the leading logarithms at 1-loop (81)–(82), and again implies the resumed results,

$$C_{0i0j}(t, \mathbf{x}) \longrightarrow C^{\text{tree}}_{0i0j}(t, \mathbf{x}) \times \Big[a(t)\Big]^{\frac{\kappa^2 H^2}{480\pi^2}}, \tag{100}$$

$$\Psi(t, r) \longrightarrow \frac{GM}{a(r)r} \times \Big[a(t)rH\Big]^{\frac{\kappa^2 H^2}{480\pi^2}}. \tag{101}$$

A loop of Dirac fermions has $\gamma_{\text{Dirac}} = 6 \times \gamma_{\text{MMCS}}$ [83] and a loop of photons gives $\gamma_{\text{EM}} = 12 \times \gamma_{\text{MMCS}}$ [82].

7 Conclusions

Alexei Starobinsky possessed a towering intellect which allowed him to see deeply hidden truths. One of these is that a simple stochastic formalism can describe many of the secular loop corrections one finds in quantum field theory on inflationary backgrounds. This is quite surprising from the context of quantum field theory because the fields which obey Starobinsky's first order Langevin equation are ultraviolet finite, commuting variables, quite unlike the original quantum fields which obey second order equations, harbor ultraviolet divergences, and do not commute on or within the light-cone.

The message of this article is that Starobinsky's stochastic formalism is neither wrong, as many QFT experts believed, nor does it supplant quantum field theory, as some cosmologists believed. What it represents is the leading secular corrections in quantum field theories of undifferentiated "active" fields which experience inflationary particle production (2). It does not directly apply to "passive" fields, which do not experience inflationary particle production, or to differentiated active fields. Those systems make nonsecular, but still nonzero, contributions which come as much from the ultraviolet as from the infrared, and from the full mode function. The correct way to deal with these systems is by integrating out the problematic passive or differentiated active fields in the presence of a constant active field background. For passive fields this typically results in a standard effective potential induced by a field-dependent mass. For differentiated active fields one encounters new sorts of effective potentials induced by field-dependent field strengths and/or field-dependent Hubble parameters.

Finally, it is best avoid dogmatic adherence to statistical mechanics, and to check against explicit computations whenever possible. Although Starobinsky's formalism does have a statistical mechanical interpretation, its derivation rests on quantum field theory, and what it does is to predict a certain class of secular corrections. One must renormalize in the presence of either passive fields or differentiated active fields. And differentiated active fields produce new types of secular corrections from renormalization (75), which cannot be represented by any stochastic formalism. This second source of secular corrections was only discovered by explicit 1-loop and 2-loop computations. It is possible that further surprises await us, which Alexei would have loved.

Acknowledgements It is a pleasure to acknowledge collaboration, conversation and correspondence on these subjects with A. Foraci, D. Glavan, S. P. Miao, T. Prokopec, N. C. Tsamis and B. Yesilyurt. This work was partially supported by NSF grant PHY-2207514 and by the Institute for Fundamental Theory at the University of Florida.

Competing Interests The author has no conflicts of interest to declare that are relevant to the content of this chapter.

References

1. A.A. Starobinsky, A new type of isotropic cosmological models without singularity. Phys. Lett. B **91**, 99–102 (1980)
2. M. Tristram et al., Improved limits on the tensor-to-scalar ratio using BICEP and Planck data. Phys. Rev. D **105**(8), 083524 (2022)
3. A.A. Starobinsky, Spectrum of relict gravitational radiation and the early state of the universe. JETP Lett. **30**, 682–685 (1979)

4. A.A. Starobinsky, Dynamics of phase transition in the new inflationary universe scenario and generation of perturbations. Phys. Lett. B **117**, 175–178 (1982)
5. A.A. Starobinsky, Stochastic De Sitter (inflationary) stage in the early universe. Lect. Notes Phys. **246**, 107–126 (1986)
6. A.A. Starobinsky, J. Yokoyama, Equilibrium state of a selfinteracting scalar field in the De Sitter background. Phys. Rev. D **50**, 6357–6368 (1994)
7. K.G. Wilson, Renormalization group and critical phenomena. 1. Renormalization group and the Kadanoff scaling picture. Phys. Rev. B **4**, 3174–3183 (1971)
8. K.G. Wilson, Renormalization group and critical phenomena. 2. Phase space cell analysis of critical behavior. Phys. Rev. B **4**, 3184–3205 (1971)
9. J.S. Schwinger, Brownian motion of a quantum oscillator. J. Math. Phys. **2**, 407–432 (1961)
10. K.T. Mahanthappa, Multiple production of photons in quantum electrodynamics. Phys. Rev. **126**, 329–340 (1962)
11. P.M. Bakshi, K.T. Mahanthappa, Expectation value formalism in quantum field theory. 1. J. Math. Phys. **4**, 1–11 (1963)
12. P.M. Bakshi, K.T. Mahanthappa, Expectation value formalism in quantum field theory. 2. J. Math. Phys. **4**, 12–16 (1963)
13. L.V. Keldysh, Diagram technique for nonequilibrium processes. Zh. Eksp. Teor. Fiz. **47**, 1515–1527 (1964)
14. V.K. Onemli, R.P. Woodard, Superacceleration from massless, minimally coupled phi**4. Class. Quant. Grav. **19**, 4607 (2002)
15. T. Prokopec, O. Tornkvist, R.P. Woodard, Photon mass from inflation. Phys. Rev. Lett. **89**, 101301 (2002)
16. T. Prokopec, O. Tornkvist, R.P. Woodard, One loop vacuum polarization in a locally de Sitter background. Annals Phys. **303**, 251–274 (2003)
17. A. Vilenkin, L.H. Ford, Gravitational effects upon cosmological phase transitions. Phys. Rev. D **26**, 1231 (1982)
18. A.D. Linde, Scalar field fluctuations in expanding universe and the new inflationary universe scenario. Phys. Lett. B **116**, 335–339 (1982)
19. B. Allen, A. Folacci, The massless minimally coupled scalar field in De Sitter space. Phys. Rev. D **35**, 3771 (1987)
20. D.N. Page, X. Wu, Massless scalar field vacuum in De Sitter spacetime. JCAP **11**, 051 (2012)
21. L.R. Abramo, R.P. Woodard, No one loop back reaction in chaotic inflation. Phys. Rev. D **65**, 063515 (2002)
22. N.C. Tsamis, R.P. Woodard, Stochastic quantum gravitational inflation. Nucl. Phys. B **724**, 295–328 (2005)
23. S.P. Miao, R.P. Woodard, The Fermion self-energy during inflation. Class. Quant. Grav. **23**, 1721–1762 (2006)
24. S.P. Miao, R.P. Woodard, Gravitons enhance fermions during inflation. Phys. Rev. D **74**, 024021 (2006)
25. T.M.C. Abbott et al., [DES], Dark Energy Survey Year 3 results: cosmological constraints from galaxy clustering and weak lensing. Phys. Rev. D **105**(2), 023520 (2022)
26. M. Kamionkowski, A.G. Riess, The Hubble tension and early dark energy. Ann. Rev. Nucl. Part. Sci. **73**, 153–180 (2023)
27. N. Aghanim, et al. [Planck], Planck 2018 results. VI. Cosmological parameters. Astron. Astrophys. 641, A6 (2020) [erratum: Astron. Astrophys. 652, C4 (2021)]
28. G. Geshnizjani, W.H. Kinney, A. Moradinezhad Dizgah, General conditions for scale-invariant perturbations in an expanding universe. JCAP 11, 049 (2011)
29. E. Lifshitz, Republication of: On the gravitational stability of the expanding universe. J. Phys. (USSR) **10**(2), 116 (1946)
30. V.F. Mukhanov, G.V. Chibisov, Quantum fluctuations and a nonsingular universe. JETP Lett. **33**, 532–535 (1981)
31. V.K. Onemli, R.P. Woodard, Quantum effects can render w < -1 on cosmological scales. Phys. Rev. D **70**, 107301 (2004)

32. E.O. Kahya, V.K. Onemli, R.P. Woodard, A completely regular quantum stress tensor with w < -1. Phys. Rev. D **81**, 023508 (2010)
33. N.A. Chernikov, E.A. Tagirov, Quantum theory of scalar fields in de Sitter space-time. Ann. Inst. H. Poincare A Phys. Theor **9**, 109 (1968)
34. C. Schomblond, P. Spindel, Unicity conditions of the scalar field propagator Delta(1) (x, y) in de Sitter universe. Ann. Inst. H. Poincare Phys. Theor. **25**, 67–78 (1976)
35. T.S. Bunch, P.C.W. Davies, Quantum field theory in de Sitter space: renormalization by point splitting. Proc. Roy. Soc. Lond. A **360**, 117–134 (1978)
36. C.L. Wang, R.P. Woodard, One-loop quantum electrodynamic correction to the gravitational potentials on de Sitter spacetime. Phys. Rev. D **92**, 084008 (2015)
37. S.R. Coleman, E.J. Weinberg, Radiative corrections as the origin of spontaneous symmetry breaking. Phys. Rev. D **7**, 1888–1910 (1973)
38. S.P. Miao, R.P. Woodard, Leading log solution for inflationary Yukawa. Phys. Rev. D **74**, 044019 (2006)
39. T. Prokopec, N.C. Tsamis, R.P. Woodard, Stochastic inflationary scalar electrodynamics. Ann. Phys. **323**, 1324–1360 (2008)
40. P. Candelas, D.J. Raine, General relativistic quantum field theory-an exactly soluble model. Phys. Rev. D **12**, 965–974 (1975)
41. N.C. Tsamis, R.P. Woodard, A maximally symmetric vector propagator. J. Math. Phys. **48**, 052306 (2007)
42. B. Allen, Phase transitions in de Sitter space. Nucl. Phys. B **226**, 228–252 (1983)
43. T. Prokopec, N.C. Tsamis, R.P. Woodard, Two loop scalar Bilinears for inflationary SQED. Class. Quant. Grav. **24**, 201–230 (2007)
44. T. Prokopec, N.C. Tsamis, R.P. Woodard, Two loop stress-energy tensor for inflationary scalar electrodynamics. Phys. Rev. D **78**, 043523 (2008)
45. N.N. Bogoliubov, O.S. Parasiuk, On the multiplication of the causal function in the quantum theory of fields. Acta Math. **97**, 227–266 (1957)
46. K. Hepp, Proof of the Bogolyubov-Parasiuk theorem on renormalization. Commun. Math. Phys. **2**, 301–326 (1966)
47. W. Zimmermann, The power counting theorem for minkowski metric. Commun. Math. Phys. **11**, 1–8 (1968)
48. W. Zimmermann, Convergence of Bogolyubov's method of renormalization in momentum space. Commun. Math. Phys. **15**, 208–234 (1969)
49. J.F. Donoghue, Leading quantum correction to the Newtonian potential. Phys. Rev. Lett. **72**, 2996–2999 (1994)
50. J.F. Donoghue, General relativity as an effective field theory: the leading quantum corrections. Phys. Rev. D **50**, 3874–3888 (1994)
51. J. Donoghue, Quantum gravity as a low energy effective field theory. Scholarpedia **12**(4), 32997 (2017)
52. T. Brunier, V.K. Onemli, R.P. Woodard, Two loop scalar self-mass during inflation. Class. Quant. Grav. **22**, 59–84 (2005)
53. N.C. Tsamis, R.P. Woodard, One loop graviton selfenergy in a locally de Sitter background. Phys. Rev. D **54**, 2621–2639 (1996)
54. S. Park, R.P. Woodard, Scalar contribution to the graviton self-energy during inflation. Phys. Rev. D **83**, 084049 (2011)
55. S.P. Miao, N.C. Tsamis, R.P. Woodard, Alternate computation of gravitational effects from a single loop of inflationary scalars. JHEP **07**, 099 (2024)
56. H. Kitamoto, Y. Kitazawa, Non-linear sigma model in de Sitter space. Phys. Rev. D **83**, 104043 (2011)
57. H. Kitamoto, Y. Kitazawa, Infra-red effects of Non-linear sigma model in de Sitter space. Phys. Rev. D **85**, 044062 (2012)
58. H. Kitamoto, Infrared resummation for derivative interactions in de Sitter space. Phys. Rev. D **100**(2), 025020 (2019)

59. S.P. Miao, N.C. Tsamis, R.P. Woodard, Summing inflationary logarithms in nonlinear sigma models. JHEP **03**, 069 (2022)
60. R.P. Woodard, B. Yesilyurt, Unfinished business in a nonlinear sigma model on de Sitter background. JHEP **06**, 206 (2023)
61. S. Basu, R.P. Woodard, Testing an ansatz for the leading secular loop corrections from quantum gravity during inflation. Class. Quant. Grav. **33**(20), 205007 (2016)
62. S. Basu, N.C. Tsamis, R.P. Woodard, Causality implies inflationary back-reaction. JHEP **07**, 037 (2017)
63. S.P. Miao, N.C. Tsamis, R.P. Woodard, Summing gravitational effects from loops of inflationary scalars. Class. Quant. Grav. **41**(21), 215007 (2024)
64. N.C. Tsamis, R.P. Woodard, B. Yesilyurt, The need to renormalize the cosmological constant. Phys. Lett. B **849**, 138472 (2024)
65. N.C. Tsamis, R.P. Woodard, The Structure of perturbative quantum gravity on a De Sitter background. Commun. Math. Phys. **162**, 217–248 (1994)
66. R.P. Woodard, de Sitter breaking in field theory. arXiv:gr-qc/0408002 [gr-qc].
67. S.P. Miao, N.C. Tsamis, R.P. Woodard, Leading logarithm quantum gravity. Universe **11**, 223 (2025). (arXiv:2409.12003 [gr-qc])
68. R.L. Arnowitt, S. Deser, C.W. Misner, The Dynamics of general relativity. Gen. Rel. Grav. **40**, 1997–2027 (2008)
69. R.P. Woodard, Resummations for inflationary quantum gravity. Int. J. Mod. Phys. D **34**, 2542002 (2025). (arXiv:2501.05077 [gr-qc])
70. N.C. Tsamis, R.P. Woodard, Dimensionally regulated graviton 1-point function in de Sitter. Annals Phys. **321**, 875–893 (2006)
71. L. Tan, N.C. Tsamis, R.P. Woodard, Graviton self-energy from gravitons in cosmology. Class. Quant. Grav. **38**(14), 145024 (2021)
72. L. Tan, N.C. Tsamis, R.P. Woodard, How inflationary gravitons affect gravitational radiation. Phil. Trans. Roy. Soc. Lond. A **380**, 0187 (2021)
73. L. Tan, N.C. Tsamis, R.P. Woodard, How inflationary gravitons affect the force of gravity. Universe **8**(7), 376 (2022)
74. S.P. Miao, R.P. Woodard, A simple operator check of the effective fermion mode function during inflation. Class. Quant. Grav. **25**, 145009 (2008)
75. E.O. Kahya, R.P. Woodard, Quantum gravity corrections to the one loop scalar self-mass during inflation. Phys. Rev. D **76**, 124005 (2007)
76. K.E. Leonard, R.P. Woodard, Graviton corrections to vacuum polarization during inflation. Class. Quant. Grav. **31**, 015010 (2014)
77. D. Glavan, S.P. Miao, T. Prokopec, R.P. Woodard, Electrodynamic effects of inflationary gravitons. Class. Quant. Grav. **31**, 175002 (2014)
78. C.L. Wang, R.P. Woodard, Excitation of photons by inflationary gravitons. Phys. Rev. D **91**(12), 124054 (2015)
79. D. Glavan, S.P. Miao, T. Prokopec, R.P. Woodard, Single graviton loop contribution to the self-mass of a massless, conformally coupled scalar on a de Sitter background. Phys. Rev. D **101**(10), 106016 (2020)
80. D. Glavan, S.P. Miao, T. Prokopec, R.P. Woodard, One-loop graviton corrections to conformal scalars on a de Sitter background. Phys. Rev. D **103**(10), 105022 (2021)
81. D. Glavan, S.P. Miao, T. Prokopec, R.P. Woodard, Large logarithms from quantum gravitational corrections to a massless, minimally coupled scalar on de Sitter. JHEP **03**, 088 (2022)
82. A.J. Foraci, R.P. Woodard, Resumming photon loops for inflationary gravity. Phys. Rev. D **111**(6), 063531 (2025)
83. A.J. Foraci, R.P. Woodard, Resumming fermion loops for inflationary gravity. Phys. Lett. B **868**, 139649 (2025). (arXiv:2501.01972 [gr-qc])
84. S. Park, T. Prokopec, R.P. Woodard, Quantum scalar corrections to the gravitational potentials on de Sitter background. JHEP **01**, 074 (2016)
85. A.F. Radkowski, Some aspects of the source description of gravitation. Ann. Phys. **56**(2), 319–354 (1970)

86. D.M. Capper, On quantum corrections to the graviton propagator. Nuovo Cim. A **25**, 29 (1975)
87. D.M. Capper, M.J. Duff, The one loop neutrino contribution to the graviton propagator. Nucl. Phys. B **82**, 147–154 (1974)
88. M.J. Duff, J.T. Liu, Complementarity of the Maldacena and Randall-Sundrum pictures. Phys. Rev. Lett. **85**, 2052–2055 (2000)
89. D.M. Capper, M.J. Duff, L. Halpern, Photon corrections to the graviton propagator. Phys. Rev. D **10**, 461–467 (1974)
90. G. 't Hooft, M.J.G. Veltman, One loop divergencies in the theory of gravitation. Ann. Inst. H. Poincare A Phys. Theor **20**, 69–94 (1974)

Dark Energy

Structure Formation in the Local Universe and the Cosmological Constant

V. G. Gurzadyan

Abstract The structure formation in the local Universe is considered within the weak-field modification of General Relativity involving the cosmological constant. This approach enables to describe the dynamics of groups and clusters of galaxies, to explain the discrepancy in the observational properties of the local (late) and the global (early) Universe, i.e. the Hubble tension as a result of two flows, local and global ones, with non-equal Hubble parameters. The kinetic analysis with the modified gravitational potential involving the cosmological constant is shown to predict semi-periodical structure of filaments in the local universe. In the local scale this complements the Zeldovich pancake theory of evolution of the primordial density perturbations and of structure formation in the cosmological scale. The role of the cosmological constant is outlined in rescaling of the physical constants from one aeon to another within the Conformal Cyclic Cosmology.

1 Introduction

During years I had the fortunate possibility to discuss with Alexei Starobinsky broad scope of cosmological and physical problems, including those resulted in joint publications dealing with the topology of the Universe [1, 2]. Below, the problem of the structure formation will be discussed which was among the topics of research by Yakov Zeldovich [3], the renowned mentor of Starobinsky. Zeldovich, along with the developing the pancake theory [3, 4] of the large-scale structure formation, outlined the efficiency of non-relativistic treatment of the structure formation in the local Universe [5].

The difference in the descriptions of the early and late Universe can be indicated by the Hubble tension, i.e. the evidence on the discrepancy of the value of the Hubble constant obtained at observations of galactic surveys [6–12] and of Cosmic Microwave Background [13].

V. G. Gurzadyan (✉)
Center for Cosmology and Astrophysics, Alikhanian National Laboratory, Yerevan State University, Yerevan, Armenia
e-mail: gurzadyan@yerphi.am

A. Barvinsky and A. Kamenshchik (eds.), *Open Issues in Gravitation and Cosmology*, Fundamental Theories of Physics 225, https://doi.org/10.1007/978-3-032-15703-4_29

The non-relativistic description of the local Universe that we consider below is based on a theorem [14] on the general function satisfying the condition of the equivalency of the gravity of the sphere and of a point mass located in its center. That function for the force has the form [14]

$$\mathbf{F}(r) = \left(-\frac{\alpha}{r^2} + \Lambda r\right)\hat{\mathbf{r}}. \quad (1)$$

The second term in the right-hand-side corresponds to the cosmological term in McCrea-Milne non-relativistic cosmological model [15], and the cosmological constant Λ enters the weak-field General Relativity. Then, within this Λ-gravity one has 2 gravity constants, G and Λ, and weak-field modified General Relativity with the metric

$$g_{00} = 1 - \frac{2Gm}{r} - \frac{\Lambda r^2}{3}; \; g_{rr} = \left(1 - \frac{2Gm}{r} - \frac{\Lambda r^2}{3}\right)^{-1}. \quad (2)$$

This metric was known before, the Schwarzschild—de Sitter metric [16], as describing a black hole type solution. While the derivation based on the theorem on sphere-point equivalency enables to describe large-scale configurations such as galaxy groups and clusters and their flows. The role of Λ in Eqs. (1), (2) within group-theoretical approach can be represented depending on the sign of Λ as three solutions for Einstein equations corresponding to the isometry groups, as shown in the Table 1 [17].

The crucial point regarding Eq. (1) is that, it satisfies the first condition of Newton's shell theorem, namely the sphere-point equivalence but not the second one, i.e. force-free field inside a spherical shell. One can recall a classical example, when Einstein, to develop the cosmological model, posed two conditions: (1) existence of non-zero mean density of space, (2) the size (radius) of the space does not depend on time [18]. When Friedmann removed the second condition, he got new remarkable solutions which later appeared to be in agreement with the observations.

Below, we deal with the non-relativistic description of the dynamics of the groups and clusters based on Eq. (1), involving the difference of the local flow of galaxies and of the global expansion of the Universe, in the context of the Hubble tension. The local structure formation is described within this non-relativistic approach with a cosmological constant in Eq. (1). Certain consequences of the consideration of Λ as a physical constant are also discussed, including in the context of Cyclic Conformal Cosmology (CCC) of Penrose [19].

Table 1 Isometry groups for different values of Λ

Sign	Spacetime	Isometry group
$\Lambda > 0$	de Sitter (dS)	O(1,4)
$\Lambda = 0$	Minkowski (M)	IO(1,3)
$\Lambda < 0$	Anti de Sitter (AdS)	O(2,3)

2 Galaxy Groups and Clusters

Turning to the application of Λ-gravity to the galaxy systems, first let us turn to the virilalized systems. For them one can derive from Eq. (1) the following formula [20]

$$\Lambda = \frac{3\sigma^2}{2c^2 R^2} \simeq 3\ 10^{-52} (\frac{\sigma}{50\, kms^{-1}})^2 (\frac{R}{300\, kpc})^{-2} m^{-2}, \quad (3)$$

where σ is the velocity dispersion and R the characteristic radius of the system. Using this formula one can test the role of the Λ-term in Eq. (1) observational data on galaxy systems. Table 2 [20], based on the data in [21] of a sample of 17 galaxy groups of Hercules—Bootes region, with *rms* galactic velocities σ and the harmonic average radii R_h of the groups, exhibits the values of Λ obtained using Eq. (3); the galaxy groups are denoted by the name of their brightest galaxy.

The value of the cosmological constant obtained from CMB Planck data yields $\Lambda = (1.09 \pm 0.028) \times 10^{-52} \mathrm{m}^{-2}$ [13]. The correspondence of the latter value of Λ with those in Table 2 is visible. In [17, 22] various samples of galaxy groups and clusters, including groups of Leo/Cancer region, clusters of CLASH, LoCuSS

Table 2 Λ obtained for the galaxy groups of the Hercules-Bootes region

Galaxy group	$\sigma(km/s^{-1})$	$R_h(kpc)$	$\Lambda(m^{-2})$
NGC4736	50	338	3.84E-52
NGC4866	58	168	2.09E-51
NGC5005	114	224	4.55E-51
NGC5117	27	424	7.12E-53
NGC5353	195	455	3.23E-51
NGC5375	47	66	8.91E-51
NGC5582	106	93	2.28E-50
NGC5600	81	275	1.52E-51
UGC9389	45	204	8.55E-52
PGC55227	14	17	1.19E-50
NGC5961	63	86	9.43E-51
NGC5962	97	60	4.59E-50
NGC5970	92	141	7.48E-51
UGC10043	67	65	1.87E-50
NGC6181	53	196	1.28E-51
UGC10445	23	230	1.76E-52
NGC6574	15	70	8.07E-52
Average			8.24E-51
St. deviation			1.15E-50

Surveys, Planck clusters, are considered and it is shown that Λ-gravity with Eq. (1) does agree with those observational data.

3 Local and Global Hubble Flows: Hubble Tension

The Λ-gravity described above enables to address the Hubble tension [6–11]. Consider the following two equations which although look similar but have rather different content [23–25]

$$H_{local}^2 = \frac{8\pi G\rho_{local}}{3} + \frac{\Lambda c^2}{3}, \tag{4}$$

$$H_{global}^2 = \frac{8\pi G\rho_{global}}{3} + \frac{\Lambda c^2}{3}. \tag{5}$$

The first equation is derived using Eq. (1) within non-relativistic description of the local Universe, while the second one is the Friedmannian equation for FLRW metric. The discrepancy in the values of the relevant Hubble parameters, H_{local} and H_{global}, due to the different values of the mean densities, is then evident, thus suggesting a natural solution to the Hubble tension.

Then, Eq. (1) enables to define a *critical radius* where the Λ-term in the r.h.s. dominates upon the first term and the test particle will run away the central body

$$r_{crit}^3 = \frac{3GM}{\Lambda c^2}. \tag{6}$$

This scale attributed to the local Hubble flow, in [24, 25] is shown to agree with the data on the galactic flows in the vicinity of the Local Group, Virgo cluster, Virgo Supercluster and Laniakea Supercluster.

One can also obtain *absolute* constraints on the lower and upper values for the local Hubble parameter [25],

$$H_{min} = \sqrt(\Lambda c^2/3) = 56.2\, km/sec\ Mpc^{-1}, \tag{7}$$

$$H_{max} = \sqrt(\Lambda c^2) = 97.3\, km/sec\ Mpc^{-1}. \tag{8}$$

The lower constraint is obtained from the condition of the validity of the weak-field limit, while the upper limit is the value of Hubble parameter at r_{crit} (for details see [25]). These constraints are in full agreement with the observational data on the galactic flows in the local Universe.

4 Structure Formation

As mentioned above, Zeldovich pancake theory [3, 4] considers the evolution of primordial fluctuations to produce the large-scale cosmological structures. On the local scale, the self-consistent gravitational interaction can be crucial to ensure the formation of structures on that scale. Hence, in [26–28] the non-relativistic (non-GR) kinetic Vlasov mechanism of the formation of cosmic voids and walls is considered. Crucial point is the use of Eq. (1) for the gravitational interaction of the particles, i.e. the existence of repulsive Λ-term along with the gravitational attraction.

Then the Vlasov–Poisson equations to describe the system of N particles is given as (see [26–28] for details)

$$\frac{\partial F(\mathbf{x}, \mathbf{v}, t)}{\partial t} + \mathrm{div}_{\mathbf{x}}(\mathbf{v}F) + \widehat{G}(F; F) = 0, \tag{9}$$

$$\widehat{G}(F; F) \equiv -\big(\nabla_{\mathbf{v}} F\big)\big(\nabla_{\mathbf{x}}(\Phi[F(\mathbf{x})]\big), \tag{10}$$

$$\Delta^{(3)}_{\mathbf{x}} \Phi[F(\mathbf{x})]\big|_{t=t_0} = AG \int F(\mathbf{x}, \mathbf{v}, t_0)\, d\mathbf{v} - \frac{c^2 \Lambda}{2}, \tag{11}$$

where $F(\mathbf{x}, \mathbf{v}, t)$ is the distribution function of particles interacting according to Eq. (1), A is a normalization factor for the particle density.

The analysis of this system of equations leads to semi-periodic solutions which for the normalized radial coordinate $u(\mathbf{x})$ have the form [27]

$$u(\mathbf{x}) = \sum_{m=0} \big(\eta \exp(-U^{(0)})\big)^{m+1/2} \Big(\sum_{\ell=1}^{m} {}_{(I)}C_{2\ell+1}^{(2m+1)} \tag{12}$$

$$\sin\big((2\ell+1)\mathbf{qx}\big) + {}_{(III)}C_{2m+1} \sin\big(\mathbf{qx}\big)\big) +$$

$$+ \sum_{m=1}^{\infty} \big(\eta \exp(-U^{(0)})\big)^{m} \Big(\sum_{\ell=1}^{m} {}_{(II)}C_{2\ell}^{(2m)} \cos\big(2m\mathbf{qx}\big) +$$

$${}_{(III)}C_{2m} \sin\big(\mathbf{qx}\big)\Big).$$

As shown in [26–28] such semi-periodic solutions of the Vlasov-Poisson equations appear due to the repulsive Λ-term in the interacting potential, thus providing a mechanism for formation of voids, walls, and 1D and 2D-filaments. The critical radius given by Eq. (6), and hence, the interplay of Λ and of the local matter density determines the scale of the voids and appears to be in agreement with the observational data in the local Universe [27, 28].

5 Λ as a Fundamental Constant

The Λ-gravity considered above based on the sphere-point equivalence theorem [14], suggests two gravity constants, G and Λ in Eq. (1). If so, then one can go further and consider the cosmological constant Λ among fundamental physical constants, along with $G, c, \hbar$. Note, that Einstein has denoted as *universal constant* the cosmological constant when he introduced it in his cosmological model [18, 29].

Then, the set of (3+1) constants $(G, \Lambda, c, \hbar)$ enables one to have the following dimensionless quantity [30].

$$I = \frac{c^{3a}}{\Lambda^a G^a \hbar^a}, \tag{13}$$

where a is a real number, while no dimensionless ratio is possible to construct from the classical 3-set $(G, c, \hbar)$.

The relation Eq. (13) of the 4 constants within a numerical factor and $a = 1$ coincides with that of the entropy of de Sitter event horizon [19, 31, 32]

$$I_{dS} = 3\pi \frac{c^3}{\Lambda G \hbar}, \tag{14}$$

and with the Bekenstein bound [33] for the information in de Sitter space

$$I_{BB} = \frac{3\pi c^3}{\Lambda G \hbar ln 2}. \tag{15}$$

The consideration of Λ as a physical constant has consequences for the Conformal Cyclic Cosmology (CCC) [19], for which the second law of thermodynamics and the positive Λ are basic conditions. Then, the dimensionless ratio Eq. (13) becomes important condition regarding the information transfer from one aeon to the next one (see [19, 34, 35]), as invariant of conformal transformation

$$\tilde{g}_{\mu\nu} = \Omega^2 g_{\mu\nu}. \tag{16}$$

Namely, as shown in [30], then the physical constants can be rescaled from one aeon to the successive one

$$c \to a_1 c, \quad \hbar \to a_2 \hbar, \quad G \to a_3 G, \quad \Lambda \to a_4 \Lambda, \quad a_i \in \mathbb{R}^+, \tag{17}$$

keeping the condition

$$\frac{a_1^3}{a_2 a_3 a_4} = 1, \tag{18}$$

Thus, considering Λ among 3 other physical constants will imply within CCC: **a** the same global cosmological dynamics in the aeons, **b** rescaling of the physical constants with modifications for physical processes from one aeon to another.

6 Conclusions

The observational indications on the peculiarities in the properties of the early and late Universe stimulate consideration of non-uniform mechanisms responsible for the relevant structure formation. The Zeldovich pancake theory [3] of structure formation at global scale deals with the evolution of the primordial density perturbations, while at local scales the self-consistent gravitational interaction can have dominating contribution.

We concisely discussed the properties of the local Universe and the formation of the matter structures based on the weak-field modification of General Relativity with the cosmological constant Λ. That approach follows from the theorem [14] on sphere-point gravity identity when the relevant general function includes a Λ-term. The Hubble tension then is naturally explained as a result of local and global flows with non-identical Hubble parameters. The analysis of Vlasov-Poisson equation set with the potential involving the Λ-term leads to the prediction of voids and walls in the matter structure of the local Universe, so that the cosmological constant, along with the local matter density, defines the scale of the voids.

Forthcoming more deep observational surveys in acquiring accurate comparative features of the early and late Universe can enable one to reveal the relevant physical processes in more details.

Acknowledgements I am thankful to A. Stepanian for the efficient joint work.

References

1. V.G. Gurzadyan, A.A. Starobinsky et al., Mod. Phys. Lett. A **22**, 2955 (2007)
2. V.G. Gurzadyan, A.A. Starobinsky et al., A&A **490**, 929 (2008)
3. Ya.B. Zeldovich, A&A **5**, 84 (1970)
4. V.I. Arnold, S.F. Shandarin, Y.B. Zeldovich, Geophys. Astrophys. Fluid Dyn. **20**, 111 (1982)
5. Ya.B. Zeldovich, *Non-relativistic cosmology*, 181, Editor's appendix 1 to Russian edition of S. Weinberg, *The First Three Minutes*, Energoizdat, Moscow (in Russian) (1981)
6. A.G. Riess, Nat. Rev. Phys. **2**, 10 (2020)
7. A.G. Riess et al., ApJ **962**, 17 (2024)
8. A.G. Riess et al., ApJ **977**, 120 (2024)
9. M.G. Dainotti et al., Galaxies **10**(1), 24 (2022)
10. M.G. Dainotti, B. De Simone, arXiv:2501.14944
11. M.G. Dainotti et al, arXiv:2501.11772
12. G. Bargiacchi, M.G. Dainotti, S. Capozziello, New Astron. Rev. **100**, 101712 (2025)
13. Planck Collaboration: N. Aghanim et al., A&A **641**, A6 (2020)
14. V.G. Gurzadyan, Observatory **105**, 42 (1985)
15. W.H. McCrea, E.A. Milne, Q. J. Math. **5**, 73 (1934)
16. W. Rindler, *Relativity, Special, General, and Cosmological* (Oxford University Press, Oxford, 2006)
17. V.G. Gurzadyan, A. Stepanian, Eur. Phys. J. C **78**, 632 (2018)
18. A. Einstein, Koniglich Preussische Akad. der Wissenschaften, 142 (1917)

19. R. Penrose, *Cycles of Time: An Extraordinary New View of the Universe* (Bodley Head, London, 2010)
20. V.G. Gurzadyan, Eur. Phys. J. Plus **134**, 14 (2019)
21. I.D. Karachentsev, O.G. Kashibadze, V.E. Karachentseva, Astr. Bull. **72**, 111 (2017)
22. V.G. Gurzadyan, A. Stepanian, Eur. Phys. J. C **79**, 169 (2019)
23. V.G. Gurzadyan, A. Stepanian, Eur. Phys. J. C **79**, 568 (2019)
24. V.G. Gurzadyan, A. Stepanyan, Eur. Phys. J. Plus **136**, 235 (2021)
25. V.G. Gurzadyan, A. Stepanian, A&A **653**, A145 (2021)
26. V.G. Gurzadyan, N.N. Fimin, V.M. Chechetkin, A&A **666**, A149 (2022)
27. V.G. Gurzadyan, N.N. Fimin, V.M. Chechetkin, A&A **672**, A95 (2023)
28. V.G. Gurzadyan, N.N. Fimin, V.M. Chechetkin, A&A **677**, A161 (2023)
29. A. Einstein, Phys. Z. **19**, 165 (1918)
30. V.G. Gurzadyan, A. Stepanian, Eur. Phys. J. Plus **134**, 98 (2019)
31. J.D. Bekenstein, Phys. Rev. D **7**, 2333 (1973)
32. G.W. Gibbons, S.W. Hawking, Phys. Rev. D **15**, 2738 (1977)
33. J.D. Bekenstein, Phys. Rev. D **23**, 287 (1981)
34. V.G. Gurzadyan, R. Penrose, Eur. Phys. J. Plus **128**, 22 (2013)
35. V.G. Gurzadyan, R. Penrose, Eur. Phys. J. Plus **131**, 11 (2016)

Scale Invariant Dark Energy

Alessandro Tronconi and Giovanni Venturi

Abstract A globally scale-invariant Dark Energy model based on Induced Gravity with the addition of a small R^2 contribution is examined. The scalar field (quintessence), playing the role of Dark Energy, has a quartic potential and generates Newton's constant with its non-minimal coupling (after introducing a suitable symmetry breaking). Even when small, the R^2 contribution significantly modifies the cosmological evolution of the matter-gravity system. The solutions to this model are obtained analytically through a perturbative expansion and oscillate with transplanckian frequency. They are then compared with similar solutions found for ΛCDM cosmology plus R^2. Finally scalar field production is perturbatively taken into account in a simple model and the resulting effects illustrated.

1 Introduction

Alexei Starobinsky was a brilliant and original theoretical physicist who made fundamental contributions to gravitation and cosmology. He visited our group in Bologna for the first time in 2003 and left a lasting scientific influence on us. This manuscript is dedicated to his memory.

Even if gravity is an attractive force, at least for the short distances we can directly probe, the current understanding of theoretical cosmology is limited by the fact that several independent observations suggest that our universe underwent, and is currently going through, a phase of cosmic expansion and acceleration. The acceleration is the consequence of a "repulsive" and long range effect which perhaps originates from some modification of General Relativity or the presence of some exotic "invisible" fluid. The inflationary phase is the period of accelerated expansion which occurred before the Hot Big Bang, and dynamically produced the initial con-

A. Tronconi · G. Venturi (✉)
Dipartimento di Fisica e Astronomia and INFN, Bologna, Italy
e-mail: giovanni.venturi@bo.infn.it

A. Tronconi
e-mail: alessandro.tronconi@bo.infn.it

A. Barvinsky and A. Kamenshchik (eds.), *Open Issues in Gravitation and Cosmology*,
Fundamental Theories of Physics 225, https://doi.org/10.1007/978-3-032-15703-4_30

ditions of the present Universe. While the exact microscopic physics at the origin of inflation is still debated, its dynamics seems well constrained by observations and we generally expect that, in the near future as such observations become more precise, we shall gain more insights on inflation and the origin of our universe. In contrast there is much uncertainty about the physics behind the present cosmic acceleration. While a cosmological constant or a mechanism similar to inflation may explain, to a certain degree, current observations, several questions related to such an accelerated evolution remain unresolved, above all the tension between the differing measured values of the present rate of expansion of the universe (H_0). This is certainly the reason behind the abundant theoretical and experimental efforts which try to improve the present understanding of the current expansion of the universe.

Several years ago a simple model based on global scale invariance was suggested for induced gravity [1, 2]. The model involved a massless scalar field coupled to the curvature scalar and having a quartic self interaction. The introduction of symmetry breaking, either through the presence of a condensate [3] or quantum effects [4], would then lead to a non-zero expectation value for the scalar field. As a consequence one would then generate a gravitational constant and a de Sitter solution associated with a cosmological constant [5]. The addition of an isentropic perfect fluid as a perturbation would then lead to a time dependence of the gravitational constant which, because of the stabilizing influence of the potential, is much smaller than that (power-law) of previous models (Brans-Dicke) [6]. Some years ago, since the observational status was compatible with an accelerating universe dominated by dark energy, we studied the cosmological evolution of the model in the presence of matter and radiation. We found that it led to Einstein gravity plus a cosmological constant as a stable attractor among homogeneous cosmologies and was therefore a viable dark energy model for a wide range of scalar field initial conditions and values for its positive coupling to the Ricci scalar [7, 8]. Subsequently the same model was also studied in the context of inflation [9] with a symmetry breaking potential for the scalar field, necessary to provide the exit from the accelerated inflationary phase.

Several years ago de Sitter solutions were also found from the homogeneous Einstein equations with quantum one-loop contributions of conformally covariant matter fields. This essentially corresponded to the addition of an R^2, scale invariant, contribution to the Einstein action obtaining the so-called Starobinsky model [10]. The Starobinsky model has been studied thoroughly as a model of inflation and is essentially indistinguishable from induced gravity with a Higgs-like potential. Modified gravity models inspired by the seminal work [10] was also proposed in order to explain the present universe acceleration. In such $f(R)$ models, Dark Energy is a consequence of the modified gravity dynamics and in general, in order to have acceleration for "small" values of the curvature scalar, one needs rather exotic actions. Moreover such models has been found to have problems which can be cured by the addition of an R^2 contribution [11].

It is then natural to ask oneself what the effect is on late time cosmological evolution of adding such a scale invariant term to the previous dark energy action [1]: this is the purpose of this note.

We shall then compare some features of the cosmological evolution of a globally scale-invariant model, with R^2 contribution added, to the evolution one finds in standard $R + R^2/m^2$ theory with a cosmological constant (we shall refer to this latter case as perturbed ΛCDM model). In particular two effects are studied: we shall first consider the classical solutions to the matter gravity system and compare such solutions with $R + R^2/m^2$ cosmology; subsequently we shall account for possible scalar particle production effects and again compare the scale invariant model with $R + R^2/m^2$. Lastly we shall study the asymptotic behaviour of the homogeneous solutions of the two models. We end by summarizing the differences between the two models and our conclusions. Let us add that the scale invariant model under consideration, has already been studied in the context of inflation [12]. As an inflationary model it belongs to the class of multi-field inflation models since the addition of R^2 introduces one more scalar degree of freedom.

2 $R + R^2$ Modified Gravity Formalism

We first consider standard cosmological fluids in an expanding universe with the following gravitational action

$$S = \frac{\mathrm{M_P}^2}{2} \int \mathrm{d}^4x \sqrt{-g} \left(R + \frac{\alpha R^2}{2\mathrm{M_P}^2} \right) \equiv \frac{\mathrm{M_P}^2}{2} \int \mathrm{d}^4x \sqrt{-g}\, f(R) \tag{1}$$

where R is the Ricci scalar. Modified gravity models, wherein the Einstein-Hilbert action is replaced by some generic function of the Ricci scalar, have been studied thoroughly in the literature in order to describe inflation and Dark Energy. Here we shall only consider the effects of the scale invariant contribution $\sim R^2$ on the ΛCDM cosmological evolution. If one considers a spatially flat FRW space-time the metric tensor is defined by

$$\mathrm{d}s^2 = g_{\mu\nu}\mathrm{d}x^\mu \mathrm{d}x^\nu = -\mathrm{d}t^2 + a(t)^2 \mathrm{d}\mathbf{x} \cdot \mathrm{d}\mathbf{x} \tag{2}$$

and t is the cosmic time. For such a homogeneous and isotropic metric the Ricci scalar takes the form $R = 6\left(2H^2 + \dot{H}\right)$. The variation of the matter-gravity action leads to the following set of equations

$$3FH^2 = \frac{FR - f}{2} - 3H\dot{F} + \frac{\rho}{\mathrm{M_P}^2}, \quad -2F\dot{H} = \ddot{F} - H\dot{F} + \frac{\rho + P}{\mathrm{M_P}^2} \tag{3}$$

where $F = \partial f / \partial R$ and ρ and P are the total energy density and pressure of the cosmological fluids, each satisfying the continuity equation

$$\dot{\rho}_i = -3H(\rho_i + P_i) = -3H\rho_i(1 + w_i) \tag{4}$$

and the index $i = r, m, \Lambda$ stands for radiation, matter and cosmological constant.

One can combine the Eqs. (3) to obtain the following equation

$$\ddot{R} + 3H\dot{R} + \frac{{M_P}^2}{3\alpha}R = \frac{\rho - 3P}{3\alpha} \tag{5}$$

having the form of the Klein Gordon equation for the Ricci scalar with a mass term and an external force term depending on trace of the energy momentum tensor of the cosmological fluids. In this case, the 00 component of the Einstein Eqs. (3) is sufficient to determine the cosmological evolution. In terms of the Hubble parameter squared it has the form

$$0 = 1 - \frac{\rho}{3{M_P}^2 H^2} + \frac{3\alpha}{{M_P}^2}\left[(H^2)_{,NN} - \frac{\left[(H^2)_{,N}\right]^2}{4H^2} + 3(H^2)_{,N}\right] \tag{6}$$

where $N \equiv \ln a$ is the e-folding number. The Friedmann equation in the context of $f(R)$ theories of gravity is a dynamical equation and its $\alpha \to 0$ limit is non-trivial. Therefore for $\alpha \ll 1$ we expect that cosmological solution can differ significantly to the standard ($\alpha = 0$) one: $3{M_P}^2 H^2 = \rho$. If we search for solutions which are close to the ΛCDM evolution then one can start from the ansatz

$$H^2 = \frac{\rho}{3{M_P}^2}(1 + \delta h_2) \tag{7}$$

and obtain the following linearised equation for δh_2, valid in the regime $\delta h_2 \ll 1$:

$$\begin{aligned} 0 = {} & \frac{(3\rho_m + 4\rho_r)^2}{\rho^2}\rho^2 \delta h_{2,\rho\rho} + \left[\frac{3(3\rho_m + 4\rho_r)^2}{2\rho^2} + 4\frac{\rho_r}{\rho}\right]\rho\,\delta h_{2,\rho} \\ & + \left[-\frac{(3\rho_m + 4\rho_r)^2}{2\rho^2} + 8\frac{\rho_r}{\rho} + \frac{{M_P}^4}{\alpha\rho}\right]\delta h_2 + 4\frac{\rho_r}{\rho} - \frac{(3\rho_m + 4\rho_r)^2}{4\rho^2} \end{aligned} \tag{8}$$

where $\rho = \rho_r + \rho_m + \rho_\Lambda$.

2.1 Radiation Domination

During the radiation domination epoch $\rho_m/\rho \ll 1$, $\rho_r/\rho \simeq 1$ and Eq. (8) simplifies and takes the form of an equation for a time dependent harmonic oscillator

$$\delta h_2'' + \frac{\delta h_2}{9x^{2/3}} = 0 \tag{9}$$

where the prime denotes a derivative with respect to $x = \left(\frac{M_P{}^4}{\alpha\rho}\right)^{3/4}$ and, for realistic models where the total energy density in many order of magnitude less than the Planck energy density $M_P{}^4$, one must consider the regime $x \gg 1$. The general solution of (9) is

$$\delta h_2(x) = \sqrt{x}\left(A_1 J_{-\frac{3}{4}}\left(\frac{x^{\frac{2}{3}}}{2}\right) + A_2 J_{\frac{3}{4}}\left(\frac{x^{\frac{2}{3}}}{2}\right)\right) \simeq \frac{x^{2/3}}{2}\left(B_1 \sin\frac{x^{2/3}}{2} + B_2 \cos\frac{x^{2/3}}{2}\right)$$
$$= \frac{M_P{}^2}{2\sqrt{\alpha\rho}}\left(B_1 \sin\frac{M_P{}^2}{2\sqrt{\alpha\rho}} + B_2 \cos\frac{M_P{}^2}{2\sqrt{\alpha\rho}}\right). \tag{10}$$

Apart from the oscillation we observe that the amplitude of the solution increases with time. The perturbative approach then fails unless the integration constants are set to tiny values and we may refer to this as fine tuning. Such a fine tuning problem can be alleviated if one takes into account the effect of perturbative particle production (see for example [14, 15]). In such a case the energy density of the oscillating field is transferred to matter fields and the amplitude of the oscillation is therefore damped.

2.2 *Matter Domination*

During matter domination $\rho_r/\rho \ll 1$, $\rho_m/\rho \simeq 1$ and Eq. (8) becomes

$$\delta h_2'' + \left(\frac{4}{9} - \frac{2}{x^2}\right)\delta h_2 - \frac{1}{x^2} = 0 \tag{11}$$

which can be solved exactly and in the large x limit has the following general solution

$$\delta h_2(x) \simeq A_1 \cos\frac{2x}{3} + A_2 \sin\frac{2x}{3} = A_1 \cos\frac{2M_P{}^2}{3\sqrt{\alpha\rho}} + A_2 \sin\frac{2M_P{}^2}{3\sqrt{\alpha\rho}} \tag{12}$$

which is an oscillation with constant amplitude. If $\delta h_2 \ll 1$ at the onset of matter domination, it remains small and the perturbative solution is valid until dark energy begins to dominate.

Let us note that the oscillating contributions in (10) and (12) have the form

$$\sin\frac{2M_P{}^2}{3(1+w_{\text{eff}})\sqrt{\alpha\rho}} \quad \text{or} \quad \cos\frac{2M_P{}^2}{3(1+w_{\text{eff}})\sqrt{\alpha\rho}} \tag{13}$$

where w_{eff} is the equation of state of dominant fluid (and is 0 for dust and 1/3 for radiation). From the continuity equation $\dot\rho = -3H\rho(1+w_{\text{eff}})$ one obtains

$$\rho^{-3/2}\dot\rho = -\frac{\sqrt{3}}{M_P}(1+w_{\text{eff}}) \Rightarrow \frac{1}{\sqrt{\rho}} = \frac{\sqrt{3}(1+w_{\text{eff}})(t-t_0)}{2M_P} \tag{14}$$

and on substituting into Eq. (13) one finds the following constant frequency for the oscillations of δh_2 (in cosmic time)

$$\omega_{m,r}^{\mathrm{GR}} = \frac{\mathrm{M_P}}{\sqrt{3\alpha}} \tag{15}$$

and the result is the same for both radiation and matter domination.

2.3 *Cosmological Constant Domination*

Let us finally consider the case when the cosmological constant dominates, one then has $\rho_m/\rho \ll 1$, $\rho_\Lambda/\rho \simeq 1$. Further, the dominant cosmological fluid (cosmological constant) does not evolve and it is convenient to search for the solutions in terms of the e-fold number. The Friedmann equation for the perturbation is

$$\delta h_{2,NN} + 3\delta h_{2,N} + \frac{\mathrm{M_P}^4}{\alpha\rho_\Lambda}\delta h_2 = 0 \tag{16}$$

and its general solution is

$$\delta h_2(N) = \frac{\bar{A}_1 \mathrm{e}^{\sqrt{\frac{9}{4}-\frac{\mathrm{M_P}^4}{\alpha\rho_\Lambda}}N} + \bar{A}_2 \mathrm{e}^{-\sqrt{\frac{9}{4}-\frac{\mathrm{M_P}^4}{\alpha\rho_\Lambda}}N}}{\mathrm{e}^{\frac{3N}{2}}} \simeq \frac{A_1 \sin\left(\frac{\mathrm{M_P}^2 N}{\sqrt{\alpha\rho_\Lambda}}\right) + A_2 \cos\left(\frac{\mathrm{M_P}^2 N}{\sqrt{\alpha\rho_\Lambda}}\right)}{\mathrm{e}^{\frac{3N}{2}}} \tag{17}$$

On calculating the frequency of oscillation in terms of the cosmic time one finds the same result as was found for both matter and radiation domination, $\omega_\Lambda^{\mathrm{GR}} = \omega_{m,r}^{\mathrm{GR}}$.

3 Scale Invariant Model

Let us now consider a simple scale invariant generalisation of the action (1) with the addition of a scalar field σ which dynamically generates Newton's constant. Therefore let us start from the following scale invariant action

$$S = \frac{1}{2}\int \mathrm{d}^4x\sqrt{-g}\left(\gamma\sigma^2 R + \frac{\alpha R^2}{2} - \frac{\lambda}{2}\sigma^4 - (\partial\sigma)^2\right). \tag{18}$$

For $\alpha = 0$ and in the presence of cosmological fluids (dust and radiation) the scalar field behaves as quintessence and, for certain initial conditions, it relaxes to a constant attractor and plays the role of Dark Energy. We are interested in the modification to the cosmological evolution associated with the presence of an R^2 contribution.

The independent equations for the matter gravity system are the continuity Eqs. (4) for the cosmological fluids, the Klein-Gordon (KG) equation for the homogeneous field and a "dynamical" Friedmann equation. It is convenient to redefine $y \equiv \sigma^2$. In terms of y the KG equation takes the form

$$\frac{y_{,NN}}{y} - \frac{y_{,N}^2}{2y^2} + \frac{(H^2)_{,N}}{2H^2}\frac{y_{,N}}{y} + 3\frac{y_{,N}}{y} - 6\gamma\frac{(H^2)_{,N}}{H^2} + 2\lambda\frac{y}{H^2} - 24\gamma = 0 \tag{19}$$

and the Friedmann equation can be rewritten as

$$1 = \frac{\rho + \frac{\lambda y^2}{4}}{3\gamma y H^2} - \frac{y_{,N}}{y} + \frac{1}{24\gamma}\left(\frac{y_{,N}}{y}\right)^2 - \frac{3\alpha\left[(H^2)_{,NN} - \left(\frac{(H^2)_{,N}^2}{2H^2}\right) + 3(H^2)_{,N}\right]}{\gamma y}. \tag{20}$$

Analogously to the GR case one can write the equation for the Ricci scalar, in the IG context, which takes the following form

$$\ddot{R} + 3H\dot{R} + \frac{\gamma(1+6\gamma)x}{3\alpha}R = \frac{\rho - 3P + (1+6\gamma)\left(\lambda x^2 - \frac{\dot{x}^2}{4x}\right)}{3\alpha} \tag{21}$$

in terms of the cosmic time.

Let us now proceed following the same approach as illustrated in the previous section. We shall then consider different cosmological eras and study the departure from the evolution with $\alpha = 0$.

3.1 Radiation Domination

During radiation domination $\rho \simeq \rho_r$ and contributions proportional to λ may be neglected. For $\alpha = 0$ one has the solution $y = \mathrm{M_P}^2/\gamma$, $H^2 \propto \mathrm{e}^{-4N}$ which exactly reproduces the standard ΛCDM evolution. When $\alpha \neq 0$ one expects perturbed solutions of the form

$$H^2 = \frac{\rho}{3\mathrm{M_P}^2}(1 + \delta h_2), \quad y = \frac{\mathrm{M_P}^2}{\gamma}(1 + \delta y) \tag{22}$$

and on linearising the dynamical equations one finds

$$\begin{cases} 6\gamma\rho^2\delta h_{2,\rho\rho} + 3\rho\delta y_{,\rho} + 4\rho^2\delta y_{,\rho\rho} = 0 \\ 4\rho^2\delta h_{2,\rho\rho} - \frac{\mathrm{M_P}^4}{\alpha}\delta y_{,\rho} + 7\rho\delta h_{2,\rho} + \frac{\mathrm{M_P}^4}{4\rho\alpha}(\delta h_2 + \delta y) = 0 \end{cases} \tag{23}$$

The first equation in (23) can be solved exactly leading to

$$\delta h_2 = c_1 + \frac{\delta y}{6\gamma} - \frac{2\rho\delta y_{,\rho}}{3\gamma}. \tag{24}$$

The second equation in (23) can be written in terms of δy and its derivatives and has the following solution in terms of Hypergeometric functions

$$\delta y = -\frac{6\gamma}{1+6\gamma}c_1 + c_2\,{}_1F_2\left(\frac{1}{4}; \frac{2}{3}, \frac{5}{6}; -\frac{x^6}{16}\right)$$
$$+c_3 x\,{}_1F_2\left(\frac{5}{12}; \frac{5}{6}, \frac{7}{6}; -\frac{x^6}{16}\right) + c_4\,{}_1F_2\left(\frac{7}{12}; \frac{7}{6}, \frac{4}{3}; -\frac{x^6}{16}\right), \qquad (25)$$

where $x = \left[\mathrm{M_P}^4/(\alpha\rho)\right]^{1/6}$ and is very large for α small and realistic values of ρ_r. In the large x limit Eq. (25) takes the compact form

$$\delta y = \bar{c}_2\left(\frac{\alpha\rho}{\mathrm{M_P}^4}\right)^{1/4} + \left[\bar{c}_3 \sin\left(\frac{\pi}{24} - \frac{\mathrm{M_P}^2}{\sqrt{4\alpha\rho}}\right) + \bar{c}_4 \cos\left(\frac{\pi}{24} - \frac{\mathrm{M_P}^2}{\sqrt{4\alpha\rho}}\right)\right]\left(\frac{\alpha\rho}{\mathrm{M_P}^4}\right)^{3/8}, \qquad (26)$$

and correspondingly one finds

$$\delta h_2 = \left(\frac{\mathrm{M_P}^4}{\alpha\rho}\right)^{1/8} \frac{-\bar{c}_3\cos\left(\frac{\pi}{24} - \sqrt{\frac{\mathrm{M_P}^4}{4\alpha\rho}}\right) + \bar{c}_4 \sin\left(\frac{\pi}{24} - \sqrt{\frac{\mathrm{M_P}^4}{4\alpha\rho}}\right)}{6\gamma}, \qquad (27)$$

and we note that the perturbed Hubble parameter oscillates around the ΛCDM solution analogously to what occurs in (10). In this latter case, however, the amplitude increases much more slowly. Finally one can calculate the oscillation frequency w.r.t. cosmic time and obtain $\omega_r = \mathrm{M_P}/\sqrt{3\alpha}$ which is the same as Eq. (15). Let us note that, as in the model studied in Sec. II, the increasing amplitude introduced a fine tuning problem related to the initial conditions of the oscillating quantities. For this specific case the effect of the perturbative production of σ field quanta will be addressed in the last section and can alleviate such a fine tuning issue.

3.2 Matter Domination

During matter domination $\rho \simeq \rho_m$, and in the $\alpha \to 0$ limit one has the following solution

$$y = y_0 \mathrm{e}^{\frac{4\gamma}{1+4\gamma}N}, \quad \rho = \rho_0 \mathrm{e}^{-3N}, \quad H^2 = \frac{1+4\gamma}{1+\frac{34}{3}\gamma + 32\gamma^2}\frac{\rho}{3\gamma y} \qquad (28)$$

with $y_0 \sim \mathrm{M_P}^2/\gamma$. Therefore for α small we expand around it:

$$y = y_0 \mathrm{e}^{\frac{4\gamma}{1+4\gamma}N}(1+\delta y), \quad H^2 = \frac{(1+4\gamma)^2}{1+\frac{34}{3}\gamma + 32\gamma^2}\frac{\rho\,\mathrm{e}^{-\frac{4\gamma}{1+4\gamma}N}}{3\gamma y_0}(1+\delta h_2). \qquad (29)$$

The linearised equations for δy and δh_2 are

$$2(1+4\gamma)\delta y_{,NN} + (3+16\gamma)\delta y_{,N} - 8\gamma(1+6\gamma)\delta h_{2,N} = 0, \tag{30}$$

$$\delta h_{2,NN} - \frac{3(1+8\gamma)}{2(1+4\gamma)} h_{2,N} + \frac{2\mathrm{M_P}^4(3+16\gamma)(1+6\gamma)^2 \mathrm{e}^{\frac{8\gamma}{1+4\gamma}N}}{9\alpha\rho(1+4\gamma)^3}\delta y_{,N}$$
$$+ \frac{(3+16\gamma)\left[\frac{\mathrm{M_P}^4}{9\alpha\rho}\frac{(3+16\gamma)(1+6\gamma)^2}{(1+4\gamma)^2}\mathrm{e}^{\frac{8\gamma}{1+4\gamma}N} - \frac{3}{4}\right]}{(1+4\gamma)^2}(\delta h_2 + \delta y) = 0, \tag{31}$$

and Eq. (30) has a solution

$$\delta h_2 = c_1 + \frac{3+16\gamma}{8\gamma(1+6\gamma)}\delta y + \frac{1+4\gamma}{4\gamma(1+6\gamma)}\delta y_{,N}, \tag{32}$$

which can be substituted into Eq. (31) obtaining a linear, third order, equation for δy with the general solution

$$\delta y = -c_1 \frac{8\gamma(1+6\gamma)}{3(1+4\gamma)^2} + c_2\, {}_1F_2\,(b;\, 1-2b, 1-b;\, -z)$$
$$+ c_3 x\, {}_1F_2\,(2b;\, 1-b, 1+b;\, -z) + c_4 x^2\, {}_1F_2\,(3b;\, 1+b, 1+2b;\, -z)\,, \tag{33}$$

where $b = \frac{4\gamma}{3(3+20\gamma)}$ and the time dependent argument of the hypergeometric functions is

$$z = \frac{(3+16\gamma)(1+6\gamma)^2}{(3+20\gamma)^2}\frac{\gamma^2 y^2}{3\alpha\rho}. \tag{34}$$

For very large (negative) values of such an argument one observes that the hypergeometric functions can be expanded and the solution takes the compact form

$$\delta y = \frac{\bar{c}_2}{z^b} + z^{2b-\frac{3}{4}}\left[\bar{c}_3 \cos\left(2\sqrt{z}\right) + \bar{c}_4 \sin\left(2\sqrt{z}\right)\right] - \frac{8\gamma(1+6\gamma)c_1}{3(1+4\gamma)^2}. \tag{35}$$

We observe that, apart from a constant contribution, the time dependent part decreases in time for γ small and the oscillating part depends on the square root of the ratio between the effective (unperturbed) Planck mass given by $\gamma^2 y^2$ and $\alpha\rho$ where ρ_m is assumed to be the dominant fluid. This is a general feature of the solutions we found so far perturbatively, and we note that the ratio is tiny leading to "wild" oscillations as N varies.

On now substituting the solution for δy in (32) and retaining the leading order contributions one finds

$$\delta h_2 = \bar{c}_0 + \frac{\bar{c}_2\,(9+40\gamma)}{24\gamma(1+6\gamma)z^{\frac{4\gamma}{3(3+20\gamma)}}} + \frac{(3+20\gamma)\left[\bar{c}_3 \sin\left(2\sqrt{z}\right) + \bar{c}_4 \cos\left(2\sqrt{z}\right)\right]}{4\gamma(1+6\gamma)z^{\frac{1}{4}-\frac{8\gamma}{3(3+20\gamma)}}}, \tag{36}$$

where, the leading order, non-constant, contribution decreases and is non-oscillating. Let us note that if one expresses the oscillating functions in terms of cosmic time for this case the frequency is not constant in contrast with the case of matter dominance discussed in the previous section. Indeed because of the time varying Newton's constant (to leading order) and the dependence of z on the ratio y^2/ρ_m (and not on $y/\rho_m \propto H^{-1} \propto t$) one has $\sqrt{z} \neq \omega_m t$. However the time dependence is tiny and proportional to γ which must be small. On neglecting γ one recovers the GR result i.e. $\omega_m = \omega_r + O(\gamma)$ where the contribution $O(\gamma)$ depends on cosmic time.

3.3 Scalar Field Domination

Let us finally examine the case in which the scalar field energy density dominates. In such a case the unperturbed solution ($\alpha \to 0$) mimics a cosmological constant (see [7]) and one has

$$y = \frac{\mathrm{M_P}^2}{\gamma}, \quad H^2 = \frac{\lambda \mathrm{M_P}^2}{12\gamma^2}. \tag{37}$$

On now considering α small the above solution is modified as follows

$$y = \frac{\mathrm{M_P}^2}{\gamma}(1+\delta y), \quad H^2 = \frac{\lambda \mathrm{M_P}^2}{12\gamma^2}(1+\delta h_2). \tag{38}$$

The linearised equations for δy and δh_2 are

$$\begin{cases} \delta h_{2,NN} + 3\delta h_{2,N} + \frac{4\gamma^2}{\alpha\lambda}\left(\delta y_N - \delta y + \delta h_2\right) = 0 \\ \delta y_{NN} + 3\delta y_{,N} + 6\gamma\left(4\delta y - \delta y_{2,N} - 4\delta y_2\right) = 0, \end{cases} \tag{39}$$

which can be solved exactly. The general solutions are

$$\begin{cases} \delta h_2 = A_2 + A_1 e^{-3N} + e^{-\frac{3}{2}N}\left(A_3 e^{\frac{\Lambda}{2\alpha\lambda}N} + A_4 e^{-\frac{\Lambda}{2\alpha\lambda}N}\right) \\ \delta y = \frac{A_1}{4}e^{-3N} + A_2 + \frac{(3\Lambda - 15\alpha\gamma)}{4\gamma(1+6\gamma)}e^{-\frac{3}{2}N}\left(A_3 e^{\frac{\Lambda}{2\alpha\lambda}N} + A_4 e^{-\frac{\Lambda}{2\alpha\lambda}N}\right), \end{cases} \tag{40}$$

where

$$\Lambda \equiv \sqrt{-\alpha\lambda\left(16\gamma^2 + 96\gamma^3 - 9\alpha\lambda + 96\alpha\gamma\lambda\right)}. \tag{41}$$

For realistic values of the parameters and taking $\alpha\lambda \ll \gamma \ll 1$, Λ is pure imaginary and approximately given by $\Lambda \simeq 4\gamma\sqrt{\alpha\lambda}\,i$. Therefore after few e-folds δh_2 and δy are oscillating functions multiplied by a damping factor

$$\begin{cases} \delta h_2 \sim \mathrm{e}^{-\frac{3}{2}N}\left[\bar{A}_3 \cos\left(\frac{2\gamma}{\sqrt{\alpha\lambda}}N\right) + \bar{A}_4 \sin\left(-\frac{2\gamma}{\sqrt{\alpha\lambda}}N\right)\right] \\ \delta y \sim -\frac{15\alpha}{4(1+6\gamma)}\mathrm{e}^{-\frac{3}{2}N}\left[\bar{A}_3 \cos\left(\frac{2\gamma}{\sqrt{\alpha\lambda}}N\right) + \bar{A}_4 \sin\left(-\frac{2\gamma}{\sqrt{\alpha\lambda}}N\right)\right]. \end{cases} \tag{42}$$

In terms of the cosmic time $N \simeq \sqrt{\frac{\lambda}{3}}\frac{\mathrm{M_P}}{2\gamma}(t - t_0)$, the oscillation frequency is the same transplanckian frequency as one finds in GR: $\omega_\Lambda = \mathrm{M_P}/\sqrt{3\alpha}$.

4 Perturbative Particle Production

The highly oscillatory behaviour of the homogeneous dynamical system will lead to particle production similarly to that occurring at the end of inflation, during the so-called reheating era [13]. In this latter case the oscillatory behaviour of the inflation condensate makes it decay into SM particles (or more exotic stuff) which ends up reheating the Universe if the inflaton couples to such particles. Gravity interacts with every particle and, for the case we are considering such a coupling certainly exists. The particle production due to an oscillating Ricci scalar in $f(R)$ gravity models has been studied in several articles [14, 15]. Here we shall limit ourselves to the study of the production of scalar particles associated with the field σ. The field is non-minimally coupled and σ field quanta production should be non-negligible, at least for large enough values of the coupling γ. A significant production of such quanta may modify cosmological evolution and dampen the oscillations of the homogenous sector, thus it is necessary to, at least, provide an estimate of such effects. Since the amplitude of the oscillations of the homogeneous sector is increasing only in a radiation dominated universe we shall restrict our analysis to that era. At later times, after radiation domination, oscillations are small, non-increasing and are quickly dampened by particle production. The effects of particle production may thus be neglected.

The equation for $\sigma(t, \mathbf{x})$ is the KG equation

$$\Box\sigma - \gamma R\sigma + \lambda\sigma^3 = 0 \tag{43}$$

where $\Box \equiv g_{\mu\nu}\nabla^\mu\nabla^\nu$. On now defining $\sigma(t, \mathbf{x}) = \sqrt{y(t)} + \delta\sigma(t, \mathbf{x})$, where $y(t)$ satisfies the homogeneous equations, one is led the following linearised equation (we neglect metric perturbations):

$$\Box\delta\sigma - 6\gamma\left(2H^2 + \dot{H}\right)\delta\sigma + 3\lambda y\,\delta\sigma = 0. \tag{44}$$

and H^2 and y are given by (22) during radiation domination, by (28) during matter domination and by (38) during scalar field domination. Thus

$$\dot{H} = \frac{1}{2}\left(H^2\right)_{,N} = \frac{1}{2}\left[\frac{\rho_{,N}}{\rho} - \frac{y_{0,N}}{y_0} + \frac{(\delta h_2)_{,N}}{1 + \delta h_2}\right]H^2. \tag{45}$$

where y_0 here denotes the unperturbed (LO) quantity and $y_{,N}/y = 0$ during radiation and scalar field domination and $y_{,N}/y = 4\gamma/(1+4\gamma)$ during matter domination.

On now inserting Eq. (45) in Eq. (44) and retaining contributions to first order in δh_2 and δy one obtains

$$\Box\delta\sigma - 3\gamma\left[4 - 3(1+w) - \frac{\dot{y}}{Hy}\right]H^2(t)\delta\sigma + 3\gamma\,\delta\dot{h}_2\,H_0(t)\delta\sigma + 3\lambda y\,\delta\sigma = 0. \tag{46}$$

where w is the equation of state of the fluid which drives the universe expansion and $H_0(t)$ is the leading order time dependence in $H(t)$ (i.e. the Hubble parameter when $\alpha = 0$). During radiation dominance $H_0(t)$ is given by the standard, ΛCDM expression $H_0 = \rho_r/(3\mathrm{M_P}^2)$ but during matter dominance it is given by the cumbersome (and non-standard) expressions in Eq. (28). Let us note that the first bracket of Eq. (46) cancels in a radiation dominated universe. Essentially Eq. (46) is an equation for a scalar field with a (slowly) time varying "effective" mass and an interaction with an "external" oscillating field. The interaction may lead to a significant production of quanta (particles) of the field $\delta\sigma$. To the leading order such particles are produced in pairs. Let us note that the typical time scale of the oscillations is much smaller than that associated with the variation of the "effective" mass which, in our rough estimates, will be taken as constant. Moreover since the oscillation is much faster than the expansion rate of the universe the decay rate can be calculated in the flat space limit. This latter assumption is generally valid also when one calculates the decay rate of the inflaton at the end of inflation. More refined treatments can be found in the literature, taking into account, for example, the tiny higher order effects which lead to the distortion of the harmonic solution, or its time dependence (which here is neglected). Essentially such diverse approaches give more refined results but coincide in the LO with ours.

4.1 Particle Production from an Oscillating Field

Let us briefly review how to estimate the particle production due to the presence of a coupled, oscillating field and consider, for simplicity, just the case of two interacting scalar fields. Let us suppose the interaction potential between such fields has the form $V = g\,m\,\chi^2\,\phi$ where m is some mass scale and $\phi(t)$ is an external, homogeneous, scalar field with $\phi(t) = \bar{\phi}\cos(\omega t)$. In flat space time the Klein-Gordon equation for χ is

$$\Box\chi + \mu^2\chi + 2g\,m\,\chi\,\phi = 0. \tag{47}$$

If one calculates the transition amplitude (in the interaction picture), then to the lowest order the amplitude for ϕ to decay into two χ particles is

$$\mathcal{A}_{\phi\to\chi\chi} \equiv \langle\mathbf{k}_1,\mathbf{k}_2|\mathrm{e}^{-i\int \mathrm{d}x^4 V_{\mathrm{int}}}|0\rangle \simeq -i(g\,m)\int \mathrm{d}t\phi(t)\int \mathrm{d}^3x\langle\mathbf{k}_1,\mathbf{k}_2|\chi^2|0\rangle. \tag{48}$$

On now expanding χ in terms of creation-annihilation operators:

$$\chi = \int \frac{d^3k}{(2\pi)^{3/2}\sqrt{2\omega_k}} \left(\hat{a}_{\mathbf{k}} e^{-i\omega_k t + i\mathbf{k}\cdot\mathbf{x}} + \text{h.c.}\right), \tag{49}$$

with $\omega_k^2 = \mathbf{k}^2 + m^2$, and integrating over space ($\mathbf{x}$) and time one finds

$$\begin{aligned} \mathcal{A}_{\phi\to\chi\chi} &\simeq -\frac{2\pi i(g\,m)}{\omega_{k_1}}\delta^{(3)}\left(\mathbf{k}_1+\mathbf{k}_2\right)\int dt \frac{1}{2\pi}\frac{\bar{\phi}\left(e^{i\omega t}+e^{-i\omega t}\right)}{2} e^{2i\omega_{k_1}t} \\ &\simeq -\frac{\pi i(g\,m)\bar{\phi}}{\omega}\delta^{(3)}\left(\mathbf{k}_1+\mathbf{k}_2\right)\delta\Big(\omega_{k_1}-\frac{\omega}{2}\Big). \end{aligned} \tag{50}$$

The probability can then be calculated from the modulus squared of (50)

$$\left|\mathcal{A}_{\phi\to\chi\chi}\right|^2 = (2\pi)^4\delta^{(4)}(0)\frac{(g\,m)^2\bar{\phi}^2}{16\pi^2\omega^2}\delta^{(3)}\left(\mathbf{k}_1+\mathbf{k}_2\right)\delta\Big(\omega_{k_1}-\frac{\omega}{2}\Big) \tag{51}$$

where, as usual, one interprets the divergent $\delta^{(4)}(0)$ as the integral over all spacetime $(2\pi)^4\delta^{(4)}(0) = VT$. On now dividing by VT one obtains the transition probability per unit time and volume:

$$dP(\mathbf{k}_1, \mathbf{k}_2) = \frac{(g\,m)^2\bar{\phi}^2}{16\pi^2\omega^2}\delta^{(3)}\left(\mathbf{k}_1+\mathbf{k}_2\right)\delta\Big(\omega_{k_1}-\frac{\omega}{2}\Big). \tag{52}$$

The decay rate of the oscillating field quanta can now be calculated using energy conservation: from each ϕ particle decay one obtains two χ particles with the same energy $\omega/2$. The number of χ particles produced with momenta $\mathbf{k}_1$ and $\mathbf{k}_2$ is proportional to Eq. (52) and the energy lost by the oscillating scalar field and transferred to the χ field is also proportional to the same quantity. In particular, the energy variation per unit time of the fluid of χ particles due to the decay of a quantum with energy ω is given by

$$\frac{dE_\chi}{dt} = \omega\, V\, dP = V\,\frac{(g\,m)^2\bar{\phi}^2}{16\pi^2\omega}\delta^{(3)}\left(\mathbf{k}_1+\mathbf{k}_2\right)\delta\Big(\omega_{k_1}-\frac{\omega}{2}\Big). \tag{53}$$

The total variation in time of the energy density $\dot{\rho}_\chi = \dot{E}_\chi/V$ is obtained by integrating the last expression over the final momenta and thus

$$\begin{aligned} \dot{\rho}_\chi = -\dot{\rho}_\phi &= \int d^3k_1 d^3k_2 \frac{(g\,m)^2\bar{\phi}^2}{16\pi^2\omega}\delta^{(3)}\left(\mathbf{k}_1+\mathbf{k}_2\right)\delta\Big(\omega_{k_1}-\frac{\omega}{2}\Big) \\ &= \sqrt{1-\frac{4\mu^2}{\omega^2}}\frac{(g\,m)^2\bar{\phi}^2\omega}{16\pi}. \end{aligned} \tag{54}$$

The energy density associated with the oscillating field ϕ, to the LO in g, is $\rho_\phi = \omega^2\bar{\phi}^2/2$ and the decay rate is finally

$$\Gamma_{\phi\to\chi\chi} = -\frac{\dot{\rho}_\phi}{\rho_\phi} = -\sqrt{1 - \frac{4\mu^2}{\omega^2}}\frac{(g\,m)^2}{8\pi\omega}. \tag{55}$$

If one now takes into account the expansion of the Universe then the amplitude of the oscillating field varies in time with a rate typically much slower that the frequency of the oscillating part. In such a case $\bar{\phi} \to \bar{\phi}(t)$. The decay rate, however, is independent of $\bar{\phi}$ and the same holds for an expanding Universe. Therefore in such an approximation $\Gamma_{\phi\to\chi\chi}$ is that found on flat space. The presence of a coupling between an oscillating function and the field $\delta\sigma$ leads to perturbative particle production with transplanckian energy. Such particles may "re-heat" the Universe significantly by adding radiation like matter to the total energy content. This mechanism has been discussed already in the literature (see [14, 15]) but not in our context, where, in particular, the non-minimal coupling between "matter" (the field σ) and gravity is a feature of the model and is not simply guessed at or arbitrarily introduced.

4.2 Radiation Domination

We now estimate the production rate and its back-reaction in a radiation dominated universe. In such a case several simplifications occur in the perturbation equation (46) which takes the form:

$$\ddot{\delta\sigma} + 3H(t)\dot{\delta\sigma} - \frac{\nabla^2}{a^2}\delta\sigma + 3\gamma\frac{\dot{y}}{y}H(t)\delta\sigma + 3\gamma\,\delta\dot{h}_2\,H_0(t)\delta\sigma + 3\lambda y\,\delta\sigma = 0, \tag{56}$$

where

$$H(t) \simeq H_0(t)\left(1 + \frac{\delta h_2}{2}\right) \text{ with } H_0(t) \simeq \sqrt{\frac{\rho_r}{3{\rm M_P}^2}} \text{ and } y \simeq \frac{{\rm M_P}^2}{\gamma}(1 + \delta y). \tag{57}$$

Let us note that δh_2 and δy are oscillating functions given by Eqs. (26) and (27). The frequency of oscillation can be easily calculated on observing that the arguments of the trigonometric functions in such expressions are inverse proportional to the square root of ρ_r. The frequency of oscillation is (15) and for α small is transplanckian. Moreover we note that the scale factor also is an oscillating function which, in principle can be obtained on integrating the Hubble constant. Thus several terms in (56) have an oscillatory behaviour and may lead to particle production. To the leading order, the $\delta\sigma$ field quanta are produced in pairs with transplanckian energies $E_{\delta\sigma} = \omega_r/2$ and may thus contribute to the total radiation content. If we just consider the interaction terms of the form $A(t)\cos(\omega_r t)\delta\sigma$ in Eq. (56), we then observe that

$$3\gamma\,\delta\dot{h}_2\,H_0(t)\delta\sigma \sim \frac{\bar{c}_4}{2}H_0(t)^2\left(\frac{{\rm M_P}^4}{\alpha\rho}\right)^{5/8}\cos\left(\omega_r t\right)\,\delta\sigma \tag{58}$$

is the contribution which has the slowest rate of decrease in an expanding universe.[1] On comparing it with Eq. (47), from Eq. (54) we have

$$\dot{\rho}_{\delta\sigma} \simeq \frac{\omega_r}{256\pi}\bar{c}_4^2 H_0(t)^4\left(\frac{{\rm M_P}^4}{\alpha\rho}\right)^{10/8} = \frac{\bar{c}_4^2}{256\pi}\left(\frac{\rm M_P}{\sqrt{3\alpha}}\right)^5\left(\frac{\alpha\rho_r}{{\rm M_P}^4}\right)^{3/4}, \tag{59}$$

where $\bar{c}_4$ is an integration constant. The energy density added by the perturbative production of the $\delta\sigma$ quanta contributes to the total radiation content in the Universe and the continuity equation for radiation must be modified accordingly into

$$\rho_r \to \rho_r + \rho_{\delta\sigma} \Rightarrow \dot{\rho}_r \equiv -4H_0\rho_r + \frac{\bar{c}_4^2}{256\pi}\left(\frac{\rm M_P}{\sqrt{3\alpha}}\right)^5\left(\frac{\alpha\rho_r}{{\rm M_P}^4}\right)^{3/4}. \tag{60}$$

Correspondingly the amplitude (and energy density) of the "driving fields" δh_2 and δy decays faster than that previously obtained without including the effect of particle production. In principle the modified decay rate can be estimated by energy conservation. Let us therefore estimate the back-reaction of particle production on the energy density associated to the oscillating homogeneous field δh_2. We first observe that the energy density associated with the oscillation of δh_2 can be estimated in analogy with that of the oscillating inflaton during reheating. In this latter case it is proportional to its mass squared and its amplitude squared (apart from dimensional factors). The mass of δh_2 is its frequency (ω_r) and thus $\rho_{\delta h_2} \sim A_{\delta h_2}^2\omega_r^2/2$. From energy conservation the continuity equation for $\rho_{\delta h_2}$ is

$$\dot{\rho}_{\delta h_2} = 2\frac{\dot{A}_{\delta h_2}}{A_{\delta h_2}}\rho_{\delta h_2} - \dot{\rho}_{\delta\sigma} \tag{61}$$

where the ratio $2\dot{A}_{\delta h_2}/A_{\delta h_2} \equiv \Gamma_{\delta h_2}^{(0)}$ is that calculated without taking into account the perturbative decay. On now using the solution (27) one has

$$A_{\delta h_2} \sim \left(\frac{{\rm M_P}^4}{\alpha\rho}\right)^{1/8} \Rightarrow \Gamma_{\delta h_2}^{(0)} = H_0 = \sqrt{\frac{\rho_r}{3{\rm M_P}^2}}. \tag{62}$$

One is then led to the continuity equation

$$\dot{\rho}_{\delta h_2} = -H_0\rho_{\delta h_2} - \frac{\bar{c}_4^2}{256\pi}\left(\frac{\rm M_P}{\sqrt{3\alpha}}\right)^5\left(\frac{\alpha\rho_r}{{\rm M_P}^4}\right)^{3/4} \tag{63}$$

[1] Observe that $\delta\dot{h}_2$ increases but H_0^2 decreases and the total amplitude is decreasing.

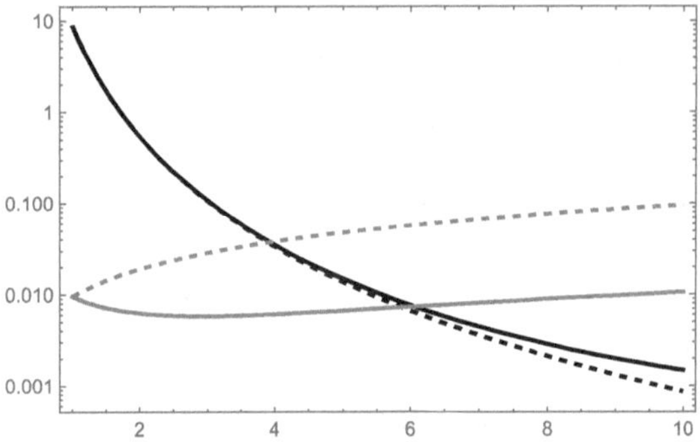

Fig. 1 The figure shows the evolution of $f_r(a)$ (black) and $f_{\delta h_2}(a)$ (gray) with $A_0 = \mathrm{e}$ and $B_0 = 10^{-3}$. The solid lines are the solutions for $C_0 = 10^{-2}$ and the dashed lines are the unperturbed solutions ($C_0 = 0$). The radiation fluid decreases slower than the usual a^{-4} behaviour and the energy density of δh_2 increases very slowly

and on solving the system of coupled Eqs. (60) and (63) one obtains the evolution of the radiation fluid and that of the time dependent amplitude $A^{\text{pert}}_{\delta h_2}$. In terms of the scale factor one has the following system

$$\begin{cases} a\frac{\mathrm{d}f_r}{\mathrm{d}a} = -4f_r + R_0 f_r^{1/4} \\ a\frac{\mathrm{d}f_{\delta h_2}}{\mathrm{d}a} = f_{\delta h_2} - R_0 f_r^{1/4} \end{cases} \tag{64}$$

where $f_r = \rho_r/\mathrm{M_P}^4$ and $f_{\delta h_2} = \rho_{\delta h_2}/\mathrm{M_P}^4$ and R_0 is a constant which depends on the integration constant $\bar{c}_4$. The system has the following general solution

$$f_r = \left(\frac{A_0^3 + C_0 a^3}{4a^3}\right)^{4/3}, \quad f_{\delta h_2} = b_0 a + \frac{C_0\left(C_0 + \frac{A_0^3}{a^3}\right)^{1/3} {}_2F_1\left(-\frac{2}{3}, -\frac{1}{3}; \frac{1}{3}; -\frac{C_0 a^3}{A^3}\right)}{2^{5/3}\left(1 + C_0 a^3/A_0^3\right)} \tag{65}$$

where A_0 and B_0 are integration constants. The general solutions Eq. (65) are illustrated in the Fig. 1, for a particular choice of the model parameters, and compared with the solutions without particle production. At $a = 1$ the initial conditions are such that the solutions coincide and the figure plots their difference as time evolves. The black line describes the energy density of radiation in Planck mass units with and without particle production (solid and dashed lines respectively) while the grey lines plot the energy density of δh_2 in Planck mass units with and without particle production (solid and dashed lines respectively). It is evident that particle production increases the radiation energy density which differs from the usual a^{-4} scaling behaviour, and lowers the energy density of δh_2. Since the frequency of δh_2 remains constant, such a decrease just affects the amplitude of its oscillations. Let us note that the decay rate decreases as the universe expands and its effects becomes less and less relevant as time evolves. The injection of very energetic particles in the plasma may also have consequences on Nucleosynthesis. The σ field, however, is coupled to SM particles indirectly and its effects might be small enough. Such a point certainly deserves further examination and we leave it to future studies.

5 Asymptotic Behaviour

In this last section we study the different asymptotic behaviour of the two models (on neglecting any particles production effect which are shown to be negligible for a large). This analysis is performed by searching for the fixed points of the two autonomous systems with different definitions of the dimensionless dynamical variables. Starting from the perturbed ΛCDM case, let us rewrite the Friedmann-like equation as follows

$$1 - x_1 - x_2 - x_3 + x_4 - \frac{\mathrm{M_P}^4}{\alpha\rho_\Lambda} x_1 x_4^2 + \frac{\alpha R_{,N}}{\mathrm{M_P}^2} = 0. \tag{66}$$

where we defined

$$x_1 = \frac{\rho_\Lambda}{3H^2\mathrm{M_P}^2},\ x_2 = \frac{\rho_m}{3H^2\mathrm{M_P}^2}\ x_3 = \frac{\rho_r}{3H^2\mathrm{M_P}^2};\ x_4 = \frac{\alpha R}{\mathrm{M_P}^2} \tag{67}$$

We can differentiate w.r.t. N the dimensionless x_i's and use the dynamical Eqs. (4) and (66). Then, on setting such derivatives to zero, one obtains the fixed points of the system. The following solutions are found

$$x_1 = x_2 = 0,\ x_3 = 1 + x_4 \quad \text{and} \quad x_1 = 1,\ x_2 = x_3 = 0,\ x_4 = \frac{4\alpha\rho_L}{\mathrm{M_P}^4} \tag{68}$$

and the first solution is $3\mathrm{M_P}^2H^2 = \rho_r$ and exists if $\rho_\Lambda = \rho_M = 0$. The second solution is the DE attractor in the future with $3\mathrm{M_P}^2H^2 = \rho_\Lambda$. These solutions have already been discussed in Sec. II during radiation domination and Dark Energy domination and also exist for the "unperturbed" system.

On instead defining $x_4 = 9\alpha(H^2)_{,N}/\mathrm{M_P}^2$ and following the same procedure as illustrated above, one also obtains the solution

$$H^2 = -\frac{\mathrm{M_P}^2}{9\alpha} N + h_2 \tag{69}$$

which is peculiar to the perturbed ΛCDM system we are considering.

5.1 Scale Invariant Cosmology

For the scale invariant model the relevant dynamical equations (apart from the continuity equations for matter and radiation) are the Friedmann-like equation and the KG equation

$$\begin{cases} 1 = \frac{\lambda y}{12\gamma H^2} + \frac{\rho_r}{3\gamma H^2 y} + \frac{\rho_m}{3\gamma H^2 y} - \frac{\alpha R}{\gamma y} + \frac{\alpha R^2}{12\gamma H^2 y} - \frac{y_{,N}}{y} + \frac{y_{,N}^2}{24\gamma y^2} - \frac{\alpha R_{,N}}{\gamma y} \\ \frac{y_{,NN}}{24\gamma y} - \frac{y_{,N}^2}{48\gamma y^2} + \frac{y_{,N}}{24\gamma y} + \frac{R y_{,N}}{144\gamma H^2 y} + \frac{\lambda y}{12\gamma H^2} - \frac{R}{12H^2} = 0 \end{cases} . \tag{70}$$

On defining the following set of dynamical variables

$$x_1 = \frac{R}{y},\ x_2 = \frac{R}{H^2},\ x_3 = \frac{y_{,N}}{y},\ x_4 = \frac{\rho_r}{3\gamma y H^2},\ x_5 = \frac{\rho_m}{3\gamma y H^2} \tag{71}$$

one obtains the following results. The first two solutions are the radiation and cosmological constant domination with $y = y_0$ and $3\gamma y_0 H^2 = \rho_r$ or $12\gamma H^2 = \lambda y_0$ respectively, which also exist in the limit $\alpha = 0$. For $\rho_r = \rho_m = 0$ one obtains

$$\gamma = -\frac{1}{6},\ y = y_0 e^{-2N}\ H^2 = h_2 e^{-4N}. \tag{72}$$

and the more cumbersome solutions

$$y = y_0 e^{-\frac{3+12(1-3\gamma)\gamma - 36\alpha\lambda \pm \tilde{\Gamma}}{2(1+6\gamma)} N},\ H^2 = \frac{y_0 (1+6\gamma)^2 e^{-\frac{3+12(1-3\gamma)\gamma - 36\alpha\lambda \pm \tilde{\Gamma}}{2(1+6\gamma)} N}}{9\alpha \left[5 + 36\left(\gamma + \gamma^2 + \alpha\lambda\right) \mp \tilde{\Gamma}\right]} \tag{73}$$

where $\tilde{\Gamma} = \sqrt{3\left[3 + 4\gamma(5+3\gamma) + 12\alpha\lambda\right]\left[(1+6\gamma)^2 + 36\alpha\lambda\right]}$. Finally the solution

$$y = y_0 e^{-2N},\ H^2 = \frac{\lambda y_0 e^{-2N}}{1+6\gamma},\ \alpha = -\frac{(1+6\gamma)^2 \left(3 + \frac{4\rho_{r,0}}{\lambda y_0^2}\right)}{108\lambda} \tag{74}$$

is found for $\rho_m = 0$ and the solution

$$y = y_0 e^{-\frac{3}{2}N},\ H^2 = \frac{8\lambda y_0 e^{-\frac{3}{2}N}}{3(3+20\gamma)},\ \alpha = -\frac{(3+20\gamma)^2 \left(\frac{3(1+6\gamma)}{3+20\gamma} + \frac{2\rho_{m,0}}{\lambda y_0^2}\right)}{360\lambda} \tag{75}$$

exists when $\rho_r = 0$. Let us note that the last two solutions have α fixed and negative.

On changing the definition of the dimensionless dynamical variables and setting $x_2 = \frac{\lambda y}{12\gamma H^2}$ one finds two new solutions for $y = y_0$ and

$$H^2 = h_2 e^{6N} - \frac{\gamma y_0}{12\alpha} \text{ or } H^2 = h_2 e^{-4N} - \frac{\gamma y_0}{12\alpha} \tag{76}$$

and the solution

$$y = y_0 \mathrm{e}^{-2N}, \; H^2 = h_2 \mathrm{e}^{-4N} - \frac{(1+6\gamma)y_0}{36\alpha}\mathrm{e}^{-2N} \tag{77}$$

valid when $\lambda = \rho_m = \rho_r = 0$. A third, possible, definition of the dynamical variables is finally considered i.e. $x_1 = \lambda y/(12\gamma H^2)$ and $x_2 = 9\alpha \left(H^2\right)_{,N}/(\gamma y)$. In such a case, when $\lambda = \rho_m = \rho_r = 0$, one finds the solution

$$y = y_0 \mathrm{e}^{-(3\pm\sqrt{9+48\gamma})N}, \; H^2 = h_2 + y_0 \frac{3+12\gamma \pm \sqrt{9+48\gamma}}{36\alpha\left(3 \pm \sqrt{9+48\gamma}\right)} \mathrm{e}^{-(3\pm\sqrt{9+48\gamma})N}. \tag{78}$$

Given the presence of the 2 more degree of freedom associated with the field σ and its derivative, the scale invariant model has more solution compared with the perturbed ΛCDM model. In particular many of them are realised when the cosmological fluids are absent or subdominant.

6 Conclusions

We studied the cosmological evolution of a scale invariant model having a quintessence scalar field non-minimally coupled to gravity and an R^2 term. The scalar field dynamically generates both Newton's constant and the cosmological constant. The model is the simplest, scale invariant, generalisation to a previously analysed dark energy model [7]. While the non-minimal coupling of the scalar field naturally arises when one considers the quantum corrections of a scalar field on a curved background, the R^2 contribution is generally introduced to describe quantum gravitational effects. This model has been already studied in the context of inflation [12] where the scalar field plays the role of the inflaton and we analyse here some of its dynamical features at lower energies, after the Hot Big Bang, as a model of quintessence. The αR^2 contribution alters significantly the cosmological evolution even for tiny values of α, adding a new scalar degree of freedom. Indeed the 00 Einstein equation becomes a dynamical equation for R and, in principle, independently of the value of α, new cosmological solutions exist. We limit our analysis to solutions which can be obtained on slightly deforming those found in the $\alpha = 0$ limit, which were already found to be viable as Dark Energy models (or at least not too different from ΛCDM) provided $\gamma \ll 1$ [7].

In Sect. 2 the cosmological evolution of ΛCDM with the addition of the "quantum corrections" αR^2 is studied by perturbing the ΛCDM model ($\alpha \ll 1$). Analytical solutions are found during radiation, matter and cosmological constant domination by expanding the dynamical equation to the first order in perturbations. Such solutions share the common feature of having an oscillating contribution with a frequency $\omega = \mathrm{M_P}/\sqrt{3\alpha}$ (transplanckian). However only during radiation domination does the oscillating contribution have an increasing amplitude which may significantly alter cosmological evolution and have consequences which are not in agreement with observations. During matter domination the amplitude is constant and at later times, when the cosmological constant begins dominating, the amplitude is decreasing. As shown in [14] the oscillatory behaviour of the solutions leads to perturbative particle production analogously to what occurs at the end on inflation, when the homogeneous inflaton transfers its energy to matter fields and reheats the Universe. Thus such an energy transfer damps the oscillatory solution which finally tends to the unperturbed ones, but with additional radiation injected into the system.

We then considered the full, scale invariant system and compared it with the former matter gravity system. On perturbing the solutions already found for $\alpha = 0$, we obtained the analytical solutions to the first order in the perturbations on assuming radiation, matter and cosmological constant domination separately. If one does not consider them separately the equations for the perturbed cosmological evolution are much more cumbersome. The solutions share some relevant features with those previously obtained for the perturbed ΛCDM model. Indeed, the perturbations oscillate with the same frequency during radiation and scalar field domination and the amplitude of such oscillations increases in time for the first case and decreases in the second. We observe, however, that the rate of increase during radiation domination is slower for the scale invariant model. During matter domination the evolution slightly changes since the "dynamical" Newton's constant varies to LO. Thus the frequency becomes slightly time dependent and the amplitude, compared with the perturbed ΛCDM model, is not constant but decreases. Finally we evaluated the effects of perturbative scalar particle production during radiation dominance. Such an analysis was already performed in the literature for the perturbed ΛCDM model in [14, 15] by considering a minimally coupled scalar field. In the scale invariant case one studies the particle production associated with the non-minimally coupled scalar field playing the role of Dark Energy and the back-reaction of such a production on the amplitude of the oscillating quantities. We just considered the LO effects of particle production on the oscillating perturbed Hubble parameter. In such a case the modified continuity equations are estimated and solved analytically. The solution shows that due to energy transfer the radiation fluid energy density has a non-standard scaling behaviour for a given time interval and correspondingly the amplitude of oscillations is damped. An efficient damping would alleviate the problem of the fine tuning of the initial conditions of the model which, otherwise, could sensibly deviate from standard cosmology and then become non-viable. A more refined analysis which takes into consideration higher order contributions to the oscillatory behaviour of the homogeneous solutions and a more complete study of the perturbative production of particle during the entire universe evolution is necessary and is left for future studies.

Finally, in the last section, we obtained some cosmological solutions, which exist for particular choices of the parameters and the universe content, for the two models being compared. This last part certainly deserves further in-depth analysis and a study of the stability of the solutions found should be also addressed.

References

1. F. Cooper, G. Venturi, Phys. Rev. D **24**, 3338 (1981)
2. G. Turchetti, G. Venturi, Nuovo Cim. A **66**, 221 (1981)
3. A. Zee, Phys. Rev. Lett. **42**, 417 (1979). (A. Zee, Phys. Rev. D **23**, 858 (1981))
4. A.D. Sakharov, Dokl. Akad. Nauk Ser. Fiz. **177**, 70–71 (1967)
5. A.Y. Kamenshchik, A. Tronconi, G. Venturi, Phys. Lett. B **713**, 358–364 (2012)
6. C. Brans, R.H. Dicke, Phys. Rev. **124**, 925–935 (1961)
7. F. Finelli, A. Tronconi, G. Venturi, Phys. Lett. B **659**, 466–470 (2008)
8. M. Ballardini, F. Finelli, C. Umiltà, D. Paoletti, JCAP **05**, 067 (2016). (C. Umiltà, M. Ballardini, F. Finelli, D. Paoletti, JCAP **08**, 017 (2015))
9. A. Cerioni, F. Finelli, A. Tronconi, G. Venturi, Phys. Lett. B **681**, 383–386 (2009). (A. Cerioni, F. Finelli, A. Tronconi, G. Venturi, Phys. Rev. D **81**, 123505 (2010))
10. A.A. Starobinsky, Phys. Lett. B **91**, 99–102 (1980)
11. A. De Felice, S. Tsujikawa, Living Rev. Rel. **13**, 3 (2010)
12. M. Rinaldi, C. Cecchini, A. Ghoshal, D. Mukherjee, J. Phys. Conf. Ser. **2531**(1), 012012 (2023). (C. Cecchini, M. De Angelis, W. Giarè, M. Rinaldi, S. Vagnozzi, JCAP **07**, 058 (2024))
13. M.S. Turner, Phys. Rev. D **28**, 1243 (1983). (F. Finelli, R.H. Brandenberger, Phys. Rev. Lett. **82**, 1362–1365 (1999))
14. E.V. Arbuzova, A.D. Dolgov, L. Reverberi, JCAP **02**, 049 (2012)
15. E.V. Arbuzova, A.D. Dolgov, R.S. Singh, JCAP **07**, 019 (2018)

Dynamical Dark Energy with AdS-dS Transitions Versus Baryon Acoustic Oscillations at $z = 2.3 - 2.4$

Özgür Akarsu, Maxim Eingorn, Leandros Perivolaropoulos, A. Emrah Yükselci, and Alexander Zhuk

Abstract In this paper, written in memory of *Alexei Starobinsky*, we discuss the observational viability of the Ph-Λ_sCDM model—a dynamical dark energy scenario based on a phantom scalar field undergoing an anti-de Sitter (AdS) to de Sitter (dS) transition—and revisit the Sahni–Shtanov braneworld model in light of updated BAO Ly-α data at $z \sim 2.3$. Both models are able to remain consistent with Planck CMB data while offering potential resolutions to the H_0 tension. In both cases, the expansion rate $H(z)$ is suppressed relative to Planck-ΛCDM at high redshift and enhanced at low redshift, while remaining consistent with the comoving distance to recombination as estimated by Planck-ΛCDM. Comparing model predictions with BAO-inferred values of $H(z)$, we find that SDSS Ly-α data at $z \approx 2.33$ mildly favor such dynamical models, whereas the recent DESI Ly-α measurements agree more closely with ΛCDM. Although current high-redshift BAO data do not decisively favor one model over another, our findings illustrate how frameworks originally developed to address earlier anomalies—such as the braneworld scenario—may gain renewed relevance in confronting today's cosmological tensions.

Ö. Akarsu · A. E. Yükselci
Department of Physics, Istanbul Technical University, Istanbul, Türkiye
e-mail: akarsuo@itu.edu.tr

A. E. Yükselci
e-mail: yukselcia@itu.edu.tr

M. Eingorn
Department of Mathematics and Physics, North Carolina Central University, Durham, NC, USA

L. Perivolaropoulos
Department of Physics, University of Ioannina, Ioannina, Greece
e-mail: leandros@uoi.gr

A. Zhuk (✉)
Center for Advanced Systems Understanding, Görlitz, Germany
e-mail: ai.zhuk2@gmail.com

Helmholtz-Zentrum Dresden-Rossendorf, Dresden, Germany

Astronomical Observatory, Odesa I.I. Mechnikov National University, Odesa, Ukraine

A. Barvinsky and A. Kamenshchik (eds.), *Open Issues in Gravitation and Cosmology*, Fundamental Theories of Physics 225, https://doi.org/10.1007/978-3-032-15703-4_31

1 Introduction

This paper, *dedicated to the memory of Alexei Starobinsky*, is inspired by the article [1], coauthored by Varun Sahni, Arman Shafieloo, and Alexei Starobinsky himself. In that work, the authors investigated the dynamics of the Universe within the braneworld model framework [2], and observed that the Hubble parameter in this model takes smaller values than in the standard ΛCDM model over an extended period in the past. A similar behavior emerged in our recent study [3], where we introduced the Ph-Λ_sCDM model, in which dark energy is modeled by a phantom scalar field with a hyperbolic tangent potential. Owing to the specific form of this potential and the phantom nature of the scalar field, our framework naturally realizes bottom-up AdS-to-dS transitions. The model is derived from first principles—grounded in a well-defined action—and offers a physical realization of the phenomenological Λ_sCDM scenario, wherein the effective cosmological constant transitions from negative to positive values as conjectured based on findings in graduated dark energy (gDE) [4–7]. We demonstrate that, on the one hand, this approach maintains consistency with the full Planck 2018 CMB data [8], and on the other, yields a present-day Hubble parameter H_0 in agreement with SH0ES measurements [9], thereby offering a potential resolution to the longstanding H_0 tension in cosmology.

We also show that the equation-of-state (EoS) parameter ω_ϕ of the phantom field evolves from $\omega_\phi \gtrsim -1$ before the transition—when its energy density is negative ($\varepsilon_\phi < 0$)—to $\omega_\phi \lesssim -1$ after the transition, when the energy density becomes positive. Notably, ω_ϕ exhibits a pole (a so-called safe singularity) at the transition point, where the total energy density of the scalar field vanishes. An analogous evolution of the effective EoS parameter is found in the braneworld model examined in Ref. [1], which is based on the framework introduced in Ref. [2]. Moreover, at redshift $z = 2.34$, this braneworld model provides a better fit to the line-of-sight baryon acoustic oscillation (BAO) signal from Ly-α forest data in SDSS DR11 [10] than the standard ΛCDM model [1]. According to Refs. [10, 11], BAO measurements at this redshift—derived from Ly-α forest auto-correlation and Ly-α–quasar cross-correlation—indicate a lower value of the Hubble parameter $H(z)$ than predicted by the standard ΛCDM model. This deviation suggests a possible preference for cosmological scenarios in which the expansion rate at high redshift is suppressed relative to ΛCDM. This is precisely the behavior exhibited by the braneworld model in Ref. [1]. In our dynamical dark energy model [3], we similarly find that $H(z)$ is lower than in ΛCDM at sufficiently high redshifts (approximately $z \gtrsim 1$). This naturally raises the question of whether our model provides a better fit to the BAO observations in this redshift regime. The present work is dedicated to addressing this question. As we will show, certain BAO measurements—particularly those based on SDSS Ly-α data—exhibit a preference for such behavior, while the most recent DESI results remain more consistent with the ΛCDM prediction.

The paper is organized as follows. In Sect. 2, we present a brief description of dynamical dark energy in the Ph-Λ_sCDM model, as well as in the generalized Sahni–Shtanov braneworld model. There, we perform numerical computations of the

Hubble parameter $H(z)$ such that both models are consistent with the CMB data on the one hand, and satisfy the SH0ES determination of H_0 on the other. In Sect. 3, we compile recent BAO Ly-α measurements and compare the corresponding values of the Hubble parameter with those predicted by the Ph-Λ_sCDM model, the braneworld model, and the standard ΛCDM model. Finally, in Sect. 4, we summarize and discuss our findings.

2 Dynamical Dark Energy with AdS-dS Transitions

In this section, we provide a brief description of the dynamical dark energy model; a detailed analysis can be found in Ref. [3]. The matter content of the Universe is taken to consist of radiation, pressureless matter—i.e., cold dark matter (CDM) and baryons—and a minimally coupled scalar field ϕ with a potential $V(\phi)$, described by the action

$$S_\phi = \int \mathrm{d}^4x\,\sqrt{-g}\left[\frac{\xi}{2}g^{ik}\partial_i\phi\partial_k\phi - V(\phi)\right]. \tag{1}$$

We consider this model at the background level, where the scalar field depends only on time, and the background metric is taken to be spatially flat Friedmann–Lemaître–Robertson–Walker (FLRW). The parameter $\xi = +1$ corresponds to quintessence (a canonical scalar field with positive kinetic energy), while $\xi = -1$ corresponds to a phantom field (a scalar field with negative kinetic energy). In what follows, we focus exclusively on the phantom case, i.e., $\xi = -1$.

The scalar field potential is taken to be

$$V(\phi) = \frac{\Lambda(\xi_1+1)}{2} - \frac{\Lambda(\xi_1-1)}{2}\tanh\left[\sqrt{\kappa}\,\nu\,(\phi-\phi_c)\right], \tag{2}$$

where $\Lambda > 0$, ν is a rapidity parameter characterizing the transition rate, and $\kappa \equiv 8\pi G_N/c^4$, with G_N being Newton's gravitational constant and c the speed of light. The parameter ϕ_c denotes the inflection point of the potential, marking the midpoint of the transition where the hyperbolic tangent changes sign. This potential exhibits the following asymptotic behavior: $V(\phi \to -\infty) \to \Lambda\xi_1$ and $V(\phi \to +\infty) \to \Lambda$. Therefore, for $\xi_1 < 0$, the phantom field—with its negative kinetic energy—evolves from a region with negative potential energy to one with positive potential energy, thereby realizing an AdS-to-dS transition. The case $\xi_1 = -1$ describes a mirror transition from $-\Lambda$ to $+\Lambda$, providing a natural dynamical realization of the Λ_sCDM scenario [4–7]. In contrast, the case $\xi = 0$, $\xi_1 = 1$ recovers the standard ΛCDM model.

The full system of equations of motion for the phantom field $\phi(t)$ and the scale factor $a(t)$ is presented in Ref. [3]. We solve these equations numerically with initial conditions $\phi_{\rm in} = \dot{\phi}_{\rm in} = 0$, where $\phi_{\rm in} < \phi_{\rm c}$. In addition, we impose the following constraints on the present-day values of the physical densities of matter (CDM + baryons):

$$\Omega_{\rm m0}h^2 = 0.14314, \tag{3}$$

and radiation:

$$\Omega_{\rm r0}h^2 = 2.469 \times 10^{-5} \times \left[1 + \frac{7}{8}\left(\frac{4}{11}\right)^{4/3} N_{\rm eff}\right], \tag{4}$$

where $h \equiv H_0/(100\ \mathrm{km\,s^{-1}\,Mpc^{-1}})$ is the dimensionless reduced Hubble constant, and $N_{\rm eff} = 3.046$, in accordance with the standard model of particle physics. The value of the physical matter density in Eq. (3) is chosen to match the Planck-CMB estimates [8]. Furthermore, the comoving angular diameter distance to the surface of last scattering, $D_M(z_*)$, with $z_* \approx 1090$, is tightly constrained by the Planck CMB power spectra within the ΛCDM framework [12]:

$$D_M(z_*) = \int_0^{z_*} \frac{c}{H(z)}\,\mathrm{d}z = 13869.57\ \mathrm{Mpc}, \tag{5}$$

which ensures that the sound horizon at recombination, r_*, remains the same as in the Planck-ΛCDM model. This is because the angular scale of the sound horizon, $\theta_* = r_*/D_M(z_*)$, is measured with high precision and is nearly model-independent. Thus, the constraints given in Eqs. (3)–(5) ensure consistency with the CMB power spectra [8]. For each solution $H(z)$, we fix the present-day Hubble parameter to $H_0 = 73.04\ \mathrm{km\,s^{-1}\,Mpc^{-1}}$, in accordance with the SH0ES measurement [9], and verify that it satisfies Eq. (5). This procedure uniquely determines the value of $\phi_{\rm c}$ and the redshift of the AdS–dS transition, $z_{\rm t}$, for each solution.

Figure 1 shows the result of the numerical solution for the comoving Hubble parameter, $\mathrm{d}a/\mathrm{d}t = H(z)/(1+z)$, for three phantom models with $\xi_1 = -1.5, -1, -0.5$ (colored curves), as well as for the standard ΛCDM model (black dashed curve). The rapidity parameter is fixed at $\nu = 100$. All phantom models are calibrated to yield $H_0 = 73.04\ \mathrm{km\,s^{-1}\,Mpc^{-1}}$, in accordance with the SH0ES measurement [9]. In contrast, the standard ΛCDM model is shown with $H_0 = 67.22\ \mathrm{km\,s^{-1}\,Mpc^{-1}}$, which represents the upper limit compatible with the constraints in Eqs. (3), (4), and (5) under the ΛCDM framework.

All phantom model curves are continuous and smooth, with clearly defined transition redshifts $z_{\rm t}$, which correspond to the inflection points of the scalar field potential given in Eq. (2). The transition redshifts are $z_{\rm t} = 2.36,\ 2.12,\ 1.83$ for $\xi_1 = -1.5,\ -1,\ -0.5$, respectively. These redshifts mark characteristic points in the evolution of the model parameters during the AdS–dS transition, as illustrated in Fig. 1. It is worth noting that another special point in the model is the redshift $z_\dagger$, defined by the condition that the energy density of the scalar field vanishes:

$$\varepsilon_\phi(z = z_\dagger) = -\frac{\dot{\phi}^2}{2c^2} + V(\phi) = 0,$$

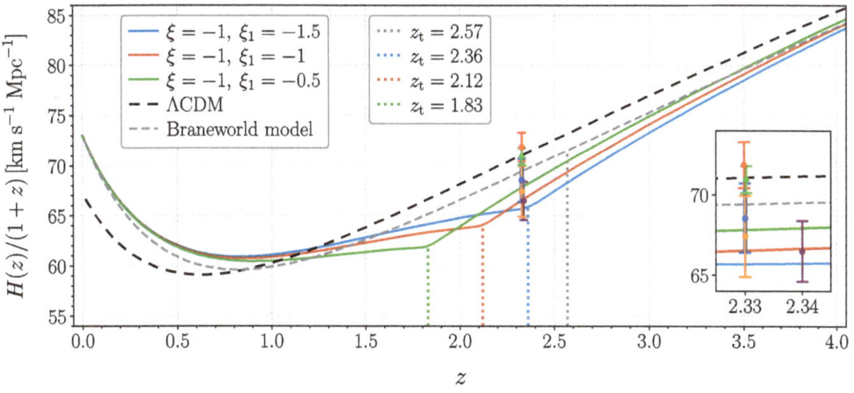

Fig. 1 The evolution of the co-moving Hubble parameter, $\dot{a} = H(z)/(1+z)$, for selected models. All curves are continuous and smooth. The vertical dotted lines determine the transition redshifts z_t. They correspond to the inflection points of potential $V(\phi)$ in the case of the phantom model and the point where the density of dark energy ε_{de} changes its sign in the case of the braneworld model. The phantom cases, i.e. $\xi = -1$, and the braneworld model satisfy the present-day value for the Hubble parameter in accordance with the SH0ES, that is 73.04 km s^{-1} Mpc^{-1}. The black dashed line corresponds to ΛCDM solution ($\xi = 0, \xi_1 = 1$) which gives $H_0 = 67.22$ km s^{-1} Mpc^{-1}. We also indicate 5 selected values of $H(z)/(1+z)$ with error bars corresponding to BAO measurements. The purple dot represents the result of [10], the orange and blue ones are for [23], whereas the bright red and green dots represent the results of [24, 25], respectively

where the overdot denotes a derivative with respect to cosmic time. The EoS parameter ω_ϕ exhibits a pole at $z_\dagger$. Numerical calculations show that $z_\dagger < z_t$, which is explained by the influence of the kinetic term.

Figure 1 demonstrates that the Hubble parameter in the considered phantom models lies below that of the ΛCDM model for $z \gtrsim 1.1$–1.2. However, in order to satisfy the fixed value of the integral in Eq. (5), the phantom model curves must lie above the ΛCDM curve for $z \lesssim 1.1$. This behavior explains why the phantom field models yield H_0 values consistent with the SH0ES measurement.

In Fig. 1, we also include the curve corresponding to the braneworld model (gray dashed line). The Hubble parameter in a spatially flat brane cosmology satisfies the equation [1, 2]:

$$\frac{H^2(z)}{H_0^2} = \Omega_{m0}(1+z)^3 + \Omega_{r0}(1+z)^4 + \tilde{\Omega}_{de}(z), \tag{6}$$

where $\Omega_{m0} = \varepsilon_{m0}/\varepsilon_{cr0}$, $\Omega_{r0} = \varepsilon_{r0}/\varepsilon_{cr0}$ (we additionally turned on radiation), and $\varepsilon_{cr0} = 3H_0^2/(\kappa c^2)$. The effective dark energy term is given by

$$\tilde{\Omega}_{de}(z) \equiv \frac{\varepsilon_{de}}{\varepsilon_{cr0}} = \Omega_\Lambda + 2\Omega_l - 2\sqrt{\Omega_l}\sqrt{\Omega_{m0}(1+z)^3 + \Omega_{r0}(1+z)^4 + \Omega_\Lambda + \Omega_l}, \tag{7}$$

where $\Omega_\Lambda = \Lambda/\varepsilon_{\mathrm{cr0}}$, and Λ is the effective brane tension associated with a three-dimensional brane embedded in a 4+1-dimensional bulk spacetime. The cosmological parameter $\Omega_l = 1/(l_c^2 H_0^2)$ is defined in terms of a characteristic length scale l_c, which is given by the ratio $l_c = M_{\mathrm{Pl}(4)}^2/M_{\mathrm{Pl}(5)}^3$. Here, $M_{\mathrm{Pl}(4)}$ and $M_{\mathrm{Pl}(5)}$ denote the four-dimensional and five-dimensional reduced Planck masses, respectively. The scale l_c characterizes the transition between four-dimensional and higher-dimensional gravitational behavior in the braneworld scenario. It is convenient to use the following consistency relation for the present-day density parameters [2]:

$$\Omega_\Lambda = 1 - \Omega_{\mathrm{m0}} - \Omega_{\mathrm{r0}} + 2\sqrt{\Omega_l}. \tag{8}$$

To integrate Eq. (6) under the constraint (8), we impose the same conditions as in Eqs. (3)–(5), and adopt the SH0ES value $H_0 = 73.04\ \mathrm{km\,s^{-1}\,Mpc^{-1}}$ [9]. These additional constraints (which were not employed in the original analysis of Ref. [1]), when combined with Eq. (8), fully determine the model parameters. For example, fixing $H_0 = 73.04\ \mathrm{km\,s^{-1}\,Mpc^{-1}}$ in accordance with the SH0ES mean value yields $\Omega_l \approx 0.023$. In turn, choosing $\Omega_l = 0.025$, as done in Ref. [1], leads to $H_0 = 73.24\ \mathrm{km\,s^{-1}\,Mpc^{-1}}$, which remains within the SH0ES error bars. As in the original work [1], we find that at high redshifts the Hubble parameter in the braneworld model lies significantly below that of the standard ΛCDM scenario. This behavior originates from the negative contribution of the dark energy density $\varepsilon_{\mathrm{de}}(z)$ in Eq. (6), which reduces the total energy density driving the expansion for $z > z_{\mathrm{t}}$. At the critical redshift $z_{\mathrm{t}} = 2.57$, $\varepsilon_{\mathrm{de}}$ vanishes and subsequently becomes positive, reversing its effect on the expansion rate. To satisfy the CMB constraint on the comoving angular diameter distance, Eq. (5), the model must compensate for this early-time suppression of $H(z)$ by producing an enhanced expansion rate at lower redshifts. As a result, the braneworld Hubble curve eventually overtakes the ΛCDM curve, reaching a present-day value of H_0 consistent with the SH0ES measurement. This transition was not captured in Ref. [1], as the condition (5) was not imposed in that analysis. In summary, much like our phantom field scenario, the inclusion of CMB-integrated distance constraints enables the braneworld model to reconcile the high-redshift suppression of the expansion rate with the high local value of H_0, thereby maintaining consistency with both Planck and SH0ES observations.

To conclude this section, we emphasize several important properties of our dynamical dark energy model. It is well known that phantom fields violate the null energy condition (NEC), since their inertial mass density—defined as $\varrho_\phi \equiv \varepsilon_\phi + p_\phi$—is negative. Moreover, the energy density ε_ϕ itself can also become negative, particularly when the scalar field potential is negative. However, in the presence of additional matter components such as cold dark matter and radiation, the total inertial mass density and total energy density can remain non-negative at all times. Our calculations [3] confirm that this is indeed the case in our model. Therefore, the weak energy condition (and consequently the NEC) is not violated in the combined system. Another common concern with phantom models is the emergence of a big rip singularity: in scenarios where the phantom field has positive energy density, the expansion typically leads to a divergence of the scale factor and energy density within finite cosmic time [13–17]. However, if the late-time geometry asymptotes to de Sitter spacetime, the big rip can be avoided [18]. It is also crucial to recognize that the negative kinetic term intrinsic to phantom fields generically leads to dynamical instabilities, unless the scalar field potential is appropriately bounded from above. Thus, both the big rip singularity and the background instability associated with negative kinetic energy can be mitigated through a carefully chosen potential. The potential given in Eq. (2) is specifically constructed to address these issues, ensuring both the stability of the model and the avoidance of singular behavior. A further potential concern is the instability of perturbations in phantom field models, particularly if the second derivative of the potential, $\mathrm{d}^2V/\mathrm{d}\phi^2$, becomes positive (convex) [19]. In our case, however, the potential remains convex only for a very limited duration [3], during which time any unstable modes do not grow appreciably.

3 Baryon Acoustic Oscillations at $z = 2.3 - 2.4$

BAO observations allow us to estimate the ratio $D_\mathrm{H}/r_\mathrm{d}$, where $D_\mathrm{H}(z) = c/H(z)$ is the Hubble distance at redshift z, and r_d is the comoving sound horizon at the drag epoch, approximately $z \approx 1060$ [8]. According to the Planck-ΛCDM results (see Table 1 in [8]), the best-fit value is $r_\mathrm{d} = 147.049 \approx 147.05\,\mathrm{Mpc}$. Given a measurement of the ratio $\frac{D_\mathrm{H}}{r_\mathrm{d}} \equiv A(z)$ at redshift z, and a known value of r_d, the Hubble parameter at that redshift can be inferred as:

$$H(z) = \frac{1}{A(z)} \frac{147.05\ \mathrm{Mpc}}{r_\mathrm{d}} \times 2038.71\ \mathrm{km\,s^{-1}\,Mpc^{-1}}, \tag{9}$$

where the prefactor arises from converting between distance and Hubble units, scaled to the reference value $r_\mathrm{d} = 147.05$ Mpc.

In Table 1, we compile measurements of the ratio $D_\mathrm{H}/r_\mathrm{d}$ from BAO observations at redshifts $z = 2.3 - 2.4$, focusing specifically on Ly-α forest-based analyses [10, 20–25] (see also [11, 26]). BAO measurements at these redshifts have had a significant impact on cosmology, as they provide a powerful and independent test of the flat ΛCDM model inferred from Planck CMB data [22]. The values of $D_\mathrm{H}/r_\mathrm{d}$ were extracted from three types of analyses: (i) Ly-α forest auto-correlation (Ly-α–Ly-α), (ii) cross-correlation between quasars and the Ly-α forest (Ly-α–QSO), and (iii) a combined analysis incorporating both auto- and cross-correlations.

As shown in Table 1, measurements of $D_\mathrm{H}/r_\mathrm{d}$ from different SDSS data releases (DR) exhibit slight variations. Even within the same data release, the results depend on the specific analysis method–auto-correlation, cross-correlation, or their combination. A more noticeable difference arises when comparing the SDSS results with

Table 1 We present the values of $D_{\rm H}/r_{\rm d}$ obtained by different methods of BAO measurements for selected numbers of redshifts z. The corresponding $H(z)$ values, the last column, are calculated via equation (9) using the Planck 2018 value of $r_{\rm d} = 147.09 \pm 0.26$ Mpc (TT,TE,EE+lowE+lensing)

Reference	Survey	$D_H/r_d(z)$	$H(z)$ km/s/Mpc
Delubac et al. 2015, A&A 574 (2015) A59	SDSS-III, DR 11	$Ly\alpha - Ly\alpha,\ z = 2.34$ 9.18 ± 0.28	222.02 ± 6.78 $z = 2.34$
Bautista et al. 2017, A&A 603(2017) A12	SDSS-III, DR 12	$Ly\alpha - Ly\alpha,\ z = 2.33$ 9.07 ± 0.31	224.71 ± 7.69 $z = 2.33$
Bourboux et al. 2017, A&A 608(2017) A130	SDSS-III, DR 12	$Ly\alpha - Ly\alpha,\ z = 2.4$ 8.94 ± 0.22	227.98 ± 5.62 $z = 2.4$
		$Ly\alpha - QSO,\ z = 2.4$ 9.01 ± 0.36	226.21 ± 9.05 $z = 2.4$
Alam et al. 2021, PRD 103(2021)083533; Bourboux et al. 2020, ApJ 901 (2020) 153	SDSS-IV	$Ly\alpha - Ly\alpha,\ z = 2.33$ 8.93 ± 0.28	228.24 ± 7.17 $z = 2.33$
		$Ly\alpha - QSO,\ z = 2.33$ 9.08 ± 0.34	224.47 ± 8.41 $z = 2.33$
		$Ly\alpha - Ly\alpha + Ly\alpha - QSO,\ z = 2.33$ 8.99 ± 0.19	226.71 ± 4.81 $z = 2.33$
Adame et al. 2024, JCAP 02 (2025) 021	DESI DR1 BAO	$Ly\alpha - QSO,\ z = 2.33$ 8.52 ± 0.17	239.22 ± 4.79 $z = 2.33$
Karim et al. 2025, Phys. Rev. D 112, (2025) 083514	DESI DR2 BAO	$Ly\alpha - Ly\alpha + Ly\alpha - QSO,\ z = 2.33$ 8.632 ± 0.101	236.12 ± 2.79 $z = 2.33$

those from the recent DESI data releases.[1] The corresponding values of the Hubble parameter $H(z)$ are calculated using Eq. (9), adopting the Planck-ΛCDM best-fit value $r_{\rm d} = 147.09 \pm 0.26$ Mpc (from TT, TE, EE + lowE + lensing) [8]. In Table 2, we also report the theoretical predictions for $H(z)$ at redshifts $z = 2.33$, 2.34, and 2.40, computed for the Ph-$\Lambda_{\rm s}$CDM model (with several choices of ξ_1), the braneworld model, and the standard ΛCDM model.

[1] We include the value $D_{\rm H}/r_{\rm d}(z = 2.330) = 8.632 \pm 0.098 \pm 0.026$ reported in Ref. [25] from DESI DR2, where the first and second uncertainties correspond to statistical and theoretical systematic errors, respectively. In our analysis, we combine these in quadrature to obtain the total uncertainty.

Table 2 The values of the Hubble parameter $H(z)$ (in km sec^{-1} Mpc^{-1}) calculated for Ph-Λ_sCDM model in three different cases of parameter ξ_1 as well as for the standard ΛCDM model and the braneworld model at three selected values of redshift z

z	Ph-Λ_sCDM model			ΛCDM	Braneworld model
	$\xi_1 = -1.5$	$\xi_1 = -1$	$\xi_1 = -0.5$	$\xi = 0, \xi_1 = 1$	
2.33	218.72	221.49	225.77	236.63	231.14
2.34	219.52	222.56	226.83	237.64	232.13
2.40	225.17	229.03	233.20	243.73	238.09

Based on the observational values summarized in Table 1 and their comparison with theoretical predictions in Table 2, we may ask: which of the models–the standard ΛCDM, our Ph-Λ_sCDM model, or the braneworld scenario–provides a better fit to the high-redshift BAO data? To address this qualitatively, we have plotted in Fig. 1 the BAO-inferred Hubble parameter values from three major surveys: SDSS-III DR11 (used in [1]), SDSS-IV, and DESI. A visual comparison suggests that the SDSS data tend to favor models in which dark energy transitions from negative to positive values during cosmic evolution, such as the Ph-Λ_sCDM and braneworld models. In contrast, the most recent DESI measurement at $z \sim 2.33$ shows better agreement with the standard ΛCDM prediction. Thus, while the current data do not yield a definitive preference for one model over another, they highlight interesting differences in model performance across surveys. Of course, a rigorous conclusion requires a full statistical analysis—such as a likelihood-based parameter estimation using the full BAO datasets and covariance matrices, along with the CMB data—which lies beyond the scope of this brief paper.

4 Conclusion

In his talk *"Future and Origin of our Universe: Modern View"* [14], Alexei Starobinsky discussed the idea of a variable cosmological constant, for instance realized through a self-interaction potential $V(\phi)$ of a scalar field. The Ph-Λ_sCDM [3] model provides a concrete realization of this concept: a phantom scalar field with a potential constructed to induce a rapid transition from an anti-de Sitter (AdS) phase to a de Sitter (dS) phase in the late Universe, thereby offering a physically well-defined embedding of the Λ_sCDM framework [4–7]—an approach that has shown promise in simultaneously addressing major cosmological tensions. Phantom fields, characterized by a negative kinetic term, violate the weak energy condition (WEC). In the same talk, Starobinsky discussed scenarios in which matter violates the WEC and emphasized that, if the equation-of-state parameter satisfies $w < -1$ and remains constant, the Universe will inevitably encounter a big rip singularity, where both the scale factor $a(t)$ and the energy density diverge within a finite cosmic time [16]. In our model, the phantom field indeed exhibits $w_\phi < -1$ after the transition. Nevertheless, the Universe asymptotically approaches a de Sitter phase, as in our model the EoS parameter satisfies $w_\phi \to -1$ in the limit $a \to \infty$, thereby evading the big rip singularity [18]. Moreover, since the total energy content includes cold dark matter and radiation, the total energy density $\varepsilon_{\rm tot}$, as well as the total inertial mass density $\varrho_{\rm tot} = \varepsilon_{\rm tot} + p_{\rm tot}$, remain positive throughout cosmic evolution. Hence, the WEC is not violated at the level of the total energy-momentum tensor, and the model remains physically viable.

Another interesting feature of our model is that the Hubble parameter $H(z)$ takes on smaller values than in the standard Planck-ΛCDM model prior to the transition. When combined with the CMB constraints given in Eqs. (3), (4), and (5), this feature enables the model to yield a Hubble constant H_0 that is in agreement with the SH0ES measurement. This outcome arises because the comoving angular diameter distance to the last scattering surface, $D_M(z_*)$, is kept fixed at the Planck-ΛCDM value. In our model, the Universe's evolution prior to recombination follows standard cosmology, ensuring that the sound horizon scale r_* remains unchanged. As a result, the integral in Eq. (5) imposes a strict constraint on the expansion history: any suppression of $H(z)$ at redshifts before the transition must be compensated by an enhancement at later times. This built-in mechanism naturally allows for a higher value of H_0 without disrupting consistency with early-universe observables.

In Ref. [1], Varun Sahni, Arman Shafieloo, and Alexei Starobinsky investigated the cosmic dynamics in a braneworld framework [2], in which the cosmological constant is dynamically screened at early times. A key prediction of their model is that the Hubble parameter $H(z)$ remains consistently lower than in the Planck-ΛCDM model over the entire redshift interval from today ($z = 0$) to recombination. This includes the epoch around $z = 2.34$, where BAO measurements from the Ly-α forest are available. In the present work, we revisit this analysis by imposing the additional constraints given in Eqs. (3)–(5), which were not considered in the original study. This refinement ensures that the braneworld model remains consistent with the full Planck-CMB dataset while simultaneously accommodating the SH0ES measurement of the Hubble constant. As shown in Fig. 1, these constraints cause the $H(z)$ curve to rise above that of ΛCDM at low redshifts ($z < 1$), enabling the model to reach $H_0 = 73.04\,\mathrm{km\,s^{-1}\,Mpc^{-1}}$, in line with local determinations.

One of the main findings of Ref. [1] was that the predicted Hubble parameter at redshift $z = 2.34$ in the braneworld model provides a better fit to the BAO measurement from Ly-α forest data [10] than the corresponding prediction from the Planck-ΛCDM model. In our paper, we set out to explore whether a similar conclusion can be drawn for our dynamical dark energy model and for the revised braneworld scenario under updated observational constraints. At the time of Ref. [1], the measurement of $H(z = 2.34)$ from [10] was the only available BAO determination in that redshift regime. Since then, several new measurements have been reported in the range $z = 2.3 - 2.4$, as compiled in Table 1. These updated results, based on both SDSS and DESI data, exhibit some variation depending on the dataset and methodology employed. In Fig. 1, we have plotted these BAO-inferred values of $H(z)$ with corresponding error bars. A visual comparison shows that the SDSS-based measurements continue to favor expansion histories in which the Hubble parameter is suppressed at high redshift—such as those predicted by our Ph-Λ_sCDM and braneworld models—whereas the most recent DESI measurement at $z = 2.33$ aligns more closely with the standard ΛCDM model. Thus, current data do not allow for a definitive conclusion regarding which model best matches the high-redshift BAO measurements.

In this brief paper, written in memory of Alexei Starobinsky, we revisited the braneworld model originally studied by Starobinsky and collaborators, which at the time was motivated by the apparent tension between Ly-α BAO data and ΛCDM. Although this anomaly appears less significant in light of more recent data, we find that the same model—once properly constrained—offers a natural resolution to the Hubble tension, a problem that gained prominence only after the original work was published. This capacity to anticipate the future relevance of ideas is a hallmark of scientific foresight—and a pattern often seen in the work of truly great minds.

Acknowledgements This work was partially supported by the Center for Advanced Systems Understanding (CASUS) which is financed by Germany's Federal Ministry of Education and Research (BMBF) and by the Saxon state government out of the State budget approved by the Saxon State Parliament. A.Z. is also grateful to ICTP, Trieste Observatory and IFPU for their hospitality during the preparation of the final version of the paper.

References

1. V. Sahni, A. Shafieloo, A.A. Starobinsky, Model independent evidence for dark energy evolution from Baryon Acoustic Oscillations. Astrophys. J. **793**, L40 (2014). arXiv:1406.2209 [astro-ph.CO]
2. V. Sahni, Y. Shtanov, Braneworld models of dark energy. JCAP **0311**, 014 (2003). arXiv:astro-ph/0202346
3. Ö. Akarsu, L. Perivolaropoulos, A. Tsikoundoura, A.E. Yükselci, A. Zhuk, Dynamical dark energy with AdS-to-dS and dS-to-dS transitions: implications for the H_0 tension. arXiv:2502.14667 [astro-ph.CO]
4. Ö. Akarsu, J.D. Barrow, L.A. Escamilla, J.A. Vazquez, Graduated dark energy: Observational hints of a spontaneous sign switch in the cosmological constant. Phys. Rev. D **101**, 063528 (2020). arXiv:2108.09239 [astro-ph.CO]
5. Ö. Akarsu, S. Kumar, E. Özülker, J.A. Vazquez, Relaxing cosmological tensions with a sign switching cosmological constant. Phys. Rev. D **104**, 123512 (2021). arXiv:2108.09239 [astro-ph.CO]
6. Ö. Akarsu, S. Kumar, E. Özülker, J.A. Vazquez, A. Yadav, Relaxing cosmological tensions with a sign switching cosmological constant: improved results with Planck, BAO, and Pantheon data. Phys. Rev. D **108**, 023513 (2023). arXiv:2211.05742 [astro-ph.CO]
7. Ö. Akarsu, E. Di Valentino, S. Kumar, R.C. Nunes, J. A. Vazquez, A. Yadav, Λ_sCDM model: a promising scenario for alleviation of cosmological tensions (2023). arXiv:2307.10899 [astro-ph.CO]
8. N. Aghanim, et al., Planck 2018 results. VI. Cosmological parameters. Astron. Astrophys. **641**, A6 (2020), [Erratum: Astron. Astrophys. **652**, C4 (2021)]; arXiv:1807.06209 [astro-ph.CO]
9. A.G. Riess et al., A comprehensive measurement of the local value of the Hubble constant with 1 km/s/Mpc uncertainty from the Hubble Space Telescope and the SH0ES team. Astrophys. J. Lett. **934**, L7 (2022). (arXiv:2112.04510 [astro-ph.CO])

10. T. Delubac et al., Baryon acoustic oscillations in the Ly-α forest of BOSS DR11 quasars. Astron. & Astrophys. **574**, A59 (2015). (arXiv:1404.1801 [astro-ph.CO])
11. É. Aubourg, et al., Cosmological implications of baryon acoustic oscillation (BAO) measurements. Phys. Rev. D **92**, 123516 (2015). arXiv:1411.1074 [astro-ph.CO]
12. Ö. Akarsu, E.Ó. Colgáin, E. Özülker, , S. Thakur, L. Yin, Inevitable manifestation of wiggles in the expansion of the late Universe. Phys. Rev. D **107**, 123526 (2023). arXiv:2207.10609
13. R.R. Caldwell, A Phantom menace? Phys. Lett. **545**, 23–29 (2002). (arXiv:astro-ph/9908168)
14. A.A. Starobinsky, Future and origin of our universe: modern view. Grav.& Cosmol. **6**, 157–163 (2000). arXiv:astro-ph/9912054
15. B. McInnes, The dS/CFT correspondence and the big smash. JHEP **0208**, 029 (2002). arXiv:hep-th/0112066
16. R.R. Caldwell, M. Kamionkowski, N.N. Weinberg, Phys. Rev. Lett. **91**, 071301 (2003). arXiv:astro-ph/0302506
17. V.B. Johri, Phantom cosmologies. Phys. Rev. D **70**, 041303 (2004). arXiv:astro-ph/0311293
18. M. Bouhmadi-Lopez, J.A. Jiménez Madrid, Escaping the Big Rip? JCAP **05**, 005 (2005). arXiv:astro-ph/0404540
19. S.M. Carroll, M. Hoffman, M. Trodden, Can the dark energy equation-of-state parameter ω be less than -1? Phys. Rev. D **68**, 023509 (2003). arXiv:astro-ph/0301273
20. J.E. Bautista, et al.: Measurement of baryon acoustic oscillation correlations at z=2.3 with SDSS DR12 Ly-α-Forests. Astron. & Astrophys. **603**, A12 (2017). arXiv:1702.00176 [astro-ph.CO]
21. H.M. Bourboux, et al., Baryon acoustic oscillations from the complete SDSS-III Ly-α -quasar cross-correlation function at z=2.4. Astron. & Astrophys. **608**, A130 (2017). arXiv:1708.02225 [astro-ph.CO]
22. H.M. Bourboux et al., The completed SDSS-IV extended Baryon oscillation spectroscopic survey: Baryon acoustic oscillations with Lyman-α forests. Astrophys. J. **901**, 153 (2020). (arXiv:2007.08995 [astro-ph.CO])
23. S. Alam et al., The completed SDSS-IV extended Baryon oscillation spectroscopic survey: cosmological implications from two decades of spectroscopic surveys at the Apache Point Observatory. Phys. Rev. D **103**, 083533 (2021). (arXiv:2007.08991 [astro-ph.CO])
24. A.G. Adame et al., DESI 2024 VI: cosmological constraints from the measurements of baryon acoustic oscillations. JCAP **02**, 021 (2025). (arXiv:2404.03002 [astro-ph.CO])
25. M.A. Karim, et al., DESI DR2 results. I. Baryon acoustic oscillations from the Lyman alpha forest. Phys. Rev. D **112**, 083514 (2025). arXiv:2503.14739 [astro-ph.CO]
26. D. Sapone, S. Nesseris, Outliers in DESI BAO: robustness and cosmological implications. Phys. Rev. D **112**, 063523 (2025). arXiv:2412.01740 [astro-ph.CO]

Memories

Remembering Alexei Starobinsky—The Gentle Giant of Cosmology

Varun Sahni

Abstract I share fond memories of my former Ph.D advisor Alexei Starobinsky with whom I was closely associated for nearly 45 years. I reflect upon my early years in Moscow when I worked with him on my thesis, and touch upon the seminal work on inflation which he did during that period. Alexei visited India often and actively interacted with Indian scientists and students on issues relating to inflation, large scale structure and dark energy. This extensive collaboration, which lasted several decades, resulted in the publication of over a dozen important papers, several Ph.D's, and the development of the Statefinder and Om diagnostics, which I briefly discuss.

1 Introduction

I first met Alexei Starobinsky on March 29, 1979 which was a Thursday. I remember this day well because it also happened to be my birthday and I was feeling quite happy and elated at having turned 23. I had arrived in Moscow from New Delhi three years earlier in 1976 to pursue my higher studies at the physics department of Moscow State University. I was admitted into the second year of the five year integrated MSc course and began studying physics in the Russian language which, at first, was not an easy task for me. As my studies advanced I became increasingly fascinated by the field of cosmology and started attending the outstanding course taught by Prof. Yakov Zeldovich to students of Moscow state university.[1] Zeldovich's academic style and personality attracted me like a magnet and I soon began working under his guidance for my 'Kursovaya rabota' – project work, which students needed to do during their intermediate years before before embarking on their MSc thesis. I thoroughly enjoyed working under Zeldovich, but when the time came for me to find a guide for my M.Sc and Ph.D Zeldovich gently informed me that he no longer

[1] See my article [1] for a summary of Zeldovich's many remarkable contributions to science.

V. Sahni (✉)
Inter-University Centre for Astronomy and Astrophysics, Pune, India
e-mail: varun@iucaa.in

A. Barvinsky and A. Kamenshchik (eds.), *Open Issues in Gravitation and Cosmology*, Fundamental Theories of Physics 225, https://doi.org/10.1007/978-3-032-15703-4_32

guided students at the advanced level. He then generously offered to introduce me to any member of his talented team with whom I might like to pursue higher studies.

I immediately thought of Alexei Starobinsky, who, at the young age of 31, was widely considered to be the rising star of Soviet cosmology. Zeldovich provided me with Alexei's contact details and so on March 29, 1979 I went to the Landau Institute on Sparrow street to meet my future guru and thesis supervisor. When I had asked Zeldovich earlier, on how I should address Dr Starobinsky, he had thought for a moment and then replied "let us simply call him Alyosha". Alyosha was the shortened nick-name with which everyone addressed Alexei Starobinsky and which was easier for me to pronounce than the more formal Alexei Alexandrovich according to Russian custom.

My association with Alyosha got off to a smooth start. He was very kind and considerate and began guiding me on how to carry out independent research in physics. At that point of time, in the 1970's, Alyosha was deeply involved in the exciting new subject area of 'quantum field theory in curved space-time'. His 1971 paper with Zeldovich on the quantum mechanical production of particles in an anisotropic universe is widely considered to be a landmark paper in this field [2]. Zeldovich and Starobinsky demonstrated that a rapidly changing gravitational field could give rise to copious particle production in the early universe. This seminal work paved the way for the development of regularization and renormalization methods which are essential for the extraction of a finite value of the vacuum energy momentum tensor, $\langle T_{\rm ik}\rangle$, from a formally divergent and infinite quantity. Of central importance was the fact that the backreaction of created matter on the background geometry, via the semi-classical Einstein equations $G_{\rm ik} = 8\pi G\langle T_{\rm ik}\rangle$, could rapidly isotropize an initially anisotropic universe [3]. This helped address the famous observation by Collins and Hawking [4] that "the set of spatially homogeneous cosmological models which approach isotropy at infinite times is of measure zero in the space of all spatially homogeneous models.... It therefore seems that there is only a small set of initial conditions that would give rise to universe models which would be isotropic to within the observed limits at the present time." The work of Zeldovich and Starobinsky [2] and Lukash and Starobinsky [3] showed that our universe could arise from a much larger set of initial conditions, if quantum effects were taken into account[2]. Decades later, Alexei once remarked to me, that his very first paper with Zeldovich [2]—written for his M.Sc—was perhaps the best paper he had ever written !

For his Ph.D Alexei explored the possibility of quantum particle production from a rotating black hole [8]. His paper "Amplification of waves reflected from a rotating black hole" published in 1973 was pathbreaking and influenced Stephen Hawking's famous work on black hole evaporation [9]. As Hawking later reminiscenced [10] "In september 1973 while I was visiting Moscow, I discussed black holes with two

---

[2] Alexei revisited this issue in [5] where he showed that a positive cosmological constant could also isotropize the universe. A similar result was obtained by Wald [6] and later generalized to an inflationary universe by Moss and Sahni [7]. These results demonstrate that inflation can iron out anisotropies and small inhomogeneities thereby considerably enlarging the set of initial conditions from which our homogeneous and isotropic universe could arise.

leading Soviet experts, Yakov Zeldovich and Alexander Starobinsky. They convinced me that according to the QM uncertainty principle, rotating black holes should create and emit particles." In 1974 Hawking showed that quantum particle production is also a key feature of non-rotating black holes which shine via the famous 'Hawking radiation' formula $T = 1/8\pi M$ [9].

I worked with Alexei from 1979 to 1985, initially for my MSc and then for my Ph.D. Most of Alexei's efforts during this period were directed towards developing inflationary cosmology. Below I list some of his seminal results that emerged during 1979-1997.

- 1979 Production of relic gravitational waves (GW) from Inflation [11].
- 1980 First model of Inflation in the framework of $R + R^2$ gravity [12].
- 1982 Determination of the primordial scalar power spectrum in Inflation [13].
- 1985 Calculation of the tensor-to-scalar ratio r in inflationary cosmology [14].
- 1986 Developed the formalism of Stochastic Inflation [15, 16].
- 1994–1997 Developed the formalism of post-inflationary preheating (with Kofman and Linde) [17–20]; also see [21].

I should mention that during my years in Moscow office space was very limited, due to which I was obliged to work from my hostel room while Alexei mostly worked from home. We agreed to meet on Thursdays at the Landau institute, provided I either had a nice result to show him, or—which was more frequent—I was fundamentally stuck in an intractable calculation. Occasionally I also visited Alexei at his stately home where his very kind wife Lyudmila made me some delicious Russian Borsch.

From the very outset Alexei struck me as a kind of *science wizard*. No matter how mathematically complicated the problem, Alexei could always guess the final answer even without performing any of the very tedious intermediate calculations (which were left to me). During my visits to the Landau institute I was occasionally accompanied by Lev Kofman, who had started working under Alexei around the same time as me. Lev and I soon became close friends. Although Lev lived in the Estonian city of Tartu, he visited Moscow often and on these occasions used to stay in my small room in the Moscow university hostel. Lev and I had a great time together. We watched movies, had awesome discussions and performed difficult calculations seated side by side next to my writing table (which also served as our dining table). Since living in the hostel required special permission, Lev's stay at my place was illegal and he was once hauled off by the police after a surprise raid. I rushed to the police station, located in the university basement, and begged and pleaded for the release of Lev. Luckily the police were lenient and let him go after a warning, after which all was forgotten and Lev began living in my room again. I defended my Ph.D in 1985 and my research papers included several collaborations with Lev. Alexei was the junior supervisor and Prof. Zeldovich the senior supervisor of my Ph.D thesis.

It is interesting that Alexei Starobinsky and Yakov Zeldovich had very contrasting science styles. Zeldovich was an intuitive physicist par excellence. He had the gift of cutting to the very essence of complex physical phenomena and then explaining them to us students using very simple language. This was a key attraction of his bi-weekly seminar series at the Shternberg institute where he invited two experts to present their results, which they sometimes did in a very cumbersome manner. At the end of the presentation Zeldovich would bound up to the podium and lucidly explain to the audience the very gist of the problem being discussed. I could never quite figure out how he did this so spontaneously, since the seminar–under the general title of 'Unified Astrophysics Seminar'—covered a wide range of topics ranging from stellar physics to quantum gravity. Zeldovich's excellent lectures to students of Moscow state university were marked by the same style: rigorous, yet at the same time intuitive. I believe that Zeldovich developed his unique scientific style because he was essentially self-taught, his formal education having ended very early—in the 8th standard.

Alexei, on the other hand, was a formidable mathematical physicist. His understanding of complicated physical problems was supported by a deep and nuanced knowledge of mathematics which he imparted to his students. Although I struggled at first during my Ph.D (partly since the text books were in Russian), those early years of exposure to complicated mathematical concepts and calculations developed in me a deep sense of confidence that stood me in good stead in the years to come.

Alexei was the recipient of numerous international awards and honours including: the Oscar Klein medal (2010), the Gruber cosmology prize (2013), the Kavli prize (2014) and the ICTP Dirac medal (2019). He was a fellow of several academic bodies including the Russian Academy of Science and the Indian National Science Academy. Below I summarize some of the key scientific achievements of Alexei Starobinsky which were a major influence on my own work in cosmology.

2 Inflation

2.1 Relic gravitational waves from Inflation

In 1979 Alexei examined an interesting cosmological scenario in which the radiative epoch was preceeded by a non-singular de Sitter universe.[3] Alexei showed that relic gravitational waves were a generic prediction of such a scenario [11]—subsequently called inflation [24–26]. It is interesting that the amplitude of relic GWs is sensitive to the value of the Hubble parameter during inflation, while the GW spectrum encodes both the inflationary and post-inflationary EOS of the universe [11, 27–33]. Since GWs interact minimally with other forms of matter they constitute one of the cleanest probes of physics of the very early universe.

---

[3] Gliner (1966–1975) originally advocated the possibility of a metastable super-dense de Sitter stage in the early universe [22]. Alexei was aware of Gliner's work when he wrote his GW paper in 1979 and cites it in [11, 23]. However Alexei was also looking for a physically better motivated scenario of a metastable early de Sitter stage which subsequently led him to develop his $R + R^2$ model of inflation in 1980 [12].

In a homogeneous and isotropic universe massless scalar fields satisfy the Klein-Gordon equation

$$(\Box + \xi R)\,\Phi = 0 \tag{1}$$

where R is the Ricci scalar and ξ characterizes the coupling to gravity, with $\xi = 1/6, 0$ corresponding to conformal and minimal coupling respectively. In a spatially flat FRW universe $\Phi = \phi_k(\tau)e^{-i\mathbf{k}.\mathbf{x}}$ where τ is the conformal time $\tau = \int dt/a(t)$. Performing the conformal transformation $\chi_k = a\phi_k$ and using $R = 6\ddot{a}/a$ leads to

$$\ddot{\chi}_k + \left[k^2 - V(\tau)\right]\chi_k = 0 \tag{2}$$

where

$$V(\tau) = (1 - 6\xi)\frac{\ddot{a}}{a}\,. \tag{3}$$

For exponential inflation $a = \tau_0/\tau \quad (|\tau| < |\tau_0|)$, whereas $a \propto \tau, \tau^2$ in a post-inflationary radiation/matter dominated universe.

Equation (2) bears an interesting resemblance to the one dimensional non-relativistic Schroedinger equation

$$\frac{d^2\Psi}{dr^2} + [E - V(r)]\,\Psi = 0. \tag{4}$$

Comparing (2) and (4) one finds that $V(\tau)$ in (2) plays the role of a 'potential barrier in *time*', which is illustrated in Fig. 1.

Two values of ξ merit special mention.

For conformal coupling ($\xi = 1/6$), $V(\tau) = 0$, which implies $\phi(\tau) \propto e^{-ik\tau}/a$ and indicates that massless conformally coupled scalars cannot be created in an expanding universe. This result applies to massless neutrino's and photons, and for a while it was felt that all massless particles may have this property [36]. However, as shown by Grishchuk [37], gravitational waves couple minimally ($\xi = 0$) to gravity and can therefore be created quantum mechanically as the universe expands.

The two polarization states of the graviton satisfy

$$h_{\times,+} = \sqrt{8\pi G}\,\frac{\chi_k(\tau)}{a}\,e^{i\mathbf{kx}}$$

where χ_k is determined from the wave equation (2) and (3).

Equations (2) and (3) inform us that for short wavelength modes with $k^2 \gg |V|$

$$\chi_k \propto e^{-ik\tau} \text{ and } \phi^+_{\text{in}} = \frac{\chi_k}{a} \propto \frac{e^{-ik\tau}}{a}\,, \tag{5}$$

which corresponds to the adiabatic decay of positive frequency modes and no particle production. However (2) and (3) also reveal that long wavelength modes freeze out once they exit the Hubble radius since

$$\phi_k \simeq \text{constant}\,, \quad \text{for } k^2 < |V|\,.$$

As a result these modes get super-adiabatically amplified once the universe stops inflating and enters a radiation or matter dominated stage, see Fig. 1. Thus infla-

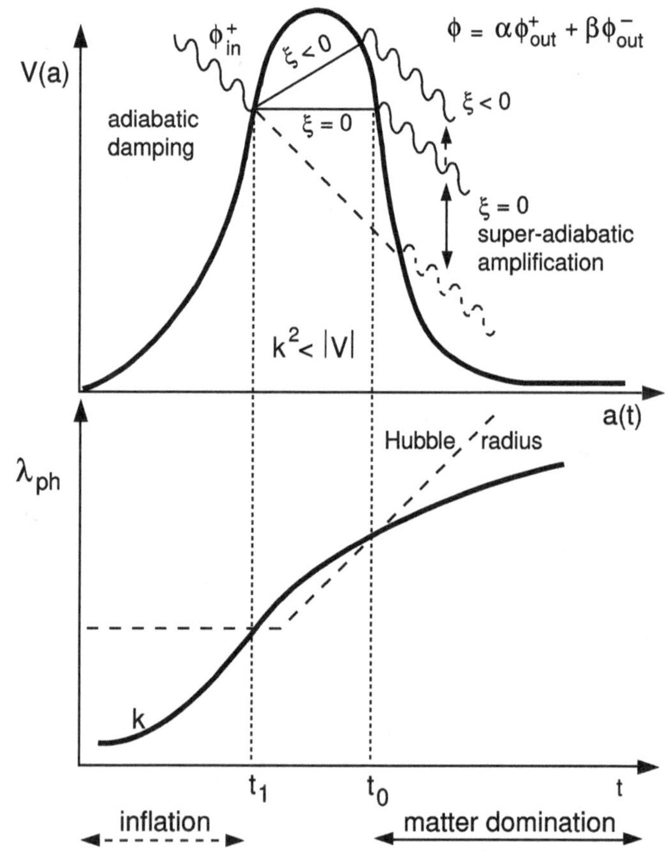

Fig. 1 **GW production** is illustrated by the time-like "potential barrier" $V(a) = (1 - 6\xi)\frac{\ddot{a}}{a}$ in inflationary cosmology (top). The evolution of the comoving Hubble radius is shown below. Short wavelength modes with $k^2 \gg |V|$ experience adiabatic damping (5) and no particle production. Long wavelength modes, on the other hand, exit the Hubble radius at t_1 during inflation and re-enter it during matter domination at t_0. The amplitude of these modes remains frozen while they are outside the Hubble radius ($k^2 < |V|$). As a result the gravitational wave ($\xi = 0$) amplitude is super-adiabatically amplified post-inflation. The amplification of fields with $\xi < 0$ can give rise to dark energy, as shown by Sahni and Habib [35]

tionary zero-point fluctuations get transformed into post-inflationary gravitational waves [11, 37].

The dimensionless amplitude of a tensor GW mode is related to the inflationary Hubble parameter $H_k^{\rm inf}$ at Hubble exit by

$$P_{\rm GW}(k) \equiv h^2_{\times,+}(k) \simeq \frac{1}{2\pi^2}\left(\frac{H_k^{\rm inf}}{m_p}\right)^2\bigg|_{k=aH}. \tag{6}$$

The spectral density of stochastic GWs, defined in terms of the critical density at the present epoch ρ_{0c}, is

$$\Omega_{\rm GW}(f) \equiv \frac{1}{\rho_{0c}}\frac{d\rho^0_{\rm GW}(f)}{d\log f}. \tag{7}$$

During the post-inflationary epoch it is described by the following set of equations [28, 32, 33]

$$\Omega^{\rm (MD)}_{\rm GW}(f) = \frac{3}{8\pi^3}P_{\rm GW}(f)\,\Omega_{0m}\left(\frac{f}{f_h}\right)^{-2}, \quad f_h < f \le f_{\rm eq} \tag{8}$$

$$\Omega^{\rm (RD)}_{\rm GW}(f) = \frac{1}{6\pi}P_{\rm GW}(f)\,\Omega_{0r}, \quad f_{\rm eq} < f \le f_{\rm re} \tag{9}$$

$$\Omega^{\rm (re)}_{\rm GW}(f) = \Omega^{\rm (RD)}_{\rm GW}\left(\frac{f}{f_{\rm re}}\right)^{2\left(\frac{w-1/3}{w+1/3}\right)}, \quad f_{\rm re} < f \le f_e\,. \tag{10}$$

The superscripts 'MD', 'RD', 're' in $\Omega_{\rm GW}$ refer to the matter dominated epoch, radiative epoch and the epoch of reheating respectively, while f_h, $f_{\rm eq}$, $f_{\rm re}$, f_e refer to the present day frequency of relic GWs corresponding to tensor modes that became sub-Hubble at: the present epoch (f_h), the epoch of matter-radiation equality ($f_{\rm eq}$), the commencement of the radiative epoch ($f_{\rm re}$) and at the end of inflation (f_e). Note that $f_{\rm re} > f_{\rm BBN} \simeq 10^{-11}$ Hz in order to satisfy the BBN bound on the reheating temperature $T_{\rm re}$. In (10) w is the post-inflationary equation of state (EOS) during the epoch of reheating.

The inflationary power spectrum $P_{\rm GW}(k)$ can be written as

$$P_{\rm GW}(k) = P_{\rm GW}(k_*)\left(\frac{k}{k_*}\right)^{n_T}, \tag{11}$$

where the tensor power at the CMB pivot scale $k_* = 0.05\ {\rm Mpc}^{-1}$ is given in terms of the scalar power spectrum A_s by

$$P_{\rm GW}(k_*) = \frac{1}{4}r\,A_s = r \times 5.25 \times 10^{-10}, \tag{12}$$

and the tensor tilt is found to be

$$n_{\text{GW}} \equiv n_T = \frac{d \log P_{\text{GW}}(k)}{d \log k} = -\frac{r}{8}, \tag{13}$$

which satisfies the consistency relation; see [33] for details.

The GW spectrum, $\Omega_{\text{GW}}(k)$, and spectral index $n_{\text{GW}} = \frac{d \log \Omega_g}{d \log k}$, therefore serve as key probes of physical processes occurring both during and after inflation. As originally shown in [11, 27, 28], and outlined in (8)–(10), the spectrum of relic gravitational radiation is exceedingly sensitive to the **post-inflationary EOS** w. From (10) one finds that the GW spectrum on small scales can have a red tilt (for $w < 1/3$), a blue tilt (for $w > 1/3$) or be scale free (if $w = 1/3$). The GW spectrum therefore provides us with a powerful signature of physics in the early universe. Some consequences of blue tilted and red tilted GW spectra are shown in Fig. 2 and a more comprehensive discussion of this issue is given in [33]. Concerning the EOS $w = w_{\text{re}}$ during the post-inflationary epoch of reheating, it is important to keep in mind the issues summarized in the next section.

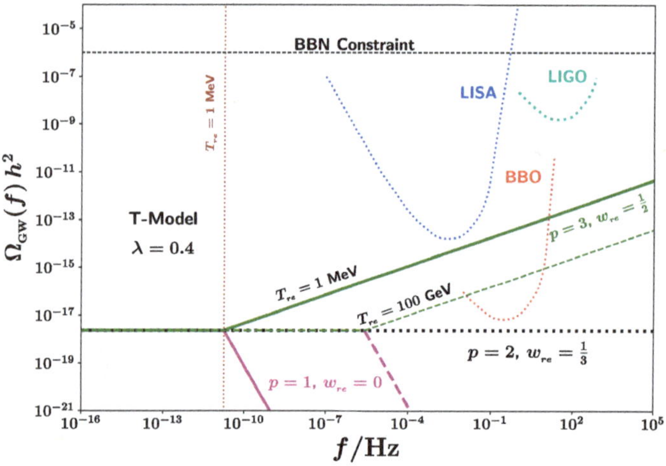

Fig. 2 The spectrum of relic gravitational waves is plotted for the T-model α-attractor potential [48] $V(\phi) = V_0 \tanh^{2p}(\frac{\lambda\phi}{m_p})$ for $\lambda = 0.4$ assuming that reheating takes place perturbatively. The dotted black curve corresponds to $p = 2$ for which the post-inflationary EOS is radiation-like, with $w_{\text{re}} = \langle w_\phi \rangle = 1/3$. The solid and dashed green curves corresponds to reheating temperatures $T_{\text{re}} = 1$ MeV and 100 GeV respectively, for the case when $p = 3$ where the post-inflationary EOS is stiff with $w_{\text{re}} = \langle w_\phi \rangle = 1/2$. We infer that blue tiled relic GWs can be detected by the future GWs observatories such as the Big Bang Observatory (BBO) [50], for a range of T_{re}. A matter-like post inflationary EOS with $w_{\text{re}} = \langle w_\phi \rangle = 0$ corresponds to $p = 1$, and is shown in red (solid, dashed). One sees that the red tilted GW amplitude is strongly suppressed in this case

2.1.1 Reheating

A key feature of inflationary cosmology is that it enables the universe to reheat by transferring the energy localized in the inflation to the matter/radiative degrees of freedom present in the universe.

In potentials possessing a minimum, reheating can occur in two distinct ways: (i) perturbatively (slowly), (ii) rapidly—via a parametric resonance. Which of these two ways is realized depends upon the nature of the coupling between the inflaton and bosons/fermions. In the case of perturbative reheating, the scalar field oscillates for a very long time gradually releasing its energy into matter/radiation. In this case the EOS during the oscillatory regime, $\langle w_\phi \rangle$, is determined primarily by the shape of the inflaton potential near its minimum value, about which the inflaton oscillates. For $V \propto \phi^{2p}$, $\langle w_\phi \rangle = \frac{p-1}{p+1}$. Perturbative reheating is expected to occur if the inflaton decays primarily into fermions (which then decay into standard model fields), its decay into bosons being strongly suppressed [19].

However if the inflaton decays into bosons, χ, through the coupling $g^2\phi^2\chi^2$ with $g \gg 10^{-3}$, then oscillations of ϕ can lead to a parametric resonance during which quanta of the field χ are produced in copious amounts. This stage is usually referred to as preheating and was studied in great detail by Alexei Starobinsky and his colleagues in [17–21]. The backreaction of χ on ϕ ends the resonance and the subsequent decay of excitations of the ϕ and χ fields into standard model (SM) fields gives rise to reheating and the subsequent thermalization of the universe at a temperature $T_{\rm re}$. The duration of the pre-radiative epoch, which includes the end of inflation, the parametric resonance, the decay of the inflaton into bosons ($\phi \to \chi\,\chi$) and fermions ($\phi \to \psi\,\psi$) and thermalization can be quite long, and it is convenient to encode its physics by means of an effective EOS parameter $w_{\rm re}$. Since $w_{\rm re}$ influences the spectrum of relic gravitational waves, observations of the GW spectrum can shed light on the complex, non-linear physics which operates during the reheating epoch (this sub-section has been adapted from [33]).

2.2 $R + R^2$ *Inflation*

In 1980 Alexei suggested an interesting model of inflation which is commonly referred to as 'Starobinsky inflation' [12]. It is based on the Jordan frame action[4]

---

[4] The quadratic R^2 term can appear as a one loop quantum correction to GR [34]. Equation (15) bears an interesting analogy with the quantum (Euler-Heizenberg) corrections to classical electrodynamics. The classical electromagnetic Lagrangian $\mathcal{L}_{\rm cl} = -\frac{1}{4}F_{\mu\nu}F^{\mu\nu} \equiv \frac{1}{2}(\mathbf{E}^2 - \mathbf{B}^2)$, in the presence of one-loop quantum corrections becomes $\mathcal{L} = \mathcal{L}_{\rm cl} + \mathcal{L}_{\rm Q}$ where the quadratic quantum correction

$$\mathcal{L}_{\rm Q} = \frac{\alpha^2}{90m_e^4}\left[\left(F_{\mu\mu}F^{\mu\nu}\right)^2 + \frac{7}{4}\left(F_{\mu\nu}\tilde{F}^{\mu\nu}\right)^2\right] \equiv \frac{2\alpha^2}{45m_e^4}\left[\left(\mathbf{E}^2 - \mathbf{B}^2\right)^2 + 7\,(\mathbf{E}.\mathbf{B})^2\right] \qquad (14)$$

leads to third order equations and non-linear electrodynamics. (α is the fine structure constant.)

$$S = \int d^4x \sqrt{-g} \frac{m_p^2}{2} \left[R + \frac{1}{6m^2} R^2 \right] \tag{15}$$

where m is a mass parameter and $m_p = 1/\sqrt{8\pi G}$. Since

$$\frac{\delta}{\delta g^{ik}} \int d^4x \sqrt{-g} \frac{R^2}{m^2} = 2g_{ik} \Box R + \cdots \tag{16}$$

this action leads to fourth order gravity, with two orders being encoded in the $\Box$ operator and another two in the Ricci scalar R.

It is sometimes convenient to write the action (15) in the Einstein frame where it has the form

$$S_E = \int d^4x \sqrt{-g} \left[\frac{m_p^2}{2} \hat{R} - \frac{1}{2} \hat{g}^{\mu\nu} \partial_\mu \phi \partial_\nu \phi - V(\phi) \right] \tag{17}$$

where the inflaton potential is [42, 43]

$$V(\phi) = \frac{3}{4} m^2 m_p^2 \left(1 - e^{-2\frac{\phi}{\sqrt{6}m_p}} \right)^2. \tag{18}$$

The CMB determined amplitude of scalar fluctuations sets $m = 1.25 \times 10^{-5} m_p$ [44]. The potential (18) is shown in Fig. 3 (left) and current CMB constraints on $R + R^2$

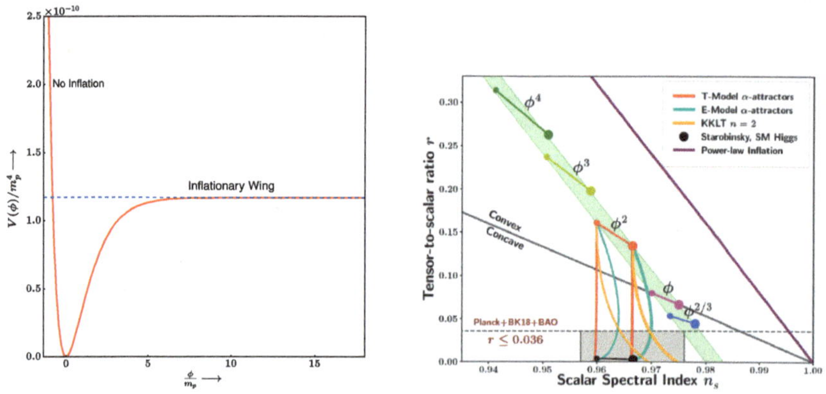

Fig. 3 **Left**: The effective potential in Starobinsky Inflation, (18), is plotted in units of m_p^4. The potential is asymmetric about the origin and has a steep left wing and plateau-like right wing. Inflation occurs along the flat right wing, the steep left wing is unable to sustain inflation. **Right**: The tensor-to-scalar ratio r and the scalar spectral index n_s from the BICEP-Keck data [49] are shown for several inflationary models. While most early models of inflation are either strongly constrained or ruled out by CMB data, the Starobinsky model provides an excellent fit. Reprinted with permission from [33].

inflation are shown in the right panel. As shown in that figure the potential for Starobinsky inflation is asymmetric about the origin. The right wing of the potential is flat and has the same functional form as the Higgs inflation potential in the Einstein frame. However the left wing of $V(\phi)$ is very steep. The slow-roll parameter for this potential is given by

$$\epsilon = \frac{m_p^2}{2}\left(\frac{V'}{V}\right)^2 = \frac{4}{3}\left[\frac{e^{-\frac{2}{\sqrt{6}}\frac{\phi}{m_p}}}{1-e^{-\frac{2}{\sqrt{6}}\frac{\phi}{m_p}}}\right]^2 .$$

Inflation ends when $\epsilon \geq 1$ which corresponds to $\phi \geq 0.937$ and indicates that no inflation can arise from the steep left wing of the potential for which $\phi < 0$ [45].

It is interesting that in formulating his $R + R^2$ model of inflation Alexei was attracted by the fact that de Sitter space can arise as a self-consistent solution to the Einstein equations with quantum corrections in the RHS

$$G_{\rm ik} = 8\pi G \langle T_{\rm ik}\rangle , \tag{19}$$

a fact that was originally noted by Dowker and Critcheley in 1976 [46]. (The R^2 term in the $R + R^2$ action generates a higher order $\Box R$ term in $\langle T_{\rm ik}\rangle$.) However Alexei went a step further and showed that de Sitter space—induced by its own vacuum polarization—was unstable and made a transition to the FRW universe. The Starobinsky model therefore successfully accommodated acceleration in the past (inflation) with deceleration at present. Self-consistent solutions to (19) were further explored in [47] and references therein.[5]

2.3 *Stochastic Inflation*

In his landmark 1986 paper [15], Alexei showed that "The dynamics of a large-scale quasi-homogeneous scalar field producing the de Sitter (inflationary) stage in the early universe is strongly affected by small-scale quantum fluctuations of the same scalar field and, in this way, becomes stochastic. The evolution of the corresponding large scale space-time metric follows that of the scalar field and is stochastic also." Alexei's formalism for stochastic inflation was very influential and paved the way for further important developments including *eternal inflation* and the *multiverse* [16, 38, 39]; see [40] for a recent review of stochasic inflation dedicated to the memory of Alexei Starobinsky.

There is an interesting parallel between stochastic inflation and classical diffusion (Brownian motion) which allows one to formulate a diffusion equation for the inflaton probability distribution $P(\phi, t)$ akin to the Fokker-Planck equation

[5] In his last conversation with me Alexei mentioned that he no longer believed that his $R + R^2$ model arose because of quantum corrections of the form (19) but might have a different origin.

$$\frac{\partial P(\phi, t)}{\partial t} = D\frac{\partial^2 P(\phi, t)}{\partial \phi^2} \tag{20}$$

where $D = H^3/8\pi^2$ is the diffusion coefficient. Solving this equation results in a Gaussian probability distribution for $P(\phi, t)$

$$P(\phi, t) = \frac{1}{\sqrt{2\pi}\sigma(t)} \exp\left(-\frac{\phi^2}{2\sigma^2(t)}\right) \tag{21}$$

where $\sigma^2 = \langle\phi^2\rangle$ is the quantum dispersion. In de Sitter space $\langle\phi^2\rangle = \frac{H^3}{4\pi^2}(t - t_0)$ where t_0 marks the start of inflation [41]. The growth in σ^2 broadens the Gaussian distribution (21) with time and arises because of the quantum diffusion of the inflaton.

In the presence of a potential $V(\phi)$ the Fokker-Planck equation gets modified to

$$\frac{\partial P(\phi)}{\partial \log a} = \frac{1}{8\pi G}\frac{\partial}{\partial \phi}\left(\frac{V'}{V}P\right) + \frac{\partial}{\partial \phi}\left(\sqrt{\Delta}\frac{\partial}{\partial \phi}\left(\sqrt{\Delta}P\right)\right), \quad \Delta = \frac{H^2}{8\pi^2}. \tag{22}$$

This equation has the solution

$$P(\phi, t) \propto \exp\left[-\frac{(\phi - \phi_{\rm cl}(t))^2}{2\sigma^2(t)}\right] \tag{23}$$

$P(\phi, t)$ describes a Gaussian which is peaked about the classical trajectory $\phi_{\rm cl}$

$$\ddot{\phi}_{\rm cl} + 3H\dot{\phi}_{\rm cl} + V' = 0. \tag{24}$$

Since $\sigma^2 \propto (t - t_0)$, the Gaussian broadens with time implying the quantum diffusion of the inflaton ϕ about its classical trajectory.

In the presence of quantum noise, and in the slow roll limit when $\ddot{\phi}_{\rm cl} \simeq 0$, the classical equation of motion (24) gets transformed into the Langevan equation for the inflaton $\phi = \phi_{\rm cl} + \Delta\phi_q$

$$\frac{d\phi}{d\log a} = -\frac{V'}{3H^2} + \frac{H}{2\pi}f(\phi, \log a) \tag{25}$$

where f is a stochastic force which describes random kicks that are imparted to ϕ as small wavelength fluctuations of quantum origin ($\Delta\phi_q$) grow in scale and cross the Hubble radius. If the quantum noise dominates over the classical drift then the universe can inflate eternally in some regions of space [38], as shown in Fig. 4.

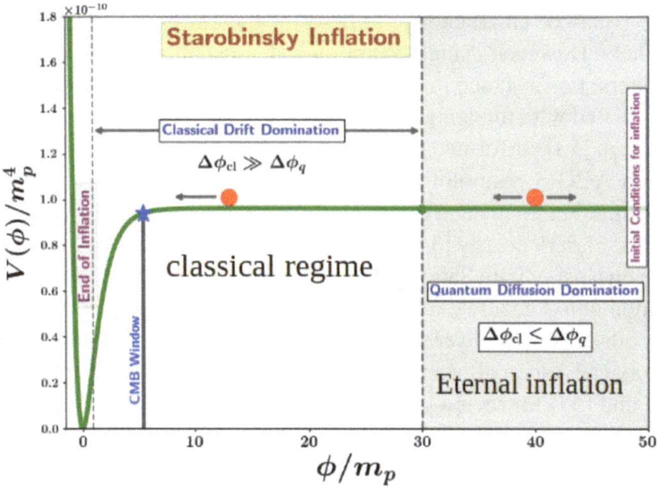

Fig. 4 An illustration of stochastic Starobinsky inflation (18)

3 Dark Energy

After completing my Ph.D in Moscow in 1985, I went to the UK and to Canada for post-doctoral work. I finally returned to India in 1991 to join a newly established institute of astrophysics in Pune called IUCAA. When I arrived the institutional campus was just being built and we were all housed in temporary quarters. What struck me immediately upon arrival were the large number of old and beautiful Banyan trees on campus and the copious abundance of snakes and other wildlife including peacocks, mongoose families and even scorpions (I was bitten by one).

At the time of my arrival the total faculty strength at IUCAA was only 6 and so we were very keen to have distinguished scientists visit our institute and interact with members of the Indian scientific community. I applied for an Indo-Russian collaboration program which allowed Alexei to visit India. He loved the ambience of IUCAA and, over the course of several visits, collaborated with a large number of IUCAA students, visitors and faculty including Tarun Souradeep, Dipak Munshi, Tarun Saini, Ujjaini Alam, Arman Shafieloo, Swagat Mishra, M. Sami, Minu Joy and Somak Raychaudhury.

3.1 Reconstructing Dark Energy

The discovery of dark energy in 1998 [51] took almost everyone by surprise. Although the existence of a cosmological constant Λ had been seriously discussed for several decades, its theoretical value was presumed to be very large (formally infinite) and so the fact that Λ may be small enough to describe the currently accelerating universe seemed unlikely. However a small cosmological constant was anticipated in several key early papers. For instance in 1967 Zeldovich suggested that a small value of Λ could be associated with fundamental physics [52] . Then in 1985 Alexei Starobinsky and Lev Kofman [53] performed the first ever calculation of large scale anisotropies in the CMB in ΛCDM cosmology (the Sachs-Wolfe effect). In 1992 I had written a paper with Hume Feldman and Albert Stebbins on a universe which *loiters* because of the presence of dark energy [54], and in 1998, at the time of the SNIa discovery, I was working on a paper with Salman Habib in which we showed that quantum effects during inflation could generate a small cosmological constant [35]. Because of this early work, both Alexei and I were psychologically prepared for Λ and began working on dark energy (DE) soon after its discovery. (See [55, 56] for other important early work on DE and [57] for reviews.)

The SNIa discovery of an accelerating universe led to many interesting papers being written which explained cosmic acceleration using new theoretical models. Alexei and I followed a parallel path and developed model independent methods which could reconstruct the nature of DE directly from observations using model agnostic techniques [58, 59].

The idea behind DE reconstruction is simple. For standard candles (SNIa) with luminosity L, the flux received by an observer is

$$\mathcal{F} = \frac{L}{4\pi d_L^2}, \tag{26}$$

where

$$d_L(z) = (1+z)\int_0^z \frac{dz'}{H(z')} \tag{27}$$

is the luminosity distance. From (27) one can obtain the expansion history

$$H(z) = \left[\frac{d}{dz}\left(\frac{d_L}{1+z}\right)\right]^{-1} \tag{28}$$

and the equation of state of DE

$$w(z) = \frac{\frac{2x}{3}\frac{d\log H}{dx} - 1}{1 - \left(\frac{H_0}{H}\right)^2 \Omega_{0m} x^3}, \quad x = 1+z\,. \tag{29}$$

Differentiating a noisy observable such as the luminosity distance $d_L(z)$ increases the noise, therefore one must smoothen $d_L(z)$ or $H(z)$ before differentiating in order to obtain $w(z)$. This can be done either by approximating $d_L(z)$ or $H(z)$ by an ansatz, or by smoothing the data directly. There are therefore two main approaches to DE reconstruction: parametric and non-parametric [59].

- In the parametric approach one approximates a key cosmological quantity such as the luminosity distance $d_L(z)$ (or angular size distance d_A), the expansion history $H(z)$ or the DE equation of state, $w(z)$, using a simple ansatz such as $w(z) = w_0 + w_1 \frac{z}{1+z}$ [60]. The free parameters w_0, w_1 are then obtained by matching with observations.
- Non-parametric reconstruction is based on the observation that one can obtain a smoothed quantity $D^S(\mathbf{x})$ from a fluctuating 'raw' quantity, $D(\mathbf{x}')$, using a low pass filter F having a characteristic scale R_f [61]

$$D^S(\mathbf{x}, R_f) = \int D(\mathbf{x}')F(|\mathbf{x} - \mathbf{x}'|; R_f)\, d\mathbf{x}'. \tag{30}$$

Commonly used filters include the Gaussian filter $F_G \propto \exp(-|\mathbf{x} - \mathbf{x}'|^2/2R_G^2)$. In studies of large scale structure D is the density fluctuation $\delta(\mathbf{x})$, whereas for DE reconstruction D could be either $H(z)$, $d_L(z)$, $d_A(z)$, etc.
Since we wish to smooth the noise and not the signal we proceed as follows [63]

$$\ln d_L(z, \Delta)^s = \ln\, d_L(z)^g + N(z) \sum_i \left[\ln d_L(z_i) - \ln\, d_L(z_i)^g\right] \times \exp\left[-\frac{\ln^2\left(\frac{1+z_i}{1+z}\right)}{2\Delta^2}\right], \tag{31}$$

$$N(z)^{-1} = \sum_i \exp\left[-\frac{\ln^2\left(\frac{1+z_i}{1+z}\right)}{2\Delta^2}\right].$$

Here, $\ln d_L(z, \Delta)^s$ is the smoothed luminosity distance at redshift z and $N(z)$ is a normalization parameter. This procedure is to be used iteratively with ΛCDM being the first 'guess' model for $d_L = d_L(z)^g$. Convergence takes place quite rapidly within only a few iterations. Using $d_L(z) \equiv d_L^s(z)\big|_{\rm final}$ one can obtain the expansion history (28) and the EOS of DE (29). A parallel approach involving cosmological reconstruction using Gaussian processes is discussed in [64].

3.2 *Null Tests of ΛCDM*

Of all Dark Energy models the cosmological constant, Λ, is singled out by its elegance and simplicity since $T_i^k = \Lambda\delta_i^k$. So, as a first step, its logical to find tests which could falsify the *cosmological constant hypothesis*. These include the Statefinder and Om diagnostics.

3.2.1 The Statefinder Diagnostic of Dark Energy

A general form for the expansion factor of the Universe is given by the Taylor series expansion

$$a(t) = a(t_0) + \dot{a}\big|_0 (t - t_0) + \frac{\ddot{a}\big|_0}{2}(t - t_0)^2 + \frac{\dddot{a}\big|_0}{6}(t - t_0)^3 + ... \tag{32}$$

In 1970 Alan Sandage described observational cosmology as being "a search for two numbers"

$$H_0 = \left(\frac{\dot{a}}{a}\right)_0 \quad \text{and} \quad q_0 = -\left(\frac{\ddot{a}}{aH^2}\right)_0 . \tag{33}$$

In 2002 Alexei and I, together with my students Ujjaini Alam and Tarun Saini, showed that a new diagnostic of dark energy—Statefinder—could be constructed using the third derivative of the expansion factor [62]. The statefinder pair $\{r, s\}$, defines two new cosmological parameters (in addition to H and q):

$$r \equiv \frac{\dddot{a}}{aH^3} = 1 + \frac{9w}{2}\Omega_{\rm DE}(1 + w) - \frac{3}{2}\Omega_{\rm DE}\frac{\dot{w}}{H} , \tag{34}$$

$$s \equiv \frac{r - 1}{3(q - 1/2)} = 1 + w - \frac{1}{3}\frac{\dot{w}}{wH} . \tag{35}$$

A key property of the Statefinder is that $\{r, s\} = \{1, 0\}$ is a **fixed point** for ΛCDM.

The statefinder therefore provides us with a *fingerprint* of DE since for all DE models excluding ΛCDM the trajectory of $\{r, s\}$ evolves with time. In Fig. 5 we show the behaviour of the statefinder pair $\{r, s\}$ for a number of DE models including

1. Quintessence models defined by $\mathcal{L} = \frac{1}{2}\dot{\phi}^2 - V(\phi)$ and $V = V_0/\phi^\alpha$.
2. DE with a constant EOS $p = w\rho$.
3. The Chaplygin gas with $\mathcal{L} = V_0\sqrt{1 - \phi_{,\mu}\phi^{,\mu}}$ and $p_c = -A/\rho_c$.

Note that null tests of ΛCDM can also be constructed using higher derivatives of $a(t)$, which results in a 'Statefinder hierarchy' of null tests [66].

3.2.2 The *Om* Diagnostic

The Om diagnostic is defined very simply as [67]

$$Om(z) \equiv \frac{h^2(z) - 1}{(1 + z)^3 - 1} , \tag{36}$$

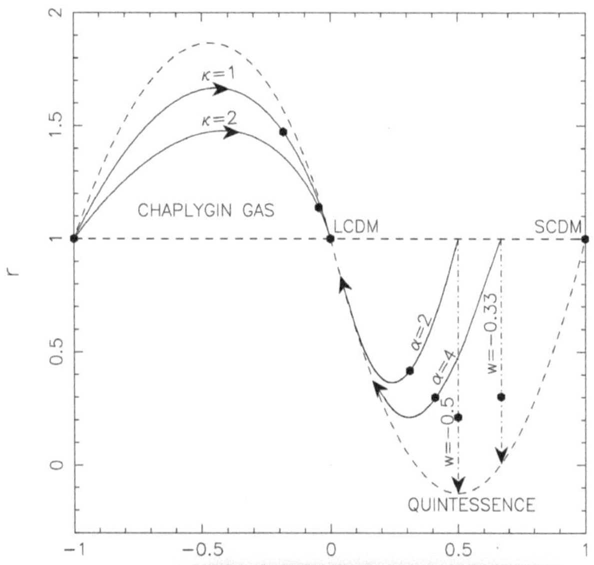

Fig. 5 The time evolution of the statefinder pair $\{r, s\}$ for quintessence models with $V = V_0/\phi^\alpha$, DE with a constant EOS w, and the Chaplygin gas. Quintessence models lie to the right of the ΛCDM fixed point ($r = 1, s = 0$) while the Chaplygin gas lies to the left. Both models approach the ΛCDM limit of $r = 1, s = 0$ in the future. For the Chaplygin gas, κ is the ratio between the matter density and the density of the Chaplygin gas at early times. The dashed curve in the lower right is the envelope of all quintessence models, while the dashed curve in the upper left is the envelope of Chaplygin gas models. In DE models with a constant EOS s remains fixed at $1 + w$ while r declines asymptotically to $1 + \frac{9}{2}w(1 + w)$. Reprinted with permission from [65].

where $h(z) = H(z)/H_0$. The possibility of using Om as a *null diagnostic* follows from the fact that for ΛCDM

$$Om(z) = \Omega_{0m}. \tag{37}$$

In other words, the value of Om is *redshift independent* for the cosmological constant, while for other models of dark energy $Om(z)$ is redshift dependent. Interestingly, $Om(z) > \Omega_{0m}$ in quintessence models while $Om(z) < \Omega_{0m}$ in phantom DE.

This important property of Om can be used to considerable advantage by defining the *difference diagnostic*

$$Om_{\text{diff}}(z_1, z_2) := Om(z_1) - Om(z_2). \tag{38}$$

From (37) one immediately finds that for the cosmological constant

$$Om_{\text{diff}}(z_1, z_2) = 0. \tag{39}$$

A departure of $Om_{\rm diff}$ from zero therefore serves as a 'smoking gun' test for ΛCDM. Since Om depends only on the Hubble parameter and not its derivatives it may be easier to determine from (noisy) observations than either $q(z)$ or $w(z)$. From (36) one finds that the Om diagnostic does not depend upon the value of Ω_{0m} and is therefore insensitive to observational uncertainties in that quantity, unlike $w(z)$ which explicitely depends upon Ω_{0m} through (29).

One can also define the two-point Om diagnostic [68]

$$Om(z_2; z_1) = \frac{h^2(z_2) - h^2(z_1)}{(1+z_2)^3 - (1+z_1)^3}\,, \quad h(z) = H(z)/H_0\,, \tag{40}$$

where $Om(z;0) = Om(z)$ and $Om(z_2;z_1) = \Omega_{0m}$ for ΛCDM. Multiplying both sides of (40) by h_{100}^2 where $h_{100} = H_0/100\,\text{km/sec/Mpc}$, results in the *improved Om diagnostic* [69]

$$\begin{aligned} Omh^2(z_2; z_1) &= \left[\frac{h^2(z_2) - h^2(z_1)}{(1+z_2)^3 - (1+z_1)^3}\right] h_{100}^2 \\ &= \frac{h_{100}^2(z_2) - h_{100}^2(z_1)}{(1+z_2)^3 - (1+z_1)^3}\,, \end{aligned} \tag{41}$$

where $h_{100}(z) = H(z)/100\text{km/sec/Mpc}$. A significant advantage of Omh^2 is that for ΛCDM:

$$Omh^2 = \Omega_{0m} h_{100}^2\,. \tag{42}$$

Since observations of the CMB inform us that [70] $\Omega_{0m} h_{100}^2 = 0.1426 \pm 0.0025$, it follows that for ΛCDM [69]

$$Omh^2(z_2; z_1) = 0.1426 \pm 0.0025. \tag{43}$$

Therefore a departure of Omh^2 from this value would signal that DE *is not* Λ.

A recent reconstruction of the Om diagnostic from DESI-BAO data suggests that cosmic acceleration may be slowing down [71]. In this context it is interesting to note that several years prior to the release of DESI data, Alexei Starobinsky drew attention to an interesting new model in which DE is metastable and can decay into dark matter (or radiation) via [72]

$$\begin{aligned} \dot{\rho}_{\rm DE} &= -\Gamma \rho_{\rm DE} \\ \dot{\rho}_m + 3H\rho_m &= \Gamma \rho_{\rm DE}\,. \end{aligned} \tag{44}$$

The decay of DE is based on its intrinsic properties and does not depend upon the external environment (expansion rate, etc.), so in this sense (44) resembles the *radioactive decay* of a substance. Equation (44) leads to transient acceleration which may be in better agreement with data than ΛCDM [71, 73]; also see [74].

4 Discussion

Alexei Starobinsky was not only an exceptional scientist, but also a very kind and generous human being, a wonderful Ph.D guide and a team player par excellence. In this latter role he helped mentor several generations of cosmologists including myself, Lev Kofman, and numerous others, many of whom have contributed articles to this memorial volume. During his visits to India, Alexei interacted with many young researchers including Tarun Souradeep, Dipak Munshi, Tarun Saini, Ujjaini Alam, Minu Joy, Arman Shafieloo and Swagat Mishra. Many of these scientists have gone on to become leaders in the international scientific community and have mentored a whole new generation of young students. I therefore firmly believe that although Alexei Starobinsky is no longer present amongst us, the legacy of the Zeldovich-Starobinsky school of cosmology will endure for many generations to come.

5 Down Memory Lane

See (Figs. 6, 7, 8, 9 and 10).

Fig. 6 With Alexei at his home in Moscow, clockwise: Alexei, Lyudmila, myself, Alexei's mother and Rohini

Fig. 7 With Alexei Starobinsky and Yakov Zeldovich during of my Ph.D defence, Moscow 1985

Fig. 8 Alexei at IUCAA. *Top row*: Alexei and myself, *middle row*: Jatush Sheth, Tarun Souradeep and Ujjaini Alam, *bottom row*: Sanjit Mitra and Amir Hajian

Fig. 9 Alexei sharing a joke with Tarun Souradeep, Amir Hajian and myself

Fig. 10 Alexei flanked by Swagat Mishra and myself, during Alexei's last visit to IUCAA in 2019

Acknowledgements I thank Swagat Mishra for a careful reading of the manuscript and for his help with the figures.

References

1. V. Sahni, Ya.B. Zeldovich, Chemist, nuclear physicist, cosmologist. IAU Symp. **308**, 25–31 (2014). arXiv:1310.8048. (1914–1987)
2. Ya.B. Zeldovich, A.A. Starobinsky, Sov. Phys. JETP **34**, 1159–1166 (1972). (Zh. Eksp. Teor. Fiz. **61**, 2161–2175, 1971)
3. V.N. Lukash, A.A. Starobinsky, Sov. Phys. JETP 39742 (1974)
4. C.B. Collins, S.W. Hawking, Astrophys. J. **180**, 317–334 (1973)
5. A.A. Starobinsky, JETP Lett. 37, 66–69 (1983)
6. R.M. Wald, Phys. Rev. D **28**, 2118–2120 (1983)
7. I.G. Moss, V. Sahni, Phys. Lett. B **178**, 159–162 (1986)
8. A.A. Starobinsky, Sov. Phys. JETP **37**, 1 (1973)
9. S.W. Hawking, Nature **248**(5443), 30–31 (1974)
10. S.W. Hawking, *A Brief History of Time* (Bantam Dell Publishing Group, UK, 1988)
11. A.A. Starobinsky, JETP Lett. **30**, 682 (1979)
12. A.A. Starobinsky, Phys. Lett. B **91**, 99–102 (1980)
13. A.A. Starobinsky, Phys. Lett. B **117**, 175–178 (1982)
14. A.A. Starobinsky, Sov. Astron. Lett. **11**, 323 (1985)
15. A.A. Starobinsky, Lect. Notes Phys. **246**, 107–126 (1986)
16. A.A. Starobinsky, J. Yokoyama, Phys. Rev. D **50**, 6357–6368 (1994). ([arXiv:astro-ph/9407016])
17. L.A. Kofman, A.D. Linde, A.A. Starobinsky, Phys. Rev. Lett. **73**, 3195 (1994)
18. L.A. Kofman, A.D. Linde, A.A. Starobinsky, Phys. Rev. Lett. **76**, 1011 (1996)
19. L. Kofman, in *Relativistic Astrophysics: A Conference in Honor of Igor Novikov's 60th Birthday, Proceedings, Copen-hagen, Denmark*, ed. by B (Marcovic Cambridge University Press, Cambridge, England, Jones and D, 1996). arXiv:astro-ph/9605155
20. L.A. Kofman, A.D. Linde, A.A. Starobinsky, Phys. Rev. D **56**, 3258 (1997)
21. Y. Shtanov, J. Traschen, R. Brandenberger, Phys. Rev. D **51**, 5438 (1995). (also see Y. Shtanov, Ukr. Fiz. Zh. **38**, 1425 (1993) (in Russian))
22. E.B. Gliner, Sov. Phys. JETP **22**, 378 (1966). arXiv:2301.13150. (E.B. Gliner, Sov. Phys. Dokl. **15**, 559 (1970); E.B. Gliner, I.G. Dymnikova, Sov. Astron. Lett. **1**, 93 (1975); I.G. Dymnikova, JETP **63**, 1111–1115 (1986); For a nice article on Gliner's important work and difficult life see D. Yakovlev, A. Kaminker, *Nearly Forgotten Cosmological Concept of E. B. Gliner*)
23. A.A. Starobinsky, Ya.B. Zeldovich, in *Spontaneous Creation of the Universe; My Universe, Selected Reviews*, ed. by Y.B. Zeldovich, M.V. Sazhin. (Harwood Academic Publishers: Chur, Switzerland, 1992), pp. 97–134
24. A.H. Guth, Phys. Rev. D **23**, 347 (1981)
25. A.D. Linde, Phys. Lett. B **108**, 389 (1982)
26. A. Albrecht, P.J. Steinhardt, Phys. Rev. Lett. **48**, 1220 (1982)
27. B. Allen, Phys. Rev. D **37**, 2078 (1988)
28. V. Sahni, Phys. Rev. D **42**, 453 (1990)
29. M. Giovannini, Phys. Rev. D **58**, 083504 (1998). (Phys. Rev. D **60**, 123511, 1999)
30. V. Sahni, M. Sami, T. Souradeep, Phys. Rev. D **65**, 023518 (2002). ([arXiv:gr-qc/0105121])
31. C. Caprini, D.G. Figueroa, Class. Quant. Grav. **35**(16), 163001 (2018). arXiv:1801.04268
32. D.G. Figueroa, E.H. Tanin, JCAP **08**, 011 (2019). arXiv:1905.11960
33. S.M. Mishra, V. Sahni, A.A. Starobinsky, JCAP **05**, 075 (2021). ([arXiv:2101.00271])
34. N.D. Birrel, P.C.W. Davies, *Quantum Fields in Curved Space* (Cambridge Monographs on Mathematical Physics, 1984)

35. V. Sahni, S. Habib, Phys. Rev. Lett. **81**, 1766–1769 (1998). ([arXiv: hep-ph/9808204])
36. L. Parker, Phys. Rev. Lett. **21**, 562 (1968)
37. L.P. Grishchuk, Sov. Phys. JETP **40**, 409 (1975)
38. A.D. Linde, Phys. Lett. B **175**(4), 395–400 (1986)
39. A.D. Linde, *Inflationary Cosmology* (Springer Berlin Heidelberg, 2008), pp. 1–54
40. R.P. Woodard, *Recent Developments in Stochastic Inflation*. arXiv:2501.15843
41. A. Vilenkin, L.H. Ford, Phys. Rev. D **26**, 1231 (1982)
42. B. Whitt, Phys. Lett. B **145**, 176 (1984)
43. K.I. Maeda, Phys. Rev. D **37**, 858 (1988)
44. D.S. Gorbunov, A.G. Panin, Phys. Lett. B **743**, 79–81 (2015). ([arXiv:1412.3407])
45. S.M. Mishra, V. Sahni, A.A. Toporensky, Phys. Rev. D **98**(8), 083538 (2018). arXiv:1801.04948
46. J.S. Dowker, R. Critchley, Phys. Rev. D **13**, 3224 (1976)
47. V. Sahni, L.A. Kofman, Phys. Lett. A **117**, 275 (1986)
48. R. Kallosh, A.D. Linde, JCAP07 002 (2013). arXiv:1306.5220
49. P.A.R. Ade et al., Phys. Rev. Lett. **127**, 151301 (2021). ([arXiv:2110.00483])
50. W.T. Ni, Int. J. Mod. Phys. D **25**(14), 1630001 (2016). arXiv:1610.01148
51. S.J. Perlmutter, et al., Nature **391**, 51 (1998). (Astroph. J. **517**, 565 (1999) . arXiv: astro-ph/9812133. A.G. Riess, et al., Astron. J. **116**, 1009 (1998). arXiv: astro-ph/9805201)
52. Ya.B. Zeldovich, Sov. Phys. JETP Lett. **6**, 316 (1967). (Sov. Phys.-Uspekhi, **11**, 381, 1968)
53. L.A. Kofman, A.A. Starobinsky, Sov. Astron. Lett. **11**, 271–274 (1985)
54. V. Sahni, H. Feldman, A. Stebbins, Astrophys. J. **385**, 1–8 (1992)
55. B. Ratra, P.J.E. Peebles, Phys. Rev. D **37**, 3406 (1988)
56. J.P. Ostriker, P.J. Steinhardt, Nature **377**, 600 (1995)
57. V. Sahni, A.A. Starobinsky, Int. J. Mod. Phys. D **9**, 373 (2000). (P.J.E. Peebles, B. Ratra, Rev. Mod. Phys. **75**, 559 (2003); T. Padmanabhan, Phys. Rep. 380, 235 (2003); V. Sahni, Lect. Notes Phys. 653, 141 (2004); V. Sahni, arXiv:astro-ph/0502032; E.J. Copeland, M. Sami and S. Tsujikawa, Int. J. Mod. Phys. D 15, 1753 (2006); J.A. Frieman, M.S. Turner, D. Huterer, Ann. Rev. Astron. Astroph. **46**, 385 (2008); S. Tsujikawa, Dark Matter and Dark Energy: a Challenge for the 21st Century, arXiv: 1004.1493; S. Nojiri, S.D. Odintsov, Phys. Rept. **505**, 59 (2011); T. Clifton, P.G. Ferreira, A. Padilla, C. Skordis, Phys. Rept. **513**, 1 (2012))
58. T.D. Saini, S. Raychaudhury, V. Sahni, A.A. Starobinsky, Phys. Rev. Lett. **85**, 1162–1165 (2000). ([arXiv:astro-ph/9910231])
59. V. Sahni, A.A. Starobinsky, Int. J. Mod. Phys. D **15**, 2105–2132 (2006). ([arXiv:astro-ph/0610026])
60. M. Chevallier, D. Polarski, IJMP D **10**, 213 (2001). (E.V. Linder, Phys. Rev. Lett. **90**, 091301, 2003)
61. P. Coles, F. Lucchin, *Cosmology, The Origin and Evolution of Large Scale Structure* (John Wiley & Sons, 1995)
62. V. Sahni, T.D. Saini, A.A. Starobinsky, U. Alam, JETP Lett. **77**, 201 (2003). ([arXiv:astro-ph/0201498])
63. A. Shafieloo, U. Alam, V. Sahni, A.A. Starobinsky, Mon. Not. Roy. Astron. Soc. **366**, 1081–1095 (2006). ([arXiv:astro-ph/0505329])
64. A. Shafieloo, A.G. Kim, E.V. Linder, Phys. Rev. D **85**, 123530 (2012). ([arXiv:1204.2272])
65. U. Alam, V. Sahni, T.D. Saini, A.A. Starobinsky, Mon. Not. Roy. Astron. Soc. **344**, 1057 (2003). arXiv: astro-ph/0303009
66. M. Arabsalmani, V. Sahni, Phys. Rev. D **83**, 043501 (2011). arXiv:1101.3436
67. V. Sahni, A. Shafieloo, A.A. Starobinsky, Phys. Rev. D **78**, 103502 (2008). arXiv:0807.3548
68. A. Shafieloo, V. Sahni, A.A. Starobinsky, Phys. Rev. D **86**, 103527 (2012). arXiv:1205.2870
69. V. Sahni, A. Shafieloo, A.A. Starobinsky, Astrophys. J. Lett. **793**(2), L40 (2014). arXiv:1406.2209
70. P. Ade et al., 2013, Planck 2013 results. XVI. Cosmol. Parameter. A&A **571**, A16 (2014). arXiv:1303.5076
71. R. Calderon, et al. JCAP10, 048 (2024). arXiv:2405.04216
72. A. Shafieloo, et al. Mon. Not. Roy. Astron. Soc. **473**(2), 2760–2770 (2018). arXiv:1610.05192
73. X. Li et al., Astrophys. J. **887**, 153 (2019). arXiv:1904.03790
74. U. Alam, V. Sahni, A.A. Starobinsky, JCAP 04 002 (2003). arXiv:astro-ph/0302302. (A. Shafieloo, V. Sahni, A.A. Starobinsky, Phys. Rev. D **80**, 101301 (2009). arXiv:0903.5141)

Starobinsky in Japan

Jun'ichi Yokoyama

Abstract Alexei A. Starobinsky visited Japan many times for 33 years and spent significant time there until the pandemic hit the world. I review some of the work he did with his colleagues in Japan and describe how he affected us as well as the outlook of inflationary cosmology.

1 Introduction

Alexei Alexandrovich Starobinsky spent significant time in Japan and wrote a number of important papers during his stay there. His first visit to Japan was back in 1990 before the collapse of the Soviet Union. He was invited to an international conference entitled "Primordial Nucleosynthesis and the Evolution of Early Universe" which was a conference endorsed by IUPAP and supported by the Japan World Exposition Commemorative Fund from the proceeds of the 1970 Osaka Expo. The conference was organized by Katsuhiko Sato, a pioneer of inflationary cosmology [1] who was an associate professor at the physics department of the University of Tokyo then, and I served as its secretariat as an assistant in Sato's lab.

At that time, it was quite rare for international conferences like this to be held in Japan. Based on his prior experience of actually managing the Yamada Conference titled "Big Bang, Active Galactic Nuclei, and Supernovae", held from March 28 to April 1, 1988, under the chairmanship of Satio Hayakawa, who was President of Nagoya University then, and with support from the Yamada Science Foundation, Sato decided to take the lead himself as organizing committee chair for this conference.

J. Yokoyama (✉)
Kavli Institute for the Physics and Mathematics of the Universe (Kavli IPMU), WPI, UTIAS, The University of Tokyo, Kashiwa, Chiba, Japan
e-mail: yokoyama@resceu.s.u-tokyo.ac.jp

Research Center for the Early Universe (RESCEU), Graduate School of Science, The University of Tokyo, Tokyo, Japan

Department of Physics, Graduate School of Science, The University of Tokyo, Tokyo, Japan

Trans-Scale Quantum Science Institute, The University of Tokyo, Tokyo, Japan

A. Barvinsky and A. Kamenshchik (eds.), *Open Issues in Gravitation and Cosmology*, Fundamental Theories of Physics 225, https://doi.org/10.1007/978-3-032-15703-4_33

The venue on both occasions was the Sanjo Conference Hall, located on the Hongo campus of the University of Tokyo. At that time, the hall had about ten guest rooms, and Starobinsky also stayed there. I still vividly remember him wandering around the conference area wearing room slippers. He had been scheduled to come as a distinguished disciple of Zel'dovich, a towering figure in Soviet physics. However, since Zel'dovich was unable to travel to Japan, only Starobinsky ended up attending.

Several years earlier, when I was still a graduate student, Andrei Linde, who was a leading authority on inflationary cosmology in the Soviet Union [2, 3], had also visited the Sato Laboratory. At the time, visitors from abroad, much less from the Soviet Union, were extremely rare, so I took Linde on a sightseeing tour of Tokyo at Sato's order. Linde wanted a toy that would fly just like the real birds that he had seen since arriving in Japan. I had no idea what it was, so I took him to the toy section of a department store, but we could not find it. It was not until much later that I learned that it was the kind of thing they sell at fairs around Ueno Park, not at a fancy department store. Compared to those days, the world has opened up considerably, although there were interruptions due to the Covid-19, which actually killed Starobinsky sadly.

2 Starobinsky in Uji and Stochastic Inflation

After that conference, I went to the USA for a year as a JSPS overseas fellow and then took a position at the Yukawa Institute for Theoretical Physics (YITP), Kyoto University. The institute has a visiting foreign professor position, which currently rotates between the fields of particle physics, nuclear physics, condensed matter physics, and astrophysics, with each position lasting up to three months. At that time, however, there were no such restrictions, and the mission of the institute was to find good people from all over the world. I emailed Starobinsky on the spur of the moment, using my acquaintance in Tokyo as leverage, and to my surprise, he was interested and agreed to stay as a visiting professor at Kyoto University for about ten months from November 1993 to September 1994. Looking back now, I am amazed that the Institute accepted the proposal from the youngest associate professor, who was not even a member of the executive committee. I would like to once again express my gratitude to Director Yosuke Nagaoka at that time for his support.

When I joined YITP, the Institute was divided into the original Research Institute for Fundamental Physics founded by Hideki Yukawa (located at the Kyoto University headquarters and called Kitashirakawa after its location) and the Uji Research Center. The Uji Research Center was a descendant of the Research Institute for Theoretical Physics of Hiroshima University, which had a long history dating back to 1943, and the staff of the Institute were transferred to the Uji Research Center. I was employed there.

I picked Starobinsky up at the train station in my own car and took him to a guesthouse that was located on the Uji campus. He said there were no towels in the room, so I took him to the Kyoto Bank, opened his account, and got a towel as a bonus. This was how he started his life in Japan.

Several weeks later, when he settled down, I told him I wanted to work with him on stochastic inflation or other topics on inflation. He suggested solving the Langevin equation iteratively to see higher-order effects. His intention was to capture loop corrections of quantum fluctuations by solving the stochastic differential equation in that way. I did the calculation partially but did not find any interesting conclusion as such higher-order effects are small in normal inflation which predicts nearly scale-invariant perturbation whose amplitude must be as small as 10^{-5} as observed by cosmic microwave background radiation [4]. Interestingly, this is the topic I have been working on recently using a completely different method based on quantum field theory of the so-called in-in formalism [5, 6] together with Jason Kristiano [7–11] who first told me the sad news of Alxei's passing on the day of Christmas Eve.

In the course of my calculations, I became interested in the spatial distribution of scalar fields and realized that by the method of stochastic inflation, it is possible to show that the probability distribution function of a field has de Sitter invariance, the fundamental symmetry of the de Sitter spacetime. Due to the high symmetry of the de Sitter background, it is natural to define the background (or "vacuum") state of any quantum field as a state vector invariant under the action of the full group of symmetries of the de Sitter spacetime O(4,1). One may also expect that this background state is a stable attractor for an open set of other quantum states in the Schrödinger picture; in other words, that it is a stable equilibrium state. In the Heisenberg picture, this means that average values of all possible operators in an open set of constant state vectors approach those in the equilibrium state as $t \to \infty$.

The investigation of a free massive scalar field in the de Sitter background has shown that this is the case, but for a massless minimally coupled field, it is well known that such an equilibrium state does not exist and the square expectation value increases in proportion to time $\langle\phi^2\rangle = H^3 t/4\pi^2$ for $Ht \gg 1$, where H is the Hubble parameter [12–16]. The square expectation value growing in proportion to time reminds us of the Brownian motion, and we are tempted to derive a Langevin equation to describe its dynamics, which is indeed done in the stochastic approach [17, 18]. So we represent the Heisenberg operator of the quantum field ϕ as

$$\phi(\mathbf{x}, t) = \overline{\phi}(\mathbf{x}, t) + \int \frac{d^3k}{(2\pi)^{3/2}} \theta(k - \epsilon a(t) H) \left[a_{\mathbf{k}} \phi_{\mathbf{k}}(t) e^{-i\mathbf{k}\mathbf{x}} + a_{\mathbf{k}}^{\dagger} \phi_{\mathbf{k}}^{*}(t) e^{i\mathbf{k}\mathbf{x}} \right]. \tag{1}$$

Here $\overline{\phi}(\mathbf{x}, t)$ is a coarse-grained or a long wavelength part of ϕ, $\theta(z)$ is the step function, and ϵ is a small constant parameter which in fact is not arbitrarily small. The short wavelength counterpart, on the other hand, is characterized by the mode functions $\phi_{\mathbf{k}}(t)$ with $a_{\mathbf{k}}$ and $a_{\mathbf{k}}^{\dagger}$ being annihilation and creation operators, respectively. It is small, so it satisfies the linear massless equation

$$\ddot{\phi}_{\mathbf{k}} + 3H\dot{\phi}_{\mathbf{k}} + \frac{\mathbf{k}^2}{a^2(t)}\phi_{\mathbf{k}} = 0, \tag{2}$$

which is solved as

$$\phi_{\mathbf{k}} = \sqrt{\frac{\pi}{4}}H(-\eta)^{3/2}H^{(1)}_{3/2}(-k\eta) = \frac{H}{\sqrt{2k}}\left(\eta - \frac{i}{k}\right)e^{-ik\eta}, \tag{3}$$

with $H^{(1)}_{3/2}(z)$ being the Hankel function of the first class, and η (< 0) is the conformal time of de Sitter space with which the scale factor, $a(t) = e^{Ht}$, is expressed as $a(\eta) = -1/(H\eta)$. With the above choice of mode functions, the constant Heisenberg quantum state annihilated by $a_{\mathbf{k}}$'s corresponds to the usual adiabatic (or Minkowski) vacuum in the limit $\eta \to -\infty$.

From the equation of motion of the scalar field with some interaction potential $V(\phi)$, $\Box\phi + V'(\phi) = 0$, we obtain the following equation for the slowly-varying coarse-grained part

$$\dot{\overline{\phi}}(\mathbf{x}, t) = -\frac{1}{3H}V'(\overline{\phi}) + f(\mathbf{x}, t), \tag{4}$$

where $f(\mathbf{x}, t)$ is given by

$$f(\mathbf{x}, t) = \epsilon a(t)H^2 \int \frac{d^3k}{(2\pi)^{3/2}}\delta(k - \epsilon a(t)H)\left[a_{\mathbf{k}}\phi_{\mathbf{k}}(t)e^{-i\mathbf{kx}} + a^{\dagger}_{\mathbf{k}}\phi^{*}_{\mathbf{k}}(t)e^{i\mathbf{kx}}\right]. \tag{5}$$

Equation (4) may be regarded as the Langevin equation for the stochastic quantity $\overline{\phi}$ with a noise term $f(\mathbf{x}, t)$ whose correlation properties are given by

$$\langle f(\mathbf{x}_1, t_1) f(\mathbf{x}_2, t_2)\rangle = \frac{H^3}{4\pi^2}\delta(t_1 - t_2)\, j_0(\epsilon a(t)H|\mathbf{x}_1 - \mathbf{x}_2|), \quad j_0(z) = \frac{\sin z}{z}. \tag{6}$$

As we see later, $j_0(z)$ or its approximate form plays a crucial role to show the de Sitter invariance of the system.

It is straightforward to derive the Fokker-Planck equation for the one-point (or one-domain) probability distribution function (PDF) $\rho_1[\overline{\phi}(\mathbf{x}, t) = \varphi] \equiv \rho_1[\varphi(\mathbf{x}, t)] \equiv \rho_1(\varphi, t)$. It has the form:

$$\frac{\partial \rho_1[\varphi(\mathbf{x}, t)]}{\partial t} = \frac{1}{3H}\frac{\partial}{\partial \varphi}\left\{V'[\varphi(\mathbf{x}, t)]\rho_1[\varphi(\mathbf{x}, t)]\right\} + \frac{H^3}{8\pi^2}\frac{\partial^2 \rho_1[\varphi(\mathbf{x}, t)]}{\partial \varphi^2} \equiv \Gamma_{\varphi}\rho_1[\varphi(\mathbf{x}, t)]. \tag{7}$$

In general, the solution of (7) can be written in the form

$$\rho_1(\varphi, t) = \exp\left(-\frac{4\pi^2 V(\varphi)}{3H^4}\right)\sum_{n=0}^{\infty} a_n \Phi_n(\varphi) e^{-\Lambda_n(t-t_0)}, \tag{8}$$

where $\Phi_n(\varphi)$ is the complete orthonormal set of eigenfunctions of the Schrödinger-type equation with the eigenvalue Λ_n:

$$\left[-\frac{1}{2}\frac{\partial^2}{\partial\varphi^2}+W(\varphi)\right]\Phi_n(\varphi)=\frac{4\pi^2\Lambda_n}{H^3}\Phi_n(\varphi), \tag{9}$$

with the effective potential

$$W(\varphi)=\frac{1}{2}\left[v'(\varphi)^2-v''(\varphi)\right], \qquad v(\varphi)\equiv\frac{4\pi^2}{3H^4}V(\varphi). \tag{10}$$

The coefficients a_n are given by an initial condition of ρ_1 at $t=t_0$ as

$$a_n=\int d\varphi\rho_1(\varphi,t)e^{v(\varphi)}\Phi_n(\varphi). \tag{11}$$

If the solution is normalizable, or

$$N\equiv\int_{-\infty}^{\infty}e^{-2v(\varphi)}d\varphi<\infty, \tag{12}$$

is satisfied, we find $\Lambda_0=0$ and the corresponding eigenfunction is given by

$$\Phi_0(\varphi)=N^{-1/2}e^{-v(\varphi)}. \tag{13}$$

Then the normalization condition $\int_{-\infty}^{\infty}d\varphi\,\rho_1(\varphi,t)=1$ results in $a_0=N^{-1/2}$, so that (8) reads

$$\rho_1(\varphi,t)=\rho_{1_0}(\varphi)+e^{-v(\varphi)}\sum_{n=1}^{\infty}a_n\Phi_n(\varphi)e^{-\Lambda_n(t-t_0)}. \tag{14}$$

Therefore, any solution (8) asymptotically approaches the static equilibrium solution

$$\rho_{1_0}(\varphi)\equiv N^{-1}\exp\left(-\frac{8\pi^2}{3H^4}V(\varphi)\right)=N^{-1}e^{-2v(\varphi)} \tag{15}$$

This PDF is significantly non-Gaussian in the presence of self-interaction. Thus (12) is the condition of the existence of an equilibrium solution.

As shown in [19], we can obtain the Fokker-Planck equation for the two-point PDF at equal time $\rho_2[\varphi_1(\mathbf{x}_1,t),\varphi_2(\mathbf{x}_2,t)]$ as

$$\frac{\partial\rho_2}{\partial t}[\varphi_1(\mathbf{x}_1,t),\varphi_2(\mathbf{x}_2,t)]=\Gamma_{\varphi_1}\rho_2+\Gamma_{\varphi_2}\rho_2+j_0(\epsilon a(t)H|\mathbf{x}_1-\mathbf{x}_2|)\frac{H^3}{4\pi^2}\frac{\partial^2\rho_2}{\partial\varphi_1\partial\varphi_2}. \tag{16}$$

Since $j_0(z)$ is a rapidly oscillating function for $|z|\gtrsim 1$ and we are concerned with coarse-grained quantities, I realized that adopting an approximation

$$j_0(\epsilon a(t)H|\mathbf{x}_1 - \mathbf{x}_2|) \simeq \theta(1 - \epsilon a(t)H|\mathbf{x}_1 - \mathbf{x}_2|), \tag{17}$$

should be adopted in the spirit of stochastic inflation which deals with the distribution of a coarse-grained field. This replacement corresponds to the approximation that two points $\mathbf{x}_1$ and $\mathbf{x}_2$ with their proper distance $a(t)|\mathbf{x}_1 - \mathbf{x}_2| \equiv a(t)r < (\epsilon H)^{-1}$ have 100% correlation while those with larger separation are mutually independent at the time t.

In fact, one can readily see that

$$\rho_2(\varphi_1, \varphi_2) \equiv \rho_{1\mathrm{eq}}(\varphi_1)\delta(\varphi_1 - \varphi_2) \tag{18}$$

constitutes a static solution of (16) with j_0 replaced by unity. Therefore one can find the equilibrium two-point PDF with $a(t)r > (\epsilon H)^{-1}$ by solving

$$\frac{\partial \rho_2}{\partial t}[\varphi_1(\mathbf{x}_1, t), \varphi_2(\mathbf{x}_2, t)] = \Gamma_{\varphi_1}\rho_2 + \Gamma_{\varphi_2}\rho_2, \tag{19}$$

with the initial condition

$$\rho_2[\varphi_1(\mathbf{x}_1, t_r), \varphi_2(\mathbf{x}_2, t_r)] = \rho_{1\mathrm{eq}}(\varphi_1)\delta(\varphi_1 - \varphi_2), \quad Ht_r = -\ln(ra_0\epsilon H). \tag{20}$$

Now we turn to the calculation of the general two-point correlation function

$$G(r, t_1, t_2) = \langle \overline{\phi}(\mathbf{x}_1, t_1)\overline{\phi}(\mathbf{x}_2, t_2)\rangle. \tag{21}$$

Following essentially the same line of arguments, we arrive at the basic expression for the general two-point PDF in the stochastic approach:

$$\rho_2[\varphi_1(\mathbf{x}_1, t_1), \varphi_2(\mathbf{x}_2, t_2)] = \int \Pi[\varphi_1(\mathbf{x}_1, t_1)|\varphi_r(\mathbf{x}_1, t_r)]$$

$$\times \Pi[\varphi_2(\mathbf{x}_2, t_2)|\varphi_r(\mathbf{x}_2, t_r)]\rho_{1\mathrm{eq}}(\varphi_r)d\varphi_r, \tag{22}$$

$$t_r = -H^{-1}\ln(ra_0\epsilon H), \quad r = |\mathbf{x}_1 - \mathbf{x}_2|, \tag{23}$$

where $\Pi[\varphi_1(\mathbf{x}, t_1)|\varphi_2(\mathbf{x}, t_2)]$ is the conditional probability to find $\overline{\phi}(\mathbf{x}, t_1) = \varphi_1$ provided that $\overline{\phi}(\mathbf{x}, t_2) = \varphi_2$. It satisfies the Fokker-Planck equation (7) with respect to both pairs of its arguments φ_1, t_1 and φ_2, t_2 [20]:

$$\frac{\partial \Pi}{\partial t_1}[\varphi_1(\mathbf{x}, t_1)|\varphi_2(\mathbf{x}, t_2)] = \Gamma_{\varphi_1}\Pi[\varphi_1(\mathbf{x}, t_1)|\varphi_2(\mathbf{x}, t_2)] \tag{24}$$

and

$$\frac{\partial \Pi}{\partial t_2}[\varphi_1(\mathbf{x}, t_1)|\varphi_2(\mathbf{x}, t_2)] = \Gamma_{\varphi_2}\Pi[\varphi_1(\mathbf{x}, t_1)|\varphi_2(\mathbf{x}, t_2)], \tag{25}$$

with the initial condition

$$\Pi[\varphi_1(\mathbf{x}, t_1)|\varphi_2(\mathbf{x}, t_1)] = \delta(\varphi_1 - \varphi_2). \tag{26}$$

Note also that the spatial points $\mathbf{x}_1$ and $\mathbf{x}_2$ were inside the same elementary averaging volume at the moment $t = t_r$, thus, they had the same one-point PDF at this time.

Returning to the case of a space-like separation between two points and using the spectral decomposition (14) for $\Pi[\varphi_1(\mathbf{x}, t_1)|\varphi_2(\mathbf{x}, t_2)]$, we obtain

$$\begin{aligned}
G(r, t_1, t_2) &= \int d\varphi_1 \varphi_1 \int d\varphi_2 \varphi_2 \rho_2[\varphi_1(\mathbf{x}_1, t_1), \varphi_2(\mathbf{x}_2, t_2)] \\
&= \iint d\varphi_1 d\varphi_2 \varphi_1 \varphi_2 e^{-v(\varphi_1)} e^{-v(\varphi_2)} \int d\varphi_r e^{2v(\varphi_r)} \rho_{1\mathrm{eq}}(\varphi_r) \\
&\quad \times \sum_m \Phi_m(\varphi_1)\Phi_m(\varphi_r) e^{-\Lambda_m(t_1 - t_r)} \sum_n \Phi_n(\varphi_2)\Phi_n(\varphi_r) e^{-\Lambda_n(t_2 - t_r)} \\
&= N \sum_m \sum_n \int d\varphi_r \Phi_m(\varphi_r)\Phi_n(\varphi_r) A_m A_n e^{-\Lambda_m(t_1 - t_r)} e^{-\Lambda_n(t_2 - t_r)} \\
&= N \sum_n A_n^2 e^{-\Lambda_n(t_1 + t_2 - 2t_r)} \\
&= N \sum_n A_n^2 e^{-\Lambda_n(t_1 + t_2)} \exp\left(-2\Lambda_n H^{-1} \ln(r a_0 \epsilon H)\right).
\end{aligned} \tag{27}$$

Here A_n is defined by

$$A_n = \int \varphi \rho_{1\mathrm{eq}}(\varphi) e^{v(\varphi)} \Phi_n(\varphi) d\varphi = N^{-1} \int \varphi e^{-v(\varphi)} \Phi_n(\varphi) d\varphi, \tag{28}$$

Now we may choose the small parameter ϵ of the coarse-graining (1) in such a way that $e^{-2\Lambda_n/H} \approx 1$ for all Λ_n which make a significant contribution to $G(t)$. This is achieved if ϵ satisfies the inequalities

$$\exp\left(-\min(H^2 m^{-2}|, \lambda^{-1/2})\right) \ll \epsilon \ll 1. \tag{29}$$

As a result, we may omit ϵ from the last line in Eq. (27). Then, from the comparison of (27) with the expression of an equal-point temporal correlation function $G(t)$, the expression for the general two-point correlation function in the equilibrium state reads

$$G(r, t_1, t_2) = G\left(t_1 + t_2 + 2H^{-1} \ln(r a_0 H)\right). \tag{30}$$

Using the de Sitter invariant separation function of $(\mathbf{x}_1, t_1)$ and $(\mathbf{x}_2, t_2)$ defined by

$$z(x_1, x_2) = \cosh H(t_1 - t_2) - \frac{H^2}{2} a_0^2 e^{Ht_1 + Ht_2} |\mathbf{x}_1 - \mathbf{x}_2|^2, \tag{31}$$

in the regime of a strongly space-like separation between two points ($z < 0, |z| \gg 1$) to which the formula (22) refers, Eq. (30) can be rewritten in the form

$$G(r, t_1, t_2) = G\left(H^{-1}\ln|2z-1|\right) \tag{32}$$

that applies both to the cases of large space-like and time-like separations. Actually, we may use this expression even in the vicinity of the point $z = 1$ where the right-hand side of (32) is simply equal to $G(0) = \langle\phi^2\rangle$. Therefore, we have proved the de Sitter invariance of the two-point correlation function.

The above is the essence of the first paper I wrote with Starobinsky [19], but we also studied an interesting case of a double well potential

$$V(\phi) = \frac{\lambda}{4}(\phi^2 - \phi_0^2)^2, \quad \phi_0^2 = \frac{|m^2|}{\lambda}. \tag{33}$$

In this model, the scalar field fluctuates around $\phi = \phi_0$ in some domains and around $\phi = -\phi_0$ in others. So there are two degenerate perturbative vacuum states.

The problem was to find the coefficient of transition amplitude (entropy factor) between degenerate ground states, and Alexei asked me to bring it because it was in Landau's Quantum Mechanics exercise. I had only the Japanese version, so I told him that "the formula is the same, so you can use the Japanese version," and when I showed him the Japanese version, he immediately opened the corresponding section and showed it to me, which surprised me. However, I still remember that he was impressed because the answer obtained by Landau's method was identical to the one I had obtained independently by a haphazard method.

In the final stages of preparing the manuscript, Starobinsky looked up relevant papers and tried to cite them. One of them was a paper I had written with Michiyasu Nagasawa [21], and Alexei asked me,

"Why didn't you cite this paper?"

Nowadays, it is common practice to cite your own papers to increase the number of citations, but I had never been educated to increase the number of citations, so I did not understand what he meant, and in the end, I did not cite it.

We submitted this paper to Physical Review D, and although the referee told us that it was a very good paper with only one comment, his request was to prove that the correlation function of a self-interacting scalar field in de Sitter spacetime, when expanded in powers of time, has zero radius of convergence. I managed to show this requirement by using a forceful method of finding the expansion coefficients one after another, and then using the inequalities satisfied by each coefficient to obtain the divergence of the series. However, he observed that the temporal correlation function in spectral expansion diverges for negative time as all the eigenvalues are non-negative, so the value at $t = 0$ is the irregular nonanalytic point of the correlation function. I was much impressed with this argument, but a long time later, I realized that the original argument was due to Langer [22]. Anyway, he rewarded my hard work by adding the sentence, "It may be verified by a direct computation, too."

I now check the database and find that the paper was rarely cited for a long time after publication and, interestingly, for the first five years it was cited only by Russians. At some point, however, the number of citations began to increase rapidly and it is now cited as the standard reference on stochastic inflation. I do not measure the value of a paper by its citation count, and when I make personnel selections, I give poor marks to candidates who flaunt their citation counts too much. I suspect that few of those who currently cite this paper have actually read it.

During his stay in Uji, his wife, daughter, and mother joined him alternately. David Polarsky also visited the Uji research center for two weeks and stayed in a guest room of the center. It was during that period they wrote a paper on isocurvature perturbations in multi-field inflation [23], in which they proposed a way to define adiabatic and isocurvature modes clearly. As an application of this paper, Alexei suggested to me to work on density perturbation in Brans-Dicke inflation which includes the Brans-Dicke scalar field in addition to the inflation field. It was not difficult to apply their formalism to this problem, and I reached a clear conclusion, actually correcting an error in the original article of soft inflation [24]. I made a presentation on this work at the 4th Workshop on General Relativity and Gravitation (JGRG4) in Japan held at Kyoto University from 28 November to 1 December in 1994, and wrote an article in its proceedings [25]. Alexei suggested to me to extend this analysis to the case when there are two scalar fields with mutually multiplicative potential and then wrote a journal paper. The calculation was easy but the paper was never published as it was difficult to communicate with him after his return to Moscow in time. Meanwhile, a paper extending our analysis was published by Wands and Garcia-Bellido [26] and what we wanted to show was discussed in [27] somewhat later.

3 First Stay at RESCEU

The Research Center for the Early Universe (RESCEU) was established in April 1999 by order of the then Ministry of Education of Japan at the Graduate School of Science, the University of Tokyo. Its predecessor, the Research Center for the Early Universe (the same name in English but the Japanese name was changed to "International Research Center for the Big Bang Universe" in 1999), was established in 1995 as an internal organization of the Graduate School of Science under the MEXT program "Basic Research for the Formation of COE (Centers of Excellence)". RESCEU is an international and comprehensive research institute that studies the creation and evolution of the universe to answer the fundamental question that humans have been asking since the beginning of time: "How did the universe begin and how did it come to be in its present form?" It has a section of visiting professorship from abroad, making use of which Alexei Starobinsky stayed there from August 2000 to February 2001.

At that time, I was an associate professor at Osaka University and Shinji Tsujikawa was a postdoc at RESCEU. Three of us worked together to write a paper entitled "Cosmological perturbations from multifield inflation in generalized Einstein theories" [28], which is in a sense an extension of our work on Brans-Dicke inflation [25]. This time, too, Tsujikawa and I had a problem communicating with Alexei after his return to Moscow to finish the paper. However, Shinji was more tactful than I, and he told Alexei that he would submit the paper to a journal if he would not send us any response by a certain deadline. And he actually did so. In this way, we have managed to publish the paper in Nuclear Physics B.

4 Stay at YITP on Kitashirakawa Main Campus

Starobinsky came to the headquarters of YITP in Kitashirakawa in 2007 as a visiting professor, this time only for three months. During that period, he completed his first work on $f(R)$ gravity, which was published in JETP Letter [29]. This paper attempts to explain the accelerated expansion in the late universe without introducing a cosmological constant nor vacuum energy. The price to pay for this is a very nontrivial functional form of the gravitational action. Thus, when I first saw the paper, I did not think it was a meaningful study. However, it was ironic that it turned out that I subsequently did research on this theory as well.

5 Starobinsky at RESCEU

I joined RESCEU in March 2005 as a professor. Since 2008, Alexei stayed with us almost every year until the Covid-19 pandemic hit the world. I list it for the sake of our record (Table 1).

RESCEU holds a summer school every year, and Starobinsky has attended as a lecturer three times: in 2009, when it was held in Ginowan, Okinawa; in 2011, when

Table 1 Starobinsky at RESCEU as a visiting professor

Start date	End date	Duration (months)
August 2008	March 2009	8
October 2010	November 2010	2
February 2011	March 2011	2
February 2012	March 2012	2
November 2012	December 2012	2
January 2013	February 2013	2
October 2013	November 2013	2
February 2015	March 2015	2
November 2015	December 2015	2
February 2017	February 2017	1
October 2018	November 2018	1
November 2019	December 2019	1

it was held at the San-ai Kogen Hotel on Mt. Aso; and in 2016, when it was held in Hida Takayama.

Looking back like this now, I checked the database and found that we wrote 10 papers together at RESCEU [30–39]. I realized that was not as many as I had expected. However, three students from my laboratory participated in these papers, and several of them were the basis for their master's thesis and doctoral dissertation.

Starobinsky conducted active research during his stay in Tokyo but also encountered several severe experiences. When he was heading to Kyoto on the Shinkansen railway, heavy rain stopped the train, where he had to spend a night. After his return to Tokyo, he kept a cool face and said, "It was quite an experience, but I was impressed that the Japanese people remained calm over the night." An even more serious incident was the East Japan great earthquake which happened on March 11, 2011, around 14:56. He went missing for several hours, alarming the people of RESCEU. They were worried that he might have been trapped in the elevator. In fact, he had gone to the Air France office in Aoyama, Tokyo that day and it took him a long time to walk back to the University. Later I asked him how he got back. He replied he first headed to the Imperial Palace, and then turned the direction to the University. In fact, I myself was on the plane to London when the earthquake happened, and I heard all the story after my return. I left Narita airport just four hours before the earthquake happened and I did not experience it myself. Anyway, Alexei was not only an excellent researcher, but also an excellent life person.

5.1 $f(R)$ Gravity

The first of such students was Hayato Motohashi, who recently became an associate professor at Tokyo Metropolitan University. He first studied the validity of the fourth-order differential equation derived by an Italian group for the evolution equation of density fluctuation in $f(R)$ gravity theory at the recommendation of Starobinsky, and found it to be inappropriate as a result of difficult analytical calculations. Although this result was good enough to be published in an academic paper, Hayato did not write the paper at this stage because of his strong determination not to end up merely criticizing the previous research. In the current climate of quantitative achievement-oriented research, such a respectful attitude toward scholarship is noteworthy. He finally succeeded in obtaining an analytical expression for the development of density fluctuation in $f(R)$ theory by using hypergeometric functions under the correct calculations, and only then did he write his paper as a positive research result [30].

When deviation from Einstein gravity is small, namely, $f(R) \cong R$ and $F(R) \equiv f'(R) \cong 1$, the evolution equations of the spatially flat, homogeneous and isotropic background universe yield the standard matter-dominated solution $a(t) = a_0(t/t_0)^{2/3}$. The differential equation describing a density perturbation in the subhorizon regime in $f(R)$ gravity in this limit reads

$$\ddot{\delta} + 2H\dot{\delta} - 4\pi G_{\rm eff}\rho\delta = 0, \tag{34}$$

where

$$G_{\text{eff}} = \frac{G}{F}\frac{1+4\frac{k^2}{a^2}\frac{F_{,R}}{F}}{1+3\frac{k^2}{a^2}\frac{F_{,R}}{F}}, \tag{35}$$

as derived by Tsujikawa [40]. Specifically, we consider a model

$$F(R) \equiv f'(R) = 1 - \left(\frac{R_0}{R}\right)^{n+1}, \quad n > -1 \tag{36}$$

This $f(R)$ corresponds to the models in [29, 41] in the regime $R \gg R_0$, and we shall use it in this regime only. Equation (34) then becomes

$$\ddot{\delta} + \frac{4}{3t}\dot{\delta} - \frac{2}{3t^2}\frac{1+4A(t/t_0)^{2n+8/3}}{1+3A(t/t_0)^{2n+8/3}}\delta = 0, \quad A \equiv A(n,k) = \frac{(n+1)k^2}{a_0^2 R_0^2}. \tag{37}$$

By changing the variable from t to $\tau = (t/t_0)^\alpha$ with $\alpha = 2n + 8/3$, Eq. (37) can be rewritten as

$$\delta'' + \left(1 + \frac{1}{3\alpha}\right)\frac{\delta'}{\tau} - \frac{2}{3\alpha^2}\frac{1+4A\tau}{1+3A\tau}\frac{\delta}{\tau^2} = 0. \tag{38}$$

Here a prime denotes derivative with respect to τ. Hayato found that this equation can be solved in terms of the hypergeometric function ${}_2F_1$. The two independent solutions are

$$\delta_{\mathbf{k}}(t) = \delta_{0\mathbf{k}}\left(\frac{t}{t_0}\right)^{\frac{-1\pm 5}{6}} {}_2F_1\left(\frac{\pm 5 - \sqrt{33}}{6\alpha}, \frac{\pm 5 + \sqrt{33}}{6\alpha}; 1 \pm \frac{5}{3\alpha}; -3A(n,k)\left(\frac{t}{t_0}\right)^\alpha\right). \tag{39}$$

The upper sign represents a growing mode while the lower sign corresponds to the decaying mode of perturbations and is singular at $t \to 0$. We can confirm that the solution has appropriate asymptotic behaviors.

As is seen in (35), in $f(R)$ model, the strength of gravity in the growth of perturbation is scale dependent. In particular, we can expect an extra growth on small scales. This observation led me to combine massive neutrinos with $f(R)$ gravity.

The possibility of the existence of light sterile neutrinos had been suggested by a number of experiments [42, 43] including reactor experiments [44] favoring sterile neutrinos of eV mass scale. On the other hand, cosmological observations of large-scale structure severely constrain the neutrino mass, and these two did not seem to be consistent with each other. It struck me an idea that extra growth of small-scale perturbations in $f(R)$ gravity may solve the tension between cosmological observations and massive neutrinos favored by experiments. I told it to Alexei and he said it was a good idea and we started working on it together with Hayato. I vividly remember it was when we were walking in the campus between the Sanshiro-pond and Law and Humanities Bldg. 2. We have successfully shown that everything is consistent in this model and the paper was published in Physical Review Letters with an editor's suggestion mark [34]. Hayato pursued this line of study and obtained generic constraints on neutrino mass in $f(R)$ gravity, which led him to get a postdoc offer from particle physics theory group in Spain rather than a cosmology group.

5.1.1 Higgs and R^2 Inflation

Observations of cosmic microwave background [45] favor the standard Higgs inflation model [46–48] and the curvature-square model, which is often called the Starobinsky model [49]. I made a simple proposal to my graduate student, Minxi He, to study the case where both the Higgs field and curvature square term in the gravitational action are at work for inflation. This will lead to a two-field inflation setting.

The action in the original Jordan frame, where the space-time metric is denoted as $\hat{g}_{\mu\nu}$, is given by

$$S_\mathrm{J} = \int d^4x\sqrt{-\hat{g}}\left[\frac{M_p^2}{2}\hat{R} + \frac{1}{2}\xi\chi^2\hat{R} + \frac{M_p^2}{12M^2}\hat{R}^2 - \frac{1}{2}\hat{g}^{\mu\nu}\hat{\nabla}_\mu\chi\hat{\nabla}_\nu\chi - \frac{\lambda}{4}\chi^4\right] \tag{40}$$

$$= \int d^4x\sqrt{-\hat{g}}\left[F(\chi,\hat{R}) - \frac{1}{2}\hat{g}^{\mu\nu}\hat{\nabla}_\mu\chi\hat{\nabla}_\nu\chi\right] \tag{41}$$

where $M_p \equiv (8\pi G)^{-1/2}$ and χ is a singlet scalar field, a simplified model of the Standard Model Higgs boson. We neglect its interaction to gauge fields. Here $F(\chi,\hat{R})$ is defined by

$$F(\chi,\hat{R}) \equiv \frac{M_p^2}{2}\hat{R} + \frac{1}{2}\xi\chi^2\hat{R} + \frac{M_p^2}{12M^2}\hat{R}^2 - \frac{\lambda}{4}\chi^4\,. \tag{42}$$

This action has also been considered in [50, 51]. χ has a non-minimal coupling term with the Ricci scalar. We take the sign of the non-minimal coupling constant ξ such that the conformal coupling corresponds to $\xi = -1/6$.

Defining the scalaron field as [52, 53]

$$\sqrt{\frac{2}{3}}\frac{\psi}{M_p} \equiv \ln\left(\frac{2}{M_p^2}\left|\frac{\partial F}{\partial\hat{R}}\right|\right)\,, \tag{43}$$

and performing a conformal transformation

$$g_{\mu\nu}(x) = e^{\sqrt{\frac{2}{3}}\frac{\psi(x)}{M_p}}\,\hat{g}_{\mu\nu}(x)\,, \tag{44}$$

we can transform the original action (40) into the one in the Einstein frame and express the new action in terms of the new scalar fields as

$$S_{\mathrm{E}} = \int d^4x\sqrt{-g}\left[\frac{M_p^2}{2}R - \frac{1}{2}g^{\mu\nu}\nabla_\mu\psi\nabla_\nu\psi - \frac{1}{2}e^{-\sqrt{\frac{2}{3}}\frac{\psi}{M_p}}g^{\mu\nu}\nabla_\mu\chi\nabla_\nu\chi - U(\psi,\chi)\right], \tag{45}$$

where the potential is expressed as

$$U(\psi,\chi) \equiv \frac{\lambda}{4}\chi^4 e^{-2\sqrt{\frac{2}{3}}\frac{\psi}{M_p}} + \frac{3}{4}M_p^2M^2e^{-2\sqrt{\frac{2}{3}}\frac{\psi}{M_p}}\left(e^{\sqrt{\frac{2}{3}}\frac{\psi}{M_p}} - 1 - \frac{1}{M_p^2}\xi\chi^2\right)^2. \tag{46}$$

In addition to the metric and Higgs field χ, ψ is the third dynamical field which originates from the R^2 term. Note that ψ shows up only in the exponent with an $\mathcal{O}(1)$ numerical factor, so that large values of ψ will significantly suppress the terms with higher orders in $\exp(-\sqrt{2/3}\psi/M_p)$.

In the Einstein frame, this is a two-field inflation model with kinetic mixing. Minxi analyzed this complicated model faithfully. However, Alexei suggested a very simple way to see the relation between ξ and the scalaron mass M without any conformal transformation by considering the relation between this two-field model and the R^2 inflation directly in the Jordan frame. From the action in this frame (40), we find that the non-minimal coupling can be partially regarded as an extra contribution to the kinetic term for the Higgs field χ proportional to ξ since there are second derivatives in the Ricci scalar so that we can realize this by using integration by parts. Then, for large ξ and ψ regime, the original kinetic term in (40) is negligible. Dropping the kinetic term for χ, the original action becomes

$$S_{\mathrm{J}} \approx \tilde{S}_{\mathrm{J}} = \int d^4x\sqrt{-\hat{g}}\left[\frac{M_p^2}{2}\hat{R} + \frac{1}{2}\xi\chi^2\hat{R} + \frac{M_p^2}{12M^2}\hat{R}^2 - \frac{\lambda}{4}\chi^4\right]. \tag{47}$$

As a result, the equation of motion for χ just becomes a constraint on χ [54] and R as $\chi^2 = \xi\hat{R}/\lambda$ which simplifies the action as

$$\tilde{S}_{\mathrm{J}} = \int d^4x\sqrt{-\hat{g}}\left[\frac{M_p^2}{2}\hat{R} + \left(\frac{M_p^2}{12M^2} + \frac{\xi^2}{4\lambda}\right)\hat{R}^2\right] \tag{48}$$

$$= \int d^4x\sqrt{-\hat{g}}\left[\frac{M_p^2}{2}\hat{R} + \frac{M_p^2}{12\tilde{M}^2}\hat{R}^2\right] \tag{49}$$

where

$$\tilde{M}^2 \equiv \frac{M^2}{1 + 3\xi^2 \frac{M^2}{\lambda M_p^2}} \tag{50}$$

is the effective mass squared of the scalaron, which should take $\tilde{M} = 1.3 \times 10^{-5} M_p$ [55, 56] to reproduce the observed amplitude of curvature perturbation. Thus, dropping the kinetic term of χ we can easily capture the essence of the theory. I was much impressed with his intuition and insight.

After this work, we focused on the reheating stage of this model to study tachyonic preheating [37, 38] and the final stage of reheating [39], which turned out to be a master's thesis of Hyun Jeong. In particular, the last paper was published as an invited article to celebrate the 20th anniversary of the Journal of Cosmology and Astroparticle Physics. We still have a follow-up project of [39] unfinished.

5.2 *Outlook of Inflationary Cosmology and Related Particle Physics*

When the BICEP collaboration claimed a detection of B-mode polarization of CMB [57] favoring tensor-to-scalar ratio $r \simeq 0.2$, many people were excited by it as evidence for large-field inflation models. I had a mixed feeling as I was just about to submit a paper on the Starobinsky inflation model in supergravity [58] which predicts much smaller tensor perturbations. In the end, however, the presentation settled on the fact that they had not found the primordial component of B-mode polarization. Later on, I asked Alexei what he thought about their original claim, as he remained silent during the boom. He said he was sure that it would not have been the detection of the primordial tensor perturbation as he put more trust in WMAP and Planck results.

For a long time, the study of inflation was to find a model which can control curvature perturbation at the observed level of 10^{-5} by introducing a small self-coupling of $\mathcal{O}(10^{-13})$, and how to stabilize such a small parameter against quantum correction was a central issue. Supersymmetry certainly helps it [59], and it was believed that model construction should be done using supergravity [60, 61].

Nowadays, however, these large-field models are disfavored by CMB observations [45], while the Higgs inflation and Starobinsky model lie in the central region of the preferred parameter space in the diagram of scalar spectral index versus tensor-to-scalar ratio. Of course, these models occupy just single points in the vast functional space of the most general single-field inflation [62]. But our recent study has revealed that quantum loop calculations of curvature perturbation significantly constrain the viable class of models [7], which indicates that we should start from simpler models that predict Gaussian fluctuations. Both Higgs inflation and the Starobinsky model possess the desired properties and can take place with the minimal extension of the standard particle physics model. On the other hand, both of them have large parameters such as $\xi \sim 10^5$ in Higgs inflation and the coefficient of R^2 term being

$\mathcal{O}(10^{10})$ in the Starobinsky model. The origin of these large parameters is the central issue of contemporary inflationary cosmology. Note that both of these parameters are related to the modification of gravity from Einstein, and the key should certainly lie in modified gravity or extra dimension.

6 Conclusion

I had the privilege of engaging with Alexei A. Starobinsky over a period of 33 years, from 1990 to 2023. His contributions to theoretical cosmology are of exceptional significance. He stands not only as one of the principal founders of inflationary cosmology—alongside Sato [1] and Guth [63]—but also as a pioneer in the theory of cosmological fluctuations, which underpins our current understanding of structure formation in the universe. His seminal work on the generation of density [64] and tensor [65] perturbations during inflation, as well as his influential studies on stochastic inflation [17, 19], dark energy, and modified theories of gravity [29], have profoundly shaped the field. If one were to ask who has made the most substantial contributions to relativistic cosmology, his name would undoubtedly be among the foremost.

Inflationary cosmology [66] provides a compelling explanation for several fundamental properties of the universe, including its large-scale homogeneity and isotropy, spatial flatness, and the generation of nearly scale-invariant curvature perturbations in remarkable agreement with observations. Nevertheless, critics of inflation have often pointed out that mechanisms other than inflation could also generate curvature perturbations, and have argued that the homogeneity, isotropy, and flatness are postdictions rather than genuine predictions. It is also frequently suggested that a Nobel Prize will not be awarded to the proponents of inflation until tensor perturbations are confirmed through the detection of B-mode polarization in the cosmic microwave background (CMB). However, I personally believe that WMAP's discovery of the negative correlation between temperature fluctuations in the cosmic microwave background radiation and E-mode polarization [67] is sufficient proof of inflation in the early universe, since it confirms the existence of fluctuations beyond the Hubble scale in the early universe. In this sense, it is unfortunate that the Nobel Prize Committee missed out on awarding the prize to the greatest contributor to inflationary cosmology.

A Korean researcher once recalled a remark made by Starobinsky: "I already know the result before I do the calculation; I perform the calculation merely to confirm it." Among the many distinguished scholars I have had the privilege to know throughout the world, I have never encountered anyone with such profound insight and such a deep and comprehensive command across so many areas of physics as he possessed.

The last time I spoke with Starobinsky was during a Zoom meeting held on August 22, 2023, to discuss our work which turned out to [39]. He mentioned that, now that the COVID-19 pandemic had subsided, he hoped to visit Japan again. Our final exchange of emails took place on November 14, when the paper was accepted for publication. That was approximately one month before his passing.

As illustrated by just a small portion of what I have shared above, I have received countless insights and inspiration from him over the years. We collaborated on the study of the distribution function of the universe within the framework of stochastic inflation, and I still have over a hundred pages of handwritten notes from our discussions, which remain unfinished.

I also wanted to ask him about his thoughts on a recent study I did with Jason Kristiano that showed that small scale fluctuations can affect larger scales that can be observed by the CMB due to higher order quantum effects, which caused a sensation among many people [9]. Being Alexei, he might well have replied, "Of course, that is to be expected."

I am profoundly grateful for the many years of intellectual companionship and friendship we shared. With deepest respect, I offer my heartfelt prayers for his peaceful repose.

Acknowledgements I am grateful to Andrei Barvinsky and Alexander Kamenshchik for inviting me to write this article.

References

1. K. Sato, First order phase transition of a vacuum and expansion of the universe. Mon. Not. Roy. Astron. Soc. **195**, 467 (1981)
2. A.D. Linde, A new inflationary universe scenario: a possible solution of the horizon, flatness, homogeneity, isotropy and primordial monopole problems. Phys. Lett. B **108**, 389 (1982)
3. A.D. Linde, Chaotic inflation. Phys. Lett. B **129**, 177–181 (1983)
4. G.F. Smoot, C.L. Bennett, A. Kogut, E.L. Wright, J. Aymon, N.W. Boggess, E.S. Cheng, G. De Amici, S. Gulkis, M.G. Hauser, G. Hinshaw, P.D. Jackson, M. Janssen, E. Kaita, T. Kelsall, P. Keegstra, C. Lineweaver, K. Loewenstein, P. Lubin, J. Mather, S.S. Meyer, H. Moseley, T. Murdock, L. Rokke, R.F. Silverberg, L. Tenorio, R. Weiss, D.T. Wilkinson, Structure in the COBE differential microwave radiometer first-year maps. Astrophys. J. Lett. **396**, L1–L5 (1992)
5. J. Schwinger, Brownian motion of a quantum oscillator. J. Math. Phys. **2**, 407–432 (1961)
6. L.V. Keldysh, Diagram technique for nonequilibrium processes. Sov. Phys. JETP **20**, 1018–1026 (1965). [Zh. Eksp. Teor. Fiz. **47**, 1515 (1964)]
7. J. Kristiano, J. Yokoyama, Why must primordial non-Gaussianity be very small? Phys. Rev. Lett. **128**, 061301 (2022)
8. J. Kristiano, J. Yokoyama, Perturbative region on non-Gaussian parameter space in single-field inflation. JCAP **07**, 007 (2022)
9. J. Kristiano, J. Yokoyama, Constraining primordial black hole formation from single-field inflation. Phys. Rev. Lett. **132**, 221003 (2024)
10. J. Kristiano, J. Yokoyama, Note on the bispectrum and one-loop corrections in single-field inflation with primordial black hole formation. Phys. Rev. D **109**, 103541 (2024)
11. J. Kristiano, J. Yokoyama, Comparing sharp and smooth transitions of the second slow-roll parameter in single-field inflation. JCAP **10**, 036 (2024)
12. N.A. Chernikov, E.A. Tagirov, Quantum theory of scalar fields in de Sitter space-time. Annales de l'Institut Henri Poincaré A **9**, 109–141 (1968)

13. T.S. Bunch, P.C.W. Davies, Quantum field theory in de Sitter space: renormalization by point-splitting. Proc. R. Soc. Lond. A **360**, 117–134 (1978)
14. A.A. Starobinsky, Dynamics of phase transition in the new inflationary universe scenario and generation of perturbations. Phys. Lett. B **117**, 175–178 (1982)
15. A. Vilenkin, L.H. Ford, Gravitational effects upon cosmological phase transitions. Phys. Rev. D **26**, 1231–1241 (1982)
16. B. Allen, Vacuum states in de Sitter space. Phys. Rev. D **32**, 3136–3149 (1985)
17. A.A. Starobinsky, Stochastic de Sitter (inflationary) stage in the early universe, in *Fundamental Interactions*, ed. by V.N. Ponomarev (MGPI Press, Moscow, 1984), pp. 55–72
18. A.A. Starobinsky, Stochastic de Sitter (inflationary) stage in the early universe, in *Field Theory, Quantum Gravity and Strings*, Lecture Notes in Physics, vol. 246, ed. by H.J. de Vega, N. Sanchez (Springer, 1986), pp. 107–126
19. A.A. Starobinsky, J. Yokoyama, Equilibrium state of a selfinteracting scalar field in the De Sitter background. Phys. Rev. D **50**, 6357–6368 (1994)
20. H. Risken, *The Fokker-Planck Equation: Methods of Solution and Applications* (Springer, 1984)
21. M. Nagasawa, J. Yokoyama, Phase transitions triggered by quantum fluctuations in the inflationary universe. Nucl. Phys. B **370**, 472–490 (1992)
22. J.S. Langer, Theory of the condensation point. Ann. Phys. **41**, 108–157 (1967)
23. D. Polarski, A.A. Starobinsky, Isocurvature perturbations in multiple inflationary models. Phys. Rev. D **50**, 6123 (1994)
24. A.L. Berkin, K.-I. Maeda, J. Yokoyama, Soft inflation. Phys. Rev. Lett. **65**, 141–144 (1990)
25. A.A. Starobinsky, J. Yokoyama, Density fluctuations in Brans-Dicke inflation, in: *4th Workshop on General Relativity and Gravitation*, vol. 381 (1994)
26. J. Garcia-Bellido, D. Wands, Constraints from inflation on scalar-tensor gravity theories. Phys. Rev. D **52**, 6739–6749 (1995)
27. V.F. Mukhanov, P.J. Steinhardt, Density perturbations in multifield inflationary models. Phys. Lett. B **422**, 52–60 (1998)
28. A.A. Starobinsky, S. Tsujikawa, J. Yokoyama, Cosmological perturbations from multifield inflation in generalized Einstein theories. Nucl. Phys. B **610**, 383–410 (2001)
29. A.A. Starobinsky, Disappearing cosmological constant in f(R) gravity. JETP Lett. **86**, 157–163 (2007)
30. H. Motohashi, A.A. Starobinsky, J. Yokoyama, Analytic solution for matter density perturbations in a class of viable cosmological f(R) models. Int. J. Mod. Phys. D **18**, 1731–1740 (2009)
31. H. Motohashi, A.A. Starobinsky, J. Yokoyama, Phantom boundary crossing and anomalous growth index of fluctuations in viable f(R) models of cosmic acceleration. Prog. Theor. Phys. **123**, 887–902 (2010)
32. H. Motohashi, A.A. Starobinsky, J. Yokoyama, Matter power spectrum in f(R) gravity with massive neutrinos. Prog. Theor. Phys. **124**, 541–546 (2010)
33. H. Motohashi, A.A. Starobinsky, J. Yokoyama, Future oscillations around phantom divide in $f(R)$ gravity. JCAP **06**, 006 (2011)
34. H. Motohashi, A.A. Starobinsky, J. Yokoyama, Cosmology based on $f(R)$ gravity admits 1 eV Sterile neutrinos. Phys. Rev. Lett. **110**, 121302 (2013)
35. H. Motohashi, A.A. Starobinsky, J. Yokoyama, Inflation with a constant rate of roll. JCAP **09**, 018 (2015)
36. M. He, A.A. Starobinsky, J. Yokoyama, Inflation in the mixed Higgs-R^2 model. JCAP **05**, 064 (2018)
37. M. He, R. Jinno, K. Kamada, S.C. Park, A.A. Starobinsky, J. Yokoyama, On the violent preheating in the mixed Higgs-R^2 inflationary model. Phys. Lett. B **791**, 36–42 (2019)
38. M. He, R. Jinno, K. Kamada, A.A. Starobinsky, J. Yokoyama, Occurrence of tachyonic preheating in the mixed Higgs-R^2 model. JCAP **01**, 066 (2021)
39. H. Jeong, K. Kamada, A.A. Starobinsky, J. Yokoyama, Reheating process in the R^2 inflationary model with the baryogenesis scenario. JCAP **11**, 023 (2023)

40. S. Tsujikawa, Matter density perturbations and effective gravitational constant in modified gravity models of dark energy. Phys. Rev. D **76**, 023514 (2007)
41. W. Hu, I. Sawicki, Models of f(R) cosmic acceleration that evade solar-system tests. Phys. Rev. D **76**, 064004 (2007)
42. A. Aguilar et al., Evidence for neutrino oscillations from the observation of $\bar{\nu}_e$ appearance in a $\bar{\nu}_\mu$ beam. Phys. Rev. D **64**, 112007 (2001)
43. A.A. Aguilar-Arevalo et al., Event excess in the MiniBooNE search for $\bar{\nu}_\mu \rightarrow \bar{\nu}_e$ oscillations. Phys. Rev. Lett. **105**, 181801 (2010)
44. G. Mention, M. Fechner, T. Lasserre, T.A. Mueller, D. Lhuillier, M. Cribier, A. Letourneau, The reactor antineutrino anomaly. Phys. Rev. D **83**, 073006 (2011)
45. Y. Akrami et al., Planck 2018 results. X. Constraints on inflation. Astron. Astrophys. **641**, A10 (2020)
46. J.L. Cervantes-Cota, H. Dehnen, Induced gravity inflation in the standard model of particle physics. Nucl. Phys. B **442**, 391 (1995)
47. F.L. Bezrukov, M. Shaposhnikov, The standard model Higgs boson as the inflaton. Phys. Lett. B **659**, 703 (2008)
48. A.O. Barvinsky, A.Y. Kamenshchik, A.A. Starobinsky, Inflation scenario via the standard model Higgs boson and LHC. JCAP **0811**, 021 (2008)
49. A.A. Starobinsky, A new type of isotropic cosmological models without singularity. Phys. Lett. B **91**, 99 (1980)
50. Y. Ema, Higgs Scalaron mixed inflation. Phys. Lett. B **770**, 403 (2017)
51. Y.-C. Wang, T. Wang, Primordial perturbations generated by Higgs field and R^2 operator. Phys. Rev. D **96**, 123506 (2017)
52. K. Ichi Maeda, Towards the Einstein-Hilbert action via conformal transformation. Phys. Rev. D **39**, 3159 (1989)
53. K. Ichi Maeda, Inflation as a transient attractor in R**2 cosmology. Phys. Rev. D **37**, 858 (1988)
54. A. Kehagias, A.M. Dizgah, A. Riotto, Remarks on the Starobinsky model of inflation and its descendants. Phys. Rev. D **89**, 043527 (2014)
55. A.A. Starobinsky, The perturbation spectrum evolving from a nonsingular initially De-Sitter cosmology and the microwave background anisotropy. Sov. Astron. Lett. **9**, 302 (1983)
56. T. Faulkner, M. Tegmark, E.F. Bunn, Y. Mao, Constraining f(R) gravity as a scalar tensor theory. Phys. Rev. D **76**, 063505 (2007)
57. P.A.R. Ade et al., Detection of B-mode polarization at degree angular scales by BICEP2. Phys. Rev. Lett. **112**, 241101 (2014)
58. T. Terada, Y. Watanabe, Y. Yamada, J. Yokoyama, Reheating processes after Starobinsky inflation in old-minimal supergravity. JHEP **02**, 105 (2015)
59. J.R. Ellis, D.V. Nanopoulos, K.A. Olive, K. Tamvakis, Cosmological inflation cries out for supersymmetry. Phys. Lett. B **118**, 335 (1982)
60. A.B. Goncharov, A.D. Linde, Chaotic inflation in supergravity. Phys. Lett. B **139**, 27–30 (1984)
61. H. Murayama, H. Suzuki, T. Yanagida, J. Yokoyama, Chaotic inflation and baryogenesis in supergravity. Phys. Rev. D **50**, R2356–R2360 (1994)
62. T. Kobayashi, M. Yamaguchi, J. Yokoyama, Generalized G-inflation: inflation with the most general second-order field equations. Prog. Theor. Phys. **126**, 511–529 (2011)
63. A.H. Guth, The inflationary universe: a possible solution to the Horizon and flatness problems. Phys. Rev. D **23**, 347 (1981)
64. A.A. Starobinsky, Dynamics of phase transition in the new inflationary universe scenario and generation of perturbations. Phys. Lett. B **117**, 175–178 (1982)
65. A.A. Starobinsky, Spectrum of relict gravitational radiation and the early state of the universe. JETP Lett. **30**, 682–685 (1979)
66. K. Sato, J. Yokoyama, Inflationary cosmology: first 30+ years. Int. J. Mod. Phys. D **24**, 1530025 (2015)
67. A. Kogut, D.N. Spergel, C. Barnes, C.L. Bennett, M. Halpern, G. Hinshaw, N. Jarosik, M. Limon, S.S. Meyer, L. Page, G.S. Tucker, E. Wollack, E.L. Wright, First-year Wilkinson microwave anisotropy probe (WMAP) observations: temperature-polarization correlation. Astrophys. J. Suppl. Ser. **148**, 161–173 (2003)

FSC
www.fsc.org
MIX
Papier | Fördert gute Waldnutzung
FSC® C083411

Kolibri 360 GmbH
Ferdinand-Jühlke-Straße 7
99095 Erfurt, Deutschland
produktsicherheit@kolibri360.de